THE ROUTLEDGE HANDBOOK ON LIVELIHOODS IN THE GLOBAL SOUTH

The Routledge Handbook on Livelihoods in the Global South presents a unique, timely, comprehensive overview of livelihoods in low- and middle-income countries. Since their widespread adoption in the 1990s, livelihoods perspectives, frameworks and methods have influenced diverse areas of research, policy and practice.

The concept of livelihoods reflects the complexity of strategies and practices used by individuals, households and communities to meet their needs and live their lives. The Handbook brings together insights and critical analysis from diverse approaches and experiences, learning from research and practice over the last 30 years. The Handbook comprises an introductory section on key concepts and frameworks, followed by five parts, on researching livelihoods, negotiating livelihoods, generating livelihoods, enabling livelihoods and contextualising livelihoods. The introduction provides readers with an appreciation of concepts researched and applied in the five parts, including chapters on vulnerability and resilience, social capital and networks, and institutions. Each part reflects the diversity of approaches taken to understanding livelihoods, whilst recognising commonalities, including the centrality of power in shaping, enabling and constraining livelihoods. The book also reflects diversity of context, including conflict, climate change and religion, as well as in generating livelihoods, through agriculture, small-scale mining and pastoralism. The aim of each chapter is to provide a critically informed introduction and overview of key concepts, issues and debates of relevance to the topic, with each chapter concluding with suggestions for further reading.

It will be an essential resource to students, researchers and practitioners of international development and related fields. Researchers and practitioners will also benefit from the book's diverse disciplinary contributions and by the wide and contemporary coverage.

Fiona Nunan is Professor of Environment and Development, International Development Department, University of Birmingham.

Clare Barnes is an Interdisciplinary Lecturer in Sustainable Livelihoods at the School of GeoSciences, University of Edinburgh.

Sukanya Krishnamurthy is a Chancellors Fellow/Senior Lecturer at the School of GeoSciences, University of Edinburgh.

THE ROUTLEDGE HANDBOOK ON LIVELIHOODS IN THE GLOBAL SOUTH

*Edited by Fiona Nunan, Clare Barnes
and Sukanya Krishnamurthy*

LONDON AND NEW YORK

Cover image: © Sriram Vittalamurthy

First published 2023
by Routledge
2 Park Square, Milton Park, Abingdon, Oxon OX14 4RN

and by Routledge
605 Third Avenue, New York, NY 10158

Routledge is an imprint of the Taylor & Francis Group, an informa business

© 2023 selection and editorial matter, Fiona Nunan, Clare Barnes and Sukanya Krishnamurthy; individual chapters, the contributors

The right of Fiona Nunan, Clare Barnes and Sukanya Krishnamurthy to be identified as the authors of the editorial material, and of the authors for their individual chapters, has been asserted in accordance with sections 77 and 78 of the Copyright, Designs and Patents Act 1988.

All rights reserved. No part of this book may be reprinted or reproduced or utilised in any form or by any electronic, mechanical, or other means, now known or hereafter invented, including photocopying and recording, or in any information storage or retrieval system, without permission in writing from the publishers.

Trademark notice: Product or corporate names may be trademarks or registered trademarks, and are used only for identification and explanation without intent to infringe.

British Library Cataloguing-in-Publication Data
A catalogue record for this book is available from the British Library

Library of Congress Cataloging-in-Publication Data
A catalog record for this book has been requested

ISBN: 978-0-367-85635-9 (hbk)
ISBN: 978-1-032-26005-1 (pbk)
ISBN: 978-1-003-01404-1 (ebk)

DOI: 10.4324/9781003014041

Typeset in Bembo
by Apex CoVantage, LLC

CONTENTS

List of figures — x
List of tables — xi
List of boxes — xii
List of contributors — xiv

Introduction — 1

1 Livelihoods in the Global South — 3
   Fiona Nunan, Clare Barnes and Sukanya Krishnamurthy

2 Livelihoods: concepts and frameworks — 10
   Fiona Nunan

3 The Capability Approach as an analytic lens for studying livelihoods — 20
   Lucy Szaboova

4 Livelihoods and institutions — 31
   Luke Whaley

5 Vulnerability and resilience — 44
   Nathan Clay

6 Social capital and social networks — 56
   Itzel San Roman Pineda

7 A rights-based approach for sustainable livelihoods — 68
   Jae-Eun Noh

PART I
Researching livelihoods: approaches and methods — 79

8 Critically understanding livelihoods in the Global South: researchers, research practices and power — 81
   Sam Staddon

9 Quantitative approaches to analyse rural livelihood strategies — 93
   Solomon Zena Walelign, Xi Jiao and Carsten Smith-Hall

10 Longitudinal research to understand the complexity of livelihoods — 104
   Joan DeJaeghere

11 The use of ethnography for livelihoods research — 114
   Thaís de Carvalho

12 Using participatory rural appraisal to research livelihoods — 124
   Klara Fischer

13 Using participatory video in researching livelihoods in the Global South: the why and wherefore — 134
   Lara Bezzina

14 Using participatory GPS methods to develop rich understandings of people's diverse and complex livelihoods in the Global South — 147
   Nathan Salvidge

PART II
Negotiating livelihoods — 157

15 Power and livelihoods — 159
   Enrico Michelutti

16 Feminist political ecology — 170
   Miriam Gay-Antaki and Ana De Luca

17 Democratic politics and livelihoods in Africa — 181
   Jeremy Seekings

18 Social accountability in Asia's livelihoods: the role of sanctions and rewards — 194
   Aries A. Arugay

19 Advocating for livelihoods through social movements — 206
   Stefan Rzedzian, Margherita Scazza and Elodie Santos Vera

20 Education and livelihoods in the Global South   217
    Vikas Maniar and Meera Chandran

21 Youth livelihoods: negotiating intergenerationality and responsibility   227
    Sukanya Krishnamurthy

22 The governance and regulation of the informal economy: implications
    for livelihoods and decent work   237
    Julian Walker, Andrea Rigon and Braima Koroma

23 Disability and sustainable livelihoods: towards inclusive
    community-based development   246
    David Cobley

PART III
Generating livelihoods   **257**

24 Environmental income and rural livelihoods   259
    Carsten Smith-Hall, Xi Jiao and Solomon Zena Walelign

25 Forests and livelihoods   271
    Clare Barnes

26 Agricultural livelihoods, rural development policy and political
    ecologies of land and water: exploring new agrarian questions   284
    Cristián Alarcón, Johanna Bergman Lodin and Flora Hajdu

27 Pastoralism and livelihoods in the Global South   302
    Lenyeletse Vincent Basupi

28 Fisheries livelihoods   312
    Deo Namwira and Fiona Nunan

29 Complexity and heterogeneity in the informal economy of waste:
    problems and prospects for organising and formalising   324
    Aman Luthra

30 Planning for sustainable urban livelihoods in Africa   335
    *Lauren Andres, Stuart Paul Denoon-Stevens, John R. Bryson, Hakeem
    Bakare and Lorena Melgaço*

31 Artisanal mining and livelihoods in the Global South   345
    Roy Maconachie

PART IV
Enabling livelihoods — 357

32 Conceptualising migration and livelihoods: perspectives from the Global South — 359
    Mariama Zaami

33 International migration and experiences of Indian women migrants: a critical analysis of the Kafala system — 368
    Jyoti Bania

34 Remittances and economic development in the Global South — 382
    Haruna Issahaku, Anthony Chiaraah and George Kwame Honya

35 Mobile money, financial inclusion and livelihoods in the Global South — 393
    Stanley Kojo Dary, Abdulai Adams and Shamsia Abdul-Wahab

36 The role of microfinance in mediating livelihoods — 403
    Karabi C. Bezboruah

37 Global markets and southern livelihoods: exploring trans-scalar connections — 415
    Thaisa Comelli

38 Contextualising urban transport systems and livelihoods in developing countries: the case of Bus Rapid Transit project — 425
    Michael Poku-Boansi, Michael Osei Asibey and Richard Apatewen Azerigyik

PART V
Contextualising livelihoods — 435

39 Livelihoods and social protection — 437
    Leo de Haan

40 Collective organisations: an introduction to their contributions to livelihoods in the Global South — 450
    Molly Atkins

41 Rebuilding livelihoods to reduce disaster vulnerabilities — 460
    Gargi Sen, Vineetha Nalla and Nihal Ranjit

42 Religion and livelihoods studies — 470
    Emma Tomalin

43 Climate change adaptation and agricultural livelihoods of
    smallholder farmers   481
    Issaka Kanton Osumanu

44 Livelihoods and disarmament, demobilisation and reintegration:
    from security to inclusive development   490
    Henry Staples

45 Livelihoods in conflict-affected settings   501
    Ibrahim Bangura

46 Land tenure transformations in the Global South: privatisation,
    marketisation and dispossession in contemporary rural Asia   510
    Lam Minh Chau

Index   *519*

FIGURES

2.1	DFID's Sustainable Livelihoods Framework	13
3.1	A socially embedded conceptualisation of the Capability Approach	22
6.1	Structural conditions for household capabilities	62
9.1	A conceptual view of the bidirectional relationships between key livelihood elements: assets, conditions, activities, strategies and outcomes	94
13.1	Participatory diagramming with disabled people in Burkina Faso	137
13.2	Training in PV with disabled people in Burkina Faso	140
14.1	Annotated GPS map of Godfrey's route (an itinerant vendor who sold oranges in Arusha)	152
15.1	Spaces for livelihoods in Mumbai non-recognised slums: Deonar dumping ground in 2010s	164
15.2	The Palestinian area C community of Khashem Ad Daraj-Hathaleen, Yatta, Hebron	166
23.1	The Community-Based Rehabilitation matrix	251
24.1	Variations in absolute and relative environmental household-level income patterns: (A) the common type: decreasing reliance and increasing absolute income with rising total income, here in the Tanzanian part of the Serengeti-Mara Ecosystem; and (B) the uncommon type: constant reliance and increasing absolute income, here in Cambodia	261
24.2	Variations in income composition (in %) across four survey quarters (Q1–Q4) in the Greater Serengeti-Mara Ecosystem in Tanzania, reflecting the seasonality and gap-filling effect of environmental income	262
24.3	Graphic illustrations of the four household-level possible movements into and out of poverty over time (adapted from Charlery and Walelign, 2015): (A) Transient poor: moving out, (B) Non-poor: staying out, (C) Poor: staying in, and (D) Transient poor: moving in. Notes: the vertical and the horizontal lines represent poverty lines in time 1 and 2, respectively. The desired movement is toward B.	264
29.1	Informal waste sector in urban India	326
31.1	Artisanal gold miners in Kono District, Sierra Leone	346
31.2	Women gold miners in Kono District, Sierra Leone	351
40.1	Conceptual framework of collective action adapted from Ibrahim's (2006) collective capabilities framework	451

TABLES

2.1	Questions to extend livelihoods analysis	17
6.1	Rural livelihood strategies models by Fierros and Avila-Foucat (2017)	63
12.1	Key PRA tools for researching rural livelihoods	126
18.1	Average indicators for health, education and WASH provision around the world	199
21.1	Summary of different definitions of what age ranges constitute 'youth' based on various UN agencies and other organisations	228
26.1	The OECD's New Rural Paradigm	291
26.2	Context and selected content of rural development projects in Chile and Ethiopia	293
31.1	A snapshot of global trends in the ASM sector	347
31.2	Employment estimates for ASM, and key minerals extracted, in selected developing countries	348
34.1	Trends in remittance flows: 2009–2021	383
36.1	Types of microfinance institutions	406
44.1	Key DDR terms	491

BOXES

2.1	Livelihood definitions	11
3.1	Key concepts in the Capability Approach	21
3.2	Power, politics and pastoralist livelihoods in Egypt and Morocco	25
3.3	Women's livelihoods and (im)mobility in Bangladesh	26
4.1	Historical fishing institution on the lakes and ponds of the Rufiji River floodplain, Tanzania	39
5.1	Resilience for whom? Engendering vulnerability in Rwanda's 'green revolution'	49
8.1	'The Danger of a Single Story'	82
8.2	Epistemologies underpinning our research (a very simplified portrayal of a complex continuum)	83
8.3	Are you 'Set Up to Listen – and Understand – Well'?	84
8.4	'So what kind of researcher are you?'	86
8.5	Feminist insights on representation and reflexivity	88
8.6	Leading on global challenges by healing our own practices	89
11.1	Example of participant observation	118
11.2	Example of document analysis embedded in ethnography	119
11.3	Disclosing a militant engagement	120
13.1	PV with recycling cooperatives in Brazil	136
13.2	PV with persons with disabilities in Burkina Faso	136
13.3	PV with marginalised women in Nepal	139
13.4	PV with community members on natural resource management in Guyana	139
15.1	Negotiating livelihoods in informal contexts: waste pickers in Mumbai non-recognised slums	163
15.2	Without spaces of negotiation: livelihoods in the occupied Palestinian territory (oPt) area C	165
17.1	Government response to drought in Botswana	188
18.1	Complaints and responsiveness in Hyderabad's water sector	200
18.2	Contracting NGOs in the delivery of primary health care (PHC)	202
19.1	Contesting neoliberalism in Bolivia: the Cochabamba water wars	211

20.1	Livelihood strategies in rural India	221
20.2	Livelihood strategies in rural South Africa	221
21.1	Youth livelihoods and future aspirations	231
21.2	Youth protagonism	232
22.1	Social governance of informal livelihoods in Freetown, Sierra Leone	241
25.1	Forest livelihoods understood as interwoven into other aspects of livelihoods	273
25.2	Sacred groves	276
25.3	Livelihood implications of defining forests as wastelands	277
29.1	Structure of the informal economy of waste in India	326
29.2	Dynamics of gender, class and caste in informal waste work in India	327
29.3	The politics of representation of waste pickers in India	328
29.4	The implications of heterogeneity for worker organising in Pune and Delhi	331
32.1	Types of migrants in South-South migration (SSM)	360
32.2	Vulnerability, gender and livelihoods	364
33.1	Meaning of the term 'migrant'	371
33.2	Definition of asylum seeker, irregular migrant and refugee	372
33.3	Origin and meaning of 'Kafala'	374
33.4	Definition of forced or compulsory labour, slavery and trafficking in persons	376
39.1	Controversies of social protection in development	439
39.2	Social protection: from protection to transformation	440
39.3	Graduation out of poverty	441
39.4	Three rationales for social protection	442
39.5	PNSP's graduation out of poverty	443
39.6	Bolsa Familia's impact on schooling: a remarkable difference between girls and boys	446
41.1	Post-disaster policies in South Asia	464
45.1	External exploitation of livelihoods as a catalyst of piracy in Somalian waters	503

CONTRIBUTORS

Shamsia Abdul-Wahab works in Development Studies and is an entrepreneur. She is a Lecturer in the Department of Economics, Simon Diedong Dombo University of Business and Integrated Development Studies (SDD-UBIDS), Wa, Ghana. She holds an MPhil from the University for Development Studies. Her research areas include development economics, development finance and enterprise development.

Abdulai Adams is an Economist and a Lecturer in the Department of Economics, Simon Diedong Dombo University of Business and Integrated Development Studies (SDD-UBIDS), Wa, Ghana. He has previously worked for Plan International Ghana, IWMI, GIZ and STPRI-CSIR in Ghana. He holds a PhD in Economics from the University of Zululand, South Africa. His research areas include microfinance, innovation economics, microeconomic analysis and applied economics.

Cristián Alarcón holds a doctoral degree from the department of Urban and Rural Development at the Swedish University of Agricultural Sciences (SLU) and was a Postdoctoral Researcher at the Department of Development Sociology at Cornell University. Currently, he is a researcher at the Division of Rural Development at SLU.

Lauren Andres is Associate Professor in Urban Planning, at the Bartlett School of Planning (University College London). Her expertise sits within the understanding of the intersectionality between people, space and temporalities in the process of urban making and living. Her research contribution spans from developing alternative models to understanding cities with key account of locality and context, to re-thinking systematically the connection between cities, planning, health and sustainability with a specific focus on the most vulnerable communities.

Aries A. Arugay is Professor of Political Science at the University of the Philippines-Diliman. He is also Editor-in-Chief of *Asian Politics & Policy*. He conducts research on civil society and democratic governance, civil-military relations, foreign and security policy, and electoral politics. He has published in *American Behavioral Scientist, Asian Perspective, Annals of the American Academy of Political and Social Science, Journal of East Asian Studies* and the *Journal of Peacebuilding and Development*, among others as well as book chapters published by Routledge and Palgrave.

Contributors

Michael Osei Asibey holds Master of Philosophy and Ph.D. Planning degrees from the Kwame Nkrumah University of Science and Technology, Ghana. Michael is an urban and environmental planner whose research is focused on the following themes of key theoretical and applied interest: transport studies; urbanisation; environmental policy, planning, management and sustainability; sustainable cities and urban containment; climate change adaptation and resilience; and urban greenery conservation and infrastructure.

Molly Atkins is a postgraduate researcher at the University of Birmingham's International Development Department. Molly is particularly interested in gender, feminist political ecology, equitable governance and collective action in small-scale fisheries. Her PhD research examines how women fish processors and traders in sub-Saharan Africa respond to shocks and stresses, such as the COVID-19 pandemic, investigating the role of collective organisations and social relations in building and sustaining resilience.

Richard Apatewen Azerigyik holds an MPhil degree in Planning from the Kwame Nkrumah University of Science and Technology, Ghana. Richard is currently a PhD candidate working on a project titled 'Land Use Planning as a Tool for Managing Transhumance-Associated Conflicts in Ghana'. The project seeks to navigate the complex relationship between competing land users and sustainable conflict management. His research interests concern urban poverty reduction, urban housing and water resource management.

Hakeem Bakare is an Environmental Human Geographer with interest in Environment-Development and Governance crossroads. He received his bachelor's and master's degrees in Geography from Olabisi Onabanjo University and University of Ibadan, Nigeria, respectively before joining the University of Birmingham to complete his PhD in Geography and Environmental Science. He is currently interested in development planning, resilience and place-based systems approaches to achieving the Sustainable Development Goals, particularly in the Global South.

Ibrahim Bangura is a senior lecturer at the Department of Peace and Conflict Studies, Fourah Bay College, University of Sierra Leone. His research interests are in the areas of youth, gender, social protection, transnational organised crimes and peacebuilding in Africa.

Jyoti Bania is a PhD research scholar at Tata Institute of Social Sciences (TISS), India. He is currently working in the area of men and masculinities for his PhD. His areas of interest include gender and development, caste and gender-based violence, international migration and sustainable development.

Clare Barnes is an Interdisciplinary Lecturer in Sustainable Livelihoods at the School of GeoSciences, University of Edinburgh, UK. She gained her PhD from Utrecht University, The Netherlands. She researches natural resource governance, especially forests in the Global South. Her research interests include critical reflective learning in civil society organisations, and the roles and influences of civil society in shaping forest governance.

Lenyeletse Vincent Basupi is a lecturer at the Department of Earth and Environmental Science; Botswana International University of Science and Technology. He holds an MSc in Environmental Science and a PhD in Environmental Sustainability from University of Leeds. His main research focuses on land use systems, land degradation, human-wildlife conflicts, climate change, community engagement, environmental policy and sustainable land management in drylands. This also include the application of Geographic Information Systems and Remote Sensing in social science research.

Karabi Bezboruah's interests are in organisational behaviour, policy analysis, cross-sector collaborations, microfinance institutions and international NGOs. She focuses on policy effects, community sustainability through policy analysis, advocacy, capacity building and networks. She has published articles on organisational behaviour, state-NGO relations and impact of microfinance on women in *Voluntas, Development in Practice, Journal of Community Practice, Health Informatics, Journal of Health Organization and Management* and the *Journal of Social Service Research*, among others.

Lara Bezzina is an independent scholar working with various NGOs in Malta and internationally. She is also a visiting lecturer at the University of Malta. Lara has facilitated participatory video and diagramming in western contexts as well as in Burkina Faso, and is currently involved in a participatory video project in West Africa.

John R. Bryson is Professor of Enterprise and Economic Geography, Birmingham Business School, University of Birmingham, UK. Professor Bryson is economic geographer whose research focusses on understanding people and organisations in place and space, and in developing an integrated approach to reading city-regions. One of his cross-cutting research interests includes understanding rapid adaptation during times of disruptive and radical change, including citizen-led end-user innovation and the development of alternative solutions in response to private and public sector failure.

Thaís de Carvalho is a PhD student in International Development at the University of East Anglia. Her work focuses on Shipibo-Konibo children's experiences amid development projects in Peruvian Amazonia. Her ethnography combines participant observation, document analysis and arts-based interviews with children.

Meera Chandran is faculty at Centre of Excellence in Teacher Education at Tata Institute of Social Sciences, Mumbai. Her research interests lie in sociology and political economy of education, schooling for social equity, ICT in education, teachers and teaching. She has worked extensively in the development and education sectors and teaches master's courses in Teacher Education, Education Development and Policy, and Research Methodology. She was research lead of Connected Learning Initiative (CLIx) and is co-PI for the government of India SPARC-CLIx project for conducting research, design workshops and developing education case studies.

Lam Minh Chau is Associate Professor of Anthropology and Chair of Social and Economic Anthropology at the College of Social Sciences and Humanities, Vietnam National University, Hanoi. He holds an MPhil (2010) and PhD (2015) in Social Anthropology from the University of Cambridge.

Anthony Chiaraah is a Lecturer at the Department of Economics, SDD University for Business and Integrated Development Studies, Ghana. He has published nine articles in refereed journals. He holds a master's degree in International Economics from the International University of Japan and is currently a PhD candidate in Economics at the Kwame Nkrumah University of Science and Technology, Ghana. He has served on various boards and committees within and outside the University. He has also undertaken various extension and advisory services.

Contributors

Nathan Clay is a Geographer, studying how global processes of social and environmental change are experienced in different places. With case studies on the interface of development and sustainability in agriculture and forest contexts, he aims to understand how rural transformation is unevenly experienced. Nathan is currently a postdoctoral fellow at the Swedish University of Agricultural Sciences. He received his PhD in 2017 from Pennsylvania State University and from 2017 to 2020 was a researcher at the University of Oxford.

David Cobley has taught and undertaken research at the International Development Department, University of Birmingham, in disability and development over the last ten years. He has conducted research and consultancy in Kenya, Uganda, Sierra Leone, India and the Philippines. He is the author of *'Disability and International Development: A Guide for Students and Practitioners'*, published by Routledge in 2018.

Thaisa Comelli is an Associate Researcher at King's College – with the Tomorrow's Cities interdisciplinary global research hub. She researches and works with community and urban development planning and has hands-on experience in Brazil, Mexico, Senegal, Spain and the UK. Her most recent research examined spaces of participatory planning and citizenship in the context of slum upgrading in Rio de Janeiro. She previously worked as lecturer and researcher at the Bartlett Development Planning Unit (University College London).

Stanley Kojo Dary is an Agricultural and Applied Economist and a Senior Lecturer in the Department of Economics, Simon Diedong Dombo University of Business and Integrated Development Studies (SDD-UBIDS), Wa, Ghana. He is a Fulbright Scholar and holds a PhD from the University of Missouri-Columbia, USA. His research areas are agricultural finance (trade credit), agro-food industry innovations, agricultural cooperatives, microfinance, non-farm employment and rural development.

Leo de Haan is Emeritus-Professor of Development Studies at the International Institute of Social Studies in The Hague. He is the former Rector of ISS and former director of the African Studies Centre Leiden and Professor of Development in Sub-Sahara Africa at Leiden University. Leo de Haan has published widely on poor people's livelihoods. His current research focuses on livelihoods, social exclusion and social protection in the context of the developmental state and on migration and the multi-locality of livelihoods.

Joan DeJaeghere is a Professor at the Department of Organizational Leadership, Policy, and Development, University of Minnesota, USA. Dr. DeJaeghere's scholarship is concerned with inequalities in education and how they affect youth's future civic engagement, livelihoods and wellbeing. She has published two books, *Educating Entrepreneurial Citizens* and *Education and Youth Agency* and numerous articles addressing gender inequalities, youth livelihoods, and longitudinal research. She has led several longitudinal research projects on youth livelihoods (East Africa), girls' education (India), and women's empowerment (Vietnam).

Stuart Paul Denoon-Stevens is a lecturer at the University of the Free State, South Africa, and a professional planner. His work spans three fields. Firstly, a key element of his research focuses on land management, covering issues such as inclusionary retail and proactive upzoning. Secondly, he has written on issues such as spatial planning, housing and public health in mining towns. Thirdly, he has started to write on planning education and practice, and the linkages between them.

Contributors

Klara Fischer is associate professor in rural development at the Swedish University of Agricultural Sciences. Her research concerns smallholder livelihoods and agriculture in the Global South, particularly in Uganda and South Africa. She has a particular interest in the interaction between local and formal knowledge systems in smallholder agriculture.

Miriam Gay-Antaki is an Assistant Professor at the University of New México in the Department of Geography and Environmental Studies. Using feminist geography and political ecology she analyses the networked connections between climate science, policy and action focusing on México.

Flora Hajdu holds a doctoral degree from the Department of Water and Environmental Studies at Linköping University, Sweden. Currently, she is Associate Professor in Rural Development at SLU.

George Kwame Honya is a Lecturer at the Department of Economics, SDD University for Business and Integrated Development Studies, Ghana. He has over two decades of teaching and research experience. He holds a Master of Philosophy in Agricultural Economics from the University of Ghana, Legon. His research interest areas are economic development, agricultural economics and rural entrepreneurship.

Haruna Issahaku is Senior Lecturer and Head, Department of Economics and Entrepreneurship Development, University for Development Studies, Ghana. He holds a PhD in Finance from the University of Ghana. Haruna is an international Visiting Scholar at Covenant University, Nigeria. His current research focuses on digital finance, financial inclusion and economic development. Over his 17 years academic career, he has to his credit 38 published academic papers. Haruna is a recipient of the 2020 Emerald Literati Awards, Highly Commended Paper category.

Xi Jiao, researcher in Development Economics, is interested in the fields of rural livelihoods and wellbeing, environmental reliance, and vulnerability and adaptive capacity to land use and climate change in developing countries, focusing on Southeast Asia and Eastern Africa. She works extensively on empirical studies, including instrument design and supervising field data collection. In addition to her academic activities, she has worked as a consultant with the Asian Development Bank in Cambodia, China, Laos, Philippines, Thailand and Vietnam.

Braima Koroma is the Director of Research and Training at SLURC and also a lecturer at Njala University, Sierra Leone. His work focuses on urban livelihoods and the city economy, resilience, vulnerability and adaptation of climate change, urban planning and development, environmental management and development impact evaluation. He has consulted with various national government ministries, department and agencies, international organisations and institutions including the World Bank, African Development Bank, UN, amongst others. Braima has published extensively, reaching both academic and non-academic audiences.

Sukanya Krishnamurthy is a Chancellors Fellow/Senior Lecturer at the School of Geosciences, University of Edinburgh. Her focus lies at the interface between urban and social geography, where her scholarship analyses how cities can use their resources and values for more sustainable development. Key research areas include: children's lived experiences in diverse urban contexts, and participatory methods for engaging children and youth in design and planning.

Contributors

Johanna Bergman Lodin holds a doctoral degree in Social and Economic Geography from Lund University, Sweden, with specialisations in development geography and feminist geography. Currently, she is a researcher at the Department of Urban and Rural Development at SLU.

Ana De Luca received her PhD in Social and Political Sciences from the National Autonomous University of Mexico. Her research expertise lies in the fields of feminist and environmental politics and discourses. She is currently editor of the environment section at Nexos magazine.

Aman Luthra is Assistant Professor in the Department of Geography at the George Washington University. His research focuses on informal workers, urban political economy and political ecology of waste management in urban India.

Roy Maconachie is Professor of Natural Resources and Development at the University of Bath. His research in sub-Saharan Africa explores the social, political and economic aspects of natural resource management, and their relationships to wider societal change. He has been carrying out field-based research in sub-Saharan Africa for over 20 years, and has spent much of this time working with artisanal gold and diamond miners.

Vikas Maniar is faculty at the School of Education, Azim Premji University, Bangalore. His research focusses on political economy of education in postcolonial contexts, education policy and administration, and ICT in school education. Dr. Maniar splits his time between teaching, research and engagement with grassroots organisations in the education sector. He is currently involved in a multi-country research on innovation diffusion and scaling of ICT-enabled capacity building of STEM teachers in the Global South.

Lorena Melgaço is a Brazilian postdoctoral researcher at the Institute for Urban Research at Malmö University, Sweden. She has an interdisciplinary and international career with collaborations in Brazil, South Africa, UK, Germany and Sweden, always working on the interface of the built environment, technological development and urban transformation, with a focus on social inclusion and spatial justice. Her areas of interest include, among others, 'bottom-bottom spatial practices' – forms of local agency within a global political structure of knowledge and technological production – and their relation to the production of space; as well as interrelations between planning education and practice.

Enrico Michelutti trained as an architect and planner (IUAV, Italy) and in sustainable development (UPC, Spain), and has carried out research projects in Brazil, Chile, India, Palestine and Tanzania, collaborating with the Economic Commission for Latin America and Caribbean (UN, ECLAC) and several international and local NGOs. Between 2014 and 2018, he was co-coordinator of the Network-Association of European Researchers on Urbanisation in the South (N-AERUS).

Vineetha Nalla is an Associate in the Practice Team at Indian Institute for Human Settlements (IIHS), Bangalore. Her research and practice spans across themes of disaster risk, recovery, climate justice and affordable housing. More recently, she has worked on the Recovery with Dignity (RwD) project wherein her research focused on representations of disasters and recovery needs in litigation. She is also PI on a study to understand low-income rental housing in India. At IIHS, Vineetha teaches courses on urban risk and resilience and affordable housing.

Deo Namwira is a Sessional Instructor teaching Environmental Sustainability and International Development, holding an MSc (Guelph 2002) and PhD in International Development (Birmingham, 2020). His MSc thesis explored poverty reduction policies and projects in the Global South, while his PhD work investigated fisheries livelihood patterns in conflict situations. Deo worked as a Research and Teaching Associate at the University of Birmingham for four years, and previously for ten years as the International Grants Manager for Mennonite Central Committee Canada.

Jae-Eun Noh works at the University of Western Australia as Postdoctoral Research Fellow. Her research and practice include human rights and development, social justice, global citizenship and emotions in development practice. Her research interests are influenced by five years of experience in international NGOs and supported by studies in a range of intersecting fields.

Fiona Nunan is a Professor of Environment and Development in the International Development Department, University of Birmingham, with over 25 years' experience in environment and development. Her research interests focus on the governance of renewable natural resources, particularly inland fisheries and coastal ecosystems, and she is the author of *Understanding Poverty and the Environment: Analytical frameworks and approaches* and editor of *Making Climate Compatible Development Happen* and *Governing Renewable Natural Resources: theories and frameworks*.

Issaka Kanton Osumanu is an associate professor of Geography and Environmental Studies in the Department of Geography, SD Dombo University of Business and Integrated Development Studies, Ghana. He combines a rich background in geography and resource development and development studies with extensive research experience in northern Ghana. His research interests straddle the following: Sustainable livelihoods; Environmental policy and planning; and Livelihood impacts of climate change and land degradation.

Itzel San Roman Pineda is a PhD researcher in the Department of Geography at the University of Sheffield. Her current doctoral research focuses on the implementation of solidarity practices in indigenous communities with touristic-driven economies as an alternative to development. Itzel has over a decade of experience conducting research in the most touristic region of Latin America, the Yucatan Peninsula. She is sponsored by the Mexican Government through a CONACYT scholarship.

Michael Poku-Boansi is a Professor in Planning at the Department of Planning, Kwame Nkrumah University of Science and Technology (KNUST) Kumasi, Ghana. He holds a PhD degree in Planning from the KNUST and has over ten years' experience in teaching and conducting research in the areas of transport planning, land use and urban studies, urbanisation, informality, climate change adaptation and urban resilience. He is a Fellow of the Ghana Institute of Planners (FGIP).

Nihal Ranjit is part of the Practice team at the Indian Institute for Human Settlements (IIHS), Bangalore. He is a key member of the Recovery with Dignity (RwD) project, which aims to understand experiences of recovery in post-disaster situations across three states in India. Currently, Nihal's research focuses on understanding the extent to which developmental plans for semi-arid regions in India are compatible with existing knowledge on climate change. At IIHS, he teaches courses on urban risk and resilience and affordable housing.

Contributors

Andrea Rigon is an Associate Professor at the Bartlett Development Planning Unit of University College London and a founder of the Sierra Leone Urban Research Centre. His work focuses on how power relations affect the participation of different people and groups in decision-making processes. He has worked to include an intersectional perspective into participatory design and to incorporate participatory approaches in the Sustainable Development Goals. His latest project was about co-designing built interventions with children affected by displacement.

Stefan Rzedzian is a University Teacher in Human Geography at the University of Edinburgh. His research cuts across three main themes: Social movements, cultural politics of the environment, and counter-hegemony. Most recently, his research has focused on the Rights of Nature movement in Ecuador.

Nathan Salvidge is a PhD candidate at the University of Reading, UK, funded by the ESRC. Broadly, his research interests are related to issues of gender, youth, informality, generation and social media/mobile phone usage, in Global South contexts.

Margherita Scazza is a PhD candidate in human geography at the University of Edinburgh. Her research combines political ecology and decoloniality to investigate Indigenous social movements and autonomy, territoriality and post-extractivism in the Ecuadorian Amazon.

Jeremy Seekings is Professor of Political Studies and Sociology at the University of Cape Town. His most recent books include *Inclusive Dualism: Labour-Intensive Development, Decent Work, and Surplus Labour in Southern Africa* (2019, with Nicoli Nattrass) and a co-edited volume on *The Politics of Social Protection in Eastern and Southern Africa* (2019).

Gargi Sen is a Senior Associate in the Practice team at the Indian Institute for Human Settlements (IIHS), Bangalore. At IIHS her work focuses on Post-Disaster Recovery, Disaster Governance and Sustainable Development Goals (SDGs). She is a key member of the team, conducting research in the disaster-affected areas, evaluating programs on disaster risk reduction, institutional aspects of risk governance and has conducted in-depth study of good practices in disaster resilience leadership across the country. Gargi also teaches courses on Urban Risk and Resilience in IIHS.

Carsten Smith-Hall is Professor at the University of Copenhagen, Department of Food and Resource Economics. His research focuses on (i) environment – livelihood relationships, including the role of environmental products in preventing and reducing poverty, (ii) forests and human health, in particular the role of forests in maintaining and improving welfare in low-income countries, and (iii) commercial utilisation of biodiversity, with emphasis on trade and conservation issues. He has extensive fieldwork experiences from Asia and Africa.

Sam Staddon is a Lecturer in Environment and Development at the University of Edinburgh in the UK. Having trained in ecology and worked in conservation around the world for ten years, Sam returned to academia to complete a PhD in the critical social sciences, exploring community-based conservation in the forests of Nepal. Her academic interests focus on the politics of conservation, critical reflective learning, and on the ethics of research.

Henry Staples is a PhD candidate in Human Geography at the University of Sheffield. His work explores the geographies of peacebuilding, focusing on the experiences and political practices of former armed actors. He holds a BA in Land Economy (University of Cambridge), and a master's in International Development, (University of Amsterdam).

Lucy Szaboova is a Research Fellow at the University of Exeter. Her research interests centre around issues of resilience, sustainability and wellbeing in the context of global change processes. Her recent research explores migration and urban governance and livelihoods.

Emma Tomalin is Professor of Religion and Public Life at the University of Leeds. She has published widely on the topic of religion and development, including the following books – *Religions and Development* (2013) and *The Handbook of Religions and Global Development* (2015). She co-edits the Routledge Research in Religion and Development book series, which now has 19 volumes. She is the co-chair of the Joint Learning Initiative on Faith and Local Community (JLI) learning hub on Anti-Human Trafficking and Modern Slavery.

Elodie Santos Vera is an independent researcher who works with marginalised communities in Ecuador to work towards communal social and political empowerment. She has worked together with various indigenous and peasant groups on projects related to extractivism, women's rights, and water rights.

Solomon Zena Walelign is Postdoctoral Fellow at the Norwegian University of Science and Technology, Visiting Scholar at University of California Berkeley, Research Fellow at the World Bank, and Adjunct Assistant Professor at University of Gondar. He completed a double PhD in Environmental and Resource Economics at University of Copenhagen and in Forest Sciences at Georg-August University of Göttingen. His main research foci are livelihoods, poverty, large-scale land transaction, and climate change resettlement. He has conducted fieldwork in Nepal, Ethiopia, Kenya, and Tanzania.

Julian Walker is an Associate Professor at Bartlett Development Planning Unit of University College London. His work focuses on social policy, diversity and citizenship, with an emphasis on gender policy and planning and disability. Key areas of his research have included urban displacement by development, including involuntary resettlement, and forced eviction, and pro-poor livelihoods and employment rights.

Luke Whaley is a Global Challenges Research Fellow in the Department of Geography at the University of Sheffield. His primary research area is environmental and natural resource governance. Luke leads a two-year research project in Eastern Uganda, exploring the link between people's worldviews and the institutions that shape access to land and water.

Mariama Zaami is a lecturer at the Department of Geography and Resource Development, University of Ghana. Her research interests include gender and poverty, migration, immigration and integration, ethnicity and livelihood change. Her research highlights the role of ethnicity and associated challenges of Black African immigrant youth in Calgary in navigating Canadian cities. In Ghana, her research focuses on gendered migration patterns from rural to urban locations, and the implications of these movements for household livelihoods.

Introduction

1
LIVELIHOODS IN THE GLOBAL SOUTH

Fiona Nunan, Clare Barnes and Sukanya Krishnamurthy

Introduction

The *Routledge Handbook on Livelihoods in the Global South* presents a unique and timely comprehensive overview of the diverse dimensions of, and perspectives on, livelihoods in low- and middle-income countries. The Handbook brings together insights and critical analysis from diverse approaches and experiences, learning from research and practice over at least the last 20 years and engaging with current and emerging themes. In doing this, the Handbook provides a comprehensive and cutting-edge overview of theories, concepts and experiences of livelihoods in diverse settings.

The aim of this Handbook is to be an essential resource for students, researchers and practitioners of international development and related fields. Students on international development programmes and related courses at undergraduate and postgraduate level will find the book invaluable due to its comprehensive approach, providing an accessible overview to key topics, issues and methods. Researchers and practitioners will benefit from the book through the diverse disciplinary contributions and the wide and contemporary coverage.

Each chapter provides a critically informed introduction and overview of key concepts, issues and debates of relevance to the topic, thereby providing a way into a topic, with pointers to key ideas and academic or grey literature for further reading.

Why livelihoods?

Livelihoods are inevitably varied and diverse yet share commonalities of being complex and having multiple dimensions. Livelihoods are far more than an economic consideration – a source of employment and/or income generation. They encompass 'the capabilities, assets (including both material and social resources) and activities required for a means of living' (Scoones, 1998: 5) and, as this volume shows, may foreground considerations of culture, rights and politics across scales. Livelihoods thinking and framing has influenced analysis of how people get by for decades, but has been particularly influential and important in development studies and practice for their holistic, multi-dimensional perspectives since the 1990s (Scoones, 2015).

The development and form of livelihoods thinking in the 1990s is described in Chapter 2, where the adoption of the approach, particularly in the form of the Sustainable Livelihoods

Framework, by the UK Government's Department for International Development (DFID), is noted for its influence on research and practice. DFID and many international NGOs appointed livelihoods advisors and rolled out projects that sought livelihood improvements, as well as integrating a livelihood focus into many other projects and programmes. The livelihoods perspective brought a holistic, multidimensional approach to poverty alleviation, recognising the complexity of people's lives and the range of constraining factors that influence and limit opportunities and livelihood strategies.

Despite the significant adoption and influence of livelihoods thinking and approaches in the 1990s and into the 2000s, prominent application has waned over the years. Instead of taking central stage and providing an overarching framework, livelihoods thinking is more often incorporated into other framings, such as resilience and climate change. However, there is support for the contention that livelihoods perspectives remain important, with Carr (2014: 110) arguing that 'livelihoods approaches remain the only broad framework of analysis that allows for the holistic investigations necessary to address issues of vulnerability and resilience at the heart of contemporary development discourse and practice'.

The diversity of contexts provided by the chapters of this volume illustrate the ongoing relevance of livelihoods framing for understanding how people live their lives, particularly in terms of making a living and sustaining their lives, but also in terms of giving meaning and purpose to their lives and negotiating how they want to live. It is our hope that this book will contribute to a reinvigoration of livelihoods perspectives, insights and approaches, in acknowledgement of how these can contribute to achieving the Sustainable Development Goals or centring 'alternative' livelihoods in the context of a changing climate and, at the time of writing, in the context of coping with, and recovering from, the global coronavirus pandemic.

Why the Global South?

Before introducing the content of the volume, rationale for the use of the phrasing 'Global South' in the title of the book is needed. As an alternative label to 'Third World', which had pejorative connotations, and 'developing countries', which was felt to be inaccurate and misleading, the UNDP helped to popularise the term Global South through their 2004 report *Forging the Global South* (Absell, 2015). 'South' referred to 'developing countries' (with the exception of Australia and New Zealand) being located to the geographical south of industrialised countries (ibid). With the continuation of a focus on geography and poverty, many have questioned whether the term does indeed mark a departure from the simplification and misportrayal of lives common under the Third World and 'developing countries' labels (Schneider, 2017). From a sociological understanding, Dados and Connell (2012) discuss the term as also providing an alternative to the belief of homogenisation across cultures contained in discussions of globalisation.

The term 'Global South' can be approached in different, often contested ways, as is evident through the chapters of this volume. Schneider (2017) presents three different readings of the Global South, which we extend here to discuss approaches to livelihoods included in this volume:

1   A geographical perspective in which the South refers to countries in Latin America, Africa and Asia which are seen as 'structurally underdeveloped and poverty-struck' (Schneider, 2017: 21). In livelihoods research this could lead to a focus on socio-economic processes, the role of trade and markets, and of geopolitics.
2   A focus on subalterns in all countries, defined as 'human beings disadvantaged by neoliberal policies who are socially, politically, and intellectually disempowered' (Schneider, 2017: 21).

Many chapters in this volume discuss the influence of neoliberal policies for livelihoods in different contexts and inequalities within countries, whilst others discuss processes of subaltern migration and solidarity between groups located in different countries.

3   As a 'flexible metaphor' which challenges ideas of there being an objective distinction between 'developed' and 'developing' countries and encompasses dynamic inequalities and political, cultural and epistemic marginalisation, as well as socio-economic aspects. Livelihoods research in this vein pays attention to the many ways political, cultural and epistemic freedoms are challenged and resisted, and includes how powerful discourses of development (emerging, perpetuating and being translated or transformed in both the Global North and South) can support or undermine livelihoods.

Many scholars point to the diversity of the Global South, with Escobar (2017: 40) arguing that it is 'powerfully plural, inhabited by radical difference of all kinds – cultural, social, ethnic and, in the last instance, ontological. To the same extent as the North, if not more so, the South is made up of multiple worlds, a pluriverse'. This challenges a purely geographical perspective (Fry, 2017) or economic classifications popular with international organisations such as the World Bank (Potter et al., 2018). This volume serves to foreground the diversity of livelihoods across and within countries.

Others point to unifying experiences of colonisation and international policies (Moore, 2018) such as structural adjustment programmes, or poverty reduction strategies. That the concept of the Global South can serve as a political tool to unite spaces of resistance and common struggles is what Schneider (2017: 28) calls its utopian appeal. Though how these processes take form and affect livelihoods reveals diversity across the Global South and can be understood in different ways. The perspectives, concepts and frameworks introduced in the Introduction and Part I of this volume highlight different aspects of these processes.

This volume considers the livelihoods of people living in systems and locations often framed as the Global South, though many chapters challenge such a framing and Eurocentric epistemologies, through drawing attention to movements, subaltern livelihoods, contestations over livelihoods within countries or systems, and the role and influence of researchers and practitioners. The volume also has utility for understanding livelihood injustices, process of marginalisation or empowerment, and the heterogeneity of livelihood experiences in the geographical Global North. There is a recognition throughout the book that livelihoods in the Global South are often shaped by processes and actors interlinking the Global South and Global North.

There are different views on capitalising the G of g/Global South depending on the position authors take towards use of the term and how they employ the term in their work. This volume allows for such diversity across chapters. As Potter et al. (2018: 46) argue, 'no matter what abstract conceptualisations are used to structure development debates – three worlds, two worlds, the South, nation states, cities or whatever – we must not forget that we are discussing human beings and their livelihoods'. Therefore, whether the Global South is approached as an academic concept, a site of interventions, a unifying rallying cry, or a powerful homogenising discourse with material consequences, the chapters in this volume centre people's lived realities and social, environmental, political, cultural and economic processes affecting them.

Contributions in this volume

The following chapters are contributions from across the world by scholars working on livelihoods in its many forms and contexts. All the authors share an interest in understanding how livelihoods are accessed, maintained and shaped. The book is organised into six parts to provide a broad understanding of livelihoods in conceptual and empirical detail. Positioning the volume is an **Introduction** section that provides an in-depth review of key concepts employed in the book. Fiona Nunan introduces the concept of livelihoods, as well as the related concepts of poverty and wellbeing. Using the Sustainable Livelihoods Framework (SLF) developed by the UK Department for International Development, Chapter 2 reviews the components of the framework and introduces broader structural constraints that impact livelihoods. Following this, Lucy Szaboova introduces readers to the Capability Approach in Chapter 3, identifying its roots and components, noting its focus on opportunities as opposed to the focus of the SLF on outcomes, and on how freedoms and agency generate 'functionings' and hence wellbeing. Chapter 4, by Luke Whaley, recognises the central place of institutions in mediating livelihoods, enabling and constraining the potential to use or benefit from livelihood assets. He draws on Mainstream and Critical Institutionalism to introduce the reader to different approaches to analysing forms and dynamics of institutions. Also key to analysing livelihoods, according to the SLF, is analysis of the 'vulnerability context'. In Chapter 5, Nathan Clay introduces and critically reviews the related concepts of vulnerability and resilience. The key livelihood asset of 'social capital' is the focus of Chapter 6, where Itzel San Roman Pineda introduces readers both to social capital and the related area of literature on social networks. The Introduction section is concluded with an introduction to Rights-based Approaches in Chapter 7 by Jae-Eun Noh, who reflects on what Rights-based Approaches can bring to sustainable livelihoods thinking and analysis.

Part I, **Researching livelihoods: approaches and methods**, identifies different approaches to researching and understanding livelihoods, drawing on different philosophies, methods and perspectives. Sam Staddon in Chapter 8 begins the part by reflecting on the positionality of researchers and practitioners in the field of international development, exploring the 'challenge of understanding' the experience of poverty by practitioners and researchers who have not shared the experience of poverty in the Global South. Chapter 9, by Solomon Zena Walelign, Xi Jiao and Carsten Smith-Hall, reviews five quantitative approaches for measuring livelihood strategies, followed by a more qualitative focus in Chapter 10 by Joan DeJaeghere, in which it is argued that qualitative longitudinal research is particularly critical to generating rich and in-depth data on livelihoods. Continuing the qualitative focus, Thaís de Carvalho introduces the potential for ethnography in researching livelihoods in Chapter 11, particularly through participant observation. The application of sustainable livelihoods in practice and research in the 1990s/2000s is particularly associated with Participatory Rural Appraisal, an approach introduced by Klara Fischer in Chapter 12. The chapter demonstrates how using Participatory Rural Appraisal tools can encourage communities to take charge of a research process and generate a more contextualised understanding of communities. The more specific method of participatory video is introduced in Chapter 13, where Lara Bezzina explains how the approach enables the research 'participants' to take over the research process. Part I concludes with Chapter 14 by Nathan Salvidge on the use of participatory Global Positioning System (GPS) methods in livelihoods research, which is particularly useful in capturing people's spatial knowledge and mobilities in relation to livelihoods.

Part II, **Negotiating livelihoods**, brings together different mechanisms for, and approaches to, enabling different aspects of livelihoods. The chapters in this part illustrate how livelihoods are shaped by multiple structural constraints and provide examples of how those constraints have sought to be addressed. Part II begins with an introduction to the core concept of power by Enrico Michelutti. Chapter 15 introduces a relational approach to understanding power and examines how power is manifested in vertical and horizontal negotiations affecting livelihoods in the Global South. Attention to the concept of power continues in Chapter 16, where Feminist Political Ecology is introduced by Ana De Luca and Miriam Gay-Antaki, providing a framework for analysing gender as a form of power relation, with implications for securing and maintaining livelihoods. Chapter 17 turns to power within national politics, as Jeremy Seekings examines connections between democracies and livelihoods in Africa. The theme of power and politics continues in Chapter 18 on social accountability, where Aries A. Arugay identifies mechanisms of sanctions and rewards in protecting and promoting livelihoods in Asia. Negotiating livelihoods inevitably concerns rights and so, in Chapter 19, Stefan Rzedzian, Margherita Scazza and Elodie Santos Vera review how social movements have facilitated the pursuit of rights and livelihoods, with examples from Latin America. The role of education in negotiating livelihoods is introduced by Vikas Maniar and Meera Chandran in Chapter 20, focusing on how Human Capital Theory underpins understanding of the education-livelihoods relationship and foregrounding alternatives to this framing, such as human rights and capabilities. In Chapter 21 on youth livelihoods, Sukanya Krishnamurthy highlights the need to better understand how youth and children access livelihoods as determined by age, gender and social responsibilities. Julian Walker, Andrea Rigon and Braima Koroma consider understandings of the informal economy in Chapter 22 and of the potential for state and social regulation to protect those deriving their livelihoods from the informal economy. David Cobley concludes Part II in Chapter 23 by reviewing literature and experience related to disability, poverty and livelihoods.

Part III, **Generating livelihoods**, focuses on strategies and mechanisms through which livelihoods are generated, focusing particularly on income and subsistence, but also identity and culture. The part begins with an introduction to the scale and economic importance of environmental income in rural livelihoods in Chapter 24 by Carsten Smith-Hall, Xi Jiao and Solomon Zena Walelign. Chapter 25 by Clare Barnes continues this focus, with specific attention on forest livelihoods, examining the nature and extent of forest use, identifying cultural and spiritual values of forests as well as ecological and economic. Keeping to a rural focus, in Chapter 26 Cristián Alarcón, Johanna Bergman Lodin and Flora Hajdu examine agricultural livelihoods, reviewing how rural development projects and agrarian change have shaped such livelihoods, informed by a local political ecology of land and water use. Lenyeletse Vincent Basupi moves the focus to pastoralism in Chapter 27, identifying factors that impact dryland ecologies, pastoralist institutions and dynamic pastoral livelihoods. Small-scale fisheries are the focus of Chapter 28, where Deo Namwira and Fiona Nunan introduce characteristics of small-scale fisheries, how actors within fisheries have been identified through a value chain and gender perspective, and the nature of poverty within fishing communities. The more urban setting of the waste economy provides the context for Chapter 29, where Aman Luthra proposes a typology for classifying and developing a better understanding of the diverse set of actors in the informal waste sector, with a focus on India. Staying with an urban focus in Chapter 30, Lauren Andres, Stuart Paul Denoon-Stevens, John R. Bryson, Hakeem Bakare and Lorena Melgaço explore the role, success and failures of spatial planning in shaping African cities and its influence on livelihoods. Roy Maconachie concludes Part III by providing an overview on artisanal and small-scale mining livelihoods in Chapter 31, examining the potential for formalisation of the sector to safeguard livelihoods and deliver greater benefits to host communities where extraction takes place.

Part IV, **Enabling livelihoods**, positions mechanisms and factors that support livelihoods from the key role of migration to remittance to rights. The part begins with two chapters on migration. In Chapter 32, Mariama Zaami introduces different forms and theories of migration, focusing on South-South migration, noting its important role in millions of livelihoods. Women's migration from India to the Gulf, under the Kafala system, in Chapter 33 by Jyoti Bania illustrates the feminisation of migration. Haruna Issahaku, Anthony Chiaraah and George Kwame Honya, in Chapter 34, investigate the role of international remittances in enabling livelihoods in the originating country, supporting entrepreneurship, education, poverty reduction, agricultural development and resilience. Staying with the subject of money, Stanley Kojo Dary, Abdulai Adams and Shamsia Abdul-Wahab in Chapter 35 explore the movement of money within countries, specifically how mobile money works. Mobile money is enabling cheaper, faster, reliable and convenient access to financial services, especially for those with limited access to formal banking services. Chapter 36 also examines how access to financial services can enable livelihoods, with Karabi Bezboruah introducing the theory and practice of microfinance, its evolution and development, reviewing its role and limitations in enabling economic activity. Global markets are the focus of Chapter 37 by Thaisa Comelli, examining the significance of global markets for livelihoods in the Global South, how livelihoods and global markets interact, and which analytical frameworks help to unpack such complex interactions. In Chapter 38, the final chapter of Part IV, Michael Poku-Boansi, Michael Osei Asibey and Richard Apatewen Azerigyik examine how bus rapid transit systems in cities of the Global South enable livelihoods through better access to markets, education and other services, and how they have affected pre-existing informal transport services.

Part V, **Contextualising livelihoods**, concludes the volume and focuses on key issues that impact the potential for, and experience of, livelihoods. The part begins with Chapter 39 by Leo de Haan, introducing the characteristics and objectives of social protection and how social protection schemes impact livelihoods. The use of social protection as a development strategy and its role in livelihood protection to livelihood transformation is discussed through theory and empirical cases. Livelihoods are often supported by and experienced within collective structures, which are introduced and analysed in Chapter 40 by Molly Atkins. The chapter focuses on producer organisations, savings and credit groups and natural resource user committees, identifying how these structures enable and constrain livelihoods. Chapter 41 presents the context of disasters, with Gargi Sen, Vineetha Nalla and Nihal Ranjit considering disaster risk and vulnerability in relation to their impact on livelihoods and economic sectors. The context of religion for livelihoods forms the focus of Chapter 42, where Emma Tomalin examines the marginalisation of religion within broader literature and implications of this for achieving improved livelihoods strategies and outcomes. The chapter examines the role that religion plays in shaping livelihoods and the impact of religion on social inequality and social exclusion. The context of climate change forms the focus of Chapter 43, where Issaka Kanton Osumanu reviews how agricultural livelihoods of smallholder farmers are affected by climate changes and the challenges faced in climate change adaptation. Two chapters follow that review how conflict interacts with and affects livelihoods. In Chapter 44, Henry Staples traces evolutions in the theory and practice of support to the reintegration of former combatants, noting the increasing focus on livelihoods in reintegration programmes. In Chapter 45, Ibrahim Bangura considers conflict itself and how conflict interacts with and affects livelihoods, including livelihood responses to conflict and programmes to support livelihoods as people cope with and recover from conflict. Lam Minh Chau concludes the volume in Chapter 46 by reviewing the contextual processes affecting land tenure in rural Asia, of privatisation, marketisation and dispossession. He observes how these processes have resulted in mixed results, with different outcomes and consequences experienced by these profound changes.

References

Absell, C.D. (2015) 'The lexicon of development: A quantitative history of the language of development studies', *Iberoamerican Journal of Development Studies*, 4(1): 4–35.

Carr, E.R. (2014) 'From description to explanation: Using the Livelihoods as Intimate Government (LIG) approach', *Applied Geography*, 52: 110–122. https://doi.org/10.1016/j.apgeog.2014.04.012

Dados, N. and Connell, R. (2012) 'The Global South', *Contexts*, 11(1): 12–13. https://doi.org/10.1177/1536504212436479

Escobar, A. (2017) 'Response: Design for/by [and from] the "Global South"', *Design Philosophy Papers*, 15(1): 39–49. https://doi.org/10.1080/14487136.2017.1301016

Fry, T. (2017) 'Design for/by "the Global South"', *Design Philosophy Papers*, 15(1): 3–37. https://doi.org/10.1080/14487136.2017.1303242

Moore, C. (2018) 'Internationalism in the Global South: The evolution of a concept', *Journal of Asian and African Studies*, 53(6): 852–865. https://doi.org/10.1177/0021909617744584

Potter, R., Binns, T. and Elliott, J. (2018) *Geographies of development: An introduction to development studies*, Oxon: Routledge

Schneider, N. (2017) 'Between promise and skepticism: The Global South and our role as engaged intellectuals', *The Global South*, 11(2): 18–38. www.muse.jhu.edu/article/696275

Scoones, I. (1998) 'Sustainable rural livelihoods: A framework for analysis', *IDS Working Paper, 72*, Brighton: Institute of Development Studies.

Scoones, I. (2015) *Sustainable livelihoods and rural development, agrarian change & peasant studies*, Rugby: Practical Action Publishing.

2
LIVELIHOODS
Concepts and frameworks

Fiona Nunan

Introduction

This chapter introduces the concept of livelihoods and related concepts of poverty and wellbeing. The Sustainable Livelihoods Framework as developed by the UK Department for International Development (DFID) provides the main focus of the chapter, with each component of the framework (vulnerability context, assets, transforming structures and processes, livelihood strategies and outcomes) being reviewed. The chapter introduces examples of political economy analysis of livelihoods in recognition of the need to consider broader structural constraints on the agency of individuals and households in their livelihood pursuits.

Livelihoods: the concept

A livelihoods perspective within international development thinking and practice is particularly associated with the Sustainable Livelihoods Approach (SLA) and Sustainable Livelihoods Framework (SLF) developed by the UK Department for International Development (DFID) and informed by the work of Ian Scoones (1998, 2009, 2015) and others at the Institute of Development Studies (IDS) in the UK. The background and chronology of the development of the livelihoods approach and framework associated with the policy and work of DFID can be found in Solesbury (2003), who traces the origins of the SLA and SLF to *Our Common Future* of the World Commission on Environment and Development (1987), in particular to the definition of sustainable development. In terms of more theoretical underpinnings, Scoones (2015) acknowledges several areas of critical foundational thinking, including Marxist political geography, political ecology and farming systems research as influencing the conceptualisation of sustainable livelihoods. These encouraged recognition of power dynamics and the need for an integrated approach encapsulating social and ecological dimensions, responding to environmental change. De Haan (2012) also provides a useful chronology of the livelihoods thinking and framework of the 1990s, tracing foundations to *genre de vie*, or 'mode of living', in French geography, and to anthropology concern with how people make a living.

A seminal paper published in 1992 by Robert Chambers and Gordon Conway developed and defined the concept of sustainable livelihoods. Box 2.1 sets out their definition. To reach this definition, Chambers and Conway drew on existing thinking within development theory and practice and the concepts of capabilities, equity and sustainability. Their focus was at least partially informed by concern about population growth and the burden of such growth for low-income countries, seeking a response that would enable larger numbers of people living in rural areas to have 'decent livelihoods in a manner which can be sustained' (Chambers and Conway, 1992: 2). Scoones (1998) adapted this definition as shown in Box 2.1, which also sets out a definition of livelihoods by Ellis (2000b). In Ellis' definition, the issue of 'access' is extracted from the list of types of assets included in Chambers and Conway's definition, recognising how access to assets is affected by institutions and social relations rather than being an asset in itself.

Box 2.1 Livelihood definitions

'A livelihood comprises the capabilities, assets (stores, resources, claims and access) and activities required for a means of living: a livelihood is sustainable which can cope with and recover from stress and shocks, maintain or enhance its capabilities and assets, and provide sustainable livelihood opportunities for the next generation; and which contributes net benefits to other livelihoods at the local and global levels and in the short and long term'.

Chambers and Conway (1992: 6)

'A livelihood comprises the capabilities, assets (including both material and social resources) and activities required for a means of living. A livelihood is sustainable when it can cope with and recover from stresses and shocks, maintain or enhance its capabilities and assets, while not undermining the natural resource base'.

Scoones (1998: 5)

'A livelihood comprises the assets (natural, physical, human, financial and social capital), the activities, and the access to these (mediated by institutions and social relations) that together determine the living gained by the individual or household'.

Ellis (2000b: 10)

Livelihoods thinking came to inform UK government aid policy and delivery in the late 1990s with the incoming Labour government and widening perceptions of poverty and development internationally beyond the domination of macro-economics, demonstrated by the publication of the first Human Development Report in 1990 and later reflected in the 2000 World Development Report on Attacking Poverty (de Haan, 2017; Solesbury, 2003). The 1997 White Paper on International Development, *Eliminating World Poverty: A Challenge for the 21st Century*, explicitly referred to sustainable development and supporting targets and policies which 'create sustainable livelihoods for poor people, promote human development and conserve the environment' (DFID, 1997: 6). Guidance notes on taking a livelihoods perspective to development policy and projects were developed and sustainable livelihoods thinking informed a wide range of development projects.

Livelihoods thinking and conceptualisation began as a rural endeavour, informed by literature on agrarian power relations, farming systems research and participatory rural appraisal (Scoones, 2015). Over time, livelihoods thinking and frameworks have been applied in many other settings and contexts, including in relation to urban areas (Rakodi, 2002), as reflected in the chapters of this volume.

Poverty and wellbeing

The purpose and focus of the livelihoods work of the 1990s was the alleviation of poverty, coinciding, as it did, with a broadening of the conceptualisation of poverty. Given this attention to poverty alleviation, the concept of livelihoods is sometimes used interchangeably with poverty alleviation or reduction and with enhancing wellbeing.

Prior to the 1990s, poverty was predominantly viewed and measured in monetary terms, for example through the International Poverty Line, measured from 2015 at US $1.90 in 2011 purchasing power parity dollars (World Bank, 2018). Since the 1990s, poverty has been recognised as having multiple dimensions, not just economic. This was reflected in the 2000/2001 World Development Report, 'Attacking Poverty', in which the World Bank defined poverty as 'pronounced deprivation in wellbeing' (2000: 15), with deprivation including vulnerability and exposure to risk, inadequate education and access to health care, and material deprivation. The report was informed by an international consultation on experiences and interpretations of poverty, recording the 'Voices of the Poor' (Narayan et al., 2000), reflecting the participatory turn in international development, closely aligned to the Sustainable Livelihoods Approach. Such multidimensionality is reflected in the Multidimensional Poverty Index (MPI), which was originally developed in 2010 for the UN's Human Development Report. The index recognises that people have multiple deprivations and experience multiple challenges at the same time, being composed of measures of the incidence and intensity of ten indicators within three themes: health (nutrition and child mortality), education (years of schooling and school attendance) and living standards (cooking fuel, sanitation, drinking water, electricity, housing and assets) (Oxford Poverty and Human Development Initiative, 2018). These multiple dimensions of poverty are reflected in the multiple components of the Sustainable Livelihoods Framework, as set out in the next section of this chapter.

Within the context of international development, however, the very concept and measure of poverty has been challenged, particularly by postdevelopment scholars. In Rahnema's (2010) scathing critique of poverty, he argued that poverty is a construct imposed from outside, Western, perspectives and that it is 'development' itself that has resulted in 'modern' poverty. Loss of traditional ties, industrialisation and urbanisation have resulted in 'poverty' as measured in Western terms of inadequate income and nutrition. Whilst there have been critiques of such postdevelopment thinking on poverty, such as Shaffer's (2012) review which questions the evidence for Rahnema's (2010) critique of poverty, such thinking is particularly informative in recognising that someone may not identify themselves as 'poor' even if their circumstances would place them under a poverty line and that poverty is experienced, defined and understood differently by different people. A livelihoods approach provides space for such thinking, enabling differing perspectives on how a livelihood is constructed and functioning to be examined.

Recognition of the subjective perspective of an individual is also found in some contemporary approaches to wellbeing. An economic interpretation of wellbeing dominated international development theory and practice until the 1990s, when Amartya Sen's capabilities approach challenged this dominance (see Chapter 3 for an overview of capabilities). McGregor (2008: 1) defined wellbeing as 'a state of being with others, where human needs are met, where one can act meaningfully to pursue one's goals, and where one enjoys a satisfactory quality of life' and White (2010: 160) simply as 'doing well – feeling good' and 'doing good – feeling well'. Within a wellbeing perspective, objective and subjective dimensions reflect an individual's priorities and perspectives. Wellbeing takes a more positive outlook than poverty, focusing on what people have rather than what they do not have, and the concept encourages a holistic perspective, taking people and their lives as a whole rather than focusing on one selected dimension or aspect.

Livelihoods: concepts and frameworks

The poverty focus of livelihoods work enables recognition of the agency that people have, agency to respond and challenge rules and situations. However, as explained below, such agency is constrained by a range of institutions.

The Sustainable Livelihoods Framework

This section explains the components and connections within DFID's 'Sustainable Livelihoods Framework', which is shown in Figure 2.1. There are other livelihoods frameworks and adaptations of the SLF, however they share similarities in components and construction.

Moving across the SLF from the left involves recognising the vulnerability context of an individual, household or community and what the sources and nature of vulnerability means for the types of assets they have. Assets are typically categorised into five types of capital in an asset pentagon, yet it is not necessarily the case that an individual, household or community can use the assets or capitals that they have. The potential to use or benefit from assets is mediated by the 'transforming structures and processes', sometimes referred to as 'policies, institutions and processes'. These formal and informal structures and processes affect the ability to benefit from assets and therefore influence the types of livelihood strategies adopted and the subsequent outcomes. Feedback loops from the livelihood outcomes affect livelihoods assets, and structures and processes can affect the context of vulnerability. In practice, the entire framework is not always used in research or to inform development projects and programmes. One or more component of the framework may be focused on.

Vulnerability context

The context of vulnerability set out in DFID's SLF diagram reflects its rural origins. The context accordingly consists of 'shocks, trends and seasonality', referring to external shocks such as rainfall variability and conflict, major trends affecting livelihoods such as price fluctuations

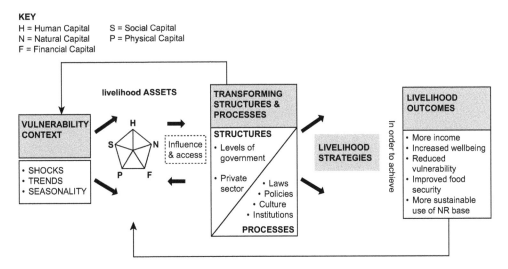

Figure 2.1 DFID's Sustainable Livelihoods Framework

Source: Ashley and Carney (1999: 47)

and changes in population density, and seasonal variations that increase or alleviate vulnerability, such as changes in prices of goods, availability of products and opportunities for employment. The source, extent and implications of vulnerability will affect people differently, particularly affected by cultural norms. Vulnerability has been defined as 'the state of susceptibility to harm, from exposure to stresses associated with environmental and social change and from the absence of capacity to adapt' (Adger, 2006: 268), with the concept examined in detail in Chapter 5 of this volume.

Sources of vulnerability include disasters, climate change and conflict, as discussed in Chapters 41, 43 and 45. Seasonality was included in the SLF due to its rural roots, recognising that seasonality in rainfall leads to associated fluctuations in food prices, agricultural wage rates and infectious diseases (Devereaux et al., 2012). The adverse consequences of seasonality are felt differently by men and women, reflecting the often-gendered division of agricultural and domestic activities, as well as gendered assets and capacity to use and benefit from those assets. Coping mechanisms may be further damaging to livelihoods, such as taking out high-interest loans, leading to longer-term consequences (Devereux et al., 2012). The impacts of climate change in the form of higher temperatures in the tropics, shorter periods of rainfall and growing seasons and more extreme weather events can only exacerbate the adverse effects of seasonality.

Livelihood assets

The asset pentagon of the SLF disaggregates assets into five types of categories, referred to as 'livelihood resources' by Scoones (1998) and as 'building blocks' by DFID (2001: Section 2.3). The five categories are natural, social, financial, human and physical capital. Natural capital reflects the rural origins of the SLF and refers to the stocks of natural capital that people have access to and can make use of, such as land, water and forests. Social capital refers to the social resources that people can draw on, such as families, networks and associations, a concept explained in more detail in Chapter 6 of this volume. Financial capital refers to monetary assets, including cash, credit/debit, savings and other economic assets. Human capital refers to the knowledge, skills, health and physical ability that people have, influenced by education and access to healthcare. And, finally, physical capital refers to the infrastructure available, including roads, transport, shelter, energy and communication.

Assets within each category are interconnected; for example, education may affect the potential to gain higher paying employment or access to land may improve food consumption and health. The endowment of assets that an individual or household may have is not static as it may change over time due to a range of sources of vulnerability, such as an extreme weather event or conflict. The institutional context encapsulated within the 'transforming structures and processes' also affects the extent and nature of assets as well as the ability to use and benefit from them.

Bebbington (1999) suggests that assets should not be seen only as a means of making a living, but as giving meaning to peoples' lives. This reflects Polanyi's view of the economy as being 'socially, culturally and historically embedded', going beyond formalist economics which seeks utility maximisation (de Haan, 2012: 348). This perspective reflects a more multi-faceted view of livelihoods, beyond an economic, income and subsistence focus, and supports the subjective dimension of wellbeing.

Arguments for additional capital assets have been made over time. Bebbington argued that culture should be recognised as a capital asset, drawing on his research in the Andes, where culture was seen to enable forms of 'action and resistance' (1999: 2034). Baumann (2000) argued that 'political capital' should be included as a distinct category and not left to inclusion within 'transforming structures and processes', mediating benefits from assets for livelihood strategies and outcomes. The issue of how politics is reflected in the framework and approach reflects wider critique of the DFID framework and application, which was seen to depoliticise livelihoods, despite Chambers and Conway's (1992) founding paper giving due recognition to 'powerlessness, discrimination, unequal distribution of assets and deprivation' (de Haan, 2017: 24). Discussion and critique on different ways of viewing capital assets indicates that the categorisation of assets in the SLF is not universally accepted and may be modified in analyses of livelihoods.

Transforming structures and processes

Central to the SLF is the box on 'transforming structures and processes'. This component of the SLF recognises that the existence of assets does not guarantee opportunity to utilise and benefit from those assets, whether that's education, land or social networks. There are many factors that influence the potential to use and benefit from assets. The framework lists these as structures, including levels of government and the private sector, and processes, which are listed as laws, policies, culture and institutions. These are found at all levels, from the household to the international level, influencing and shaping assets from multiple perspectives, from international trade agreements, agricultural subsidies in another country and the design of agricultural product certification schemes to gender relations within households and social norms associated with age and wealth status. The influence of the policies, institutions and processes within this component on livelihood strategies and outcomes is considerable. Several chapters of this volume provide further insight into the components of the box, particularly Chapter 4 on institutions and Chapter 15 on power. Scoones (2015: 60) describes this box as representing 'attention to power and politics, and the social and political relations that under pin them', thereby refuting concerns that the framework does not adequately recognise the place and influence of politics in livelihoods.

Livelihood strategies

Given the assets of an individual or household, and how the use of these is mediated by a range of policies, institutions and processes, livelihood strategies are then pursued, with differing degrees of choice in what those strategies are. Livelihood strategies may change over time, including over seasons of a year, for example a dry season leading to urban migration, and may differ between members of a household. Livelihood strategies may consist of income and subsistence from multiple sources, including crops, livestock, casual employment and remittances. These sources of income, employment, culture and identity are explored in-depth in Parts 3 and 4 of this volume, examining the nature and context of livelihoods related to forests (Chapter 25), pastoralism (Chapter 27) and small-scale fisheries (Chapter 28), and how the capacity to benefit from natural resources, such as minerals (Chapter 31 on artisanal mining), is intricately linked to questions of access and control, and how diverse experiences of migration, in terms of duration, location and outcomes, impact on the livelihoods of those migrating and those left behind (Chapter 32).

Whilst the chapters examine the experience and context of 'sectoral' livelihoods, it must be stressed that having access to diverse sources of livelihood and the potential for diversification of livelihoods is recognised as critical for livelihood security (Ellis, 2000a). Ellis (2000b: 14) explains the difference between there being *diverse* sources of livelihoods at any point in time and *diversification*, as an 'ongoing social and economic process'. Diversification itself is distinguished by Alobo Loison (2015) between diversification of the rural economy, away from farm to non-farm activities, and diversification of livelihoods within households and by individuals, increasing the number of livelihood-related activities. Reasons for households pursuing diversification of livelihoods have been categorised as 'necessity or choice' (Ellis, 2000a: 291) and as push and pull factors (Alobo Loison, 2015). Necessity or push factors are involuntary, meaning that individuals and households have no choice but to seek out alternative livelihood sources and strategies. Such factors may include natural disasters, policy change and seasonal variation, affecting agricultural activities and productivity in particular. Choice or pull factors include commercialisation of agriculture, improved infrastructure and access to technology (Ellis, 2000a; Alobo Loison, 2015).

Whether by necessity or choice, the opportunities and processes for diversifying livelihoods are influenced by 'transforming structures and processes', that is a wide range of policies, norms and institutions which mediate how people can utilise and benefit from their assets. The potential for and experience of diversification differs within and between households as policies, norms and institutions very often affect people differently by age, gender and ethnicity. Alobo Loison (2019) suggests that there is little in literature on livelihood diversification which addresses how opportunities and experiences differ by gender. She goes on to observe that women's opportunities are limited by a range of 'social, economic, physical and cultural barriers', being involved in less lucrative food processing and household-based cottage industries (Alobo Loison, 2019: 157).

Livelihood outcomes

Given the livelihood strategies pursued, these result in a range of outcomes, including income and subsistence, improved wellbeing, reduced vulnerability and increased food security. The aim is to at least sustain livelihoods, if not improve them, or to alleviate or reduce poverty. Changes in outcomes may be difficult to measure and certainly to attribute to certain measures or interventions.

The political economy of livelihoods

One of the debates concerning the use of a livelihoods perspective is how much it focuses on actors' agency rather than the broader structures, or political economy, that constrain and enable opportunities. This critique is linked to the concern that the approach, and particularly the framework, pays insufficient attention to power relations (de Haan, 2012). Understanding the nature and influence of the political economy of livelihoods is recognised as being essential, articulated by Scoones (2015: 74) as encapsulating the 'long-term, historical patterns of structurally defined relations of power between social groups, of processes of economic and political control by the state and other powerful actors, and of differential patterns of production, accumulation, investment and reproduction across society'. The 'transforming structures and processes' component of the SLF provides incentive and a mechanism for encouraging such analysis but does not explicitly call attention to analysis of how dominant political and economic actors and relations mediate the endowment and capacity to utilise and benefit from livelihood assets. This suggests that greater attention is needed to the wider political economy in terms of the opportunities and constraints it brings to livelihoods.

Scoones (2015: 82–83) puts forward an 'extended livelihoods approach' that enables investigation of the political economy of livelihoods. The approach includes asking four core and two supplementary questions, as shown in Table 2.1.

Taking these questions, Scoones (2015) goes on to suggest that political economy analysis of livelihoods may include class analysis and examination of relationships between the state, markets and citizens, building on critical agrarian studies. However, he does not advocate that these questions are always appropriate for an extended livelihood analysis, but encourages the development and application of similarly critical questions relevant to the political economy of the context and situation under study. This suggests that a range of theoretical and analytical

Table 2.1 Questions to extend livelihoods analysis

Question	Elaboration
1. 'Who owns what (or who has access to what)?'	Analysis of property rights and ownership of livelihood assets and resources.
2. 'Who does what?'	Recognising social divisions of labour, employment relations and divisions in activities shaped by gender norms and relations
3. 'Who gets what?'	Analysis of who benefits from income and assets and accumulation of these over time, influenced by social and economic differentiation.
4. 'What do they do with it?'	Analysis of livelihood strategies and outcomes.
5. 'How do social classes and groups in society and within the state interact with each other?'	Analysis of social relations, institutions and forms of domination in society that affect livelihoods.
6. 'How do changes in politics get shaped by dynamic ecologies and vice versa?'	How power and environmental dynamics affect, and are affected by, livelihoods through resource access and entitlements.

Source: Author, adapted from Scoones (2015: 82–83).

approaches can be applied, as further illustrated by Banks (2015), who analysed the social networks and patron-client relations of urban livelihoods in Dhaka, Bangladesh, to investigate the local political economy.

Chapters in Part II of this volume provide further examples of analytical approaches that can be taken to investigate the political economy of livelihoods. In particular, the nature and experience of power (Chapter 15), the political context of livelihoods in sub-Saharan Africa (Chapter 17) and the political and policy context of the informal sector (Chapter 22) provide tools and evidence for understanding how the multi-level context of politics and power interact with and inform livelihood vulnerability, assets, strategies and outcomes.

Conclusion

The development and application of livelihoods thinking is strongly linked to the framework and approach adopted by DFID in the 1990s and supported theoretically by Scoones and others at IDS. Multiple foundational roots to the approach have been identified and the DFID framework and approach has been adapted over time and between contexts, as demonstrated in the chapters that follow in this volume. Critique of livelihoods analysis, in particular SLA and the SLF, has particularly centred on there being insufficient attention to power and politics, leading to examples of analysis of the political economy of livelihoods.

Key points

- Livelihoods analysis in international development is strongly associated with the Sustainable Livelihoods Approach and Sustainable Livelihoods Framework of DFID and has roots in Marxist political geography, political ecology and farming systems research.
- The concept of livelihoods is often used interchangeably with poverty and wellbeing, reflecting the focus of early livelihoods approaches on poverty alleviation and the broad, holistic and multidimensional understanding of all three concepts.

- Critique of livelihoods approaches have particularly centred on insufficient attention to power and politics, addressed by Scoones in 2015 through the development of 'extended livelihoods analysis', incorporating political economy analysis.

Further reading

Scoones, I. (1998) 'Sustainable rural livelihoods: A framework for analysis', *IDS Working Paper, 72*, Brighton: Institute of Development Studies.
This is the foundational paper on Sustainable Rural Livelihoods, making for essential reading on the rationale for the structure and content of the SLF, though Scoones' version differs slightly from that adopted by DFID.

Scoones, I. (2015) *Sustainable livelihoods and rural development, Agrarian change & Peasant studies*, Rugby: Practical Action Publishing.
This text provides an invaluable overview into livelihoods, livelihoods analysis and the Sustainable Livelihoods Framework. It provides a useful review of foundational theory and concepts as well as takes the framework and approach forward through proposing an 'extended livelihoods analysis' that incorporates political economy analysis.

References

Adger, W.N. (2006) 'Vulnerability', *Global Environmental Change*, 16(3): 268–281. https://doi.org/10.1016/j.gloenvcha.2006.02.006

Alobo Loison, S. (2015) 'Rural livelihood diversification in sub-Saharan Africa: A literature review', *The Journal of Development Studies*, 51(9): 1125–1138. https://doi.org/10.1080/00220388.2015.1046445

Alobo Loison, S. (2019) 'Household livelihood diversification and gender: Panel evidence from rural Kenya', *Journal of Rural Studies*, 69: 156–172. https://doi.org/10.1016/j.jrurstud.2019.03.001

Ashley, C. and Carney, D. (1999) *Sustainable livelihoods: Lessons from early experience*, London: Department for International Development.

Banks, N. (2015) 'Livelihoods limitations: The political economy of urban poverty in Dhaka, Bangladesh', *Development and Change*, 47(2): 266–292. https://doi.org/10.1111/dech.12219

Baumann, P. (2000) 'Sustainable livelihoods and political capital: Arguments and evidence from decentralisation and natural resource management in India', *Working Paper, 136*, London: Overseas Development Institute.

Bebbington, A. (1999) 'Capitals and capabilities: A framework for analyzing peasant viability, rural livelihoods and poverty', *World Development*, 27(12): 2021–2044. https://doi.org/10.1016/S0305-750X(99)00104-7

Chambers, R. and Conway, G. (1992) 'Sustainable rural livelihoods: practical concepts for the 21st century', *IDS Discussion Paper 296*, Brighton: Institute of Development Studies.

De Haan, L.J. (2012) 'The livelihood approach: A critical exploration', *Erdkunde*, 66(4): 345–357. https://doi.org/10.3112/erdkunde.2012.04.05

de Haan, L.J. (2017) 'Livelihoods in development', *Canadian Journal of Development Studies/Revue canadienne d'études du développement*, 38(1): 22–38. https://doi.org/10.1080/02255189.2016.1171748

Devereux, S., Sabates-Wheeler, R. and Longhurst, R. (2012) 'Seasonality revisited: New perspectives on seasonal poverty', in Devereaux, S., Sabates-Wheeler, R. and Longhurst, R. (eds.), *Seasonality, rural livelihoods and development*, Abingdon: Earthscan, pp. 1–21.

DFID (1997) 'Eliminating world poverty: A challenges for the 21st century', *White Paper on International Development*, London: Department for International Development.

DFID (2001) *Sustainable livelihood guidance sheets*, London: Department for International Development.

Ellis, F. (2000a) 'The determinants of rural livelihood diversification in developing countries', *Journal of Agricultural Economics*, 51(2): 289–302. https://doi.org/10.1111/j.1477-9552.2000.tb01229.x

Ellis, F. (2000b) *Rural livelihoods and diversity in developing countries*, Oxford: Oxford University Press.

McGregor, J.A. (2008) 'Wellbeing, poverty and conflict', *WeD Briefing Paper 01/08*, WeD Research Group, Bath: University of Bath.

Narayan, D., Chambers, R., Shah, M.K. and Petesch, P. (2000) *Voices of the Poor: Crying out for change*, Oxford: Oxford University Press.

Oxford Poverty and Human Development Initiative (2018) *Global Multidimensional Poverty Index 2018: The most detailed picture to date of the world's poorest people*, Oxford: University of Oxford.

Rahnema, M. (2010) 'Poverty', in Sachs, W. (ed.), *The development dictionary*, London: Zed Books, pp. 174–194.

Rakodi, C., with Lloyd-Jones, T. (2002) *Urban livelihoods: A people-centred approach to reducing poverty*, Abingdon: Earthscan.

Scoones, I. (1998) 'Sustainable rural livelihoods: A framework for analysis', *IDS Working Paper, 72*, Brighton: Institute of Development Studies.

Scoones, I. (2009) 'Livelihoods perspectives and rural development', *Journal of Peasant Studies*, 36(1): 171–196. https://doi.org/10.1080/03066150902820503

Scoones, I. (2015) *Sustainable livelihoods and rural development, agrarian change & peasant studies*, Rugby: Practical Action Publishing.

Shaffer, P. (2012) 'Post-development and poverty: An assessment', *Third World Quarterly*, 33(10): 1767–1782. https://doi.org/10.1080/01436597.2012.728314

Solesbury, W. (2003) 'Sustainable livelihoods: A case study of the evolution of DFID policy', *Working Paper 217*, London: Overseas Development Institute.

White, S. (2010) 'Analysing wellbeing: A framework for development practice', *Development in Practice*, 20(2): 158–172. https://doi.org/10.1080/09614520903564199

World Bank (2000) *World development report 2000/01: Attacking poverty*, Washington, DC: The World Bank.

World Bank (2018) *Poverty and shared prosperity 2018: Piecing together the poverty puzzle*, Washington, DC: The World Bank.

World Commission on the Environment and Development (WCED) (1987) *Our common future*, Oxford: Oxford University Press.

3
THE CAPABILITY APPROACH AS AN ANALYTIC LENS FOR STUDYING LIVELIHOODS

Lucy Szaboova

Introduction

The chapter explores the potential of the Capability Approach as an analytic lens for analysing rural and urban livelihoods in the Global South. It is not the intention of the chapter to propose the Capability Approach as an alternative framework, but rather as an additional conceptual layer that can help understand the nuances between structure and people's agency, and their implication for the range of outcomes that are possible for different actors.

The chapter first provides a brief introduction to the Capability Approach, its concepts, discusses key critiques of the approach that are also relevant for livelihoods analyses and situates the concept of capability within existing livelihood approaches. The remainder of the chapter attends to two questions: what a Capability Approach can offer to livelihood analyses and how can insights from such analyses inform debates about livelihood adaptation.

Capability theories: a brief overview

Nobel laureate Amartya Sen formulated the Capability Approach (CA) as a normative framework for the evaluation of wellbeing which emphasises opportunities over outcomes. As such, the CA challenges conventional utilitarian conceptualisations of wellbeing, which focus on the maximisation of utility via incomes, consumption or the ownership of commodities (Alkire, 2008; Fukuda-Parr, 2003). Sen highlights that there is an important qualitative difference between the idea of wellbeing and being well-off: '[t]he value of the living standard lies in the living, and not in the possessing of commodities' (Sen, 1987: 25). Therefore, wellbeing should be thought of in terms of the opportunities – referred to as *capabilities* – people have and the outcomes – termed as *functionings* – they are able to achieve through these (Sen, 1985, 1999, 2008). In CA analyses, people's *freedom* and *agency* to convert capabilities into valued functionings take centre stage in the evaluation of wellbeing (e.g. Sen, 1985, 2008) (see Box 3.1).

> **Box 3.1 Key concepts in the Capability Approach**
>
> **Functionings.** Represent the alternative 'beings' and 'doings' a person has reason to value. These may include activities (eating, reading, resting) and states of existence or being (being well-nourished, being able to appear in public without shame, being free from disease) (Sen, 1985).
>
> **Capabilities.** Also referred to as people's capability set, that is the range of opportunities that a person can choose from. The capability set reflects the freedom to achieve valuable functionings (Sen, 1990).
>
> **Freedom.** There are two aspects to freedom: opportunity, or the (cap)ability to achieve what a person values, and process, or the degree of agency a person is able to exercise when choosing. Choices are affected by people's preferences and ideals of a good life, which are shaped by the broader social and cultural context (Nussbaum, 2000; Sen, 1990). Hence, two people with identical sets of capabilities will likely achieve different functionings based on the choices they make (Robeyns, 2005).
>
> **Agency.** The ability to make choices, that is to exercise freedom to pursue the goals and aspirations that a person has reason to value (Sen, 1985). According to Sen, an agent can bring about change, while a person lacking agency is someone who is oppressed, forced or passive (Sen, 1999).

Sen (1992) used the example of going without food to illustrate why we need to focus on capabilities or opportunities instead of outcomes, and why the notions of freedom and agency are important entry points for understanding how well a person is doing. In his example, two people, one of whom is starving because they lack access to food and the other because they are fasting for religious reasons, both achieve the same functioning of being hungry. However, one exercises agency when choosing to go without food while the other is deprived of the capability to be well-nourished due to not having access to food. Wellbeing, therefore, is best understood not in terms of achievements but in terms of capabilities and freedoms that allow people to live a flourishing life (Sen, 1999).

Before considering how the CA might enrich livelihood analyses, it is important to highlight two common critiques of the approach, because they have implications for applying the CA as an analytic lens in livelihood research. They have also shaped the evolution of capability thinking and theories over time. These two common criticisms concern the CA's incomplete and individualistic nature (Robeyns, 2005).

The CA is intentionally left incomplete, in that Sen refused to provide a set list of capabilities, due to his belief that these should be produced through a democratic process of discussion and deliberation to generate a list applicable to a given context or situation (Sen, 2004). However, following calls for a universally applicable set of capabilities, Nussbaum (2000, 2003) proposed an alternative iteration of the framework which endorses a list of ten capabilities as a benchmark of basic requirements for a dignified life. These include life, bodily health, bodily integrity, senses, imagination and thought, emotions, practical reason, affiliation, other species, play and, finally, control over one's environment (Nussbaum, 2003). Some believe that the list makes Nussbaum's version of the CA easier to implement in an empirical setting, and this version has also been used in some livelihoods analyses (e.g. Briones, 2009; Lienert and Burger, 2015).

Although the CA is often critiqued for being too individualistic, not engaging sufficiently with the collective and social context, both Sen's and Robeyns' articulation of the approach explicitly recognises the role of social context in shaping an individual's agency and choice. This is evident in Sen's discussion of freedom and agency:

> the freedom of agency that we individually have is inescapably qualified and constrained by the social, political and economic opportunities that are available to us. There is a deep complimentary between individual agency and social arrangements.
>
> (Sen, 1999: xi–xii)

Sen also considered the wider social setting when he elaborated on examples of valued functionings, listing a number of 'social functionings', such as taking part in the life of a community, communicating with others, being well-integrated into society, or being able to appear in public without shame (Gore, 1997).

Robeyns (2005) further clarifies that in addition to recognising that choices regarding transforming capabilities into functionings are socially mediated, the CA also captures the role of social structures by acknowledging that social conditions, alongside other factors, mediate the conversion of goods and resources into functionings. Robeyns describes three types of conversion factors: personal characteristics, the social conditions surrounding the person, and environmental factors stemming from both built and natural environments (Robeyns, 2005, 2017) (see Figure 3.1). The example of the bicycle is often used to illustrate the practical implications of different conversion factors. Owning or having access to a bike can *potentially* enable the functioning of mobility. However, different people will have different scope to achieve that functioning depending on their physical condition such as fitness or (dis)ability (personal conversion factor), the prevailing social norms that condition who can ride a bike and may prohibit women from doing so (social conversion factor), and the availability of roads or cycling paths (environmental conversion factor) (Robeyns, 2017).

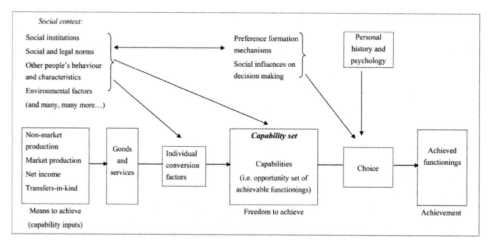

Figure 3.1 A socially embedded conceptualisation of the Capability Approach

Source: Robeyns (2005)

We can therefore say that the CA subscribes to ethical individualism, which implies that the role of social structures and institutions is evaluated with regard to the wellbeing of individuals (Gore, 1997; Robeyns, 2005). Sen sees this as a strength rather than shortcoming, because by making the individual the unit of analysis, we are able to engage with inequalities and deprivations that may exist in larger social units such as households, families or communities, within which individuals are embedded. For example, the unequal position of women alluded to in the above example of the bicycle. The CA's focus on the individual enables critical engagement with the socially conditioned experience of translating the same goods and resources into different functionings. For example, researching the contribution of ecosystem services to wellbeing, Dawson and Martin (2015) showcase the usefulness of the CA for disaggregated analyses that recognise that different people with similar resources achieve different outcomes due to differences in capabilities and subjective preferences regarding valued outcomes.

Capabilities in existing livelihood approaches

In situating capabilities within existing livelihood frameworks, two related questions appear to be particularly relevant: first, how do livelihood theories conceptualise capabilities, and second, how are capabilities interpreted and operationalised in practical or empirical applications of livelihood frameworks?

The concept of capability, alongside equity and sustainability, constitutes the normative basis of sustainable livelihoods. Chambers and Conway's (1991) seminal definition of livelihoods has been widely adopted (and adapted) by bilateral, multilateral and non-governmental organisations working in the Global South on issues of poverty alleviation, rural development and food security (Carney et al., 1999; Hussein, 2002), as well as it underpins Scoones' (1998) Sustainable Livelihoods Framework (SLF). This definition regards capabilities as both inputs (means) and outcomes (ends) of livelihoods:

> a livelihood comprises the capabilities, assets (stores, resources, claims and access) and activities required for a means of living: a livelihood is sustainable which can cope with and recover from stress and shocks, maintain or enhance its capabilities and assets.
> *(Chambers and Conway, 1991: 6)*

Existing livelihood theories, however, only provide a brief explanation of what capabilities are and very limited guidance on how the concept might be applied in a practical setting. Frediani (2010) notes that assets, capitals and capabilities are sometimes used interchangeably, leading to a rather narrow interpretation of the role of capabilities within livelihoods. While Chambers and Conway (1991) recognise that livelihoods can enhance capabilities by providing conditions and opportunities for widening choices, diminishing powerlessness, and improving quality of life, they go on to describe capabilities in a rather instrumental way. They view 'livelihood capabilities' as the ability to foresee and respond to changes and uncertainties and highlight the role of human and financial capitals (such as skills, knowledge, education, inheritance or credit) for strengthening such capabilities. Other than the role of capabilities in supporting livelihood adaptation, the authors offer little insight on the less tangible links between capabilities, livelihoods and wellbeing outcomes.

Scoones (1998) acknowledges that the idea of capabilities broadens the scope of the livelihoods concept and takes it beyond material concerns about income or food intake to also consider intrinsically valued capabilities. As such, intangible, non-material and intrinsic wellbeing outcomes, such as security, stress, happiness, can also be considered as the goal of sustainable livelihoods. Building on ideas of environmental entitlements (Leach et al., 1999) and access to resources (Ribot, 1998), Bebbington (1999) develops a more nuanced understanding of the pathway between capitals and outcomes and thus offers a more politicised take on livelihoods. According to him, capitals or assets underpin people's capability not only to make a living, but to make living meaningful and to exercise agency to reproduce, challenge or change the structures that govern livelihoods.

Despite conceptual advances on the role of capabilities within livelihoods, most organisations working with communities in a livelihood context continue to emphasise capitals and assets, perhaps because they are more conducive to being used as a checklist and are easier to measure to track progress (Scoones, 2009). A number of organisations retain the idea of capabilities in their approach to livelihoods but use simplified language and terminology to communicate it (Carney et al., 1999). By not addressing capabilities in explicit terms, however, they run the risk of these being overlooked in the design and implementation of projects and programmes.

Assets and capabilities

The Department for International Development (DFID) livelihoods approach (see Chapter 2 of this volume) links improved access to assets with people's increased ability to shape structures and processes. This resonates with Bebbington's (1999) proposition that assets can promote the capability to take emancipatory action and change or challenge existing structures. This idea is most clearly articulated in the International Fund for Development (IFAD) strategic framework, which states that by strengthening poor people's capabilities to access assets, we can enable them to shape the institutions, policies and decisions that affect their lives and livelihoods (Hussein, 2002).

However, through her work in urban informal settlements of Dhaka, Bangladesh, Banks (2016) draws attention to the wider political economy of poverty and the limits it poses to the agency of low-income groups to improve livelihoods. Gains are only ever small, incremental and subject to reversal. Dominant social structures and hierarchies within settlements restrict people's access to assets, services and livelihoods. As assets are embedded within social relations and political structures, asset-building alone is unlikely to suffice (Banks, 2016). Instead, enhancing the capability of actors to challenge prevailing systems of inequality and precarity needs to be the focus of interventions. However, power, politics and agency are neither sufficiently articulated nor operationalised in most livelihood approaches and analyses, a shortcoming also highlighted by Scoones (2015).

What can a capability lens add to livelihood analyses?

The CA can complement existing livelihood theories and can help create a holistic analytic space that situates actors within the wider context and considers the role of structures, institutions and processes in shaping livelihood choices and decisions. As such, the CA can make a useful contribution to addressing key gaps in existing livelihoods approaches concerning the structure-agency relationship and the role of power and politics (Lienert and Burger, 2015; Sakdapolrak, 2014; Scoones, 2009; van Dijk, 2011).

Although livelihood frameworks acknowledge the role of politics and power, these are less prominent in their practical implementations. Power is represented differently across frameworks, as context, access to capitals, institutions, and choices and strategies. Some attempt to make power more explicit by highlighting that power relations are embedded in social institutions (e.g. SLF) or by adding political capital to the asset pentagon (e.g. United Nations Development Programme) (Scoones, 2009). Sakdapolrak (2014) observes that the SLF does not do justice to the issue of power, because it does not consider how structures such as institutions or policies shape livelihoods and vice versa. The impact of policy on the ability of nomadic pastoralists to continue practicing their traditional livelihoods illustrates the interaction between politics of power and agency (Box 3.2).

Box 3.2 Power, politics and pastoralist livelihoods in Egypt and Morocco

Pastoralist groups who rely on communal lands and informal tenure for access to grazing land are often disproportionately affected by policy decisions. As some of the poorest and most vulnerable, pastoralists lack political representation and voice in decision making processes and become increasingly marginalised as a result of changes in tenure systems (La Rovere et al., 2009; Young et al., 2009). In Egypt and Morocco, government ambitions to align national economies with the global market system have resulted in macroeconomic policy incentives and interventions (e.g. land reclamation, subsidies that incentivised sedentarisation) that eroded long-standing informal tenure arrangements and led to conflict and competition between previously complementary livelihood practices (Kmoch et al., 2018; Rignall, 2016) and in some cases resulted in unsustainable land use (Rignall and Kusunose, 2018). Due to their vulnerable and marginalised status, pastoralist groups lack the capability to challenge policy decisions. Their power to act is diminished due to their low social standing.

Despite this recognition, existing livelihood frameworks offer an incomplete account of the role of structures and institutions in shaping livelihoods, because they do not directly engage with or analyse agency. Tiwari (2014) argues that the notion of agency is the missing link in existing livelihood approaches, because they fail to elaborate on how one might transform capabilities into valued ends. Indeed, livelihood approaches tend to emphasise the influence of structures and institutions on access to capitals, assets or resources that underpin livelihoods (e.g. Scoones, 1998). However, access alone does not guarantee the achievement of livelihood aspirations, because actors' agency might be constrained by the structural context within which decisions and choices regarding livelihoods are made (Pritchard et al., 2013; Sakdapolrak, 2014). For example, in conservative societies, women's ability to pursue certain livelihoods is conditioned by prevailing norms that render certain activities inappropriate for women (Box 3.3).

The example of women in the garment industry in Bangladesh highlights the need to engage with the concept of agency in livelihood analyses in order to unpack how structures and institutions shape people's choices with regard to pursuing different livelihoods. The CA recognises that choices regarding converting capabilities into functionings are socially mediated and considers the degree of agency people are able to exercise as a central concern of analysis (Robeyns, 2005; Sen, 1999). Focusing on agency is key to understanding why two people with equal capabilities (e.g. financial means, skills, educational attainment) achieve different outcomes. Using this angle in livelihood analyses has the potential to expose the deep-rooted structural causes of undesired outcomes and helps identify entry points for strengthening the capacity of people to translate capabilities into valued outcomes. It can also help shed light on structures and processes that perpetuate or even exacerbate vulnerability by constraining people's agency. For example, women's constrained agency to migrate or return could mean that women are compelled to persist in places that are highly exposed to environmental or security risks such as floods or harassment.

> **Box 3.3 Women's livelihoods and (im)mobility in Bangladesh**
>
> Women's migration in search of work still carries social costs in rural parts of Bangladesh. Women working in cities as domestic workers or in garment factories are stigmatised and accused of violating their *purdah* (Ayeb-Karlsson, 2021; Evertsen and van der Geest, 2020). As well as mediating women's ability to migrate, religious gender norms and social expectations also shape women's ability to return. Because city life is perceived to represent a normative shift in women's behaviour (e.g. by mixing with unrelated males or not covering their head), women who chose to work in garment factories are often seen as dishonourable in rural societies, making it difficult or even impossible for women to return home (Ayeb-Karlsson, 2021). Thus, women's agency to migrate and return is mediated by strict social norms and understandings of gender-appropriate livelihoods and conduct, highlighting women's unequal position in society with regard to men.

The experience of women's constrained agency to choose certain livelihood paths also highlights the importance of intrahousehold dynamics and inequalities. Yet, households are often the unit of analysis in livelihood studies (Sakdapolrak, 2014; van Dijk, 2011). This approach risks overlooking important trade-offs in the wellbeing outcomes of different individuals within households. The CA's analytic focus on the individual can facilitate critical engagement with prevailing inequalities in larger social units such as communities, households or families. Applying a CA analytic lens to livelihoods thus promotes disaggregated analyses that can uncover the ways in which social structures and institutions produce and reproduce inequality. Explicitly recognising trade-offs between different social actors is important, as these can have ramifications for the long-term sustainability of livelihoods.

How can the CA inform debates about livelihood adaptation?

A central tenet of livelihood approaches is to explore how can livelihoods remain sustainable in the context of differentiated vulnerability to shocks and risks. Adaptation is key to sustaining livelihoods over time and it might entail a range of reactive coping and proactive resilience-building strategies. Traditionally, the focus of livelihood frameworks has been on rural livelihoods and local adaptation, but resilience and sustainability cannot always be achieved only through local adaptation, especially where conditions of extreme vulnerability coincide with climate change impacts (Scoones, 2009). Migration and translocality, manifest through interlinkages between rural and urban livelihood spaces, have gained prominence and call for the reconceptualisation of livelihood frameworks to reflect these changes. Bebbington (1999) spotlights the role of migration and points out the limitations of livelihood frameworks that constrain their analysis to agriculture and natural resources. In his study of livelihoods in the high Andes, he finds that while projects have tended to invest in technology and assets related to agricultural production, the primary income source of beneficiaries were in fact migrant remittances.

Livelihood analyses should, therefore, consider livelihoods at different spatial scales, including the interconnectedness between different geographic locations and social actors. A CA lens can provide a platform for exploring whether translocal livelihoods contribute to building capabilities that enable adaptation, reduce vulnerability and enhance development (Porst and Sakdapolrak, 2018). While migration has the potential to provide a pathway out of poverty and can improve the adaptive capacity of rural households, it also entails trade-offs between the wellbeing and resilience of the rural household and the individual who migrates (Singh, 2019). Work with rural-to-urban migrants in India, Bangladesh and Ghana highlights the social costs of migration and the implications these have for migrants' multidimensional wellbeing (Adger et al., 2021; Siddiqui et al., 2021). Interviews with migrants in Chattogram, Bangladesh, revealed that migrants' agency to challenge the urban structures that create precarious conditions is curtailed by their poverty, informality and lack of citizenship rights. Some also perceive a strong sense of caring obligation towards their family and kin left behind in the village, which further entraps them in hazardous work and living environments (Siddiqui et al., 2021). The CA recognises that people may exercise their agency to also pursue 'other-regarding objectives' (Sen, 2008), for example helping their family or community. The achievement of other-regarding goals, however, does not necessarily coincide with improved personal wellbeing and at times may even diminish that. This can have important implications for livelihood adaptation. The high vulnerability and low adaptive capacity of migrants in urban and international destinations has been exposed by the COVID-19 pandemic, particularly in the context of Asia, where the impact of public health measures on livelihoods and gaps in social protection prompted many migrants to attempt returning home, bearing implications for future livelihoods (Suhardiman et al., 2021).

Conclusion

While the CA can evidently enrich analyses of livelihoods, it has received criticism for only presenting an incomplete theory due to not providing a universally applicable set of capabilities – unless one uses Nussbaum's (2003) version of the approach. This is seen as a key challenge to operationalising the capabilities in a practical empirical setting. However, as noted earlier, the CA is intentionally under specified. This means that when implementing the CA, the first step will be establishing what are valued capabilities in a specific context. This can be done using a range of qualitative participatory and deliberative approaches, however one should be aware of the respective limitations of each (see e.g. Robeyns, 2006).

It is precisely the incomplete nature of the CA that lends it applicability to a range of problems and in a diversity of geographic and disciplinary contexts. As a normative framework, it can be applied in conjunction with other frameworks and theories for a variety of purposes (e.g. evaluation, project or intervention design, comparison). In the case of livelihoods, the CA adds an additional conceptual layer that can help us unpack the structures, institutions and processes that shape how people convert capabilities into achievements. This complements existing livelihoods approaches and analyses that focus on the instrumental role of assets in underpinning livelihoods or the livelihood outcomes that are achieved. The CA directs our attention to the opportunities people have and the agency they are able to exercise in order to transform these into valued outcomes or aspirations.

Key points

- The CA emphasises the opportunities people have and the agency they are able to exercise in order to transform these into valued outcomes. Yet, agency is the missing link in many livelihood approaches, which tend to emphasise access to capitals, assets or resources that underpin livelihoods.
- Focusing on agency in livelihood analyses is key to understanding why two people with equal capabilities achieve different outcomes and can expose the structures and processes that perpetuate or even exacerbate vulnerability by constraining people's agency.

- While people's agency might be constrained by the structural and institutional context within which decisions and choices regarding livelihoods are made, power, politics and agency are not sufficiently articulated and operationalised in most livelihood approaches and analyses.
- The CA can inform debates about livelihood adaptation by providing an analytic lens for exploring whether translocal livelihoods contribute to building capabilities that enable adaptation, reduce vulnerability and enhance development at interconnected spatial and social scales.

Further reading

Chiappero-Martinetti, E., Osmani, S. and Qizilbash, M. (2021) *The Cambridge handbook of the capability approach*, Cambridge: Cambridge University Press.

Deneulin, S. and Shahani, L. (2009) *An introduction to the human development and capability approach: Freedom and agency*, London: Earthscan.

Nussbaum, M.C. (2003) 'Capabilities as fundamental entitlements: Sen and social justice', *Feminist Economics*, 9(2–3): 33–59. https://doi.org/10.1080/1354570022000077926

Robeyns, I. (2005) 'The capability approach: A theoretical survey', *Journal of Human Development*, 6(1): 93–117. https://doi.org/10.1080/146498805200034266

Robeyns, I. (2017) *Wellbeing, freedom and social justice: The capability approach re-examined*, Cambridge: Open Book Publishers.

Sen, A. (1999) *Commodities and capabilities*, 2nd ed., Oxford: Oxford University Press.

Sen, A. (1999) *Development as freedom*, Oxford: Oxford University Press.

References

Adger, W.N., Safra de Campos, R., Siddiqui, T., Franco Gavonel, M., Szaboova, L., Rocky, M.H. and Billah, T. (2021) 'Human security of urban migrant populations affected by length of residence and environmental hazards', *Journal of Peace Research*, 58(1): 50–66. https://doi.org/10.1177/0022343320973717

Alkire, S. (2008) 'The capability approach to the quality of life', *Oxford Poverty & Human Development Initiative Working Paper*, Oxford: University of Oxford.

Ayeb-Karlsson, S. (2021) '"When we were children we had dreams, then we came to Dhaka to survive": Urban stories connecting loss of wellbeing, displacement and (im)mobility', *Climate and Development*, 13(4): 348–359, 1–13. https://doi.org/10.1080/17565529.2020.1777078

Banks, N. (2016) 'Livelihoods limitations: The political economy of urban poverty in Dhaka, Bangladesh', *Development and Change*, 47(2): 266–292. https://doi.org/10.1111/dech.12219

Bebbington, A. (1999) 'Capitals and capabilities: A framework for analyzing peasant viability, rural livelihoods and poverty', *World Development*, 27(12): 2021–2044. https://doi.org/10.1016/S0305-750X(99)00104-7

Briones, L. (2009) 'Reconsidering the migration-development link: Capability and livelihood in Filipina experiences of domestic work in Paris', *Population, Space and Place*, 15: 133–145. https://doi.org/10.1002/psp.532

Carney, D., Drinkwater, M., Rusinow, T., Neefjes, K., Wanmali, S. and Singh, N. (1999) *Livelihood approaches compared: A brief comparison of the livelihoods approaches of the UK Department for International Development (DfID), CARE, Oxfam and the United Nations Development Programme (UNDP)*, London: DFID.

Chambers, R. and Conway, G.R. (1991) 'Sustainable rural livelihoods: Practical concepts for the 21st century', *IDS Discussion Paper No. 296*, Brighton: Institute of Development Studies. https://doi.org/ISBN 0 903715 58 9

Dawson, N. and Martin, A. (2015) 'Assessing the contribution of ecosystem services to human wellbeing: A disaggregated study in western Rwanda', *Ecological Economics*, 117: 62–72. https://doi.org/10.1016/j.ecolecon.2015.06.018

Evertsen, K.F. and van der Geest, K. (2020) 'Gender, environment and migration in Bangladesh', *Climate and Development*, 12(1): 12–22. https://doi.org/10.1080/17565529.2019.1596059

Frediani, A.A. (2010) 'Sen's capability approach as a framework to the practice of development', *Development in Practice*, 20(2): 173–187. https://doi.org/10.1080/09614520903564181

Fukuda-Parr, S. (2003) 'The human development paradigm: Operationalizing Sen's ideas on capabilities', *Feminist Economics*, 9(2–3): 301–317. https://doi.org/10.1080/1354570022000077980

Gore, C. (1997) 'Irreducibly social goods and the informational basis of Amartya Sen's capability approach', *Journal of International Development*, 9(2): 235–250. https://doi.org/10.1002/(SICI)1099-1328(199703)9:2<235::AID-JID436>3.0.CO;2-J

Hussein, K. (2002) *Livelihoods approaches compared: A multi-agency review of current practice*, London: Overseas Development Institute.

Kmoch, L., Pagella, T., Palm, M. and Sinclair, F. (2018) 'Using local agroecological knowledge in climate change adaptation: A study of tree-based options in Northern Morocco', *Sustainability*, 10(3719): 17. https://doi.org/10.3390/su10103719

La Rovere, R., Bruggeman, A., Turkelboom, F., Aw-Hassan, A., Thomas, R. and Al-Ahmad, K. (2009) 'Options to improve livelihoods and protect natural resources in dry environments', *The Journal of Environment and Development*, 18(2): 107–129. https://doi.org/10.1177/1070496509333564

Leach, M., Mearns, R. and Scoones, I. (1999) 'Environmental entitlements: Dynamics and institutions in community-based natural resource management', *World Development*, 27(2): 225–247. https://doi.org/10.1016/S0305-750X(98)00141-7

Lienert, J. and Burger, P. (2015) 'Merging capabilities and livelihoods: Analyzing the use of biological resources to improve well-being', *Ecology and Society*, 20(2). https://doi.org/10.5751/ES-07405-200220

Nussbaum, M.C. (2000) *Women and human development: The capabilities approach*, Cambridge: Cambridge University Press.

Nussbaum, M.C. (2003) 'Capabilities as fundamental entitlements: Sen and social justice', *Feminist Economics*, 9(2–3): 33–59. https://doi.org/10.1080/1354570022000077926

Porst, L. and Sakdapolrak, P. (2018) 'Advancing adaptation or producing precarity? The role of rural-urban migration and translocal embeddedness in navigating household resilience in Thailand', *Geoforum*, 97: 35–45. https://doi.org/10.1016/j.geoforum.2018.10.011

Pritchard, B., Rammohan, A., Sekher, M., Parasuraman, S. and Choithani, C. (2013) *Feeding India: Livelihoods, entitlements and capabilities*, London: Routledge.

Ribot, J.C. (1998) 'Theorizing access : Forest profits along Senegal's charcoal commodity chain', *Development and Change*, 29: 307–341. https://doi.org/10.1111/1467-7660.00080

Rignall, K. (2016) 'The labor of agrodiversity in a Moroccan oasis', *Journal of Peasant Studies*, 43(3): 711–730. https://doi.org/10.1080/03066150.2015.1034112

Rignall, K. and Kusunose, Y. (2018) 'Governing livelihood and land use transitions: The role of customary tenure in southeastern Morocco', *Land Use Policy*, 78: 91–103. https://doi.org/10.1016/j.landusepol.2018.03.035

Robeyns, I. (2005) 'The capability approach: A theoretical survey', *Journal of Human Development*, 6(1): 93–117. https://doi.org/10.1080/146498805200034266

Robeyns, I. (2006) 'The capability approach in practice', *Journal of Political Philosophy*, 14(3): 351–376. https://doi.org/10.1111/j.1467-9760.2006.00263.x

Robeyns, I. (2017) *Wellbeing, freedom and social justice: The capability approach re-examined*, Cambridge: Open Book Publishers.

Sakdapolrak, P. (2014) 'Livelihoods as social practices – re-energising livelihoods research with bourdieu's theory of practice', *Geographica Helvetica*, 69(1): 19–28. https://doi.org/10.5194/gh-69-19-2014

Scoones, I. (1998) 'Sustainable rural livelihoods: A framework for analysis', *IDS Working Paper 72*, Brighton: Institute of Development Studies.

Scoones, I. (2009) 'Livelihoods perspectives and rural development', *Journal of Peasant Studies*, 36(1): 171–196. https://doi.org/10.1080/03066150902820503

Scoones, I. (2015) *Sustainable livelihoods and rural development*, Rugby: Practical Action Publishing.

Sen, A. (1985) 'Well-being, agency and freedom: The Dewey lectures 1984', *The Journal of Philosophy*, 82(4): 169–221. https://doi.org/10.2307/2026184

Sen, A. (1987) 'The standard of living: Lecture II, lives and capabilities', in Hawthorn, G. (ed.), *The standard of living*, Cambridge: Cambridge University Press, pp. 20–38.

Sen, A. (1990) 'Justice: Means versus freedoms', *Philosophy & Public Affairs*, 19(2): 111–121.

Sen, A. (1992) *Inequality reexamined*, New York, Cambridge, MA: Harward University Press.

Sen, A. (1999) *Development as freedom*, Oxford: Oxford University Press.

Sen, A. (2004) 'Capabilities, lists, and public reason: Continuing the conversation', *Feminist Economics*, 10(3): 77–80. https://doi.org/10.1080/1354570042000315163

Sen, A. (2008) 'Capability and well-being', in Hausman, D.M. (ed.), *The philosophy of economics: An anthology*, 3rd ed., Cambridge: Cambridge University Press, pp. 270–295.

Siddiqui, T., Szaboova, L., Adger, W.N., Safra de Campos, R., Bhuiyan, M.R.A. and Billah, T. (2021) 'Policy opportunities and constraints for addressing urban precarity of migrant populations', *Global Policy*, 12(S2): 91–105. https://doi.org/10.1111/1758-5899.12855

Singh, C. (2019) 'Migration as a driver of changing household structures: Implications for local livelihoods and adaptation', *Migration and Development*, 8(3): 301–319. https://doi.org/10.1080/21632324.2019.1589073

Suhardiman, D., Rigg, J., Bandur, M., Marschke, M., Miller, M.A., Pheuangsavanh, N., . . . Taylor, D. (2021) 'On the Coattails of globalization: Migration, migrants and COVID-19 in Asia', *Journal of Ethnic and Migration Studies*, 47(1): 88–109. https://doi.org/10.1080/1369183X.2020.1844561

Tiwari, M. (2014) 'Capability approach, livelihoods and social inclusion: Agents of change in rural India', in Ibrahim, S. and Tiwari, M. (eds.), *The capability approach: From theory to practice*, London: Palgrave Macmillan, pp. 29–51. https://doi.org/10.1057/9781137001436

van Dijk, T. (2011) 'Livelihoods, capitals and livelihood trajectories: A more sociological conceptualization', *Progress in Development Studies*, 11(2): 101–117. https://doi.org/10.1177/146499341001100202

Young, H., Osman, A.M., Abusin, A.M., Asher, M. and Egemi, O. (2009) *Livelihoods, power, and choice: The vulnerability of the northern Rizaygat of Darfur, Sudan*, Massachusetts: Feinstein International Center, Tufts University.

4
LIVELIHOODS AND INSTITUTIONS

Luke Whaley

Introduction

People's livelihoods involve entering into relationships with others. These relationships typically establish themselves as relatively stable arrangements; rule-structured interactions that we inherit, adopt and adapt. Examples include the rules and norms for owning and accessing land, selling one's produce in the market, or running a taxi service. These arrangements, durable across time and space but always subject to change, are institutions. Yet exactly what institutions are and how they function differs according to the theoretical lens through which you view them.

This chapter begins by providing answers from the wider literature to three fundamental questions about institutions. It proceeds to outline two schools of thought – Mainstream and Critical Institutionalism – that draw variously on these answers to develop their own distinctive approach. The chapter then considers livelihoods and institutions from the perspectives of the two schools. The discussion throws up questions about the nature of power, meaning and human agency, pointing to the centrality of institutions for understanding livelihoods in the Global South.[1]

Three questions

What are institutions?

An oft-cited answer to this question is institutions are 'durable systems of established and embedded social rules that structure social interactions' (Hodgson, 2006: 13). Hodgson gives examples that include language, money, law, systems of weights and measures, table manners and organisations. This definition and the examples provided are expedient not least because they cover in general terms several key debates about institutions that will be worked through in this section.

A useful starting point for delving further into the nature of institutions is to focus on the concept of a rule. A common interpretation of rules is that they regulate human behaviour. For example, there may be a rule that says women are prohibited from owning land. The metaphor associated with institutions comprising regulatory rules is a cage-like structure that constrains how people behave. A second interpretation is that rules are constitutive; they give form to social situations by stating what should or must be the case. For example, there may be a rule that says in order to sell your plot of land a neighbour must be present to witness the transaction. On closer examination, the constraint of behaviour and the constitution of behaviour are in fact two features of rules rather than two separate types of rules. That is, rules are both regulatory and constitutive (Giddens, 1984). The rule that prohibits women from owning land both regulates behaviour and partly constitutes how humans interact with each other and with the land.

It follows from this interpretation that as systems of social rules, institutions both constrain and enable human behaviour. A related idea is institutions reduce transaction costs (Coase, 1960; Gabre-Madhin, 2001; Williamson, 2002). Imagine a situation where every time you wanted to communicate with someone there was no language for doing so. Without the institution of language – with its rules of syntax, phonology, and the like – each interaction would require inventing a new system of shared meanings. It is clear in this case that transaction costs would be prohibitively high and the social world impossible. North (1990) described institutions as 'the rules of the game'. To the extent that people play by the rules (or are forced to play by them), institutions confer a degree of predictability to human behaviour, allowing one to move on in the world.

The rule stating that women cannot own land is controversial because it discriminates against women. One may find that when the men and women in question are asked, they say that all people are equally free to own land. People often say one thing and do another (De Herdt and Olivier De Sardan, 2015). It may even be the case that these people are not particularly aware of the rule at all. Yet by and large they adhere to it. This tells us that rules need not be formal, explicit or codified (e.g. written down). Rather, many of the rules that make up institutions are implicit social norms and conventions that are enacted with little conscious thought. This is the 'established and embedded' in Hodgson's definition of institutions above.

Institutions typically have associated sanctions designed to enforce them. When rules are implicit and shared widely, it may be that only weak sanctions are necessary. For some scholars, these weak sanctions relate to feelings such as shame or guilt on the part of the rule transgressor (Ostrom, 2005). On the other hand, rules that are explicit and formal, including a country's legal rules (i.e. laws), likely involve stronger sanctions. Included here is the threat of a fine, jail time, violence or even death. Of course, a rule may be formally codified whilst also taking on the character of an implicit norm or convention – think of the traffic law in a country that states on what side of the road people must drive.

How do institutions emerge and evolve?

A common answer to this question is that people design institutions to achieve certain goals or overcome particular problems. The institution of the village meeting, for example, allows the community to come together and raise public concerns, resolve disputes, and make decisions. To the extent that institutions do not achieve their purpose, or as circumstances change, rules can be devised or altered accordingly. However, making or changing rules is often costly in terms of time and energy. As a result, institutions tend to exhibit path dependence whereby their present and future form is contingent on what has come before. This perspective on institutional emergence and evolution reflects influential developments in economics, sociology and political science during the 20th century. At its heart is the assumption that institutions would not be necessary at all but for the need to overcome transaction costs, provide stability and predictability, and limit opportunistic behaviour (North, 1990). This last point relates to 'free riding', whereby certain individuals attempt to enjoy the benefits accrued through the collective action of a group but without contributing anything themselves (Olson, 1965). An example is a person who jumps the queue.

Previously we considered the distinction between explicit and implicit rules. This distinction, which infers a continuum rather than a binary, is important for understanding the evolution of durable institutions. To understand how rules move along this continuum from explicit to implicit requires a temporal dimension. It is only over time – an individual lifetime or the course of generations – that institutions become established and embedded. A new rule that prohibits women from speaking during village meetings may face considerable challenge and require strong sanctions to enforce it. Over time, however, the rule comes to form part of the common sense of a population – it seems natural. This tells us something very important about institutions: they reflect historically situated power relations (Kenny, 2007). Embedded in their past may be processes of negotiation, contestation, discipline and violence that have shaped present-day arrangements. Institutional path dependence therefore relates not simply to transaction costs, in the economic sense. It is about power and the ongoing, often unquestioned reproduction of interests, biases and privilege (ibid; Mosse, 1997).

Whilst the institutional order one is born into appears self-evident, ultimately the legitimacy of institutions rests on knowledge that both justifies and explains their existence. Included here is the very vocabulary available to speak about an institution in the first place; relevant proverbs, moral maxims and wise sayings; and explicit theories about the institution in question (Berger and Luckmann, 1966). Thus the institution of the village meeting is legitimated by the fact it has a name at all and a vocabulary for the different roles and responsibilities of attendees; by proverbs like 'teamwork makes the dream work' or 'unity is strength'; and by explicit theories that explain, say, the purpose of village meetings in terms of the benefits for individuals and the community. Members of a new generation learn this legitimising knowledge during the same process that socialises them into the institutional order (ibid). This would suggest a self-reinforcing dynamic whereby institutions shape human thought and behaviour, which in turn helps to legitimise institutions.

At the broadest level of legitimation, a person's worldview encompasses and makes sense of the entire institutional order. Douglas (1986: 46) writes that 'most institutions, if challenged, are able to rest their claims to legitimacy on their fit with the nature of the universe. She further argued that to get off the ground, any new institution must accord with people's worldview beliefs, with their common-sense understanding of the way things are. One mechanism that achieves this is analogy. Meanings and logics from different domains of the social world are applied to the new institutional arrangement, often unconsciously, making it appear natural. A new community land assembly may explicitly borrow from the form and logic of the traditional village meeting. In doing so, it bestows the new institution with legitimacy by fitting with existing understandings and ways of doing things.

The answer we have arrived at is different to the one we started with. It suggests that institutions do not emerge and evolve only through the action of individuals consciously designing rules to overcome instrumental challenges and achieve particular goals. Institutions also tend to reflect and reproduce power relations, including relations of dominance and dependence. Explicit attempts to change institutions must necessarily challenge or entrench these power dynamics. Moreover, humans think and act largely within the bounds provided by institutions, suggesting limits to the possibility of change.

What is human behaviour, what is society?

The answer to this question rests on the longstanding debate in the social sciences about the relationship between human agency and social structure. On the one end of the spectrum, humans are information processors who act in their own self-interest to maximise what economists call their 'utility'. Each situation an individual is confronted with presents a range of choices about how to act. The individual calculates the course of action that will best maximise their utility by comparing the choices available to them against a set of internal preferences (Becker, 1976). Society, then, is nothing but the aggregate behaviour of individuals following their own rational self-interest. The concept of 'bounded rationality' circumscribes this model of human behaviour by recognising that time and cognitive constraints limit an individual's ability to process information and make decisions (Simon, 1957). Noted previously, from this rational choice perspective the whole purpose of institutions is to overcome transaction costs, provide stability and predictability, and limit opportunistic behaviour.

On the other end of the spectrum, humans are 'social dupes' who act entirely in accordance with the structures of society; that is, within institutional rules and norms (Granovetter, 1985). Processes of socialisation are total. The norms pertaining to a person's class or gender, for example, are rigidly adhered to. Proponents of both the under-socialised individual of 'rational choice theory' and the over-socialised individual of 'structuralism' argue that these are only models used to analyse social life. They do not necessarily imply how humans actually are. In practice, however, these models hold considerable sway over the imaginations of many. They lie at the heart of assumptions about society upon which real-world decisions are made.

A major preoccupation in sociology has been the attempt to rethink the relationship between human agency and social structure and move beyond the approaches discussed so far. A basic premise common to a number of prominent scholars is that social structure is both the medium and outcome of human behaviour (Archer, 1995; Bhaskar, 1979; Giddens, 1984). That is, humans are not social dupes, they do have agency. However, to act they must necessarily draw upon the structures of society. In so doing, they invariably reproduce it, and sometimes change it. Giddens (1984) defines social structure as material resources (e.g. technology, natural resources) and non-material resources (e.g. ideas, symbols, discourses). These resources structure relationships between people of different social positions. For example, relationships between men and women are structured by access to material resources such as land and by non-material resources such as discourses about what it means to be a man or woman.

From this perspective, institutions are something different to social structure. Social structure (material and non-material resources) shapes institutions, whilst institutions (systems of rules and norms) in turn emerge from and condition human behaviour (Fleetwood, 2008). Socialisation into the institutional order results in individuals with a particular dispensation, a tendency to think and act in certain ways (Bourdieu, 1977). Human behaviour is therefore largely habitual and takes the form of routinised social practices. For example, the practices of attending village meetings, debating issues, and making communal decisions are learned during socialisation. Yet humans also exercise agency, the capacity to think and act otherwise. However, they do so whilst engaged in social practices, and always in relation to the institutions and structures of society. Human behaviour is a case of creativity within limits.

Two schools of thought

In this section, we look at two schools of institutional thought: Mainstream Institutionalism (MI) and Critical Institutionalism (CI). They have been chosen because they draw differently on the answers to the three questions in the previous section, in the process developing their own distinctive approach to institutions. These two schools have been proposed by Cleaver (2012) and represent one of a number of different ways of grouping institutional traditions.[2] Both MI and CI have been concerned primarily with the governance of natural resources (e.g. fisheries, forests, wildlife, water). This focus is apt given that the development of livelihoods thinking itself has historical ties to the relationship between humans and natural resources (Scoones, 2009).

Mainstream institutionalism

MI draws upon influential rational choice assumptions of humans as goal-oriented and self-interested utility maximisers. In the case of natural resources, the model is of people as resource users acting strategically to generate outcomes that maximise individual gain. These deliberative and calculating aspects of human behaviour are generally curtailed somewhat by applying the notion of bounded rationality. Ostrom (2005: 99–109) went to some length to articulate a model of bounded rationality that incorporates, 1) the way that people acquire, process, retain, and use information, 2) the valuation that people assign to possible actions and outcomes, and 3) the process people use for selecting particular actions or sequences of actions in relation to others and the resource in question. Repeated interactions between boundedly rational individuals helps to build trust and enhance the information they have about the situation in question.

A good deal of MI thinking focuses on the public spaces that represent an important part of institutional life, whether formal or informal. Rational choice assumptions underpin the idea that institutions are consciously designed by individuals to address specific issues through processes of deliberative 'crafting' (Ostrom, 1999). Much emphasis has been given to local institutions and to sets of design principles that appear to characterise robust institutional arrangements (Agrawal, 2001). These design principles have generated considerable interest, not least because they provide a coherent approach to assessing and intervening in institutions that otherwise vary considerably in space and time. Beyond the local, the concept of polycentric governance attempts to capture how many centres of decision making that are formally independent of each other are nested across scales and interact within a general system of rules (Carlisle and Gruby, 2019; V. Ostrom et al., 1961).

MI has been accompanied by attempts to develop a framework for analysing institutions. This endeavour was pursued in particular by Ostrom and colleagues and resulted in the Institutional Analysis and Development (IAD) Framework (Blomquist and DeLeon, 2011; Imperial, 1999; Kiser and Ostrom, 1982; Ostrom, 1990, 2005). In more recent times, MI has informed resilience research on social-ecological systems, which recognises change, uncertainty, and the need for multi-scalar and adaptive systems of governance. This in turn has led to the development of a Social-Ecological Systems (SES) Framework to analyse more comprehensively the interrelated social and ecological dimensions of human-environment relations (McGinnis and Ostrom, 2014). Taken together, MI has had a notable influence on development policy and planning in the Global South.

Critical institutionalism

A central premise of CI is that institutions are both complex and embedded. This 'complex-embeddedness' (Peters, 1987) stems from the fact that institutions are situated within broader political and economic structures. Furthermore, they enmesh with and emerge out of people's systems of meaning and culturally accepted ways of doing things. As a result, institutions tend to reflect, and often entrench, historically specific power relations (Mosse, 1997; Nightingale, 2011). Where attempts are made to create or impose new institutions, this is never undertaken upon a blank slate but instead must contend with these existing social relations and cultural paradigms, often leading to unintended outcomes (de Koning, 2011).

We have seen that the rational choice assumptions of MI result in a tendency to focus on public spaces and the deliberative designing or crafting of institutions. CI revises this framing to incorporate the ways in which institutions are entwined in people's everyday practices, and emerge and evolve through dynamic processes of 'institutional bricolage' (Cleaver, 2001, 2002). Institutional bricolage is a concept that attempts to capture how people both consciously and non-consciously patch together institutional arrangements from the social and cultural resources available to them. Elaborating on the concept, Cleaver (2012: 34) writes that by 'imbu[ing] configurations of rules, traditions, norms, and relationships with meaning and authority . . . people modify old arrangements and invent new ones'. Furthermore, 'innovations are always linked authoritatively to acceptable ways of doing things' where 'these refurbishments are everyday responses to changing circumstances' (ibid). As a result, institutions tend to elude design (and design principles).

Critical institutionalism is founded on a model of humans with complex identities and multiple forms of belonging (Nightingale, 2011). Institutions formed through bricolage are characteristically plural in their functioning, where arrangements that purportedly serve one purpose are adapted to serve others as circumstances dictate (Cleaver, 2007; Jones, 2015; Schnegg and Linke, 2015). This also implies a degree of indeterminacy, overlap and potential contestation when it comes to the functional remit across different institutions (Lund, 2006). Furthermore, institutions seldom perform consistently over time but instead operate intermittently in relation to changing conditions (Berry, 1994; Chase Smith et al., 2001).

CI has not had anything like the same influence as MI on policy and planning. In part, this is because it embraces the complex-embeddedness of real-world institutions and so does not easily succumb to the simple narratives that are the grist of policymaking. Proposals for 'facilitated institutional bricolage' (Haapala et al., 2016; Hassenforder et al., 2015; Merrey and Cook, 2012) and 'working with the grain' (Whaley et al., 2021) seek to bridge this gap. Moreover, Whaley (2018) developed the Critical Institutional Analysis and Development (CIAD) Framework, which attempts to structure CI analyses in ways that capture the complex-embeddedness of institutions, foregrounding the workings of power and meaning whilst rendering CI research more amenable to policymaking.

Livelihoods and institutions in the Global South

The previous sections highlight the significance of institutions for understanding the social world. It is not surprising therefore that theory and practice recognise the central place of institutions in mediating livelihoods. In this section, I draw upon examples from the literature to consider livelihoods institutions from the perspectives of MI and CI. At the outset, it is worth recognising the incredible breadth of livelihoods in the Global South and the dynamism and ingenuity often exhibited by those engaged in them. This fact makes analysing institutions challenging but ultimately rewarding for understanding livelihood dynamics.

The idea that institutions mediate livelihoods has been most concretely and influentially formulated by the Sustainable Livelihoods Framework (SLF) (see Chapter 2 of this volume). The SLF is a heuristic for conceptualising how means, or livelihood assets, are converted to ends, or livelihood outcomes, through institutions and the strategic actions of individuals (Scoones, 1998). MI has lent itself well to this framing, helping the SLF gain a powerful hold over different communities of practice since the late 1990s. Whilst the SLF can be married to a variety of theories and understandings of institutions, the allure of MI has tended to orient it in straightforwardly economic terms (Jakimow, 2013).

From an MI perspective, livelihood institutions serve clear purposes related to owning and accessing resources, the constitution of relevant groups and organisations, and more generally the incentives and sanctions that shape markets, labour relations and so forth. MI tends to render these institutions public, visible and clearly bounded. For example, much livelihoods research in the Global South examines local collective arrangements such as marketing cooperatives or micro-finance groups (e.g. Getnet and Anullo, 2012; Kumar et al., 2015; Mazumder and Lu, 2015; Sivachithappa, 2013). Other research examines how these local institutions nest within broader systems of polycentric governance (e.g. Baldwin et al., 2016; Ros-Tonen et al., 2014).

In keeping with much economic theory, it is generally assumed that livelihood institutions reduce transactions costs and engender efficient resource management and use (Di Gregorio et al., 2008; Kydd, 2002; Lambini and Nguyen, 2014). People's identities tend to be economic and singular, captured by terms such as farmer, traffic warden, or trader. Natural resources too are conceived in narrowly economic ways. For example, in relation to farming livelihoods, land and water become 'inputs' or 'factors of production'. At the same time, biophysical and material conditions, including forms of technology and the properties of natural resources, strongly influence livelihood institutions (Tucker et al., 2007; Wang et al., 2013).

The MI focus on rules, incentives and sanctions underpinned a drive in the early 2000s to 'get institutions right' in order to facilitate the development of more sustainable livelihoods.[3] MI's model of the rational or boundedly rational actor puts much stock in the agency of individuals to develop appropriate livelihood strategies or to consciously change the institutions that structure livelihood opportunities and constraints. The same logic provided a mandate for governments and development organisations by suggesting where and how they may usefully intervene to change livelihood institutions (Saunders, 2014). Allied to this are the various sets of design principles, proposed by Ostrom and others, that allow researchers and practitioners to assess institutional performance.

The effect of MI on livelihoods research and practice has therefore been to promote a largely instrumental and functionalist agenda. This has provided scope for analysing and understanding how institutions mediate livelihoods. It has also helped to make legible livelihood institutions in ways that fit with the prerogatives of development agencies. At the same time, the narrow remit has meant that MI tends to downplay or mask the relationship between livelihoods and questions of complexity, power and meaning (de Haan and Zoomers, 2005; Rigg, 2007; Scoones, 2009). Moreover, many of MI's assumptions about the nature of institutions and human behaviour appear dubious when applied to a diversity of real-world situations. We now turn to consider how CI understands livelihood institutions.

From a CI perspective, key principles of livelihood institutions are pluralism, dynamism and relationship. Institutions are complex and embedded in relations of power spanning the global to the local. It is not the case that a livelihood institution serves narrowly instrumental functions alone. Instead, these institutions are often plural and patchy in their functioning, contingent on the situation. For example, Whaley et al. (2021) show how a village water committee in Ethiopia also contributes labour and funds to the repair of roads and construction of a school. At the same time, the committee's role in water supply is intermittent, becoming more prominent only in dry periods. The committee itself does not exist in isolation. Its form and function are shaped by the existing institutional landscape. This includes a close relationship with the village *iddir*, an informal community mutual aid organisation.

CI draws attention to broader political and economic processes. This framing reveals how social relations, and especially power relations, operating within and across different scales come to bear on livelihood institutions. It also provides critical insights that are beyond the scope of MI. For example, without a focus on history and political economy it would be difficult to recognise that the institution of the police exists 'primarily as a system for managing and even producing social inequality', with its origins in 18th century slave patrols, colonialism and the control of the new industrial working class (Vitale, 2018: 34). This provides a different complexion to the livelihoods of police officers in the Global South. One that is lost when only analysing the rules of the game and the strategic behaviour of boundedly rational individuals.

Through processes of bricolage, people with complex identities consciously and non-consciously adopt and adapt livelihood institutions in relation to changing circumstances. de Koning (2014) details attempts to introduce new formal institutions for forest governance in the Amazon. She shows how these institutions were transformed by local farmers and forest users. The institutional blueprint was reconfigured and repurposed in combination with existing traditions, community norms, previous legislation and ideas introduced by NGOs.

CI also recognises that people are embedded in social networks and have multiple allegiances and obligations. These may include, for example, the sharing of water, labour, food and ancestries. These social ties result in 'institutional multiplexity' whereby different livelihood institutions hang together, entwine and overlap in relation to a spectrum of individual and shared needs (Schnegg, 2018). This again is illustrated by de Koning's study, which reveals how attempts to implement forest legislation 'affects more aspects of local livelihoods than just those related to the forest' (de Koning, 2014: 367).

Whilst livelihood institutions may be durable and even appear unchanging, in practice they are in constant negotiation. People exercise power and agency to work with and within rules and norms. Those with more power typically have more room for manoeuvre and are better placed to bend or break rules without facing sanction. Yet it is more complex than this. Those with less power may tactically weaponise the structural inequalities they face to create room for manoeuvre. Zwarteveen and Neupane's (1996) study of a farming irrigation scheme in Nepal reveals how women drew upon prevailing gender discourses (e.g. women as weak and in need of protection) to strengthen their position in negotiations for more water and reduced labour commitments for maintenance. This allowed women to take more water than they were entitled to and contribute less labour to maintenance without punishment.

At the same time, livelihood institutions endure precisely because they reflect existing power relations and are invested with meaning and authority. This is well illustrated by Moreau's (2014) study of fishing institutions in Tanzania since the colonial period, on a series of lakes and ponds on the Rufiji River floodplain (see Box 4.1). Moreau shows how the durability of this livelihood institution resulted from its embeddedness in social relations (gender, generational, kinship) and in the natural and supernatural world. For example, community fishing norms developed in relation to the type of fishing technology employed and seasonal fluctuations in water levels. Supernatural beliefs imbued the environment with profound significance, provided cause-and-effect explanations, and conferred legitimacy on the social roles and relations underpinning the fishing institution.

It was only the forced displacement of local populations after Tanzanian independence that upended this local fishing institution. Moreau (2014) discusses how vestiges of the former institution live on alongside attempts to modernise state fisheries management and the claims of 'outsiders' and neighbouring villages to fish the lakes and ponds. There now exists an arena of contending discourses, different explanations of environmental change, a mix of fishing technologies, policies, disputed village boundaries, bylaws, norms and threats of state violence that jostle alongside one another. Within this arena, different actors seek to legitimise their claims to access or manage access to the fishing grounds. Fishing institutions are highly contested and in flux.

> **Box 4.1 Historical fishing institution on the lakes and ponds of the Rufiji River floodplain, Tanzania**
>
> Prior to the 1960s, local fishing communities believed the system of lakes and ponds associated with the Rufiji River were gifts from God, with each waterbody also having its own water spirit. Different clans had responsibility for specific ponds, based on territoriality. Ultimate responsibility for guardianship of a pond rested with the clan elder, a man, who would grant permission to other clans to access the pond for fishing. The clan elder or clan witchdoctor – often the same person – in turn sought permission from the pond's water spirit. Norms dictated that a fishing event would only catch as much as was needed. Collective fishing took place during the dry season to overcome hunger, by catching fish both to eat directly and to exchange for cassava with highland farmers. The technology used was the *nyando*, a long reed fence pushed through the water over several days or weeks to corral the fish. Wading through the water was dangerous. Crocodiles, hippopotamuses, pythons and poisonous fish all posed a threat. For the fishers, safety and courage was sought through appeasement of the water spirit and medicine prepared by the witchdoctor.
>
> Only married men with children were allowed to join the *nyando*, after receiving permission from their father. A *nyando* event would typically involve members of a lineage group and their neighbours. Women had no direct role in fishing but would help to smoke the catch. The wives of the *fundi*, the expedition leader, along with the witchdoctor, followed certain taboos such as leaving their hair unbraided for the duration of the expedition. Widows, the elderly, and those otherwise unable to participate were given fish from the collective catch. It was expected that access granted to a lake or pond by one clan would be reciprocated. Up until the present day, local fishers have understood fishing in moral terms, where its purpose is to 'heal hunger'. A need to eat infers a right to fish. Attempts by the state to restrict fishing practices is therefore widely perceived as morally wrong.

Conclusion

The relevance of institutions for understanding livelihoods cannot be understated. This chapter has outlined different answers to three key questions about institutions. It has discussed how these answers have been selectively drawn upon by two schools of institutional thought. The nature of livelihood institutions varies considerably depending on the perspective of each one.

Mainstream Institutionalism draws upon influential developments in economics and political science. From this perspective, livelihood institutions serve relatively clear purposes and tend to be public and visible. As rational goal-oriented individuals, people consciously adopt, design, and craft institutions to enhance their livelihood opportunities and outcomes. This framing has helped to make legible livelihood institutions in ways that fit with the prerogatives of policymakers and development agencies. At the same time, its narrow remit has meant that Mainstream Institutionalism downplays the relationship between livelihoods and questions of complexity, power and meaning.

Critical Institutionalism foregrounds the workings of power and meaning. Livelihood institutions are embedded in the broader political economy, everyday social relations and culturally accepted ways of doing things. People's identities are complex and multifaceted. Human agency is exercised through routinised social practices, informed by a disposition learned during socialisation into the institutional order. Livelihood institutions emerge and evolve as people consciously and non-consciously piece together arrangements from the social and cultural resources to hand. This framing renders Critical Institutionalism less amenable to livelihoods policy and development planning. More recently, however, attempts have been made to bridge this gap.

To end on a reflective note, the thinking and theory outlined in this chapter exhibits its own institutional path dependence. As an earlier section pointed out, path dependence is in part about power and the ongoing, often unquestioned reproduction of interests, biases and privilege. This can be observed, for example, by the preponderance of Western and Western-based scholars cited throughout the chapter. In finishing, I would like to recognise this point. As work on institutions becomes increasingly diverse, both geographically and epistemologically, it promises to create new and important insights for understanding livelihoods across the world, not least in the Global South.

Key points

- Institutions are central to understanding livelihoods in the Global South. Exactly what institutions are and how they function differs according to the theoretical lens through which you view them.
- One way of grouping institutional theory to understand livelihoods is into two schools of thought: Mainstream Institutionalism and Critical Institutionalism.
- Mainstream Institutionalism is founded on prominent developments in economics and political science. It draws attention to how institutions can be consciously designed or crafted to bring about more desirable livelihood outcomes.
- Critical Institutionalism is grounded in more critical social sciences traditions. It draws attention to the role of power, meaning and complexity in shaping livelihood processes and outcomes.

Further reading

- Cleaver, F. (2012) *Development through Bricolage: Rethinking institutions for natural resource management*, Oxon, UK: Earthscan.
- Jakimow, T. (2013) 'Unlocking the black box of institutions in livelihoods analysis: Case Study From Andhra Pradesh, India', *Oxford Development Studies*, 41: 493–516. https://doi.org/10.1080/13600818.2013.847078
- Ostrom, E. (2005) *Understanding institutional diversity*, New Jersey, USA: Princeton University Press.

Notes

1. It is worth recognising both the promise and peril of the term Global South. See for example: Dados and Connell (2012), Joshi (2015), Schneider (2017)
2. See for example DiMaggio (1998), Hall and Taylor (1996), Jakimow (2013), Mehta et al. (1999)
3. See for example the series of World Development Reports from the late 1990s into the early 2000s. Institutions are proposed as key to ensuring the flourishing of markets, sustainable environmental management and good governance, in the interests of pro-poor development.

References

Agrawal, A. (2001) 'Common property institutions and sustainable governance of resources', *World Development*, 29(10): 1649–1672. https://doi.org/10.1016/S0305-750X(01)00063-8

Archer, M. (1995) *Realist social theory: The morphogenetic approach*, Cambridge, UK: Cambridge University Press.

Baldwin, E., Washington-Ottombre, C., Dell'Angelo, J., Cole, D. and Evans, T. (2016) 'Polycentric governance and irrigation reform in Kenya', *Governance*, 29(2): 207–225. https://doi.org/10.1111/gove.12160

Becker, G.S. (1976) *The economic approach to human behaviour*, Chicago: The University of Chicago Press.

Berger, P.L. and Luckmann, T. (1966) *The social construction of reality*, Harmondsworth, UK: Penguin Books Ltd.

Berry, S. (1994) 'Resource access and management as historical process: Conceptual and methodological issues', in Lund, C. and Marcussen, H.S. (eds.), *Access, control and management of natural resources in sub-Saharan Africa: Methodological considerations*, No.13/1994, Roskilde, Denmark: International Development Studies, Roskilde University, pp. 24–45.

Bhaskar, R. (1979) *The possibility of naturalism: A philosophical critique of contemporary human science*, Brighton, UK: Harvester.

Blomquist, W. and DeLeon, P. (2011) 'The design and promise of the institutional analysis and development framework', *Policy Studies Journal*, 39(1): 1–7.

Bourdieu, P. (1977) *Outline of a theory of practice*, Cambridge, UK: Cambridge University Press.

Carlisle, K. and Gruby, R.L. (2019) 'Polycentric systems of governance: A theoretical model for the commons', *Policy Studies Journal*, 47(4): 921–946. https://doi.org/10.1111/psj.12212

Chase Smith, R., Pinedo, D., Summers, P.M. and Almeyda, A. (2001) 'Tropical rhythms and collective action: Community-based fisheries management in the face of Amazonian unpredictability', *IDS Bulletin*, 32(4): 36–46. https://doi.org/10.1111/j.1759-5436.2001.mp32004005.x

Cleaver, F. (2001) 'Institutional bricolage, conflict and cooperation in Usangu, Tanzania', *IDS Bulletin*, 32(4): 26–35. https://doi.org/10.1111/j.1759-5436.2001.mp32004004.x

Cleaver, F. (2002) 'Reinventing institutions: Bricolage and the social embeddedness of natural resource management', *European Journal of Development Research*, 14(2): 11–30.

Cleaver, F. (2007) 'Understanding agency in collective action', *Journal of Human Development*, 8(2): 37–41. https://doi.org/10.1080/14649880701371067

Cleaver, F. (2012) *Development through bricolage: Rethinking institutions for natural resource management*, Oxon, UK: Earthscan.

Coase, R.H. (1960) 'The problem of social cost', *The Journal of Law and Economics*, 3(1): 1–44. https://doi.org/10.1057/9780230523210_6

Dados, N. and Connell, R. (2012) 'The global South', *Contexts*, 11(1): 12–13. https://doi.org/10.1177/1536504212436479

de Haan, L. and Zoomers, A. (2005) 'Exploring the frontier of livelihoods research', *Development and Change*, 36(1): 27–47. https://doi.org/10.1111/j.0012-155X.2005.00401.x

De Herdt, T. and Olivier De Sardan, J.-P. (2015) *Real governance and practical norms in sub-Saharan Africa: The game of the rules*, Oxon, UK: Routledge.

de Koning, J. (2011) *Reshaping institutions: Bricolage processes in smallholder forestry in the Amazon*, PhD Thesis, Wageningen: Wageningen University & Research.

de Koning, J. (2014) 'Unpredictable outcomes in forestry-governance institutions in practice', *Society and Natural Resources*, 27(4): 358–371. https://doi.org/10.1080/08941920.2013.861557

Di Gregorio, M., Science, P. and Hagedorn, K. (2008) *Property rights, collective action, and poverty: The role of institutions for poverty reduction*, Washington, DC: International Food Policy Institute.

DiMaggio, P. (1998) 'The new institutionalisms: Avenues of collaboration', *Journal of Institutional and Theoretical Economics*, 154(4): 696–705. www.jstor.org/stable/40752104

Douglas, M. (1986) *How institutions think*, New York: Syracuse University Press.

Fleetwood, S. (2008) 'Institutions and social structure', *Journal for Theory of Social Behaviour*, 28(3): 241–265. https://doi.org/10.1111/j.1468-5914.2008.00370.x

Gabre-Madhin, E.Z. (2001) 'Market institutions, transaction costs, and social capital in the Ethiopian grain market', in *Research report 124*, Washington, DC: International Food Policy Research Institute.

Getnet, K. and Anullo, T. (2012) 'Agricultural cooperatives and rural livelihoods: Evidence from Ethiopia', *Annals of Public and Cooperative Economics*, 83(2): 181–198. https://doi.org/10.1111/j.1467-8292.2012.00460.x

Giddens, A. (1984) *The constitution of society: Outline of the theory of structuration*, Cambridge: Polity Press.

Granovetter, M. (1985) 'Economic action and social structure: The problem of embeddedness', *American Journal of Sociology*, 91(3): 481–510. www.jstor.org/stable/2780199

Haapala, J., Rautanen, S.L., White, P., Keskinen, M. and Varis, O. (2016) 'Facilitating bricolage through more organic institutional Designs? the case of water users' associations in rural Nepal', *International Journal of the Commons*, 10(2): 1172–1201. http://doi.org/10.18352/ijc.688

Hall, P.A. and Taylor, R.C.R. (1996) 'Political science and the three new institutionalisms', *Political Studies*, 44(5): 936–957. https://doi.org/10.1111/j.1467-9248.1996.tb00343.x

Hassenforder, E., Ferrand, N., Pittock, J., Daniell, K.A. and Barreteau, O. (2015) 'A participatory planning process as an arena for facilitating institutional bricolage: example from the Rwenzori region, Uganda', *Society and Natural Resources*, 28(9): 995–1012. https://doi.org/10.1080/08941920.2015.1054977

Hodgson, G.M. (2006) 'What are institutions?' *Journal of Economic Issues*, XL(1): 1–25. https://doi.org/10.1080/00213624.2006.11506879

Imperial, M.T. (1999) 'Institutional analysis and ecosystem-based management: The Institutional analysis and development framework', *Environmental Management*, 24(4): 449–465. https://doi.org/10.1007/s002679900246

Jakimow, T. (2013) 'Unlocking the Black box of institutions in livelihoods analysis: Case study from Andhra Pradesh, India', *Oxford Development Studies*, 41(4): 493–516. https://doi.org/10.1080/13600818.2013.847078

Jones, S. (2015) 'Bridging political economy analysis and critical institutionalism: An approach to help analyse institutional change for rural water services', *International Journal of the Commons*, 9(1): 65–86. http://doi.org/10.18352/ijc.520

Joshi, S. (2015) 'Postcoloniality and the north-south Binary revisited: The case of India's climate politics', in Bryant, R.L. (ed.), *The international handbook of political ecology*, Cheltenham, UK: Edward Elgar Publishing, pp. 117–130.

Kenny, M. (2007) 'Gender, institutions and power: A critical review', *Politics*, 27(2): 91–100. https://doi.org/10.1111/j.1467-9256.2007.00284.x

Kiser, L.L. and Ostrom, E. (1982) 'The three worlds of action: A metatheoretical synthesis of institutional approaches', in Ostrom, E. (ed.), *Strategies of political enquiry*, Beverly Hills, USA: Sage Publications, pp. 179–222.

Kumar, V., Wankhede, K.G. and Gena, H.C. (2015) 'Role of cooperatives in improving livelihood of farmers on sustainable basis', *American Journal of Educational Research*, 3(10): 1258–1266. https://doi.org/10.12691/education-3-10-8

Kydd, J. (2002) 'Agriculture and rural livelihoods: Is globalization opening or blocking paths out of rural poverty', *AgREN Network Paper No. 121*, London: Overseas Development Institute.

Lambini, C.K. and Nguyen, T.T. (2014) 'A comparative analysis of the effects of institutional property rights on forest livelihoods and forest conditions: Evidence from Ghana and Vietnam', *Forest Policy and Economics*, 38: 178–190. https://doi.org/10.1016/j.forpol.2013.09.006

Lund, C. (2006) 'Twilight institutions, public authority and political culture in local contexts: An introduction', *Development & Change*, 37(4): 1–27. https://doi.org/10.1111/j.1467-7660.2006.00497.x

Mazumder, M.S.U. and Lu, W. (2015) 'What impact does microfinance have on rural livelihood? A comparison of governmental and non-governmental microfinance programs in Bangladesh', *World Development*, 68: 336–354. https://doi.org/10.1016/j.worlddev.2014.12.002

McGinnis, M.D. and Ostrom, E. (2014) 'Social-ecological system framework: Initial changes and continuing challenges', *Ecology and Society*, 19(2): 30. http://doi.org/10.5751/ES-06387-190230

Mehta, L., Leach, M., Newell, P., Scoones, I., Sivaramakrishnan, K. and Way, S. (1999) 'Exploring understandings of institutions and uncertainty: New directions in natural resource management', *IDS Discussion Paper 372*, Sussex: Institute of Development Studies.

Merrey, D.J. and Cook, S. (2012) 'Fostering institutional creativity at multiple levels: Towards facilitated institutional bricolage', *Water Alternatives*, 5(1): 1–19.

Moreau, M.-A. (2014) *'The lake is our office': Fisheries resources in rural livelihoods and local governance on the Rufiji River floodplain, Tanzania*, PhD Thesis, London: University College London.

Mosse, D. (1997) 'The symbolic making of a common property resource: History, ecology and locality in a tank-irrigated landscape in South India', *Development and Change*, 28(3): 467–504. https://doi.org/10.1111/1467-7660.00051

Nightingale, A.J. (2011) 'Beyond design principles: Subjectivity, emotion, and the (Ir)Rational commons', *Society & Natural Resources*, 24(2): 119–132. https://doi.org/10.1080/08941920903278160

North, D. (1990) *Institutions, institutional change, and economic performance*, Cambridge, UK: Cambridge University Press.

Olson, M. (1965) *The logic of collective action: Public goods and the theory of groups*, Harvard: Harvard University Press.

Ostrom, E. (1990) *Governing the commons: The evolution of institutions for collective action*, Cambridge, UK: Cambridge University Press.

Ostrom, E. (1999) 'Coping with tragedies of the commons', *Annual Review of Political Science*, 2: 493–535. https://doi.org/10.1146/annurev.polisci.2.1.493

Ostrom, E. (2005) *Understanding institutional diversity*, New Jersey: Princeton University Press.

Ostrom, V., Tiebout, C.M. and Warren, R. (1961) 'The organization of government in metropolitan areas: A theoretical inquiry', *The American Political Science Review*, 55(4): 831–842. https://doi.org/10.2307/1952530

Peters, P. (1987) 'Embedded systems and rooted models', in McCay, B.J. and Acheson, J. (eds.), *The question of the commons: The culture and ecology of communal resources*, Arizona, USA: University of Arizona Press.

Rigg, J. (2007) *An everyday geography of the Global South*, London: Routledge.

Ros-Tonen, M.A.F., Derkyi, M. and Insaidoo, T.F.G. (2014) 'From co-management to landscape governance: Whither Ghana's modified Taungya system?', *Forests*, 5(12): 2996–3021. https://doi.org/10.3390/f5122996

Saunders, F. (2014) 'The promise of common pool resource theory and the reality of commons projects', *International Journal of the Commons*, 8(2): 636–656. http://doi.org/10.18352/ijc.477

Schnegg, M. (2018) 'Institutional multiplexity: Social networks and community-based natural resource management', *Sustainability Science*, 13: 1017–1030. https://doi.org/10.1007/s11625-018-0549-2

Schnegg, M. and Linke, T. (2015) 'Living institutions: Sharing and sanctioning water among pastoralists in Namibia', *World Development*, 68: 205–214. https://doi.org/10.1016/j.worlddev.2014.11.024

Schneider, N. (2017) 'Between promise and skepticism: The Global South and our role as engaged intellectuals', *The Global South*, 11(2): 18–38. https://doi.org/10.2979/globalsouth.11.2.02

Scoones, I. (1998) 'Sustainable rural livelihoods: A framework for analysis', *IDS Working Paper 72*, Sussex: Institute of Development Studies.

Scoones, I. (2009) 'Livelihoods perspectives and rural development', *Journal of Peasant Studies*, 36(1): 171–196. https://doi.org/10.1080/03066150902820503

Simon, H. (1957) *Models of man*, New York: John Wiley.

Sivachithappa, K. (2013) 'Impact of micro finance on income generation and livelihood of members of self help groups – A case study of Mandya district, India', *Procedia – Social and Behavioral Sciences*, 91: 228–240. https://doi.org/10.1016/j.sbspro.2013.08.421

Tucker, C.M., Randolph, J.C. and Castellanos, E.J. (2007) 'Institutions, biophysical factors and history: An integrative analysis of private and common property forests in Guatemala and Honduras', *Human Ecology*, 35(3): 259–274. https://doi.org/10.1007/s10745-006-9087-0

Vitale, A.S. (2018) *The end of policing*, New York: Verso.

Wang, J., Brown, D.G. and Agrawal, A. (2013) 'Climate adaptation, local institutions, and rural livelihoods: A comparative study of herder communities in Mongolia and Inner Mongolia, China', *Global Environmental Change*, 23(6): 1673–1683. https://doi.org/10.1016/j.gloenvcha.2013.08.014

Whaley, L. (2018) 'The Critical Institutional Analysis and Development (CIAD) framework', *International Journal of the Commons*, 12(2): 137–161. http://doi.org/10.18352/ijc.848

Whaley, L., Cleaver, F. and Mwathunga, E. (2021) 'Flesh and bones: Working with the grain to improve community management of water', *World Development*, 138: 105286. https://doi.org/10.1016/j.worlddev.2020.105286

Williamson, O.E. (2002) 'The theory of the firm as governance structure: From choice to contract', *Journal of Economic Perspectives*, 16(3): 171–195. https://doi.org/10.1257/089533002760278776

Zwarteveen, M. and Neupane, N. (1996) 'Free-riders or victims: Women's nonparticipation in irrigation management in Nepal's Chhattis Mauja irrigation scheme', *Research Report 7*, Colombo: International Water Management Institute. https://doi.org/10.22004/ag.econ.52731

5
VULNERABILITY AND RESILIENCE

Nathan Clay

Introduction

Throughout the Global South, people make livelihood decisions in the context of innumerable social and environmental stressors. Market uncertainties, civil conflict, climate change and communicable disease, to name just a few. The concepts of vulnerability and resilience therefore feature prominently in livelihoods analysis. Understanding how people make a living and make life meaningful (Bebbington, 1999) requires examining how people respond to chronic and acute stressors and how they manage uncertainty.

Vulnerability refers to the risk that harm may befall an individual, a community or a broader social group that is unable to either absorb the impacts of stressors or adapt to reduce exposure and sensitivity to stressors (Eakin and Luers, 2006). *Resilience* refers to the capacity of a social-environmental system (e.g. a forest ecosystem and indigenous hunter-gatherer populations living there) to cope with and adapt to changing risks and opportunities (Adger and Brown, 2009). Central to both vulnerability and resilience is the concept of *adaptive capacity* – the ability of individuals and systems to adapt in order to mitigate social and environmental stressors (Engle, 2011).

This chapter unpacks vulnerability and resilience as they pertain to livelihoods in the Global South. It reviews how vulnerability and resilience have featured in livelihoods research over the past 20 years, introducing key approaches and methods along with critiques and debates. The chapter's central argument is that livelihoods analysis can provide an incisive lens to consider how vulnerability and resilience are *socially differentiated*. That is: how capacities to absorb and respond to stressors and opportunities vary among individuals, groups, and societies and how these differences result from social structures and unequal power relations that are imbued in class, race, gender and ethnicity.

This focus on the social processes and power relations that create and maintain vulnerability and resilience sets livelihoods analysis apart from development discourse and practice. Development organisations and state institutions consistently equate vulnerability with poverty (Eakin and Lemos, 2006; Ribot, 2014) and have tended to regard resilience as the persistence of existing systems rather than the capacity for systems to rearrange (Carr, 2019). This framing, coupled with a need to generate standardised and quantifiable results, has meant that development programmes and policies often overlook the complex historical social processes and power dynamics that give rise to vulnerability and peoples' uneven capacities to adapt (Shinn, 2017;

Turner, 2016). Livelihoods analysis could help correct this by identifying how vulnerability and resilience emerge in places over time through intersecting political, economic, cultural and environmental elements (Carr, 2020; Clay, 2018).

The increased attention of development communities to resilience and vulnerability begs reflection on how these terms are framed, assessed, and governed. With crises such as climate change and the COVID-19 pandemic stimulating global conversation about vulnerability and resilience, there is an opportunity to re-envision what these concepts mean with respect to livelihoods.

Towards this end, this chapter reviews two decades of thinking about livelihoods, vulnerability, and resilience. It first traces the linked origins of livelihoods analysis and vulnerability assessment. It then introduces several key ways in which livelihoods analysis has helped to identify hidden processes of vulnerability. Next, it identifies a paradox where the increased attention to adaptation and 'livelihood resilience' in development programmes and policies has begun to obscure the social processes that make certain livelihoods vulnerable. The chapter concludes by suggesting how three sets of emerging insights can guide future research, policy and practice on livelihoods, vulnerability and resilience.

Linked origins of livelihoods and vulnerability

The roots of livelihoods analysis and vulnerability assessment are deep and intertwined. Livelihoods approaches (see Chapter 2 of this volume) took shape in the 1990s in an effort to better understand the lived experiences of people in the Global South who were the subjects of development interventions that were largely designed by the Global North (Chambers and Conway, 1992). The emergence of livelihoods analysis coincided with a similar intervention in the 'natural hazards' field. Natural hazards research had long employed quantitative assessments to catalogue the impacts of droughts, hurricanes and earthquakes. However, the framing of these events as 'natural disasters' was challenged by innovative research (Sen, 1982; Watts, 1983) that demonstrated how political economic structures and cultural institutions influence people's capacities to cope with biophysical shocks. Offering detailed explorations of the political economy of poverty and famine, Amartya Sen (1982) and Michael Watts (1983) paved the way for a new wave of vulnerability studies by illustrating how famine is differentially experienced. Their theories of how social and cultural institutions enable people to access material resources that are necessary to cope with drought-induced crop losses was foundational to vulnerability studies, which began to examine how social difference and resource entitlements influence people's ability to manage environmental risk (Blaikie et al., 1994; Bohle et al., 1994; Downing, 1991; Watts and Bohle, 1993).

The notion of unequal abilities to navigate change, buffer uncertainty and mitigate social and environmental stressors was also strongly present in early livelihoods approaches. For instance, Chambers and Conway's (1992: 4) 'livelihood capabilities' concept aimed to identify who is able to cope with stress and shocks and how people respond to change in 'proactive and dynamically adaptable' ways. Likewise, Ian Scoones' Sustainable Rural Livelihoods Framework (1998: 6) called attention to 'livelihood adaptation, vulnerability, and resilience' and enabled 'analysis of a range of factors, including an evaluation of historical experiences of responses to various shocks and stressors'. These foundational texts on livelihoods analysis aligned closely with the aforementioned work on the social dimensions of vulnerability. Specifically, livelihoods and vulnerability analysis shared a focus on 1) peasant agency and cultural ecological processes that comprise how people make a living, and 2) political economic and institutional structures that shape resource entitlements.

With this shared focus, vulnerability studies and livelihoods frameworks were two pieces in a broader effort to reorient development studies and practice beyond material-economic measures such as economic growth, food availability or employment. A core motive for this re-orientation was the increasingly recognised failure of top-down, technology-driven interventions that characterised development practice in the 1990s (Ferguson, 1990). For example, development projects often sought to address food insecurity by getting farmers to adopt high-yielding crop varieties that can generate more income. One symptom of the stranglehold that economics had on both development and the concept of vulnerability was that development projects commonly conflated 'vulnerability' and 'poverty'. Yet, as Chambers (1989: 1) pointed out, poverty does not encompass vulnerability because vulnerability 'means not lack or want but defenselessness, insecurity, and exposure to risk'. In turn, defenselessness is not merely a product of economic deficiency but the inability to cope with stressors without accruing physical, economic, social or psychological harm. Moreover, efforts to reduce poverty, which often involve borrowing and investing resources, can actually increase vulnerability by creating cycles of debt (Chambers, 1989).

While the evaluation of development programmes and natural hazards impacts relied on quantitative, economic indicators, livelihoods analysis and vulnerability research sought to re-centre peasant lives and agency (Chambers and Conway, 1992). These approaches attempted to understand the world from a plurality of local perspectives, focusing on the diversity of interconnected social-environmental factors that constrain and enable how people make a living (Scoones, 1998). In doing so, livelihood approaches emphasised not poverty but wellbeing and capability (Bebbington, 1999; Leach et al., 1999; Ribot and Peluso, 2003). To avoid losing sight of the political economic structures that can constrain and enable decision-making, these approaches advanced holistic measures such as wellbeing and capability (De Haan and Zoomers, 2005). In turn, this required widening the scope of analysis beyond household-level variables (such as assets and income, which characterised traditional development assessments) to include power differentials, social norms and emotional experiences (Chambers, 1987; Sen, 1982).

Livelihoods approaches drew attention to uneven resource access regimes, entitlements and capabilities, in large part to understand vulnerability beyond the realm of the economic and the quantitative. In this sense, the project of reorienting development studies through livelihoods perspectives might be thought of as an effort to unmask vulnerability. In other words, to ensure that the term does not sustain stereotypes of undifferentiated poverty in the Global South (Chambers, 1989) and to generate understanding of the complex processes that make some people vulnerable and others not (Ribot et al., 1996). To summarise, livelihoods analysis built on vulnerability studies' insights to conceptualise vulnerability not as an innate feature of individuals or societies, but as a cross-scalar process that is mediated by local institutions at the interface with national and regional political economy (Ellis and Mdoe, 2003). The concept of vulnerability thus played a foundational role in the very concept of livelihoods. The next section explores the recursive contributions that livelihoods perspectives have offered to thinking about vulnerability.

Unmasking vulnerability with livelihoods analysis

Over the past 20 years, livelihoods analysis has proved vital to identifying patterns and underlying processes of social vulnerability. Countless studies have employed livelihoods analysis as a platform to assess how the unequal distribution of resource entitlements shapes peoples' differential experiences with social and environmental stressors. Livelihoods analysis has been key in moving beyond the simple cataloguing of who or what is vulnerable, instead offering a 'thick description' (Geertz, 1973) that assesses how uneven power relations shape who gets to be resilient and who remains vulnerable.

One of the most enduring contributions of livelihoods analysis is how social and biophysical factors interrelate in ways that can exacerbate or mitigate vulnerability (McCusker and Carr, 2006; Scoones, 1998). A second strength that livelihoods analysis brings to considering vulnerability is the focus on institutional processes. Institutions – the formal and informal rules and norms for resource governance and use (see Chapter 4 of this volume) – are seen as mediators of political economic structures and cultural ecological agency, playing a key role in shaping entitlements, access and capacities (Bebbington, 2000; Ellis and Mdoe, 2003; Scoones, 1998). In these two ways, livelihoods approaches offer analytical potential for studies aiming to go beyond narrow economistic interpretations of how shocks and stressors affect people's lives to understand how political and social processes create and maintain vulnerability.

Over the past two decades, hundreds of case studies have employed livelihoods analysis to explore how intersecting political, economic, social and environmental processes shape vulnerability to social and environmental stressors (Adger, 2006; Eakin and Luers, 2006). An important contribution of this livelihoods and vulnerability research is how social and environmental stressors intersect – what O'Brien and Leichenko (2000) termed 'double exposure' – to influence peoples' lived experiences in complex ways that more conventional development assessments would almost certainly overlook. For example, climate shocks have been demonstrated to affect societies in conjunction with economic globalisation and market liberalisation, including trade agreements (Adger, 1999; Eakin, 2005; O'Brien et al., 2004). Similarly, institutions of climate risk management commonly work through livelihoods because risk management is built into the daily activities of making a living. For instance, land use strategies and seed sharing (e.g. maintenance of agrobiodiversity among smallholder communities in the Andes who optimise diverse potato cultivars across variegated agro-ecological conditions of the mountain landscape [Zimmerer, 1999]) and community-level governance (e.g. local authorities allow spatial adjustments of resource collection and agricultural activities in response to variable precipitation and flooding [King et al., 2016]). These studies involved detailed assessment of livelihood systems and the multi-scalar institutions that shape peoples' uneven capacities to respond to change.

A hallmark of livelihoods analysis has been its focus on the lived experience of vulnerability. Reid and Vogel (2006: 204), for example, employed a livelihoods approach to consider how rural communities in South Africa live with and respond to multiple stressors. They find that 'the limitations of "normal" or "daily" life impose certain constraints on community livelihoods and these cannot be easily disentangled from processes that make the same people vulnerable to a range of additional risks'. Another example is Eakin's (2005) study of smallholder farmers in Mexico, which demonstrates that climate variability factors directly into farmers' livelihood strategies, yet institutional change is more prominent in shaping farmers' vulnerability as it enables and constrains certain livelihood strategies. As Eakin (2005: 1923) puts it: 'farmers' capacities to manage climatic risk are circumscribed by the ways in which they are able to negotiate changes in agricultural policy'. As a third example, Tschakert (2007) employs a similar research design to consider how smallholder farmers in Senegal perceive their own vulnerability and what their adaptive actions are to do something about it. By broadening the focus to include the possibility of adaptive decision making, Tschakert adds a dimension of agency to research on historically marginalised societies; a vital step to identifying possible adaptation options. This offers an important counter to impact assessments, which have tended to minimise the perspectives and the agency of smallholder farmers (Tschakert, 2007).

Hiding vulnerability behind adaptation and resilience

Although livelihoods analysis – as reviewed above – enables attention to longer-term, complex processes of change that operate across scales to shape vulnerability and capacity to adapt, such studies have been more of an exception than the norm (Bassett and Fogelman, 2013). The complexity of livelihoods approaches and challenges of analysing change over time has meant that development impact assessments tend to focus on shorter-term coping despite persistent challenges from development scholars that such snapshots overlook longer-term trajectories and how vulnerability is unevenly produced (Carr, 2019). While livelihoods approaches have been widely applied to assess development interventions, in practice they have still biased economic dimensions while giving far less attention to social and political processes (Scoones, 2009). Thus, despite the promising elements of livelihoods approaches for exploring complex dimensions of vulnerability, these insights failed to gain much traction in mainstream development.

Perhaps it is unsurprising, then, that even while development policy and practice have focused more and more on adaptation and resilience as explicit goals, the tendency towards more superficial applications of livelihoods analysis has continued (Carr, 2020). This section examines how an increasing focus of development on goals of adaptation and resilience has paradoxically diminished attention to vulnerability.

Largely due to the growing recognition of global climate change, enthusiasm about adaptation (that is, adjusting to better prepare for and cope with expected changes) surged in development communities at the beginning of the 21st century. The application of adaptation has met with strong criticism from development scholars, who have argued that the resurgent attention to adaptation has obstructed efforts to reduce inequality. Some have argued that a focus on adaptation overemphasises acute external stressors while undercounting the multiple chronic stressors – and their origins in processes of social inequality – that people confront every day (Bassett and Fogelman, 2013). Others challenge that development strategies aiming to ensure that populations can adapt, meaning that the possibility of non-adaptation is excluded from consideration, even when that option could reduce vulnerability for some people (Orlove, 2009).

Perhaps most problematically, development projects have tended to place the onus of adapting on those with the least capacity to adapt (Burnham and Ma, 2017). As Shinn (2017) demonstrates about ethnic minorities in Botswana, requiring the most marginalised and vulnerable to adopt what are often sets of pre-defined technologically-based adaptations, such as setting up tented villages to avoid flooding, can further marginalise the livelihoods of the most vulnerable. Even among households in a single community, differential asset bases (including material, financial, social, and cultural capitals) can mean that households take vastly divergent adaptation strategies (Jain et al., 2015). In short, the emphasis on adaptation can serve to mask vital questions about the social processes that create vulnerability (Ribot, 2014).

Comparable levels of enthusiasm and scepticism have accompanied 'resilience', which has similarly surged into development discourse over the past few years (Carr, 2019). The widespread adoption of the term resilience likely owes to its intuitive sense of 'bouncing back' to normal following disruption (Leitch and Bohensky, 2014). Resilience now features in a range of international platforms, such as the 2015 United Nations Sustainable Development Goals, and in programmes operated by the World Bank and other development agencies, where the term 'resilient livelihoods' often appears. Critics have argued that the vagueness of the resilience concept enables the word to serve as a vehicle for myriad agendas. Béné and co-authors (2012) noted that the use of resilience in policy discourse denotes a colloquial sense of self-improvement that seems at odds with the goals of pro-poor development programs. In reviewing how resilience is employed in development, Cretney (2014) finds that despite the contested meanings of the term, its use is seldom accompanied by reflection on power, agency and inequality.

A risk is that superficial applications of resilience could facilitate the reorganisation of social services in ways absolve states of responsibilities in favour of building individuals' capacities to adapt and cope (MacKinnon and Derickson, 2012). Resilience, in this sense, can take shape as a rather conservative effort to stabilise power for the powerful (Nelson, 2014). Some have begun to interrogate the discourse and specific projects of resilience, asking what is being maintained and transformed, under the work and direction of whom, and who this benefits (Matin et al., 2018). Others point specifically to the need for researchers and practitioners to reflect on how they frame resilience, asking whether it is 'an individual capacity, an inherent part of the lives and livelihoods of the global poor, or something the poorest and most vulnerable lack?' (Carr, 2019: 70).

Overall, the surge of development programmes now aiming to create 'resilient livelihoods', decrease 'livelihood vulnerability', and facilitate 'livelihood adaptation', have attracted substantial scepticism. Many have challenged that these terms recreate top-down programmes that fail to assess and address the root causes of vulnerability (Bassett and Fogelman, 2013; Carr, 2019; Clay and Zimmerer, 2020; Ribot, 2014; Taylor, 2015). Vulnerability, resilience and adaptation, these authors argue, have become powerful buzzwords that have resuscitated the aspirations of conventional economic development. As an example, Box 5.1 describe how Climate Smart Agriculture has assembled the ample development funding for climate-related projects through vague promises of smallholder resilience. In Rwanda, this galvanised the outdated paradigm of agriculture-led growth by enabling top-down application of a standardised agricultural intensification technology packet.

Box 5.1 Resilience for whom? Engendering vulnerability in Rwanda's 'green revolution'

Under the heading of 'A New Green Revolution for Africa', development interventions promote technologies (hybrid seeds, agronomic engineering, market linkages and chemical fertilisers) to drive economic growth while making agriculture 'climate smart' (Patel, 2013). Rwanda's experience with 'Climate Smart Agriculture' (CSA) highlights crucial limitations in how development discourse and practice currently approach resilience. Since 2010, prominent development institutions have claimed that CSA offers a 'triple win': increasing crop yields, building climate resilience, and reducing greenhouse gas emissions (FAO, 2013). Yet, we must ask *who gains and who loses* from the forms of resilience provided through CSA?

Research in Rwanda (Clay and Zimmerer, 2020) shows how Green Revolution agendas of increasing crop yields can override pro-poor claims of CSA in ways that exacerbate the vulnerability of women and the poor to the negative impacts of climatic shocks. While CSA is advanced as a pro-poor development strategy, in practice 'resilience' is equated with crop productivity and technology adoption. This is exemplified in the World Bank's description of its success building 'climate-smart productive landscapes', which it claims 'substantially increased food security and the climate resilience of target populations':

'Drawing upon IDA and Donor funding and expertise in climate-smart agriculture since 2010, Rwanda has increased the productivity of hillside agriculture tenfold for target irrigated areas, five-fold for targeted non-irrigated areas, doubled the share of commercialised products, reaching over

> 292,000 beneficiaries (over 142,000 women), and minimizing erosion and sediment load by 76%' (World Bank, 2018).
>
> Applying a livelihoods analysis, we can see how CSA overlooks the social dimensions of vulnerability and resilience in favour of adjusting technologies and management structures. In Rwanda, CSA functions as top-down management that serves to reduce community resilience to droughts and uncertain rainfall by limiting sovereignty over land use and decreasing livelihood flexibility (Clay and Zimmerer, 2020). The social groups able to adapt livelihoods and land use to cope with climatic shocks are wealthy men with land and non-farm income. For the many landless rural poor and women, who depend on working in others' fields, adaptive capacity is much lower. These individuals experienced heightened vulnerability to climate shocks because they are forced to pursue land use and livelihood strategies that further drain their already scarce resources (Clay and King, 2019).

In short, CSA has granted further authority to expert technical knowledge and international organisations to decide what futures and forms of resilience are valued. This serves to hide the social differentiated nature of vulnerability.

Moving forward: contested livelihood pathways

Intersecting social and environmental crises – including climate change, racial and gender inequities, and the COVID-19 pandemic – underscore the need to examine the structural processes that give rise to uneven vulnerabilities. It is also essential to identify seeds of equitable resilience: instances where emancipatory relationships might be nurtured in pursuit of more just outcomes. However, both of these aspirations will require reframing how vulnerability and resilience are conceptualised and approached in research, policy and practice.

Development discourse and practice still tend to frame vulnerability and resilience as *internal* characteristics that are challenged by *external* stressors (Brown, 2014). This framing de-politicises the social, political and cultural processes that create socially differentiated vulnerability. As Box 5.1 illustrates, development practice can normalise technical and managerial solutions in the name of resilience even while these solutions result in exacerbated vulnerability for many people. This misses key opportunities to address historical inequities and power imbalances that are at the heart of vulnerability. The paradox is that the urgency of addressing crises has further entrenched a predisposition towards simplified economistic analysis that frames vulnerability as a quantifiable outcome that can be reduced through specific adaptation measures.

To examine how vulnerability, resilience and adaptive capacity are unevenly produced in places over time, future research can build on three insights: 1) vulnerability and resilience are relational and reproduced through social, environmental, cultural and political economic processes, 2) competing knowledge and value systems shape how vulnerability and resilience are perceived, experienced and managed, and 3) attention to pathways of change and response can help us understand how vulnerability and resilience are negotiated over time. Together, these insights can help expose the interworking of power and agency, enabling us to see more clearly processes of marginalisation and emancipation.

First, recent writing has encouraged us to see vulnerability, adaptive capacity and resilience as *relational*. That is: adaptation or resilience for one individual or group may result in vulnerability for another (Clay, 2018; Shinn, 2016; Turner, 2016). The idea that increased adaptive capacity or decreased vulnerability for one person or community entails a corollary decrease for others may seem like common sense. Yet, attention to trade-offs remains conspicuously absent from development and adaptation programs, which continue to peddle hopes of win-win solutions through techno-scientific approaches such as Climate Smart Agriculture (Taylor, 2018). Thus, even while identifying patterns of vulnerability is difficult to do with standardised instruments (Eakin and Luers, 2006), less attention has been given to how multiple intersecting dimensions of vulnerability emerge in contexts of social-environmental change (Shinn, 2017; Turner, 2016). One reason for this is that such work is time-intensive. Conducting analyses that position the 'local' (community, household, individual) in dialogue with broader structural and institutional processes to assess how vulnerability is co-constituted (and overcome) through networks of actors and institutions (c.f. Birkenholtz, 2012; Yeh et al., 2014) takes significant time and requires extensive networks and detailed understanding of places. Yet, such work is essential if we are to account for dynamic social, ecological and institutional processes within which power struggles are enmeshed (Turner, 2014).

Secondly, to see how vulnerability, resilience and adaptation are relational, it helps to take a step back and examine how these concepts expose competing values and knowledge systems. Development scholars have consistently called for attention to how resilience and adaptation are contested processes that expose competing value systems (Cote and Nightingale, 2012). Development programmes and policies that aim to enhance resilience and adaptive capacity implicitly value certain futures over others, which in turn makes certain livelihoods viable and others not (Clay and Zimmerer, 2020). Negotiations about whose resilience and what adaptations should be prioritised are therefore opportunities for either reworking or maintaining power differentials (Burnham and Ma, 2016; Goldman and Riosmena, 2013). Moreover, the discourse about vulnerability and resilience often takes on gendered dimensions. In labelling women in the Global South as the most vulnerable, Arora-Jonsson (2011) argues that their agency is stripped away, blinding assessments of decision-making processes within which power dynamics and competing value systems are enmeshed. In short, the question of *resilience for whom* (Matin et al., 2018) is invariably a political question about who gets to define resilience, vulnerability, and adaptation for a particular place or group of people. Claims about the 'successes' or 'failures' of development programs expose the values of the social groups doing the assessment (Nightingale et al., 2020). Such negotiations about the meaning of vulnerability, resilience and adaptation therefore constitute an important empirical focus for future research.

Bringing the above insights together, a third promising focus for livelihoods analysis lies in thinking about pathways of change and response. The emerging 'adaptation pathways' approach examines how specific responses to social and environmental change are embedded in broader socio-cultural and environmental shifts (Fazey et al., 2016; Wise et al., 2014). This approach closely aligns with livelihoods perspectives in a number of ways. For one, it rejects temporally static framings that end the conversation by assessing which adaptations worked or did not work. Instead, it visualises how and why particular adaptations (including *maladaptations*, or adaptations for the worse) come to stabilise over time. Moreover, this approach attends to competing knowledge systems that make claims about what resilience is, while also attending to the uneven effects of adaptation efforts. Studies in this vein have demonstrated how people take up or reject development interventions that superimpose technological-based adaptation schemes such as drip irrigation upon pre-existing trajectories of adaptation (Burnham and Ma, 2017). Importantly, pathways thinking overlaps with livelihoods insight that individuals and groups can rework political economic structures through daily activities, effectively exercising their agency and their entitlements and capabilities to change the world (c.f. Bebbington, 2000; King, 2011; McSweeney, 2004). A pathways approach can therefore enhance livelihoods analysis, which has long endeavoured to assess change over time, yet has consistently struggled to do so (Scoones, 2009).

Conclusion

In an uncertain world filled with uneven risks and socially embedded vulnerabilities, many people are marginalised and beholden to the powerful. This chapter has considered how livelihoods perspectives align with and enhance efforts to dissect this vulnerability. It has drawn attention to how the concepts of adaptation and resilience can help us understand the social-environmental roots of vulnerability and uneven capacity to adapt to change. A robust framework for such assessment is already found in livelihoods analysis, provided that adequate attention is given to 1) how patterns of vulnerability, adaptation and resilience are dynamic, relational, contested and 2) how livelihood trajectories are shaped through processes of continuous negotiation that are used to refine them in particular places. Together, these insights can help procure more analytically powerful assessments that can balance the need for maintaining security with desires for and realities of social and environmental transformation. Doing so is essential to engendering a more *equitable resilience* (Matin et al., 2018) where the most vulnerable also have the capacity to reorganise the systems within which they have long been marginalised.

Key points

- Livelihoods analysis can help identify the social processes (political economic structures, cultural norms and power relations) that create and maintain vulnerability and resilience.
- By focusing on the interface of people's agency to adapt and the structural and systemic inequalities that make them vulnerable to social and environmental stress, livelihoods approaches offer a valuable counterpoint to conventional vulnerability assessments, which tend to conflate vulnerability with poverty.
- While livelihoods analysis has produced abundant data on how vulnerability and resilience are composed of interlinked social, political, economic and environmental dimensions, development policy and practice have consistently employed a narrower focus on economic drivers and outcomes.
- To develop more equitable development policies, it is essential to understand the historical social-environmental processes that give rise to uneven vulnerability, adaptive capacity and resilience.

Further reading

Carr, E. (2019) 'Properties and projects: Reconciling resilience and transformation for adaptation and development', *World Development*, 122: 70–84. https://doi.org/10.1016/j.worlddev.2019.05.011

Clay, N. (2018) 'Integrating livelihoods approaches with research on development and climate change adaptation', *Progress in Development Studies*, 18(1): 1–17. https://doi.org/10.1177/1464993417735923

Matin, N., Forrester, J. and Ensor, J. (2018) 'What is equitable resilience?', *World Development*, 109: 197–205. https://doi.org/10.1016/j.worlddev.2018.04.020

References

Adger, N. (1999) 'Social vulnerability to climate change and extremes in coastal Vietnam', *World Development*, 27(2): 249–269. https://doi.org/10.1016/S0305-750X(98)00136-3

Adger, N. and Brown, K. (2009) 'Vulnerability and resilience to environmental change: Ecological and social perspectives', in Castree, N., Demeritt, D., Liverman, D. and Rhoads, B. (eds.), *A companion to environmental geography*, Oxford: Wiley-Blackwell. https://doi.org/10.1002/9781444305722.ch8

Adger, W.N. (2006) 'Vulnerability', *Global Environmental Change*, 16(3): 268–281. https://doi.org/10.1016/j.gloenvcha.2006.02.006

Arora-Jonsson, S. (2011) 'Virtue and vulnerability: Discourses on women, gender and climate change', *Global Environmental Change*, 21(2): 744–751. https://doi.org/10.1016/j.gloenvcha.2011.01.005

Bassett, T.J. and Fogelman, C. (2013) 'Déjà vu or something new? The adaptation concept in the climate change literature', *Geoforum*, 48: 42–53. https://doi.org/10.1016/j.geoforum.2013.04.010

Bebbington, A. (1999) 'Capitals and capabilities: A framework for analyzing peasant viability, rural livelihoods and poverty', *World Development*, 27(12): 2012–2044. https://doi.org/10.1016/S0305-750X(99)00104-7

Bebbington, A. (2000) 'Reencountering development: Livelihood transitions and place transformations in the Andes', *Annals of the Association of American Geographers*, 90(3): 495–520. https://doi.org/10.1111/0004-5608.00206

Béné, C., Godfrey Wood, R., Newsham, A. and Davies, M. (2012) 'Resilience: New Utopia or New Tyranny? Reflection about the potentials and limits of the concept of resilience in relation to vulnerability reduction programmes', *IDS Working Paper 405*, Brighton: Institute of Development Studies. https://doi.org/10.1111/j.2040-0209.2012.00405.x

Birkenholtz, T. (2012) 'Network political ecology: Method and theory in climate change vulnerability and adaptation research', *Progress in Human Geography*, 36(3): 295–315. https://doi.org/10.1177/0309132511421532

Blaikie, P., Cannon, T., Davis, I. and Wisner, B. (1994) *At risk: Natural hazards, people's vulnerability and disasters*, Routledge: London.

Bohle, H.G., Downing, T.E. and Watts, M.J. (1994) 'Climate change and social vulnerability: Toward a sociology and geography of food insecurity', *Global Environmental Change*, 4: 37–48. https://doi.org/10.1016/0959-3780(94)90020-5

Brown, K. (2014) 'Global environmental change I: A social turn for resilience?', *Progress in Human Geography*, 38(1): 107–117. https://doi.org/10.1177/0309132513498837

Burnham, M. and Ma, Z. (2016) 'Linking smallholder farmer climate change adaptation decisions to development', *Climate and Development*, 8(4): 289–311. https://doi.org/10.1080/17565529.2015.1067180

Burnham, M. and Ma, Z. (2017) 'Multi-scalar pathways to smallholder adaptation', *World Development*, 108: 249–262. https://doi.org/10.1016/j.worlddev.2017.08.005

Carr, E. (2019) 'Properties and projects: Reconciling resilience and transformation for adaptation and development', *World Development*, 122: 70–84. https://doi.org/10.1016/j.worlddev.2019.05.011

Carr, E. (2020) 'Resilient livelihoods in an era of global transformation', *Global Environmental Change*, 64: 102155. https://doi.org/10.1016/j.gloenvcha.2020.102155

Chambers, R. (1987) 'Sustainable rural livelihoods. A strategy for people, environment and development', *An overview paper for only one earth: Conference on sustainable development, 28–30 April 1987*, London: International Institute for Environment and Development.

Chambers, R. (1989) 'Editorial introduction: Vulnerability, coping and policy', *Institute of Development Bulletin*, 20(2): 1–7. https://doi.org/10.1111/j.1759-5436.1989.mp20002001.x

Chambers, R. and Conway, G. (1992) 'Sustainable rural livelihoods: Practical concepts for the 21st century', *IDS Discussion Paper 296*, Brighton: Institute for Development Studies.

Clay, N. (2018) 'Integrating livelihoods approaches with research on development and climate change adaptation', *Progress in Development Studies*, 18(1): 1–17. https://doi.org/10.1177/1464993417735923

Clay, N. and King, B. (2019) 'Smallholders' uneven capacities to adapt to climate change amid Africa's "green revolution": Case study of Rwanda's Crop Intensification Program', *World Development*, 116: 1–14. https://doi.org/10.1016/j.worlddev.2018.11.022

Clay, N. and Zimmerer, K. (2020) 'Who is resilient in Africa's Green Revolution? Sustainable intensification and Climate Smart Agriculture in Rwanda', *Land Use Policy*, 97: 104558. https://doi.org/10.1016/j.landusepol.2020.104558

Cote, M. and Nightingale, A.J. (2012) 'Resilience thinking meets social theory: Situating social change in socio-ecological systems (SES) research', *Progress in Human Geography*, 36(4): 475–489. https://doi.org/10.1177/0309132511425708

Cretney, R. (2014) 'Resilience for whom? Emerging critical geographies of socio-ecological resilience', *Geography Compass*, 8(9): 627–640. https://doi.org/10.1111/gec3.12154

De Haan, L. and Zoomers, A. (2005) 'Exploring the frontier of livelihoods research', *Development and Change*, 36(1): 27–47. https://doi.org/10.1111/j.0012-155X.2005.00401.x

Downing, T.E. (1991) 'Vulnerability to hunger in Africa: A climate change perspective', *Global Environmental Change*, 1(5): 365–380. https://doi.org/10.1016/0959-3780(91)90003-C

Eakin, H. (2005) 'Institutional change, climate risk, and rural vulnerability: Cases from Central Mexico', *World Development*, 33(11): 1923–1938. https://doi.org/10.1016/j.worlddev.2005.06.005

Eakin, H. and Lemos, M.C. (2006) 'Adaptation and the state: Latin America and the challenge of capacity building under globalization', *Global Environmental Change*, 16(1): 7–18. https://doi.org/10.1016/j.gloenvcha.2005.10.004

Eakin, H. and Luers, A.L. (2006) 'Assessing the vulnerability of social-environmental systems', *Annual Review of Environment and Resources*, 31: 365–394. https://doi.org/10.1146/annurev.energy.30.050504.144352

Ellis, F. and Mdoe, N. (2003) 'Livelihoods and rural poverty reduction in Tanzania', *World Development*, 31(8): 1367–1384. https://doi.org/10.1016/S0305-750X(03)00100-1

Engle, N.L. (2011) 'Adaptive capacity and its assessment', *Global Environmental Change*, 21(2): 647–656. https://doi.org/10.1016/j.gloenvcha.2011.01.019

Fazey, I., Wise, R.M., Lyon, C., Câmpeanu, C., Moug, P. and Davies, T.E. (2016) 'Past and future adaptation pathways', *Climate and Development*, 8(1): 26–44. https://doi.org/10.1080/17565529.2014.989192

Ferguson, J. (1990) *The anti-politics machine: 'Development', depoliticization and bureaucratic power in Lesotho*, Cambridge: Cambridge University Press.

Food and Agricultural Organization (FAO) (2013) *Climate smart agriculture sourcebook*, Rome: Food and Agriculture Organization.

Geertz, C. (1973) 'Thick description: Toward an interpretive theory of culture', in *The interpretation of cultures: Selected essays*, New York: Basic Books, pp. 3–30.

Goldman, M.J. and Riosmena, F. (2013) 'Adaptive capacity in Tanzanian Maasailand: Changing strategies to cope with drought in fragmented landscapes', *Global Environmental Change*, 23: 588–597. https://doi.org/10.1016/j.gloenvcha.2013.02.010

Jain, M., Naeem, S., Orlove, B., Modi, V. and DeFries, R.S. (2015) 'Understanding the causes and consequences of differential decision-making in adaptation research: Adapting to a delayed monsoon onset in Gujarat, India', *Global Environmental Change*, 31: 98–109. https://doi.org/10.1016/j.gloenvcha.2014.12.008

King, B. (2011) 'Spatializing livelihoods: Resource access and livelihood spaces in South Africa', *Transactions of the Institute of British Geographers*, 36(2): 297–313. https://doi.org/10.1111/j.1475-5661.2010.00423.x

King, B., Shinn, J.S., Crews, K.A. and Young, K.R. (2016) 'Fluid waters and rigid livelihoods in the Okavango Delta of Botswana', *Land*, 5(2): 1–76. https://doi.org/10.3390/land5020016

Leach, M., Mearns, R. and Scoones, I. (1999) 'Environmental entitlements: Dynamics and institutions in community-based natural resource management', *World Development*, 27(2): 225–247. https://doi.org/10.1016/S0305-750X(98)00141-7

Leitch, A.M. and Bohensky, E.L. (2014) 'Return to 'a new normal': Discourses of resilience to natural disasters in Australian newspapers 2006–2010', *Global Environmental Change*, 26: 14–26. https://doi.org/10.1016/j.gloenvcha.2014.03.006

MacKinnon, D. and Derickson, K.D. (2012) 'From resilience to resourcefulness: A critique of resilience policy and activism', *Progress in Human Geography*, 37(2): 253–270. https://doi.org/10.1177/0309132512454775

Matin, N., Forrester, J. and Ensor, J. (2018) 'What is equitable resilience?', *World Development*, 109: 197–205. https://doi.org/10.1016/j.worlddev.2018.04.020

McCusker, B. and Carr, E.R. (2006) 'The co-production of livelihoods and land use change: Case studies from South Africa and Ghana', *Geoforum*, 37(5): 790–804. https://doi.org/10.1016/j.geoforum.2005.09.007

McSweeney, K. (2004) 'The dugout canoe trade in Central America's Mosquitia: Approaching rural livelihoods through systems of exchange', *Annals of the Association of American Geographers*, 94(3): 638–661. https://doi.org/10.1111/j.1467-8306.2004.00418.x

Nelson, S.H. (2014) 'Resilience and the neoliberal counter-revolution: From ecologies of control to production of the common', *Resilience*, 2(1): 1–17. https://doi.org/10.1080/21693293.2014.872456

Nightingale, A.J., Eriksen, S., Taylor, M., Forsyth, T., Pelling, M., Newsham, A., Boyd, E., Brown, K., Harvey, B., Jones, L., Bezner Kerr, R., Mehta, L., Naes, L.O., Ockwell, D., Scoones, I., Tanner, T. and Whitfield, S. (2020) 'Beyond technical fixes: Climate solutions and the great derangement', *Climate and Development*, 12(4): 343–352. https://doi.org/10.1080/17565529.2019.1624495

O'Brien, K., Leichenko, R., Kelkar, U., Venema, H., Aandahl, G., Tompkins, H., Javed, A., Bhadwal, S., Barg, S., Nygaard, L. and West, J. (2004) 'Mapping vulnerability to multiple stressors: Climate change and globalization in India', *Global Environmental Change*, 14(4): 303–313. https://doi.org/10.1016/j.gloenvcha.2004.01.001

O'Brien, K.L. and Leichenko, R.M. (2000) 'Double exposure: Assessing the impacts of climate change within the context of economic globalization', *Global Environmental Change*, 10: 221–232. https://doi.org/10.1016/S0959-3780(00)00021-2

Orlove, B. (2009) 'The past, the present and some possible futures of adaptation', in Adger, W.N., Lorenzoni, I. and O'Brien, K. (eds.), *Adaptation to climate change: Thresholds, values, governance*, Cambridge: Cambridge University Press, pp. 131–163.

Patel, R. (2013) 'The long green revolution', *The Journal of Peasant Studies*, 40(1): 1–63. http://doi.org/10.1080/03066150.2012.719224

Reid, P. and Vogel, C. (2006) 'Living and responding to multiple stressors in South Africa – Glimpses from KwaZulu-Natal', *Global Environmental Change*, 16(2): 195–206. https://doi.org/10.1016/j.gloenvcha.2006.01.003

Ribot, J. (2014) 'Cause and response: Vulnerability and climate in the Anthropocene', *The Journal of Peasant Studies*, 41(5): 667–705. https://doi.org/10.1080/03066150.2014.894911

Ribot, J., Magalhães, A.R. and Panagides, S.S. (eds.) (1996) *Climate variability, climate change and social vulnerability in the semi-arid tropics*, Cambridge: Cambridge University Press.

Ribot, J. and Peluso, N.L. (2003) 'A theory of access', *Rural Sociology*, 68(2): 153–181. https://doi.org/10.1111/j.1549-0831.2003.tb00133.x

Scoones, I. (1998) 'Sustainable rural livelihoods: A framework for analysis', *IDS Working Paper 72*, Brighton: Institute of Development Studies.

Scoones, I. (2009) 'Livelihoods perspectives and rural development', *The Journal of Peasant Studies*, 36(1): 171–196. https://doi.org/10.1080/03066150902820503

Sen, A. (1982) *Poverty and famines: An essay on entitlement and deprivation*, Oxford: Oxford University Press.

Shinn, J.E. (2016) 'Adaptive environmental governance of changing social-ecological systems: Empirical insights from the Okavango Delta, Botswana', *Global Environmental Change*, 40: 50–59. https://doi.org/10.1016/j.gloenvcha.2016.06.011

Shinn, J.E. (2017) 'Toward anticipatory adaptation: Transforming social-ecological vulnerabilities in the Okavango Delta, Botswana', *The Geographical Journal*, 184: 179–191. https://doi.org/10.1111/geoj.12244

Taylor, M. (2015) *The political ecology of climate change adaptation: Livelihoods, agrarian change and the conflicts of development*, London: Routledge

Taylor, M. (2018) 'Climate-smart agriculture: What is it good for?', *The Journal of Peasant Studies*, 45(1): 89–107. https://doi.org/10.1080/03066150.2017.1312355

Tschakert, P. (2007) 'Views from the vulnerable: Understanding climatic and other stressors in the Sahel', *Global Environmental Change*, 17(3–4): 381–396. https://doi.org/10.1016/j.gloenvcha.2006.11.008

Turner, M.D. (2014) 'Political ecology I: An alliance with resilience?', *Progress in Human Geography*, 38(4): 616–623. https://doi.org/10.1177/0309132513502770

Turner, M.D. (2016) 'Climate vulnerability as a relational concept', *Geoforum*, 68: 29–38. https://doi.org/10.1016/j.geoforum.2015.11.006

Watts, M. (1983) *Silent violence: Food, famine, and peasantry in northern Nigeria*, Berkeley: University of California Press.

Watts, M.J. and Bohle, H.G. (1993) 'The space of vulnerability: The causal structure of hunger and famine', *Progress in Human Geography*, 17: 43–67. https://doi.org/10.1177/030913259301700103

Wise, R.M., Fazey, I., Stafford Smith, M., Park, S.E., Eakin, H.C., Archer Van Garderen, E.R.M. and Campbell, B. (2014) 'Reconceptualising adaptation to climate change as part of pathways of change and response', *Global Environmental Change*, 28: 325–336. https://doi.org/10.1016/j.gloenvcha.2013.12.002

World Bank (2018) *Climate-smart productive landscapes increase incomes and combat climate change: Hillside agriculture intensification in Rwanda*. Available at: www.worldbank.org/en/results/2018/09/10/climate-smart-productive-landscapes-increase-incomes-and-combat-climate-change-hillside-agriculture-intensification-in-rwanda (Accessed 25 January 2021).

Yeh, E.T., Nyima, Y., Hopping, K.A. and Klein, J.A. (2014) 'Tibetan pastoralists' vulnerability to climate change: A political ecology analysis of snowstorm coping capacity', *Human Ecology*, 42(1): 61–74. https://doi.org/10.1007/s10745-013-9625-5

Zimmerer, K.S. (1999) 'Overlapping patchworks of mountain agriculture in Peru and Bolivia: Toward a regional-global landscape model', *Human Ecology*, 27(1): 135–165. https://doi.org/10.1023/A:1018761418477

6
SOCIAL CAPITAL AND SOCIAL NETWORKS

Itzel San Roman Pineda

Introduction

Individuals and households obtain diverse resources from their social linkages, referred to as networks of social exchange (NSE), that are not easily quantified because they can include exchanges of intangible benefits such as favours or information that help achieve a livelihood. In an attempt to try to acknowledge these gains unrecognised by economic metrics, Pierre Bourdieu (2001) coined the term 'social capital', later popularised by English-speaking academics such as Robert Putnam (1995) and James Coleman (1988).

Understanding the complexities of social capital and NSE is integral in the study of livelihood strategies. It is particularly important for marginalised individuals [those excluded by the hegemonic economic system (Maya Jariego, 2003)] and the chronically poor [those deemed to be vulnerable and unable to reverse the downward trajectory of their livelihoods (Cleaver, 2005)]. In particular, NSE appears to be instrumental for attaining livelihoods of marginalised individuals and their households, as it allows access to the accumulated resources of the members of the social structures they belong to (Pretty and Ward, 2001).

The World Bank popularised the usage of social capital in international development with its Social Capital Initiative (1998–2001), which declared that social capital was the missing link for sustainable development, as economic growth and development hinges not only on natural, physical and human capital, but also on the way actors interact and organise themselves for production (World Bank, 1998). Further research concluded that social capital was equally as important as human capital in reducing poverty at a household level (Bebbington et al., 2004; Grootaert and Narayan, 2001). The broad acceptance of these observations led to the popularisation of what Gonzalez de la Rocha (2007) referred to as the 'myth of survival', which implies that the poor are capable of coping with stresses and shocks on their own through the usage of their labour and NSE (González De La Rocha, 2007; Thomas et al., 2010).

Nevertheless, participating in NSE and having access to social capital is restricted to members that can afford to be reciprocal, which relies on their capacity to contribute and repay any debt to their networks. Therefore, reciprocity is bounded to the existence of a critical mass of resources, such as material, financial, and human resources (health and labour) (Cleaver, 2005; Thomas et al., 2010). Without the availability of these resources, individuals will not always have the capacity to participate in their NSE and, therefore, they will often face the cost of social isolation (Gonzalez de la Rocha, 1994a).

The first part of this chapter is dedicated to reviewing social capital. The term social capital is defined according to its main proponents Pierre Bourdieu (2001), James Coleman (1988), and Robert Putnam (1995). This is followed by an introduction to the levels of analysis of social capital, description of the forms of social capital and identification of the key components of social capital. The second part of the chapter expands on one of the main components of social capital, NSE, and their importance for the achievement of livelihoods in the Global South. Networks are described as one of the structural capabilities for households to achieve livelihoods and the myth of survival is unveiled as a panacea for achieving livelihoods. Lastly, the chapter outlines key points that summarise the chapter and further readings that may be of interest.

Social capital

Social capital is an umbrella term broadly discussed by academics in social sciences in an attempt to describe the benefits that socialisation brings to individuals, such as access to economic and non-economic resources. Social capital, and its many components, like norms, reciprocity and networks of social exchange (NSE), are critical for the poorest population groups to achieve livelihoods, as well as their ability to pursue collective goals and objectives.

Academics have theorised social capital from different perspectives. For example, sociologists focus on the applications of social capital, while economists try to measure the benefits that social capital brings to individuals. To understand the concept of social capital, the definitions of its first proponents need to be integrated. The first pioneer was Pierre Bourdieu, who coined the term and emphasised its instrumental value, as well as the resources individuals can access through their NSE. Bourdieu defined social capital as

> the aggregate of the actual or potential resources linked to possession of a durable network of more or less institutionalised relationships of mutual acquaintance and recognition – to membership in a group – which provides each of its members with the backing of the collective-owned capital, a 'credential' which entitles them to credit, in the various senses of the word.
>
> *(2001: 102–103)*

According to Bourdieu, the quantity of social capital that an individual owns depends on the size of the networks they belong to, and the amount of the different capitals possessed by its members.

Conversely, James Coleman focused on its functions, and Robert Putman theorised about social capital from its cognitive social components – trust, reciprocity and norms, among others (Johnson, 2016; Portes, 1998; Portes and Landolt, 1996). Specifically, Coleman defined social capital by indicating that 'it is not a single entity but a variety of different entities, with two elements in common: they all consist of some aspect of social structures, and they facilitate certain actions of actors-whether persons or corporate actors-within the structure' (1988: 98). Coleman's theorisation of social capital recognises the potential negative effects of social capital, an often neglected perspective by academics, as he considered that 'a given form of social capital that is valuable in facilitating certain actions may be useless or even harmful for others' (Coleman, 1988: 98). For Putnam, social capital is 'features of social life-networks, norms, and trust that enable participants to act together more effectively to pursue shared objectives' (1995: 664–665). This definition implies that the components of social capital – networks, norms, and trust – foster social cohesion ['the extent of connectedness and solidarity among groups in a society' (Kawachi and Berkman, 2000: 175)] for its members to achieve common goals.

These three definitions are useful in the study of livelihood strategies of marginalised groups, as they highlight key elements of social capital that enable, and potentially constrain, access to this collective capital. While Coleman (1988) emphasises that social structures – networks – are enablers of actions, Bourdieu (2001) focuses on the utilitarian aspect of social capital -access to the collective-owned resources of the network. Putnam (1995) highlights that trust and norms are key components of social capital because members of a network must act according to the norms established by the social structure, thereby generating trust to access its collective resources. Putnam (1995) indicates that social capital enables the pursuit of 'collective goals', which can encompass improvement of the economy, labour rights, or access to land of the members of a network, among others. However, Bourdieu (2001) and Coleman (1988) do not constrain the use of social capital for the sake of the collectivity and consider its usage for the accomplishment of individual goals – like achieving livelihoods – or the pursuit of activities with adverse consequences.

Levels of social capital

Social capital operates at different levels and understanding its components and actors at each level allows acknowledgement of their interconnections and their effects on the social capital available at each level (Gelderblom, 2018). Therefore, social capital can be studied from micro-, meso- and macro-level perspectives.

At the micro-level, social capital comprises individuals, households, kin, friendships and neighbourhoods, as well as the norms, values and networks of horizontal linkages that enable their interactions (Bhandari and Yasunobu, 2009; Halpern, 2005; Johnson, 2016). At the meso-level, social capital is formed by community linkages, associational organisations, and institutions. It also comprehends the interactions between groups, communities and NGOs that occur in networks of associations and networks of vertical linkages (Bhandari and Yasunobu, 2009; Halpern, 2005). At the macro-level, social capital includes formal institutional relationships that govern the political regime, civil society, rule of law and government (Bhandari and Yasunobu, 2009). Social capital at the macro-level also encompasses state, national and international connections such as language, customs, and laws.

While most scholars suggest that social capital at the macro-level is an aggregation of social capital in the levels below, Halpern (2005) and Gelderblom (2018) suggest that it is interdependent because variations in the quantity and quality of social capital at one level influence the social capital available at other levels. Halpern exemplifies this, explaining 'if people in a society begin to have weaker ties to their family (declining micro-level social capital), this loss could be functionally offset by an increase in participation in community organizations (meso-level) or more fervent nationalism (macro-level)' (2005: 19).

Gelderblom explores the power relations between micro and macro actors and indicates that the 'agency of the former is often subsumed under that of the latter' (Gelderblom, 2018: 1315). Strategies and actions implemented through institutional structures at the macro-level can, intentional or not, either decrease or increase the amount of social capital at the meso- and micro-levels. For example, decisions of macro actors can improve or reduce access to resources that can alleviate or generate conflicts among the population affected by the implementation of laws and regulations that incentivise competition or collaboration and actions that affect political and economic stability necessary to foster trust (Gelderblom, 2018). The consequences of these actions can impose broader identities to substitute existing ones, such as favouring national over indigenous identities. These actions can also generate relations of submission, fostering clientelism through social programmes – or favour projects of macro actors, jeopardising ventures of micro or meso actors (Gelderblom, 2018).

Less powerful individuals are aided or restricted by the projects and strategies of the most powerful actors at the macro-level. Nevertheless, this does not mean that micro actors cannot influence social capital at the macro-level since they can organise and mobilise to protect their own interests. Falk and Kilpatrick (2000) indicate that interactions at the micro-level can influence social, civic, and economic results at the meso- and macro-levels, although according to Gelderblom (2018), their influence is limited.

Forms of social capital

The academic literature presents different classifications of social capital. According to the nature of the indicators used to measure it, social capital is defined as structural if it refers to objectively measurable activities such as affiliation to religious or political groups, whereas if it refers to subjective indicators, like trust or social support, it is defined as cognitive social capital (Uphoff et al., 2013). From the way it's applied, it can be defined as collective, such as collective property, or individual, referring to access to information or support (Johnson, 2016; Bhandari and Yasunobu, 2009). It can also be created by highly formal networks (e.g. members of a union or a religious group), or it can develop through informal interactions, such as meeting friends in a pub, or be formed in networks that are highly dense (Gannon and Roberts, 2020). However, the most recognised forms of social capital are defined according to the similarity of the characteristics of the individuals or groups interacting, as well as the direction of the connections, referring to vertical or horizontal linkages. These forms are bonding, bridging, and linking social capital.

Bonding capital

Bonding capital refers to the connections built among homogenous groups that share a common identity or have similar objectives. Onyx and Leonard (2010) indicate that bonding social capital comprises of densely interconnected networks with a high degree of trust. This form of capital normally consists of connections and NSE among family members and friends, neighbours, work colleagues, and community members, among others, that allow reciprocity to be mobilised.

Bonding social capital mainly serves to achieve day-to-day livelihoods, as normally the collective resources available through these networks are homogenous among the members (Baird and Gray, 2014; Gannon and Roberts, 2020; Siisiainen, 2003). Thomas et al. (2010) exemplify the importance of bonding social capital for poor households in Kerala, India, to help the family cope when the main breadwinner suffers an incapacitating illness. In these situations, social capital becomes crucial for the household to compensate the costs of illness, including expenses related to treatment and loss of work. Through NSE, these individuals can obtain financial help, gifts, and labour that enable their survival (Thomas et al., 2010).

Bridging and linking capital

Bridging capital comprises the connections among groups with socio-demographic differences (identity, ethnicity, age, etc.). According to Onyx and Leonard (2010) this form of capital is formed by weak ties, as defined by Granovetter (1973), and impersonal trust in strangers. One form of bridging social capital is linking capital, which comprises the connections among individuals and groups with different levels of authority and power, such as the connections between community associations and government authorities.

Unlike bonding capital that enables livelihoods, bridging and linking social capital have the potential to alleviate poverty and improve the social structure of the poor because these connections enable access to external resources and information that can significantly improve their livelihoods. However, scholars like Putnam et al. (1994) and Gonzalez de la Rocha (1994b) argue that vertical linkages can destroy the social capital of disadvantaged groups or actors, due to power relations and the interdependent nature of social capital. Gonzalez de la Rocha (1994a) gives an example from interviewing domestic servants living with their employers in the city of Guadalajara, Mexico. These employees are geographically isolated from their NSE formed by family, friends, and neighbours, and depend on their linkages with their patrons to access social capital. Through these vertical connections, domestic servants receive 'gifts' (clothes, old furniture, medicines, financial help, etc.) from their patrons when needed. However, due to the unequal power relations, employees' reciprocity requires subordination, fidelity and provision of additional services. This dependency destroys social capital by generating mistrust from the employee seeking help of the patron. Additionally, the patron can leave the relation without any sanction due to the power differences among these actors (Gelderblom, 2018; Maya Jariego, 2003), curtailing the possibilities of the chronically poor achieving a livelihood.

Key components of social capital

To reiterate, most academics refer to the key components of social capital in their definitions, although each scholar emphasises different attributes. Pretty and Ward (2001) summarise the key components of social capital as these five elements: relations of trust, reciprocity and exchanges, common rules, norms and sanctions, and connectedness, networks and groups. However, separating relations of trust into membership to a group as Bourdieu (2001) indicates and trust, another part of social capital that academics address separately, is useful to describe each component. Therefore, the key elements of social capital can be identified as: membership, trust, reciprocity and exchange, norms and sanctions, and connectedness.

Membership

The first requirement for an individual to access social capital is to be part of a group, i.e. membership in a network or social structure. Membership is ratified by the usage of a common name – family, class, ethnic group, school, neighbourhood, place of work, among many others – and by the institutionalised rites of the group, which are enhanced by the occurrence of material exchanges and reciprocity, which act as symbolic exchanges. The incidence of exchanges restricts networks to proximal relations that occur in the same geographical, social, or economic spaces (Bourdieu, 2001).

Trust

Coleman (1988) proposes that social capital depends on trustworthiness in the social structure. This value is essential to foster cooperation among the members of a network. Pretty and Ward (2001) state that there are two types of trust: trust in the individuals one knows and trust in individuals because of the social structure they belong to. Trusting someone generates reciprocal trust because trust can be defined as the anticipation that each member of the social structure will do what is expected of them. Therefore, trust enables reduction of cost and time, as monitoring the contributions of its members becomes unnecessary because the network can rely on everyone fulfilling their obligations (Coleman, 1988). Contributing to a network generates 'obligation credits', that is, access to the social capital, at the time and in the form that the previous contributors' needs.

Reciprocity and exchange

Reciprocity originates in trustworthiness among members of a social structure (Portes, 2003). In his essay, 'The Gift: the form and reason for exchange in archaic societies', Mauss (2002) indicates that reciprocity is the capacity to give, receive, and return. However, reciprocity must be distinguished from exchange, which is defined as the mere circulation of objects. Instead, reciprocity compels the relationship between individuals making an exchange (Coraggio, 2014). It is the recognition of the other and the social structure, therefore, that makes the value of the redistributed object secondary to the act of interchanging it (Lopez Cordova, 2014). Reciprocity is a crucial component of social capital, as it is a requirement of the members of a network to continue to access the 'collective resources', while exchange is an act that can happen in the market without the need for reciprocity, trust, or membership of a network, and can happen both vertically and horizontally between individuals with high or low degrees of commonalities (Lopez Cordova, 2014).

Norms and sanctions

According to Polanyi (2001), reciprocity and exchange depend on the value systems (values, norms, and expectations) shared by a given society because the economic system is embedded in its social structure. Norms are what is considered standard or usual in a specific social group, and these are time-period specific or limited to various factors such as particular geographical or socio-economic spaces, political situation, religion and ethnicity (Johnson, 2016).

Norms delimitate the boundaries between the individual and the collective, as these are a consensus of what each member of a group accepts as favouring the common good (Pretty and Ward, 2001). Norms can be formal – laws and regulations established by authorities – or informal – customs or common rules that are implicit or agreed by a specific social group. When norms are violated, sanctions are imposed that can be formal or informal forms of punishments and can encompass social isolation (Coleman, 1988; Gonzalez de la Rocha, 1994a).

Connectedness (social structures/networks/groups)

Connectedness refers to the semi-institutionalised 'relationships of mutual acquaintance and recognition' of members of a group or network (Bourdieu, 2001: 103). According to Pretty and Ward (2001), there are various types of connections that can be one-way or two-way, long-term established, or those that require constant updating. However, Bourdieu (2001) suggests that NSE requires constant effort at the institutional level (e.g. family, religious group or neighbourhood) to produce and reproduce long-lasting relationships that assure access to social capital. This element of social capital and its importance to the livelihood strategies of marginalised groups is detailed in the next section.

Networks for livelihoods in the Global South

Networks of social exchange (NSE) are important for marginalised households in achieving livelihoods, particularly in the Global South. These social structures seem to enable resilience because social capital found through the NSE allows households to manage risks and cope with external shocks, as well as to foster community cohesion, and thereby, collective action for their development (Baird and Gray, 2014; Rockenbauch and Sakdapolrak, 2017). In their NSE, poor households find the social security (e.g. healthcare and pensions) denied to them by the hegemonic system, access to resources, information, contacts, and favours and collective development of survival strategies (González de la Rocha and Escobar Latapí, 2008; Maya Jariego, 2003). However, these are not a panacea for marginalised households as participating in the NSE requires reciprocity that is only possible when the household has a critical mass of material, human and financial resources (Thomas et al., 2010).

Networks as one of the household's capabilities for survival in Mexico

Gonzalez de la Rocha (1994a) observed urban-poor households in a Mexican city and identified four capabilities that enable households to achieve livelihood or mere survival. These household capabilities are shown in Figure 6.1 and consist of: the possibility of obtaining a fixed wage either in the formal or informal economic sectors, the production of goods and services for household consumption, production for petty trade and participation in their NSE.

While Mexican poor-rural households present a greater diversification of their livelihood strategies as a result of greater access to a broader variety of natural resources than poor-urban households, it is possible to observe that the same capabilities elaborated by Gonzalez de la Rocha (1994b) enable their livelihoods. Through a cluster analysis, Fierros and Avila-Foucat

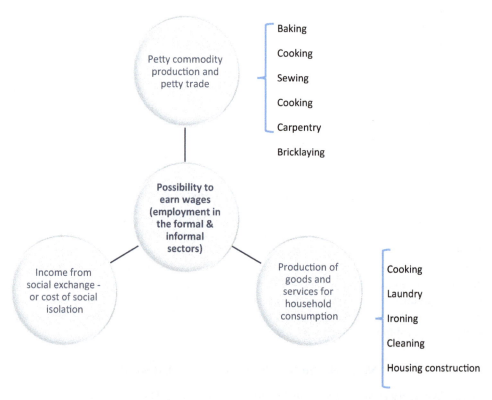

Figure 6.1 Structural conditions for household capabilities

Source: Based on Gonzalez de la Rocha (1994b, 2007)

(2017) characterised the livelihood strategies followed by Mexican rural households and determined that the main sources of income of these domestic units are wages from agricultural and non-agricultural activities, equivalent to fixed wages. These include traditional economic activities, agriculture, cattle raising and exploitation of natural resources and self-employment, which are comparable to the production of goods and services for self-consumption and petty trade, as well as remittances comparable to resources received through NSE.

This investigation defined four models of rural livelihood strategies; 'Small producers', 'Agricultural Wage-earners', 'Non-agricultural Wage-earners' and 'Family business owners', and their share of income per economic activity is shown in Table 6.1. This chart demonstrates that for the chronically poor, the 'Small Producers', remittances (e.g. resources obtained from their NSE) represents almost 33% of their income, while this source of income is less important for the groups with higher incomes, contributing only 16–18% to their livelihoods (Fierros and Avila-Foucat, 2017). This indicates that accumulation of capital allows rural households to diversify their livelihood strategies towards non-agricultural activities. This also suggest that the income obtained from their alternative economic activities substitutes, to a certain extent, their NSE (Avila-Foucat and Rodriguez-Robayo, 2018; Baird and Gray, 2014; Rockenbauch and Sakdapolrak, 2017).

Demystifying networks and social capital as the panacea of the poor

Mainstream understandings of social capital are based on the belief that the poor can access additional resources through their NSE unconditionally, disregarding that participation in these networks depend on the capacity of their members to reciprocate. Networks do not happen naturally, as they are the result of continuous, conscious and unconscious, collective and individual efforts, exchanges, and rites that allow the production and reproduction of long-lasting relations that facilitate the accumulation and procurement of resources (González De La Rocha, 2004, 2006, 1994a).

Therefore, membership of a network demands reciprocity, which depends on the availability of a critical mass of human, financial, and material resources (Thomas et al., 2010). This critical mass of resources consists of human capital in the form of labour, an overused resource by the chronically poor, as well as material resources such as a property or land and a minimum of

Table 6.1 Rural livelihood strategies models by Fierros and Avila-Foucat (2017)

Type of household		Small producers	Agricultural wage-earning	Non-agricultural wage-earning	Family business owners
Yearly income MXN		3,390	19,676	59,521	191,354
Source of income %	Agriculture	5.4	5.2	5.1	<u>20.7</u>
	Cattle raising	7.8	2.2	4.7	7.7
	Use of natural resources	<u>15.8</u>	4.9	3.3	3.1
	Agricultural wages	<u>21.3</u>	**38.5**	<u>18.3</u>	4.5
	Non-agricultural wages	10.4	23.5	**41.0**	**28.6**
	Self-employment	6.7	7.0	10.4	<u>19.7</u>
	Remittances USA and MEX	**32.6**	18.6	<u>17.1</u>	15.7

Source: Fierros and Avila-Foucat (2017)

financial resources. In case that the poor cannot cope with stresses and shocks, like chronic illness of the main breadwinner or deprivation of land for the creation of natural protected areas, they will see erosion of this critical mass of resources. This erosion can diminish the poor's capabilities to achieve a livelihood, leading to a loss of fixed wages, reduction of production of goods and services for household consumption and petty trade, and the penalty of social isolation (Gonzalez De La Rocha, 2004). This unveils the 'myth of survival' that suggests that the poor can access an infinite pool of resources through their NSE (González De La Rocha, 2007).

Additionally, academic literature on social capital overlooks the fact that the chronically poor and vulnerable have weak connections, characterised by unequal exchanges, and that investing in their NSE bears high costs with diffuse long-term benefits. Cleaver (2005) observed these inequalities in the linkages between the chronically poor and the richer households in Usangu, Tanzania, as these connections functioned as patron-client relationships. The poor had to exchange labour for food and credit in time of need;

> one poor person who still retained a few cattle grazed his animals with the much larger herd of his pastoralist neighbour. . . . The same wealthy neighbour sometimes hired the poor man for agricultural work and provided him with small amounts of credit in times of difficulty.
>
> *(Cleaver, 2005: 899).*

The recognition of the limits of social capital can prevent the implementation of public policies that weaken the livelihoods of the poor. Thomas et al. (2010) suggest that governments can play a key role in enabling the critical mass of resources that allows the poor to achieve livelihood or mere survival. This can be achieved with transferences and endowments of financial and human capital, for the creation of safety nets by covering basic needs of the chronically poor and facilitating resources during crisis. These strategies can lead to an increase of social capital. However, for these policies to work, the state must transform the structural conditions that reproduce the exclusion of the poor in their social relations and at the institutional level (Cleaver, 2005; Thomas et al., 2010).

Conclusion

Social capital is a key resource in the achievement of livelihoods or mere survival of the poor. Research in the Global South demonstrates that networks that enable access to social capital are one of the structural conditions for households to achieve livelihoods. The resources facilitated through these networks of social exchange (NSE) are the collective resources owned by their members, who share a common name, kin, neighbourhood, community, school or workplace, and who maintain their connections through continuous participation in rituals, exchanges and reciprocity.

Through NSE, members access resources such as financial capital or non-monetary resources that enable livelihoods. However, participation in these networks requires reciprocity, a value enabled by the existence of a fixed income. Therefore, social capital cannot be considered a panacea of the chronically poor, as their access to social capital is restricted to their capacity to give, receive, and return to the network, and it is bounded to the existence of a critical mass of resources.

Key Points

- Social capital is the aggregate of collective resources of a network that can contribute to the social, civic or economic wellbeing of individual members or for the collective. The

- key elements of social capital are membership of a group, trust, reciprocity and exchanges, norms and sanctions, and connectedness or networks.
- Social capital can be analysed from micro-, meso- and macro-level perspectives, and acknowledging the way it functions at each level allows understanding of the interdependence of the available social capital at each level. It is also studied from its function. Bonding social capital enables the survival or achievement of livelihoods, while vertical linkages of social capital, like bridging and linking social capital, can enable the improvement of the social structures of the poor. However, vertical linkages can diminish social capital due to power relations that potentially reduce trust.
- Networks of social exchange (NSE) enable livelihoods of the poor. However, access to the collective resources is constrained by reciprocity. The capability of being reciprocal depends on the availability of a critical mass of human, material and financial resources, which constrain the probabilities of survival of the chronically poor.
- Acknowledging the constraints of social capital allows the implementation of the public policies that create the safety nets – creation and maintenance of the critical mass of resources – necessary for the poor to achieve their livelihoods. However, to achieve these, the state must address the structural conditions that reproduce the exclusion of the chronically poor.

Further readings

1 Gannon, B. and Roberts, J. (2020) 'Social capital: exploring the theory and empirical divide', *Empirical Economics*, 58: 899–919. https://doi.org/10.1007/s00181-018-1556-y Social capital remains a controversial concept as more academics highlight its constraints and the mismatch between the theoretical and the empirical scholarly work. Brenda Gannon and Jennifer Roberts explore this, and demonstrate the importance of acknowledging the multi-dimensional nature of social capital in this article.
2 Portes, A. (1998) 'Social capital: its origins and applications in modern sociology', *Annual Review of Sociology*, 24: 1–24. http://doi.org/10.1146/annurev.soc.24.1.1 Literature on social capital is interdisciplinary and imprecise, highlighted by Alejandro Portes in this article.

References

Avila-Foucat, V.S. and Rodriguez-Robayo, K.J. (2018) 'Determinants of livelihood diversification: The case wildlife tourism in four coastal communities in Oaxaca, Mexico', *Tourism Management*, 69: 223–231. https://doi.org/10.1016/j.tourman.2018.06.021

Baird, T.D. and Gray, C.L. (2014) 'Livelihood diversification and shifting social networks of exchange: A social network transition?', *World Development*, 60: 14–30. https://doi.org/10.1016/j.worlddev.2014.02.002

Bebbington, A., Guggenheim, S., Olson, E. and Woolcock, M. (2004) 'Exploring social capital debates at the world bank', *The Journal of Development Studies*, 40(5): 33–64. https://doi.org/10.1080/0022038042000218134

Bhandari, H. and Yasunobu, K. (2009) 'What is social capital? A comprehensive review of the concept', *Asian Journal of Social Science*, 37: 480–510. https://doi.org/10.1163/156853109X436847

Bourdieu, P. (2001) 'The forms of capital', in Granovetter, M. and Swedberg, R. (eds.), *The sociology of economic life*, Boulder: Westview Press Books, pp. 96–111.

Cleaver, F. (2005) 'The inequality of social capital and the reproduction of chronic poverty', *World Development*, 33(6): 893–906. https://doi.org/10.1016/j.worlddev.2004.09.015

Coleman, J. (1988) 'Social capital in the creation of human capital', *American Journal of Sociology*, 94: S95–S120. www.jstor.org/stable/2780243

Coraggio, J.L. (2014) 'Una lectura de Polanyi desde la economía social y solidaria en América Latina', *Cadernos Metrópole*, 16(31): 17–35. https://doi.org/10.1590/2236-9996.2014-3101

Falk, I. and Kilpatrick, S. (2000) 'What is social capital? A study of interaction in a rural community', *Sociologia Ruralis*, 40(1): 87–110. https://doi.org/10.1111/1467-9523.00133

Fierros, I. and Avila-Foucat, V.S. (2017) 'Medios de Vida Sustentables y Contexto de Vulnerabilidad de los Hogares Rurales de Mexico', *Revista Problemas del Desarrollo*, 48(191): 107–131. https://doi.org/10.22201/iiec.20078951e.2017.191.58747

Gannon, B. and Roberts, J. (2020) 'Social capital: Exploring the theory and empirical divide', *Empirical Economics*, 58: 899–919. https://doi.org/10.1007/s00181-018-1556-y

Gelderblom, D. (2018) 'The limits to bridging social capital: Power, social context and the theory of Robert Putnam', *The Sociological Review*, 66(6): 1309–1324. https://doi.org/10.1177/0038026118765360

Gonzalez De La Rocha, M. (1994a) 'The penalties of social isolation', in Gonzalez de la Rocha, M. (ed.), *The resources of poverty women and survival in a Mexican City*, Cambridge, Massachusetts: Blackwell Publishers, pp. 213–228.

Gonzalez De La Rocha, M. (1994b) *The resources of poverty women and survival in a Mexican City*, Cambridge, Massachusetts: Blackwell Publishers.

Gonzalez De La Rocha, M. (2004) 'De los "Recursos de la Pobreza" a la "Pobreza de Recursos" y a las "Desventajas Acumuladas"', *Latin American Research Review*, 39: 192–195. https://doi.org/10.1353/lar.2004.0006

González De La Rocha, M. (2006) 'Vanishing assets: Cumulative disadvantage among the urban poor', *The Annals of the American Academy of Political and Social Science*, 606: 68–94. https://doi.org/10.1177/0002716206288779

González De La Rocha, M. (2007) 'The construction of the myth of survival', *Development and Change*, 38(1): 45–66. https://doi.org/10.1111/j.1467-7660.2007.00402.x

González De La Rocha, M. and Escobar Latapí, A. (2008) 'Choices or constraints? Informality, labour market and poverty in Mexico', *IDS Bulletin*, 39: 37–47. https://doi.org/10.1111/j.1759-5436.2008.tb00443.x

Granovetter, M.S. (1973) 'The strength of weak ties', *American Journal of Sociology*, 78(6): 1360–1380. https://doi.org/10.1086/225469

Grootaert, C. and Narayan, D. (2001) 'Local institutions, poverty, and household welfare in Bolivia', *World Development*, 32(7): 1179–1198. https://doi.org/10.1016/j.worlddev.2004.02.001

Halpern, D. (2005) *Social capital*, Malden, MA: Polity Press.

Johnson, L. (2016) 'What is social capital?', in Greenberg, A.G., Gullotta, T.P. and Bloom, M. (eds.), *Social capital and community well-being*, New York: Springer, pp. 53–66.

Kawachi, I. and Berkman, L. (2000) 'Social cohesion, social capital, and health', in Berkman, L.F., Kawachi, I. and Glymour, M.M. (eds.), *Social epidemiology*, Oxford: Oxford University Press.

Lopez Cordova, D. (2014) 'La reciprocidad como lazo social fundamental entre las personas y con la naturaleza en una propuesta de transformacion societal', in Marañón Pimentel, B. (ed.), *Buen Vivir y descolonialidad. Crítica al desarrollo y la racionalidad instrumentales*, Mexico: UNAM, Instituto de Investigaciones Económicas, pp. 98–118.

Mauss, M. (2002) *The gift: The form and reason for exchange in archaic societies*, London: Routledge.

Maya Jariego, I. (2003) 'Larissa Adler de Lomnitz: Categorías, redes y cadenas (Por qué se mantiene la desigualdad)', *Araucaria*, 5(10): 238–242. https://revistascientificas.us.es/index.php/araucaria/article/view/3241.

Onyx, J. and Leonard, R. (2010) 'The conversion of social capital into community development: An intervention in Australia's Outback', *International Journal of Urban and Regional Research*, 34(2): 381–397. https://doi.org/10.1111/j.1468-2427.2009.00897.x

Polanyi, K. (2001) *The great transformation: The political and economic origins of our time*, Boston: Beacon Press.

Portes, A. (1998) 'Social capital: Its origins and applications in modern sociology', *Annual Review of Sociology*, 24: 1–24. http://doi.org/10.1146/annurev.soc.24.1.1

Portes, A. (2003) 'Unsolved mysteries: The Tocqueville files II', *The American Prospect* [Online]. Available at: https://prospect.org/culture/unsolved-mysteries-tocqueville-files-ii-d3/ (Accessed 04 August 2020).

Portes, A. and Landolt, P. (1996) 'The downside of social capital', *The American Prospect*, pp. 18–94. http://hdl.handle.net/10919/67453 (Accessed 03 August 2020).

Pretty, J. and Ward, H. (2001) 'Social capital and the environment', *World Development*, 29(2): 209–227. https://doi.org/10.1016/S0305-750X(00)00098-X

Putnam, R. (1995) 'Tuning in, tuning out: The strange disappearance of social capital in America', *PS: Political Science and Politics*, 28(40): 664–683. https://doi.org/10.2307/420517

Putnam, R.D., Leonardi, R. and Nanetti, R.Y. (1994) *Making democracy work: Civic traditions in modern Italy*, Princeton: Princeton University Press.

Rockenbauch, T. and Sakdapolrak, P. (2017) 'Social networks and the resilience of rural communities in the Global South a critical review and conceptual reflections', *Ecology and Society*, 22(1): 10. https://doi.org/10.5751/ES-09009-220110

Siisiainen, M. (2003) 'Two concepts of social capital: Bourdieu vs. Putnam', Paper presented at *The International Society for Third Sector Research, 4th International Conference on 'The Third Sector: For What and For Whom'*, Trinity College, Dublin, 5–8 July.

Thomas, B.K., Muradian Sarache, R.P., Groot, G.A.D. and Ruijter, A.D. (2010) 'Resilient and resourceful?: A case study on how the poor cope in Kerala, India', *Journal of Asian and African Studies (Leiden)*, 45: 29–45. https://doi.org/10.1177/0021909610353580

Uphoff, E.P., Pickett, K.E., Cabieses, B., Small, N. and Wright, J. (2013) 'A systematic review of the relationships between social capital and socioeconomic inequalities in health: A contribution to understanding the psychosocial pathway of health inequalities', *International Journal for Equity in Health*, 12: 54. https://doi.org/10.1186/1475-9276-12-54

The World Bank (1998) *The initiative on defining, monitoring and measuring social capital*, Washington, DC: World Bank.

7
A RIGHTS-BASED APPROACH FOR SUSTAINABLE LIVELIHOODS

Jae-Eun Noh

Introduction

Development discourses have evolved through the addition of new concepts to specify and improve development practice. 'Human rights' is one of the concepts which emerged in development discourses in the late 1990s. The link between development and human rights has developed in the name of a 'human rights-based approach to development', more commonly in the shortened form, a 'rights-based approach' (RBA).

The objective of this chapter is to explore the potential contributions of a RBA to advancing sustainable livelihoods. This chapter examines the relationship between a RBA and a sustainable livelihood approach (SLA) and the implications of that relationship for livelihood-related strategies and outcomes. It is structured in three parts: 1) Part 1 is dedicated to a reflection on RBA in relation to its emergence, theoretical background, key features, operational implications and major critiques, 2) Part 2 explores how a RBA and a SLA supplement each other based on the similarities and differences between the two, 3) Part 3 describes ongoing attempts to incorporate a RBA into strategies for sustainable livelihoods drawing on a case study.

A rights-based approach

Why a rights-based approach to development?

Human rights and development have long been regarded as separate areas (Cornwall and Nyamu-Musembi, 2004; Uvin, 2004). The importance of human rights in development was acknowledged following historical events such as the end of the Cold War, the declaration of the rights to development in 1986 and the Vienna Conference on human rights in 1993 (Cornwall and Nyamu-Musembi, 2005; Hamm, 2001). The mainstreaming of human rights by the United Nations (UN) in the late 1990s is one of the most significant driving forces behind the rise of a RBA (Slim, 2002). As a result, a RBA gained wide recognition, particularly in UN agencies, some bilateral aid agencies and renowned international non-governmental organisations (NGOs) (Nelson and Dorsey, 2003). Although development organisations have developed their understandings, a RBA is widely defined as a framework 'for the process of human development that is normatively based on international human rights standards and operationally directed to promoting and protecting human rights' (OHCHR, 2006: 15) and 'for integrating the norms, principles, standards and goals of the international human rights system into the plans and processes of development' (Boesen and Martin, 2007: 9).

A RBA has been promoted in response to growing questions about the legitimacy and ineffectiveness of international aid (Cornwall and Nyamu-Musembi, 2005; Gready and Ensor, 2005; Slim, 2002). Firstly, a RBA can legitimise a development intervention by reframing it in terms of internationally agreed human rights and pushing the boundary of duty bearer from the state to various development players (Hamm, 2001; Nelson and Dorsey, 2003). Secondly, a RBA is expected to be sustainable in terms of impacts through tackling underlying causes of poverty (De Greiff and Duthie, 2009; Uvin, 2004). Accordingly, a RBA is regarded as improving legitimacy and effectiveness.

Theories and thinking behind it

Sen (1999) contributed towards the theory of a RBA by expanding the meaning of poverty from solely one of economic status to include considerations of deprivation of opportunities and power. Although Sen's work does not centre on human rights, but on capability, he draws attention to the importance of human rights realisation as the means and the ends of development. The interdependency between human rights and development challenged economy-centred development that lacked concern for process. In this sense, despite differences between the open concept of capability and the universal norm of a RBA, Sen's ideas are considerably reflected in a RBA, as acknowledged in the UNDP's report in 2000 (Frediani, 2010; Uvin, 2007).

Before Sen, Shue (1980) discussed 'basic rights' and included non-state actors as duty bearers as well as the state by listing three duty types: 'duties to avoid depriving, to protect from deprivation and to aid the deprived' (1980: 52). However, Shue's 'basic rights' invalidates the indivisibility of human rights by attaching instrumental values to certain types of rights and claims these rights based on morality. Overcoming Shue's limitations, Sen's argument reconfirms the equal importance of different rights. His study shows that the realisation of civil and political rights enables the protection of economic and social rights and vice versa. In addition, collective rights such as cultural rights, environmental rights and rights to sustainable development are increasingly recognised as essential for full development (Frezzo, 2015).

Key features and implications for development practices

The following quote from the UN Secretary-General (1997) (cited in DMCDD, 2017: 1; original italics) suggests how a RBA is different from existing approaches:

> *A rights-based approach to development describes situations not simply in terms of human needs, or developmental requirements, but in terms of society's obligations to respond to the inalienable rights of individuals, empowers people to demand justice as a right, not as a charity, and gives communities a moral basis from which to claim international assistance when needed.*

As suggested above, distinctions between a RBA and other approaches can be drawn by how focal problems are identified, goals are set, and relationships are built with aid recipients.

In a RBA, development is understood as a process of realising human rights (Conway, 2001). Its goals are set in terms of protecting, promoting and fulfilling human rights (Jonsson, 2003) and the process should be in line with principles such as participation, empowerment, accountability, non-discrimination and linkage to international human rights instruments (Boesen and

Martin, 2007; OHCHR, 2006). Rather than providing resources to meet identified needs, a RBA focuses on structural issues that prevent human rights' realisation and aims to foster power in people (OHCHR, 2006). In tackling structural issues, a RBA emphasises the state's accountability, which has shrunk in the era of neo-liberalism, and the politicisation of development, which has been challenged by managerialism and instrumentalism (Gready, 2008). Uvin holds that a RBA should be about 'the way the interactions between citizens, states, and corporations are structured, and how they affect the most marginalised and weakest in society' (2007: 600). In a RBA, people are rights holders, not pity-seekers (Slim, 2002). To summarise, a RBA shifts development discourse from charity to entitlement, from alleviating symptoms to addressing root causes of poverty, and from immediate outputs to both ultimate outcomes and process.

Critiques

Conceptually, talk of human rights has provoked debates about the universalism of human rights and cultural relativism (Donnelly, 1984), individual and collective rights (Chabal, 2012; Offenheiser and Holcombe, 2003) and rights and responsibilities (Griffin, 2008). These debates have involved consideration of differences between the West and the non-West. For example, Westerners are regarded as being familiar with human rights with their tradition of valuing equality, individuals and rights, while non-Westerners focus on socio-cultural norms, community and responsibility (Hugman, 2010). Examples of critiques from a non-Western perspective are the African Charter on Human and People's Rights and the Asian Value debate.

Adopted by 49 countries in 1986, the African charter emphasises traditional values and collective rights (Steiner et al., 2008). A set of 'Asian values' was pitched by Asian leaders in the 1993 Bangkok Declaration, responding to arguments that the human rights regime is blind to particular cultural values (Engle, 1999). A counter-argument is that Asian culture is not homogeneous and does not have particular 'Asian values' (Barr, 1998). Such differences between Western and non-Western countries raise the question of what constitutes human rights in non-Western countries (Chabal, 2012). This is why concerns are expressed regarding the legacy of this Western philosophy, even among those who hold hope for a RBA (Nyamu-Musembi and Musyoki, 2004) and more reference to local perspectives is required (Gready and Ensor, 2005).

When it comes to development practice, a gap between discourse and practice has been noted in empirical studies. In the first place, a RBA has not been comprehensively embraced by development organisations. This fragmented adoption of a RBA has provoked criticism that rights language is used to repackage old practices (Cornwall and Nyamu-Musembi, 2004; Uvin, 2004) and that RBA elements are selectively accepted (Frankovits, 2008). Some studies note that a RBA can be inappropriate for some communities because of its limited focus on service delivery (Broberg and Sano, 2018) or its possibility of endangering development workers and community members (Slim, 2002). A RBA's confrontational strategies can be counter-productive (Hickey and Mitlin, 2009).

Another critique concerns its lack of enforceability. Scarcity of resources or institutional incapability has often been the excuse for failing to protect rights, in particular economic, social and cultural rights (Hickey and Mitlin, 2009). This is why Kreimer (1984) expressed concerns about difficulties in telling the difference between failure to act and violation of human rights. In addition, a RBA is questioned as it is not so practical for non-state actors, including corporations, which are not signatories of international human rights conventions (Onazi, 2013). Although there are some examples that human rights NGOs brought justice to transnational corporations' wrongdoings abroad using the national legal framework in the USA (Karp, 2014), it is still challenging to make corporations accountable to the poor. Corporations are reported to lack awareness of human rights around their business practice and willingness to work on them (Van der Ploeg and Vanclay, 2018). Corporations tend to avoid human rights violations only when stakeholder pressure exists (Rathert, 2016).

Complementarities between a RBA and a SLA

Chapter 2 introduced livelihoods as 'the means of gaining a living, including livelihood capabilities, tangible assets and intangible assets' (Chambers and Conway, 1992: 9). A SLA recognises the importance of assets for poor people's livelihoods, both material and social resources, those they own and those they do not own. It also provides a framework for analysing stresses and shocks and developing policies relevant to the poor (DFID, 1999).

Similarities

Both approaches commonly highlight the importance of founding values such as empowerment and participation. Both a SLA and a RBA oppose dominant discourses which centre on economic growth and technical intervention. Both approaches seek to achieve empowerment of the most vulnerable by strengthening their capabilities. A SLA empowers people by creating opportunities and conditions for sustainable livelihoods and facilitates people's participation in planning. However, some development agencies understand a SLA as a linear framework based on the logic of cause and effect (Morse and McNamara, 2013). A RBA is also suggested to have the potential for restoring authentic values of participation and empowerment, which have been misused by the neoliberal notion of development (Ife, 2010).

In addition to sharing values underpinning the alternative discourses as discussed above, a SLA and a RBA commonly suggest the critical roles of policies and institutions (Farrington, 2001). Both approaches take policy analyses and institutional appraisals as a starting point to understand how people's lives are shaped. Gender and power are also analysed to uncover the dynamics across various issues of access, control, vulnerability, exclusion and social relations. A RBA places additional emphasis on the accountability of the state, which is generally the primary duty bearer in the human rights framework.

In practice, both take multi-sectoral approaches to poverty eradication (Farrington, 2001; Thomas et al., 2015). This is related to their understanding of poverty. Considering the multidimensional nature of poverty (see Chapter 2), both do not run sectoral programmes in isolation. Instead, each approach provides an overarching framework for embracing multi-sectoral strategies and programmes. Such frameworks lead to 'multi-sectoral mandates, multiple skill sets and multiple levels of working' (Gready, 2013: 1341).

Differences

Despite many similarities, there are differences between the two approaches. Firstly, a RBA tends to give primacy to individual rights although collective rights are also recognised in the RBA, while a SLA engages with a community for collective activities. This difference can yield tensions: a SLA highlights local norms and knowledge, while a RBA refers to universal human rights (Nyamu-Musembi and Musyoki, 2004); a community-centred SLA can retain inherent tensions between individuals and sub-groups in a community (Farrington, 2001) and a RBA can fuel the tensions through making them visible and siding with the most marginalised.

Secondly, a RBA seeks articulation of 'what people's entitlements are' while a SLA seeks an understanding of 'what impact the absence of certain entitlements has on people's livelihoods' and responding to its influences on the lives of the poor (Farrington, 2001: 3). Further, a RBA is seen as not necessarily helping environmental sustainability as it gives priority to rights over responsibility and to present generations over future generations (Conway et al., 2002). Thirdly, both approaches commonly pay attention to institutional contexts. However, institutions are regarded as a mechanism to improve livelihoods in a SLA whereas institutional capacity building is a critical goal to make duty bearers accountable in a RBA (Conway et al., 2002). Lastly, a RBA aims to bring out structural and societal changes (Thomas et al., 2015). Therefore, it is often seen as normative and highly critical, distinguishing it from other pragmatic and non-confrontational approaches (Gready, 2015). A SLA highlights wellbeing and sustainability rather than fundamental changes (Foresti et al., 2007). Its ameliorative orientation toward sustainable livelihoods keeps it distanced from any transformative measures, which can involve conflicts (Moser et al., 2001). However, the SLA's limited attention to political power has been critically noted, and its addition of political capital to the asset pentagon was suggested to promote action to enhance livelihoods (Baumann and Sinha, 2001).

Mutual reinforcement

Despite notable differences between a SLA and a RBA, they can be complementary, both theoretically and practically. Combining a SLA and a RBA can provide a comprehensive view of people's lives and an analytical and operational tool for sustainable development (Moser et al., 2001). Selznick (1995) maintains the importance of engaging with communities to understand and operationalise human rights in a specific context. Onazi (2013) suggests that the pursuit of human rights realisation can be reinforced by incorporating the theory of 'community', which is less confrontational in pursuing human rights with a focus on obligation and love.

In practice, a SLA enables prioritisation for the effective use of limited resources, while a RBA regards every right as equally important (Farrington, 2001). RBA's lack of concern about environmental sustainability can also be complemented by a SLA (Foresti et al., 2007). A RBA can help identify threats and opportunities for livelihood strategies (Hickey and Mitlin, 2009). A RBA also provides common language and tools for civil society to urge the state's accountability for the livelihoods of the vulnerable (Conway et al., 2002). Therefore, embracing a RBA is supported in the expectation of the long-term sustainability of pro-poor policies and programs (Conway et al., 2002). In short, a RBA's insight into power dynamics and a SLA's prioritisation of interventions can be complementary (Conway et al., 2002).

The nexus of SLA and RBA: sustainable livelihoods as rights in the Global South

Some development organisations have incorporated a RBA in their livelihood programmes. Rights are regarded as essential to secure sustainable livelihoods, as they involve access to and control of resources (Brown et al., 2006). Existing studies demonstrate how a RBA has been incorporated to strengthen livelihood opportunities and assets for the most marginalised, including street children (Trent and von Kotze, 2009), slum dwellers (Power and Wanner, 2017) and sex workers (Rosenberg and Bakomeza, 2017). In a comparative multiple case study of livelihood approaches, CARE International adopted RBA elements such as accountability and social justice in its livelihood programmes (Hussein, 2002).

This section introduces an example of combining a SLA with a RBA to identify the implications for development practices. The case project selected is the 'Women's Rights to Sustainable Livelihoods Project' implemented by ActionAid. ActionAid is a development NGO that has aligned its work with a RBA since its adoption in 1998 (Archer, 2011). This case project served as a pilot project for blending women's rights with livelihoods in ActionAid. An external review of this project captures well both strengths and weaknesses of a RBA (see Nordic Consulting Group and Policy Research Institute, 2016).

As denoted in the title of ActionAid's project, livelihoods are understood as an entitled right. There is no explicit expression of 'right to livelihoods' in international human rights laws and conventions. However, relevant ideas were recognised as 'right to work' and 'right to adequate food' in the Universal Declaration of Human Rights, 'right to decent living' and 'right to safe and healthy working conditions' in the International Covenant on Economic, Social and Cultural Rights, and 'collective right to land and resources' in the UN Declaration on the Rights of Indigenous Peoples (PDHRE, n.d.).

ActionAid in Ghana and Rwanda aimed to improve food security through climate-resilient and sustainable agriculture interventions. Food security is greatly influenced by socio-economic and political conditions, gender, and power relations (Mehta et al., 2019). Given that women are disproportionately influenced by food insecurity despite their essential roles in food production and caregiving, ActionAid put a clear focus on female farmers. As many RBA programmes are centred around awareness-raising and rights claiming, this project also involved sensitising to women's rights around food security and unpaid care work. Acknowledging that raised awareness is not enough to realise women's rights, ActionAid highlighted the importance of policy changes to support women's participation in food production and advocated for an alternative approach to conventional agriculture to be included in the agenda of the African Union. Mobilised female smallholder farmers made claims for land and water and held duty bearers such as traditional community leaders and landlords accountable. Although this project did not show strong advocacy for nationwide changes in laws and policies, this project was evaluated to be well aligned with a RBA by making duty bearers accountable and rights holders empowered.

This project shows how some shortcomings of a RBA could be addressed: its lack of commitment to sustainability was complemented by promoting sustainable agriculture; its failure to meet economic needs was addressed by diversifying foods to be resilient to climate change. In contrast to other studies that observed conflict between environmental sustainability and people's rights to livelihoods, this project framed food security as directly associated with adopting climate-resilient agriculture to reduce vulnerability. The project acknowledged that a right to livelihoods could be effectively promoted when rights to safe water and land are secured. Further, the evaluation report confirmed the necessity for expanding the project to include income and nutrition security, technology literacy, entrepreneurship training and microfinance. These recommendations are primarily related to livelihood strategies, which seek alternatives rather than stopping at making claims (Hickey and Mitlin, 2009). This case study also illustrates that sustainable livelihoods can be attained through multi-sectoral approaches: analysing vulnerabilities and working on them at domestic, community, national and regional levels, encompassing livelihood-related rights issues, and employing diverse strategies from training to advocacy.

Translation of a SLA and a RBA is greatly influenced by organisational and national contexts (Brocklesby and Fisher, 2003; Noh, 2017). A comparative study suggests that ActionAid's RBA is centred on grassroots movements, unlike other RBA NGOs which focus on campaigning or referring to conventions (Plipat, 2005). This difference implies that other organisations can employ different strategies (e.g. national legislation, the establishment of a network or coalition) when incorporating them in sustainable livelihoods programmes. This case study project captures differences as shaped by national contexts within the same organisation. For example, land rights and access to services were identified as critical in both countries. However, land for female farmers was restricted due to the limited size of land available to the family in Rwanda. In Ghana, it was due to the denial of women's access to land. Highly needed services were also different:

tractors in Rwanda and water in Ghana. Therefore, each programme should be assessed by considering its specific contexts.

Ways forward

This chapter attempted to show how a RBA can be understood and practiced concerning sustainable livelihoods. A RBA has been suggested to have added value to improving people's livelihoods. Local communities have developed their livelihoods strategies based on their local knowledge and lived experiences (Banerjee and Duflo, 2011). Incorporating a human rights perspective into livelihoods programmes can strengthen their identification and analysis of the structural causes of poverty and vulnerability. In this way, a RBA can provide analytical and operational tools to secure sustainable livelihoods, which involve access to and control of resources (Brown et al., 2006). Ultimately, enhancing human rights can be a foundation for advancing sustainable livelihoods by reducing risks and building the capabilities of both right holders and duty bearers.

However, the reality makes it very complicated because of diverse interpretations and selective adoption of human rights (Cornwall and Molyneux, 2006). Challenging existing power relationships is essential for a RBA to be transformative (Foresti et al., 2007; Nyamu-Musembi and Cornwall, 2004). This can create divisions among community members (Cornwall and Molyneux, 2006). In addition, some studies reported tensions over which rights or whose rights matter. For example, studies on community forestry in Thailand depicted tensions between environmental conservation and livelihoods rights of the poor (Sato, 2003), commercial interests of the private sector and collective interests (Johnson and Forsyth, 2002), and urban elites and rural farmers (Wittayapak and Baird, 2018).

To summarise, a RBA can be powerful for concerting development actors' efforts to improve sustainable livelihoods with its focus on structural issues and engagement with local, national, regional and international levels. At the same time, a SLA can be helpful to prioritise and operationalise rights-guided interventions and develop alternative ways of realising rights to livelihoods. However, combining a RBA and a SLA is neither easy nor necessarily desirable because of their innate differences and the complicated reality facing the poor in the Global South (Farrington, 2001). Rather than mixing the end-product of two approaches, reflecting on where each approach came from can guide us to improve development practice.

Key points

- A rights-based approach (RBA) can contribute to designing and implementing sustainable livelihood programmes.
- Sustainable livelihood is recognised as a right and pursued with other relevant rights in a RBA.
- A RBA offers an insight into structural power issues, identifies rights-holders and corresponding duty-bearers, and strengthens the capabilities of both parties.

Key readings

Farrington, J. (2001) *Sustainable livelihoods, rights and the new architecture of aid*, London: Overseas Development Institute.
Hickey, S. and Mitlin, D. (eds.) (2009) *Rights-based approaches to development: Exploring the potential and pitfalls*, Sterling: Kumarian Press.
Uvin, P. (2004) *Human rights and development*, Bloomfield: Kumarian Press.

References

Archer, D. (2011) 'Developing a new strategy for ActionAid to advance a human rights-based approach to development', *Journal of Human Rights Practice*, 3(3): 332–354. https://doi.org/10.1093/jhuman/hur018

Banerjee, A.V. and Duflo, E. (2011) *Poor economics: A radical rethinking of the way to fight global poverty*, New York: Public Affairs.

Barr, T. (1998) *Human rights conflict and the 'Engaged Australian': The challenge of culture in Australia's relations with Indonesia: 1983–1006*, PhD Thesis, Brisbane: University of Queensland.

Baumann, P. and Sinha, S. (2001) *Linking development with democratic processes in India: Political capital and sustainable livelihoods analysis*, London: Overseas Development Institute.

Boesen, J.K. and Martin, T. (2007) *Applying a rights-based approach: An inspirational guide for civil society*, Copenhagen: The Danish Institute for Human Rights.

Broberg, M. and Sano, H.O. (2018) 'Strengths and weaknesses in a human rights-based approach to international development – an analysis of a rights-based approach to development assistance based on practical experiences', *The International Journal of Human Rights*, 22(5): 664–680. https://doi.org/10.1080/13642987.2017.1408591

Brocklesby, M.A. and Fisher, E. (2003) 'Community development in sustainable livelihoods approaches – an introduction', *Community Development Journal*, 38(3): 185–198.

Brown, D., Wells, A., Luttrell, C. and Bird, N. (2006) 'Public goods and private rights: The illegal logging debate and the rights of the poor', in Bird, N. (ed.), *Human rights and poverty reduction: Realities, controversies and strategies*, London: Overseas Development Institute, pp. 85–90. https://cdn.odi.org/media/documents/2398.pdf#page=92

Chabal, P. (2012) *The end of conceit: Western rationality after postcolonialism*, New York: Zed Books.

Chambers, R. and Conway, G. (1992) *Sustainable rural livelihoods: Practical concepts for the 21st century*, Brighton: Institute of Development Studies.

Conway, T. (ed.) (2001) *Case studies on livelihood security, human rights and sustainable development*, Prepared for the workshop on human rights, assets and livelihood security, and sustainable development, 19–20 June 2001, London: ODI.

Conway, T., Moser, C., Norton, A. and Farrington, J. (2002) 'Rights and livelihoods approaches: Exploring policy dimensions', *Natural Resource Perspectives*, 78: 1–6.

Cornwall, A. and Molyneux, M. (2006) 'The politics of rights – dilemmas for feminist praxis: An introduction', *Third World Quarterly*, 27(7): 1175–1191. https://doi.org/10.1080/01436590600933255

Cornwall, A. and Nyamu-Musembi, C. (2004) 'Putting the "rights-based approach" to development into perspective', *Third World Quarterly*, 25(8): 1415–1437. https://doi.org/10.1080/0143659042000308447

Cornwall, A. and Nyamu-Musembi, C. (2005) 'Why rights, why now? Reflections on the rise of rights in international development discourse', *IDS Bulletin*, 36(1): 9–18. https://doi.org/10.1111/j.1759-5436.2005.tb00174.x

Danish Mission Council Development Department (DMCDD) (2017) *Position paper on a rights-based approach to development*, Frederiksberg: Danish Mission Council Development Department. www.dmcdd.org/fileadmin/Filer/Dokumenter/Position_paper_on_RBA.pdf

De Greiff, P. and Duthie, R. (2009) *Transitional justice and development: Making connections*, New York: Social Science Research Council.

Department for International Development (DFID) (1999) *Sustainable livelihoods guidance sheets*, London: DFID.

Donnelly, J. (1984) 'Cultural relativism and universal human rights', *Human Rights Quarterly*, 6(4): 400–419.

Engle, K. (1999) 'Culture and human rights: The Asian values debate in context', *International Law and Politics*, 32: 291–333.

Farrington, J. (2001) *Sustainable livelihoods, rights and the new architecture of aid*, London: Overseas Development Institute.

Foresti, M., Ludi, E. and Griffiths, R. (2007) *Human rights and livelihood approaches for poverty reduction*, London: Overseas Development Institute.

Frankovits, A. (2008) 'The efforts of the international community to integrate human rights and development cooperation', presented at *BASPIA Conference*, Seoul: Human Rights Council of Korea.

Frediani, A.A. (2010) 'Sen's capability approach as a framework to the practice of development', *Development in Practice*, 20(2): 173–187. https://doi.org/10.1080/09614520903564181

Frezzo, M. (2015) *The sociology of human rights*, Bristol: Polity Press.

Gready, P. (2008) 'Rights-based approaches to development: What is the value-added?', *Development in Practice*, 18(6): 735–747. https://doi.org/10.1080/09614520802386454

Gready, P. (2013) 'Organisational theories of change in the era of organisational cosmopolitanism: Lessons from ActionAid's human rights-based approach', *Third World Quarterly*, 34(8): 1339–1360. https://doi.org/10.1080/01436597.2013.831535

Gready, P. (2015) 'Theories of change for human rights and for development', in Lennox, C. (ed.), *Contemporary challenges in securing human rights*, London: Human Rights Consortium.

Gready, P. and Ensor, J. (2005) *Reinventing development?: Translating rights-based approaches from theory into practice*, London: Zed Books.

Griffin, J. (2008) *On human rights*, Oxford: Oxford University Press.

Hamm, B.I. (2001) 'A human rights approach to development', *Human Rights Quarterly*, 23(4): 1005–1031.

Hickey, S. and Mitlin, D. (ed.) (2009) *Rights-based approaches to development: Exploring the potential and pitfalls*, Sterling: Kumarian Press.

Hugman, R. (2010) *Understanding international social work: A critical analysis*, London: Palgrave Macmillan.

Hussein, K. (2002) *Livelihoods approaches compared: A multi-agency review of current practice*, London: DFID and ODI.

Ife, J. (2010) *Human rights from below: Achieving rights through community development*, Melbourne: Cambridge University Press.

Johnson, C. and Forsyth, T. (2002) 'In the eyes of the state: Negotiating a "rights-based approach" to forest conservation in Thailand', *World Development*, 30(9): 1591–1605. https://doi.org/10.1016/S0305-750X(02)00057-8

Jonsson, U. (2003) *Human rights approach to development programming*, Nairobi: UNICEF.

Karp, D.J. (2014) *Responsibility for human rights: Transnational corporations in imperfect states*, Cambridge: Cambridge University Press.

Kreimer, S.F. (1984) 'Allocational sanctions: The problem of negative rights in a positive state', *University of Pennsylvania Law Review*, 1293–1397. https://doi.org/10.2307/3311891

Mehta, L., Oweis, T., Ringler, C., Schreiner, B. and Varghese, S. (2019) *Water for food security, nutrition and social justice*, London: Routledge.

Morse, S. and McNamara, N. (2013) *Sustainable livelihood approach: A critique of theory and practice*, Berlin: Springer Science & Business Media.

Moser, C., Norton, A., Conway, T., Ferguson, C. and Vizard, P. (2001) *To claim our rights: Livelihood security, human rights and sustainable development*, London: Overseas Development Institute.

Nelson, P. and Dorsey, E. (2003) 'At the nexus of human rights and development: New methods and strategies of global NGOs', *World Development*, 31(12): 2013–2026. https://doi.org/10.1016/j.worlddev.2003.06.009

Noh, J.E. (2017) 'Contextualisation of human rights discourse by NGO workers in the context of Bangladesh', *Journal of International Development*, 29(8): 1106–1122. https://doi.org/10.1002/jid.3274

Nordic Consulting Group and Policy Research Institute (2016) *Women's rights to sustainable livelihoods project. End-of-project evaluation report*, London: ActionAid.

Nyamu-Musembi, C. and Cornwall, A. (2004) *What is the "rights-based approach" all about?: Perspectives from international development agencies*, Brighton: Institute of Development Studies.

Nyamu-Musembi, C. and Musyoki, S. (2004) *Kenyan civil society perspectives on rights, rights-based approaches to development, and participation*, Brighton: Institute of Development Studies.

Offenheiser, R.C. and Holcombe, S.H. (2003) 'Challenges and opportunities in implementing a rights-based approach to development: An Oxfam America perspective', *Nonprofit and Voluntary Sector Quarterly*, 32(2): 268–301. https://doi.org/10.1177/0899764003032002006

Office of the United Nations High Commissioner for Human Rights (OHCHR) (2006) *Frequently asked questions on a human rights-based approach to development cooperation*, New York and Geneva: UN.

Onazi, O. (2013) *Human rights from community*, Edinburgh: Edinburgh University Press.

People's Movement for Human Rights Education (PDHRE) (n.d.). *The human rights to livelihood and land*. Available at: www.pdhre.org/rights/land.html (Accessed 24 August 2020).

Plipat, S. (2005) *Developmentizing human rights: How development NGOs interpret and implement a human rights-based approach to development policy*, PhD Thesis. University of Pittsburgh, Pittsburgh.

Power, S.L. and Wanner, T.K. (2017) 'Improving sanitation in the slums of Mumbai: An analysis of human rights-based approaches for NGOs', *Asian Studies Review*, 41(2): 209–226. https://doi.org/10.1080/10357823.2017.1298566

Rathert, N. (2016) 'Strategies of legitimation: MNEs and the adoption of CSR in response to host-country institutions', *Journal of International Business Studies*, 47(7): 858–879. https://doi.org/10.1057/jibs.2016.19

Rosenberg, J.S. and Bakomeza, D. (2017) 'Let's talk about sex work in humanitarian settings: Piloting a rights-based approach to working with refugee women selling sex in Kampala', *Reproductive Health Matters*, 25(51): 95–102. https://doi.org/10.1080/09688080.2017.1405674

Sato, J. (2003) 'Public land for the people: The institutional basis of community forestry in Thailand', *Journal of Southeast Asian Studies*, 34(2): 329–346. https://doi.org/10.1017/S0022463403000286

Selznick, P. (1995) 'Thinking about community: Ten theses', *Society*, 32(5): 33–37. https://doi.org/10.1007/BF02693335

Sen, A. (1999) *Development as freedom*, Oxford: Oxford University Press.

Shue, H. (1980) *Basic rights*, Princeton: Princeton University Press Princeton.

Slim, H. (2002) 'Not philanthropy but rights', *International Journal of Human Rights*, 6(2): 1–22. https://doi.org/10.1080/714003759

Steiner, J., Alston, P. and Goodman, R. (2008) *International human rights in context: Law, politics, morals: Text and materials*, New York: Oxford University Press.

Thomas, R., Kuruvilla, S., Hinton, R., Jensen, S.L., Magar, V. and Bustreo, F. (2015) 'Assessing the impact of a human rights-based approach across a spectrum of change for women's, children's, and adolescents' health', *Health & Human Rights: An International Journal*, 17(2).

Trent, J. and von Kotze, A. (2009) 'A place in society? Strengthening livelihood opportunities for street children – A rights-based approach', *Social Work/Maatskaplike Werk*, 45(2): 182–197. https://doi.org/10.15270/45-2-214

Uvin, P. (2004) *Human rights and development*, Bloomfield: Kumarian Press.

Uvin, P. (2007) 'From the right to development to the rights-based approach: How "human rights" entered development', *Development in Practice*, 17(4/5): 597–606. https://doi.org/10.1080/09614520701469617

Van der Ploeg, L. and Vanclay, F. (2018) 'Challenges in implementing the corporate responsibility to respect human rights in the context of project-induced displacement and resettlement', *Resources Policy*, 55: 210–222. https://doi.org/10.1016/j.resourpol.2017.12.001

Wittayapak, C. and Baird, I.G. (2018) 'Communal land titling dilemmas in northern Thailand: From community forestry to beneficial yet risky and uncertain options', *Land Use Policy*, 71: 320–328. https://doi.org/10.1016/j.landusepol.2017.12.019

PART I

Researching livelihoods

Approaches and methods

PART I

8
CRITICALLY UNDERSTANDING LIVELIHOODS IN THE GLOBAL SOUTH

Researchers, research practices and power

Sam Staddon

Introduction: why critical understanding is important for livelihoods research in the Global South

> *Even the best-trained, most experienced, and seemingly impartial professionals can make systematically biased decisions . . . [the] most pressing concern is whether professionals adequately <u>understand</u> the circumstances in which the beneficiaries of their policies actually live . . . [given they] have never been poor and thus have never personally experienced . . . poverty*
>
> *(World Bank, 2014: 180/186, emphasis added)*

Taken from a World Bank publication written for its staff on 'The biases of development professionals', this quote suggests that despite training and experience, professionals can still make 'biased' decisions due to the fact that they have never themselves been poor, and thus that they cannot truly 'understand' the livelihoods of those they seek to help. This chapter explores this assertion and expands upon it, introducing ideas of 'positionality' and engaging with the politics of our research practices, so as to offer a more nuanced account of 'understanding' in livelihoods research. Rather than focusing on livelihoods in the Global South in order to understand 'poverty' alone, this chapter instead adopts a 'critical science' approach (Box 8.1) to understanding livelihoods, doing so in order to contribute to more transformatory goals of global justice. The chapter therefore draws attention to the need in livelihoods research to understand relations of power and responsibility – including our own as researchers and development practitioners.

Adopting a critical science approach to 'understanding' in livelihoods research, this chapter aligns with long-held demands to attend to the imperial and colonial histories of development, the 'Othering' of those in the Global South, and the cognitive injustice and cultural imperialism of 'expert' development knowledge (Escobar, 1995; de Sousa Santos, 2008; Kothari, 2005; Said, 1978). Global anti-racist movements, such as Black Lives Matter, have brought renewed attention to the need to de-centre the 'white gaze' of development, with scholars arguing that 'to oppose racism one must notice race . . . transformative development requires speaking and writing race into existence, because it does exist' (Pailey, 2019: 743). This need to 'notice' inequalities and injustices is echoed in Sultana's (2019) proposition, that 'decolonizing development requires recognizing, understanding and addressing . . . historical theft and exploitation that built the West at the expense of the rest' (Sultana, 2019: 32). Such demands on the researcher to notice, to recognise, and to understand, are the focus of Smith's (1999) classic text *Decolonising Methodologies. Research and Indigenous Peoples*, in which she reminds researchers that the 'ways in which scientific research is implicated in the worst excesses of colonialism remains a powerful remembered history for many of the world's colonized' (Smith, 1999: 1).

The chapter explores what a commitment to these agendas – of de-centring, decolonising and pluralising development (Kothari et al., 2019) – might mean for the ways in which we conduct our livelihoods research in the Global South. Whilst other parts of this book provide accounts of theories, concepts and methods for researching and exploring livelihoods, this chapter aims to provide something of an over-arching framework for attending to the political dimensions and ethical imperatives of our research practices. It focuses attention explicitly on ourselves, as researchers, considering how we think about our livelihoods research, our positionality and relationships within it, our emotional and embodied engagements through it, and our attempts to represent the livelihoods we research and to reflect on those attempts. The chapter ultimately argues that before we can understand the livelihoods of others, we first need to understand ourselves, as researchers, and the politics and power of our research practices.

Researchers, research practices and power in livelihoods research

How, as researchers, we think about our livelihoods research

> No story can be told nor any theory proposed that is not responding to prior (implicit or explicit) questions, and our questions are always the products of our situated selves
>
> (Somers, 2008: cited in Ribot, 2014: 695)

In order to avoid recreating the extractive histories of research, Smith (1999) ends her classic text, introduced above, with a plea for 'Getting the Story Right, Telling the Story Well'. Given the importance of stories, narratives and framings of 'development' in the Global South (Box 8.2) – and the need to pluralise those and promote cognitive justice – we begin our exploration of critically understanding livelihoods, by focusing on how the stories we tell about livelihoods through our research are the products of our thinking as researchers (i.e. of our situated selves). This may sound frustratingly obvious, but given that 'ways of thinking and ways of doing are inseparable' (Kabeer, 1994), it is imperative to understand where our thinking (i.e. our knowledge and epistemology) arises from and what it leads us to do in our research. Epistemologies have been described as 'understandings of what can constitute knowledge, or what can be known' (Evely et al., 2008: 1) (Box 8.3). Whether explicitly acknowledged or not, epistemologies frame the ways we think – and are trained to think in our particular disciplines – and the ways in which we therefore conduct our research, our methodological choices, and the stories we tell as a result.

Box 8.1 'The Danger of a Single Story'

In her hugely eloquent Ted Talk, Nigerian author Chimamanda Ngozi Adiche, explores the power of stories and the danger of 'single stories', which she argues create stereotypes and are used to dispossess and malign. Reflecting deeply on herself and her own thinking in response for example to her upbringing, Adiche emphasises the need to unpack stereotypes (not because '*they are untrue, but that they are incomplete*'), and to search for stories which emphasise similarities and connections, which can be used to empower and humanise. She importantly draws attention to the significance of where people choose to start the stories they tell; '*Start the story with the failure of the African state, and not with the colonial creation of the African state, and you have an entirely different story*'. Whilst intended for a general public audience, the talk is a powerful articulation of the significance of stories, discourses, narratives and framings within international development, but also how, as researchers, we may push back against 'single stories' and stereotypes through careful, self-reflection.

www.ted.com/talks/chimamanda_adichie_the_danger_of_a_single_story?language=en

Box 8.2 Epistemologies underpinning our research (a very simplified portrayal of a complex continuum)

Objectivism and positivism: '*To a positivist, science provides the observer with an objective account of the world as a concrete entity. . . . The senses are used to accumulate data that are objective, discernible, and measurable, thus methods are chosen to obtain estimates of the truth, using data and estimators that are both unbiased and . . . precise*' (Evely et al., 2008: 2)

Methodological choices: Include standardised surveys and questionnaires gathering quantitative data on people's actions, behaviour and perceptions; 'control' communities may be studied as part of a comparative research design

Researcher role: Research is conducted by 'objective' or neutral enumerators, who are assumed not to influence the research findings, although they may seek to reduce 'bias' in the research based on aspects of their identity

Constructivism and Subjectivism: '*rejects the idea that objective "truth" is waiting to be discovered . . . assumes that different individuals construct meaning of the same object or phenomenon in different ways; how an individual engages with and understands their world is based on their cultural, historical, and social perspectives. . . . People impose meaning and value on the world and interpret it in a way that makes sense to them*' (Moon and Blackman, 2014: 6)

Methodological choices: Include interviews, focus groups, participant observation and other methods which generate qualitative data and research material on people's values, emotions, worldviews and the meanings associated with the research topic

Researcher role: Researcher is seen as influential in the research process, based on their positionality, and they will seek to reflect on the inevitable impact of that on their research findings

Critical science approaches (described above as those focusing on power and justice), are more typically aligned with constructivist and subjectivist epistemologies, and their associated emancipatory theories including feminism, Marxism and post-structuralism (Moon and Blackman, 2014). Critical science approaches to livelihoods research would thus reject the idea that there is an objective 'truth' waiting to be discovered through our research, and rather draw attention to the role of the researcher as central to the creation of research findings and the stories told about those whose livelihoods we research. We next explore the centrality of the researcher – and their positionalities and research relationships – in livelihoods research.

Our positionalities and relationships in livelihoods research

> *our representations, especially of marginalised Third World groups, are intimately linked to our positioning (socioeconomic, gendered, cultural, geographic, historical, institutional) . . . we cannot encounter the Third World today without carrying a lot of baggage . . . our interaction with, and representations of, the subaltern are inevitably loaded*
>
> *(Kapoor, 2005: 627–628)*

Engaging the work of development scholar Gayatri Spivak, this quote articulates why positionality is so important in our livelihoods research in the Global South – whether or not this 'baggage' and the 'loaded' nature of our research is acknowledged in and accounted for by the research's epistemology. Moving beyond an objectivist or positivist conception of 'bias', whereby attempts are made to reduce the influence of the researcher (Box 8.3), engaging with ideas of positionality (Box 8.4) forces us to reject narrow definitions of ourselves simply as professionals i.e. as researchers, and rather to embrace our intersectional identities and the personal.

Sharing in common an aspect of our identities, e.g. language or gender, can afford connections with research participants, providing us some form of 'insider' status, based on the idea that we may inherently understand them better than a researcher would who does not share that aspect of their identity. Considering the assertion above for example, that in international development 'to oppose racism one must notice race' (Pailey, 2019: 743), we might imagine that a researcher of colour might 'automatically' notice racism as operating within the lives of those they research, more than a white researcher might. Caution is required with naïve notions of 'insider' or 'outsider' however given these may simply reflect stereotypes, whilst in fact our identities intersect across multiple axes and our relationships with participants (and others) are situated in specific spaces and times, meaning they are fluid, dynamic and not predictable (Katz, 1994; Sultana, 2007) (Box 8.5). Within livelihoods research it is important to reflect upon our positionalities and to see them as part of complex and contingent relationships. For example, if interviews in research touch on sensitive issues of caste and caste-based discrimination and are conducted by a researcher of the same caste as the interviewee, we might expect a certain set of responses, however if the interview takes place in the house of a so-called 'higher' caste family, then we might anticipate a slightly, but potentially significantly, different response. Much livelihoods research involves a team of researchers, and thus their individual positionalities and the relationships within the team, are of obvious relevance here.

Box 8.3 Are you 'Set Up to Listen – and Understand – Well'?

The following excerpt is taken from a podcast in 2019 on 'The Darker Side of Development' in which two senior development professionals share their views on connections between knowledge and power within the international development sector. Their conversation articulates some of the nuances of positionality (for example going beyond a shared nationality to think also about class) and its connection to being able to listen to others, and ultimately to be able to understand them.

> It also draws attention to the speakers' own positionalities (as white men from the Global North), and how well, or not, they feel that sets them up to listen, and to understand development issues:
>
> A: *'So, there's a knowledge gap* [in international development], *often, and it's because we have folks on the ground that may not be deep, they may not know the context, they may not speak the language, or they may not speak the dialect, and so on and so forth'*.
>
> B: *'I think, more and more, that's an obsolete criticism, because what I see from both international NGOs and local NGOs is the people they have on the ground are people from that society, from that region. . . . One of the knowledge gaps that I've experienced working overseas is where you have local leaders who are a part of an educated elite. And they've grown up in the city. They're often the children of government officials and of university professors and the elite of that society. And it's not uncommon, in my experience, for that group to have less knowledge about conditions outside of the capital city or the urban center than some of the expatriates who have been living out in the rural areas for 5 or 10 years . . . who've immersed themselves in the culture and who have a real personal understanding of the dynamics of community life'*.
>
> A: *'I'm living now in a world where . . . checking one's own privilege, understanding the blinders that I certainly have as somebody who checks a lot of the privilege boxes. [Laughter] I'm white. I'm male. I'm heterosexual. I have enough money to-to not worry about falling asleep hungry. All of those things –* <u>*I'm not set up to listen well. I'm not set up to understand that well'*</u>
>
> https://degrees.fhi360.org/2019/06/the-darker-side-of-development-addressing-power-dynamics-within-development/ (emphasis added)

Returning to the quote from Kapoor (2005) that opened this section, we can see that thinking about positionality offers us a way to consider the 'baggage' we carry as researchers, and the 'inevitably loaded' nature of our relationships with, and representation of, those whose livelihoods we explore through our research. This demands that we remain attentive to the politics of representation in livelihoods research and that we be reflexive in our research practice (Haraway, 1998). Representation and reflexivity are explored further below, but for now, attention is turned to how we might build research relationships by paying careful attention to the emotional and embodied engagements inherent in livelihoods research.

Emotional and embodied engagements in livelihoods research

> *Emotions are fundamental to the ways we comprehend the world and to our experiences of it. To understand what it means to be deprived, to experience the world differently and to be motivated to change it means attending to the way we are affected by it and how we affect others*
>
> *(Wright, 2012: 1114)*

Through this quote, Wright (2012) argues for the centrality of emotions both to processes of development and to our responses to those processes as researchers. Emotions such as hope, despair, confidence and defiance are seen as central to development in that they are capable of producing both regressive racist politics but also resistance and progressive ethics of care (Wright, 2012). In research settings emotions are an intrinsic part of relationship-building, e.g. feelings of friendship and care towards collaborators (Blazek and Askins, 2019), as well as part of specific research methods, e.g. feelings of dislike towards particular interviewees (Smith, 2014). Emotions are thus seen to matter due to their relationality i.e. their capacity to circulate, and to create connections (or disconnects) between people, places and processes (Bondi, 2005). The artificial separation of thinking and emotions i.e. 'the head from the heart' (Gender At Work, 2021) is argued against by both practitioners and academics, who foreground *'emotions as connected to thought . . . emotions matter . . . from developing a topic or research approach, to later stages of knowledge circulation'* (Askins and Blazek, 2017: 1089).

Within research, emotions can emerge in response to and can be expressed through what Sultana (2007) describes as embodied 'everyday acts' of fieldwork. When conducting livelihoods research, which necessarily involves daily physical interactions with those whose livelihoods are of interest, everyday acts and decisions include those around what to wear, what to eat and how to address people. Reflecting on her own fieldwork, Sultana (2007: 379) writes that 'such little actions, however mundane, are not insignificant I believe, and speak to the embodied situatedness of me as the researcher that I had to constantly keep in mind'. Such acts not only produce emotional responses but are significant ethically, and hold potential to build relationships of reciprocity and responsibility through a 'politics of care'; as Blazek and Askins (2019: 465) write,

> relations initially entrenched in professional codes and expectations (of the organisation and academic research frameworks) extend through emotional and physical engagements with others into something else: complex, reciprocal relations that demand care-full responses and surpass our positions as researchers.

This statement reflects the importance of looking beyond our identities simply as professional researchers, to our intersectional identities and the personal (as discussed above), as well as the temporal dimensions of our research relationships, for example by looking beyond time-bound periods of fieldwork to consider relationships and engagement in the longer-term (Box 8.6).

Box 8.4 'So what kind of researcher are you?'

This was a question posed to me by a participant during my PhD research, which engaged with the livelihoods of rural communities involved in Nepal's devolved forest management (Staddon et al., 2014: Staddon et al., 2015). Before asking the question, the participant had relayed a story of two other PhD students who'd come to the village to conduct research in the past – one who'd helped raise funds to build a school for the village, the other who they'd never heard of again – they clearly thought more of the former than the latter. I welcomed the question, as it got to the heart of concerns I'd harboured throughout my PhD, of how I might adequately and meaningfully 'give back' to the communities I engaged with. I have written in detail about this ethical dilemma and the emotional and embodied challenges it raised, in terms of what I might 'give back', to whom, and when (Staddon, 2014). In that work I also shared;

> *I'd had many discussions, particularly with women, about my [perceived] lack of children given my age (33) and the fact that I'd already been married for five years, so I promised to bring the girls I told them I hoped I'd have to visit them in the future. Funnily enough, writing this four years later, I now have two girls. . . . I hope one day to be able to demonstrate to the villagers that I have gone on remembering them, despite the time and space that separates us. Whether I will find that they care about the fact that I have continued to think of them, should I return there, remains to be seen.*
>
> *(Staddon, 2014: 258/9)*

> Since that time, I have had the huge privilege of taking my two girls to Nepal, aged 6 and 8, and the villagers really *did* care that I'd taken them to meet them, and we received a very warm welcome. I had always had something of a distant relationship with the wife of one of my village Research Assistants, partly reflecting the norms of her caste, but partly, I presumed, due to my some-what strange identity as a foreign woman, who in effect occupied more of a man's role whilst in rural Nepal, given my freedom and independence to move around and do as I wished. Our relationship changed the instant I took my children to meet her; we were closer, more relaxed and we couldn't stop smiling, I presume as she could now see me and relate to me in a different way, as a woman and a mother – as she was. She plaited my girls' hair and invited them into her kitchen to cook with her and her daughter. It was a day I will never forget, and one I will cherish as a researcher, even though it took place long after research fieldwork and arguably had nothing to do with my professional identity, rather being all about the personal. The embodied act of travelling with my children, my friend plaiting my girls' hair and cooking with them, and the emotional responses it generated in all of us, to me demonstrate the centrality and beauty of relationships in research. Reflecting on these things has also helped me to further understand the dynamic and situated nature of positionality, and to appreciate the complexity of research ethics.

Emotions and embodied everyday acts during fieldwork help to build relationships between us as researchers and our research participants and research contexts, and ultimately allow us to more fully understand livelihoods. As articulated by Wright (2012) above, this comprehension can help motivate us to intervene in and change those livelihoods for the better. Our attempts to intervene in and represent the livelihoods we research, and the necessity of reflecting on these attempts, forms the next and final part of this section.

Representation and reflexivity in livelihoods research

> *Though the speaker may be trying to materially improve the situation of some lesser-privileged group, the effects of her discourse is to reinforce racist, imperialist conceptions and perhaps also to further silence the lesser-privileged group's own ability to speak and be heard*
>
> (Kapoor, 2005: 631–632)

The ethical and political challenges articulated here by Kapoor, express what has been referred to since the 1980s as a 'crisis of representation' in academic work led by those in the Global North or the elite global professional class (Katz, 1994; Nast, 1994). This 'crisis' clearly demonstrates how our positionalities as researchers are significant well beyond the fieldwork component of our research (as discussed above), influencing decisions over who is deemed eligible to represent the views of others, such as those whose livelihoods we research. Whilst there is no easy answer to this challenge; given that it will depend in part on our ethical and political perspectives, as well as the contexts of our research, scholars urge us not to give up on the possibilities offered by research to tackle global inequalities (Sultana, 2007). In order to avoid extractive and imperialist research, and to promote research which contributes to decolonial agendas, Smith (1999) (as we heard above), urges us to take care in our research to 'get the story right' and to 'tell it well'. The critical science perspective on livelihoods research expounded in this chapter aims, in a small way, to contribute to that agenda, by bringing attention to the politics of our research – but also its' potential power for good.

In their promotion of a 'relationship-centred approach to research', Blazek and Askins (2019), argue for the '*more than data*' aspects of research relations, proposing that insights, skills and emotional sensitivities acquired through research can be useful beyond the specific research project or representation of its findings, for example within University teaching, student supervision, scholar-activism and academic collegiality. This potential wider application of research insights is evident in Sultana's (2019) writing on the need to decolonise development higher education, in which she offers suggestions for how researchers may engage in that process, for example by adopting 'critical pedagogies of hope' and by 'nurturing global citizens who are critically self-reflexive, ethical, aware, and committed to building substantive solidarities and alliances' (Sultana, 2019: 37). This hopeful and nurturing work extends well beyond higher education institutions and students, given we all have colleagues and peers who we can reach out to and share our livelihoods research insights and reflections with; and who we can listen to and learn from in return.

'Deep listening' has been offered as one way of practicing 'intellectual humility' during fieldwork (Koch, 2020: 56), in which 'Humility is nothing more than being modest about our own knowledge claims' (ibid). Koch (2020) goes on to explain how such a reflective approach not only inspires her own research, but how it can be used as a teaching tool to inspire others; 'I consistently tell my students that what I love most about fieldwork – in places quite different from the United States and with individuals having backgrounds quite distinct from myself – is that *every single day I change my mind*' (Koch, 2020: 56, emphasis in original). Whilst changing our minds can cause confusion, such confusion and discomfort are considered a source of creativity in research (Idahosa and Bradbury, 2020); as articulated by Roy et al. (2016: 10), 'Doubt and contradiction rather than certainty are seen as generative of social change'. Similarly, within the international development sector, 'reflective learning' has been offered as a way of 'making a better world', through its promotion of a more considered and transformatory development practice; it is described as 'a deliberate process of becoming unsettled about what is normal . . . and recognising that . . . how [one] personally understand[s] and act[s] in the world is shaped by the interplay of history, power and relationships' (Eyben, 2014: 20). Other fields of work, such as feminist thought (see Chapter 16 in this volume), also have much to offer to further our understandings of representation and reflexivity in livelihoods research (Box 8.5) and on how we might engage with and lead it (Box 8.6).

Box 8.5 Feminist insights on representation and reflexivity

Feminist thought and action (i.e. praxis) articulates and embodies a critical, subjectivist approach to knowing and engaging with the world – as such, it has much to contribute to understandings of representation and reflexivity (and more) in our livelihoods research. Two examples are shared here as a glimpse into such work:

In *Living a Feminist Life* (2017), Sara Ahmed provides insights into positionality and self-reflection when writing that '*an embodied experience of power provides the basis of knowledge*', and thus that '*theory can do more the closer it gets to the skin*' (Ahmed, 2017: 10). This intimate and insightful relationship between experience and knowledge, points to the potential to understand those in our livelihoods research based on experiences shared in common, such as those of racism or sexism. This is not to suggest that all experiences are equal or experienced equally, rather that embodied (and emotional) experiences can offer a source of understanding which academic theory or writing alone cannot. As such we should be cautious in our livelihoods research of claims to represent knowledge or 'truth' which don't attend to the embodied, everyday engagements that engender all livelihoods, and our interpretations of them.

In their book, *Troubling gender expertise in environment and development. Voices from feminist political ecology* (2020), Bernadette Resurrección and Rebecca Elmhirst draw on feminist insights to unpack the professional livelihoods of 'gender experts', now ubiquitous in current day international development. Having occupied the role of gender expert themselves, they seek to reflect on and represent 'our' work, significantly with a crucial recognition that 'as experts, power also works through us: thus, we are not separate from the hegemonic institutions we are trying to transform' (2020: 224). By sharing conversations with a range of practitioners, the book finds that past training in feminist thinking and embodied engagements with feminist activists provide 'ethical moorings' for experts' diverse strategies, intended not as markers of professional standing, but rather 'to ensure and uphold a transformative mission' (2020: 227). In the book's foreword, Professor Melissa Leach draws attention to such careful feminist praxis, writing that 'often the best and most useful expertise is also the most negotiated, reflexive, and humble. These are hugely valuable qualities in people and professionals involved with environment and development, and especially so in an era where sustainability needs plural, politicised transformations. Gender expertise that is self-aware about its troubles could actually be part of the politics of hope that the world now needs' (Resurrección and Elmhirst, 2020: xxv). The same can be said of livelihoods researchers, with self-aware researchers' part of a politics of hope for a better world.

Box 8.6 Leading on global challenges by healing our own practices

Delivering a Plenary Lecture at the Development Studies Association Conference 2020, Srilatha Batliwala provided a powerful and persuasive account of what is required for 'New Leadership for Global Challenges', which was the theme of the conference. Batliwala spoke to the importance of three 'deep challenges' in development;

1   *The discursive* i.e. how we frame development challenges
2   *The relational* i.e. how we relate to each other, in order to shift power and address challenges
3   *The psychic* i.e. ourselves, our thinking, our emotions and our wellness

Exploring new ways of engaging in and leading development, Batliwala talked about the need to see ourselves as sites of change, and the *'need to heal our own practices'*. She argued for a 'dispersed' and 'un-authorised' form of leadership, directing attention and agency to all involved in development, and promoted an approach to leadership that is about caring for and releasing the power of others. This chapter has been an attempt to promote such an approach to engaging in and leading livelihoods research, in order that it contribute to a more transformatory global justice; touching on the power of framings and stories, of relationships and care, and of ourselves as researchers.

www.devstud.org.uk/past-conferences/2020-new-leadership-for-global-challenges/

Conclusion: why critical understanding is important for livelihoods research in the Global South

As livelihoods researchers, we obviously wish to 'understand' the livelihoods of those our research engages with, and to seek no doubt, in some small way at least, to transform those livelihoods either directly or indirectly. This chapter opened with a World Bank quote that drew attention to the 'bias' of development professionals, given, it argues, that they have never themselves been poor, meaning they struggle to understand those whose lives they seek to intervene in. Whilst a useful starting point for reflection as researchers, this chapter has critiqued that objectivist and positivist view of 'bias' and has adopted a critical science approach in order to complicate and deepen how we understand ourselves as researchers and our abilities to represent those who livelihoods we research. It has done so by exploring our thinking and epistemologies, our positionalities and research relationships, emotions and embodied engagements in our research, and the opportunities for more transformatory research practices afforded through reflexivity. Seeing ourselves and our research as imbricated and inculcated in an unequal world, forces us to face the politics and responsibilities associated with our livelihoods research; it also helps us to 'notice' better (cf. Pailey, 2019) the power of the stories we tell through our research, and thus helps us to tell the 'right stories' and to 'tell them well' (cf. Smith, 1999). The chapter ultimately argues that before we can understand the livelihoods of others, we first need to understand ourselves, as researchers, and the politics and power of our research practices.

Key points

- As livelihoods researchers, before we seek to understand the livelihoods of others, we should seek to understand ourselves as researchers, and our influence on our research practices and the understanding it generates.
- We should seek to reflect upon the influence of our thinking and epistemologies on the direction and intention of our research.
- We should consider how our unique and situated positionalities influence the process of our research and the relationships established through it.
- We should seek to engage with the emotional and embodied aspects of our research
- We should pay attention to the politics of our attempts to represent those whose livelihoods we research, and reflect on our attempts to do so.
- We should maintain a 'politics of hope' that our efforts and engagements as livelihoods researchers can contribute to more transformatory goals of global justice (Sultana, 2019).

Further reading

- Katz, C. (1994) 'Playing the field – Questions of fieldwork in geography', *Professional Geographer*, 46(1): 67–72. https://doi.org/10.1111/j.0033-0124.1994.00067.x
- Koch, N. (2020) 'Deep listening: Practicing intellectual humility in geographic fieldwork', *Geographical Review*, 110(1–2): 52–64. https://doi.org/10.1111/gere.12334
- Smith, L. (1999) *Decolonizing methodologies. Research and indigenous peoples*, Otago: Zed Books/Otago University Press.
- Sultana, F. (2007) '*Reflexivity, positionality and participatory ethics: Negotiating fieldwork dilemmas in international research*', ACME, 6(3): 374–385. https://acme-journal.org/index.php/acme/article/view/786
- Sultana, F. (2019) 'Decolonizing development education and the pursuit of social justice', *Human Geography*, 12(3): 31–46. https://doi.org/10.1177/194277861901200305

References

Ahmed, S. (2017) *Living a feminist life*, Durham and London: Duke University Press.
Askins, K. and Blazek, M. (2017) 'Feeling our way: Academia, emotions and a politics of care', *Social and Cultural Geography*, 18(8): 1086–1105. https://doi.org/10.1080/14649365.2016.1240224
Blazek, M. and Askins, K. (2019) 'For a relationship perspective on geographical ethics', *Area*, 52(3): 464–471. https://doi.org/10.1111/area.12561
Bondi, L. (2005) 'The place of emotions in research', in Davidson, J., Bondi, L. and Smith, M. (eds.), *Emotional geographies*, London: Routledge.
de Sousa Santos, B. (2008) *Another knowledge is possible: Beyond northern epistemologies*, London: Verso.
Escobar, A. (1995) *Encountering development: The making and unmaking of the Third World*, Princeton: Princeton University Press.
Evely, A., Fazey, I., Pinard, M. and Lambin, X. (2008) 'the influence of philosophical perspectives in integrative research: A conservation case study in the Cairngorms National Park', *Ecology and Society*, 13(52). www.ecologyandsociety.org/vol13/iss2/art52/
Eyben, R. (2014) *International aid and the making of a better world*, London: Routledge.
Gender At Work (2021) *Troubling gender expertise. Voices from feminist political ecology*, Webinar 15 April 2021. Available at: www.youtube.com/watch?v=N-ms5g1B0do (Accessed 13 May 2021).
Haraway, D. (1998) 'Situated knowledges: The science question in feminism and the privilege of partial perspective', *Feminist Studies*, 14(3): 575–599. https://doi.org/10.2307/3178066
Idahosa, G.E.-O. and Bradbury, V. (2020) 'Challenging the way we know the world: Overcoming paralysis and utilising discomfort through critical reflexive thought', *Acta Academica*, 52(1): 31–53. http://doi.org/10.18820/24150479/aa52i1/SP3
Kabeer, N. (1994) *Reversed realities. Gender hierarchies in development thought*, New York: Verso.
Kapoor, I. (2005) 'Participatory development, complicity and desire', *Third World Quarterly*, 26(8): 1203–1220. www.jstor.org/stable/4017712
Katz, C. (1994) 'Playing the field – questions of fieldwork in Geography', *Professional Geographer*, 46(1): 67–72. https://doi.org/10.1111/j.0033-0124.1994.00067.x
Koch, N. (2020) 'Deep listening: Practicing intellectual humility in geographic fieldwork', *Geographical Review*, 110(1–2): 52–64. https://doi.org/10.1111/gere.12334
Kothari, A., Salleh, A., Escobar, A., Demaria, F. and Acosta, A. (2019) *Pluriverse. A post-development dictionary*, Chennai, Indianna: Tulike Books and Authors Upfront.
Kothari, U. (2005) 'Authority and expertise: The professionalisation of international development and the ordering of dissent', *Antipode*, 37(3): 425–446. https://doi.org/10.1111/j.0066-4812.2005.00505.x
Moon, K. and Blackman, D. (2014) 'A guide to understanding social science research for natural scientists', *Conservation Biology*, 28(5): 1167–1177. https://doi.org/10.1111/cobi.12326
Nast, H.J. (1994) 'Opening remarks on "women in the field"', *Professional Geographer*, 46(1): 54–66. https://doi.org/10.1111/j.0033-0124.1994.00054.x
Pailey, R.N. (2019) 'De-centring the "White Gaze" of development', *Development and Change*, 51(3): 729–745. https://doi.org/10.1111/dech.12550
Resurrección, B. and Elmhirst, R. (2020) *Troubling gender expertise in environment and development. Voices from feminist political ecology*, London: Routledge.
Ribot, J. (2014) 'Cause and response: Vulnerability and climate in the Anthropocene', *The Journal of Peasant Studies*, 41(5): 667–705. https://doi.org/10.1080/03066150.2014.894911
Roy, A., Negrón-Gonzales, G., Opoku-Agyemang, K. and Talwalker, C. (2016) *Encountering poverty. Thinking and acting in an unequal world*, Berkeley: University of California Press.
Said, E. (1978) *Orientalism*, New York: Pantheon Books.
Smith, L. (1999) *Decolonizing methodologies. Research and indigenous peoples*, Otago: Zed Books/Otago University Press.
Smith, T.A. (2014) 'Unsettling the ethical interviewer: Emotions, personality, and the interview', in Lunn, J. (ed.) *Fieldwork in the Global South: Ethical challenges and dilemmas*, London: Routledge.
Somers, M. (2008) *Genealogies of citizenship: Markets, statelessness, and the right to have rights*, Cambridge: Cambridge University Press.
Staddon, S. (2014) 'So what kind of student are you? The ethics of 'giving back' to research participants: Experiences from fieldwork in the community forests of Nepal', in Lunn, J., (ed.), *Fieldwork in the Global South: Ethical challenges and dilemmas*, London: Routledge.

Staddon, S., Nightingale, A. and Shrestha, S. (2014) 'The social nature of participatory ecological monitoring', *Society & Natural Resources*, 27(9): 899–914. https://doi.org/10.1080/08941920.2014.905897

Staddon, S., Nightingale, A. and Shrestha, S. (2015) 'Exploring participation in ecological monitoring in Nepal's community forests', *Environmental Conservation*, 42(3): 268–277. https://doi.org/10.1017/S037689291500003X

Sultana, F. (2007) 'Reflexivity, positionality and participatory ethics: Negotiating fieldwork dilemmas in international research', *ACME*, 6(3): 374–385. https://acme-journal.org/index.php/acme/article/view/786

Sultana, F. (2019) 'Decolonizing development education and the pursuit of social justice', *Human Geography*, 12(3): 31–46. https://doi.org/10.1177/194277861901200305

World Bank (2014) 'The biases of development professionals', in World Bank (ed.), *World development report 2015: Mind, society, and behavior*, Washington: World Bank. https://elibrary.worldbank.org/doi/abs/10.1596/978-1-4648-0342-0_ch10

Wright, S. (2012) 'Emotional geographies of development', *Third World Quarterly*, 33(6): 1113–1127. https://doi.org/10.1080/01436597.2012.681500

9
QUANTITATIVE APPROACHES TO ANALYSE RURAL LIVELIHOOD STRATEGIES

Solomon Zena Walelign, Xi Jiao and Carsten Smith-Hall

Introduction

Livelihood strategies describe what people do for a living (Ellis, 2000). Understanding these strategies thus entails investigation of what assets households own or have access to, and how these assets are allocated to different livelihood activities. A particular asset may have many different uses, e.g. a plot of agricultural land can be used to grow wheat or be leased out, resulting in a number of possible activities and thus strategies. Understanding livelihood strategies is important to gain insights into what households do, given their assets and enabling and constraining contextual factors such as policies regulating environmental product harvesting. Households are far from homogenous and their problems cannot be solved with uniform interventions. However, some rural households exhibit common features in the way they access and utilise assets to make a living and this can form the basis for grouping households into similar livelihood strategies, allowing development of strategy-targeted policy interventions based on what households do at a particular point in time and what they lack, rather than targeting households solely based on livelihood outcomes (e.g. income). As household members tend to have a shared livelihood strategy, e.g. as they share pooled assets, the household is a relevant unit of analysis. It should be noted, however, that individual household members may make different contributions and that the methods presented here can also be applied to analyse livelihood strategies at the level of individual household members with some adaptations wherever necessary.

A livelihood strategy can be defined as the composition of activities that a household undertakes to sustain or improve its livelihood (Ellis, 2000). Specifically, given the context and modifying factors (e.g. prices, institutions, shocks), a household adopts a specific livelihood strategy by allocating available assets to specific activities resulting in livelihood outcomes (Figure 9.1). For instance, in a study of 576 rural households in Bolivia, Mozambique and Nepal, Nielsen et al. (2013) distinguished five livelihood strategies: small-scale farmers, large-scale farmers, off-farm workers, livestock producers and off-farm workers, and off-farm workers and business operators. These strategies differed in how assets were used in livelihood activities, with some strategies being more remunerative than others (measured in average total household income). In general, livelihood strategies are dynamic, with households adapting to changes in resources and contextual conditions over time (Davis et al., 2010; Ellis, 2000; Jiao et al., 2017; Scoones, 2015; Walelign et al., 2017). For instance, when a household experiences a crop failure (a shock,

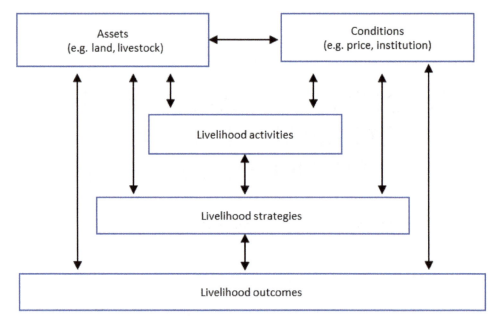

Figure 9.1 A conceptual view of the bidirectional relationships between key livelihood elements: assets, conditions, activities, strategies and outcomes

Source: Authors' own, drawing on Ellis (2000)

as a modifying factor), agricultural income is reduced and the household may try to compensate by shifting from farming to off-farm wage employment, obtaining loans or selling assets. Disposing of assets may have long-term negative consequences for entering into more remunerative livelihood strategies (or being trapped in lower remunerative livelihood strategies).

Rural livelihood strategies in the Global South are usually heterogeneous. Diversification is the norm for three reasons (Davis et al., 2010; Ellis, 2000). First, household activities are often spatially and temporally interrelated and mutually supportive. For instance, livestock may feed on crop residues in a field where they later supply draught power for ploughing and manure that is mixed with leaves from forests and used as fertiliser to maintain soil fertility (Walelign, 2016). Second, many rural markets are weak, fragile or even absent, making it difficult for households to obtain what they do not produce. This simultaneously increases the need for self-sufficiency in production and the importance of activities generating cash income (e.g. for medicine or school fees) such as wage labour. Third, as most rural households engage in rainfed agriculture, susceptible to climate and weather shocks, they pursue a number of other activities as part of risk management. Co-production is common, e.g. collecting fuelwood while herding livestock (Wunder et al., 2011), and poses a challenge to allocate inputs such as labour to different activities. Measuring the allocation of inputs to livelihood activities as accurately as possible is essential to analyse livelihood strategies, often involving analyses using quantitative data, and to design evidence-based interventions to meet (sustainable) development goals.

How can livelihood strategies be quantitatively determined? There are many approaches, with different strengths and weaknesses, making the choice of approach dependent on individual study objectives, context and availability of data. Central to livelihood strategy determination are choice variables – the chosen set of attributes used to measure livelihood strategies. The aims of this chapter are to present approaches to (i) measure choice variables, which allow the subsequent (ii) analysis to determine livelihood strategies, including assigning study units (e.g. households) to strategies. This includes discussing the advantages and disadvantages of each approach, their data requirements and contextual considerations that may guide in approach selection. We also provide examples of studies applying the various approaches.

Approaches to measuring choice variables

Quantitative identification of livelihood strategies involves five steps: 1) selection of choice variables, 2) development of data collection instruments (or identifying existing datasets), 3) data collection and quality control, 4) data analysis of choice variables, and 5) determination of livelihood strategies (and subsequent analyses). Choice variables are a set of attributes from the selected livelihood element for measuring and classifying livelihood strategies; selection of choice variables depends on the objective and chosen focus of a particular study. For instance, Nielsen et al. (2013) emphasised variables reflecting asset allocation for income generating activities, and selected variables such as purchased inputs in agricultural activities and transfer payments (such as pensions) to account for income from non-productive assets. Existing datasets should be used together with an in-depth understanding of how data was generated, including contextual limitations and qualitative information to guide explanation of findings. Excellent advice is available on development of livelihood data collection instruments (Angelsen and Lund, 2011; Lund et al., 2011), environmental product valuation techniques (Wunder et al., 2011), conducting high quality field work (Jagger et al., 2011a, 2011b) and managing data (Babigumira, 2011). In this section, we focus on discussing selection and analysis of choice variables.

Choice variables often comprise multiple attributes of a single livelihood element (see Figure 9.1), e.g. household income (a livelihood outcome) can be disaggregated into the attributes consisting of environmental income, farm income and nonfarm income (Timmermann and Smith-Hall, 2019). It should be noted that the nature of the selected variables is closely related to the subsequent possibilities for livelihood strategy analysis, e.g. the dominant income source approach (see below) requires income data. Ideally, the selection of choice variables and approach to livelihood strategy analysis should be decided before data collection to optimise data collection instruments.

There are five common approaches to measure choice variables in the literature: the income, asset, a combination of income and asset, activity choice, and choice experiment approaches. We briefly present and discuss each of these approaches.

The income approach

This approach uses income from livelihood activities (see Figure 9.1) as choice variables. Income can be defined in many ways but is, in quantitative livelihoods studies, typically counted as the sum of the value of all the products and services from the income source minus the cost of all inputs used (Angelsen et al., 2011, 2014). Income is often estimated from the entire spectrum of household activities and includes both cash (also from remittances and pensions) and subsistence incomes (e.g. if a household collects firewood for use in food preparation in their own home, the value of the firewood is counted as subsistence income). While there are no standard headings for aggregating income types, common groupings are crop income, livestock income, environmental income, business income, wage and salary employment, and other income. Either per activity absolute or relative income (the share of the income source in total household income), but not both, can be used as choice variable. Tesfaye et al. (2011) and Zenteno et al. (2013)

applied the income approach in Southern Ethiopia and in the Bolivian Amazon, respectively. Both used relative income from different sources as choice variables, since this directly reflected the importance of each income source in the households' livelihoods and particularly forest income that is the focus of both studies; Tesfaye et al. (2011) applied broader income categories (e.g. crop, livestock, forest, business) for classifying livelihood strategies, while Zenteno et al. (2013) derived strategies from a matrix of 12 more specific income sources (e.g. Brazil nut, round wood commercial, hunting and fishing).

Advantages: Availability of income data. Most livelihood focused (e.g. Poverty Environment Network, PEN) and welfare focused (e.g. World Bank Living Standard Measurement Surveys, LSMS) surveys contain income data from different livelihood activities, although welfare-based datasets do not contain income from forest and non-forest environmental resources. Another attractive advantage of this approach is the presence of standard methods and protocols for measuring income in straightforward ways (e.g. Angelsen et al., 2011; PEN, 2007) that are now also being implemented as part of larger-scale standardised surveys (FAO, 2016).

Disadvantages: Income data does not contain information regarding allocations of inputs and hence does not reflect use of assets across livelihood activities (Walelign et al., 2017). Another disadvantage is the stochastic nature of income: it is highly volatile over time due to exogenous factors, e.g. crop income in rainfed agriculture varies with rainfall and may even be negative in certain years. In such years, households whose main strategy is farm-based may appear to adopt non-farming strategies as they pursue alternative sources of income, making the income approach less accurate in identifying livelihood strategies, which are often long-term in nature. This can be addressed using panel data, i.e. repeated observation of the same study units over time. The income approach may also provide inaccurate estimates of livelihood strategies when these are characterised by (i) income elements that are difficult to value (e.g. some forms of environmental income, Angelsen et al., 2014), and (ii) elements that cannot be income estimated, such as sacred goods (Wunder et al., 2011). Lastly, as income estimates a livelihood outcome, and not livelihood activities, it may often be better suited to measure welfare than to determine livelihood strategies.

The asset approach

This approach uses assets (see Figure 9.1) as choice variables. Assets are typically divided into natural, physical, human, social and financial categories (Ellis, 2000) and the approach is often implemented across the entire spectrum of categories. Assets can also be categorised as productive (e.g. land, education) and non-productive (e.g. jewellery), or tangible (e.g. land, jewellery) and intangible (e.g. education, experience). Ansoms and McKay (2010) applied an asset approach to analyse changes in livelihood strategies in Rwanda, combining assets with a few additional variables that reflect livelihood conditions (e.g. remoteness) and outcomes (e.g. income) as choice variables.

Advantages: Availability of asset data. As above, many livelihood and welfare surveys contain data on asset types. Another attractive advantage is that assets are less stochastic (compared to income) and more closely aligned to households' choices of livelihood strategies, e.g. households with a high number of cattle pursuing livestock-based strategies. The approach also permits inclusion of intangible assets (e.g. social capital, education), that may be difficult to assign to specific income generating activities, as choice variables (Ansoms and McKay, 2010; Walelign et al., 2017).

Disadvantages: The approach focuses on assets owned rather than assets allocated to individual activities, weakening the link between household resource investment and livelihood activities.

Another disadvantage is that assets are measured in different units (e.g. land in square meters, jewellery in monetary terms, bicycles in numbers) necessitating conversion to a common unit (typically monetary value) before comparisons can be made across the asset choice variables (Walelign et al., 2017).

The combined income and asset approach

This approach uses both income and assets as choice variables. As they are measured in different units, they can be related using a multivariate statistical model where all asset types (using the actual values of the assets that each household owns) are regressed against the income generated from each livelihood activity (e.g. grouped under headings such as environmental resource extraction, crop production, livestock rearing, business operation and wage employment). This creates a composite asset index that allows predictions of each household's income from each livelihood activity, e.g. the predicted crop income is derived from the composite asset index used in crop production. Both the dependent variable (income) and the choice variables (the predicted values) are expressed in monetary units. Walelign et al. (2017) and Walelign and Jiao (2017) provide examples of application of the combined income and asset approach in Nepal.

Advantages: This approach shares all the advantages of the asset approach. And, in addition, it establishes linkages between owned household assets and income from each source, by indirectly inferring how much of each asset is invested in each livelihood activity. The approach also allows for capturing of nonlinearities and interaction effects of asset use on income generation (Walelign et al., 2017).

Disadvantages: This approach overcomes both limitations of the asset approach but retains the disadvantage in the income approach related to the stochasticity of income (e.g. that farm-based strategy households may appear to adopt non-farming strategies: the predicted composite asset indices may be underestimated if income from individual activities have been affected by shocks). It should also be noted that this approach is computationally demanding (running appropriate statistical models to create the choice variables) (Walelign et al., 2017).

The activity choice approach

The activity choice approach is a common approach to determine livelihood strategies. It classifies household livelihood strategies according to households' allocation of assets (including labour and other inputs) into different income generating activities (Barrett et al., 2001; Jiao et al., 2017; Nielsen et al., 2013). To capture the ways in which households allocate their assets into income generating activities, the approach uses activity-based variables as proxies. Labour allocation and input costs are arguably the most direct measures of asset allocation to livelihood activities (Nielsen et al., 2013). Labour allocation in employment activities are commonly included in surveys, e.g. the number of wage labour days for each household member in agricultural and non-farm activities. Labour inputs in self-employed activities, such as the amount of time spent collecting firewood and mushrooms while also herding livestock, are difficult and time-consuming to measure (and hence not included as an input cost in standard income estimates). Instead, input costs (such as for raw materials) are often selected as activity variables for self-employed activities (Nielsen et al., 2013). In cases where self-employed natural resource-based activities require minimal input cost, indirect activity variables may be employed; for instance, the number of times that environmental products are collected can be a proxy variable for environmental activities (Jiao et al., 2017). Transfers, such as pensions and remittances, and business inputs and capital can be included as activity variables to reflect incomes generated from non-productive and other financial assets (Jiao et al., 2017; Nielsen et al., 2013). Jiao et al. (2017) and Nielsen et al. (2013) provide examples of applications of the activity choice approach in Cambodia, and Bolivia, Mozambique and Nepal, respectively.

Advantages: The activity choice approach reflects the characteristics of asset investments in each livelihood activity and overcomes the problems related to the stochastic nature of income (Nielsen et al., 2013). The approach also enables analysis of relationships between livelihood strategies and welfare outcomes, e.g. allowing identification of possible push and pull factors that influence household choices of livelihood strategies (Jiao et al., 2017).

Disadvantages: Households in rural areas employ a significant part of their own labour in productive activities, a cost that is difficult to estimate due to lack of accounting of labour use (and sometimes difficulties in estimating appropriate local wage rates because of thin labour markets) and widespread multitasking/co-production. This makes it difficult to assign human asset uses to specific livelihood activities. This approach also overlooks the importance of non-productive assets, such as social capital, in households' livelihood choice.

The choice experiment approach

The choice experiment approach uses stated livelihood strategies. The study units (often households) are expected to consider the assets they have and their enabling and constraining contexts before making their stated choice. Unlike the approaches presented above, the choice variables (called attributes, and often comprising a set of livelihood activities, assets, and contextual factors) are selected and defined by the researcher based on literature and focus group discussions. The choice variables are combined through shuffling their attribute levels (using an experimental design) to generate a number of choice sets, each representing a couple of alternative livelihood strategies. Typically, several choice sets are generated and presented to the study units that select their most preferred livelihood strategy for each choice set. This approach is mainly used to assess (i) responses to changes that affect livelihood strategies, and (ii) the relative value placed on different livelihood activities. Walelign et al. (2019a) applied this approach to examine preferences for different livelihood activities in connection to the construction of an all-weather road in the Greater Serengeti Ecosystem in Tanzania. Nielsen et al. (2014) used it to assess bushmeat hunters' preferences between bushmeat hunting and wage employment in the Kilombero Valley of Tanzania.

Advantages: The choice experiment approach is useful to uncover people's likely responses to interventions prior to implementation. The approach is also useful to estimate trade-offs – what are people's willingness to forgo one activity to get one unit of another activity. It is also relevant to infer people's preferences for different livelihood activities, particularly sensitive (illegal) activities (Nielsen et al., 2014).

Disadvantages: This approach is highly hypothetical – the choice sets represent hypothetical scenarios of livelihood strategies and people make hypothetical choices. The approach thus involves numerous assumptions, while not directly measuring livelihood strategies. It also involves sophisticated and complicated statistical models for constructing the choice sets and inferring trade-offs and preferences (Nielsen et al., 2014). Lastly, it is time and resource consuming to generate data for designing the choice sets and to undertake the piloting and final analysis.

Approaches to identify livelihood strategies

When the data on choice variables is ready, it can be used to identify livelihood strategies, i.e. determining the characteristics of a definite number of livelihood strategies and assigning households to the strategies. This requires selection of an appropriate analysis technique. In this section, we present three such techniques to identifying livelihood strategies: the occupation-based, the dominant income source and the statistical clustering approaches.

The occupation-based approach

This focuses on the primary livelihood activity in which a household is engaged. Identification of livelihood strategies and assignment of study units into strategies is done on the basis of (i) the primary occupation of the household or more typically the household head, or (ii) scores of the primary and secondary occupations of the household head derived from principal component analysis, a technique to transform a large set of variables into a smaller one retaining most of the original information. Jacobs and Makaudze (2012) and Pender et al. (2004) applied this approach in South Africa and Uganda, respectively.

The advantages of this approach are that it is simple to apply and does not required detailed livelihood investigation, only data on households' primary and secondary occupation (dominant livelihood activities) are required. *The major disadvantages* are that (i) the determination of (the number of) livelihood strategies is subjective, mainly based on the researcher's contextual understanding, and (ii) the approach often does not consider the wider spectrum of income generating activities (Walelign, 2016).

The dominant income source approach

This approach focuses on each household's dominant income source. The share of each income source in total household income is estimated and households are then assigned to the livelihood strategy corresponding to their dominant income source. For instance, a household with its largest share of income from crop production is assigned to the crop production-based livelihood strategy. Eastwood et al. (2006) applied this approach in Limpopo province in South Africa.

The advantage of this approach is that it is relatively simple to apply, only requiring estimation of the share of each income component in total household income, and uses the entire spectrum of income components. *The disadvantages* are that the determination of the number of livelihood strategies remains subjective, mainly based on contextual understanding of the researcher, and that the approach often does not use the entire spectrum of income components when assigning households to strategies (Walelign, 2016).

Statistical clustering approaches

This is the most common way to identify rural household livelihood strategies. Each of the clustering approaches use multiple variables (the choice variables) to produce a single nominal variable (representing livelihood strategy clusters). The approaches are data driven, allowing the data to speak for itself, thus (i) determining the optimal distinct number of livelihood strategies using statistical methods, and (ii) ensuring that study units in the same livelihood strategy are more similar to each other than to units in other strategies.

Before applying clustering approaches, the choice variables must be adjusted for household size to enable comparison across study units (e.g. Angelsen et al., 2014; Cavendish, 2002) and the correlation among the choice variables must be reduced using data reduction methods (such as principal component analysis or factor analysis) to minimise distortion during cluster analysis. There are several statistical clustering methods. Here, we present two common methods: k-means and latent class clustering.

K-means clustering

K-means clustering is an iterative algorithm that assigns the study units to a predefined number of distinct livelihood strategies (clusters). The number of livelihood strategies is typically determined by the researcher's subjective judgement, and/or visual inspection of a graph called a dendrogram, a tree-like diagram showing the arrangement of each study unit in the clusters. The assignment of individual study units is based on the distance between it and the centre of each livelihood strategy – each study unit's sum of squared distances to each livelihood strategy centre is calculated and the unit is assigned to the closest strategy (where the sum of squared distances to the centre is minimum). Brown et al. (2006) operationalised this approach in the Kenyan highlands to identify distinct livelihood strategies that could be ordered according to mean per capita income (with statistical testing documenting significant differences between strategies).

The advantage of this approach is that the assignment of study units to livelihood strategies is objective, based on distance. *The disadvantage* is the lack of statistical tests for determining the optimal number of livelihood strategies, which remains subjective to the interpretation of the dendrogram which may have consequences for subsequent data analysis (Nielsen et al., 2013). For instance, if an analyst decides on a lower number of livelihood strategies than the actual number of clusters, less similar study units may end up in the same strategy.

Latent class clustering

This method has been employed in a number of recent studies to identify livelihood strategies at a particular point in time (e.g. Nielsen et al., 2013) or over time (e.g. Jiao et al., 2017; Walelign et al., 2017). Unlike other methods, this approach provides statistical tests (Haughton et al., 2009) for determining the optimal number of clusters and allows estimation of the probability that a study unit belongs to a particular livelihood strategy (cluster) thus enabling assignment of study units to the livelihood strategies that they are most likely to belong to.

The advantages of this approach are that the determination of the optimal number of livelihood strategies and the assignment of households to strategies is objective, based on statistical testing and membership probability. *The disadvantage* is that the statistical optimal number of strategies may be too high to be useful in further analysis (theoretically, each household could represent a unique livelihood strategy).

Moving forward

Once livelihood strategies have been identified, they should be labelled (named). The strategies are often labelled based on the characteristics of choice variables (e.g. Walelign and Jiao, 2017; Walelign et al., 2017), i.e. to reflect dominant livelihood activities in a strategy. For instance, if a strategy is characterised by asset investment in farming combined with a number of other minor livelihood activities such as raising a limited number of livestock and engaging in petty trade of wild collected plants, it could be labelled as 'diversified small-scale farmers'.

Labelled strategies can be further analysed in three ways. Dominance analysis can be used to investigate whether some strategies are more remunerative than others. The cumulative proportion of study units is mapped against the yearly income per capita for each livelihood strategy (e.g. Nielsen et al., 2013). If the cumulative proportions for a particular livelihood strategy are higher than in all other strategies, the strategy is dominant and the most remunerative (and vice versa for low remunerative livelihood strategies). An alternative is to calculate the average total household income for each livelihood strategy and test for significant differences between strategies.

A common second step is identification of covariates (e.g. Jiao et al., 2017; Nielsen et al., 2013; Rayamajhi et al., 2012; Walelign et al., 2017) – which factors determine why households pursue their livelihood strategies. This is often done using appropriate regression models (e.g. multinomial logit model or ordered probit model) using livelihood strategies as the dependent variable and a set of relevant variables (such as labour, land size and livestock holding) as the independent variables. A third step is to look into how livelihood strategies change over time, both within a year and across years. The latter requires panel data and is not widespread. From the few available studies undertaken in Nepal, we know that asset accumulation is both positively and negatively associated with agricultural income, wage and business, and that environmental income does not appear to constitute a pathway out of poverty (Walelign et al., 2019b). This corresponds with findings that households in the lower remunerative strategies have higher environmental income reliance (Walelign and Jiao, 2017).

Key points

- Available techniques allow quantitative investigation and identification of rural household livelihood strategies in the Global South.
- Each approach to determining livelihood strategies has its own advantages and disadvantages. The choice of technique depends on study objectives, contexts and available resources.
- The latent class cluster analysis has the distinct advantages of providing statistical tests for determining the optimal number of livelihood strategies and allowing assignment of households to the livelihood strategies that they are statistically most likely to belong to.

Further readings

The book 'Measuring Livelihoods and Environmental Dependence: Methods for Research and Fieldwork' (Angelsen et al., 2011) provides an overview of the research process from planning to reporting, including advice on data collection instrument design, data collection, and data management.

Readers wishing to become more familiar with applying the quantitative approaches to determine livelihood strategies mentioned in this chapter could consult a book like 'Stata – A Practical Introduction' (Sønderskov, 2015) and Latent Gold user guides (www.statisticalinnovations.com/user-guides/). For examples of applications of the techniques, see studies in the list of References below.

References

Angelsen, A., Jagger, P., Babigumira, R., Belcher, B., Hogarth, N.J., Bauch, S., Börner, J., Smith-Hall, C. and Wunder, S. (2014) 'Environmental income and rural livelihoods: A global-comparative analysis', *World Development*, 64: S12–S28. https://doi.org/10.1016/j.worlddev.2014.03.006

Angelsen, A., Larsen, H.O., Lund, J.F., Smith-Hall, C. and Wunder, S. (eds.) (2011) *Measuring livelihoods and environmental dependence – methods for research and fieldwork*, London: Earthscan.

Angelsen, A. and Lund, J.F. (2011) 'Designing the household questionnaire', in Angelsen, A., Larsen, H.O., Lund, J.F., Smith-Hall, C. and Wunder, S. (eds.), *Measuring livelihoods and environmental dependence*, London: Earthscan, pp. 107–126.

Ansoms, A. and McKay, A. (2010) 'A quantitative analysis of poverty and livelihood profiles: The case of rural Rwanda', *Food Policy*, 35: 584–598. https://doi.org/10.1016/j.foodpol.2010.06.006

Babigumira, R. (2011) 'Data entry and quality checking', in Angelsen, A., Larsen, H.O., Lund, J.F., Smith-Hall, C. and Wunder, S. (eds.), *Measuring livelihoods and environmental dependence*, London: Earthscan, pp. 191–208.

Barrett, C.B., Reardon, T. and Webb, P. (2001) 'Nonfarm income diversification and household livelihood strategies in rural Africa: Concepts, dynamics, and policy implications', *Food Policy*, 26: 315–331. https://doi.org/10.1016/S0306-9192(01)00014-8

Brown, D.R., Stephens, E.C., Ouma, J.O., Murithi, F.M. and Barrett, C.B. (2006) 'Livelihood strategies in the rural Kenyan highlands', *African Journal of Agricultural and Resource Economics*, 1: 21–35. https://doi.org/10.22004/ag.econ.57019

Cavendish, W. (2002) 'Quantitative methods for estimating the economic value of resource use to rural households', in Campbell, B.M. and Luckert, M.K. (eds.), *Uncovering the hidden harvest: Valuation methods for woodland and forest resources*, London: Earthscan, pp. 17–66.

Davis, B., Winters, P., Carletto, G., Covarrubias, K., Quiñones, E.J., Zezza, A., Stamoulis, K., Azzarri, C. and DiGiuseppe, S. (2010) 'A cross-country comparison of rural income generating activities', *World Development*, 38: 48–63. https://doi.org/10.1016/j.worlddev.2009.01.003

Eastwood, R., Kirsten, J. and Lipton, M. (2006) 'Premature deagriculturalisation? Land inequality and rural dependency in Limpopo province, South Africa', *Journal of Development Studies*, 42: 1325–1349. https://doi.org/10.1080/00220380600930614

Ellis, F. (2000) *Rural livelihoods and diversity in developing countries*, Oxford: Oxford University Press.

Food and Agricultural Organization (FAO) (2016) 'National socioeconomic surveys in forestry', *FAO Forestry Paper 179*, Rome: FAO.

Haughton, D., Legrand, P. and Woolford, S. (2009) 'Review of three latent class cluster analysis packages: Latent gold, poLCA, and MCLUST', *The American Statistician*, 63: 81–91. https://doi.org/10.1198/tast.2009.0016

Jacobs, P. and Makaudze, E. (2012) 'Understanding rural livelihoods in the West Coast district, South Africa', *Development Southern Africa*, 29: 574–587. https://doi.org/10.1080/0376835X.2012.715443

Jagger, P., Duchelle, A., Dutt, S. and Wyman, M. (2011a) 'Preparing for the field: Managing and enjoying fieldwork', in Angelsen, A., Larsen, H.O., Lund, J.F., Smith-Hall, C. and Wunder, S. (eds.), *Measuring livelihoods and environmental dependence*, London: Earthscan, pp. 147–162.

Jagger, P., Duchelle, A., Larsen, H.O. and Nielsen, Ø.J. (2011b) 'Hiring training and managing a field team', in Angelsen, A., Larsen, H.O., Lund, J.F., Smith-Hall, C. and Wunder, S. (eds.), *Measuring livelihoods and environmental dependence*, London: Earthscan, pp. 163–174.

Jiao, X., Pouliot, M. and Walelign, S.Z. (2017) 'Livelihood strategies and dynamics in rural Cambodia', *World Development*, 97: 266–278. https://doi.org/10.1016/j.worlddev.2017.04.019

Lund, J.F., Shackleton, S. and Luckert, M. (2011) 'Getting quality data', in Angelsen, A., Larsen, H.O., Lund, J.F., Smith-Hall, C. and Wunder, S. (eds.), *Measuring livelihoods and environmental dependence*, London: Earthscan, pp. 175–190.

Nielsen, M.R., Jacobsen, J.B. and Thorsen, B.J. (2014) 'Factors determining the choice of hunting and trading bushmeat in the Kilombero Valley, Tanzania: Choice of hunting and trading bushmeat', *Conservation Biology*, 28: 382–391. https://doi.org/10.1111/cobi.12197

Nielsen, Ø.J., Rayamajhi, S., Uberhuaga, P., Meilby, H. and Smith-Hall, C. (2013) 'Quantifying rural livelihood strategies in developing countries using an activity choice approach', *Agricultural Economics*, 44: 57–71. https://doi.org/10.1111/j.1574-0862.2012.00632.x

Pender, J., Jagger, P., Nkonya, E. and Sserunkuuma, D. (2004) 'Development pathways and land management in Uganda', *World Development*, 32: 767–792. https://doi.org/10.1016/j.worlddev.2003.11.003

Poverty Environment Network (2007) *PEN technical guidelines version 4*, Poverty Environment Network. Available at: 13 May 2021. https://www2.cifor.org/pen/the-pen-technical-guidelines/.

Rayamajhi, S., Smith-Hall, C. and Helles, F. (2012) 'Empirical evidence of the economic importance of Central Himalayan forests to rural households', *Forest Policy and Economics*, 20: 25–35. https://doi.org/10.1016/j.forpol.2012.02.007

Scoones, I. (2015) *Sustainable livelihoods and rural development*, Rugby: Practical Action Publishing.

Sønderskov, K.M. (2015) *Stata: A practical introduction*, Copenhagen: Hans Reitzel.

Tesfaye, Y., Roos, A., Campbell, B.M. and Bohlin, F. (2011) 'Livelihood strategies and the role of forest income in participatory-managed forests of Dodola area in the Bale highlands, southern Ethiopia', *Forest Policy and Economics*, 13: 258–265. https://doi.org/10.1016/j.forpol.2011.01.002

Timmermann, L. and Smith-Hall, C. (2019) 'Commercial medicinal plant collection is transforming high-altitude livelihoods in the Himalayas', *Mountain Research and Development*, 39(3): R13–R21. https://doi.org/10.1659/MRD-JOURNAL-D-18-00103.1

Walelign, S. and Jiao, X. (2017) 'Dynamics of rural livelihoods and environmental reliance: Empirical evidence from Nepal', *Forest Policy and Economics*, 83: 199–209. https://doi.org/10.1016/j.forpol.2017.04.008

Walelign, S.Z. (2016) 'Livelihood strategies, environmental dependency and rural poverty: The case of two villages in rural Mozambique', *Environment, Development and Sustainability*, 18: 593–613. https://doi.org/10.1007/s10668-015-9658-6

Walelign, S.Z., Nielsen, M.R. and Jacobsen, J.B. (2019b) 'Roads and livelihood activity choices in the Greater Serengeti Ecosystem, Tanzania', *PLoS One*, 14: e0213089. https://doi.org/10.1371/journal.pone.0213089

Walelign, S.Z., Nielsen, M.R. and Larsen, H.O. (2019a) 'Environmental income as a pathway out of poverty? Empirical evidence on asset accumulation in Nepal', *Journal of Development Studies*, 55: 1508–1526. https://doi.org/10.1080/00220388.2017.1408796

Walelign, S.Z., Pouliot, M., Larsen, H.O. and Smith-Hall, C. (2017) 'Combining household income and asset data to identify livelihood strategies and their dynamics', *Journal of Development Studies*, 53: 769–787. https://doi.org/10.1080/00220388.2016.1199856

Wunder, S., Luckert, M. and Smith-Hall, C. (2011) 'Valuing the priceless: What are non-marketed products worth?', in Angelsen, A., Larsen, H.O., Lund, J.F., Smith-Hall, C. and Wunder, S. (eds.), *Measuring livelihoods and environmental dependence*, London: Earthscan, pp. 127–145.

Zenteno, M., Zuidema, P.A., de Jong, W. and Boot, R.G.A. (2013) 'Livelihood strategies and forest dependence: New insights from Bolivian forest communities', *Forest Policy and Economics*, 26: 12–21. https://doi.org/10.1016/j.forpol.2012.09.011

10
LONGITUDINAL RESEARCH TO UNDERSTAND THE COMPLEXITY OF LIVELIHOODS

Joan DeJaeghere

Introduction

Livelihoods, as used in this chapter, builds on the definition from Chambers and Conway (1992) that focuses on an individual's capabilities, assets and agency, along with the benefits from livelihoods, and adds a critical perspective by giving attention to the social, environmental, cultural and economic conditions that influence long-term wellbeing. These conditions include local as well as global influences on ecologies, economies and climate. Furthermore, this conceptualisation of livelihoods gives particular attention to what people desire for their wellbeing in these environments. In this sense, livelihoods are not themselves an outcome but a means to wellbeing. This perspective draws specifically on Sen's (2001) capability approach to examine how an individual's livelihood is embedded within social and ecological environments that influence if and how one can achieve wellbeing. Wellbeing, from a capability approach, is multidimensional and includes material, social, and physical/emotional outcomes. The perspective used here builds on Chapter 2 on livelihoods and wellbeing and Chapter 3 on the capabilities approach to discuss the implications for longitudinal research designs.

While many studies of livelihoods address how they relate to economic development, and particularly to overcoming poverty in low and middle-income countries (De Haan, 2012), the concept of livelihoods is not limited to these contexts, nor to the idea of getting out of poverty. The interaction of local livelihoods with globalisation through trade, supply chains, migration and cultural exchange requires us to think about livelihoods for wellbeing in any context, including poorer households in high income countries. Murray (2002) also pointed out that a relational approach should be used to examine the livelihoods of those who are and are not poor in specific contexts to understand the social and structural relations that influence these different trajectories. This dynamic and multidimensional perspective on livelihoods has implications for how practitioners and researchers study livelihood programs and outcomes, with longitudinal research being a useful set of approaches to understand multidimensional outcomes through time.

Why conduct longitudinal studies

The livelihood literature has long noted a few critical shortcomings that can be addressed by longitudinal research, including both quantitative and qualitative methods and data. These shortcomings include: 1) a focus on outcomes achieved at a specific point in time, what Murray (2002: 490) calls the 'circumspect approach', but these outcomes often quickly change, 2) an over-reliance on quantitative data of income and assets, which can be unreliable and lack validity (Ansoms et al., 2017), and 3) analyses of the many variations of local livelihoods, rather than on the effects of livelihoods on wellbeing (De Haan, 2012).

Longitudinal research can provide several advantages to attend to these shortcomings because it uses a conceptualisation that livelihoods are dynamic, influenced by social and structural relations, and change through time. First, gathering research over an extended period of time and at multiple time points can reveal and explain multidimensional outcomes, including longer-term wellbeing. Researchers have noted the importance of more longitudinal research over several time points because many studies include only one time point (post-programme) or a pre- and post-programme analysis of change during a short time period. For instance, Murray (2002) argued for a focus on trajectories, or changes through time, in all forms of livelihood research. Second, longitudinal research can examine the complex conditions, including the social, environmental, political and other conditions, that influence livelihood change and outcomes through time. Time, a key concept of longitudinal research, is critical for understanding these complex conditions. Finally, a longitudinal approach can address the third shortcoming, that studies have produced an endless variation of local livelihoods, by explaining the complex nature of these variations and what they mean for wellbeing. By focusing on longer-term outcomes, longitudinal studies can contribute to a better understanding of wellbeing, beyond the snapshot of economic or social outcomes that tend to be measured in many studies.

Longitudinal studies come in many variations, including exclusively quantitative, completely qualitative, and mixed-methods, but time and change are key concepts in the analysis of all types (Saldaña, 2003). Murray (2002) states that the contribution of quantitative studies are to identify what is achieved and the representativeness of the livelihood outcome. Qualitative studies explain the how and why of livelihood outcomes, while also problematising assumptions and contextualising findings. Other livelihood scholars have argued for attending to macro-level conditions and influences, often referred to as the political economy approach, along with micro-level livelihood strategies, changes, and effects (De Haan, 2012; Neves, 2017). Such an approach requires the use of mixed methods, both quantitative and qualitative, that can examine historically and retrospectively as well as prospectively the present and the future. Many mixed-methods studies, however, do not adequately detail the nature of the qualitative data or analysis. In particular, many mixed-methods studies use qualitative data but often at one time point – not longitudinally, or they do not show how they analyse qualitative data through time.

A key concept for all longitudinal studies is the measure of time and how time relates to changes of desired livelihood outcomes. Critical questions to ask in designing and analysing longitudinal studies are: Is the timeframe for the study based on a program, on life course changes, or on events that occur for participants? Is time considered retrospectively, prospectively or both? What duration of time is necessary for the data to explain livelihood wellbeing? For example, some studies explicitly measure changes in livelihood outcomes immediately following a program and then again one year later to see if the patterns hold (e.g. Alcid, 2014). Some studies measure livelihood outcomes based on key events, such as migration, examining data before and after the migration in which households are considered settled (Li et al., 2018; Wilmsen and Hulten, 2017). Panel studies can also be used to follow people from youth to adulthood to explain life course changes, or to examine how households in a community are affected by 'development' over decades (Östberg et al., 2018). Time is not equal to change so study designers need to consider how they conceptualise and measure or examine change. For example, is change regarded as an increase or decrease in the outcome variable? How is no change or consistency in the outcome through time interpreted? How are trajectories of change conceptualised? Is change examined through turning points, stability, or tipping points?

Finally, a consideration for longitudinal studies of livelihoods is to assess relevant multidimensional outcomes at the time points accounted for in the design. Yet, many studies assess singular outcomes, such as income or assets, usually at two or more time points. A wellbeing perspective, however, regards income as an intermediate outcome, and a longitudinal design allow us to see how a livelihood activity and related income are used to achieve certain wellbeing outcomes that a person, household or community desires. Therefore, the relationship between different outcomes and their feedback or connection to other activities and outcomes should be examined (see, for example, Jiao et al., 2017, for a dynamic model illustrating these various outcomes and feedback loops). The chapter now turns to discussing the specific features and examples of quantitative longitudinal studies.

Quantitative longitudinal research on livelihoods

Quantitative longitudinal studies are more common in the livelihood literature possibly because of the dominance of the disciplines of economics and sociology that informed the early livelihoods research and practice. Quantitative longitudinal studies are generally referred to as panel studies in which data are collected from individuals or households over a series of time points, often years (see, for example, Jiao et al., 2017; Wilmsen and Hulten, 2017). Some quantitative studies also use cohort data, rather than tracking individuals, to look at changes through generations, such as examining the livelihoods of 30-year-olds during the 1990s and again in 2010. Repeated measures analysis, growth curves, and linear mixed effects are different analyses used to show change and the factors influencing change.

Quantitative longitudinal designs and data on livelihoods are commonly used in two ways. The first is as pre- and post-survey for evaluating programs or a policy intervention (see, for example, Krause et al., 2016). Livelihood programmes include many different interventions suited to the 'target' population. The different interventions used, such as technical skills for self-employment or financial literacy to improve savings or investment, determine the outcomes measured in quantitative designs. But many programs, particularly as they target women and youth in conditions of poverty, are multi-pronged and need to attend to finance, skills and self-employment simultaneously (Cho and Honorati, 2013; Grimm and Paffhausen, 2015; Shehu and Nilsson, 2014). The age, experience and context of the participants and the desired outcomes of the program influence the measures used in the evaluation (more on this below).

A pre- and post-design, or a design including multiple time periods, to assess an intervention's outcomes or effectiveness usually includes an assessment of types of livelihood activities, income, and other measures relevant to the program, such as health outcomes, asset acquisition or loan repayment. Quantitative studies often assess outcomes that are most easily measured, such as income or assets, but the interpretation of these data must be contextualised within localities (e.g. DeJaeghere et al., 2020; McCabe et al., 2014). These studies, therefore, need to consider the program theory of change and desired outcomes, and they must be designed to be relevant to the context and experiences of the participants. Quantitative livelihood studies are often measured at the household level, but that measure can be problematic for programs targeting groups across households, such as women and youth (Ansoms et al., 2017).

Assessing longitudinal livelihood outcomes for programs targeting women poses its own unique challenges. Many studies that use household level data assume men as the main contributor to the livelihood, and that men's income earning and strategies differ from those of women. The following considerations need to be taken into account when assessing women's livelihoods and outcomes longitudinally: 1) gendered norms influence access to types of services, loans, supply chains, and labour markets, and, in turn, the amount of income earned and how it is used; 2) gender relations within the household and community shift when programs are introduced, or due to other external factors, and these effects on livelihood outcomes need to be assessed; and 3) women and men bring different experiences, values and knowledge to their livelihoods and these need to be considered in how they affect livelihood outcomes (Anderson et al., 2021; Badstue et al., 2020; Oestreicher et al., 2014). These three considerations are illustrated by Östberg et al. (2018) who argue that a shortcoming of household surveys is that they do not adequately account for gendered and generational dynamics within households as they affect livelihoods and earnings. For instance, household income is not earned equally, nor is it necessarily shared or used by both men and women household members. In some households, the elder generation may control land or income and the younger generation does not. Other studies show that commodity chains in which women are engaged as producers and consumers have deeply gendered inequalities (Ramamurthy, 2014). Certain kinds of production, including many agricultural products, are highly gendered, with women and girls more engaged in their production and often at lower commodity prices. Therefore, longitudinal studies of women's (and households') livelihoods need to account for the gendered dynamics within the household as well as in relation to markets. The Women's Empowerment in Agriculture Index (WEAI) is one quantitative measure that has been used to assess change in gendered participation in and decision-making about agricultural production (Alkire et al., 2013; Anderson et al., 2021).

Many livelihood programmes focus on youth at critical life points when they are completing education or transitioning to employment. Yet, longitudinal research of youth's livelihoods also has some unique features because their lives are more transitional or dynamic. For example, a study on youth livelihoods in Rwanda (Alcid, 2014) and another on youth in Tanzania, Uganda and Kenya (Lefebvre et al., 2018) found that youth engaged in multiple livelihoods, and they are also likely to change their forms of earning and livelihoods many times even within a year. Therefore, determining how often and how to assess these livelihood changes depends on the dynamics of their life course, which do not follow a universal model of development (e.g. whether they are married, settled in a home, living with family, transitioning between education and work, etc.) (see, for example, Morrow, 2013). Furthermore, youth's access to assets, finance, land and other physical, financial and human capitals also differ from older adults, or adults who are a head of household. Cultural and social processes related to youthhood, such as when and how they complete education, and acquire a home or family assets are important conditions to examine with regard to livelihood outcomes (see, for example, Crivello, 2011). While longitudinal studies offer the possibility to capture changes in youth's livelihoods through key transitional years and events, many surveys are not designed to do so, in part because they use measures from surveys with older adults.

A second type of quantitative longitudinal design includes household survey data to assess changes in a specific site due to policy, environmental or other social changes through time, often a decade or more (e.g. Betcherman and Marschke, 2016). In this approach, the outcomes assessed are livelihood activities, as well as income, wealth or assets, which can be easily measured in household surveys. Data from surveys can also measure other factors, including physical, human, financial, and social capital factors (Li et al., 2018). Quantitative longitudinal studies that analyse household survey data are particularly useful to researchers and policymakers because they are often nationally representative and they are collected on a regular basis (decadal).

However, longitudinal quantitative studies drawing on household surveys have considerable shortcomings to understanding the complex and dynamic nature of livelihoods and wellbeing of diverse household members. Ansoms et al. (2017) found that household survey data may misrepresent poverty and income because of notions of a static household. For instance, they found that Rwanda's household survey data may not adequately capture household members who have migrated, and their forms of livelihoods and contributions to income and household wellbeing. Furthermore, households may not report types of income and forms of livelihoods if they fall outside regulations defining forms of formal livelihoods or homesteads. The authors argue that survey data collected in relation to government projects in Rwanda (including census surveys) may mis-represent poverty or livelihoods because household members have learned to provide responses that align with government targets.

In summary, quantitative longitudinal studies of livelihoods are most often used to assess measurable outcomes, such as income, assets and changes in kinds of livelihoods, so that change over time points can be assessed and multi-factorial analyses of related factors can be used. These studies are less able to capture relational and subjective outcomes of wellbeing, which are an important feature of longitudinal studies because they can illustrate how change in financial and material outcomes affect other aspects of life. These kinds of multidimensional outcomes and the conditions affecting them are best examined through qualitative studies.

Qualitative longitudinal research on livelihoods

Qualitative longitudinal data are often used as a complement or supplement to survey data in mixed methods designs. Complementing survey data, some studies use qualitative data to understand local meanings of wealth, assets, and livelihoods and how these meanings change over time and place (e.g. Östberg et al., 2018). Other studies use qualitative data to supplement quantitative data, particularly to explain outcomes that are not achieved, or why they were or were not achieved, though often these qualitative data are not explicitly longitudinal nor analysed as such (e.g. Kura et al., 2017). For example, in a study of youth employability in Rwanda, youth in the program were found to not have increased their employment over the non-intervention group, and particularly in certain communities. Therefore, a qualitative study was designed after the initial quantitative study to explain why there were not significant differences between groups of youth (Alcid, 2014). The qualitative data also added another dimension of time by following the youth for a period after the initial program and post-survey collection.

Another important use of qualitative longitudinal studies is to examine underlying assumptions and to show the dynamic nature of livelihoods (Murray, 2002). Studies that have a strong qualitative longitudinal component can offer important insights into the social meanings of livelihoods and the dynamics that affect them. For example, governments and practitioners make assumptions that certain policies and programs will improve livelihoods, but they may not. For instance, Li et al. (2018: 110) used qualitative interviews, though not repeated over time, along with quantitative survey data, to 'explore household's less visible experiences of livelihood change' and to better understand the social, material and financial factors that affected positive and negative livelihood outcomes after land acquisition in China. The government assumed that by providing certain material support, farmers would improve their livelihoods over time, but this was not the case for many of them. They also found that external support, such as job training, was inadequate. They argue for policies and research that take a more nuanced approach to understanding and prioritising human conditions, or wellbeing, over development.

Qualitative longitudinal studies utilise various methods; most commonly, repeat interviews are used to follow participants prospectively through the timeframe of a study. This form of qualitative data allows for analysing change as it occurs in relation to a program, specific context and life course events of participants. While the design of repeat interviews should follow a logic of time and change that has been hypothesised as relevant, often the multiple time points for conducting interviews (and related observations) are more practical. For example, DeJaeghere et al. (2020) designed a five-year study of youth's livelihoods that included annual interviews alongside some quantitative data. Because the interviews were conducted annually, they were both retrospective, asking youth to reflect on their livelihood strategies and outcomes over the past year, and prospective, in that it revealed changes in their lives as they were occurring from one year to the next. Analysis of qualitative data from this study problematised some of the findings from the quantitative survey, as youths' engagement with work, financial lending, and even further training was often more dynamic and complex than captured in quantitative data points (see DeJaeghere et al., 2020).

Retrospective interviews or life histories is another method used to understand change through time. This method relies on participant memories and reflections of particular life events. While relying on memories of livelihoods (events, times, activities) has been critiqued for its 'reliability', or accuracy of what actually happened in a person's life, the use of reflective narration of one's life and livelihoods is particularly useful in understanding how participants saw their wellbeing at different points in time, as a comparative (past and present) analysis of wellbeing outcomes. The Timescapes studies provide an excellent example of qualitative longitudinal research, using life histories as a method (Neale et al., 2012). From these studies, Tarrant and Hughes (2019) bring together a re-analysis of two datasets of men, one of young fathers and another of mid-life grandfathers, to understand inter- and intra-generational dynamics of vulnerability (the studies are not specifically about livelihoods but they examine effects of poverty). The men in each study were interviewed five and four times, respectively. The interviews were conducted as life histories to gather an in-depth understanding of men's lives through time, including capturing these men's lives in the present time. The analysis of the data was both retrospective and prospective, looking back through time at the data and forward to explain the different pathways and factors affecting men's life course in contexts of poverty.

Finally, ethnographic research constitutes another form of qualitative longitudinal research, particularly when it aims to explain how time and change matter in the context of specific sites. Ethnography (see Chapter 11) is used to explain complex contextual and cultural dynamics of livelihoods, and it also incorporates time (time in the field), a key condition of longitudinal studies. But not all ethnographies examine time and change explicitly. A qualitative longitudinal study using ethnography gives explicit attention to how time matters in fostering change in livelihoods, whether that is calendar or seasonal time, or timing of particular events or life course changes. For example, Ursin (2016) engages in a 10-year study of young men living on the street and their livelihoods in Brazil, using multiple rounds of interviews with 10 participants, as well as interviews with residents, business owners and other neighbourhood stakeholders. Her analysis shows the complex relationships these young men developed through time in order to have a livelihood. The study offers important findings for the engagement between mid-income residents and street youth, and for public policy on how to foster livelihoods that are inclusive and promote youth wellbeing.

A slightly different variation of qualitative longitudinal studies of communities and community change are case studies. Case studies can be used to follow individuals as well as groups of people within a community to understand livelihood changes. McCabe et al.'s (2014) case studies draw on extensive fieldwork over two decades that included individual interviews, focus groups and survey data to understand livelihood diversification among pastoral groups in Tanzania. Individual interviews allow for examining changes through time of specific families/individuals, while the group interviews provide contextualisation of these changes and their variations within the community. These data can also show the dynamics of individual decisions and outcomes in relation to family dynamics or outcomes. For example, migration in one community was often the decision of young men, without consulting family members, while in another community, the decision was made by the family. The qualitative data (in contrast to the survey data) also show how poverty is understood differently among participants and communities.

Finally, another form of qualitative ethnographic or case studies is a 'restudy' methodology that examines change over a long period of time based on data collected for a prior study and then gathering new data from households in the original study community. A restudy includes historical analyses to explain past understandings of livelihoods; it also provides data in the present to analyse how social and political-economic conditions have shifted and affect livelihoods. Östberg et al. (2018) used this methodology to identify households in villages where there had been a study in past decades (1990s) of assets and income of villagers to understand changes in local meanings of wealth. Qualitative data were used to understand past and present meanings of wealth. For example, wealth was not solely measured by land owned, but used; and livestock ownership was an important indicator of wealth. These meanings of wealth remained fairly consistent through the decades and generations of households. However, casual labour was not regarded a sign of poverty in the current study as it had been in the 1990s. Rather it was a strategy for increasing wealth (in addition to their other livelihoods). Changes in assets and livelihood strategies were not deemed necessarily as positive by community members as they expressed concerns about the future, and the change in land use and the climate. Finally, the idea of wealth – or being strong – as a community was illustrated through different 'indicators' in 2016 than those of the 1990s, including having health care and educating children.

As discussed above, qualitative longitudinal studies are particularly useful to capture dynamic changes in gendered relations or in youth trajectories that cannot be easily studied through quantitative data. Qualitative data can reveal local meanings of livelihoods and wellbeing; they can also explore the nature of political, economic and social structures that influence livelihood outcomes, including the forms of patriarchy, racism or economic inequalities within local and global systems. Qualitative longitudinal studies are both flexible and locally contextualised, drawing on various forms of data in their analysis, including historical documents, participatory approaches, ethnographic observations, and participant interviews.

Conclusion

Longitudinal studies of livelihoods are a specific kind of research design that can show livelihood trajectories, change and long-term outcomes. This design allows for engaging conceptually with the dynamic conditions that affect livelihood outcomes through time. As Murray (2002) argued in the early years of livelihoods research, all livelihoods studies should have a longitudinal component. As an approach (rather than a method), longitudinal studies are particularly adaptive and complex. They draw on a combination of research methods discussed in this book: surveys, interviews, ethnography, life histories, participatory methods, case studies and historical documents.

Quantitative longitudinal studies, usually using pre-post program or household surveys, are helpful:

- to examine changes in livelihood activities and whether and how they improve income, assets and other established measures of outcomes;
- to evaluate programs, using an experimental design or randomised control trial, which can attribute outcomes to a program; and
- to be representative of the population and livelihood changes through time.

But survey data have several shortcomings, including:

1. the kinds of indicators measured, in that certain aspects of wellbeing are not easily quantifiable;
2. relational conditions, such as gender inequalities, that affect livelihood outcomes are not easy to capture;
3. difficulty in appropriately contextualizing and interpreting the data without qualitative data;
4. difficulty in adapting to changes in the program design or conditions or events in the context.

Qualitative longitudinal studies are best used:

- to understand local meanings of livelihood outcomes through time;
- to combine multiple methods and sources of data, including repeat narrative interviews, life histories, ethnography, and other participatory methods; and
- to capture important contextual and local meanings of livelihoods and wellbeing; they can also examine the conditions and mechanisms that influence diverse outcomes.

But qualitative studies also have several challenges in order to be carried out robustly.

1. They require extensive data, planning and analysis. Often funding organisations, programs and researchers do not adequately account for the time and resources needed for longitudinal analysis of qualitative data (Miller, 2015).
2. They must consider the ethical implications when engaging with the same participants over time to ensure their confidentiality (Warin, 2011).
3. There is also a responsibility that long-term studies contribute to the participants and community in ways that are meaningful to them rather than being extractive. Qualitative longitudinal research can be designed with communities to make important contributions to policy and practice.

Bringing quantitative survey data together with qualitative studies in a mixed-method longitudinal design can examine the complexities of livelihood activities and processes for achieving wellbeing (Murray, 2002; Davis and Baulch, 2011). Large scale survey data can provide a broader picture of livelihoods across a population and time. Household surveys are also used to compare poor and non-poor participants to understand divergent trajectories. Qualitative methods can capture the micro-level changes among individuals, households and communities, as well as historicise and contextualise livelihood dynamics through time. More livelihood research needs to explicitly include a longitudinal design of both qualitative and quantitative data so these studies can contribute to researchers', practitioners' and policymakers' understandings of livelihood trajectories, changes and long-term outcomes.

Recommended additional readings

Bamattre, R., Schowengerdt, B., Nikoi, A. and DeJaeghere, J. (2019) 'Time matters: The potentials and pitfalls in longitudinal mixed methods in social research', *International Journal of Social Research Methods*, 22(4): 335–349. https://doi.org/10.1080/13645579.2018.1553674

Davis, P. and Baulch, B. (2011) 'Parallel realities: Exploring poverty dynamics using mixed methods in rural Bangladesh', *The Journal of Development Studies*, 47(1): 118–142. https://doi.org/10.1080/00220388.2010.492860

Murray, C. (2002) 'Livelihoods research: Transcending boundaries of time and space', *Journal of Southern African Studies*, 28: 489–509. https://doi.org/10.1080/0305707022000006486

References

Alcid, A. (2014) *A randomized control trial of Akazi Kanoze youth in rural Rwanda*, Washington, DC: Education Development Center, Inc.

Alkire, S., Meinzen-Dick, R., Peterman, A., Quisumbing, A., Seymour, G. and Vaz, A. (2013) 'The women's empowerment in agriculture index', *World Development*, 52: 71–91. https://doi.org/10.1016/j.worlddev.2013.06.007

Anderson, L.C., Reynolds, T.W., Biscaye, P., Patwardhan, V. and Schmidt, C. (2021) 'Economic benefits of empowering women in agriculture: Assumptions and evidence', *The Journal of Development Studies*, 57(2): 193–208. https://doi.org/10.1080/00220388.2020.1769071

Ansoms, A., Marijnen, E., Cioffo, G. and Murison, J. (2017) 'Statistics versus livelihoods: Questioning Rwanda's pathway out of poverty', *Review of African Political Economy*, 44(151): 47–65. https://doi.org/10.1080/03056244.2016.1214119

Badstue, L., Elias, M., Kommerell, V., Petesch, P., Prain, G., Pyburn, R. and Umantseva, A. (2020) 'Making room for manoeuvre: Addressing gender norms to strengthen the enabling environment for agricultural innovation', *Development in Practice*, 30(4): 541–547. https://doi.org/10.1080/09614524.2020.1757624

Betcherman, G. and Marschke, M. (2016) 'Coastal livelihoods in transition: How are Vietnamese households responding to changes in the fisheries and in the economy?', *Journal of Rural Studies*, 45: 24–33. https://doi.org/10.1016/j.jrurstud.2016.02.012

Chambers, R. and Conway, G. (1992) 'Sustainable rural livelihoods: Practical concepts for the 21st century', *IDS Discussion Paper 296*, Brighton: Institute of Development Studies.

Cho, Y. and Honorati, M. (2013) 'Entrepreneurship programs in developing countries: A meta regression analysis', *Labour Economics*, 28: 110–130. https://doi.org/10.1016/j.labeco.2014.03.011

Crivello, G. (2011) '"Becoming somebody": Youth transitions through education and migration in Peru', *Journal of Youth Studies*, 14(4): 395–411. https://doi.org/10.1080/13676261.2010.538043

Davis, P. and Baulch, B. (2011) 'Parallel realities: Exploring poverty dynamics using mixed methods in rural Bangladesh', *The Journal of Development Studies*, 47(1): 118–142. https://doi.org/10.1080/00220388.2010.492860

De Haan, L.J. (2012) 'The livelihood approach: A critical exploration', *Erdkunde*, 66(4): 345–357. https://doi.org/10.3112/erdkunde.2012.04.05

DeJaeghere, J., Morris, E. and Bamattre, R. (2020) 'Moving beyond employment and earnings: Reframing how youth livelihoods and wellbeing are evaluated in East Africa', *Journal of Youth Studies*, 23(5): 667–685. https://doi.org/10.1080/13676261.2019.1636013

Grimm, M. and Paffhausen, A.L. (2015) 'Do interventions targeted at micro-entrepreneurs and small and medium-sized firms create jobs? A systematic review of the evidence for low and middle income countries', *Labour Economics*, 32: 67–85. https://doi.org/10.1016/j.labeco.2015.01.003

Jiao, X., Pouliot, M. and Walelign, S.Z. (2017) 'Livelihood strategies and dynamics in rural Cambodia', *World Development*, 97: 266–278. https://doi.org/10.1016/j.worlddev.2017.04.019

Krause, B.L., McCarthy, A.S. and Chapman, D. (2016) 'Fuelling financial literacy: Estimating the impact of youth entrepreneurship training in Tanzania', *Journal of Development Effectiveness*, 8(2): 234–256. https://doi.org/10.1080/19439342.2015.1092463

Kura, Y., Joffre, O., Laplante, B. and Sengvilaykham, B. (2017) 'Coping with resettlement: A livelihood adaptation analysis in the Mekong River basin', *Land Use Policy*, 60: 139–149. https://doi.org/10.1016/j.landusepol.2016.10.017

Lefebvre, E., Nikoi, A., Bamattre, R., Jafaar, A., Morris, E., Chapman, D. and DeJaeghere, J. (2018) *Getting ahead and getting by in youth livelihoods and employment*, Toronto, CA: The Mastercard Foundation.

Li, C., Wang, M. and Song, Y. (2018) 'Vulnerability and livelihood restoration of landless households after land acquisition: Evidence from peri-urban China', *Habitat International*, 79: 109–115. https://doi.org/10.1016/j.habitatint.2018.08.003

McCabe, J.T., Smith, N.M., Leslie, P.W. and Telligman, A.L. (2014) 'Livelihood diversification through migration among a pastoral people: Contrasting case studies of Maasai in Northern Tanzania', *Human Organization*, 73(4): 389–400. https://doi.org/10.17730/humo.73.4. vkr10nhr65g18400

Miller, T. (2015) 'Going back: "Stalking", talking and researcher responsibilities in qualitative longitudinal research', *International Journal of Social Research Methodology*, 18(3): 293–305. https://doi.org/10.1080/13645579.2015.1017902

Morrow, V. (2013) 'Troubling transitions: Young people's experiences of growing up in poverty in rural Andhra Pradesh, India', *Journal of Youth Studies*,16(1): 86–100. https://doi.org/10.1080/13676261.2012.704986

Murray, C. (2002) 'Livelihoods research: Transcending boundaries of time and space', *Journal of Southern African Studies*, 28(3): 489–509. https://doi.org/10.1080/0305707022000006486

Neale, B., Henwood, K. and Holland, J. (2012) 'Researching lives through time: An introduction to the Timescapes approach', *Qualitative Research*, 12(1): 4–15. https://doi.org/10.1177/1468794111426229

Neves, D. (2017) *Reconsidering rural development: Using livelihood analysis to examine rural development in the former homelands of South Africa*, Cape Town: PLAAS, University of Western Cape.

Oestreicher, J.S., Farella, N., Paquet, S., Davidson, R., Lucotte, M., Mertens, F. and Saint-Charles, J. (2014) 'Livelihood activities and land-use at a riparian frontier of the Brazilian Amazon: Quantitative characterization and qualitative insights into the influence of knowledge, values, and beliefs', *Human Ecology*, 42: 521–540. https://doi.org/10.1007/s10745-014-9667-3

Östberg, W., Howland, O., Mduma, J. and Brockington, D. (2018) 'Tracing improving livelihoods in rural Africa using local measures of wealth: A case study from central Tanzania, 1991–2016', *Land*, 7(2): 44. https://doi.org/10.3390/land7020044

Ramamurthy, P. (2014) 'A feminist commodity chain analysis of rural transformation in contemporary India', in Fernandes, L. (ed.), *Routledge handbook of gender in South Asia*, London and New York: Routledge, pp. 247–259.

Saldaña, J. (2003) *Longitudinal qualitative research: Analyzing change through time*, Altamira: Rowman.

Sen, A. (2001) *Development as freedom*, Oxford: Oxford University Press.

Shehu, E. and Nilsson, B. (2014) *Informal employment among youth: Evidence from 20 school-to-work transition surveys*, Geneva: Youth Employment Programme, Employment Policy Department, International Labour Office.

Tarrant, A. and Hughes, K. (2019) 'Qualitative secondary analysis: Building longitudinal samples to understand men's generational identities in low income contexts', *Sociology*, 53: 538–553. https://doi.org/10.1177/0038038518772743

Ursin, M. (2016) 'Contradictory and intersecting patterns of inclusion and exclusion of street youth in Salvador, Brazil', *Social Inclusion*, 4(4): 39–50. https://doi.org/10.17645/si.v4i4.667

Warin, J. (2011) 'Ethical mindfulness and reflexivity: Managing a research relationship with children and young people in a 14-year qualitative longitudinal research (QLR) study', *Qualitative Inquiry*, 17(9): 805–814. https://doi.org/10.1177/1077800411423196

Wilmsen, B. and Hulten, A. (2017) 'Following resettled people over time: The value of longitudinal data collection for understanding the livelihood impacts of the Three Gorges Dam, China', *Impact Assessment and Project Appraisal*, 35(1): 94–105. https://doi.org/10.1080/14615517.2016.1271542

11
THE USE OF ETHNOGRAPHY FOR LIVELIHOODS RESEARCH

Thaís de Carvalho

Introduction

Poverty is not exclusively an economic condition (see Chapter 2 of this volume). After Sen's (1983) work on entitlement, development studies embraced a perspective of scarcity as a relational and multidimensional phenomenon (see Chapter 3 of this volume). It was then understood that the circumstances of deprivation do not result solely from individual choices. Rather, they depend on access to basic assets for survival, and from social structures that rule over and limit one's strategies of making a living. Moreover, postcolonial scholars have critiqued the representations of the Global South as 'underdeveloped', signalling residual colonial practices that sustain economic dependency and the stigma of certain cultures as inferior (Escobar, 2011; Harrison, 1997).

The concept of livelihoods has arisen in a search to understand the lives of impoverished peoples taking into account the structures that fabricate their social exclusion (de Haan, 2017). Instead of proposing universal interventions of poverty alleviation, livelihoods research works with a contextualised lens (de Haan, 2012). The concept is rooted in the notion that scarcity is relational and deeply connected to inequality. Thus, livelihoods research must pay attention to social networks and cultural values (Geiser, 2017).

Ethnography is a useful method to gather holistic data on people's lives in their everyday contexts. Originating from Anthropology, the method was used to research small communities' livelihood strategies long before this concept was incorporated into the lexicon of International Development (e.g. Malinowski, 2014 [1922]). Ethnographies describe a people's way of living, with attention to cultural and socioenvironmental factors that shape local strategies of survival. The method sheds light on diverse lifeways through case studies wherein the researcher is embedded in the studied community. The resulting data interprets and compares the local social organisation and culture in relation to theoretical debates in the social sciences.

Ethnography remains the main research tool of anthropologists and has been incorporated by several other disciplines in the social sciences. This is because, instead of bringing participants into experiments with highly structured interviews, the method places the researcher inside the participants' social environments, where they will observe and interact with people on a daily basis (Atkinson and Hammersley, 2007). The immersive case study offers a new standpoint from which to reflect on the original research questions, which is convenient when looking for an in-depth understanding of a certain context.

Ethnographies uncover the flaws of top-down models of intervention by examining complex aspects of local life experiences that could not be predicted without an intimate engagement with people and their context (Crawford et al., 2017; Mosse, 2005). They also contribute to humanising people's survival strategies in the face of normalised scarcity, untangling the multiplex network of cultural, political and environmental factors that push people to hazardous livelihoods. For instance, Scheper-Hughes' (1993) work unveiled an apparent maternal indifference in contexts of high infant mortality in rural Brazil, showing that mothers developed an emotional resilience to the recurrent loss of children. Apart from being scientifically revealing, the human content of ethnographic accounts incites empathy and attracts the attention of donors and policymakers (Biehl and Petryna, 2013).

This chapter offers an overview of the method and its applications and is divided into four sections. Firstly, it introduces ethnography and its potential for livelihoods research. Secondly, it discusses fieldwork, focusing mostly on participant observation but also presenting other common strategies of data collection. Finally, it discusses the process of writing about culture, drawing on anthropological reflections about ethics, bias, and validity.

What is ethnography?

'Ethnography' is, quite literally, a written description (*graphos*) of a people (*ethnos*). In the 19th century, ethnographies had the aim to catalogue and salvage the memory of what were viewed as vanishing cultures (namely, the Indigenous peoples of the New World). Despite the colonial roots of this research dynamic,[1] the method and its findings have largely contributed to challenging Western thought. A classic example is a cross-disciplinary debate about the gift economy, initiated with discussions arising from Malinowski's (2014[1922]) description of reciprocity in the Trobriand Islands and later explored in Mauss' (2002 [1954]) theory of gift exchange.

The first outlines of ethnographic praxis are also found in Malinowski's (2014) work. The author claimed that ethnographers must immerse themselves in a culture to understand local realities. In order to obtain enough data to comprehend a community, the researcher must live with the people they are studying, learn their language and modes of living. The researcher should look for the reasons why people behave as they do and maintain a vigilant posture throughout fieldwork. This often demands a long period of data collection, ideally of more than a year, for people to get acquainted to the researcher and vice-versa. Through detailed notes of the realities observed, the ethnographer records both concrete data (e.g. demographics) and abstract reasoning (e.g. cultural values).

These principles still guide participant observation, but the method has responded to the demands of the globalised world. Depending on the researcher's relationship with the studied context, ethnography can be pursued in shorter periods of time, in multi-sited (Marcus, 1995) and even multi-country case studies (e.g. Narotzky and Goddard, 2017), or in digital format (Hjorth et al., 2016). The research design will depend on the questions that the project aims to answer, but limitations of funding and external deadlines impact the amount of time that can be devoted to fieldwork. Overall, ethnographies focus on few delimited communities to build an in-depth account of a context.

While ethnography can take many forms, there is some common ground about what it does (Atkinson and Hammersley, 2007). The method relies on an open-ended mode of inquiry, developed in an intimate engagement with the studied community (Shah, 2017). Its data is qualitative and results from the researcher's impressions of the local context and their dialogues with the people in the study. Researchers may (and often do) employ several strategies of data collection. Fieldwork can be enhanced with activities that help the researcher to understand the context or to bond with the people they are studying, such as play and drawing (e.g. de Carvalho, 2021), photos and videos (Briggs and Mantini-Briggs, 2016; Pink, 2013). However, participant observation and its unstructured interviews are the core of an ethnographic case study.

A case study produces local explanations to social phenomena, rather than testing hypotheses. As a relational and exploratory mode of inquiry, the method thrives in open-endedness, in adjusting oneself (and the scope of research) to the realities of lived experience. Well-sketched plans often change during fieldwork. While this is expected, it poses a challenge when research proposals are expected to specify quantities, durations and hypotheses. The initial design guides the researcher's attention to certain aspects of social life (e.g. gender relations), but fieldwork experience will likely refine the conceptual framework and questions that were originally envisioned.

Immersed in the studied context, the ethnographer searches for an understanding of how people view the realities they face, their community and their own identity (Atkinson and Hammersley, 2007). The intimacy with people's daily routines, developed during months of shared interactions, gradually reveals a new perspective on local modes of living. In-depth case studies often bring innovative understandings to mainstream concepts, as illustrated by anthropological works on wealth and wellbeing that explore the different meanings of these notions to various communities in the Global South (Humphrey and Hugh-Jones, 1992; Parry and Bloch, 1989; Santos-Granero, 2015). Thus, the method illuminates the intricacy and diversity of human livelihoods.

Doing fieldwork

The researcher's immersion in a studied community is the core element of ethnography. The process to obtain consent for this varies at each context. Access to the field may rely on extensive bureaucracy and negotiation with gatekeepers, in a process that demands assertiveness about the researcher's boundaries and ethics. To accept you as a member of their community, even temporarily, people may ask for political and financial engagement with communal issues. The researcher has to reflect on the level of involvement that they feel comfortable in offering in reciprocity, and this will be constantly negotiated with informants (Scheper-Hughes, 1995).

Where fieldworkers face limitations to enter a community or if there are safety concerns, there is the possibility of collaborative research (Hoffman and Tarawalley, 2014). This means hiring a local research assistant who can collect part of the data and facilitate the ethnographic process. Gatekeepers and experienced researchers can instruct the ethnographer to opt for collaborative fieldwork. Research assistants may be crucial to gain access to certain spaces and demographics or act as translators (Middleton and Cons, 2014). Mediators are also handy if the ethnography aims to evaluate a project or support advocacy and policymaking, albeit the challenge of setting boundaries will be extended to the those relationships as well (Mosse, 2005).

Collaborative research brings up issues of power, authorship and even ethnographic validity (Gupta, 2014). The researcher's identity and social fluency influences data collection, and thus an assistant often has a different field experience. Assistants may also have a method that differs from the main ethnographer's practice (Middleton and Pradhan, 2014). This is both the value and challenge of this approach. Since the researcher has the main prerogative of ethnography, they will have to grapple with these dual perspectives on the case study during data analysis. The researcher also has to negotiate with the assistant the level of involvement expected in the production of the manuscript. This ranges from a written acknowledgment of the assistant's contributions to a fully collaborative authorship. When disclosing the work of a research assistant, it is good practice to add a reflection on ethics, particularly considering the innate power imbalances of this relationship.

Fieldnotes

The beginning of fieldwork is a treasured stage of ethnography. In the process of entering the research site, the researcher will observe how their relationships unfold and what the community makes of a person inquiring about their lifestyles. This offers valuable clues about local sociopolitical dynamics and the community's links to outsiders. Fieldnotes keep track of one's impressions and experiences during data collection. In retrospect, they express how the researcher's view of the community changed with time, and how the relationships in the field have developed. Without the practice of journaling, the ethnographer risks forgetting the initially intriguing aspects of daily life, which will become trivial with their growing familiarity to the context.

There is no universal guideline to record field experience. Notes could be taken through audio recording, on a phone, with pen and paper, through images, or even using a combination of these (Sanjek, 1990). Opting for a written, digital format is simpler if the intention is to do software-based analysis. The habit of taking notes may affect relationships with informants as they reinforce the researcher's role and exogenous origin (Atkinson and Hammersley, 2007). While bullet points may be a quick solution to this, without a more descriptive journaling practice the ethnography risks relying solely on memory. Part of the challenge of fieldwork is finding adequate time and a format for note-taking that preserves people's privacy and minimises disruptions in interactions with others. It is important to mark the direct speech of informants and the researcher's recollection of conversations, as this needs to be acknowledged in the manuscript.

The practice of journaling incites a preliminary data analysis. Adopting a broader scope for daily annotations, especially during the initial months of fieldwork, allows the research to gain shape. Through writing, the researcher identifies key concepts and phenomena, along with new angles from which to approach the research. Later in analysis, these key elements serve as basic categories for coding. Descriptive notes leave ground for the researcher to revisit fieldwork memory in detail and reassess underplayed data with hindsight. In the manuscript, different excerpts will merge to convey the richness of fieldwork through thick description (example in Box 11.1).

Participant observation

Ethnography has a dialogic condition. It engages the perspectives of the observer (etic) with those of the local people (emic). This builds an account of a culture that is both innovative and theoretically informed. The attempt to become a part of a community is fundamental to acquire, as much as possible, an emic point of view. The embeddedness of ethnography incites the researcher to re-imagine lives from the standpoint of local people and their relations to their environment, based on the premise that dwelling in the world is essential to theorise about it:

> The scientist may indeed think himself to be an isolated, rational subject confronting the world as a spectacle, yet were he in reality so removed from worldly existence he could not think the thoughts he does.
>
> *(Ingold, 2000: 433)*

In a fluid arrangement with their hosts, the ethnographer aims to transform the initial estrangement provoked by their presence into a sense of familiarity. This will depend on the researcher's capacity to surrender to the local mode of living (Shah, 2017). When researchers are foreigners, this means embracing the discomfort of being alien to a culture. Engaged with communal chores, the researcher produces knowledge by 'being and action' (Shah, 2017:45). This is also exemplified in Box 11.1. They nurture relationships of rapport with local people and attempt to incorporate the necessary skills to survive in the context. These skills are social (e.g. etiquette) and bodily (e.g. chopping wood), as both are a worthy source of information about the cultural values of a community. The etiquette and the rituals that allow integration into a new community will vary in each context and will need to be observed. During this process, the researcher's body is, simultaneously, a tool that makes culture and the product of symbolic significance (Csordas, 1994).

The community's welcome will depend on their perception of the researcher's identity. Markers of difference, such as gender, influence a researcher's access to certain aspects of local life, as well as their perception of risk (Bell et al., 1993). For self-protection and social integration, the ethnographer negotiates (and performs) local gender norms (Huggins and Glebbeek, 2009). Other aspects of one's identity, such as race and class, may also interfere in building rapport with interlocutors, especially in contexts of greater inequality (e.g. Baird, 2018). The researcher should carefully reflect on the effects of their identities at all stages of fieldwork, being attentive to both limitations and privileges. When this is disclosed in the final report, it gives information about local power imbalances and social dynamics from the ethnographer's own embodied experience.

Box 11.1 Example of participant observation

In my fieldwork in Peruvian Amazonia, learning to sew became a source of visual, verbal, and embodied information about Shipibo-Konibo livelihoods. At first, I only learned the craft because women spend long hours sewing with kin. However, the habit served as a practical visualisation of the concept of 'connectedness', central to local descriptions of a good life. *Kenés* are an intricate pattern of interconnected elements, eluding to Amazonian fauna and flora. They are often stitched in cloths tinted with ink made from cooked tree barks and mud. The elaborate tapestry is an easy sell to European tourists and the main source of independent income for women. The importance of *kené* for the household economy also explained why young girls would undergo a painful treatment of *waste*, the application of plant-based eye drops that made them envision those patterns.

Interviews

The broad description of the ethnographic method as 'participant observation' conceals the importance of other forms of inquiry that are inherent to fieldwork. The most common one is the informal interview, which naturally happens as the researcher interacts with others (Hockey and Forsey, 2012). Open-ended and unstructured questions surge as the researcher notices activities and interactions that spark an interest: a simple walk in the forest in Amazonia may lead to a conversation about hallucinogenic plants that can predict a person's career prospects, and that offers a perfect opportunity of inquiry about aspirations (e.g. Buitron, 2020).

Participatory methods

In contemporary ethnography, participatory methods are often used to elicit conversations, especially when the available time for fieldwork is short (see Chapter 13 of this volume). These methods also have the potential to discover functional approaches of intervention from the perspective of local people, engaging research participants in the design of alternatives (Crawford et al., 2017; de Haan, 2017). The familiarity with the local culture and the power dynamics of the community, developed during participant observation, help the researcher to shape the best strategy for the participatory stage, later in the fieldwork. The approach should be appealing and functional to the group of people that are the focus of study.

Participatory approaches can also enrich the study of groups with different levels of literacy. A classic example is the use of the arts in research with children, as illustrated by Hunleth's (2017) research with children that took care of ill family members in Zambia. By observing the dynamics of role-playing and the aesthetic choices in representations of daily life, the researcher elicited conversations about their topics of interest and reduced the risk of distress to research participants. It is also through a combined analysis of ethnographic data, drawings and interviews that Pires (2014) envisions the role of children as family sponsors in a conditional cash transfer programme in Brazil.

The interpretation and writing of cultures

Data analysis

Each culture is full of interwoven symbolic meanings that shape people's beliefs and influence their life choices (Geertz, 2007). The role of the ethnographer is to decipher the complex nuances of social life through the critical analysis of lived experiences. Since ethnography is a reflexive method, analysis is somewhat an ongoing process. During fieldwork, the researcher is constantly reflecting on their interactions with others, the potential meaning of everyday rituals, the structures of social organisation and power dynamics. Nevertheless, the major chunk of analysis is done in preparation for writing.

Ethnographers may opt to use a software (e.g. NVivo) or to code data manually. In both cases, they will have to revisit their fieldnotes to recognise the patterns and unforeseen events that shaped their experience in the field. The development of categories that classify fieldnotes tentatively marks the key themes in an ethnography. In this process, fieldnotes can be combined with document analysis to provide a historical and political context to an ethnography. In writing, this conveys an elucidative synthesis of how overarching structures affects local livelihoods. Box 11.2 is an example of this.

Box 11.2 Example of document analysis embedded in ethnography

In multi-ethnic Peru, the State determined the form of Indigenous communities in a law from 1978. The law established that, in order to have collective land titles, all Indigenous peoples should settle in delimited territories designed by the State. Their land would take the form of a rural village, to which the government would provide a school and a medical unit. This law radically altered the livelihood strategies of Amazonian peoples. In 2014, the government also published a resolution through the Ministry of Development and Social Inclusion that considered indigenous peoples in Amazonia to be 'extremely poor' and reliant on a non-monetary economy.

Ethnographies in the region make explicit the social changes faced by Indigenous peoples as an effect of such development initiatives in the rainforest (e.g. Santos-Granero, 2015). They contest the Peruvian State's notions of wealth and poverty as being quite distant from Indigenous realities.

> Labour is a fundamental and demanding part of life in Amazonia, and a good character is often defined in terms of energy and strength. Although money becomes increasingly more necessary to survive in settled communities, the purpose of accumulation is seldom individual profit. Amazonian peoples perceive wealth as a multifaceted combination of vital energy, communal assets and harmonious behaviour (ibid).

Writing culture

The act of selecting which stories will appear in the manuscript and how to construe their meaning manifests the subjective character of ethnography. Authors have power over the narrative that will be told about a community. The ethical implications of this should be considered in the process of writing – after all, the method is based mostly on an 'interpretation of cultures' (Geertz, 2007). The ethnographic account engages with relevant social theory that supports the researcher's interpretations.

Social scientists once read ethnographies as accurate portraits of a culture, but this vision is old-fashioned. Since the publication of *Writing Culture* (Clifford and Marcus, 2011) in the 1980s, ethnographic compositions are recognised as the writer's subjective perception of fieldwork. The researcher's identity, their cultural values and political inclinations influence social interactions during the fieldwork and even data analysis. Hence, the importance of an exercise of self-reflection, known as reflexivity (Harrison, 1997). Acknowledgement of positionality is an ethical concern of contemporary ethnography. In other words, ethnographers are expected to disclose their identity along with any research bias in writing.

Since ethnography's neutrality was debunked, a politically engaged account is acceptable if presented as such. It may even be considered necessary if the researcher is immersed in a setting of severe social injustice and deprivation (Scheper-Hughes, 1995; Biehl and Petryna, 2013). Disclosure statements can be embedded in the description of fieldwork, as shown in Box 11.3.

> **Box 11.3 Disclosing a militant engagement**
>
> My Cofán friends asked me to take pictures of the disaster [the oil spill in Ecuadorian Amazonia]. They wanted to record and publicise the event so that other people, in other lands, could see what was happening. I sent photos and reports to contacts in the United States and Ecuador's capital, Quito. They passed the information along social media and activist networks. I filmed a Cofán leader's impromptu remarks beside the oil-covered river. In A'ingae, he talked about how often such things had happened to his people. He said the spill meant they would have nowhere to bathe, fish, or wash clothes and dishes (Cepek, 2018: 4).

When ethnographers opt to be involved, they can use emotion to convey empathy (Behar, 1996). This compelling writing strategy attracts attention from policymakers with compelling human stories (Biehl and Petryna, 2013). The mention of one's emotional responses in the field does not threaten validity. Montgomery's (2001) work about child prostitution in Thailand is an example of this. Her book starts by describing the misconceptions and simplistic views associated with the practice, as well as her own internal conflict in studying such a sensitive topic. However, despite these enormous ethical challenges, Montgomery managed to uncover the cultural values and fraught legislations that sustain international child prostitution.

Regardless of the aesthetics choices for the final manuscript, ethnographers must continuously ponder the ethics of their research practices. The preservation of integrity of any communities in which studies were based is paramount throughout the process of outreach and publication. This prevents the access of potentially harmful actors to the community while also keeping people's secrets safe. Research ethics is safeguarded with confidentiality and anonymity of research participants.

Conclusion

This chapter has shown that ethnography is an effective method for the in-depth study of livelihoods. Through the open-endedness of participant observation, the researcher can develop rapport with informants and gradually access intimate knowledge of people's lives. Thus, the method has the potential to reveal aspects of local livelihoods that could not be known without an intimate mode of inquiry. Moreover, case studies that examine people in their real environments allow the researcher to gain a comprehensive view of local livelihood strategies.

The chapter details all stages of an ethnography, from fieldwork to the final manuscript. It describes various methods of data collection, focusing on the main ethnographic strategy of participant observation and detailed fieldnotes. It also presents strategies of data analysis, showing how the researcher can combine fieldnotes with document analysis to provide a political and historical context to a community. Finally, it discusses strategies of writing that can boost policy impact.

When situated in relation to broader structures that govern people's lives, ethnographic accounts offer a holistic view of local modes of living. Its data results in an in-depth account of local livelihoods and contextual elements (e.g. laws and power relations) that may affect them. Through the interesting dialectics of estrangement and intimacy (Shah, 2017), the researcher searches for an insider's perspective. Later, they make sense of fieldnotes in dialogue with existing social theory. This help contest misconception about a community, and shape contextualised strategies of intervention in International Development.

Being embedded in the context is a fruitful standpoint to reflect on how people's choices are shaped by their environment, and their customs and beliefs, which is a fundamental concern of livelihoods research. A profound understanding of people's culture engaged with relevant theory offers a new outlook on livelihood strategies in the Global South. Since ethnography has the capacity to illustrate the intricacy and diversity of people's livelihood strategies, it contributes to shaping contextualised and respectful policy and practice in Development.

Key points

- Ethnography is a qualitative and open-ended method that relies on the researcher's immersion in the context of the people they are studying. It gives information about people's ways of living in their context.
- In ethnographies, information is gathered in daily life interactions and through observation of cultural patterns. Participant observation is its core strategy of data collection, but it is often combined with other methods such as open-ended and unstructured interviews, and document analysis.
- Participatory methods can be handy when working with groups with different level of literacy (e.g. children) or to elicit specific conversations.
- The benefit of ethnography for livelihoods research is its comprehensive view of a community. The method pays attention to social relations and cultural influence in people's lives, contributing to an understanding of *how* local livelihoods strategies came to be.

- The method counters top-down approaches through a contextualised approach to local realities. Furthermore, ethnographic accounts provide human stories and promote empathy with different livelihoods.

Further reading

Atkinson, P. and Hammersley, M. (2007) *Ethnography: Principles in practice*, London: Routledge.
An excellent starting point for future ethnographers. Offers a step-by-step to the method, detailing different kinds of ethnography and their uses.

Shah, A. (2017) 'Ethnography? Participant observation, a potentially revolutionary praxis', *HAU: Journal of Ethnographic Theory*, 7(1): 45–59. https://doi.org/10.14318/hau7.1.008.
Presents a dense argument about the importance of ethnography for knowledge production, focusing on the method's potential to confront research biases.

Notes

1 For more about the decolonisation of Anthropology and its methodologies, see Harrison (1997) and Smith (2012).

References

Atkinson, P. and Hammersley, M. (2007) *Ethnography: Principles in practice*, London: Routledge.
Baird, A. (2018) 'Dancing with danger: Ethnographic safety, male bravado and gang research in Colombia', *Qualitative Research*, 18(3): 342–360. https://doi.org/10.1177/1468794117722194
Behar, R. (1996) *The vulnerable observer: Anthropology that breaks your heart*, Boston: Beacon Press.
Bell, D., Caplan, P. and Karim, W.-J.B. (1993) *Gendered fields: Women, men, and ethnography*, London: Routledge.
Biehl, J.G. and Petryna, A. (2013) *When people come first: Critical studies in global health*, New Jersey: Princeton University Press.
Briggs, C.L. and Mantini-Briggs, C. (2016) *Tell me why my children died: Rabies, indigenous knowledge, and communicative justice*, Durham: Duke University Press.
Buitron, N. (2020) 'Autonomy, productiveness, and community: The rise of inequality in an Amazonian society', *Journal of the Royal Anthropological Institute*, 26(1): 48–66. https://doi.org/10.1111/1467-9655.13180
Cepek, M. (2018) *Life in oil. Cofan survival in the petroleum fields of Amazonia*, Austin: University of Texas Press.
Clifford, J. and Marcus, G.E. (2011) *Writing culture: The poetics and politics of ethnography*, Berkeley: University of California Press.
Crawford, G., Kruckenberg, L.J., Loubere, N. and Morgan, R. (2017) *Understanding global development research: Fieldwork issues, experiences and reflections*, London: Sage Publications. https://doi.org/10.4135/9781473983236
Csordas, T.J. (1994) *Embodiment and experience: The existential ground of culture and self*, Cambridge: Cambridge University Press.
de Carvalho, T. (2021) 'White men and electric guns: Analysing the Amazonian dystopia through Shipibo-Konibo children's drawings', *Global Studies of Childhood*, 11(1): 40–53. https://doi.org/10.1177/2043610621995837
de Haan, L. (2017) *Livelihoods and development: New perspectives*, Leiden: Brill.
de Haan, L.J. (2012) 'The livelihood approach: A critical exploration', *Erdkunde*, 66(4): 345–357. www.jstor.org/stable/41759104
Escobar, A. (2011) *Encountering development: The making and unmaking of the third world*, Princeton: Princeton University Press.
Geertz, C. (2007) *The interpretation of cultures: Selected essays*, New York: Basic Books.
Geiser, U. (2017) 'Understanding poverty, defining interventions: Why social relations need more attention in livelihoods analyses and why this complicates development practice', in de Haan, L.J. (ed.), *Livelihoods and development: New perspectives*, Leiden: Brill, pp. 13–43. https://doi.org/10.1163/9789004347182_003

Gupta, A. (2014) 'Authorship, research assistants and the ethnographic field', *Ethnography*, 15(3): 394–400. https://doi.org/10.1177/1466138114533460

Harrison, F.V. (ed.) (1997) *Decolonizing anthropology: Moving further toward an anthropology of liberation*, Arlington: American Anthropological Association.

Hjorth, L., Horst, H., Gallowat, A. and Bell, G. (2016) *The Routledge Companion to digital ethnography*, London: Taylor & Francis Group.

Hockey, J. and Forsey, M. (2012) 'Ethnography is not participant observation: Reflections on the interview as participatory qualitative research', in Skinner, J. (ed.), *The interview: An ethnographic approach*, London: Bloomsbury Academic, pp. 68–87.

Hoffman, D. and Tarawalley, M. (2014) 'Frontline collaborations: The research relationship in unstable places', *Ethnography*, 15(3): 291–310. https://doi.org/10.1177/1466138114533463

Huggins, M.K. and Glebbeek, M.-L. (2009) *Women fielding danger: Negotiating ethnographic identities in field research*, London: Rowman & Littlefield.

Humphrey, C. and Hugh-Jones, S. (1992) *Barter, exchange, and value: An anthropological approach*, Cambridge: Cambridge University Press.

Hunleth, J. (2017) *Children as caregivers: The global fight against tuberculosis and HIV in Zambia*, New Brunswick: Rutgers University Press.

Ingold, T. (2000) *The perception of the environment: Essays on livelihood, dwelling and skill*, London: Routledge.

Malinowski, B. (2014) *Argonauts of the western Pacific: An account of native enterprise and adventure in the archipelagoes of Melanesian New Guinea*, Hoboken: Taylor & Francis.

Marcus, G.E. (1995) 'Ethnography in/of the world system: The emergence of multi-sited ethnography', *Annual Review of Anthropology*, 24: 95–117. https://doi.org/10.1146/annurev.an.24.100195.000523

Mauss, M. (2002) *The gift*, London: Routledge Classics.

Middleton, T. and Cons, J. (2014) 'Coming to terms: Reinserting research assistants into ethnography's past and present', *Ethnography*, 15(3): 279–290. https://doi.org/10.1177/1466138114533466

Middleton, T. and Pradhan, E. (2014) 'Dynamic duos: On partnership and the possibilities of postcolonial ethnography', *Ethnography*, 15(3): 355–374. https://doi.org/10.1177/1466138114533451

Montgomery, H. (2001) *Modern babylon? Prostituting children in Thailand*, Oxford: Berghahn

Mosse, D. (2005) *Cultivating development: An ethnography of aid policy and practice*, London: Pluto Press. https://doi.org/10.2307/j.ctt18fs4st

Narotzky, S. and Goddard, V. (2017) *Work and livelihoods : History, ethnography and models in times of crisis*, London: Routledge.

Parry, J. and Bloch, M. (1989) *Money and the morality of exchange*, Cambridge: Cambridge University Press.

Pink, S. (2013) *Doing visual ethnography*, Los Angeles: SAGE.

Pires, F. (2014) 'Child as family sponsor: An unforeseen effect of Programa Bolsa Família in northeastern Brazil', *Childhood*, 21(1): 134–147. https://doi.org/10.1177/0907568213484341

Sanjek, R. (1990) *Fieldnotes: The makings of anthropology*, Ithaca: Cornell University Press.

Santos-Granero, F. (2015) *Images of public wealth or the anatomy of well-being in indigenous Amazonia*, Tucson: University of Arizona Press. www.jstor.org/stable/j.ctt183gxkn

Scheper-Hughes, N. (1993) *Death without weeping: The violence of everyday life in Brazil*, Berkeley: University of California Press.

Scheper-Hughes, N. (1995) 'The primacy of the ethical: Propositions for a militant anthropology', *Current Anthropology*, 36(3): 409–440. https://doi.org/10.1086/204378

Sen, A.K. (1983) *Poverty and famines: An essay on entitlement and deprivation*, Oxford: Oxford University Press.

Shah, A. (2017) 'Ethnography? Participant observation, a potentially revolutionary praxis', *HAU: Journal of Ethnographic Theory*, 7(1): 45–59. https://doi.org/10.14318/hau7.1.008

Smith, L.T. (2012) *Decolonizing methodologies: Research and indigenous peoples*, New York: Zed Books.

12
USING PARTICIPATORY RURAL APPRAISAL TO RESEARCH LIVELIHOODS

Klara Fischer

Introduction

This chapter provides an introduction to participatory methods, focusing in particular on Participatory Rural Appraisal (PRA, also known as participatory reflection and action) in the tradition of the work of Robert Chambers (Chambers, 1983a, 1992a, 1994a, 1997). It examines relevant issues when using PRA to research livelihoods, as well as key critiques and developments.

PRA evolved in the 1980s and 1990s, greatly inspired by Chambers' own experiences from development work in East Africa and India, and in response to top-down development and natural resource management projects that had failed to achieve their intended outcomes (Chambers, 1983b). The methods were critiqued and developed in several books and academic papers in the early 2000s (Campbell, 2002; Cooke and Kothari, 2001; Hickey and Mohan, 2004; Kapoor, 2002; Kesby, 2005; Mosse, 1994) and have since been used in many livelihood studies about and with people in the Global South (e.g. Angelsen et al., 2011; Ansell et al., 2012; Cramb et al., 2004; Ellis and Mdoe, 2003; Jacobson, 2013; Lunt et al., 2018; Radel, 2012).

The development of PRA stems from two key insights. The first is that people targeted by development need to be consulted if development projects are to achieve sustainable change. In an interview, Robert Chambers discusses how his own top-down interventions in natural resource management, which failed to lead to sustained change in local practices, was an important eye-opener and inspiration for the work he went on to do with PRA:

> My time working as a manager in the Samburu district [in Kenya] was very formative in the sense that I worked flat-out, with a certain missionary zeal, on grazing projects for the Samburu. We thought they were destroying their environment, which to some extent they were. We thought we had to do something very rapidly, as the erosion was spectacular; it was a tragedy of the commons type of situation through over-grazing. That attempt was disastrous: as soon as I left, the grazing schemes I had been involved in collapsed.
>
> *(Interview with Robert Chambers in Biekart and Gasper, 2013: 706)*

This led Chambers to appreciate the need to pay attention to local people's resource use and priorities and start from local perspectives in order to bring about lasting development.

The second key insight is that poverty is context-specific and multidimensional, and is best understood by listening to the people who are affected by it. This insight became more widespread in the work that led to the 'World Development Report 2000/2001: Attacking Poverty', in which PRA was extensively used, as reported in the three-part series 'Voices of the poor' (Narayan and Petesch, 2002). This broader perspective on poverty emanated from a reaction to the heavy focus on the monetary aspects of poverty often seen in policy and research, and large-scale survey studies aiming to objectively measure and standardise poverty measures across contexts. It was noted that such studies were costly and time-consuming, but nevertheless did not sufficiently aid our understanding of what poverty means or how to solve it (Chambers, 2007). As such, there is a common genealogy between livelihoods thinking and PRA in their critique of conventional methods of researching the situations of the poor and their common aim of describing these more accurately and holistically by embracing the diversity of livelihoods of the (rural) poor from their own perspectives (Chambers, 1992b, 1997, 2007; Chambers and Conway, 1992; Scoones, 2009, 2015).

PRA as a methodology offers a valuable perspective of listening to people as experts about their own situations, and the vast list of PRA methods can serve as a useful toolbox when researching rural livelihoods. In this context, it should be noted that to a great extent poverty is caused and maintained by larger processes beyond the control of individual people or households, and that such dimensions are only minimally captured by participatory research (Cooke and Kothari, 2001).[1] For such an understanding we must place our analyses within a broader historical and political context, which makes visible the structural constraints and dynamics that lead to local poverty (Scoones, 2015).

The next section provides a brief genealogy of PRA, followed by a description of the key PRA methods useful in livelihoods research. The two subsequent sections delve in more detail into two important aspects to consider when researching livelihoods using PRA: how to use PRA in ways that embrace heterogeneous livelihood possibilities, and the role of language in accessing people's perspectives through PRA. The chapter ends with a summary of the key points.

The genealogy of PRA

Academic inspiration for PRA came from, among others, farming systems research (Chambers et al., 1989), anthropology (Brokensha et al., 1980), theories of adult learning and activist participatory research (Freire, 1996). The idea was that PRA would lead to more accurate data and consequently better-informed development interventions, would be quicker than both long-term anthropological fieldwork and questionnaire surveys, and would empower people by engaging them in data collection and analysis, giving them control over describing their own situations and voicing their concerns to development practitioners and researchers (Campbell, 2002; Chambers, 1994a). PRA developed out of the earlier, less participatory Rapid Rural Appraisal (RRA) (Chambers, 1983a; McCracken, 1988), and has branched off into, inspired or developed in parallel with, other methods and concepts of participatory research and development, such as Participatory Learning and Action (PLA) (Chambers, 2007; Pretty et al., 1995) and (Participatory) Action Research ((P)AR) (Reason and Bradbury, 2008).

Early PRA writings place significant emphasis on participation, i.e. the engagement of local people in the research process, and in suggesting directions for action (Chambers, 1994b, 1997; Radel, 2012). Despite this, much work has been labelled PRA that '*is elicitive or extractive rather than participatory*' (Chambers, 1994b: 958). Chambers early on suggested that this extractive work should be referred to as RRA rather than PRA as '[a]n RRA is intended for learning by outsiders [whereas a] PRA is intended to enable local people to conduct their own analysis, and often to plan and take action' (Chambers, 1994b: 958). He also noted that what he termed RRA was more frequently used by researchers and published in academic journals under the name of PRA, whereas 'true' PRA was the focus mainly of NGOs aiming for local empowerment and action. Despite Chambers' early critique of this way of terming non-participatory work PRA, the term RRA never really took off in academia, and instead PRA has remained a more popular term in research, including for describing more extractive forms of data collection.[2]

PRA as a package of methods for researching livelihoods from local perspectives

PRA has been used in a diversity of livelihood studies and is largely known for its valuable set of tools aimed at joint learning in a group and enabling local participants to lead data collection and determine associated categorisations and classifications of their local situations. This section describes and summarises key methods that can be useful when researching livelihoods through PRA.

In early PRA work, there was a reluctance by some to provide practical PRA manuals for fear that such manuals would take attention away from one of the core principles of PRA, that of letting the context and local participants guide the choice of activities (Chambers, 1994c). There was, and remains, a concern that manuals with lists of PRA tools would lead to their top-down or instrumental use by researchers without reflection (Kapoor, 2002; Chambers, 1994b), which would undermine efforts within PRA to embrace local and alternative ways of understanding the world than those of the researcher (Chambers, 1997). Nevertheless, many useful manuals on PRA have been produced that can be of significant value when planning and performing livelihoods research (e.g. Chambers, 2002a; Mukherjee, 1993; Narayanasamy, 2009; Pretty et al., 1995).

The wealth of tools described in these manuals allows the researcher to select and adapt a wide range of tools and approaches, enabling adaptation to the context and methodological triangulation. A list of some commonly used PRA tools is given in Table 12.1.

Table 12.1 Key PRA tools for researching rural livelihoods

Tool	Description	Read more
Mapping (e.g. village mapping, social mapping, resource mapping)	Participants in a group jointly draw maps of e.g. settlement areas, locally important resources, distribution of households of different wealth etc. The map can subsequently be used for further interviews on emerging issues.	Mukherjee, 1993; Chambers, 2002a; Narayanasamy, 2009; Gentle and Maraseni, 2012
Transects/transect walks	Participants walk with the researcher(s) along a transect and describe what can be seen; a transect across a particular section of a village is drawn and described, pointing out important landscape features.	Pretty et al., 1995; Mukherjee, 1993; Narayanasamy, 2009
Historical transect/timeline	Elderly participants draw and describe important historical events, landscape features, environmental or agricultural changes over time etc. and map these along a timeline or transect.	Mukherjee, 1993;

Tool	Description	Read more
Scoring and ranking (e.g. pairwise ranking, matrix scoring, wealth ranking)	Participants rank and score different items or issues by their relative importance, e.g. usefulness of different tree species for firewood, tolerance of different local crop varieties to drought, categorisation of households in a village based on local classifications of relative poverty and wealth etc.	Scoring and ranking: Mukherjee, 1993; Pretty et al., 1995 Wealth ranking: Lunt et al., 2018; Jacobson, 2013; Ellis and Mdoe, 2003
Seasonal calendar	Participants list and describe important issues such as celebrations, key agricultural events, seasonal weather changes etc., e.g. along a line or a circle representing one year.	Mukherjee, 1993; Pretty et al., 1995; Narayanasamy, 2009; Gentle and Maraseni, 2012
Diagramming (Venn diagram)	Circles of different size and proximity to each other are used for visualising e.g. access to key institutions or resources, relationships between social groups etc.	Pretty et al., 1995; Narayanasamy, 2009; Angelsen et al., 2011; Gentle and Maraseni, 2012

Accessing heterogeneity and different local perspectives through PRA: pitfalls and solutions

A key tenet of livelihoods research is the diversity of livelihoods, not only within but also between households, implying heterogeneity among a community's residents in terms of their access to different livelihood opportunities (de Haan and Zoomers, 2005; Ellis, 2000). Thus, an important part of understanding rural livelihoods is understanding this diversity. This section discusses the possibility of capturing the breadth of voices from heterogeneous local communities through PRA.

Early PRA texts emphasised the importance of researchers making sure that they do not just talk to the more privileged members of rural communities. These are often referred to by Chambers in terms of biases, e.g. spatial and person biases, i.e. only speaking to people who are easy to reach by car or who are more willing to speak, who are often local elites (Chambers, 1983a, 1994b). It was emphasised that spending a limited time in the field (referred to by Chambers as rural development tourism) would enhance the risk of the researcher falling for such biases (Chambers, 1994c). Despite this, early PRA writing has been criticised for being unclear about how to deal with power and heterogeneity in local communities (Cooke and Kothari, 2001; Gaventa and Cornwall, 2006; Kapoor, 2002). This is because despite highlighting local heterogeneity, much of the early literature still spoke of rural people as one group. Terms such as 'rural people', 'local experts' or 'the poor' are concepts used in the PRA literature about those targeted by research or development (Chambers, 1994a). At times this has been interpreted as a lack of appreciation by early PRA researchers of the heterogeneity in rural communities (Fischer and Chenais, 2019).

Related to this, there is criticism of the ability of PRA tools to grasp this heterogeneity. While the broad spectrum of PRA tools as described in the previous section can be helpful in exploring different dimensions of livelihoods and is able to facilitate different ways of expression, many of these tools are designed for use in group meetings with the aim of reaching consensual answers (Mubaya et al., 2012; Mukherjee, 1993). A frequently expressed critique of this idea is that group activities that focus on reaching consensus might actually marginalise non-dominant perspectives (Campbell, 2002; Chenais and Fischer, 2018; Fischer et al., 2020). Some will find it easier than others to express themselves freely in a group and might even dominate the discussion at the expense of other participants. Structural cultural factors might also lead certain participants (e.g. women, younger people or ethnic minorities) to refrain from voicing their discontent, or even express support for dominant perspectives voiced in a community meeting even though they might express disagreement if they were in another group constellation (Cooke and Kothari, 2001; Gaventa and Cornwall, 2006; Kesby, 2005). The methodological focus on consensus might potentially both reinforce, and be reinforced by, the absence of any mention of heterogeneity in early PRA texts (Kapoor, 2002).

There are generally two types of strategies outlined in the literature for ensuring that a greater diversity of perspectives are embraced when using PRA:

1 Strategies employed within a group to ensure that different participants are heard, and
2 Dividing groups to increase their homogeneity and reduce the risk of certain actors dominating the discussion.

Chambers places a significant emphasis on work within groups and gives a key role to facilitators in PRA exercises to ensure that everyone is engaged and heard in discussions (see also Chambers, 1994a; Narayanasamy, 2009). Others have suggested that Chambers placed too much emphasis on the personality of the facilitators, and that a more structured and theoretically informed facilitation of group activities would be better at ensuring participation across power differentials (Kapoor, 2002).

It has also been emphasised that reliance on facilitators from the local community does not automatically mean that diverse local perspectives are taken into account. Many countries in the Global South are highly unequal (Seery et al., 2019). Extreme gaps in educational levels and hierarchical education systems jointly create a significant gap between textbook knowledge and local practical knowledges (Chenais and Fischer, 2018; Hebinck et al., 2011). There are examples of researchers experiencing difficulties in getting local (but often comparatively well-educated) facilitators and translators to establish a rapport between researchers and the local community or to accurately interpret local activities without correcting or even ignoring less educated participants (Berreman, 2012; Chenais and Fischer, 2018).

The second strategy is to divide groups for the purpose of increasing their homogeneity. This aims to ensure that people within the group feel comfortable sharing their experiences. In particular the importance of holding separate groups with women has been emphasised in the PRA literature (Djoudi and Brockhaus, 2011; Mubaya et al., 2012; Mukherjee, 1993). More recent PRA literature also discusses how the division of groups along other axes of local relevance can be important (Ansell et al., 2012; Chenais and Fischer, 2018; Fischer et al., 2020).

A strategy mentioned less frequently for embracing diverse perspectives is to move beyond group activities and seek out individuals who do not come to meetings or are uncomfortable at expressing themselves in groups, and to interview these people individually (Bedelian et al., 2007; Chambers, 2002b; Fischer et al., 2020). Meeting people individually is also more accommodating of their day-to-day priorities.

More generally, there is a wealth of literature on participatory research that deals with heterogeneity and power in rural communities that can be constructively consulted when planning and executing research (Ebata et al., 2020; Gaventa and Cornwall, 2006; Hickey and Mohan, 2004; Kapoor, 2002; Kesby, 2005; Mosse, 1994).

Embracing local perspectives by paying attention to language

Two related dimensions of language use mentioned here are also of importance in PRA: the relationship between language and power, and the constructive aspect of language.

Language is closely related to power and knowledge. Different languages, but also different styles and dialects have an impact on how people are perceived by others. Those with a limited formal education might not be as skilful at formulating their arguments because they do not have the same vocabulary as their better-educated peers, or might be intimidated if the researcher uses formal language or specialist terminology (Fraser, 1990; Hagendijk and Irwin, 2006). They might also censor themselves, or not trust their own judgement when asked to provide their views in a group with more educated peers (Gaventa and Cornwall, 2006). In part, this aspect of language is clearly connected to issues of group dynamics discussed in the previous section.

Much PRA research is conducted in situations where the researcher does not speak the same language as local participants (Chambers, 2002a; Fischer et al., 2020; Mariner and Paskin, 2000). This means that significant responsibility, and influence over the participatory process, lies with the local interpreters and assistants who facilitate the process. However, the role of language and the need for interpretation are rarely mentioned in participatory methodology texts (e.g. it is left out of the following key texts on PRA (Chambers, 1992a, 2007; Narayanasamy, 2009; Pretty et al., 1995)). It is possible that this absence of methodological reflections on language and the use of interpreters in PRA is a result of the wider absence of these aspects from a great deal of anthropological and other social science research (Borchgrevink, 2003; Skjelsbæk, 2016).

In contrast to much traditional PRA work, participatory epidemiology (PE) and the related participatory disease surveillance (PDS) have engaged more with the role of language. Thus, looking at developments in this field could be important for the development of PRA. PE is a field of participatory research within (veterinary) medicine and epidemiology that has its roots in PRA and RRA, and aims to engage local communities in resource-constrained settings to co-research human and animal health problems and suggest measures for prevention, control and surveillance (Allepuz et al., 2017). PE work has emphasised the importance of collecting data on diseases and health problems in local terminology (Bedelian et al., 2007; Jost et al., 2007; Mariner and Paskin, 2000). Equally important, but less frequently acknowledged in the PE literature, is the analysis of data collected in local terminology with the insight that there is not necessarily a 1:1 match between local and formal classifications (Fischer et al., 2020). This means that a translation that focuses on explaining the local *meaning* of words and phrases is preferable, rather than trying to make local terms fit with formal classifications or the researcher's language (Fischer et al., 2020). For example, when describing damage to crops and illness in livestock, smallholder farmers in the Global South often refer to symptoms (e.g. wilted, becoming powder, fever) rather than causes (e.g. specific insects or pathogens). Furthermore, dialects of the same language and local knowledge of seemingly culturally similar neighbouring communities can vary significantly. For example, Jost et al. (2007: 539) describe how

> the Somalis have a very detailed grasp of disease vectors and have local names for most species important in disease transmission. Like the Somalis, the Karamojong of Uganda are pastoralists, but do not associate insects or arthropods with specific diseases. Understanding these factors is essential to carrying out disease investigations and designing control programmes that work.

This research indicates the relevance of PRA work that engages more closely with local knowledge systems, interpretation, and language.

While spoken language is rarely discussed in the PRA literature, PRA tools are designed to be flexible in accommodating different types of classification, communication, and non-literacy, with an explicit aim to avoid top-down steering by researchers or facilitators. Many PRA tools are visual and/or can be performed without pen and paper (Table 12.1). The idea is that exercises such as ranking and scoring, which do not rely mainly on writing, make it easier and less threatening for participants with poor literacy skills to engage in the activity (Mukherjee, 1993). Likewise, the suggestion of using local materials such as stones or beans and writing with a stick on the ground rather than with a pen and paper is based on the idea that this will make it easier for local participants to take charge of the process, avoiding a sense of a school-like situation in which the researcher and facilitators will easily be seen as having more competence (Narayanasamy, 2009). Letting local people steer the research process in this way is also expected to make it easier for the researcher to understand local categorisations (Chambers, 1994a). Chambers has spoken about this as 'handing over the stick' (1994a: 2), emphasising the core idea of letting local participants steer data collection and analysis.

In summary, the use of PRA tools, encouraging local people to take charge of the research process, and to be open and curious about local classifications and explanations, will help researchers acquire a better understanding of how local participants perceive their world. Paying attention to dimensions of power and differences within rural communities, and designing data collection with this in mind, will further facilitate a nuanced and locally relevant understanding of diverse rural livelihoods.

Key points

- PRA tools should be modified to fit local participants and contexts. PRA books offer a plethora of different tools to be used when researching local livelihoods. These should be used as a source of inspiration and be modified with local participants so that they make sense to them and their situations.
- Local communities are heterogeneous. The researcher needs to bear in mind how group dynamics and different modes of expression offered by different PRA tools might favour certain voices over others. This chapter provides examples and suggestions for how to embrace heterogeneous local livelihoods when using PRA.
- PRA is often performed in situations where the researcher is not fluent in the local language. There is seldom a straightforward relationship between terminologies in different languages, therefore it is important to pay attention to translation. Many PRA tools facilitate an openness to local classifications and categorisations, nevertheless it is also important to pay attention to verbal forms of translation as well. This chapter provides examples and suggests strategies and further reading for dealing appropriately with language and translation in PRA.
- PRA can be used and is useful in a variety of types of participation by local communities. While initially developed to empower local people, PRA tools have also been found of value in extractive forms of research where the main purpose is to learn about local communities. For ethical reasons, it is important that researchers and local participants have a common understanding about the expected level and nature of local involvement.

Recommended reading

Borchgrevink, A. (2003) 'Silencing language: Of anthropologists and interpreters', *Ethnography*, 4(1): 95–121. https://doi.org/10.1177/1466138103004001005

Cooke, B. and Kothari, U. (2001) *Participation: The new tyranny?* London and New York: Zed books.

Fischer, K., Schulz, K. and Chenais, E. (2020) '"Can we agree on that?" Plurality, power and language in participatory research', *Preventive Veterinary Medicine*, 180: 104991. https://doi.org/10.1016/j.prevetmed.2020.104991

Gaventa, J. and Cornwall, A. (2006) 'Challenging the boundaries of the possible: Participation, knowledge and power', *IDS Bulletin* 37(6): 122–128. https://doi.org/10.1111/j.1759-5436.2006.tb00329.x

Hickey, S. and Mohan, G. (2004) *Participation – from tyranny to transformation? Exploring new approaches to participation in development*, London: Zed books.

Mukherjee, N. (1993) *Participatory rural appraisal: Methodology and applications*, New Delhi: Concept Publishing Company.

Narayanasamy, N. (2009) *Participatory rural appraisal: Principles, methods and application*, New Delhi: SAGE.

Pretty, J.N., Guijt, I., Thompson, J. and Scoones, I. (1995) *Participatory learning and action: A trainer's guide*, London: International Institute for Environment and Development (IIED).

Notes

1. Although there are exceptions. See, for example, Kesby (2005) and Radel (2012) in the list of references.
2. A search of the Web of Science core collection from 1980–2020 for the topic 'Rapid Rural Appraisal' revealed a peak in the term's use in 1995 with 62 publications, subsequently reducing to eight publications in 2019. A similar search for 'Participatory Rural Appraisal' shows that between the mid-1990s and 2019, the concept has been used in 40–90 new publications per year.

References

Allepuz, A., De Balogh, K., Aguanno, R., Heilmamm, M. and Beltraon-Alcrudo, D. (2017) 'Review of participatory epidemiology practices in animal health (1980–2015) and future practice directions', *PLoS One*, 12(1): e0169198. https://doi.org/10.1371/journal.pone.0169198

Angelsen, A., Overgaard Larsen, H. and Friis Lund, J. (2011) *Measuring livelihoods and environmental dependence: Methods for research and fieldwork*, London: Earthscan.

Ansell, N., Robson, E., Hajdu, F. and van Blerk, L. (2012) 'Learning from young people about their lives: Using participatory methods to research the impacts of AIDS in southern Africa', *Children's Geographies*, 10(2): 169–186. https://doi.org/10.1080/14733285.2012.667918

Bedelian, C., Nkedianye, D. and Herrero, M. (2007) 'Maasai perception of the impact and incidence of malignant catarrhal fever (MCF) in southern Kenya', *Preventive Veterinary Medicine*, 78(3–4): 296–316. https://doi.org/10.1016/j.prevetmed.2006.10.012

Berreman, G.D. (2012) 'Behind many masks: Ethnography and impression management', in Robben, A.C.G.M. and Sluka, J.A. (eds.), *Ethnographic fieldwork: An anthropological reader*, Chichester: Wiley, pp. 152–174.

Biekart, K. and Gasper, D. (2013) 'Robert Chambers', *Development and Change*, 44(3): 705–725. https://doi.org/10.1111/dech.12025

Borchgrevink, A. (2003) 'Silencing language: Of anthropologists and interpreters', *Ethnography*, 4(1): 95–121. https://doi.org/10.1177/1466138103004001005

Brokensha, D., Warren, D.M. and Werner, O. (1980) *Indigenous knowledge systems and development*, Lanham, Maryland: University Press of America.

Campbell, J. (2002) 'A critical appraisal of participatory methods in development research', *International Journal of Social Research Methodology*, 5(1): 19–29. https://doi.org/10.1080/13645570110098046

Chambers, R. (1983a) *Putting the last first*, New York: Routledge.

Chambers, R. (1983b) *Rural development: Putting the last first*, Essex: Longman.

Chambers, R. (1992a) 'Relaxed and participatory rural appraisal, notes on practical approaches and methods', *Participatory Rural Appraisal Workshop Recommended Reading*, Pietermaritzburg: Midnat.

Chambers, R. (1992b) *Rural Appraisal: Rapid, Relaxed and Participatory*, Brighton: Institute of Development Studies.

Chambers, R. (1994a) 'Participating Rural Appraisal (PRA): Analysis of experience', *World Development*, 22(9): 1253–1268. https://doi.org/10.1016/0305-750X(94)90003-5

Chambers, R. (1994b) 'The origins and practice of participatory rural appraisal', *World Development*, 22(7): 953–969. https://doi.org/10.1016/0305-750X(94)90141-4

Chambers, R. (1994c) 'Participatory Rural Appraisal (PRA): Challenges, potentials and paradigm', *World Development*, 22(10): 1437–1454. https://doi.org/10.1016/0305-750X(94)90030-2

Chambers, R. (1997) *Whose reality counts? Putting the first last*, London: Intermediate Technology Publications.
Chambers, R. (2002a) *Participatory workshops: A sourcebook of 21 sets of ideas and activities*, New York: Earthscan.
Chambers, R. (2002b) *Relaxed and participatory appraisal: Notes on practical approaches and methods for participants in PRA/PLA-related familiarisation workshops: Participation group, institute of development studies, University of Sussex*, Brighton: Eldis.
Chambers, R. (2007) 'Participation and poverty', *Development*, 50: 20–25. https://doi.org/10.1057/palgrave.development.1100382
Chambers, R. and Conway, G. (1992) 'Sustainable rural livelihoods: Practical concepts for the 21st century', *IDS Discussion Paper 296*, Brighton: Institute for Development Studies.
Chambers, R., Pacey, A. and Thrupp, L.A. (1989) *Farmer first: Farmer innovation and agricultural research*, London: Intermediate Technology Publications Ltd.
Chenais, E. and Fischer, K. (2018) 'Increasing the local relevance of epidemiological research: Situated knowledge of cattle disease among Basongora pastoralists in Uganda', *Frontiers in Veterinary Science*, 5: 119. https://doi.org/10.3389/fvets.2018.00119
Cooke, B. and Kothari, U. (2001) *Participation: The new tyranny?* New York: Zed books.
Cramb, R., Purcell, T. and Ho, T. (2004) 'Participatory assessment of rural livelihoods in the Central Highlands of Vietnam', *Agricultural Systems*, 81(3): 255–272. https://doi.org/10.1016/j.agsy.2003.11.005
de Haan, L. and Zoomers, A. (2005) 'Exploring the frontier of livelihoods research', *Development and Change*, 36(1): 27–47. https://doi.org/10.1111/j.0012-155X.2005.00401.x
Djoudi, H. and Brockhaus, M. (2011) 'Is adaptation to climate change gender neutral? Lessons from communities dependent on livestock and forests in northern Mali', *International Forestry Review*, 13(2): 123–135. https://doi.org/10.1505/146554811797406606
Ebata, A., Hodge, C., Braam, D., Waldman, L., Sharp, J., MacGregor, H. and Moore, H. (2020) 'Power, participation and their problems: A consideration of power dynamics in the use of participatory epidemiology for one health and zoonoses research', *Preventive Veterinary Medicine*, 177: 104940. https://doi.org/10.1016/j.prevetmed.2020.104940
Ellis, F. (2000) *Rural livelihoods and diversity in developing countries*, Oxford: Oxford University Press.
Ellis, F. and Mdoe, N. (2003) 'Livelihoods and rural poverty reduction in Tanzania', *World Development*, 31(8): 1367–1384. https://doi.org/10.1016/S0305-750X(03)00100-1
Fischer, K. and Chenais, E. (2019) 'What's in a name: Participatory epidemiology', *Preventive Veterinary Medicine*, 165(1): 34–35. https://doi.org/10.1016/j.prevetmed.2019.01.009
Fischer, K., Schulz, K. and Chenais, E. (2020) '"Can we agree on that"? Plurality, power and language in participatory research', *Preventive Veterinary Medicine*, 180: 104991. https://doi.org/10.1016/j.prevetmed.2020.104991
Fraser, N. (1990) 'Rethinking the public sphere: A contribution to the critique of actually existing democracy', *Social Text*, 25/26: 56–80. https://doi.org/10.2307/466240
Freire, P. (1996) *Pedagogy of the oppressed (revised)*, New York: Continuum.
Gaventa, J. and Cornwall, A. (2006) 'Challenging the boundaries of the possible: Participation, knowledge and power', *IDS Bulletin*, 37(6): 122–128. https://doi.org/10.1111/j.1759-5436.2006.tb00329.x
Gentle, P. and Maraseni, T.N. (2012) 'Climate change, poverty and livelihoods: Adaptation practices by rural mountain communities in Nepal', *Environmental Science & Policy*, 21: 24–34. https://doi.org/10.1016/j.envsci.2012.03.007
Hagendijk, R. and Irwin, A. (2006) 'Public deliberation and governance: Engaging with science and technology in contemporary Europe', *Minerva*, 44: 167–184. https://doi.org/10.1007/s11024-006-0012-x
Hebinck, P., Fay, D. and Kondlo, K. (2011) 'Land and agrarian reform in South Africa's Eastern Cape Province: Caught by Continuities', *Journal of Agrarian Change*, 11(2): 220–240. https://doi.org/10.1111/j.1471-0366.2010.00297.x
Hickey, S. and Mohan, G. (2004) *Participation – from tyranny to transformation? Exploring new approaches to participation in development*, New York: Zed Books.
Jacobson, K. (2013) *From betterment to Bt maize: Agricultural development and the introduction of genetically modified Maize to South African smallholders department of urban and rural development*, Uppsala: Swedish University of Agricultural Sciences, pp. 155.
Jost, C.C., Mariner, J.C., Roeder, P.L., Sawitri, E. and Macgregor-Skinner, G.J. (2007) 'Participatory epidemiology in disease surveillance and research', *Revue Scientifique Et Technique-Office International Des Epizooties*, 26: 537–549. https://doi.org/10.3389/fvets.2020.532763

Kapoor, I. (2002) 'The devil's in the theory: A critical assessment of Robert Chambers' work on participatory development', *Third World Quarterly*, 23(1): 101–117. https://doi.org/10.1080/01436590220108199

Kesby, M. (2005) 'Retheorizing empowerment-through-participation as a performance in space: Beyond tyranny to transformation', *Signs: Journal of Women in Culture and Society*, 30(4): 2037–2065. https://doi.org/10.1086/428422

Lunt, T., Ellis-Jones, J., Mekonnen, K., Schulz, S., Thorne, P., Schulte-Geldermann, E. and Sharma, K. (2018) 'Participatory community analysis: Identifying and addressing challenges to Ethiopian smallholder livelihoods', *Development in Practice*, 28(2): 208–226. https://doi.org/10.1080/09614524.2018.1417354

Mariner, J. and Paskin, R. (2000) *FAO animal health manual 10 – manual on participatory epidemiology – method for the collection of action-oriented epidemiological intelligence*, Rome: Food and Agriculture Organization of the United Nations.

McCracken, J. (1988) *Participatory rapid rural appraisal in Gujarat: A trial model for the Aga Khan Rural Support Programme (India)*, London: International Institute for Environment and Development.

Mosse, D. (1994) 'Authority, gender and knowledge: Theoretical reflections on the practice of participatory rural appraisal', *Development and Change*, 25(3): 497–526. https://doi.org/10.1111/j.1467-7660.1994.tb00524.x

Mubaya, C.P., Njuki, J., Mutsvangwa, E.P., Migabe, F.T. and Nanja, D. (2012) 'Climate variability and change or multiple stressors? Farmer perceptions regarding threats to livelihoods in Zimbabwe and Zambia', *Journal of Environmental Management*, 102: 9–17. https://doi.org/10.1016/j.jenvman.2012.02.005

Mukherjee, N. (1993) *Participatory rural appraisal: Methodology and applications*, New Delhi: Concept Publishing Company.

Narayan, D. and Petesch, P. (2002) *Voices of the poor: From many lands*, Washington, DC: World Bank Publications.

Narayanasamy, N. (2009) *Participatory rural appraisal: Principles, methods and application*, New Delhi: SAGE.

Pretty, J.N., Guijt, I., Thompson J, Scoones, I. (1995) *Participatory learning and action: A trainer's guide*, London: International Institute for Environment and Development (IIED).

Radel, C. (2012) 'Gendered livelihoods and the politics of socio-environmental identity: Women's participation in conservation projects in Calakmul, Mexico', *Gender, Place & Culture*, 19(1): 61–82. https://doi.org/10.1080/0966369X.2011.617905

Reason, P. and Bradbury, H. (2008) *The Sage handbook of action research: Participative inquiry and practice*, 2nd ed., New Delhi: Sage

Scoones, I. (2009) 'Livelihoods perspectives and rural development', *Journal of Peasant Studies*, 36: 171–196. https://doi.org/10.1080/03066150902820503

Scoones, I. (2015) *Sustainable livelihoods and rural development*, Rugby: Practical Action Publishing.

Seery, E., Okanda, J. and Lawson, M. (2019) *A tale of two continents: Fighting inequality in Africa. Oxfam briefing paper- September 2019*, London: Oxfam International.

Skjelsbæk, I. (2016) 'Interpreting the interpreter: Navigating translation, interpretation, and mediation', *Culture & Psychology*, 22(4): 502–519. https://doi.org/10.1177/1354067X16650830

13
USING PARTICIPATORY VIDEO IN RESEARCHING LIVELIHOODS IN THE GLOBAL SOUTH

The why and wherefore

Lara Bezzina

Introduction

The practice of participatory video (PV) has grown in popularity as a research method and tool used by both academics and practitioners (Milne et al., 2012). It involves a group of people designing and making their own film (Lunch and Lunch, 2006), enabling them to convey their ideas without the limitations that may be set by the participant's societal status or the fear of possible adverse consequences one may face from communicating controversial views (White, 2003). While PV is not unique in creating spaces for participants' voices to be heard or in contributing towards action for development (Roberts and Lunch, 2015), Johansson et al. (1999) argue that one would be hard-pressed to envisage a method that is more effective in understanding people's perspectives than producing, watching, discussing, and analysing the PV with them. Within this framework, PV has been widely used as a method to research livelihoods in the Global South, varying in scope from enabling marginalised women in Nepal to formulate the support they need to adapt to climate change (Khamis et al., 2009) and researching disabled people's income-generating activities (IGAs) in Burkina Faso (Bezzina, 2020), to exploring the role of social memory for natural resource management in Guyana (Mistry et al., 2014) and recognising the importance of recycling cooperatives in Brazil (Tremblay and de Oliveira Jayme, 2015). These cases, among others, illustrate the importance of PV – and participation – in social transformation and, consequently, in researching livelihoods in contexts where marginalised people's voices are not always heard. Participatory action research (PAR) approaches – of which PV forms part – create spaces where the participants take charge of the research process and reflect upon the action needed to improve their situations (McTaggart, 1997). Nonetheless, PV is not without its limitations, and its critiques need to be taken into account in order to maximise its potential and ensure that it benefits everyone involved.

This chapter begins by describing the participatory video process, its history, and current use. The second part then focuses on PV in researching and understanding livelihoods in Global South contexts, including critiques of such approaches. Lastly, the chapter reflects on the future of PV as a useful tool in the hands of those who are usually 'invisible', in influencing policy and decision-making processes that affect their means of survival.

Why participatory video?

Participatory video traces some of its roots back to the 1970s' community arts movement and Paolo Freire's 'praxis of reflection upon action' in enabling the oppressed to reflect upon and transform their reality (Roberts and Lunch, 2015: 2), and became broadly accessible with the launch of the consumer camcorder in the 1980s (Varghese et al., 2020). However, the origins of PV go further back, with one "beginning"[1] attributed to the Fogo Process. In the late 1960s, the National Film Board of Canada and Memorial University of Newfoundland's Extension Service filmed a series of short films with fishing communities on Fogo island, which they then screened to the island's residents. This stimulated dialogues across communities who recognised the similarity of their experiences, and the fact that some of their problems could be solved by working together (Newhook, 2009).

In order to understand what participatory video is, it is essential to understand what it is not. Specifically, PV differs from other types of video media, such as documentaries, where a filmmaker – usually western – decides what, who and where to film. In the PV process, the film emerges from grassroots groups (Johansson et al., 1999), providing the opportunity to look '"alongside" rather than "at" research subjects' (Kindon, 2003: 142). Thus, PV can be beneficial to both academic researchers (or PV facilitators) and those who are usually researched (in PV, the participants or co-researchers). It is worthwhile noting that PV can also be used for purposes other than research, including advocacy, (project/impact) evaluation, and development.

The PV process is much more important than the finished product (Lunch and Lunch, 2006) because the process also educates participants (White, 2003) in various aspects. Keeping in mind that the implementation of PV varies according to the facilitator, community, and context, the process can take more or less the following shape: participants are trained in using video equipment, following which the facilitator(s) support(s) the group in identifying and analysing significant matters in their community (Lunch and Lunch, 2006), enabling personal stories to emerge and ultimately identifying the broad topic that the participants will select as their collective story (Plush and Shahrokh, 2015). This 'personal story' stage can involve the use of other participatory methods such as mapping, where participants draw a mind map of the community according to how they perceive their environment. Such maps can not only be used to instigate ideas and conversation, but also to identify community members to be interviewed for the film (Lunch and Lunch, 2006).

Next, the participants design and film short videos and messages (Lunch and Lunch, 2006). This process is iterative (Johansson et al., 1999), because the participants review the scenes they have shot before continuing to film or re-film the same scenes. The PV process can be as participatory as the facilitators and the participants choose. For example, an important phase following the filming may include group (participatory) editing, where the participants decide which parts of the film are to be included (or excluded), and in what order.

Once – or before – the video is edited, the film can be screened to the wider community, generating a 'dynamic process of community-led learning, sharing and exchange' (Lunch and Lunch, 2006: 12). Screening it to other communities enables 'horizontal communication', that is, the exchanging of ideas between communities. In addition, screening the video(s) to decision-makers, policy-makers, development workers, and others who affect the community members' lives, enables 'vertical communication' (Ibid: 13).

PV also allows the often forgotten but vital, part of the research process: the dissemination of findings. Rather than disseminating the knowledge gathered solely in academic circles through, for example, monographs which are generally inaccessible to most of the research participants (Fals Borda, 1988), in PV, the research "results" are accessible at a community level. Besides being produced by the community members themselves, the visual medium is also accessible to a wide range of audiences, including, for example, persons who are illiterate. The participants, who own the film, are also able to easily reproduce and transport it (Lunch and Lunch, 2006), enabling them to further disseminate the research findings after completion of the research project (and presumably the departure of the academic researcher). In this way, PV can also help transcend barriers posed by academic publishing fees, rendering research dissemination financially accessible to both non-western researchers as well as audiences.

Despite the numerous advantages PV holds over traditional methods, it still has a number of limitations. For example, working with what is an essentially a western medium in isolated communities could reinforce the illusion of the 'cultural superiority' of the west (Lunch and Lunch, 2006: 7). Furthermore, amid pertinent critiques of participatory techniques (see, for example Cooke and Kothari, 2001; Rahnema, 1992), Cleaver (2001) warns against the presumption of the existence of solidarity between community members, or of one detectable community in a particular location. Some of the setbacks of using PV are further discussed in the following section, which explores this approach in relation to understanding diverse livelihoods.

Participatory video in researching livelihoods in the Global South

PV has often been hailed as a powerful tool for empowering communities (see, for example, Haynes and Tanner, 2013; Varghese et al., 2020). White (2003) asserts that its primary goal is to foster respect to oneself and others, and to create a space where all persons' views are equally valued, regardless of their standing in society. It is perhaps evident, then, that PV is a valuable tool in researching and understanding livelihoods in Global South contexts. Examples of such use of PV can be found in Boxes 13.1 to 13.4.

Box 13.1 PV with recycling cooperatives in Brazil

After training leaders from recycling cooperatives in the São Paulo region – which are vulnerable due to limited government support – in filmmaking, four short (between 12 and 15 minutes) films[2] were co-produced between 2009 and 2011. The main themes included occupational health, validation and acknowledgment of service, and gender equality. A post-production group editing session was also held. Upon the completion of the final films, focus sessions were held with each group to review the video and make changes. The next step involved focus groups with leaders of cooperatives and the municipal governments, which enabled participant-led discussions and reflections on needed action such as positive changes in policy.

Source: Tremblay and de Oliveira Jayme (2015)

Box 13.2 PV with persons with disabilities in Burkina Faso

A participatory diagramming exercise – where participants created a diagram conveying their experiences and ideas (see Figure 13.1) – was facilitated with a group of seven disabled people in the Est region in 2014, which highlighted the issues the participants wanted to explore in their film: how personal commitment is essential to combating discrimination in such areas as employment (Bezzina, 2020). Following training in using the filming equipment, the participants designed, filmed, and partially edited the video.[3] Subtitles in French (and later in English) were added, and the film was then screened to a group of disabled people in the Cascades region, where the audience included representatives from the ministry responsible for disabled people's welfare (Bezzina, 2017). The screening sparked a discussion on the similarities in experiences, as well as with the authorities on disabled people's livelihoods and the struggles they face to find employment (Bezzina, 2018).

Figure 13.1 Participatory diagramming with disabled people in Burkina Faso

Figure 13.1 *Continued*

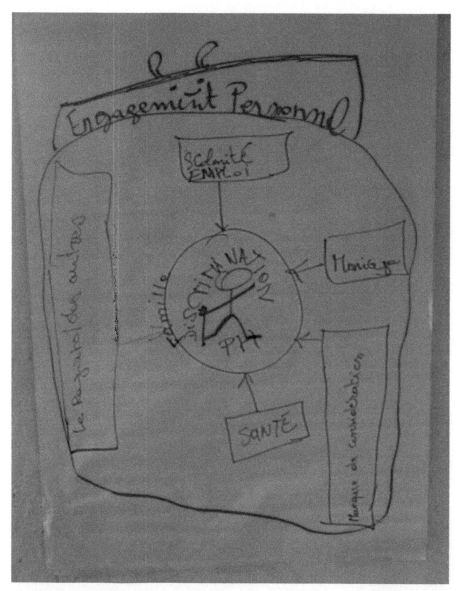

Figure 13.1 Continued

A group of persons with different disabilities in Fada N' Gourma, Burkina Faso developing a diagramme – in preparation for the PV – reflecting the discrimination a disabled person (portrayed in the circle in the middle) encounters. Close to the circle is the family, whose role is extremely influential in a disabled person's life. Surrounding this are the areas of life in which persons with disabilities in Burkina Faso encounter discrimination, including education, employment, marriage and health. This is manifested through lack of consideration by others and in the perceptions of disabled people held by non-disabled ones.

(Source: author)

> **Box 13.3 PV with marginalised women in Nepal**
>
> In 2008, a workshop training participants in holding video interviews, was held with the staff of a local environmental NGO and community members from Bageshwori and Matehiya. In the ensuing stage, the participants conducted video interviews with women about their experiences of climate change, their coping strategies and the possibilities of adaptation. Following screenings of these interviews and the ensuing discussions, the women members of the community selected the most urgent adaptation issue, which became the theme of the film.[4] After developing the storyboard, the women then created a video that was shown to the community, the local government, and local and international NGOs, in order to raise awareness on the topic and generate the needed support.
>
> *Source*: Khamis et al. (2009)

> **Box 13.4 PV with community members on natural resource management in Guyana**
>
> During a conservation and development project linking local indigenous livelihoods with biodiversity conservation in the North Rupununi in south-west Guyana, between 2006 and 2008, training on PV was opened to all communities of the region. After this, eight individuals were selected as community researchers. Using PV, these researchers selected, met, and interviewed community members and representatives of the local development board (Mistry et al., 2014). Narratives from the interviews with the members of the community were then used to explore the effect of social memory on the resilience of the local communities in the face of natural resource management challenges, the range of current challenges, and crises events. Upon the editing of the films and their translation into English, these were screened to communities, enabling the group to reflect on the emerging findings.
>
> *Source*: Mistry and Berardi (2012)

These examples demonstrate the diversity of the stages from one PV process to the next. The length of time required for training in the use of film equipment varies from two days (Bezzina, 2020) to seven or nine days (Mistry and Berardi, 2012; Tremblay and de Oliveira Jayme, 2015), where the latter also involved training in participatory methods and ethics. In this case, training was not only open to the PV participants, but also to 18 trainees who would then relate their experiences to their communities. However, this training was non-residential, unlike the training for disabled participants in Burkina Faso, which was residential because participants hailed from different municipalities which were too far away for them to return to every evening (Bezzina, 2017). At times, training can extend even beyond nine days, as in the case of Guyana, where the project lasted 18 months (Mistry and Berardi, 2012). These aspects require consideration when planning and implementing a PV process, especially because residential training increases costs for the researcher or organisation facilitating the research if they are the ones responsible for providing lodging and refreshments.

Challenges in PV process

Deciding on whether PV is the right method to employ involves considering various aspects of the process. The expenses of creating and using PV include those discussed above, as well as the purchasing of film equipment, which, while having become less expensive and easier to purchase, can still amount to a sum[5] much larger than needed for traditional research methods. One way in which this can be mitigated is by inviting the participants to use their mobile phones to film, capitalising on the increasingly cheaper and accessible smartphones in the Global South. While this is a frequently used practice (see, for example, Boone, 2014; Plush and Shahrokh, 2015; Schleser, 2012), inequalities (and the ensuing power dynamics) concerning ownership of, and access to, smartphones and related technologies (Rani et al., 2020) are important considerations and require caution when employing such practices. Furthermore, the available smartphones may not necessarily be suitable to produce films, and one might still need external equipment such as tripods and microphones to ensure the films are of accessible quality.

PV also tends to be more time consuming, both in planning and carrying out, than many other research methods. Furthermore, the process may have to be adapted to the participants' needs (Bezzina, 2018). For instance, when facilitating PV with disabled persons in Burkina Faso, two of the participants with physical impairments could not efficiently use their right arms to hold the video camera (which had a handle on the right), and instead, used the tripod for balance (see Figure 13.2). A running commentary was also kept by the participants during the diagramming process to enable a blind participant to follow what was being depicted. This, however, also meant that this person's participation was not as thorough as that of the other participants, also because during the filming, the same participant took more of a role in front of the camera rather than behind it.

Figure 13.2　Training in PV with disabled people in Burkina Faso
Source: Author

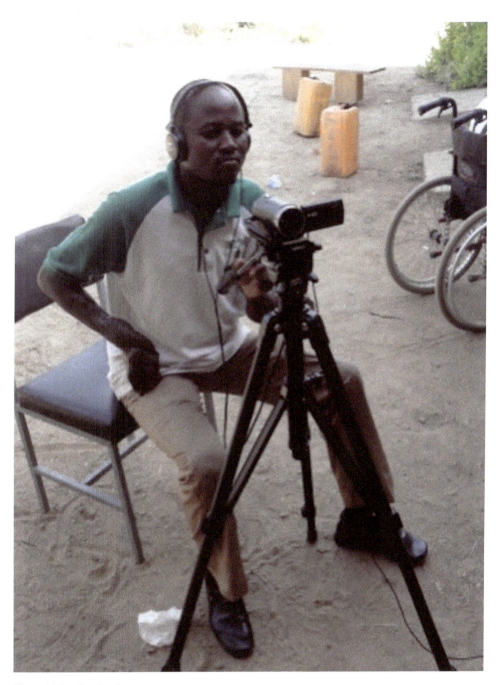

Figure 13.2 *Continued*

Furthermore, PV is not always necessarily conducted with the most marginalised people. In the case of Burkina Faso – both because of time and financial constraints – the PV was facilitated with what Chambers (1994: 12) calls 'the local elite'. The participants have 'less severe' disabilities, a relatively high level of education, and knowledge of the official language, French. These factors contributed to a "smoother" PV process than if it had been facilitated with those who are more marginalised, such as people with severe intellectual disabilities. Therefore, the importance of reflexivity in such processes is essential. While analysis of the "research findings" is held during the PV process itself with the participants, in order to reflect on the way forward, it is also necessary to keep in mind that the participants' situations might not necessarily be generalised to all persons – in this case, disabled individuals. Such challenges can be partially mitigated if resources allow for the production of more than one film, as was the case with recycling cooperatives in Brazil (see Box 13.1).

While all the examples given above involved participants behind the camera, there are projects where, while the participants did not film the videos themselves, the process was participatory in the sense that those filmed could say whatever they chose to. An example of this was the Forests Trees and People (FTP) project in the Ngorongoro Conservation Area (NCA) in Tanzania. In this case, people's views on a draft management plan regarding natural resource management were filmed without the use of structured interviews or selection of whose comments were to be recorded. The filmed participants took the opportunity to make their voices heard, knowing that the recordings would be shown to authorities. Since the views of the participants were negative, the authorities were not pleased with the video, and prohibited further video recordings by the NCA residents (Arora-Jonsson, 1996). Nonetheless, the fact that the participants were able to express their opinions to the authorities highlights one of the important aspects of PV, that of enabling vertical communication between the community and authorities.

This example is significant in highlighting the fact that PV does not always reach the aims/audiences it sets out to achieve/target. During the PV process outlined in Box 13.2, for instance, the envisioned audience for another screening of the film comprised international NGOs (INGOs) that intervene with disabled people in Burkina Faso. However, this screening did not take place due to INGOs' busy schedules. An important part of the PV process thus went missing.

PV in improving livelihoods policy and practice

Challenges notwithstanding, PV provides a unique opportunity to lobby for action, an often much-needed element in improving livelihoods. PV can be the medium through which action (or at least reflection upon action) kicks off, especially when policymakers are involved in viewings and discussions, as seen in the examples – of which there are many more[6] – provided earlier in this chapter. This can be seen in one PV process that took place in Nepal,[7] which was a collaborative effort between the Children in a Changing Climate (CCC) coalition[8] and ActionAid Nepal,[9] where children made short films about the effects of climate change in their communities. The children identified the challenges to their families' livelihoods (among others), the actions that their families were taking to cope, and the obstacles they encountered in their coping mechanisms. Through this, they spoke on the support they needed to adapt, including reforestation programmes, better agricultural technologies and infrastructure, more knowledge on climate change impacts, and efficient disaster prevention. Such PV processes not only enable children – who are often not invited to voice their opinions on climate change or livelihoods – to speak out, but also to reflect on what strategies the community members were already implementing in order to save their livelihoods, including what support they needed to better do so (Gautam and Oswald, 2008).

In the same vein, PV also offers the possibility of allowing participants to define for themselves the meaning of "livelihoods", beyond predetermined categories found in established frameworks. In the case of Burkina Faso (see Box 13.2), one of the issues emerging during the 'personal story' stage – which often results in the gathering of extremely rich data in itself – was that development actors intervening with disabled people tend to assume that persons with specific impairments want to engage in a particular IGA, and that persons with similar impairments want to engage in the same

IGAs as each other. Such reflections, which made their way into the PV, served to challenge preconceptions of the type of livelihoods that disabled persons want to engage in, as well as for disabled persons to speak out on the most beneficial ways to secure their means of survival (Bezzina, 2020).

In the same PV process, the participants also wished to challenge misconceptions surrounding disabled people in Burkina Faso. The filmmakers endeavoured to demonstrate to other disabled people that having a disability is not necessarily a tragedy. They also aimed to communicate to the wider community that disabled persons are not the burden they are often thought to be, but can work and fend for themselves and their families – an important issue which links to their status in society. Thus, being gainfully employed has multiple positive outcomes for disabled persons (see Chapter 23 of this volume for more on livelihoods and disability), which the PV serves to bring to the forefront (Bezzina, 2020).

Finally, the PV process can serve to instil a sense of pride in the filmmakers. In the case of the recycling cooperatives in Brazil (see Box 13.1), the PV process not only served to mobilise the community's knowledge of the recycling sector, but the increased self-confidence of the participants also linked to their pride in being recyclers. For example, one of the participants expressed that the PV project had increased 'her motivation and confidence as a recycler, and in recognizing her power and to *"fight and defend a work that is beautiful and important"*' (Tremblay and de Oliveira, 2015: 305, italics in original).

Concluding reflections

This chapter has provided an insight into the use and utility of PV in researching and understanding livelihoods in Global South contexts. While the discussion covered the details of PV processes, it is only, to use a well-worn phrase, the "tip of the iceberg", and, as suggested earlier, only touches upon a very small number of examples of the multitude of PV processes investigating communities' and different groups' livelihoods in Global South contexts and elsewhere.

The significance of PV in such contexts is undisputed: PV not only involves the research participants as co-researchers, but creates spaces where participants, who are experts on their own lives, can speak on the challenges they routinely encounter and reflect on the resources and skills they have in ameliorating their situation. Thus, PV not only enables participants to create their own film but also to disseminate the "research results" to other communities and policymakers through film screenings and ensuing discussions, without relying on the presence of the academic researcher or facilitating organisation. Such processes also provide the academic researcher with invaluable insights into the research participants' lived experiences, while engaging in research processes that enable groups of typically invisible persons to reach out to those who affect their means of survival.

Nonetheless, the academic researcher needs to employ reflexivity throughout the research process, and recognise the fact that while PV might provide a more ethical and decolonising means to investigate and understand lived experiences in Global South contexts, such participatory processes are often employed by international organisations such as the World Bank in order to legitimise decisions that have already been made (Hildyard et al., 2001). Instances such as these are called out by Rahnema (1992: 126), who cautions against reducing participation to a slogan – or simply a methodology – and argues that '[r]educed to such trivialities, not only does it cease to be a boon, but it runs the risk of acting as a deceptive myth or a dangerous tool for manipulation'. Considering this, such approaches need to be taken with care, making sure to avoid falling into the trap of using them as the new 'tyranny' (Cooke and Kothari, 2001: 3).

It is also important to recognise that researchers and communities in many Global South contexts, especially rural areas, may not have access to resources to initiate research processes which are recognised and approved of by western academia. Scholars who are trained, and working in, the Global South rarely have the opportunity to publish in western peer-reviewed journals or to participate in western-based conferences. Thus, rather than deciding on the topics to be researched and then implementing them in Global South contexts, research needs to involve scholars and researchers from Global South contexts whose knowledge and experience need to be highlighted and respected in international arenas. It is such processes in particular that will benefit from the use of PV to research, understand, and ultimately improve livelihoods in the Global South.

Key points

- The participatory video process consists of a group of people or community creating and filming their own film, in which they represent themselves as they deem fit. It is much used in researching and understanding people's livelihoods in Global South contexts.
- PV, like other participatory research approaches, brings in those who are usually 'researched' as co-researchers; and creates a space where the voices of those who are often unheard can be heard and those who are usually invisible can be seen. It can thus be an ethical means of conducting and disseminating research in contexts which have been – and are still – colonised.
- PV processes vary from one context to the other, and it can be used to speak out to other communities as well as to policy-makers and organisations influencing and intervening in the area of local people's livelihoods.
- Nonetheless, PV, like other participatory approaches, should be treated with caution in terms of not using it simply to tick a box: reflexivity and cultural sensitivity need to be employed throughout.
- Meanwhile, paving the way for researchers from Global South contexts to start their own research processes – while also recognising that research might already be taking place, albeit not in forms recognised by western academic institutions and western-based organisations (Swartz, 2009) – and supporting them in researching the aspects which they deem of importance, should be given priority.

Further reading

Kindon, S. (2003) 'Participatory video in geographic research: A feminist practice of looking?', *Area*, 35(2): 142–153. https://doi.org/10.1111/1475-4762.00236

Lunch, N. and Lunch, C. (2006) *Insights into participatory video: A handbook for the field* [Electronic]. Available at https://insightshare.org/resources/insights-into-participatory-video-a-handbook-for-the-field (Accessed 5 May 2020).

Roberts, T. and Lunch, C. (2015) 'Participatory video', in Mansell, R. and Ang, P.H. (eds.), *The international encyclopedia of digital communication and society*, New York: John Wiley & Sons Inc.

White, S.A. (ed.) (2003) *Participatory video: Images that transform and empower*, New Delhi: Sage Publications India Pvt Ltd.

Notes

1 Other scholars document earlier, or other, origins and histories of PV or alternative film productions (see, for example, Mistry and Berardi, 2012; Turner, 1991).
2 One of the films can be found here: www.youtube.com/watch?v=mB-oMNV0YlI
3 The long version of the film with English subtitles can be found here: www.youtube.com/watch?v=pNygwowI4xY&t=73s
4 The films facilitated with women in Nepal can be found here: www.youtube.com/user/idsclimatechange
5 For an example of PV costs, see Plush (2009).
6 See, for example, the work of Cobra Collective: https://communityownedsolutions.org; Insightshare: https://insightshare.org; Daldali-Dhansiri Community Video: www.youtube.com/channel/UC_GBfFyNtghd8CRpZtYo8Rw; Indigenous Community Video Collective - NE India: www.youtube.com/channel/UCyii0joHJVmMuFxL1uMjuew/videos

7 More information, including links to the video, can be found here: www.tamaraplush.com/participatoryvideo
8 The CCC coalition is made up of five child-centred development and humanitarian organisations. More information on CCC can be found here: www.childreninachangingclimate.org
9 ActionAid Nepal, a branch of Action Aid, focuses on ensuring the resilience of livelihoods against disasters as one of its foci. More information can be found here: https://nepal.actionaid.org

References

Arora-Jonsson, S. (1996) *Forests trees and people networking*, Unpublished.
Bezzina, L. (2017) *Disabled voices in development? The implications of listening to disabled people in Burkina Faso*, PhD Thesis, Durham: Durham University.
Bezzina, L. (2018) 'The role of indigenous and external knowledge in development interventions with disabled people in Burkina Faso: The implications of engaging with lived experiences', *Disability and the Global South*, 5(2): 1488–1507. https://dgsjournal.org/vol-5-no-2
Bezzina, L. (2020) *Disability and development in Burkina Faso: Critical perspectives*, Cham: Palgrave Macmillan.
Boone, K. (2014) 'Recalling and remembering community: Cellphone Diaries', in Bose, M., Horrigan, P., Doble, C. and Shipp, S. (eds.), *Community matters: Service-learning in engaged design and planning*, London: Routledge.
Chambers, R. (1994) *Paradigm shifts and the practice of participatory research and development*, Brighton: Institute of Development Studies.
Cleaver, F. (2001) 'Institutions, agency and the limitations of participatory approaches to development', in Cooke, B. and Kothari, U. (eds.), *Participation: The New Tyranny?* London: Zed Books Ltd.
Cooke, B. and Kothari, U. (eds.) (2001) *Participation: The New Tyranny?* London: Zed Books Ltd.
Fals Borda, O. (1988) *Knowledge and people's power: Lessons with peasants in Nicaragua, Mexico and Colombia*, New Delhi: Indian Social Institute.
Gautam, D. and Oswald, K. (2008) *Child voices: Children of Nepal speak out on climate change adaptation*, Brighton: Children in a Changing Climate, Institute of Development Studies; London: Action Aid. www.childreninachangingclimate.org/uploads/6/3/1/1/63116409/child_voices_np.pdf
Haynes, K. and Tanner, T.M. (2013) 'Empowering young people and strengthening resilience: Youth-centred participatory video as a tool for climate change adaptation and disaster risk reduction', *Children's Geographies*, 13(3): 357–371. https://doi.org/10.1080/14733285.2013.848599
Hildyard, N., Hedge, P., Wolvekamp, P. and Reddy, S. (2001) 'Pluralism, participation and power: Joint forest management in India', in Cooke, B. and Kothari, U. (ed.), *Participation: The New Tyranny?* London: Zed Books Ltd.
Johansson, L., Knippel, V., de Waal, D. and Nyamachumbe, F. (1999) 'Questions and answers about participatory video', *Forests, Trees and People Newsletter*, 40/41: 21–23.
Khamis, M., Plush, T. and Sepúlveda Zelaya, C. (2009) 'Women's rights in climate change: Using video as a tool for empowerment in Nepal', *Gender & Development*, 17(1): 125–135. https://doi.org/10.1080/13552070802697001
Kindon, S. (2003) 'Participatory video in geographic research: A feminist practice of looking?', *Area*, 35(2): 142–153. https://doi.org/10.1111/1475-4762.00236
Lunch, N. and Lunch, C. (2006) *Insights into participatory video: A handbook for the field* [Electronic]. Available at: https://insightshare.org/resources/insights-into-participatory-video-a-handbook-for-the-field (Accessed 5 May 2020).
McTaggart, R. (1997) 'Guiding principles for participatory action research', in McTaggart, R. (ed.), *Participatory action research: International contexts and consequences*, Albany: New York Press.
Milne, E-J., Mitchell, C. and de Lange, N. (ed.) (2012) *Handbook of participatory video*, Lanham: Altamira Press.
Mistry, J. and Berardi, A. (2012) 'The challenges and opportunities of participatory video in geographical research: Exploring collaboration with indigenous communities in the North Rupununi, Guyana', *Area*, 44(1): 110–116. https://doi.org/10.1111/j.1475-4762.2011.01064.x
Mistry, J., Berardi, A., Haynes, L., Davis, D., Xavier, R. and Andries, J. (2014) 'The role of social memory in natural resource management: Insights from participatory video', *Transactions of the Institute of British Geographers*, 39(1): 115–127. https://doi.org/10.1111/tran.12010
Newhook, S. (2009) 'The godfathers of Fogo: Donald Snowden, Fred Earle and the roots of the Fogo Island Films, 1964–1967', *Newfoundland and Labrador Studies*, 23(20): 1719–1726.
Plush, T. (2009) *Video and voice: How participatory video can support marginalized groups in their efforts to adapt to a changing climate*, Brighton: Institute of Development Studies.

Plush, T. and Shahrokh, T. (2015) *Story-driven participatory video for mobile technologies facilitation*, South Africa: Sonke Gender Justice

Rahnema, M. (1992) 'Participation', in Sachs, W. (ed.), *The development dictionary: A guide to knowledge as power*, London: Zed Books Ltd.

Rani, P., Naik, M.G. and Agrawal, B.C. (2020) 'Widening the wedge: Digital inequalities and social media in India', in Ragnedda, M. and Gladkova, A. (ed.), *Inequalities in the Global South*, Cham: Palgrave Macmillan.

Roberts, T. and Lunch, C. (2015) 'Participatory video', in Mansell, R. and Ang, P.H. (ed.), *The international encyclopedia of digital communication and society*, New York: John Wiley & Sons Inc.

Schleser, M.R.C. (2012) 'Collaborative mobile phone filmmaking', in Milne, E-J., Mitchell, C. and de Lange, N. (ed.), *Handbook of participatory video*, Lanham: Altamira Press.

Swartz, L. (2009) 'Building disability research capacity in low-income contexts: Possibilities and challenges', in MacLachlan, M. and Swartz, L. (ed.), *Disability & international development: Towards inclusive global health*, New York: Springer.

Tremblay, C. and de Oliveira Jayme, B. (2015) 'Community knowledge co-creation through participatory video', *Action Research*, 13(3): 298–314. https://doi.org/10.1177/1476750315572158

Turner, T. (1991) 'The social dynamics of video media in an indigenous society: The cultural meaning and the personal politics of video-making in Kayapo communities', *Visual Anthropology Review*, 7(2): 68–76. https://doi.org/10.1525/var.1991.7.2.68

Varghese, D., Olivier, P. and Bartindale, T. (2020) 'Towards participatory video 2.0', *CHI Conference on Human Factors in Computing Systems* (CHI '20), 25–30 April 2020, Honolulu, USA, [Online]. Available at: https://doi.org/10.1145/3313831.3376694 (Accessed 5 May 2020).

White, S.A. (ed.) (2003) *Participatory video: Images that transform and empower*, New Delhi: Sage Publications India Pvt Ltd.

14
USING PARTICIPATORY GPS METHODS TO DEVELOP RICH UNDERSTANDINGS OF PEOPLE'S DIVERSE AND COMPLEX LIVELIHOODS IN THE GLOBAL SOUTH

Nathan Salvidge

Introduction

Recent advancements in GPS technology have presented exciting opportunities for social scientists to expand the methods they use in their research. This chapter provides a critical overview of literature employing Global Positioning System (GPS) method approaches to develop insights into livelihoods and mobility. Opportunities and challenges of using GPS techniques in research are reviewed, as well as ethical considerations that need to be assessed prior to their employment. In response to an uptake of participatory approaches in development-related research, some, although not all, social scientists have begun to innovatively adapt these methods to promote the input(s) of participants in research processes who rely on mobility to undertake their livelihoods. Discussions in literature highlight how detailed insights into people's spatial experiences, knowledge and strategies in relation to their livelihoods can be gained through application of participatory GPS approaches. This is important so that theoretical and practical interpretations of the multifaceted nature of people's livelihoods can continue to be advanced. Moreover, as contemporary challenges such as high rates of urbanisation and climatic changes continue to impact populations across the Global South, it will be important to understand how livelihoods are being sustained and adapted in response to these vicissitudes; the chapter explores how participatory GPS approaches can be used to develop these understandings.

GPS use and developments in research practice

Originally used for military purposes, GPS technology is now widely available for public use and has been extensively employed in empirical studies over the last few decades (Shoval et al., 2018). This technology has enabled researchers and members of the public to produce high resolution spatial and temporal data of both human and object movements (Goodchild, 2007; Shoval,

DOI: 10.4324/9781003014041-16

2008). Understanding time-space patterns has been of particular interest to social scientists, and the employment of GPS methods has been especially beneficial to geographical and mobility research (Janusz et al., 2019). These methods have allowed for locational information of local environments to be produced (Oloo, 2018), and for people's individual patterns of movements to be greater understood (Oyana, 2017). As time has passed, GPS technology has advanced considerably and has become increasingly more affordable and easier to use. This has increased possibilities for research.

Contemporary GPS tracking

Dedicated GPS loggers have been used to track people's movements in a vast number of studies. However, these devices have been critiqued for being overly specialised and expensive (Lee et al., 2015), which has limited their use within social science research (Jones et al., 2011; Shoval et al., 2014). In recent years, advancements in smartphone technology have shown potential in being able to address these abovementioned drawbacks. Compared to stand alone GPS devices, contemporary smartphones are relatively inexpensive (Korpilo et al., 2017) and they have the capacity to run numerous mapping applications which can be repaired, replaced and updated with little cost and effort (Lee et al., 2015). GPS apps can also record multiple data in relation to movement such as the actual distances travelled, the time it took to complete a journey, the route taken and the destinations that were visited (Joseph et al., 2019). This is possible because modern smartphones have several embedded sensors that include technologies such as GPS, Wi-Fi positioning, accelerometers, gyroscopes, magnetometer, microphones and cameras (Birenboim and Shoval, 2016). These additions have also presented opportunities for real-time data to be collected through GPS enabled smartphones (ibid). In turn, this can afford deeper understandings into people's mobile lives (Evans et al., 2018; Shoval et al., 2018).

In terms of accuracy and sampling rate, it had been identified that dedicated GPS devices had more superior GPS capabilities than smartphones (Birenboim and Shoval, 2016). Yet, this gap has narrowed over time, and recent research has shown that smartphone-based GPS apps are able to produce accurate and detailed spatial and temporal information (Korpilo et al., 2017). Using these smartphone-based applications also requires less expertise than is required to use a GPS device, which makes it more accessible for both researchers and participants to engage in spatially contextualised research (Jones et al., 2011). However, researchers must also be mindful not to exclude participants by using approaches which demand smartphone ownership (Birenboim and Shoval, 2016; Joseph et al., 2019). Although smartphones are becoming cheaper throughout Global South contexts, there are still many who do not have access to such devices. The later sections of this chapter detail the ways smartphone-based GPS methods can be adapted so that participation can be promoted and sustained.

Using GPS technology to develop livelihood mobility research

The development of GPS tracking methods in research has coincided with a conceptual turn within the social sciences, the 'new mobilities paradigm', which emphasise the need to move beyond an over reliance on sedentary research practices within academia (Kusenbach, 2012). Previous approaches were critiqued for being unable to understand people's diverse and complex movements, and the connections that people have with local environments (Sheller and Urry, 2006). Consequently, social scientists have innovated their methodologies so that participants' mobilities and their engagements with different spaces can be better understood (Hein et al., 2008). Innovative GPS approaches, along with developments in mobile methods, have proliferated allowing social researchers to advance insights into people's diverse lived experiences as they perform different types of mobility across time and space (Jones and Evans, 2012).

Yet, contemporary mobilities research has been slow to investigate the interrelationship between livelihoods and mobility. As Esson et al. (2016: 187) claim, 'the new mobilities paradigm itself neglects the essential livelihood dimensions underpinning much mobility'. This is surprising, given that almost two decades ago research recognised that livelihoods and mobility intersect in multiple ways (Bryceson et al., 2003). It has also been widely acknowledged that mobility supports and defines livelihoods, especially across the Global South (Gough, 2008; Rigg, 2007). Within these contexts, there has been a growing number of contemporary studies which have begun to advance insights into people's livelihood mobilities through using GPS methods. These have identified the importance of mobility in relation to livelihood strategies, as well as the role of mobility in shaping people's work opportunities and challenges (Janusz et al., 2019; Naybor et al., 2016). Other research which has used tracking technology has also highlighted that mobile livelihood practices, such as transportation services, are a crucial component to the survival and development of many communities (Evans et al., 2018).

Nevertheless, more research into people's mobile livelihoods across Global South contexts is required (Lund, 2014). To date, much mobility theory has been founded on understandings developed in the Global North (Esson et al., 2016), which cannot adequately represent mobility experiences and practices in the Global South. Mobility is socially and culturally constructed, meaning that it is experienced differently depending on factors such as a person's age, gender, class, ethnicity and geographical location (Lund, 2014). As such, the mobilities people undertake in relation to their livelihoods will differ enormously depending on who they are and where in the world they are. The following discussion explains how GPS methods can be used to highlight the heterogenous and context specific nature of people's mobile livelihood practices.

Contextualising GPS data

Although GPS methods can allow for accurate and high-quality data to be produced, this technology does not ascribe meaning(s) to the data (Christensen et al., 2011). If an itinerant vendor's route is being tracked for example, a geographical pattern of this will be produced. However, this alone cannot reveal why this person chose to walk the route they did, nor will it be able to detail the environments that they walked through and their experiences of these (Christensen et al., 2011). For this reason, researchers have been using methods such as participant observation and go-along interviews in combination with GPS tracking to contextualise maps. Situating themselves within the environments through which participants navigate can afford researchers richer insights into livelihood mobilities and practices (Janusz et al., 2019). Naybor et al. (2016) integrated GPS tracking with several methods including ethnographic observation and participation to contextualise women's livelihood patterns in rural Uganda. Through this approach, they observed that it was common for these women to carry on foot the crops and other goods they wanted to sell. The women were physically limited by the amount of stock they were able to carry, therefore they had to make multiple trips which resulted in them tiring thus restricting the distances they were able to walk. The authors note that this forced them to sell their goods to nearby friends and neighbours at lower prices than they would have been able to get if they could travel further distances to markets. Used alone, GPS tracks would not have been able to produce this level of detail.

However, researchers do not necessarily need to be present for a detailed context of people's spatial journeys and experiences to be gained. Birenboim and Shoval (2016) claim that smartphones offer exciting opportunities through which contextual information and subjective insights into spatial journeys can be obtained. Self-reporting on smartphones has been recognised as a useful technique in combination with GPS tracking, which enables participants to report what they see, smell, hear, as well as how they feel, as they move through certain places (Birenboim and Shoval, 2016). This method has shown to be useful in understanding mobile experiences which are marked by irregularity and changeability (Sugie, 2018). Additionally, surveys can also be used on a smartphone device with participants, which can be triggered by location and/or time of day (Shoval et al., 2018). Advancing technology is giving researchers opportunities to be creative and innovative with GPS methods, which is furthering understandings into how people engage with, and interpret, the spaces they traverse (Jones et al., 2011).

Participatory GPS approaches

Participatory research places emphasis on working 'alongside' people and communities so that their knowledge and experiences, as well as the issues that are of concern to them, can be understood (Kesby et al., 2005). According to Oyana (2017), GPS methods offer unique opportunities for participants to actively engage and contribute to research processes. Little expertise is required for participants to be involved in these approaches, which can be beneficial for promoting high levels of participation within research (Jones et al., 2011).

In Global South contexts, participatory GPS approaches are starting to be used more frequently to understand people's livelihoods. Navarrete et al. (2017) undertook research around the Spermonde Archipelago, Indonesia, where they equipped artisanal fishermen with GPS devices so that they could map their spatial fishing practices. The authors note that the fishermen were keen to play an active role in this study, because a lack of regulation in this area had led to ecosystem degradation which threatened the future of their livelihoods and others who lived in surrounding local communities. The data which the fishermen played a vital role in collecting, produced findings which could then be used to inform policies and/or management strategies in the region.

In Brazil, Offenhuber and Lee (2012) employed participatory GPS mapping with informal recyclers, locally known as 'catadores', so that they could understand the detailed knowledge of the city which these workers possessed. This approach allowed the tacit knowledge of the 'catadores' to be documented and mapped in ways which was explicit and accessible to 'everyone', including the workers themselves, and to leaders and policymakers who it was hoped would be able to offer them support. Participatory mapping helped to validate the work of these informal recyclers by demonstrating the vital services they provide throughout the city. This was crucial, because often these workers are criminalised as they do not fit the modern image that many cities are trying to create (Offenhuber and Lee, 2012), thus mapping was able to show the work the 'catadores' do from a different perspective, marking their place in the city.

Similarly, Lee et al. (2015) undertook research with informal waste pickers in Mombasa, Kenya, who used their own mobile phones to map the routes they undertook. Through visualising waste pickers' routes, it was revealed that in certain areas of the city they had to walk long distances to reach open dumping points. Thus, this information suggested that open dumping points could be more equally spread throughout the city, which would save the pickers both time and effort whilst undertaking their work. The visualisation of these journeys also promoted participatory dialogue between several actors (waste collectors, municipal authorities and communities) in this study, regarding current practices and challenges of waste collection, as well as the future directions that should be taken in relation to waste management in the city.

In the above examples, participatory GPS was used in ways which can promote dialogue regarding context specific issues which people face in relation to their livelihoods. In turn, this can lead to interventions which can help to tackle issues faced by certain groups of people and communities (Oyana, 2017). Policymakers and other actors are becoming increasingly interested in the meaning and value of spaces (Hein et al., 2008), and GPS applications are one way through which the realities that people face can be better understood and engaged with (Joseph et al., 2019).

Increasing participation: a case study from Tanzania

I undertook ethnographic research in Dar es Salaam and Arusha, Tanzania, for one year from August 2018. In this study I utilised a range of methods including participant observation, participatory diagramming, semi-structured interviews and mobile-based GPS tracking. I employed GPS mapping with 8 youth participants (aged 15–35) across both cities. This method was used to develop understanding into young vendors' experiences, knowledge and practices of itinerant street vending. However, because of funding constraints (Evans, 2016), it was not possible to equip participants with their own GPS devices and none of these 8 participants owned their own smartphones. I had to use my own personal smartphone device, which meant that I had sole control of mapping. Because of this, I would accompany participants during their working days, which typically lasted 6–8 hours, but could last up to 12 hours. I used Strava, a free mobile-based fitness tracking application, which I was familiar with prior to using it in the field. This application was also chosen as it was easy to use and provided a simple yet effective visualisation of participants' routes. However, there are now an abundance of other free-to-download applications that are available to researchers for tracking purposes.

Mindful of participants' lack of participation in mapping their routes, I adapted the GPS method at later stages of the research process, to promote greater participant involvement. During follow-up interviews, participants were invited to annotate the GPS maps that had been created of their routes. Through prompts, I would ask participants whether their movements were strategised, why they chose to walk the routes they did and what parts of their journeys they found easiest/hardest, and why. This allowed subjective experiences into participants' livelihood practices and experiences to be gained (Birenboim and Shoval, 2016). The maps that were created were clear and appealing (Hansson and Roulston, 2017), which assisted participants in interpreting their routes and assigning meaning to parts of the city they navigated through (Offenhuber and Lee, 2012).

Figure 14.1 shows a map annotated by Godfrey (pseudonym), an itinerant vendor in Arusha, who sold oranges using a metal cart. In his map interpretations, Godfrey highlighted where he would expect to find customers whilst undertaking his route. Unsurprisingly, locating customers was a crucial aspect of his livelihood activities, and his knowledge of where they were most likely to be found determined which part of the city he would navigate through. In this instance, the location of customers gave meaning to his livelihood mobilities (Langevang and Gough, 2009). Furthermore, the value vendors assign to city spaces differs. Areas of the city which are beneficial for one business will not necessarily be good for another. GPS mapping and annotations can be used to capture this heterogeneity.

Godfrey's annotations also highlighted an area of the city which he found physically difficult to navigate, due to the steep hilly terrain he was confronted with. Godfrey's inclusion of these embodied encounters in his annotations, suggests that his interactions with urban environments, which evoked specific feelings, were an important aspect of his livelihood practices and experiences. Again, these insights are not generalisable and people's reactions and interpretations to space will vary (Porter et al., 2010). Yet, this further demonstrates the capability of GPS methods in being able to gain subjective insights into people's diverse experiences as they perform livelihood mobilities.

Figure 14.1 Annotated GPS map of Godfrey's route (an itinerant vendor who sold oranges in Arusha)

Source: Author

This case study demonstrates the versatility of GPS methods in that they can be adapted to include participation at later stages of a research process. Through using these approaches, detailed insights can also be gained into the knowledge and strategies which people employ to maintain their livelihoods.

Practicalities and ethics

Mapping the routes of participants using GPS devices/applications does not demand very much of participants during the data collection process (Shoval, 2008). As discussed previously, other methods can be used to contextualise participants' routes whilst they are 'on the move'. However, the suitably of including these with GPS tracking depends on the purpose of the research and the activities that participants are performing. Through my own experiences, I found that when undertaking research with participants who moved around in search of income it was not possible for them to stop, complete surveys, write down their thoughts, and/or have regular and detailed conversations with me during their working day. Thus, contextualising maps after they were created was the most suitable approach to take during my research. It is imperative that researchers consider how they will implement GPS methods in a research process, ensuring that participants do not become overburdened and/or disadvantaged because of the use of these techniques.

As noted previously, a growing number of development-related studies are adopting approaches whereby participants are involved in most, if not all, stages of a research process (Chambers, 2008; Kesby et al., 2005). Participants are often given possession of GPS devices to track their own routes. From an ethical perspective, giving participants control over mapping allows them to choose when they record their journeys. Research by Naybor et al. (2016) found that this approach afforded their participants with privacy as they were able to stop recording whenever they did not want something to be mapped, such as when they travelled to a HIV/AIDS clinic or to an illegal alcohol vendor. Yet even when participants have control over tracking, researchers must be mindful that participants' routes can still reveal locations such as their homes or places of work, which could expose their identity (Korpilo et al., 2017). As Taylor (2016) states, human mobility is becoming more legible and detailed through using technology such as GPS. Researchers should thoroughly consider the use of these methods before employing them to ensure that participant anonymity can be maintained. Participants also need to be informed as to why this data is being collected and how it will be used; will maps of their routes feature within a thesis, publications or/and reports? How will the data be stored, and who will be able to access this data? In relation to the latter point, Taylor (2016) states that such data can be used to survey and control unwanted movement. This can result in the movements of certain groups of people being restricted and/or stigmatised. Because data can be accessed and used in 'unknown ways', different from its intended purpose(s), it could be argued that gaining informed consent is not possible with research and data which involves the use of tracking technologies. However, participants can be made aware that the full consequences of their participation cannot be fully known and with this knowledge still agree to partake in research processes.

Tracking technology may also lead to the increased illegibility of certain people's mobilities. Actors who have and maintain power over smartphone devices will be able to use the technology to map their mobilities, which may be taken as representative of the community/society they are in. This can overlook the heterogeneity of movements within a specific location, and may work to further marginalise and underrepresent the mobilities of those who may be in less powerful positions due to their (dis)ability, age, class, ethnicity and/or gender, among other attributes. For example, Alozie and Akpan-Obong (2017) identify that throughout much of Africa, women are less likely to own and use ICT technology, compared to men, due to having lower levels of education and economic status.

Moreover, the act of carrying a GPS device may be unusual for participants. It has been suggested that the presence of recorders may influence participants' movements, resulting in them walking further distances, for longer lengths of time, or to different places than they would have done without possession of such a device (Christensen et al., 2011). Yet, Birenboim and Shoval (2016) argue that the growing popularity of mobile-based GPS applications to track movements is reducing these influences because people are increasingly used to carrying a mobile device with them everywhere.

Although GPS technology has advanced considerably in recent years (Korpilo et al., 2017), there are still environments, such as dense urban areas, where the GPS signal can be blocked (Shoval, 2008). During fieldwork in Arusha in 2019, I had trouble obtaining GPS signal in an area of the city where I was surrounded by three storey buildings. On this occasion, after some perseverance and constant refreshing of the tracking application, I was able to acquire signal further down the road and begin tracking. However, participants cannot constantly monitor the GPS application to ensure that it is working properly, as they will be engaged in their own activities. Moreover, I have also had situations whereby I have not been able to acquire a GPS signal at all, meaning that the routes of some participants were not tracked. Research contexts vary enormously, and through my own experiences, a lack of GPS signal cannot always be attributed to 'obvious reasons' such as high building density or tree cover; it can sometimes be unclear as to why GPS signal cannot be obtained, which can make it difficult to resolve such issues. Thus, for studies which rely heavily on GPS tracking, piloting the technology where possible is crucial to ensure that it works within the environment(s) in which research is taking place.

Conclusion

As this chapter has demonstrated, participatory GPS methods are becoming an important way through which insights into people's livelihoods can be developed. These approaches are particularly beneficial in developing understandings of the complex and diverse mobilities which many people across Global South contexts perform in relation to their livelihoods (Gough, 2008; Rigg, 2007). The knowledge, strategies and practices which people employ in relation to managing and sustaining their livelihoods is important to understand given that people across Global South contexts are having to confront contemporary challenges such as rapid urbanisation and climatic changes. Gaining insight into people's spatial knowledge(s) and subjective insights of the spaces they frequent can also shed light on how (if at all) they are adapting and managing their livelihoods. In turn, these understandings can be used to develop theoretical understandings of contemporary livelihoods, which can also be of importance to both policymakers and development agencies alike.

Of course, GPS methods do not provide the only means through which understandings into people's livelihoods can be developed. These methods have shown promise in being able to interrogate the complexity and diversity of livelihoods through providing accurate and robust spatial and temporal data which does not require high levels of expertise to produce.

Key points

- GPS methods can be used and adapted in multiple ways to complement studies which aim to develop understandings into people's livelihoods – the appropriateness of these approaches depends on aims and of a research project, and the participants that will be involved.
- GPS technology is relatively easy to operate and it is becoming more affordable over time.
- Participatory GPS enables participants to have greater input into research processes which can afford studies with subjective insights into people's livelihood experiences, knowledge, strategies and practices.

- Participatory mapping can detail and visualise people's livelihood mobility practices and it can highlight the challenges they face. This can lead to targeted policy interventions and support.
- Consideration is required prior to GPS use, because these approaches can increase participants' visibility, which can raise several ethical dilemmas.

Further readings

Birenboim, A. and Shoval, N. (2016) 'Mobility research in the age of the smartphone', *Annals of the American Association of Geographers*, 106(2): 283–291. https://doi.org/10.1080/00045608.2015.1100058

Hansson, U. and Roulston, S. (2017) 'Evaluations of diaries and GPS-enabled trackers to plot young peoples' geographies – asking the participants what they think', *Children's Geographies*, 15: 517–530. https://doi.org/10.1080/14733285.2016.1272915

Joseph, L., Neven, A., Martens, K., Kweka, O., Wets, G. and Janssens, D. (2019) 'Measuring individuals' travel behaviour by use of a GPS-based smartphone application in Dar es Salaam, Tanzania', *Journal of Transport Geography*, 88: 102477. https://doi.org/10.1016/j.jtrangeo.2019.102477

Shoval, N., Schvimer, Y. and Tamir, M. (2018) 'Tracking technologies and urban analysis: Adding the emotional dimension', *Cities*, 72: 34–42. https://doi.org/10.1016/j.cities.2017.08.005

References

Alozie, N.O. and Akpan-Obong, P. (2017) 'The digital gender divide: Confronting obstacles to women's development in Africa', *Development Policy Review*, 35(2): 137–160. https://doi.org/10.1111/dpr.12204

Birenboim, A. and Shoval, N. (2016) 'Mobility research in the age of the smartphone', *Annals of the American Association of Geographers*, 106(2): 283–291. https://doi.org/10.1080/00045608.2015.1100058

Bryceson, D.F., Mbara, T.C. and Maunder, D. (2003) 'Livelihoods, daily mobility and poverty in sub-Saharan Africa', *Transport Reviews*, 23(2): 177–196. https://doi.org/10.1080/01441640309891

Chambers, R. (2008) *Revolutions in development inquiry*, London: Routledge. https://doi.org/10.4324/9781849772266

Christensen, P., Mikkelsen, M.R., Nielson, T.A.S. and Harder, H. (2011) 'Children, mobility, and space: Using GPS and mobile phone technologies in ethnographic research', *Journal of Mixed Methods Research*, 5(3): 227–246. https://doi.org/10.1177%2F1558689811406121

Esson, J., Gough, K.V., Simon, D., Amankwaa, E.F., Ninot, O. and Yankson, P.K.W. (2016) 'Livelihoods in motion: Linking transport, mobility and income-generating activities', *Journal of Transport Geography*, 55: 182–188. https://doi.org/10.1016/j.jtrangeo.2016.06.020

Evans, J., O'Brien, J. and Ng, B.C. (2018) 'Towards a geography of informal transport: Mobility, infrastructure and urban sustainability from the back of a motorbike', *Transactions of the Institute of British Geographers*, 43(4): 674–688. https://doi.org/10.1111/tran.12239

Evans, R. (2016) 'Achieving and evidencing research "impact"? Tensions and dilemmas from an ethic of care perspective', *Area*, 48(2): 213–221. https://doi.org/10.1111/area.12256

Goodchild, M.F. (2007) 'Citizens as sensors: The world of volunteered geography', *GeoJournal*, 69: 211–221. https://doi.org/10.1007/s10708-007-9111-y

Gough, K.V. (2008) '"Moving around": The social and spatial mobility of youth in Lusaka', *Geografiska Annaler: Series B, Human Geography*, 90(3): 243–255. https://doi.org/10.1111/j.1468-0467.2008.290.x

Hansson, U. and Roulston, S. (2017) 'Evaluations of diaries and GPS-enabled trackers to plot young peoples' geographies – Asking the participants what they think', *Children's Geographies*, 15(5): 517–530. https://doi.org/10.1080/14733285.2016.1272915

Hein, J.R., Evans, J. and Jones, P. (2008) 'Mobile methodologies: Theory, technology and practice', *Geography Compass*, 2(5): 1266–1285. https://doi.org/10.1111/j.1749-8198.2008.00139.x

Janusz, K., Kesteloot, C., Vermeiren, K. and Rompaey, A.V. (2019) 'Daily mobility, livelihoods and transport politics in Kampala, Uganda: A Hagerstrandian analysis', *Tijdschrift Voor Economische en Sociale Geografie, Royal Dutch Geographical Society KNAG*, 110(4): 412–427. https://doi.org/10.1111/tesg.12349

Jones, P., Drury, R. and McBeath, J. (2011) 'Using GPS-enabled mobile computing to Augment qualitative interviewing: Two case studies', *Field Methods*, 23(2): 173–187. https://doi.org/10.1177%2F1525822X10388467

Jones, P. and Evans, J. (2012) 'The spatial transcript: Analysing mobilities through qualitative GIS', *Area*, 44(1): 92–99. https://doi.org/10.1111/j.1475-4762.2011.01058.x

Joseph, L., Neven, A., Martens, K., Kweka, O., Wets, G. and Janssens, D. (2019) 'Measuring individuals' travel behaviour by use of a GPS-based smartphone application in Dar es Salaam, Tanzania', *Journal of Transport Geography*, 88: 102477. https://doi.org/10.1016/j.jtrangeo.2019.102477

Kesby, M., Kindon, S. and Pain, R. (2005) 'Participatory approaches and diagramming techniques', in Flowerdew, R. and Martin, D. (eds.), *Methods in human geography*, London: Pearson Publishing.

Korpilo, S., Virtanen, T. and Lehvavirta, S. (2017) 'Smartphone GPS tracking – Inexpensive and efficient data collection on recreational movement', *Landscape and Urban Planning*, 157: 608–617. https://doi.org/10.1016/j.landurbplan.2016.08.005

Kusenbach, M. (2012) 'Mobile methods', in Delamont, S. (eds.), *Handbook of qualitative research in education*, Cheltenham: Edward Elgar, pp. 252–264.

Langevang, T. and Gough, K.V. (2009) 'Surviving through movement: The mobility of urban youth in Ghana', *Social & Cultural Geography*, 10(7): 741–756. https://doi.org/10.1080/14649360903205116

Lee, D., Kung, K. and Ratti, C. (2015) 'Mapping the waste handling dynamics in Mombasa using mobile phone GPS', *The 14th International Conference on Computers in Urban Planning and Urban Management*, CUPIM 2015.

Lund, R. (2014) 'Gender, mobilities and livelihood transformations: An introduction', in Lund, R., Kusakabe, K., Panda, S.M. and Wang, Y. (eds.), *Gender, mobilities, and livelihood transformations: Comparing indigenous people in China, India, and Laos*, New York: Routledge, pp. 1–20.

Navarrete Forero, G.N., Miñarro, S., Mildenberger, T.K., Breckwoldt, A., Sudirman, S. and Reuter, H. (2017) 'Participatory boat tracking reveals spatial fishing patterns in an Indonesian artisanal fishery', *Frontiers in Marine Science*, 4: 409. https://doi.org/10.3389/fmars.2017.00409

Naybor, D., Poon, J.P.H. and Casas, I. (2016) 'Mobility disadvantage and livelihood opportunities of marginalized widowed women in rural Uganda', *Annals of the American Association of Geographers*, 106(2): 404–412. https://doi.org/10.1080/00045608.2015.1113110

Offenhuber, D. and Lee, D. (2012) 'Putting the informal on the map: Tools for participatory waste management', *Proceedings of the 12th Participatory Design Conference: Exploratory Papers, Workshop Descriptions, Industry Cases*, 2, PDC '12, New York: ACM, pp. 13–16.

Oloo, F. (2018) 'Mapping rural road networks from Global Positioning System (GPS) trajectories of motorcycle taxis in Sigomre Area, Siaya County, Kenya', *International Journal of Geo-Information*, 7(8): 309–324. https://doi.org/10.3390/ijgi7080309

Oyana, T.J. (2017) 'The use of GIS/GPS and spatial analyses in community-based participatory research', in Coughlin, S.S., Smith, S. and Fernandez, M.E. (eds.), *Handbook of community-based participatory research*, Oxford: Oxford University Press. https://doi.org/10.1093/acprof:oso/9780190652234.003.0004

Porter, G., Hampshire, K., Abane, A., Munthali, A., Robson, E., Mashiri, M. and Maponya, G. (2010) 'Where dogs, ghosts and lions roam: Learning from mobile ethnographies on the journey from school', *Children's Geographies*, 8(2): 91–105. http://doi.org/10.1080/14733281003691343

Rigg, J. (2007) 'Moving lives: Migration and livelihoods in the Lao PDR', *Population, Space and Place*, 13(3): 163–178. https://doi.org/10.1002/psp.438

Sheller, M. and Urry, J. (2006) 'The new mobilities paradigm', *Environment and Planning A*, 38(2): 207–226. https://doi.org/10.1068%2Fa37268

Shoval, N. (2008) 'Tracking technologies and urban analysis', *Cities*, 25(1): 21–28. https://doi.org/10.1016/j.cities.2007.07.005

Shoval, N., Kwan, M.-P., Reinau, K.H. and Harder, H. (2014) 'The shoemaker's son always goes barefoot: Implementations of GPS and other tracking technologies for geographic research', *Geoforum*, 51: 1–5. https://doi.org/10.1016/j.geoforum.2013.09.016

Shoval, N., Schvimer, Y. and Tamir, M. (2018) 'Tracking technologies and urban analysis: Adding the emotional dimension', *Cities*, 72(Part A): 34–42. https://doi.org/10.1016/j.cities.2017.08.005

Sugie, N.F. (2018) 'Utilizing smartphones to study disadvantaged and hard-to-reach groups', *Sociological Methods and Research*, 47(3): 458–491. https://doi.org/10.1177%2F0049124115626176

Taylor, L. (2016) 'No place to hide? The ethics and analytics of tracking mobility using mobile phone data', *Environment and Planning D: Society and Space*, 34(2): 219–336. https://doi.org/10.1177%2F0263775815608851

PART II
Negotiating livelihoods

15
POWER AND LIVELIHOODS

Enrico Michelutti

Introduction

Power and livelihoods are interwoven in complex relations, which have a polysemic, multi-dimensional nature. The multiple configurations assumed by these relations lead to an exploration of power and livelihoods in a transdisciplinary way, recognising that the structures of power for livelihoods and the role of livelihoods in power set-ups are context-dependent. An in-depth understanding of power and livelihoods in the Global South means reconstructing these relational fabrics and rethinking their political function and role.

In the section 'Power and Negotiation: Definitions', the chapter introduces categories for the definition of power relations and negotiation processes, understanding negotiations for livelihoods as dynamic processes, which involve a plurality of rationalities, different forms of authority and sets of institutional solutions. In the section 'Power and Negotiation: Processes', 'vertical' negotiation processes (connecting formal and informal systems, and focusing on the action of the State and political-economic elites), and 'horizontal' types of negotiation (embedded in informal networks with diverse institutional ramifications) are presented through a relational approach, enabling the framing of empowerment dynamics on an individual and community level in a radical process of change. In Boxes 15.1 and 15.2, two 'narratives of power' show the plural roles of livelihoods in power interplays shaped by specific contextual conditions. In the section 'Further Readings', philosophical and political readings are provided as key instruments to deepen the exploration between power and livelihoods.

Power and negotiation: definitions

In the Global South, livelihoods exist as complex webs of power relations connecting individuals, social groups and communities in negotiation processes that take place at all levels and scales. Power is plural and the approach to power relations changes according to the chosen theoretical/philosophical framework and the set of ethical values embedded within it. Considering livelihoods in their interactions with socio-economic, political-institutional and cultural structures, a relational approach to power (Koggel, 2013), based on recognition of the actors' networks and inter-dependent tactics, offers analytical tools to explore negotiation processes for livelihoods within different strategic interplays. In this fabric of relationships, specific typologies of power can be identified and can involve different conceptualisations, rationalities and logics of action (from a micro-politics point of view, see Smeed et al., 2009; from a planning and development perspective, see Healey, 2007):

- *Power over*. Based on a command-and-control logic, 'power over' has an authoritative shape, with individuals or groups imposing actions and behaviours on others by using persuasion, seduction, etc. (Allen, 2003; Ball and Peters, 2005). 'Power over' includes decision-making mechanisms and choices imposed by established power systems on livelihoods at any level.
- *Power to*. Understood as a vital force, a potentiality, 'power to' refers to the capacity of generating energy to act, to engage people in an action, to project or to create on a personal/individual and/or collective level (Dyrberg, 1997; Giddens, 1984). 'Power to' involves power as a strategic device generating livelihoods and sets livelihoods as a factor enabling socio-political change and alternative power configurations.
- *Power through*. Seen as an instrument for facilitation, 'power through' deals with the ability to enable other subjects to achieve objectives through power sharing mechanisms (Lukes, 2005). 'Power through' includes the transfer of power allowing the creation of new institutional solutions and utilisation of assets for livelihoods.
- *Power with*. Related to mutual capacity, 'power with' refers to the capacity to create connections among actors/stakeholders, fostering empowerment through collective actions (Smeed et al., 2009). 'Power with' involves forms of collaborative power management in livelihoods and the role/use of livelihoods in (re-)structuring political groups, social movements and grassroots institutions.

In its multiple forms, power is a key driver for setting-up institutions and policies for livelihoods. These processes involve the potentiality and use of power (power to/power through), and the individual-collective relationships between actors/stakeholders and power (power over/power with). In these institutional/political dynamics, power is both an 'objective' (holding power to control organisations and obtain advantage from specific political arrangements for livelihoods, see Box 15.1) and an 'instrument' (wielding power on livelihoods to have access to other power domains, see Box 15.2). For their role in the configuration and development of society, livelihoods extend the 'territories of power', shaping organisations and mental models (on power and organisations, Vallas and Hill, 2012) and setting-up governance mechanisms on all levels (for a critical approach, Kooiman, 2003). Through the definition of socio-economic initiatives and political agendas, power plays an important role for the Sustainable Livelihoods Approach. In the Sustainable Livelihoods Framework, power relations are part of livelihoods processes underpinning the strategies to achieve livelihood outcomes. At the same time, power systems affect the structures that determine the use of livelihood assets: power defines the conditions for the agency people have in livelihood processes (see Chapters 2 and 4 of this volume).

Power is managed through different rationalities and power sharing is actualised according to various interests and objectives, and by means of diverse actions and strategies. The set-ups of power balances include different types and levels of negotiation, which depend on the nature of the actors/stakeholders involved in the processes, their relational webs and changing internal-external conditions. Negotiation processes are not linear: negotiations aiming to achieve specific objectives (access to resources, control of the territory and/or markets, etc.) and determine stakeholders' actions (leading to decision-making mechanisms, codification of behaviours and routines, establishment of practices, etc.) are dynamic. Negotiations can be (re-)built continuously, depending on contingent interests and political momentum, beyond ideologies, fixed agenda or planned programs (Healey, 2007). The 'institutionalisation' of negotiation processes

(Hajer, 1997) consists of an incessant redefinition of the authority (Crocker, 2017) through which these processes become stable, including the consolidation of consensus around specific power systems (e.g. the acceptance of subordination practices, including clientelism-based relationships, illegal recruitment and payments, in exchange for access to livelihoods, see Box 15.1). These power configurations are based on a 'myriad of small details' (Mintzberg, 1994), which shape the complexity underpinning power mechanisms set-up and show the conflictive nature of power dynamics (the use of minimal normative changes to wielding power in incidental situations is an example of these 'details', see Box 15.2).

In the Global South, communities are a crucial socio-spatial element and institutional space where conflicts for livelihoods take shape and power balances are reconfigured (on the concept of 'community' in a political and power perspective, see Meade et al., 2016). An exploration of power relations for livelihoods from a community perspective, 'from the margins of power' (Riddle, 2018), offers instruments to understand power and livelihoods as parts of community self-definition processes, in a participatory development framework. From this point of view, access to livelihoods has a key political value for marginal individuals, excluded social groups and disadvantaged minorities. It becomes a device of empowerment, enabling community members and organisations to gain space in decision-making processes at all levels (on power as individuals' capacity to make decisions, see Drydyk, 2013), according to specific means and ends (Cronin-Furman et al., 2017). As a process of change, empowerment has a contextual and relational character: through livelihoods, empowerment dynamics can foster a larger process of human development, both in terms of agency, understood as the ability to achieve and actually achieving, and capability (Nussbaum, 2000; Sen, 1999).

Power and negotiation: processes

Power types show the evolving nature of power relations, which are structured in *ad hoc* power balances for specific contexts and sector/area of interest. Livelihoods are a strategic part of these power dynamics and a key object/space of negotiation in setting power-systems. Processes of negotiation for livelihoods involve individuals, social groups, and communities in complex relational webs, where formal/informal stakeholders and intermediate actors work on different institutional levels. The object of the negotiation can include all types and conditions for livelihoods (e.g. the 'right' to develop activities in specific territories, the 'rules of the game' through which economic activities are implemented in certain sectors, etc.). In the Global South, where the weak institutional capacity of the State and local authorities (Grindle, 1996) underpins a wide range of informal arrangements among different actors in the arena, a 'liquid' legal framework makes power relations decisive in shaping negotiation processes. In the relationships between formal and informal systems and systemic drivers (i.e. macroeconomics trends, legal regulatory environments, value chain dynamics), power relations assume a central function in defining policies and shaping practices for livelihoods, inside and outside the existing legal frameworks (Chen et al., 2016; Rao and Diwadkar, 2015); livelihoods are a compromise resulting from opposite interests (Watson, 2009). Reaching a definition of power balances allows conflictive conditions and tensions to be overcome at all scales and levels. At the same time, negotiations for livelihoods are spaces for reconfiguration and reproduction of power relations in cultural, socio-economic and political terms (see Box 15.2).

Moving from established power set-ups to new power balances, dynamic processes of power redefinition are based on context-dependent rationalities and change in the course of time. Rationalities entail actors/stakeholders' choices and decisional mechanisms, structuring mental models and affecting organisational arrangements on different planes. The interconnections between individual and collective rationalities become visible on the community level. In fact, familiar and clan relationships (for a perspective in terms of social security conditions, see Mtetwa, 2018), religious and ethnic solidarity practices (e.g. solidarity mechanisms in waste picking, Gutberlet et al., 2017), the transfer of traditions from rural to urban environments (Jenkins, 2013), and the respect of culture-based social hierarchies play a decisive role in structuring rationalities and logics of action in negotiation processes. Despite the great variety of processes and lack of transparency in many types of negotiation, formal/informal and informal/informal relationships can be identified as key interpretative tools in exploring negotiation processes (on formal/informal dichotomies, see Lombard, 2015). These processes can be multiple and mixed, and frequently overlap at different levels; specific 'vertical' and 'horizontal' negotiation types are presented here as instrumental in a power-oriented approach to the multi-layered fabric of negotiation processes involving livelihoods.

'Vertical' processes

The first level/type of negotiation processes involves power relations between formal and informal actors, where the State defines the institutional framework of the negotiation and the socio-spatial web of relationships between stakeholders, (re-)configuring power systems. By providing categories of legitimacy and illegitimacy (i.e. 'formal', 'informal', 'illegal', 'criminal') in legal/political and socio-economic terms (Chen and Skinner, 2014; Herrle and Fokdal, 2011), the state reproduces its power (Roy, 2005). In this context, 'vertical' power relations shape informality 'from above' (Chen et al., 2016). Informal arrangements and decision-making processes are underpinned by recognised power systems, depending mainly on negotiations between the state and socio-political elites/economic powers. Normative exceptions, quasi-legal solutions and negotiations outside regulatory procedures result from these processes.

Without a real recognition of informal actors and practices, legal/political redefinitions of informality facilitate the actualisation and representation of the power systems in place (for a conceptualisation of informality, see Marx and Kelling (2019) and Roger (2020) on the formal/informal divide through a case-study approach, see Schindler (2014)). This condition of power balance is not given *a priori*, it is the result of conflicting rationalities and cities are the space where the conflict is particularly evident; urban institutional conditions increase the dynamic character of power relations and the clash of rationalities involving formal and informal systems (Alfaro d'Alençon et al., 2018; Watson, 2009). Access to urban services and the control of public spaces and goods, and procurement processes at the local and regional level, become fields of action for conflicting rationalities, which determine decision-making mechanisms for urban livelihoods.

'Horizontal' processes

The contextual webs of relations among informal actors direct a second level/type of negotiation process that has the community as the main institutional space. 'Horizontal' power relations underpin informality 'from below' and context-dependent institutional set-ups take place at the margins of the State, working on the gaps of formal systems and/or creating alternative organisational structures (Chen et al., 2016). At a local level, an articulated set of mutual (re)actions, institutional solutions and informal arrangements characterise access to livelihoods for workers who occupy the fringes of the formal economy or generate economies outside the formal system (e.g. street vendors, waste pickers, poor self-employed, workers on daily wages, etc.). In urban areas, local processes can frequently expand, involving other horizontal/vertical formal networks, with consequences at regional levels (e.g. metropolitan informal markets). In this framework, relational geographies are made more complex by the plurality of roles played by each stakeholder. Community leaders and 'big men', criminal brokers, and political or religious authorities can act as separated subjects with opposite interests or can overlap in one single 'player', who merges divergent objectives in the same personal strategy. The presence of multi-level conflicts (e.g. infra-clan tensions, hidden power imbalances within community organisations, etc.) is an additional factor of complexity. Being a space of conflict in the configuration of informal economic fabrics, livelihoods are an immediate representation of the instrumental use of informality as a practice of power (see Box 15.1).

Box 15.1 Negotiating livelihoods in informal contexts: waste pickers in Mumbai non-recognised slums

Waste picking is one of the main sources of livelihoods in Mumbai slum areas. Having a central role in the Mumbai urban economy, waste picking influences the formal processes which structure city waste management. The high specialisation of the work (and workers) has increasingly shaped Mumbai's spatial environment at the city/ward scale, with areas and communities dedicated to specific recycled products, and at the slum/home scale, with *ad hoc* typological and distributive solutions (Shannon and Gosseye, 2009). In this context, the dumping grounds become key nodes for informal livelihoods, 'permeable' zones where formal and informal mechanisms of access to livelihoods take shape, change, and re-assume different configurations according to the power balances in place. The Deonar dumping ground, located in the north-east of the Mumbai metropolitan region, is the biggest area for waste storage in the municipality. Expected to close in 2019 but likely to remain in service at least until 2023, it is a space of negotiation in which formal and informal stakeholders interact, following specific 'rules of the game'; public officers, private companies, slum dwellers, community institutions and criminal organisations create a complex web of relationships and determine waste management practices.

For the communities around Deonar livelihoods depend on access to the dumping ground resources. In this sense, the legal conditions of the communities are a key factor. Besides national and state regulations, 'being formally recognised as slum' opens opportunity of negotiation for livelihoods at different levels. 'Recognised' slum communities can negotiate access and terms of use of the dumping ground with formal authorities and other informal subjects and develop their activities in waste recycling processes with a certain security and continuity (e.g. recognised communities in Shivaji Nagar; for an overview, see Chatterjee, 2019); 'non-recognised' slum communities, who suffer from demolitions and eviction processes due to their (il)legal status, can only develop precarious and unstable arrangements with formal stakeholders and are frequently forced to set-up 'no-choice' agreements with criminal organisations in a general condition of deep socio-economic vulnerability (e.g. non-recognised communities in Rafi Nagar; for a specific focus on waste pickers groups, see Sinha, 2010).

In non-recognised slums, embedded in a 'vertical' power dynamic, communities and individual households build a complex fabric of negotiations with municipality and state officers controlling the dumping ground areas. Besides 'formal' arrangements, negotiations include 'informal' payments and recurrent corruption practices. Additional negotiations involve informal and criminal actors managing the recycling processes and distribution markets (recycling activities take place outside any legal frame, including violent assaults to workers, discrimination for gender, ethnic and religious

> reasons, etc.). Despite being a key asset in city recycling processes, non-recognised slum communities are in a position of complete dependency and construct their negotiations with limited chances to get sustainable solutions. As a response, aiming to achieve better working conditions, communities are mainly engaged in widening their relational networks towards more 'horizontal' power arrangements. Slum dwellers' actions seek to strengthen the involvement of political parties and local and international NGOs in livelihoods negotiation processes, lobbying with other non-recognised workers, unions, and communities (Michelutti and Smith, 2014).

Negotiations can go through fragmentation processes, with a breakdown of organisational structures, political/economic solutions, and relational webs, or periods of suspension, being in place only *pro forma* due to a change in relation to the initial conditions, objectives, and interests. In these phases, a transformation of power equilibria takes place, leading to an increase of authoritative relationships among stakeholders (i.e. no compromise between power holders and subaltern counterparts, see Box 15.1), or to emerging conflicts (e.g. practices of resistance against existing power equilibria developed by excluded/marginalised subjects, etc.). These processes can go beyond the redefinition of power balances in specific socio-spatial environments, involving alternative ideological positions and ideas of development, thus becoming radical political actions (for an introduction on the relationships between power systems and ideas of development, see Nilsen, 2016). The consequential condition of instability/crisis and the creation of spaces of change include not only types of livelihoods and internal mechanisms but also the political use of livelihoods as an operational instrument in the conflict. In this sense,

Figure 15.1 Spaces for livelihoods in Mumbai non-recognised slums: Deonar dumping ground in 2010s
Source: © Enrico Michelutti

the politicisation of livelihoods finds a specific space in informal environments (Kamete, 2010), scaling-up from personal/individual resilient behaviours to collective actions of unions and social movements, practices of resistance contribute to the generation of an integral change with a global redefinition of institutional structures.

Box 15.2 Without spaces of negotiation: livelihoods in the occupied Palestinian territory (oPt) area C

Covering 60% of the West Bank, the oPt area C includes 300,000 Palestinian residents distributed in more than 500 communities and approximately 400,000 Israeli settlers living in 230 settlements. Unlike areas A and B, where Palestinian authorities have full administrative control or share authority with Israeli counterparts, in area C, Israel retains almost complete power over the territory, involving law enforcement, access, movement, planning, and construction (OCHA, 2017). Israeli control is even stricter in the 'firing zones' for military training, which accounts for nearly 30% of area C. Here, a hidden state of conflict is permanent and Palestinian residents suffer from Israeli direct interventions, demolitions, and evictions (OCHA, 2014). The legal framework of area C and its specific environmental conditions (e.g. recurrent drought crisis, GVC and UNICEF, 2010) make Palestinian residents vulnerable to the climate change dynamics, with severe limitations in the development of sustainable responses. These factors radically affect access to livelihoods for the area C population.

In this context, livelihoods mainly consist of farming and agricultural activities in peri-urban and rural areas, along with services/commercial activities and related daily wages jobs in urban areas, involving cities located in both areas A and B. In area C, livelihoods are deeply interconnected with the provision of basic services, especially in regard to access to water and mobility (within the Palestinian territory, but also in relation to Israeli areas and roads). Livelihoods are the object of a multi-layered web of negotiations involving a wide range of territorial contentions. Through high-level political negotiations among international subjects/mediators, Palestinian and Israeli authorities set-up the general conditions for access to livelihoods through 'vertical' power mechanisms, while on the local level, 'underlying' informal negotiations take place between Israeli settlers and Palestinian communities. This places contextual livelihoods practices in a complex horizontal power frame.

Water services represent an ideal example to illustrate the nature of the process. The lack of rainwater harvesting due to the environmental crisis and the restraints imposed by Israeli authorities on Palestinian water tanks' mobility facilitate technical/political arrangements that involve the entire water sector, including pipeline design and the organisation of water tanking routes. Long-term, sustainable solutions are limited due to the Palestinian restrictions in planning and construction, which make maintenance and extension of water networks very problematic or impossible. Living conditions and livelihoods are determined by these infrastructures. By controlling infrastructural/service assets, Israeli powers 'use' Palestinian forced choices for livelihoods to obtain political control on contested territories.

In this framework, the results of negotiation processes consist mainly of temporary concessions to Palestinian residents, precarious/short-term informal agreements or tacit consensus for the implementation of local projects. These institutional solutions do not impinge on the actions of the Israeli army, which can operate for military reasons in area C without considering any 'local agreement'. Palestinian independent works are usually demolished and, in areas close to Israeli settlements, demolitions and evictions recurrently increase in periods of political tension. These practices led Palestinian, Israeli, and international activists to argue for a global redefinition of oPt area C legal status (for a legal-historical review on area C from an Israeli perspective, see BIMKOM, 2008).

Figure 15.2 The Palestinian area C community of Khashem Ad Daraj-Hathaleen

Source: © Enrico Michelutti

Where practices of resistance and socio-spatial change overcome the contextual interplay of interests in power configuration (on the implications of 'rebel' actions in urban contexts, see Harvey, 2012), the struggle for livelihoods emerges as a vehicle for empowerment on all levels. Being a factor against exclusion and inequality, livelihoods address empowerment towards a socio-economic, political, and cultural re-definition of individuals, communities, and social structures. Religious, class, and gender relations, understood as power relations, are re-thought in the empowerment process (on power and politics in a gendered perspective, see Runyan and Spike Peterson, 2013). At the same time, the consciousness of the role of livelihoods in changing power balances leads to self-recognition processes for community actors, which become meaningful 'through livelihoods'. This appropriation of the political meaning of livelihoods as a source and device of power becomes integral for the future of communities in the Global South.

Key points

In approaching livelihoods from a power perspective, there is the need:

- To consider the plurality of power conceptualisations, which underpin different uses of power as an analytical tool in understanding mechanisms and practices for livelihoods.
- To recognise livelihoods as the result of specific power equilibria, which have an unstable, dynamic character, implying continuous power reconfigurations (power definition, authority consolidation, crisis/conflict, power redefinition).

- To include a relational approach in the analysis of power balances, enabling a socio-spatial reading of power relations and an understanding of institutional set-ups for livelihoods.
- To understand the role of livelihoods for a political, rights-oriented change, where individuals, communities and social structures take part in a global empowerment process 'through livelihoods'.

In an exploration of negotiation processes for livelihoods, understood as power dynamics, it is necessary:

- To consider political and normative processes for the definition of informality as forms of power reproduction affecting livelihoods.
- To analyse 'vertical' and 'horizontal' power practices, which shape negotiation mechanisms and relational geographies for livelihoods at all scales/levels.
- To understand the plurality of socio-spatial consequences brought by livelihoods negotiation processes and the political nature of livelihoods as cultural-institutional driver of change.
- To see negotiation processes for livelihoods through an empowerment perspective at the individual and community level, understanding the re-appropriation of livelihoods as a grassroots force for human development.

Further reading

The complexity of the relationships between power and livelihoods suggests opting for transdisciplinary approaches, which are based on different literatures. Further reading may include:

- A philosophical focus, considering the 'negative thought' (from its roots in Schopenahuer and Nietzsche to further developments in Wittgenstein and Heidegger; see Heidegger, 1979) and the 'revolutionary thought' (from Feuerbach to Marx; for a synthetic introduction, Singer, 2018) to construct a plural philosophical framework for power understanding and the ethical implications for livelihoods. The study of the left Hegelian wing offers bases for the analysis of neo-Marxist and post-Marxist works, which are still a key reference for alternative approaches to livelihoods. For an in-depth study on the concept of authority, see Kojève, 2014; for a philosophical/theological introduction to power in political systems, see Cacciari, 2018.
- A contextual focus, exploring the changing power balances for livelihoods in specific regional contexts and political environments that shape different institutional solutions/arrangements and set-ups. For context-dependent perspectives in different global regions, see Vervisch et al., 2013; Kapadia, 2015; Parizeau, 2015; for dynamics in urban areas of the Global South, see Ness et al., 2017.
- A political focus, identifying the relationships between power and political practices of change towards individual and community empowerment through livelihoods (see Friedson-Ridenour et al., 2019; Lapeyre, 2013).

References

Alfaro d'Alençon, P., Smith, H., Álvarez de Andrés, E., Cabrera, C., Fokdal, J., Lombard, M., Mazzolini, A., Michelutti, E., Moretto, L. and Spire, A. (2018) 'Interrogating informality: Conceptualisations, practices and policies in the light of the New Urban Agenda', *Habitat International*, 75: 59–66. https://doi.org/10.1016/j.habitatint.2018.04.007

Allen, J. (2003) *Lost geographies of power*, Oxford: Blackwell Publishing.
Ball, A. and Peters, B.G. (2005) *Modern politics and government*, London: Red Globe Press.
BIMKOM (2008) *The prohibited zone. Israeli planning policy in the Palestinian villages in area C*, Jerusalem: BIMKOM.
Cacciari, M. (2018) *The withholding power. An essay on political theology*, London: Oxford Bloomsbury Academic.
Chatterjee, S. (2019) 'The labors of failure: Labor, toxicity, and belonging in Mumbai', *International Labor and Working-Class History*, 95: 49–75. https://doi.org/10.1017/S0147547919000073
Chen, M., Roever, S. and Skinner, C. (2016) 'Editorial: Urban livelihoods: Reframing theory and policy', *Environment & Urbanization*, 28(2): 331–342. https://doi.org/10.1177/0956247816662405
Chen, M. and Skinner, C. (2014) 'The urban informal economy: Enhanced knowledge, appropriate policies and effective organization', in Parnell, S. and Oldfield, S. (eds.), *Routledge handbook of cities of the Global South*, New York: Routledge, pp. 219–235.
Crocker, D. (2017) 'Confronting inequality and corruption: Agency, empowerment and democratic development', in Otto, H.-U., Pantazis, S., Ziegler, H. and Potsi, A. (eds.), *Human development in times of crisis: Renegotiating social justice*, New York: Palgrave McMillan, pp. 291–302.
Cronin-Furman, K., Gowrinathan, N. and Zakaria, R. (2017) *Emissaries of empowerment*, New York: Colin Powell School for Civic and Global Leadership, The City College of New York.
Drydyk, J. (2013) 'Empowerment, agency and power', *Journal of Global Ethics*, 9(3): 249–262. https://doi.org/10.1080/17449626.2013.818374
Dyrberg, T.B. (1997) *The circular structure of power*, London: Verso.
Friedson-Ridenour, S., Clark-Barol, M., Wilson, K., Shrestha, S. and Ofori, C.M. (2019) 'The limitations of market-based approaches to empowerment: Lessons from a case study in Northern Ghana', *Development in Practice*, 29(6): 774–785. https://doi.org/10.1080/09614524.2019.1577802
Giddens, A. (1984) *The constitution of society*, Cambridge: Polity Press.
Grindle, M.S. (1996) *Challenging the state. Crisis and innovation in Latin America and Africa*, Cambridge: Cambridge University Press.
Gutberlet, J., Carenzo, S., Kain, J. and Mantovani Martiniano de Azevedo, A. (2017) 'Waste picker organizations and their contribution to the circular economy: Two case studies from a Global South perspective', *Resources*, 6(4): 52. https://doi.org/10.3390/resources6040052
GVC, Gruppo di Volontariato Civile and UNICEF, United Nations International Children's Emergency Fund (2010) *Assessment of water availability and access in the areas vulnerable to drought in the Jordan Valley*, New York: UNICEF and GVC.
Hajer, M. (1997) *The politics of environmental discourse: Ecological modernization and the policy process*, Oxford: Oxford University Press.
Harvey, D. (2012) *Rebel cities: From the right to the city to the urban revolution*, New York-London: Verso.
Healey, P. (2007) *Urban complexity and spatial strategies. Towards a relational planning for our times*, London-New York: Routledge.
Heidegger, M. (1979) *Nietzsche*, 4th vol., New York: Harper & Row.
Herrle, P. and Fokdal, J. (2011) 'Beyond the urban informality discourse: Negotiating power, legitimacy and resources', *Geographische Zeitschrift*, 99(1): 3–15. www.jstor.org/stable/23226577
Jenkins, P. (2013) *Urbanization, urbanism, and urbanity in an African city. Home spaces and house cultures*, London: Palgrave Macmillan.
Kamete, A.Y. (2010) 'Defending illicit livelihoods: Youth resistance in Harare's contested spaces', *International Journal of Urban and Regional Research*, 34(1): 55–75. https://doi.org/10.1111/j.1468-2427.2009.00854.x
Kapadia, K. (2015) 'Sri Lankan livelihoods after the tsunami: Searching for entrepreneurs, unveiling relations of power', *Disasters*, 39(1): 23–50. https://doi.org/10.1111/disa.12090
Koggel, C. (2013) 'Is the capability approach a sufficient challenge to distributive accounts of global justice?', *Journal of Global Ethics*, 9(2): 145–157. https://doi.org/10.1080/17449626.2013.818458
Kojève, A. (2014) *The notion of authority. A brief presentation*, London, New York: Verso Books.
Kooiman, J. (2003) *Governing as governance*, London: Sage.
Lapeyre, F. (2013) 'Securing livelihoods in Africa: Towards multi-scalar policy frameworks', *European Journal of Development Research*, 25(5): 659–679. https://doi.org/10.1057/ejdr.2013.5
Lombard, M. (2015) 'Discursive construction of low-income neighbourhoods', *Geography Compass*, 9(12): 648–659. https://doi.org/10.1111/gec3.12251
Lukes, S. (2005) *Power: A radical view*, London: Palgrave Macmillan.
Marx, C. and Kelling, E. (2019) 'Knowing urban informalities', *Urban Studies*, 56(3): 494–509.

Meade, R., Shaw, M., Banks, S. (eds.) (2016) *Politics, power and community development*, Bristol: Policy Press.

Michelutti, E. and Smith, H. (2014) 'The realpolitik of the informal city governance. The interplay of powers in Mumbai's un-recognized settlements', *Habitat International*, 44(4): 367–374. https://doi.org/10.1016/j.habitatint.2014.07.010

Mintzberg, H. (1994) *The rise and fall of strategic planning*, Edinburgh: Pearson Education Limited.

Mtetwa, E. (2018) 'Rural social security for Zimbabwe: Challenges and opportunities for state and non-state actors', *Journal of Social Development in Africa*, 33(1): 131–156.

Ness, I., Ngwane, T., Sinwell, L. (2017) *Urban revolt: State power and the rise of people's movements in the Global South*, Chicago: Haymarket Books.

Nilsen, A. (2016) 'Power, resistance and development in the Global South: Notes towards a critical research agenda', *International Journal of Politics, Culture, and Society*, 29(3): 269–287. https://doi.org/10.1007/s10767-016-9224-8

Nussbaum, N. (2000) *Women and human development*, Cambridge, MA: Harvard University Press.

OCHA, United Nations Office for the Coordination of Humanitarian Affairs (2014) *Area C vulnerability profile*, East Jerusalem: United Nations Office for the Coordination of Humanitarian Affairs occupied Palestinian territory.

OCHA, United Nations Office for the Coordination of Humanitarian Affairs (2017) *West bank area C: Key humanitarian concerns*. Available at: www.ochaopt.org/content/west-bank-area-c-key-humanitarian-concerns(Accessed 2 July 2021).

Parizeau, K. (2015) 'When assets are vulnerabilities: An assessment of informal recyclers' livelihood strategies in Buenos Aires, Argentina', *World Development*, 67: 161–173. https://doi.org/10.1016/j.worlddev.2014.10.012

Rao, V. and Diwadkar, V. (2015) 'From informality to parametrics and back again', in Mörtenböck, P., Mooshammer, H., Cruz, T. and Forman, F. (eds.), *Informal market worlds reader*, Rotterdam: Nai010 publishers, pp. 161–184.

Riddle, K.C. (2018) 'Empowerment. Participatory development and the problem of cooptation', in Drydyk, J. and Keleher, L. (eds.), *Routledge handbook of development ethics*, London: Routledge, pp. 171–182.

Roger, C.B. (2020) *The origins of informality: Why the legal foundations of global governance are shifting, and why it matters*, Oxford: Oxford University Press.

Roy, A. (2005) 'Urban informality: Towards an epistemological of planning', *Journal of the American Planning Association*, 71(2): 147–158. https://doi.org/10.1080/01944360508976689

Runyan, A.S. and Spike Peterson, V. (2013) *Global gender issues in the New Millennium*, New York: Hachette Book Group.

Schindler, S. (2014) 'Producing and contesting the formal/informal divide: Regulating street hawking in Delhi, India', *Urban Studies*, 51(12): 2596–2612. https://doi.org/10.1177/0042098013510566

Sen, A. (1999) *Development as freedom*, New York: Anchor Books.

Shannon, K. and Gosseye, J. (eds.) (2009) *Reclaiming (the urbanism of) Mumbai*, Amsterdam: Sun Academia.

Singer, P. (2018) *Marx. A very short introduction*, Oxford: Oxford University Press.

Sinha, R. (2010) *Violence and health: a study among an unorganised labour group in the solid waste disposal system in Mumbai*, Ph.D. Thesis, New Delhi: Jawaharlal Nehru University.

Smeed, J.L., Kimber, M., Millwater, J. and Ehrich, L.C. (2009) 'Power over, with and through: Another look at micropolitics', *Leading & Managing*, 15(1): 26–41.

Vallas, S.P. and Hill, A. (2012) 'Conceptualizing power in organizations', in Golsorkhi, D., Courpasson, D. and Sallaz, J. (eds.), *Rethinking power in organizations, institutions, and markets*, Bingley: Emerald Group Publishing, pp. 165–198.

Vervisch, T.G.A., Vlassenroot, K. and Braeckman, J. (2013) 'Livelihoods, power, and food insecurity: Adaptation of social capital portfolios in protracted crises-case study Burundi', *Disasters*, 37(2): 267–292. https://doi.org/10.1111/j.1467-7717.2012.01301.x

Watson, V. (2009) 'Seeing from the South: Refocusing urban planning on the globe's central urban issue', *Urban Studies*, 46(11): 2259–2275. https://doi.org/10.1177/0042098009342598

16
FEMINIST POLITICAL ECOLOGY

Miriam Gay-Antaki and Ana De Luca

Introduction

Environmental issues have often been characterised as being the result of mismanagement of natural resources by local populations, without any consideration of geopolitical forces, particularly in the Global South. In this way, apolitical ecology understandings of environmental problems neglect to consider how the global political economy is affecting and reconfiguring local livelihoods. Political Ecology (PE) was born as an attempt to give a "political" dimension to environmental studies by centring power relations, thus stressing the role of political economy in environmental management, alongside globalisation, capitalism, and neoliberalism in understanding localised environmental challenges. Nevertheless, early PE work did not consider gender relations in its analysis leading to a mischaracterisation of grounded environmental issues. Feminist Political Ecology (FPE) addressed this gap by centring gender alongside other forms of oppression to better understand access to, and control over, natural resources and other assets such as education or wage labour that determine how people make a living.

A focus on FPE in the context of livelihoods in the Global South helps to better understand the micropolitics of everyday life with a focus on broader historical processes. Latin American Feminist Political Ecology (LAFPE) highlights the role of colonialism and patriarchy in shaping access and control over natural resources in the Global South, as well as their role in creating the climate crisis. As Global South livelihoods will be the most affected by environmental change, FPE provides an ethical framework to improve the living conditions of those who will be the most affected. As Carr (2015) points out, livelihood frameworks that dialogue with political ecology produce a more robust representation of people's lives, while helping us understand paths towards sustainability. We begin with a brief political ecology review to then examine its development and applications in Latin America. We then introduce FPE and bring Latin American insights to better understand the globalised ways in which gender is interwoven with other socio-political categories in nature and society relations. We close by underscoring the importance of LAFPE to better understand the intersections of oppressive structures and the embodied impact of urbanisation and climate change on Global South livelihoods.

Notes on political ecology

PE was born as an approach to politicise apolitical environmental studies from the 1960s and 1970s and challenge the purely positivist approaches toward human-environment issues (Delgado, 2013; Perreault et al., 2015). Particularly critical of works from Hardin (1968), Ehrlich (1968) and publications such as The *Limits to Growth* by Meadows et al. (1972) positing 'over-population' as the biggest threat to the Earth's resources (Perreault et al., 2015). Importantly as Perreault et al. (2015: 5) have stressed, what political ecology has set out to do as an epistemological project has been to 'shatter comfortable and simplistic 'truths' about the relationship between society and its natural environment'. PE is a useful approach to analyse the complex relationships between resource use, access, management, control, rights and environmental decision making with political, social, and economic structures (Harris et al., 2017; Resurrección, 2017). More profoundly, PE offers a robust criticism of modernity and the nature-society relationship it was built upon, in which the only possibility of human survival resides in the submission and exploitation of nature (Lezama, 2019).

Approaching the study of livelihoods through a PE lens allows us to interrogate the impact that environmental degradation has on livelihoods, the way in which people respond to environmental shocks and stresses and determine who is most vulnerable (Cruz Torres and McElwee, 2017). PE approaches the concept of livelihoods by highlighting that livelihoods are much more than describing how people make a living. Livelihoods underscore the diverse roles of social actors and civil society in transforming mainstream environmental strategies, while highlighting the relational aspects of the embodied and the political. PE approaches to livelihoods elucidate the power dynamics that give shape to the differentiated and unequal access, use, management, control, and rights and decision-making process over resources at the local, regional, and global scales (Resurrección, 2017). Power dynamics shape how we conceptualise nature, how we think about what is natural, what we value and should be preserved, the knowledge that we need to preserve it and who the producer of this knowledge is (Robbins, 2012).

Latin American political ecology

Even though PE has a strong European and American foundation it became a prolific intellectual field in Latin America, and a broad social and political movement demanding environmental justice in the region (Alimonda, 2015). Latin American Political Ecology (LAPE) has been interested in the epistemic, economic, cultural and political processes of domination at the national, continental and global levels (Alimonda, 2011; Escobar, 2011; Porto-Gonçalves and Leff, 2015). LAPE has deepened our understanding of the links between environmental and social injustice with a focus on the process of ethnocide and ecocide in the region, thus making strong contributions around ethics and environmental justice (Delgado, 2013). LAPE seeks to respond to what Leff calls 'a crisis of civilization', that is, the unsustainability of life that humanity has inadvertently produced (Leff, 2015).

LAPE theorists have been critical of development efforts seeking to 'modernise' nonwestern spaces because these hegemonic practices imposed on the rest of the world assume the West as the standard, erasing other ways of living and existing in this world (Escobar, 1995; Esteva, 2010; Ferguson, 1990; Sachs, 1992; Ziai, 2007). According to these authors, development is rooted in an ideology of 'modernisation' seeking to sustain and strengthen western economic structures. As Kabeer (1998) points out, development is a way in which rich countries define the development agenda, including the debates and discussions around priorities and the problems that need to be addressed.

These critics argue that the problems that development is wishing to tackle not only have failed to bring the promised benefits but have caused problems of their own, and development has become a means by which countries from the Global North constantly intervene in countries of the Global South. In that sense, Latin American PE engages with a diverse range of ecological and alternative debates as well as different proposals for civilising transitions that challenge Western hegemonic practices of modernisation, that push for interconnected relationships with nature and other humans rather than exploiting both for profit, validating and reviving non-Western ways of life (Escobar, 1995). As in the case of 'el buen vivir' (good living), which values a harmonious life with nature and other people, seeking both dignified life and death, prioritising the collective rather than the individual, and pursuing the flourishing and prolongation of distinct human cultures. As Boaventura De Sousa Santos (2006) explains, Latin American alternatives to development aim to create a new epistemology of the South that seeks to liberate people from the metaphorical South as made up of the historically discriminated sectors and also from the metaphorical North including the capitalist (global and local) elites.

Feminist political ecology: beyond women

Just as PE addressed the political aspect of environmental degradation, FPE attends to gendered power dynamics that have often been invisible. FPE has highlighted the importance of local agency and studying the transformative potential at the scale of our bodies, exploring the relationships between the symbolic and the material, and importantly how social signifiers such as gender, class and race serve to limit or advance control over natural resources (Carey et al., 2016; Doshi, 2017; Vaz-Jones, 2018).

Centring women in environmental and development work has forced us to re-evaluate much of what we have come to learn, regarding our environmental commons and around knowledge more broadly. Knowledge produced without women and other marginalised groups has proved to be incomplete time and again (Maynard, 1997). By writing women into environmental narratives as agents of change, as bearers of knowledge with power to transform their surroundings, rather than as passive victims of change, Feminist Political Ecology (FPE) has pushed PE to consider additional epistemological directions. As Sundberg (2017: 4) maintains 'to disrupt conventional assumptions about men as the primary environmental actors is to ask fundamental epistemological questions about how knowledge is produced and legitimized'.

FPE emerged in the 1990s and was first coined in the seminal book by Rocheleau et al. (1996) 'Feminist Political Ecology: Global Issues and Local Experiences' where it became evident, as the title suggested, that a rescaling of the impact of global environmental issues was necessary to attend to grounded social relations. Learning about local responses shifts our attention from imagining communities as traditional and monolithic entities toward diverse, fluid and adaptive entities. A grounded focus underscores the differential access and control over natural resources and how it is gendered. The rescaling of global challenges to attend to gender relations has brought more attention to the embodied experiences of these changes. It is at the scale of the body that FPE scholarship finds that identities are fluid and that social categories such as race, gender, ethnicity are negotiated and articulated in accordance with specific social, political and environmental contexts (Gay-Antaki, 2020a; Harris, 2016; Nightingale, 2006; Sundberg, 2004).

FPE challenges essentialist ecofeminist tropes that assume women are naturally closer to nature, common in Women and Development (WED) approaches. The classic study by Ester Boserup's (1970) 'Women's Role in Economic Development' was critical in the emergence of women as a key concern in development. Development interventions began incorporating women solely in terms of their delivery capacity and ability to extend their working day. Under global capitalism, the state and the corporate sector shift responsibilities for social reproduction onto poor women and men who effectively subsidise this system of production (Di Chiro, 2008; Nagar et al., 2002). The way that women have been incorporated in development has been widely critiqued in FPE literature because of the hyper-focus on patriarchy and dismissal of colonialism, its essentialisation of gender roles, the homogenising of women in the South and their portrayal as natural caretakers, and how these framings have continued exploitation of women and the environment (Arora-Jonsson, 2009; Arora-Jonsson, 2011; Gay-Antaki, 2016; Leach, 2007; Simon-Kumar et al., 2018; Sturgeon, 1999). To disengage from development projects, Cruz-Torres and McElwee (2017) suggest talking about livelihoods and the necessary conditions that are required to dignify the lives of the population of the Global South.

Challenging monolithic representations of communities in the Global South has been accompanied by a growing body of work that challenges the neat divide between the Global South and the Global North. Simon-Kumar et al.'s (2018) work highlights how globalised capitalism is challenging the North/South binary as we find increasing wealth in the Global South and increasing poverty in the Global North. Much as our understanding of the Global South and North evolves, so does our understanding of gender. It is important to note that gendered differences are not natural and a product of biology but rather are due to social and cultural roles, norms, institutions, and the sexual division of labor. Gendered social roles effectively differentiate how men and women relate to their environment and these roles have historically assigned more power and control to men, for example in the form of property rights (Gay-Antaki, 2016). Rather than accepting women's disproportionate vulnerability to environmental changes, FPE work asks why women are disproportionately vulnerable and underscores the relationship between systematic oppression based on gender, race, class and sexuality and the domination of nature (Mollet, 2017; Mollet and Faria, 2013; Resurrección, 2017; Sundberg, 2017).

Latin American FPE

Latin American FPE (LAFPE) has understood patriarchy and colonialism as systems of domination that encompass oppressive and exploitative relationships that affect women, nature and other feminised bodies, resulting in distinct gendered and racialised systems influencing access and control over resources (Lugones, 2007; Mies, 1970). LAFPE has had to come to terms with different oppressions beyond gender and how the matrix of oppression – racism, heterosexuality, colonialism and classism has acted over their own lives. Latin American FPE understands lived realities and the processes by which certain bodies in the Global South come to be known as brown, poor and female (Bolados and Sánchez Cuevas, 2017; Curiel Pichardo, 2015; De Luca et al., 2020; Gay-Antaki, 2020b; Hill Collins, 1986; Merlinsky, 2017; Susial-Martín, 2017; Ulloa, 2014, 2016; Vázquez et al., 2016). To better understand the situation of colonised women vis-a-vis white women, a decolonial approach provides unique insights. Maldonado-Torres (2007: 243) underscores that coloniality survives colonialism, remaining alive in what we read and aspire to, in academic criteria, in our common sense, in our self-image and how 'as modern subjects we breathe coloniality all day and every day'. Any adequate understanding of livelihoods in the Global South must attend to the intersections of patriarchy and colonialism (Lugones, 2007). Decolonial and feminist theorising have allowed FPE to deepen our understanding of situated knowledges, epistemic privilege and authority, challenging universalising claims and rethinking the world from the Global South (Elmhirst, 2018). LAFPE's strength comes from a diversity of resisting voices from autonomous, decolonial and community feminisms, and those of indigenous and Afro-descendant women that criticise modernity and coloniality (Rivera Cusicanqui et al., 2016).

An important and recurring theme is the multi-faceted forms of violence experienced by women in Latin America which includes colonial violence, living in postcolonial conditions, representational violence, the violence brought on by neoliberal policies that further displace and marginalise them, as well as physical and sexual violence (Segato, 2014; Segato, 2011; Svampa, 2019). In that sense, extractive practices, such as mining, are detrimental to the environment but also reconfigure the lives of indigenous peoples, Afro-descendants and peasants, and produce changes both in local economic relations and in gender relations that enhance structures of male domination and exacerbate cruelty against women and children (Salazar and Rodríguez, 2015; Ulloa, 2014, 2016; Vázquez et al., 2020). Care-related activities in the extractive production process are rendered invisible, as well as the collateral environmental risks and damages that mining imposes on women, such as water and land contamination. However, these practices are actively challenged through what Ulloa (2016) describes as 'territorial feminisms', movements led by women who defend life, their bodies and territory offering alternative proposals to the modernity/colonial relationship, to extractivism and to patriarchy.

Women in the Global South have always played an active role defending life, territory and the health of their community. Agroecological practices play a fundamental role in the search for food security in Latin America (Zuluaga, 2013). Local agriculture has meant an economic income) for households in addition to accessing various nutritious foods (Vázquez, 2007). However, the introduction of monocrops such as soy, have encroached into women's livelihoods. Not only are these efforts displacing women's livelihoods but as Chiappe (2016) elucidates have been shown to be detrimental to both the environment and human health.

The agency of poor women in the Global South has long been threatened by overpopulation narratives. This persistent linkage between overpopulation in the Global South and environmental degradation has led to a myriad of family planning programmes from the Global North implemented in the Global South under the guise of development. These programmes use discourses of women's human rights, and the legitimacy of the environmental crisis to present family planning as the panacea to resolve the environmental crisis rather than pointing out consumption patterns and the intrinsic inequality derived from the capitalist system (Hartmann, 1995; Hendrixon et al., 2019). These types of practices end up significantly altering the livelihoods of women in the Global South. This is the case of the Population, Health and Environment (PHE) programs that bring family planning services to ecological hotspots mainly in the Global South. By using a reproductive rights discourse they seek to perpetuate a Malthusian approach to environmental programs disguised as progressive programs that help both the environment and women (De Luca, 2020).

Urban FPE

Most FPE scholarship has studied gender dynamics around the causes, experiences and management of environmental problems in rural settings. There is a lack of FPE literature on urban spaces, particularly in the Global South, with few exceptions (Casas et al., 2018; De Luca and Gay-Antaki, 2017; Navarro et al., 2015) even though most of today's world population lives in cities. The Global South is urbanising rapidly; Latin America is one of the most urbanised regions in the world with 81% of its population living in cities, Asia's population is now approximating 50% urban, while Africa still remains mostly rural with 43% of its population living in cities (UN, 2018). Most of the wealth and consumption in the world is generated in cities and they are responsible for absorbing most of the material and energy extracted for its renovation and expansion (Delgado, 2017). Any adequate analysis of environmental challenges must also consider the socio-economic forces changing the urban landscape, and their differential impacts across social strata. Some of these challenges include the lack of clean and potable water, lack of proper urban infrastructure causing flooding, extreme overcrowding in informal settlements with very few services, air pollution, lack of adequate transportation, food insecurity, lack of appropriate land tenure, dependency on the informal market and the lack of financial services, lack of building regulations, and many more impacts (Khosla and Masaud, 2010).

Going forward, FPE literature on Global South livelihoods can benefit from feminist literature that analyses the experiences and perceptions of women in the city, including security, mobility, access to health, education and economic and political participation, among other things (Barraza, 2006; Bolos, 2008; Falú, 2009; Moser, 2012; Nash et al., 2005; Sabsay, 2011). The literature has examined the way in which women deal with oppression, inequality, violence, exclusion and poverty in the cities, but at that same time have considered these spaces as drivers of positive change as they favour the recognition and exercise of women's rights, including the right to social mobilisation and protests. Future research on FPE should further explore themes around mobility and the obstacles towards achieving sustainable urban mobility with gender equality, including the design of a non-motorised mobility infrastructure, oriented towards spatial, climatic and gender justice (Casas et al., 2018; Chávez-Rodriguez et al., 2018). In this sense FPE could trace sustainable pathways for livelihoods in the Global South that could help deal with poverty in urban areas, increase resilience towards environmental shocks and stresses, and improve livelihood strategies for women's lives.

FPE in the context of climate change

There is tremendous opportunity for LAFPE perspectives to enhance understanding and policies around climate change on livelihoods. Gender relations are implicated within our analysis and understanding of the causes and consequences of climate change, and in the design and effectiveness of policies ranging from mitigation to adaptation and geoengineering.

While climate impacts will be embodied, plans and funding for adaptation emerge at international sites that end up reifying uneven power dynamics when it comes to accessing climate science, policy and action (Gay-Antaki, 2020a, 2020b; Gay-Antaki and Liverman, 2018). International interventions that do not consider social inequalities due to gender, race, class, ethnicity, in their effort to address climate, further exacerbate their vulnerability to its impacts (Westholm and Arora-Jonsson, 2018). For instance, because access to land is a prerequisite for participating in many development schemes, it is vital that we understand how their implementation may adversely impact women and other landless individuals (Gay-Antaki, 2016). The partial or total dislocation of women's agricultural activities towards ever more marginal lands, have had negative repercussions for women and landless individual's ability to meet basic household obligations such as food provision and food security (Rossi and Lambrou, 2008).

Methodologically, FPE is a useful tool to understand patterns and processes of climate governance and the multiple ways that gender and other socially constructed categories and identities intersect with the causes and consequences of climate change. FPE's attention to the intersections of oppressive structures and their embodied impact makes it uniquely suited to understand the differential impacts of climate change on livelihoods. By paying close attention to the spaces of everyday life, and keeping women's roles visible, we gain important insight into not only the impacts of climate change, but strategies deployed to adapt and mitigate climate change (Bee, 2013). According to Bee (2013) because the housing, health and livelihoods of marginalised people are already being threatened by climate change-related impacts worldwide, they have already gained the know-how in addressing climate change. As such, their experiences, knowledge and views must be part of local, regional, national and international governance.

Feminist approaches counter narratives that portray women as vulnerable, powerless victims of a changing climate but rather look to them as diverse and having different interests and priorities, to learn about strategies to adapt to, and mitigate, climate change. Intersectionality, a term coined by Crenshaw (1989) underscores that social categories such as, gender, race and ethnicity are dynamic and operate simultaneously by including or excluding people from actively participating in society. Intersectionality reminds us that a narrow focus on gender may constrain our understanding of climate change on livelihoods when other forms of social difference and oppression such as class, race and sexuality are simultaneously important in responsibility, vulnerability, or governance of environment and climate change (Kaijser and Kronsell, 2013; Nightingale, 2011).

FPE can offer insight into the processes by which some people and regions are more vulnerable to climatic changes than others. This differentiation is achieved through everyday practices from local to global sites that maintain unequal power relations. These practices might include the silencing of non-English speaking voices from climate negotiations and climate assessments or the token acknowledgement of gender in climate discourses without addressing material gender inequities. Contesting these everyday inequities requires counter narratives, solidarity, and struggles to destabilise the structures and discourses that maintain control over the resources, political decisions and peoples that create vulnerability, foster high carbon emissions, or overlook power in climate policy. The most popular approaches to tackle climate change do not seriously incorporate a diversity of voices and their experiences and as a result dampen the effectiveness of policies and programs.

Climate change reveals what our modern society has forsaken in the name of growth, modernisation and development. Rather than just identifying peoples as vulnerable to climate change, FPE lays bare the structures and mechanisms that have made certain people disproportionately vulnerable to climate change. Structures embedded in epistemic hierarchies, paternalistic approaches, modernisation and marketisation ideologies, that continue to exploit those from the periphery even in the struggle to address climate change as we can see from climate science, action and policy. FPE offers important epistemological and methodological contributions to conceptualise livelihoods in the Global South to better understand micropolitical and broader historical processes. A focus on FPE in the context of livelihoods in the Global South looks to local residents as agents of change helping envision solutions to climate change beyond purely technocratic fixes that attempt to resolve climate change through market mechanisms or other neoliberal interventions.

Key points

- Political ecology emerged to politicise environmental studies.
- PE in LA is a prolific intellectual field and movement demanding environmental justice and an ethical path towards sustainability.
- FPE understands that gender, class, race and ethnicity are critical in shaping access and control over natural resources.
- Latin American FPE encompasses autonomous, decolonial, and community feminisms, including those from indigenous and Afro-descendant women.
- FPE's attention to the intersections of oppressive structures and their embodied impact is suited to understand the uneven impacts of urbanisation and climate change in the Global South.

Further reading

Carney, J.A. (1998) 'Women's land rights in Gambian irrigated rice schemes: Constraints and opportunities', *Agriculture and Human Values*, 15: 325–336. https://doi.org/10.1023/A:1007580801416

De Luca, A., Fosado, E. and Velázquez, M. (2020) 'Feminismo Socioambiental. Revitalizando El Debate Desde América Latina', in De Luca, A., Fosado, E. and Gutiérrez, M. (eds.), *Centro Regional de Investigaciones Multidisciplinarias*, Mexico City: CRIM-UNAM.

Escobar, A. (1995) *Encountering development: The making and unmaking of the third world*, Princeton: Princeton University Press.

Hart, G. (1991) 'Engendering everyday resistance: Gender, patronage and production politics in rural Malaysia', *Journal of Peasant Studies*, 19: 93–121. https://doi.org/10.1080/03066159108438472

MacGregor, S. (ed.) (2017) *Routledge handbook of gender and environment*, London and New York: Routledge.

Mollett, S. and Faria, C. (2013) 'Messing with gender in feminist political ecology', *Geoforum* 45: 116–125. https://doi.org/10.1016/j.geoforum.2012.10.009

Rocheleau, D.E., Thomas-Slayter, B. and Wangari, E. (1996) *Feminist political ecology: Global issues and local experience, international studies of women and place*, New York & London: Routledge.

Sturgeon, N. (1997) *Ecofeminist natures: Race, gender, feminist theory and political action*, New York & London: Routledge

References

Alimonda, H. (2011) 'La colonialidad de la naturaleza. Una aproximación a la Ecología Política Latinoamericana', in Alimonda, H. (ed.), *La naturaleza colonizada. Ecología política y minería en America Latina*, Lima: Tejiendo Saberes PDTG, pp. 21–60.

Alimonda, H. (2015) 'Ecologia política latinoamericana y pensamiento crítico: Vanguardias arraigadas', *Desenvolvimento e Meio Ambiente*, 35: 161–168. http://doi.org/10.5380/dma.v35i0.44757

Arora-Jonsson, S. (2009) 'Discordant connections: Discourses on gender and grassroots activism in two forest communities in India and Sweden', *Signs: Journal of Women in Culture and Society*, 35(1): 213–240. https://doi.org/10.1086/599259

Arora-Jonsson, S. (2011) 'Virtue and vulnerability: Discourses on women, gender and climate change', *Global Environmental Change*, 21(2): 744–751. https://doi.org/10.1016/j.gloenvcha.2011.01.005

Barraza, S. (2006) *Mujeres en la construcción de ciudad, en Ciudadanas*, Barcelona: Red 12 del Programa URB-AL.

Bee, B.A. (2013) 'Who reaps what is sown? A feminist inquiry into climate change adaptation in two Mexican ejidos', *Acme*, 12(1): 131–154.

Bolados, G. and Sánchez Cuevas, P. (2017) 'Una ecología política feminista en construcción: El caso de las "Mujeres de zonas de sacrificio en resistencia", Región de Valparaíso, Chile', *Psicoperspectivas*, 16(2): 33–42. http://doi.org/10.5027/psicoperspectivas-vol16-issue2-fulltext-977

Bolos, S. (2008) *Mujeres y espacio público. Construcción y ejercicio de la ciudadanía*, México: Universidad Iberoamericana.

Boserup, E. (1970) *Woman's role in economic development*, New York: St. Martin's Press

Carey, M., Jackson, M., Antonello, A. and Rushing, J. (2016) 'Glaciers, gender, and science: A feminist glaciology framework for global environmental change research', *Progress in Human Geography*, 40: 770–793. https://doi.org/10.1177/0309132515623368

Carr, E. (2015) 'Political ecology and livelihoods', in Perreault, T., Bridge, G. and McCarthy, J. (eds.), *The Routledge handbook of political ecology*, New York: Routledge, pp. 332–342.

Casas, R., De Luca, A. and Velázquez, M. (2018) 'Rutas hacia la desigualdad: La movilidad de las mujeres en la Ciudad de México ante el cambio climático', in De Luca, A., Vázquez, V., Bose, P. and Velazquez, M. (eds.), *Género, energía y sustentabilidad. Aproximaciones desde la academia*, Cuernavaca: CRIM, UNAM

Chávez-Rodriguez, L., Treviño, R. and Curry, L. (2018) 'Canarios en la mina de asfalto. Vulnerabilidades y privilegios de género en la movilidad alternativa de Monterrey', in De Luca, A., Vázquez, V., Bose, P. and Velazquez, M. (eds.), *Género, energía y sustentabilidad. Aproximaciones desde la academia*, Cuernavaca: CRIM, UNAM

Chiappe, M. (2016) 'Rompiendo el siliencio. Las mujeres rurales ante la expansión de uso de agroquímicos en Uruguay', in Velázquez, M., Vázquez, V., De Luca, A. and Sosa, D. (eds.), *Transformaciones ambientales e igualdad de género en América Latina*, Cuernavaca: CRIM, UNAM.

Crenshaw, K. (1989) 'Demarginalizing the intersection of race and sex: A Black feminist critique of antidiscrimination doctrine, feminist theory and antiracist policies', *The University of Chicago Legal Forum*, 1989(1): 139–167. https://doi.org/10.1525/sp.2007.54.1.23

Cruz-Torres, M. and McElwee, P. (2017) 'Gender, livelihoods, and sustainability: Anthropological research', in MacGregor, S. (ed.), *Routledge handbook of gender and environment*, London: Routledge, pp. 133–145. https://doi.org/10.4324/9781315886572

Curiel, O. (2015) 'Construyendo Metodologías Feministas Desde El Feminismo Decolonial', in Mendia, I., Luxan, M., Legarreta, M., Guzman, G., Zirion, I. and Azpoazu, J. (eds.), *Otras Formas de (Re) Conocer: Reflexiones, Herramientas y Aplicaciones Desde La Investigación Feminista*, Bilbao: Universidad del Pais Vasco, Hegoa.

De Luca, A., Fosado, E. and Velázquez, M. (eds.) (2020) *Feminismo Socioambiental. Revitalizando El Debate Desde América Latina*, Mexico City: Centro Regional de Investigaciones Multidisciplinarias, UNAM.

De Luca, A. and Gay-Antaki, M. (2017) 'Gender, climate change and cities: A case study of gendered climate policy in Mexico City', in Delgado Ramos, G. (ed.), *Climate change-sensitive cities*, Mexico City: PINCC-UNAM, pp. 307–326.

De Sousa Santos, B. (2006) *Conocer desde el Sur – Para una cultura política emancipatoria*, Lima: Universidad Nacional Mayor de San Marcos.

Delgado Ramos, G. (2013) '¿Porqué es importante la ecología política?', *Nueva Sociedad*, 244: 47–60.

Delgado Ramos, G. (2017) 'Climate change-sensitive cities. Editor's introduction', in Delgado Ramos, G. (ed.), *Climate sensitive cities*, México: PINCC/UNAM.

Di Chiro, G. (2008) 'Living environmentalisms: Coalition politics, social reproduction, and environmental justice', *Environmental Politics*, 17(2): 276–298. https://doi.org/10.1080/09644010801936230

Doshi, S. (2017) 'Embodied urban political ecology: Five propositions', *Area*, 49(1): 125–128. https://doi.org/10.1111/area.12293

Ehrlich, P.R. (1968) *The population bomb*, New York: Ballantine Books.

Elmhirst, R. (2018) *Ecologías políticas feministas: perspectivas situadas y abordajes emergentes*. Available at: www.ecologiapolitica.info/?p=10162 (Accessed 22 November 2020).

Escobar, A. (1995) *Encountering development: The making and unmaking of the third world*, Princeton: Princeton University Press.

Escobar, A. (2011) 'Ecología política de la globalidad y la diferencia', in Alimonda, H. (ed.), *La naturaleza colonizada. Ecología política y minería en America Latina*, Lima, Peru: Tejiendo Saberes PDTG, pp. 61–92.

Esteva, G. (2010) 'Development', in Sachs, W. (ed.), *The development dictionary: A guide to knowledge as power*, London: Zed Books, pp. 1–23.

Falú, A. (2009) *Mujeres en la ciudad. De violencias y derechos*, Santiago de Chile: Sur.

Ferguson, J. (1990) *The anti-politics machine. 'Development', depoliticization and bureaucratic power in Lesotho*, Minnesota: University of Minnesota Press.

Gay-Antaki, M. (2016) '"Now we have equality": A feminist political ecology analysis of carbon markets in Oaxaca, México', *Journal of Latin American Geography*, 15(3): 49–66. https://doi.org/10.1353/lag.2016.0030

Gay-Antaki, M. (2020a) 'Grounding climate governance through women's stories in Oaxaca, México', *Gender, Place & Culture*, online first. https://doi.org/10.1080/0966369X.2020.1789563

Gay-Antaki, M. (2020b) 'Feminist geographies of climate change: Negotiating gender at climate talks', *Geoforum*, 115: 1–10. https://doi.org/10.1016/j.geoforum.2020.06.012

Gay-Antaki, M. and Liverman, M. (2018) 'Climate for women in climate science: Women scientists and the intergovernmental panel on climate change', *Proceedings of the National Academy of Sciences of the United States of America*, 115(9): 2060–2065. https://doi.org/10.1073/pnas.1710271115

Hardin, G. (1968) 'The tragedy of the commons', *Science*, 162(3859): 1243–1248. https://doi.org/10.1126/SCIENCE.162.3859.1243

Harris, L. (2016) 'Irrigation, gender, and social geographies of the changing waterscapes of southeastern Anatolia', *Environment and Planning D: Society and Space*, 24(2): 187–213.

Harris, L., Kleiber, D., Goldin, J., Darkwah, A. and Morinville, C. (2017) 'Intersections of gender and water: Comparative approaches to everyday gendered negotiations of water access in underserved areas of accra, Ghana and Cape Town, South Africa', *Journal of Gender Studies*, 26(5): 561–582. https://doi.org/10.1080/09589236.2016.1150819

Hartmann, B. (1995) *Reproductive rights and wrongs*, Washington: South End Press.

Hendrixson, A., Ojeda, D., Sasser, J., Foley, E. and Bhatia, R. (2019) 'Confronting populationism: Feminist challenges to population control in an era of climate change', *Gender, Place and Culture*, 27(3): 307–315. https://doi.org/10.1080/0966369X.2019.1639634

Hill Collins, P. (1986) 'Learning from the outsider within: The sociological significance of Black feminist thought', *Social Problems*, 33(6): S14–32.

Kabeer, N. (1998) *Realidades trastocadas. Jeraquías de género en el pensamiento del desarrollo*, Ciudad de México: Paidós.

Kaijser, A. and Kronsell, A. (2014) 'Climate change through the lens of intersectionality', *Environmental Politics*, 23(3): 417–433. https://doi.org/10.1080/09644016.2013.835203

Khosla, P. and Masaud, A. (2010) 'Cities and gender. A brief overview', in Dankelman, I. (ed.), *Gender and climate change. An introduction*, London: Earthscan.

Leach, M. (2007) 'Earth mother myths and other ecofeminist fables: How a strategic notion rose and fell', *Development and Change*, 38(1): 67–85. https://doi.org/10.1111/j.1467-7660.2007.00403.x

Leff, E. (2015) 'The powerful distribution of knowledge in political ecology: A view from the South', in Perreault, T., Bridge, G. and McCarthy, J. (eds.), *Routledge handbook of political ecology*, London & New York: Routledge, pp. 64–75.

Lezama, J. (2019) *La naturaleza ante la tríada divina. Marx, Durkheim, Weber*, México City: El Colegio de México.

Lugones, M. (2007) 'Heterosexualism and the Colonial/modern gender system', *Hypatia*, 22(1): 186–219. https://doi.org/10.1111/j.1527-2001.2007.tb01156.x

Maldonado-Torres, N. (2007) 'On the coloniality of being', *Cultural Studies*, 21(2–3): 240–270. doi: 10.1080/09502380601162548

Maynard, M. (1997) *Science and the construction of women*, London: UCL Press.

Meadows, D.H., Meadows, D.L., Randers, J. and Behrens, W. (1972) *The limits to growth; a report for the Club of Rome's project on the predicament of mankind*, New York: Universe Books.

Merlinsky, G. (2017) 'Los movimientos de justicia ambiental y la defensa de lo común en América Latina. Cinco tesis en elaboración', in Héctor Alimonda, Catalina Toro Pérez y Facundo Martín (eds.), *Ecología política latinoamericana: pensamiento crítico, diferencia latinoamericana y rearticulación epistémica Volume II*, Buenos Aires: CLACSO, pp. 241–264.

Mies, M. (1970) *Patriarchy and accumulation on a world scale: Women in the international division of labour*, London: Zed Books.

Mollet, S. (2017) 'Postcolonial intersectionality, and the coupling of race and gender', in MacGregor, S. (ed.), *Routledge handbook of gender and environment. Feminist political ecology, postcolonial intersectionality, and the coupling of race and gender*, New York: Routledge, pp. 146–158. https://doi.org/10.4324/9781315886572

Mollet, S. and Faria, C. (2013) 'Messing with gender in feminist political ecology', *Geoforum*, 45: 116–125. https://doi.org/10.1016/j.geoforum.2012.10.009

Moser, C. (2012) 'Mainstreaming women's safety in cities into gender-based policy and programmes', *Gender and Development*, 20(3): 435–452. https://doi.org/10.1080/13552074.2012.731742

Nagar, R., Lawson, V., Mcdowell, L. and Hanson, S. (2002) 'Locating globalization: Feminist (re)readings of the subjects and spaces of globalization', *Economic Geography*, 78(3): 257–284. https://doi.org/10.2307/4140810

Nash, M., Benach, N. and Tello, R. (2005) *Inmigración, género y espacios urbanos. Los retos de la diversidad*, Barcelona: Bellaterra.

Navarro-Mantas, L., Velásquez, M. and López-Megías, J. (2015) *Violencia contra las mujeres en El Salvador: Estudio Poblacional 2014*, El Salvador: Editorial Universidad Tecnológica.

Nightingale, A. (2006) 'The nature of gender: Work, gender, and environment', *Environment and Planning D: Society and Space*, 24(2): 165–185. https://doi.org/10.1068/d01k

Nightingale, A.J. (2011) 'Bounding difference: Intersectionality and the material production of gender, caste, class and environment in Nepal', *Geoforum*, 42(2): 153–162. https://doi.org/10.1016/j.geoforum.2010.03.004

Perreault, T., Bridge, G. and McCarthy, J. (2015) *The Routledge handbook of political ecology*, New York: Routledge.

Porto-Gonçalves, W. and Leff, E. (2015) 'Political ecology in Latin America: The social re-appropriation of nature, the reinvention of territories and the construction of an environmental rationality', *Desenvolvimento e Meio Ambiente*, 35.

Resurrección, B. (2017) 'Gender and environment in the Global South. From women, environment and development to feminist political ecology', in MacGregor, S. (ed.), *Routledge handbook on gender and environment*, New York: Routledge, pp. 71–85. https://doi.org/10.4324/9781315886572

Rivera Cusicanqui, S., Domingues, M., Escobar, A. and Leff, E. (2016) 'Debate Sobre El Colonialismo Intelectual y Los Dilemas de La Teoría Social Latinoamericana', *Cuestiones de Sociología*, 14: 9.

Robbins, P. (2012) *Political ecology: A critical introduction*, 2nd ed., New York: John Wiley & Sons.

Rocheleau, D.E, Thomas-Slayter, B. and Wangari, E. (1996) *Feminist political ecology: Global issues and local experience*, New York: Routledge.

Rossi, A. and Lambrou, Y. (2008) *Gender and equity issues in liquid biofuels production: Minimizing the risks to maximize the opportunities*, Rome: FAO.

Sabsay, L. (2011) *Fronteras sexuales. Espacio urbano, cuerpos y ciudadanía*, Buenos Aires: Paidós.

Sachs, W. (1992) *The development dictionary. Pluriverse: A post-development dictionary*, London: Zed Books.
Salazar, H. and Rodríguez, M. (2015) *Miradas en el territorio, cómo mujeres y hombres enfrentan la minería. Aproximaciones a tres comunidades mineras en México*, Ciudad de México: Fundación Heinrich Böll Stiftung.
Segato, R. (2011) 'Género y colonialidad: en busca de claves de lectura y de un vocabulario estratégico descolonial', in Bidaseca, K. and Vázquez, V. (eds.), *Feminismos y poscolonialidad: descolonizando el feminismo desde y en América Latina*, Buenos Aires: Godot, pp. 17–48.
Segato, R. (2014) 'Las nuevas formas de la guerra y el cuerpo de las mujeres', *Sociedade e Estado*, 29(2): 341–371. https://doi.org/10.1590/S0102-69922014000200003
Simon-Kumar, R., MacBride-Stewart, S., Baker, S. and Patnaik Saxena, L. (2018) 'Towards north-south interconnectedness: A critique of gender dualities in sustainable development, the environment and women's health', *Gender, Work & Organization*, 25(3): 246–263. https://doi.org/10.1111/gwao.12193
Sturgeon, N. (1999) 'Ecofeminist appropriations and transnational environmentalisms', *Identities*, 6(2–3): 255–279. https://doi.org/10.1080/1070289X.1999.9962645
Sundberg, J. (2004) 'Identities in the making: Conservation, gender and race in the Maya Biosphere Reserve, Guatemala', *Gender, Place and Culture: A Journal of Feminist Geography*, 11(1): 43–66. https://doi.org/10.1080/0966369042000188549
Sundberg, J. (2017) 'Feminist political ecology', in Richardson, D., Castree, N., Goodchild, M.F., Kobayashi, A., Liu, W. and Martson, R.A. (eds.), *International encyclopedia of geography: People, the earth, environment and technology*, New York: John Wiley & Sons, Ltd, pp. 1–12.
Susial-Martín, P. (2017) 'Apuntes para una AgroEcología Política Feminista del siglo XXI', *Conference Paper, Seminario Agroecología y Sociedad*, El Colegio de la Frontera Sur. https://doi.org/10.13140/rg.2.2.15037.36326
Svampa, M. (2019) *Las fronteras del neoextractivismo en Abya Yala. Conflictos socioambientales, giro ecoterritorial y nuevas dependencias*, Bielefeld: Bielefeld University Press. https://doi.org/10.14361/9783839445266
Ulloa, A. (2014) 'Escenarios de creación, extracción, apropiación y globalización de las naturalezas: emergencia de desigualdades socioambientales', in Göbel, B., Góngora-Mera, M. and Ulloa, A. (eds.), *Desigualdades socioambientales en América Latina*, Bogotá: Universidad Nacional de Colombia/Ibero-Amerikanisches Institut, pp. 139–166.
Ulloa, A. (2016) 'Feminismos territoriales en América Latina: Defensas de la vida frente a los extractivismos', *Nómadas (Col)*, 45: 123–139.
United Nations (2018) *World urbanization prospects, the 2018 revision. The population division of the department of economic and social affairs of the United Nations*. Available at: www.un.org/development/desa/publications/2018-revision-of-world-urbanization-prospects.html (Accessed 15 January 2021).
Vaz-Jones, L. (2018) 'Struggles over land, livelihood, and future possibilities: Reframing displacement through feminist political ecology', *Signs: Journal of Women in Culture and Society*, 43(3): 711–735.
Vázquez García, V., Castañeda Salgado, M., Jazíbi Cárcamo Toalá, N. and Santos Tapia, A. (2016) *Género y medio ambiente en México: una antología*, Cuernavaca, Morelos: Universidad Nacional Autónoma de México, CRIM.
Vázquez, V. (2007) 'La recolección de plantas y la construcción genérica del espacio. Un estudio de Veracruz, México', *Ra Ximhai*, 3(3): 805–825. https://doi.org/10.35197/rx.03.03.2007.17.vv
Vázquez, V., Sosa, D. and Martínez, R. (2020) *Megaminería: sentires y luchas de las mujeres*, Nexos. Available at: https://medioambiente.nexos.com.mx/megamineria-sentires-y-luchas-de-las-mujeres/ (Accessed 15 September 2020).
Westholm, L. and Arora-Jonsson, S. (2018) 'What room for politics and change in global climate governance? Addressing gender in co-benefits and safeguards', *Environmental Politics*, 27(5): 917–938. https://doi.org/10.1080/09644016.2018.1479115
Ziai, A. (2007) 'Development discourse and its critics: An introduction to post-development', in Ziai, A. (ed.), *Exploring post-development theory and practice, problem and perspectives*, London and New York: Routledge.
Zuluaga, G. and Arango, C. (2013) 'Mujeres campesinas: Resistencia, organización y agroecología en medio del conflicto armado', *Cuadernos de Desarrollo Rural*, 10(72): 159–180.

17
DEMOCRATIC POLITICS AND LIVELIHOODS IN AFRICA

Jeremy Seekings

Introduction

In country after country across Africa, over a single generation, the political landscape of one-party states, life presidents, authoritarian regimes and military coups has given way to a new world of competitive multi-party democracy, characterised by term limits, multi-party elections and even the reality of presidents and parties losing power through electoral defeat (Carbone, 2013). Both Ghana and Malawi have had their third turnover of power since redemocratisation (in 2016 and 2020 respectively). Whilst elections and democracy across much of Africa remain flawed in many respects, the institutionalisation of regular, multiparty elections represents a 'major shift in African politics' and a 'political and institutional revolution' (Bleck and van de Walle, 2019: 3–6). Even when incumbents hold onto power, as is usually the case, the possibility of defeat can push them to be more responsive to voters' concerns and priorities.

Poverty persists in Africa despite these political changes – and in contrast to most other parts of the world. Between 1990 and 2015 the proportion of the *global* population in extreme income poverty fell by two-thirds, from 36% to 10%, whilst the estimated total number of people living in extreme income poverty fell from just under 1.9 billion people in 1990 to 735 million by 2015. In *Africa*, however, the overall income poverty rate declined much more modestly, from 55% to 41%, whilst the absolute number of people living in extreme income poverty in Africa actually rose from 280 million to 413 million.[1] Poverty in Africa has persisted despite economic growth, because too few of the benefits of growth have trickled down to the poor. The growth elasticity of income poverty in Africa between 1990 and 2015 was only −0.7, compared with −2 for other regions, meaning that every 1% of economic growth in Africa reduced poverty by only 0.7%, whereas the same growth elsewhere reduced poverty by 2% (Bicaba et al., 2015: 10). Across Africa as a whole the income poverty gap had fallen (although only to 16%, using the extreme poverty line of US $1.90/day as of 2011),[2] but even countries with strong economic growth such as Nigeria, Tanzania and Zambia (until the 2010s) had failed to reduce poverty rates significantly (Arndt et al., 2016). Income poverty persisted amidst economic growth because income inequality rose. In the early 2000s, Africa overtook Latin America as the region with the highest overall income inequality.

The lived experience behind these cold statistics is one of constrained, vulnerable and insecure livelihoods for many people. Across much of East and Southern Africa (except South Africa and its smaller neighbours), three-quarters of the working population is engaged in smallholder or peasant agriculture. Informal self-employment in 'household enterprises' accounts for much of the other quarter, with formal employment accounting for a very small percentage of the working population. Yet productivity in smallholder agriculture has rarely risen significantly. Economic growth has often been driven by mining, with weak linkages to other sectors. Africa has not experienced either the rise in agricultural productivity or the expansion of labour-intensive manufacturing that were integral to the much more rapid and inclusive economic growth in East Asia and parts of South-east and South Asia (Henley, 2012; Newman et al., 2016).

The persistence of low-productivity livelihoods and widespread income poverty may in part reflect environmental conditions (including the dependence on rainfall and production of only one crop per year across much of Africa) but it also reflects the policies pursued by governments and non-government organisations. Prior to redemocratisation in the 1990s, 'distributional regimes' across much of the region were 'neo-patrimonial', in that public office was used for private gain by political elites. Insofar as they needed to take into account broader constituencies, they either sought to deploy patronage or were biased towards urban minorities or rural elites at the expense of rural majorities. 'Urban bias' often entailed a specific bias towards skilled labour in the name of modernisation and development. By 'distributional regime', we mean the economic growth path that determines, in the first instance, who gets what, as well as the policies that shape this and whatever policies serve to redistribute resources, perhaps but not necessarily to poor people left behind as the economy grows along its particular pathway. Most African governments failed to steer their countries' economies down a growth path that expanded opportunities for their poor, rural populations to improve their productivity, whether in agriculture or in other sectors (Seekings, 2019c; Williams, 1976). In their efforts to control peasant farmers (Scott, 1998), state elites perpetuated rural poverty (Hyden, 1980).

Has the expansion of representative democracy changed distributional regimes or 'who gets what' in Africa? Has it pushed political elites and states to provide more consequential support for their poorer citizens' livelihoods? This chapter reviews how the institutionalisation of (qualified) democracy has affected the politics of livelihoods in Africa. The chapter examines whether and how contested elections have transformed governance and policy-making with respect to selected areas of public policy with direct implications for popular livelihoods: agricultural policies, policies towards formal and informal sectors, policies that substitute for livelihoods when people are unable to provide for themselves (including drought relief and social protection) and public services that underpin future livelihoods (especially schooling and health care). Most of the examples are drawn from Anglophone East and Southern Africa.

Flawed democracies and elite-oriented political settlements

The institutionalisation of regular, multiparty elections did not entail a complete break with the prior decades of undemocratic politics. Incumbents enjoy considerable advantages through their access to state resources both to strengthen their own campaigns and to inhibit and undermine their opponents'. If necessary, incumbents may subvert the electoral process. Only about one in six elections resulted in turnovers in government, compared to a global average of about one in three. In many countries, the 'same political class' continued to dominate politics and govern, with a limited circulation of elites (Bleck and van de Walle, 2019).

Such continuities in governance have fuelled a growing literature on 'political settlements' (Behuria et al., 2017). The underlying premise of the 'political settlements' approach is that formal political institutions themselves do not have general effects on development (or, therefore, distribution). The ways in which formal political institutions operate (and hence their consequences) depend on underlying power relations within the elite and between them and non-elite groups, i.e. on the underlying 'political settlement'. This approach first examines how the 'ruling coalition' is constituted, i.e. how concentrated (or dispersed) is the 'horizontal' distribution of power within the elite. The political settlement approach then examines the ruling coalition's social foundations, i.e. how power is distributed 'vertically' in terms of breadth and depth. This affects whether the ruling coalition can ensure that its preferences are implemented at lower levels. The approach also examines how the political settlement is financed, i.e. the relations between the ruling coalition and economic actors (especially capitalists/business) (Behuria et al., 2017; Kelsall, 2018a). Political settlements thus entail broadly stable and largely non-violent 'rules of the game' governing the distribution of resources, including especially rents. In the African context, the achievement of democracy in South Africa in the early 1990s represented the clearest case of a new political settlement in that the rules of the game after 1994 were fundamentally different to those under apartheid.

Political settlements in Africa have generally been characterised as 'competitive-clientelist'. This points to the prevalence of political clientelism – i.e. political support is constructed through patron-client networks, with political elites competing to buy the support of local leaders. This has some similarities with previous work focused on neo-patrimonialism. The concept of competitive clientelism has been applied to the democratic cases of Ghana (Abdulai and Hickey, 2016; Whitfield, 2011) and Kenya (Tyce, 2019). More authoritarian regimes have been assessed to have 'dominant party' or 'weak dominant' political settlements. This allows governments in countries such as Uganda to adopt a longer time-horizon than their election-oriented democratic neighbours (see, for example, Hickey and Izama, 2017. Kelsall (2018b) prefers a more precise specification of political settlements according to the distribution of power vertically and horizontally. In the case of Tanzania, for example, the political settlement has shifted to and fro according to whether power was (a) broad-based (vertically) and dispersed (horizontally) or (b) narrowly based and concentrated. These two types of political settlement alternated both before and after the reintroduction of competitive, multi-party elections. Cammack and Kelsall (2011) provide a similar analysis for Malawi.

The political settlement literature pays little attention to elections. Pedersen and Jacob (2019), in their study of Tanzania, emphasise the role of elections in incentivising the ruling elite to deliver more to voters so as to legitimate their rule. The implication of Pedersen and Jacob's (2019) account is that Tanzania's settlement has been broadened across society even if power has become more concentrated within the elite. In their account, the political settlement was 'reshaped', moreover, not replaced. Similarly, in their study of Ghana, Abdulai and Hickey (2016) refer to changes in the ruling coalition – with successive NPP and NDC governments having different regional bases – operating within an apparently stable 'competitive-clientelist' political settlement.

The political settlement approach reminds us not to assume that elections ensure that governments are accountable or that elites compete by offering voters clear alternatives. The approach is a blunt tool for understanding precisely when, how and why governments adopt or implement policies that benefit particular groups of citizens (or voters), and specifically when, how and why they adopt policies that expand rather than restrict the opportunities for poorer citizens to shift into higher-productivity livelihoods. As Bleck and van de Walle (2019: 16–17) put it, elections open up 'political moments' with the possibility of change and 'heightened citizenship'.

One reason why elections might not matter is that competing elites share similar beliefs and values about development. Mkandawire (2005) emphasised that developmentalism was the hegemonic ideology across most of Africa for most of the period from independence (see also Aerni-Flessner, 2018). Elites embraced an ideology of modernisation that tended to view rural smallholders as backward, serving as a brake on development that required forceful action by the state. As was clear in the case of 'socialist' Tanzania in the 1970s, peasants had to be forced to modernise, including through villagisation (Schneider, 2007; Scott, 1998). These beliefs remain widely held, including among more market-friendly elites. Kalebe-Nyamongo and Marquette (2014) document the deep prejudice against the poor among Malawian elites in the 2000s. Poverty was attributed (at least in part) to laziness and the preferred solution was hard work. Viewing the poor as 'passive, dependent, and fatalistic', Malawian elites worried that social cash transfers to the poor would not encourage self-reliance and entrepreneurial initiative but rather would encourage laziness and 'dependency' (Kalebe-Nyamongo and Marquette, 2014). At the same time, some conservative elites hold to an ideology of paternalistic responsibility. The ideology of the Botswana Democratic Party – which has won every multi-party election since independence – accepts inequality, emphasises self-help and denounces 'dependency', but also emphasises social justice. As the country's first president, Seretse Khama (himself the country's premier chief by birth and owner of many cattle) put it, quoting a Tswana proverb: 'A lean cow cannot climb out of the mud, but a good cattleman does not leave it to perish' (Seekings, 2016b: 24). Furthermore, electoral self-interest has often pushed political elites into adopting or implementing pro-poor policies, in Botswana (Seekings, 2019a) and elsewhere.

Agricultural policies

Poverty persists in Africa in part because productivity in smallholder agriculture has generally remained stagnant, in contrast to some other previously poor and agrarian societies across the world. Average output per worker was, in 2005, only just above the standard extreme poverty line (at the time, US $1/day) (McMillan and Headey, 2014). This was despite the shift, caused by population growth, from an abundance of land and labour constraints to a shortage of land and surplus labour.

Generally, stagnant productivity has not been primarily due to any lack of understanding of what works, but rather to political constraints on policymaking and implementation. 'The experience has shown that policies that support small farms by correcting for the market failures inherent in smallholder agriculture, especially in the early phases of agricultural development, are a particularly promising strategy to achieve pro-poor growth', write Birner and Resnick; 'the empirical evidence also shows that it is politically difficult to implement such policies' (Birner and Resnick, 2010: 1450). Bates, in his classic *Markets and States in Tropical Africa* (1981), contrasted the powerlessness of rural populations with the power of urban populations. The result, often, was the regulation of crop prices and management of exchange rates to ensure low consumer prices in towns but low producer prices for farmers. This urban bias was largely brought to an end by the combination of liberalisation that formed part of structural adjustment in the 1980s and 1990s and the enfranchisement of largely rural electorates under the restoration of multi-party democracy (Bates and Block, 2013). Incumbent parties are disproportionately likely to have rural strongholds (Boone and Wahman, 2015).

Overall, democratisation reduced agricultural taxation and increased expenditure on agriculture (Bates and Block, 2013). The most dramatic change in agricultural policies was the proliferation of programmes that subsidise fertiliser and seeds for farmers (Jayne and Rashid, 2013; Poulton, 2014). Moreover, in the case of Ghana, subsidies that had previously been captured by wealthier farmers (mostly growing export crops) were diverted to smaller farmers (growing maize) (Banful, 2011).

Rural politics typically continues to revolve around patronage (Booth, 2014, generally; Poulton and Kanyinga, 2014, on Kenya). National programmes such as fertiliser subsidies can sustain patronage on a massive scale at the same time as providing a party or president with a national brand that appears to transcend regional or ethnic appeals and thus helps to build broad support among voters. The outstanding example of this was in Malawi in the late 2000s. President Bingu wa Mutharika branded himself and his party in terms of their commitment to maize production, with the FISP as their flagship programme. In 2009, for the first and only time, the winner of the presidential election (Mutharika) won majorities in all regions (Ferree and Horowitz, 2010). Fertiliser subsidies appeal for another reason also. In Malawi, the Mutharika government used contracts to provide or transport fertilisers to recruit regional elites from opposition-supporting areas into the regional coalition (Chinsinga and Poulton, 2014). In Zambia, the Lungu government's belated enthusiasm for fertiliser subsidies was reportedly linked to kickbacks from contractors to fund the ruling party's election campaign (Africa Confidential, 2020).

Elected governments may direct expenditure towards districts where they enjoy solid support as a reward (e.g. Leiderer, 2014) or to 'swing' districts as an inducement (e.g. Banful, 2011). Districts or regions can sometimes be targeted through crop-specific price support or other polices as either reward or inducement. In Kenya, for example, governments allowed institutions supporting cotton farmers (most of whom are in opposition-supporting areas) to atrophy whilst dairy farmers (mostly in government-supporting areas) were strengthened (Poulton and Kanyinga, 2014).

Fertiliser subsidies may be popular but it is not clear that they are the best use of scarce public resources. Public investment in public goods – including agricultural extension and transport infrastructure – may be more consequential in terms of improving agricultural livelihoods (Poulton, 2014), but they may have little appeal to political leaders. Agricultural extension offers few opportunities for patronage or publicity. Governments do sometimes invest in roads, but tend to prefer high-profile, expensive projects rather than the more routine investment in improving and maintaining local roads that are of most benefit to poorer people (e.g. Whitworth, 2012). There is often too little of the technocratic support required to ensure that programmes are pro-poor (including through containing corruption). Some authoritarian regimes have demonstrated what can be achieved with well-designed programmes, but the objective might not simply be to support agricultural livelihoods. Ethiopia's agricultural extension programme, for example, was the vehicle for the intensification of party-state power across the countryside, with party loyalists recruited as extension agents and using extension to reward loyal farmers and to promote the party (including during election campaigns) (Berhanu and Poulton, 2014).

Structural change and higher-productivity livelihoods

Improved livelihoods generally require structural economic change, with 'surplus' labour shifting into higher-productivity sectors. Typically this means that labour moves from subsistence agriculture into informal commerce, then into labour-intensive but more formal sectors such as clothing manufacturing, commercial agriculture and tourism services, and finally into higher-productivity manufacturing and service sectors. In practice, economic growth and development in Africa has often been driven by high-productivity economic enclaves (especially mining) without the kind of broad-based structural change that underpinned rapid economic growth in East Asia and elsewhere. Across much of Southern Africa, state support to formal sectors together with land dispossession resulted in the destruction of smallholder agriculture, constrained informal commerce and widespread open unemployment – and hence severe income inequality and high levels of income poverty relative to GDP per capita.

Across much of Africa, informal employment provides a crucial safety net of opportunity for the poor. African cities and towns are worlds in which people hustle and jostle to earn their living (Simone, 2004), often in the face of repressive state regulation. State hostility to informal commerce is most evident across Southern Africa, where both colonial and post-colonial states tend to view the urban poor as posing a political threat. The solution both before and after independence was often to repatriate delinquent urban populations to rural areas. In Zimbabwe, in 2005, the government launched a military-style 'Operation Murambatsvina' (Operation Restore Order) against informal producers and street traders (as well as informal housing) in the country's cities (Potts, 2006). Whilst the operation was no doubt politically instrumental, asserting the power of the ruling party in opposition-supporting areas, it also reflected a deep-rooted ideological antipathy towards the urban poor as unproductive and delinquent (Dorman, 2016) – an attitude reflected more widely in discussions of Africa's 'urban youth' at the African Union. Traders lacked the capacity to resist the destruction of their livelihoods collectively and directly, but instead practiced 'adaptive resistance', resuming commercial activity in safer, less visible locations (Musoni, 2010). This Zimbabwean case was extreme, but its core features were common across African cities. Ruling elites and states sometimes tolerated informality – practising 'forbearance', as Holland (2017) has discussed in the Latin American context – but have often engaged in highly repressive episodes. Informal producers and traders can rarely organise effectively (Brown et al., 2010), but instead take advantage of the limits to everyday state power and quickly resume activity.

The urban poor have votes even if they lack collective organisation. In many areas, urban politics remains largely clientelistic. In some, opposition parties mobilise discontent among the urban poor and win municipal elections, although few municipalities across Africa have significant resources or power. More rarely, opposition parties win a national election and take state power. In Zambia, the Patriotic Front won the 2011 election on the basis of a populist appeal to urban voters (as well as a more ethnic or regional appeal to rural voters in the country's northern provinces). The party promised 'lower taxes, more jobs, more money in your pockets', improved urban services and an end to state harassment (Resnick, 2013).

Faced with 'backward' smallholder farmers and the delinquent (and threatening) urban poor, political elites aspire to grow formal and 'modern' employment. Their track record has been poor. Unlike a series of countries in East, Southeast and South Asia, African countries failed to expand the contribution of manufacturing to GDP between the 1970s and 2010s. Africa's share of global manufacturing actually declined over this period (Newman et al., 2016). Crucially, they failed to build substantial export-oriented manufacturing sectors such as clothing manufacturing. More capital- and skill-intensive sectors often rely on state subsidies or – like mining – have few linkages to the rest of the economy.

State-led import-substituting industrialisation allowed for rapid industrial growth in the 1960s but inefficient production for local markers meant that this strategy reached its limits by the 1980s. Structural adjustment programmes fuelled deindustrialisation. Manufacturing growth picked up in the 2000s, but slowly, with the notable exception of Nigeria (and, in North Africa, Tunisia). The continuing cause of stagnation in manufacturing was and remains the failure to export, itself in large part the result of public policies that directly and indirectly discouraged rather than facilitated export production (Newman et al., 2016).

African states have failed to set rules that allow or encourage private (capitalist) businesses to invest profits (or rents) productively and to ensure that state bureaucrats enforce these rules uniformly. African capitalists have in general remained too weak and dependent on the state to be able to push the government to adopt and implement such rules. At the same time, democratisation has strengthened somewhat their hands in that political parties required funding to compete in elections. Whitfield et al. (2015) employ the 'political settlement' framework to explain the equilibrium that results under 'competitive clientelism': politicians adopt short time horizons, preferring policies that strengthen their factional patronage structures in the short-term and neglect policies that would strengthen manufacturing in the longer-term, whilst individual businesses extend financial support to parties or factions in return for privileged access to government contracts and patronage. Rents are generated but are used to improve the prospects of short-term political survival rather than for economically productive investment.

An additional factor in some of the 'surplus labour' economies of Southern Africa – especially South Africa – is that labour costs in the more labour-intensive sectors are high relative to productivity. In the South African case, this is in part the consequence of the government choosing to regulate (i.e. raise) wages to promote 'decent work' at the cost of job destruction and the forestalling of job creation. This choice was framed by both ideological beliefs and pressure from both trade unions and larger, more capital- and skill-intensive employers (Nattrass and Seekings, 2019).

The economic crisis that followed the global COVID-19 pandemic in 2020 revealed the distributional politics of the most affected African countries. Whilst international agencies and commentators pointed out that the burden of economic crisis fell especially heavily on the urban poor – who lacked the safety net of subsistence agriculture – governments prioritised relief for employers and workers in the formal sector.

Famine relief and social protection

Given growing populations but stagnant agricultural and non-agricultural livelihoods, African governments have faced intensified pressure to expand social protection as a substitute for independent livelihoods. Whilst social protection has expanded across much of Africa, the region remains a laggard in comparison to most other regions (ILO, 2017) with the very important exception of drought relief.

Perhaps the most prevalent risk facing rural African societies was and remains drought (and less often, flooding). On a continent where agriculture generally depends upon rainfall and rural populations depend on agriculture, droughts and flooding can lead to famine, impoverishment and death. Amartya Sen (2005: 188), reflecting primarily on the Indian and Chinese experiences, famously suggested that 'major famines do not occur in democracies' (see further Drèze and Sen, 1991). Sen argued that famines only occur if they are not prevented, and democracies prevent them because a free press publicised the problem and pluralist politics serves to hold governments to account for inaction. De Waal (2000) reframed this argument in terms of 'anti-famine political contracts' between society and state.

Drought relief policies in Africa were pioneered in Botswana, where independence coincided with terrible drought (see Box 17.1). Since the early 1960s, famine has been confined to countries with undemocratic governments. There has been only one clear exception: Malawi, in 2001–02. Devereux and Tiba (2007) suggest that an interconnected set of private maize traders, politicians and civil servants stood to gain from the rising food prices that resulted from shortage. Material interest outweighed political interest. Rubin (2008) argues that the famine was the result of a weak government that was dependent on foreign donors: 'The strained relationship between international donors and the Malawi government led to famine responses that were belated and inadequate' (Rubin, 2008: 61).

Box 17.1 Government response to drought in Botswana

At independence in 1966, Botswana was one of the poorest countries in the world, with a population divided between a small minority of cattle-owners and salaried workers, and a large number of very poor households dependent on agricultural production that was difficult even in years of good rains. But independence coincided with the worst drought in living memory. Whereas the colonial government had done little to mitigate the devastating effects of drought on either the human or cattle populations, Seretse Khama's new independent government quickly moved to organise a massive drought relief operation, perhaps without precedent in Africa. Combining with the newly established World Food Programme, the new government organised relief food and even stockfeed across the entire country. The operation comprised three pillars: school and preschool feeding schemes; workfare, to support entire households; and food aid provided directly to individuals living in households unable to participate in workfare. These programmes continued not only through the 1970s and 1980s – when the country experienced further devastating droughts – but were thereafter slowly repackaged as regular social protection programmes (including pensions for the elderly). This expanding welfare state was required because of the high level of inequality in the distribution of incomes. Initially, the cattle-owning elite captured most of the benefits of public investments (in water supplies and education). Later, with the mining sector driving rapid economic growth, opportunities improved for many but not all citizens. Whilst agricultural livelihoods generally remained stagnant, non-agricultural livelihoods improved rapidly and social protection provided for those households and individuals who remained poor.

In Botswana, Ethiopia and elsewhere, 'emergency' drought relief policies evolved into permanent social protection programmes (Lavers, 2019; Seekings, 2019a). Social protection in Africa involves a distinctive emphasis on social assistance (i.e. tax- or donor-financed programmes) rather than social insurance (tied to formal employment and financed at least in part through contributions from employers and employees). Social assistance is most extensive in South Africa, where its origins lay in the country's racial history. Elsewhere, most social assistance dates from the 2000s, when it was promoted strongly by international agencies and aid donors (von Gliszczynski and Leisering, 2016; Seekings, 2021; see also Chapter 40 in this volume).

The expansion of social assistance has been far from even. In general, national governments do far less than international organisations recommend, in part because national political elites do not share the enthusiasm for extending social assistance to the poor as a right. The prejudice against the poor and veneration of work documented in the case of Malawi by Kalebe-Nyamongo and Marquette (2014) leads to a widespread preference for workfare (i.e. public employment programmes), feeding schemes and supposedly 'developmental' programmes (such as microfinance) over cash transfers. In a number of countries – including Tanzania (Ulriksen, 2019) and Zambia (Siachiwena, 2020) – governments have allowed donors to initiate and largely operate social protection programmes as long as the donors contribute most or all of the funding.

Conservative elites have sometimes been pushed into reform by pressure from within the country. Civil society played an important role in South Africa's expansion of social assistance in the 2000s, but this was an exception. Electoral competition has also provided an impetus to reform. In Botswana, welfare programmes were crucial to preserving support for the ruling party in rural areas (Gulbrandsen, 2012). The government's political vulnerability led it later to extend programmes to urban areas also (Seekings, 2019a). It is not clear, however, that African voters favour social protection over other government programmes. In Malawi, the incumbent president campaigned in 2014 on the platform of expanding social protection but lost to the opposition candidate who championed maize subsidies instead (Hamer and Seekings, 2019). In South Africa, receipt of a grant seems to have little direct effect on voting behaviour (Graham et al., 2016; Seekings, 2019b); even unemployed young men prefer job creation to grants for people like themselves (Dawson and Fouksman, 2020).

The muted response of most African governments to the COVID-19 crisis reflected continuing ambivalence towards social protection. South Africa did expand its already broad social assistance system – at least temporarily – but most countries continued to drag their heels as they continued along the path of slow reform promoted by international organisations (Gronbach and Seekings, 2021).

Conclusion

The 'political settlements' literature emphasises the importance of factional competition and compromise within the political elite, the general political weaknesses of the poor and the continued semi-dependence of domestic capitalists on the government and state. The approach does, however, tend to underestimate the importance of political factors – including ideology and partisan competition – that are important to understanding both similarities and differences between African states in their policies towards the livelihoods of their poor populations. Elites' beliefs about the causes of poverty (and success) sustain the status quo in many areas of public policy, whilst electoral competition has provided impetus to reform of some policies – especially fertiliser subsidies (and, in another policy area, the elimination of school fees – see Harding and Stasavage, 2013) – but not the more profound reforms that would restructure the economy and open up new opportunities for improved livelihoods.

Key points

- Redemocratisation in the 1990s meant that governments across much of Africa became more formally accountable to their citizens, but the quality of democracy remained uneven and poverty remains widespread even during periods of economic growth, prompting scholars to question how much the underlying 'rules of the game' have changed.
- Across most of Africa, a majority of households depend at least in part on smallholder or peasant agriculture, but political elites have generally either neglected small farmers or have actively inhibited any improvement in their livelihoods.
- Case-studies of policy 'sectors' suggest that redemocratisation contributed to some livelihood-improving public policies but not others.
- Livelihood-improving reforms include the provision of state-subsidised fertiliser and other 'inputs' to small farmers, drought relief (preventing famine) and some forms of social protection (mitigating poverty).
- At the same time, many governments have been hostile to the urban informal sector, unable to adopt the long-term time horizons that enables the kind of stable policy environment conducive to private investment, are especially reluctant to support (or even permit) labour-intensive industrialisation and oppose any large-scale expansion of social protection funded from their own domestic revenues.

Suggestions for further reading

The literature on the politics of livelihoods in Africa is generally fragmented between political dynamics on the one hand and public policy on the other, with the latter further fragmented into different areas of public policy. Bates (1981) was the classic study that combined both sides of this literature, prior to redemocratisation in the 1990s. For more recent integrations of the politics of policy and hence livelihoods, see: Poulton (2014) on agricultural extension; Whitfield et al. (2015), Newman et al. (2016) and Nattrass and Seekings (2019) on industrial policy; and Hickey et al. (2019) on social protection. The fast-growing political settlements literature attempts to link politics and policy, inspiring or informing many of the studies cited above. This literature has been usefully summarised by Behuria et al. (2017). Bleck and van de Walle (2019) provide the most recent review of parties, voters and elections in Africa. Simone (2004) is an outstanding ethnographic study of urban livelihood politics. Recent ethnographic studies of rural areas include Phillips (2018), on Tanzania. On the case of Botswana, see especially Peters (1994), Gulbrandsen (2012), Selolwane (2012) and Seekings (2016a, 2016b, 2019a).

Notes

1 Data from PovCalNet: http://iresearch.worldbank.org/PovcalNet/home.aspx.
2 Ibid.

References

Abdulai, A.-G. and Hickey, S. (2016) 'The politics of development under competitive clientelism: Insights from Ghana's education sector', *African Affairs*, 115(458): 44–72. https://doi.org/10.1093/afraf/adv071

Aerni-Flessner, J. (2018) *Dreams for lesotho*, Notre Dame: University of Notre Dame Press.

Africa Confidential (2020) 'Default hits election plan', *Africa Confidential*, 61(19), 24th September: 1–2.

Arndt, C., McKay, A. and Tarp, F. (2016) 'Synthesis: Two cheers for the African growth renaissance (but not three)', in Arndt, C., McKay, A. and Tarp, F. (eds.), *Growth and poverty in sub-Saharan Africa*, Oxford: Oxford University Press and UNU-WIDER, pp. 11–39. https://doi.org/10.1093/acprof:oso/9780198744795.003.0002

Banful, A.B. (2011) 'Old problems in the new solutions? Politically motivated allocation of program benefits and the "new" fertilizer subsidies', *World Development*, 39(7): 1166–1176. https://doi.org/10.1016/j.worlddev.2010.11.004

Bates, R. (1981) *Markets and states in tropical Africa: The political basis of agricultural policies*, Berkeley: University of California Press.

Bates, R. and Block, S. (2013) 'Revisiting African agriculture: Institutional change and productivity growth', *Journal of Politics*, 75(2): 372–384. https://doi.org/10.1017/S0022381613000078

Behuria, P., Buur, L. and Gray, H. (2017) 'Studying political settlements in Africa', *African Affairs*, 116(464): 508–525. https://doi.org/10.1093/afraf/adx019

Berhanu, K. and Poulton, C. (2014) 'The political economy of agricultural extension policy in Ethiopia: Economic growth and political control', *Development Policy Review*, 32(s2): s199–s216. https://doi.org/10.1111/dpr.12082

Bicaba, Z., Brixiová, Z. and Ncube, M. (2015) 'Eliminating extreme poverty in Africa: Trends, policies and the roles of international organisations', *Working Paper 223*, Abidjan: African Development Bank.

Birner, R. and Resnick, D. (2010) 'The political economy of policies for smallholder agriculture', *World Development*, 38(10): 1442–1452. https://doi.org/10.1016/j.worlddev.2010.06.001

Bleck, J. and van de Walle, N. (2019) *Electoral politics in Africa since 1990*, Cambridge: Cambridge University Press.

Boone, C. and Wahman, M. (2015) 'Rural bias in African electoral systems: Legacies of unequal representation in African democracies', *Electoral Studies*, 40: 335–346. https://doi.org/10.1016/j.electstud.2015.10.004.

Booth, D. (2014) 'Agricultural policy choice: Interests, ideas and the scope of reform', *Working Paper*, London: International Institute for Environment and Development.

Brown, A., Lyons, M. and Dankoco, I. (2010) 'Street traders and the emerging spaces for urban voice and citizenship in African cities', *Urban Studies*, 47(3): 666–683. https://doi.org/10.1177/0042098009351187

Cammack, D. and Kelsall, T. (2011) 'Neo-patrimonialism, institutions and economic growth: The case of Malawi, 1964–2009', *IDS Bulletin*, 42(2): 88–96. https://doi.org/10.1111/j.1759-5436.2011.00214.x

Carbone, G. (2013) 'Leadership Turnovers in Sub-Saharan Africa: From Violence and Coups to Peaceful Elections?', *ISPI Analysis 192*, Milan: Italian Institute for International Political Studies.

Chinsinga, B. and Poulton, C. (2014) 'Beyond technocratic debates: The significance and transience of political incentives in the Malawi Farm Input Subsidy Programme (FISP)', *Development Policy Review*, 32(s2): s123–s150. https://doi.org/10.1111/dpr.12079

Dawson, H. and Fouksman, E. (2020) 'Labour, laziness and distribution: Work imaginaries among the South African unemployed', *Africa*, 90(2): 229–251. https://doi.org/10.1017/S0001972019001037

De Waal, A. (2000) 'Democratic political process and the fight against famine', *IDS Working Paper, 107*, Brighton: Institute of Development Studies.

Devereux, S. and Tiba, Z. (2007) 'Malawi's first famine, 2001–2002', in Devereux, S. (ed.), *The new famines: Why famines persist in an era of globalization*, Abingdon: Routledge, pp. 143–177.

Dorman, S.R. (2016) '"We have not made anybody homeless": Regulation and control of urban life in Zimbabwe', *Citizenship Studies*, 20(1): 84–98. https://doi.org/10.1080/13621025.2015.1054791

Drèze, J. and Sen, A. (1991) *Hunger and public action*, Oxford: Oxford University Press.

Ferree, K. and Horowitz, J. (2010) 'Ties that bind? The rise and decline of ethno-regional partisanship in Malawi, 1994–2009', *Democratization*, 17(3): 534–563. https://doi.org/10.1080/13510341003700394

Graham, V., Sadie, Y. and Patel, L. (2016) 'Social grants, food parcels and voting behavior: A case study of three South African communities', *Transformation*, 91: 106–135. https://doi.org/10.1353/trn.2016.0020

Gronbach, L. and Seekings, J. (2021) 'Pandemic, lockdown and the stalled urbanization of welfare regimes in Southern Africa', *Global Social Policy*, online first. https://doi.org/10.1177/14680181211013725

Gulbrandsen, Ø. (2012) *The state and the social: State formation in Botswana and its precolonial and colonial genealogies*, New York: Berghahn.

Hamer, S. and Seekings, J. (2019) 'Social assistance, electoral competition, and political branding in Malawi', in Hickey, S., Lavers, T., Niño-Zarazúa, M. and Seekings, J. (eds.), *The politics of social protection in eastern and southern Africa*, Oxford: Oxford University Press, pp. 225–248.

Harding, R. and Stasavage, D. (2013) 'What democracy does (and doesn't do) for basic services: School fees, school inputs, and African elections', *Journal of Politics*, 76(1): 229–245. https://doi.org/10.1017/S0022381613001254

Henley, D. (2012) 'The agrarian roots of industrial growth: Rural development in South-East Asia and Sub-Saharan Africa', *Development Policy Review*, 30(s1): s25–s47. https://doi.org/10.1111/j.1467-7679.2012.00564.x

Hickey, S. and Izama, A. (2017) 'The politics of governing oil in Uganda: Going against the grain?', *African Affairs*, 116(463): 163–185. https://doi.org/10.1093/afraf/adw048

Hickey, S., Lavers, T., Niño-Zarazúa, M. and Seekings, J. (eds.) (2019) *The politics of social protection in east and southern Africa*, Oxford: Oxford University Press.

Holland, A. (2017) *Forbearance as Redistribution: The politics of informal welfare in Latin America*, Cambridge: Cambridge University Press.

Hyden, G. (1980) *Beyond Ujamaa in Tanzania: Underdevelopment and an Uncaptured Peasantry*, London: Heinemann.

ILO (2017) *World social protection report 2017–2019*, Geneva: International Labour Organization.

Jayne, T.S. and Rashid, S. (2013) 'Input subsidy programs in sub-Saharan Africa: A synthesis of recent evidence', *Agricultural Economics*, 44: 547–562. https://doi.org/10.1111/agec.12073

Kalebe-Nyamongo, C. and Marquette, H. (2014) 'Elite attitudes towards cash transfers and the poor in Malawi', *DLP Research Paper 30*, Birmingham: Developmental Leadership Programme, University of Birmingham.

Kelsall, T. (2018a) 'Towards a universal political settlement concept: A response to Mushtaq Khan', *African Affairs*, 117(469): 656–669. https://doi.org/10.1093/afraf/ady018

Kelsall, T. (2018b) 'Thinking and working with political settlements: The case of Tanzania', *ODI Working Paper 541*, London: Overseas Development Institute.

Lavers, T. (2019) 'Distributional concerns, the "developmental state" and the agrarian origins of social assistance in Ethiopia', in Hickey, S., Lavers, T., Niño-Zarazúa, M. and Seekings, J. (eds.), *The Politics of social protection in eastern and southern Africa*, Oxford: Oxford University Press, pp. 68–94.

Leiderer, S. (2014) 'Who gets the schools? Political targeting of economic and social infrastructure provision in Zambia', *DIE Discussion Paper 27/2014*, Bonn: Deutsches Institut für Entwicklungspolitik.

McMillan, M. and Headey, D. (2014) 'Introduction: Understanding structural transformation in Africa', *World Development*, 63: 1–10. https://doi.org/10.1016/j.worlddev.2014.02.007

Mkandawire, T. (2005) 'African intellectuals and nationalism', in Mkandawire, T. (ed.), *African intellectuals: Rethinking politics, language, gender and development*, Dakar: Codesria and London: Zed, pp. 10–55.

Musoni, F. (2010) 'Operation Murambatsvina and the politics of street vendors in Zimbabwe', *Journal of Southern African Studies*, 36(2): 301–317. https://doi.org/10.1080/03057070.2010.485786

Nattrass, N. and Seekings, J. (2019) *Inclusive dualism: Labour-intensive development, decent work, and surplus labour in southern Africa*, Oxford: Oxford University Press.

Newman, C., Page, J., Rand, J., Shimeles, A., Söderbom, M. and Tarp, F. (2016) *Manufacturing transformation: Comparative studies of industrial development in Africa and emerging Asia*, Oxford: Oxford University Press.

Pedersen, R. and Jacob, T. (2019) 'Political settlement and the politics of legitimation in countries undergoing democratisation: Insights from Tanzania', *ESID Working Paper 124*, Manchester: Effective States and Inclusive Development, University of Manchester.

Peters, P. (1994) *Dividing the commons: Politics, policy and culture in Botswana*, Charlottesville: University of Virginia Press.

Phillips, K. (2018) *An ethnography of hunger: Politics, subsistence, and the unpredictable grace of the sun*, Bloomington: Indiana University Press.

Potts, D. (2006) '"Restoring order"? Operation Murambatsvina and the urban crisis in Zimbabwe', *Journal of Southern African Studies*, 32(2): 273–291. https://doi.org/10.1080/03057070600656200

Poulton, C. (2014) 'Democratisation and the political incentives for agricultural policy in Africa', *Development Policy Review*, 32(s2): s101–s122. https://doi.org/10.1111/dpr.12078

Poulton, C. and Kanyinga, K. (2014) 'The politics of revitalising agriculture in Kenya', *Development Policy Review*, 32(s2): s151–s172. https://doi.org/10.1111/dpr.12080

Resnick, D. (2013) *Urban poverty and party populism in African democracies*, Cambridge: Cambridge University Press.

Rubin, O. (2008) 'The Malawi 2002 famine: Destitution, democracy and donors', *Nordic Journal of African Studies*, 17(1): 47–65.

Schneider, L. (2007) 'High on modernity? Explaining the failings of Tanzanian villagisation', *African Studies*, 66(1): 9–38. https://doi.org/10.1080/00020180701275931

Scott, J.C. (1998) *Seeing like a state: How certain schemes to improve the human condition have failed*, New Haven: Yale University Press.

Seekings, J. (2016a) 'Drought relief and the origins of a conservative welfare state in Botswana, 1965–1980', *CSSR Working Paper 37*, Cape Town: Centre for Social Science Research, University of Cape Town.

Seekings, J. (2016b) '"A lean cow cannot climb out of the mud, but a good cattleman does not leave it to perish": The origins of a conservative welfare doctrine in Botswana under Seretse Khama, 1966–1980', *CSSR Working Paper 387*, Cape Town: Centre for Social Science Research, University of Cape Town.

Seekings, J. (2019a) 'Building a conservative welfare state in Botswana', in Hickey, S., Lavers, T., Niño-Zarazúa, M. and Seekings, J. (eds.), *The politics of social protection in eastern and southern Africa*, Oxford: Oxford University Press, pp. 42–67.

Seekings, J. (2019b) 'Social grants and voting in south Africa', *CSSR Working Paper 436*, Cape Town: Centre for Social Science Research, University of Cape Town.

Seekings, J. (2019c) 'Poverty and inequality: south Africa in a continental context', in Soudien, C., Woolard, I. and Reddy, V. (eds.), *State of the nation 2019: Poverty and inequality*, Cape Town: HSRC Press, pp. 42–56.

Seekings, J. (2021) 'International actors and social protection in Africa, 2000–2020', in Schüring, E. and Loewe, M. (eds.), *Handbook of social protection systems*, Aldershot: Edward Elgar, pp. 491–506.

Selolwane, O. (ed.) (2012) *Poverty reduction and changing policy regimes in Botswana*, London: Palgrave Macmillan with UNRISD.

Sen, A. (2005) *The argumentative Indian: Writings on Indian history, culture and identity*, London: Allen Lane.

Siachiwena, H. (2020) *The politics of welfare policy reforms: A comparative study of how and why changes of government affect policy making on social cash transfer programmes in Zambia and Malawi*, PhD thesis, Cape Town: University of Cape Town.

Simone, A.-M. (2004) *For the city yet to come: Changing African life in four cities*, Durham, NC: Duke University Press.

Tyce, M. (2019) 'The politics of industrial policy in a context of competitive clientelism: The case of Kenya's garment export sector', *African Affairs*, 118(472): 553–579. https://doi.org/10.1093/afraf/ady059

Ulriksen, M. (2019) 'Pushing for policy innovation: The framing of social protection policies in Tanzania', in Hickey, S., Lavers, T., Niño-Zarazúa, M. and Seekings, J. (eds.), *The politics of social protection in eastern and southern Africa*, Oxford: Oxford University Press, pp. 122–147.

Von Gliszczynski, M. and Leisering, L. (2016) 'Constructing new global models of social security: How international organizations defined the field of social cash transfers in the 2000s', *Journal of Social Policy*, 45(2): 325–343. https://doi.org/10.1017/S0047279415000720

Whitfield, L. (2011) 'Competitive clientelism, easy financing and weak capitalists: The contemporary political settlement in Ghana', *DIIS Working Paper 2011/27*, Copenhagen: Danish Institute for International Studies.

Whitfield, L., Therkildsen, O., Buur, L. and Kjær, A.-M. (2015) *The politics of African industrial policy: A comparative perspective*, Cambridge, UK: Cambridge University Press.

Whitworth, A. (2012) 'Creating and wasting fiscal space: Zambian fiscal performance, 2002–11', *ZIPAR Working Paper 6*, Lusaka: Zambia Institute for Policy Analysis and Research.

Williams, G. (1976) 'Taking the part of peasants: Rural development in Nigeria and Tanzania', in Gutkind, P. (ed.), *The political economy of contemporary Africa*, London: Sage, pp. 131–154.

18
SOCIAL ACCOUNTABILITY IN ASIA'S LIVELIHOODS

The role of sanctions and rewards

Aries A. Arugay

Introduction

Under what conditions are societal actors able to hold political leaders accountable in protecting and promoting the livelihoods of communities they serve? How do they assert their rights through social accountability to safeguard the means of their livelihoods? Using cases from Asia, this chapter analyses the mechanisms of sanctions (enforcing accountability) and rewards (incentivising accountability) directed at governments and the private sector in livelihood areas such as education, health, and water, sanitation, and hygiene (WASH). Social accountability is defined as the ability of societal actors – nongovernmental organisations, people's associations, citizen groups, and other members of civil society, to make political leaders answerable for their decisions and/or sanction them for wrongful behaviour (Smulovitz and Peruzzotti, 2000). The chapter analyses the various ways societal actors articulate demands, negotiate terms, and/or implement effective arrangements that ensure inclusive participation, openness and transparency, and democratic accountability in public policies.

Accountability is a critical pillar in sustaining livelihoods as it directly contributes to the responsiveness of actors and institutions mandated to provide, protect and enhance the livelihoods of communities. An accountability relationship between the providers and beneficiaries of livelihoods entails not only the ability of the latter to monitor and evaluate the performance of the former. It also includes active participation in the entire process, information sharing, and intervention in the actual management of the many aspects of their livelihoods (Serrat, 2017).

The first part of the chapter discusses how the social and political mechanisms of sanctions and rewards improve social accountability in the livelihoods of communities. *Sanctions* refer to the threat and/or actual punishment because of poor performance, abuse of discretion and other errant behaviour. Holding power to account becomes a means of institutionalising distrust over the exercise of power. On the other hand, *rewards* refer to the motivations, interests, and values of those who are supposed to be held accountable for their policies and decisions. In particular, appealing to the political and social interests of politicians or bureaucrats can incentivise them to recognise social accountability efforts but also to participate in these processes that wilfully control their power (Behn, 2001; Day and Klein, 1987).

Due to the variety in political, cultural, historical and economic conditions, Asia is an appropriate site to examine social accountability efforts to ensure the livelihoods of communities. This chapter defines *livelihoods* as comprising the capabilities, assets (including both material and social resources) and activities required for a means of living. A livelihood is sustainable when it can cope with and recover from stress and shocks and maintain or enhance its capabilities and assets both now and in the future, while not undermining the natural resource base.

(Chambers and Conway, 1992: 6)

Livelihoods are at the nexus of human rights, good governance, as well as sustainability and resilience (Davies et al., 2013; Islam et al., 2014). Among the significant aspects of livelihoods in the developing world that this chapter covers are the areas of health, education and water, sanitation, and hygiene, or WASH. Numerous studies reveal that these three public services are important in the livelihoods of Asian countries (Biggs et al., 2015; Midorikawa et al., 2010; Ramesh, 2004). Moreover, many of the challenges in service delivery necessary for the livelihoods of individuals and communities are exposed through focusing on health, education and WASH. Unlike other service sectors, they simultaneously operate at multiple levels of governance and are often supported by international aid programs. All three services also entail long-term investments that makes democratic accountability challenging (Joshi, 2013).

The second part of this chapter serves as its empirical backbone. It provides a discussion of selected case studies in which societal actors have successfully sanctioned or incentivised duty bearers such as local governments and the national bureaucracy in Asia such as in the Philippines, India, Cambodia and Myanmar, among others. It identifies the modalities of accountability through sanctions and accountability through reward and highlights the issues, challenges, and lessons from their application to livelihoods of individuals and communities. By way of conclusion, the chapter identifies key lessons by carefully teasing out the contextual nuances of the experiences of social accountability for safeguarding livelihood in this part of the world.

Mechanisms of social accountability: sanctions and rewards

Scholars and practitioners alike note the close linkage between effective accountability and public services necessary for livelihood security (Lynch et al., 2013). This relationship, however, is often complex, less-than direct, multifarious, and/or difficult to observe. It becomes even more complicated if we add democratic governance to the mix, since accountability could theoretically be exercised even in authoritarian political settings (Schedler, 2013). The critical difference lies in the belief that the interface between democracy and accountability in providing public goods such as health, education and WASH offers the highest possible standard of government performance to the people. When public officials are held accountable through democratic principles, there are more chances that service provision will be faster, better, fairer, more inclusive and sustainable (Keefer, 2007).

Societal actors such as civil society organisations could play a critical role in fostering norms of accountability in the developing world (Arugay, 2005). There are four dimensions of social accountability, namely: 1) *standards:* definition of the rules, 2) *answerability:* the duty to explain and justify decisions, 3) *responsiveness:* the duty to consult people or their representatives, and 4) *enforceability:* formal and informal consequences that duty bearers may face, such as sanctions or rewards, for their actions (International IDEA, 2014). The success of any initiative by societal actors can be gauged by the extent to which it adheres to these dimensions. The complexity behind accountability relationships is partly due to the involvement of two sets of actors. *Duty bearers* are 'elected or unelected officials or private-sector service providers with the power and responsibility to fulfil a mandate and a duty to explain and justify their actions' (International IDEA, 2014: 18). *Claim holders* are citizens, civil society groups or representative institutions 'the right or the mandate to check on and question duty bearers, pass judgments on them, and impose necessary consequences when required' (International IDEA, 2014: 18).

The inadequacies and limitations of formal and institutional mechanisms of accountability were the impetus for the emergence of social accountability initiatives around the world (Fox, 2015). Through social accountability, the most vulnerable groups in society and those discriminated on the basis of identity, age, income, disability, power and sexual orientation could claim their human rights to health, education, water and other public services just like other more powerful citizens. Social accountability therefore not only denotes that there is a chance for popular control over decision-making. It also means ensuring inclusiveness in the entire process with safeguards that marginalised and disempowered groups are given equal access, opportunities, and resources to voice their demands, and participate in holding power to account (Mansuri and Rao, 2013). Apart from the possibility of more inclusive outcomes, another possible benefit of empowered societal actors demanding accountability is the habituation of exercising active and social citizenship (Gaventa and Barrett, 2012).

Sanctions: enforcing accountability

Sanctions are accountability mechanisms *par excellence*. The promise of punishment because of poor performance, abuse of discretion and other errant behaviour is one of the reasons why accountability is desired by citizens and at the same time feared by public officials. Holding power to account becomes a means of institutionalising distrust over the exercise of power (O'Donnell, 2003). Sanctions are considered to be at the culmination of any accountability process that includes successful attempts at making duty bearers *answerable* for their past decisions and actions. When government officials and service providers are required to explain and justify their behaviour, it helps citizens and political institutions acquire the necessary information to make decisions on whether to activate sanctions (Schedler, 2013).

What are the sanctions that typically revolve around accountability relations? It depends on the standards and rules set when the relationship of accountability was established. They range from the *legal* (censure, dismissal, indictment, conviction) to *political* (exit, withdrawal of vote) but can also be *social* (loss of reputation, shaming, etc.). Scholars and advocates of accountability have stressed the limitations of these types of accountability, particularly the inability of claim holders to enforce sanctions (Przeworski et al., 1999).

In many democracies, social accountability has emerged as an informal type of accountability. This vertical yet non-electoral form of accountability revolves around collective action of groups often belonging to civil society. While its efforts do not lie in the realm of formal institutions, this bottom-up approach can exact huge reputational costs through social mobilisation. Demands from this collective action could lead to the activation of political and legal accountability processes (Smulovitz and Peruzzotti, 2000). The existing literature also notes that several factors explain the sustainability of, and commitment to pursue social accountability such as regular and symmetric information exchanges, time and attention span of social actors, and a concrete action plan that involves legal-institutional venues for accountability (Ringold et al., 2012).

In many low and middle-income countries, however, rules and standards of public service delivery are either non-existent or rarely implemented. This holds true especially in fragile, post-conflict and transitioning societies (Baird, 2010). The reliance on orthodox 'Weberian' approaches that privilege formal institutions do not necessarily work in these contexts since power has shifted away from the state to unofficial non-state actors (Carpenter et al., 2012). With lack of legitimacy, institutions often fail to exercise effective accountability. In this regard, the power to sanction is heavily diminished within the formal realm. However, there is an increasing realisation that informal institutions, often dismissed as undesirable, may act as substitutes to failing authority (Unsworth, 2010). Thus, it is important to view accountability mechanisms as arrangements where formal and informal elements do not necessarily clash but in fact could co-exist and complement each other while the process of building public authority is underway (Helmke and Levitsky, 2006).

In sum, sanctions are vital mechanisms of democratic accountability. The threat of punishment for wrongdoing, inefficiency and abuse of power related to service provision has a deterring effect for governments and service providers. In order to work, however, sanctions must be embedded within a clear legal framework understood by all actors engaged in accountability relationships as well a network of functioning institutions. This is a challenge in countries with unfavourable socio-political and economic conditions.

Rewards: inducing accountability

Rewards and incentives are another means by which actor motivations, decisions and behaviour could be influenced. This realisation has become more salient in governance reforms in the past two decades as the threat of sanctions has proven to be inadequate in guaranteeing the delivery of services (UNDP, 2006). Rewards can come from different sources.

Elections as a vertical mechanism of accountability cannot be purely seen as a tool for sanctioning officials for poor service delivery. Those entrusted with the duty to safeguard the public interest can also be nudged to deliver better services through appropriate incentives. After all, voting in a democracy, in theory, also provides the opportunity to reward excellent performance through the re-election of the candidate, political party or coalition. Apart from a political environment where policy issues and incumbent performance are the driving forces behind electoral choices, voters should have adequate information on these matters in any given electoral cycle. Information about lines of responsibility for service delivery outcomes needs to be shared equally to all groups in society, especially to minority and marginalised groups. This in turn creates powerful incentives for politicians to build a credible reputation for inclusive and effective service provision, one that they may use to help them remain in their political positions (Keefer and Khemani, 2005). It is questionable though whether this set of dynamics is applicable to all kinds of public services. For example, targeted goods that are easy to implement in the short-run, such as roads, water wells, etc., might deliver more immediate electoral rewards to politicians than investment-intensive and highly transactional goods such as health care and primary education (Wild et al., 2012). This discrepancy within public services will be further discussed in the next section of this chapter.

There is more than one way for this incentive structure to fall apart. Aside from the lack of access to information and the absence of institutions that allow longer time horizons for political actors, incentives could be shaped toward allocating services in a clientelistic manner. If politicians cannot build a reputation as a comprehensive and credible provider, they can be motivated to target specific groups instead (Bell, 2011). Based on their calculations of political returns, they might also resort to allocating easily identifiable goods such as jobs for their constituents, or contracts for favoured suppliers rather than actually providing services (Booth, 2011). This is more relevant in contexts where widespread poverty and pervasive inequality significantly influences electoral outcomes. When politicians depend on low-income populations for their political careers, they are more disposed to provide dole-outs and clientelistic goods rather than provide public services that empower people to lift themselves from poverty, such as delivering universal health care and education (Mcloughlin and Batley, 2012). Democratic accountability then suffers since elections do not necessarily generate incentives for government to provide effective and equitable service delivery.

Elections are not the only way to reward politicians for effective accountability. Their cyclical nature limits dynamic interaction between democratic actors to a specific time frame but at the same time privileges an *individual* pursuit of accountability since elections simply aggregate voter preferences. Democratic accountability should strive to balance elections with other mechanisms that focus on collective decision-making and deliberative group processes among communities (El Arifeen et al., 2013).

Instead of relying on the uncoordinated actions of individual voters, political leaders mobilise their own networks when implementing public policies. These groups could potentially become supporters of accountability reform and could radically alter the dynamics between politicians whether during election campaigns or within policy processes (Fung and Wright, 2001). Instead of appealing to individuals, they align with the advocacies of existing civil society groups representing popular sectors. For example, some local leaders in Brazil opted to implement health and education programs, not because of electoral incentives, but rather because they saw these programs as capable of broadening social citizenship, a progressive goal they mutually shared with civil society (Sugiyama, 2008).

The different types of rewards are increasingly recognised in both academic and policy circles as a mechanism of democratic accountability. As a more proactive approach to generating accountability, it appeals to the interests of politicians or service providers rather than threatening them for poor performance. In a rewards framework, actors will be willing to be held accountable since it generates incentives related to their careers, reputations, material wellbeing and for private service contractors, the promise of profit and increase in income.

Social accountability and livelihoods: select Asian cases

Low- and middle-income states around the world continue to struggle to provide public services in the areas of health, education, and WASH. Moreover, the gaps in service provision between them and higher income countries continue to be wide. It can also be observed that while there are gains over the past two decades as measured in the progress in meeting the targets set by the United Nations Sustainable Development Goals (SDGs), not all countries improved equitably. The most relevant SDGs for public service provision are: 1) Goal 3: ensure healthy lives and promote wellbeing for all at all ages, 2) Goal 4: ensure inclusive and equitable quality education and promote lifelong learning opportunities for all, and 3) Goal 6: ensure availability and sustainable management of water and sanitation for all. Table 18.1 summarises the gains in these three SDGs in 2015.

Why is it difficult to exact accountability for delivering services related to health, education, and WASH? There are sector-specific reasons behind the lack of accountability in the provision of health and education compared to WASH, such as their universal nature and the relatively large investments in resources and time required. Politicians who are driven by electoral incentives but operate on shorter time horizons are also prone to relegate health and education priorities in favour of more targeted and clientelistic goods such as infrastructure and dole-out programs (Keefer and Khemani, 2005). Together with more material public goods, they are easier to provide, highly visible to the public and could be directly attributed to the politicians themselves (Nelson, 2007; Wild et al., 2012).

Some scholars argue that health and education (social services) are different from other public goods such as infrastructure since they by nature are universal goods that require huge investments in time, resources and collective effort. On the contrary, the provision of WASH is somewhat similar to infrastructure and other narrow targetable goods (Pande, 2003). In other words, health and education are transaction-intensive services that require longer time horizons and

Table 18.1 Average indicators for health, education and WASH provision around the world

UN Sustainable Development Goal	Country					
	Low income		Middle income		High income	
	1994–2003	2004–2014	1994–2003	2004–2014	1994–2003	2004–2014
Goal 3: Ensure healthy lives and promote wellbeing for all at all ages						
Life expectancy at birth, total (years)	51	57	66	69	76	78
Mortality rate, infant (per 1,000 live births)	96	66	56	39	9	7
Immunisation, measles (% of children ages 12–23 months)	53	71	73	82	90	93
Adolescent fertility rate (births per 1,000 women ages 15–19)	129	110	56	43	29	24
Maternal mortality ratio (modelled estimate, per 100,000 live births)	935	593	300	197	24	23
Goal 4: Ensure inclusive and equitable quality education and promote lifelong learning opportunities for all						
School enrolment, primary (% net)	55	76	84	90	95	96
Literacy rate, adult total (% of people ages 15 and above)	54	57	79	83	98	99
Primary completion rate, total (% of relevant age group)	43	62	83	93	95	98
Goal 6: Ensure availability and sustainable management of water and sanitation for all						
Improved water source (% of population with access)	50	60	80	88	98	99
Improved sanitation facilities (% of population with access)	19	25	51	60	95	96

Source: Author's own compilation from World Development Indicators 2015 (World Bank, 2021)

sustained resource mobilisation. These conditions might not exist in political contexts where institutions are weak, societies are divided, public order is fragile and elites are unresponsive to popular demands.

This select group of cases organised around the mechanisms of accountability are assessed on the basis of three criteria. The first is the extent to which it is *successful* in exacting accountability through either of the three mechanisms. The second is whether the case had an open and *inclusive* process that involved all relevant stakeholders. Finally, it is evaluated on whether the case is or could potentially be a *sustainable* accountability endeavour. In general, the cases offer rich qualitative evidence that balances context-specific lessons as well as comparable insights of how democratic accountability could be pursued through different mechanisms in these service sectors.

The cases display significant differences between and within Asian countries as well as in the mechanisms employed in the three public services. They also vary considerably in outcomes as different accountability arrangements could provide efficient services only for some sectors in society. Some cases might result in democratic accountability but with mediocre quality of services. Also, in some Asian cases countries, efforts to improve accountability were met with success, albeit temporary, underscoring the importance of institutionalising these initiatives. Finally, the cases reveal both the complementarity between formal and informal institutions (see Chapter 4 of this volume), traditional and contemporary, and local and external sources of accountability in some and their conflictive relationship in others.

Sanctions: responsiveness and enforceability

Sanctions cannot be an effective mechanism of accountability without accessible and reliable information. Transparency and openness of government officials and service providers are necessary for the process chain of accountability to begin. In some cases, the availability of information itself could be an adequate tool for increasing the exposure of wrongdoing or inefficiencies in service delivery that could lay the basis for possible sanctions.

Box 18.1 Complaints and responsiveness in Hyderabad's water sector

The Metropolitan Water Supply and Sewerage Board (Metro Water) in Hyderabad, India presents an example where complaint mechanisms resulted in improved service delivery. Reforms at Metro Water were of two types – reforms in customer service relations and reforms in the organisation. Prior to these reforms, customers were highly dissatisfied with Metro Water's services, as there were bottlenecks in addressing customer complaints in the repair of broken water and sewerage lines. To respond to this, Metro Water provided hotlines where these complaints can be lodged. This proved efficient and effective, as bottlenecks in addressing complaints were reduced. Citizens no longer needed to show up at their local water districts to file service complaints, and Metro Water personnel can easily be deployed to repair broken water and sewerage lines based solely on information relayed through the hotlines.

The ability to sanction bad performance of frontline workers was also transferred from local politicians to Metro Water's district managers. This resulted in better working and less politicised relationships between local politicians and frontline staff as well as stronger organisational cohesion within Metro Water.

Source: Author's own, drawing from Caseley (2003, 2006)

Dissatisfaction with service delivery could be channelled through grievance or complaints mechanisms. This tool for amplifying voice directly linked citizens with service providers. High income countries tend to have multiple service providers, and this allows complaints from users to result in compensation, better service, and/or exit toward a new provider. These options are normally unavailable in low-income countries. In Hyderabad, India for example, this complaint mechanism set up by a private water resources company helped curb corruption and improved the performance of its own workers (Caseley, 2003; see Box 18.1). In other cities in India, information technology was used to monitor the speed of action on complaints and compared them to usual turnover periods (Sirker and Cosic, 2007).

Reputational costs also prominently figure in sanction-oriented mechanisms of accountability. In India, public hearings are conducted as a means of acquiring information on the disbursement of funds for healthcare. These informal meetings, also called social audits, are capable of increasing the costs of wrongdoing and poor performance through naming and shaming campaigns. Some scholars though question the sustainability of this initiative (Singh and Vutukuru, 2010) since they rely on collective action premised on 'rude accountability' as in the case of Bangladesh (Hossain, 2009). Shame tactics lead to more adversarial accountability relations which are difficult in services that require frequent interaction.

There is a mixed record of how effective sanctions are as mechanisms of accountability. There is considerable evidence that it leads to the improvement of service delivery (Deininger and Mpuga, 2005) but more research is needed on whether these gains are sustainable and inclusive in a way that it empowers minorities and under-represented sectors of society. Governments and service providers fear sanctions if they are credibly imposed by a clear legal framework coupled with the presence of effective and impartial institutions. Inoperative sanctions tend to provide a pretence of accountability that undermines service delivery for all.

Rewards: standards and answerability

One of the strategies that the national government can provide is to foster an environment where local governments are incentivised to meet certain standards of service delivery. For example, citizen report cards are a means to determine the satisfaction of individual users carried out by independent actors such as communities and civil society organisations. This strategy hopes to expose poor performance but because of its comparative nature, it can also stimulate lagging officials and providers to improve their services (Svensson and Bjorkman, 2007). However, studies showed that their effectiveness depends on the specific indicators used for evaluation, as well as the existence of post-assessment activities such as lobbying and discussion with providers. For example, a workshop with providers of sanitation services organised by donors and nongovernmental organisations in Bangladesh helped in identifying specific performance indicators to monitor improvement (Cavill and Sohail, 2004). Incentives also change when results of citizen scorecards diverge from the self-evaluation of providers. In Andhra Pradesh, India, discussion of these differences in evaluation between service providers and recipients opened channels for collaboration, coordination and other ways of interaction (Misra, 2007). The mixed record of this accountability mechanism implies that its potential is contingent on several factors such as openness, social mobilisation and the existence of follow-up schemes (Hernández et al., 2019).

Partnerships between the government, private sector and civil society can induce rewards in terms of answerability. The case of Cambodia's primary health care sector is an example where health outcomes were improved because of the state's role of synergising linkages between profit-oriented health companies and nongovernmental organisations (NGOs) that advocate for universal health care (see Box 18.2). By contracting health services, the state utilised the private sector's greater flexibility to improve services and responsiveness to consumers. It also allowed the government to focus less on service delivery and more on roles that it is uniquely placed to carry out – such as planning, financing, and regulation (Palmer, 2000).

> **Box 18.2 Contracting NGOs in the delivery of primary health care (PHC)**
>
> The Ministry of Health devised a coverage plan to restructure and broaden the system by constructing health centres and merging smaller administrative districts into operational districts. By involving NGOs and the private sector to improve the delivery of PHC services, three approaches are used to assess how well contracting for health services works in developing countries: 1) contracting out, in which contractors have complete authority for hiring, firing, and paying staff as well as procuring drugs and supplies, 2) contracting in, where contractors provide management services within the existing district health structure and 3) comparison/control, where the existing district health management teams receive a budget supplement (as do 'contracting in' districts).
>
> Involving NGOs and the private sector was seen as a way to quickly improve services, manage the transition to the coverage plan and make up for weak district-level management. Contracting for the delivery of PHC services using a competitive bidding approach was found to be feasible, and was carried out efficiently and transparently. While the contracts and contract extensions depict rewards for the private sector, the involvement of NGOs induces a level of oversight as they can function as watchdogs incorporating a perspective outside the for-profit modus operandi.
>
> *Source*: Author's own, drawing from Loevinsohn (2000)

Myanmar's WASH sector provides another case of accountability through rewards. Cognizant of Myanmar's status as one of the least-developed countries, and its lack of sufficient resources to provide for hygiene and sanitation services, the government opted to adopt the National Sanitation Week and SocMob for hygiene through information campaigns on the importance of handwashing and the construction of sanitary latrines in their households. The primary actors are village officials and authorities, NGOs and households. A major part of the campaign is to reward 'seals of compliance' to villages that had the best sanitation and hygienic practices. In the end, survey data showed that access to sanitary means of household waste ('excreta') disposal increased from 45% in 1997 to 67% in 2001. Handwashing also increased significantly – from a mere 18% in 1996 to 43% in 2001. Through information and incentivising compliance, the citizens can report whether they met the standards set by the government (Bajracharya, 2003).

Conclusion

This chapter argued that social accountability could be implemented through the mechanisms of sanctions and rewards. It discussed this in relation to the potential of societal-led accountability initiatives in protecting and promoting livelihoods of individuals and communities. Specifically, it provided the various ways social accountability can enhance the provision of three public services – education, health, and WASH – as core components of livelihood security and resilience. Empirically, this chapter examined several cases in Asia and the successful application of various social accountability programs such as in Bangladesh.

This chapter makes analytical distinctions between the two mechanisms of social accountability, but a probe of existing case experiences shows that their boundaries are porous. The realities on the ground undoubtedly exhibit the complicated interplay between sanctions and rewards. The challenge for researchers is to capture this complexity through more longitudinal rather than single and isolated studies. Policy interventions, guided by research, should also appreciate the mutual interaction between these mechanisms. This means that accountability relationships could entail the simultaneous implementation of sanctions and the granting of rewards across multiple actors at different points in time for the protection and promotion of livelihoods of individuals and communities.

Key Points

This chapter can be summarised into the following key points:

- Sanctions are the most common accountability mechanisms in Asia. Apart from the availability of accurate and reliable information, the presence of durable and independent institutions that impose standards of behaviour is critical in enforcing sanctions. In other words, sanctions work more effectively when they have a deterrent effect, generated when formal institutions can credibly commit to enforce sanctions and if they are deemed legitimate by the citizenry.
- Cases from Asia however showed that stakeholders are implementing innovative ways to enforce sanctions despite the absence of strong institutions. However, relying on purely sanction-based mechanisms limits accountability, as it maintains a merely adversarial relationship especially when the engagements between these actors are continuous in the long-term.
- Rewards as a means of social accountability refer to the role of incentives that could shape the interests of public officials to alter their actions in favour of more efficient and inclusive service delivery outcomes. Apart from electoral incentives, some cases also highlighted the utilisation of market-oriented approaches in improving service delivery through relying on private contractors and service providers.
- Democratic accountability through rewards seems to be more salient if the profit motivation of private actors is balanced by the need of the government to ensure that services will be delivered efficiently, inclusively, and fairly to citizens.

Recommended reading

Caseley, J. (2006) 'Multiple accountability relationships and improved service delivery performance in Hyderabad City, Southern India', *International Review of Administrative Sciences*, 72(4): 531–546. https://doi.org/10.1177/0020852306070082

Fox, J.A. (2015) 'Social accountability: What does the evidence really say?', *World Development*, 72: 346–361. https://doi.org/10.1016/j.worlddev.2015.03.011

Joshi, A. (2013) 'Do they work? Assessing the impact of transparency and accountability initiatives in service delivery', *Development Policy Review*, 31(S1): 29–48. https://doi.org/10.1111/dpr.12018

References

Arugay, A.A. (2005) 'The accountability deficit in the Philippines: Implications and prospects for democratic consolidation', *Philippine Political Science Journal*, 26(49): 63–88. https://doi.org/10.1080/01154451.2005.9723491

Baird, M. (2010) 'Service delivery in fragile and conflict-affected states', *World Development Report 2011 Background Paper*, Washington, D.C: World Bank.

Bajracharya, D. (2003) 'Myanmar experiences in sanitation and hygiene promotion: Lessons learned and future directions', *International Journal of Environmental Health Research*, 13(1): S141–S152. https://doi.org/10.1080/0960312031000102903

Behn, R. (2001) *Rethinking democratic accountability*, Washington, DC: Brookings Institution Press.

Bell, C. (2011) 'Buying support and buying time: The effect of regime consolidation on public goods provision', *International Studies Quarterly*, 55(3): 625–646. https://doi.org/10.1111/j.1468-2478.2011.00664.x

Biggs, E.M., Bruce, E., Boruff, B., Duncan, J.M., Horsley, J., Pauli, McNeill, K., Neef, A., Van Ogtrop, F., Curnow, J., Haworth, B., Duce, S. and Imanari, Y. (2015) 'Sustainable development and the water–energy–food nexus: A perspective on livelihoods', *Environmental Science & Policy*, 54: 389–397. https://doi.org/10.1016/j.envsci.2015.08.002.

Booth, D. (2011) 'Towards a theory of local governance and public goods provision', *IDS Bulletin*, 42(2): 11–21. https://doi.org/10.1111/j.1759-5436.2011.00207.x

Carpenter, S., Slater, R. and Mallett, R. (2012) 'Social protection and basic services in fragile and conflict affected situations: A global review of the evidence', *Secure Livelihoods Research Consortium Briefing Paper 8*, London: Overseas Development Institute.

Caseley, J. (2003) 'Blocked drains and open minds: Multiple accountability relationships and improved service delivery performance in an Indian city', *IDS Working Paper 211*, Brighton: Institute of Development Studies.

Caseley, J. (2006) 'Multiple accountability relationships and improved service delivery performance in Hyderabad City, Southern India', *International Review of Administrative Sciences*, 72(4): 531–546. https://doi.org/10.1177%2F0020852306070082

Cavill, S. and Sohail, M. (2004) 'Strengthening accountability for urban services', *Environment & Urbanization*, 16(1): 155–170. https://doi.org/10.1177%2F095624780401600113

Chambers, R. and Conway, G. (1992) *Sustainable rural livelihoods: Practical concepts for the 21st century*, Brighton: Institute of Development Studies.

Davies, M., Béné, C., Arnall, A., Tanner, T., Newsham, A. and Coirolo, C. (2013) 'Promoting resilient livelihoods through adaptive social protection: Lessons from 124 programmes in South Asia', *Development Policy Review*, 31(1): 27–58. https://doi.org/10.1111/j.1467-7679.2013.00600.x

Day, P. and Klein, R. (1987) *Accountabilities: Five public services*, London: Tavistock.

Deininger, K. and Mpuga, P. (2005) 'Does greater accountability improve the quality of public service delivery? Evidence from Uganda', *World Development*, 33(1): 171–191. https://doi.org/10.1016/j.worlddev.2004.09.002

El Arifeen, S., Christou, A., Reichenbach, L., Osman, F.A., Azad, K., Islam, K.S., . . . & Peters, D.H. (2013) 'Community-based approaches and partnerships: Innovations in health-service delivery in Bangladesh', *The Lancet*, 382(9909): 2012–2026. https://doi.org/10.1016/S0140-6736(13)62149-2

Fox, J.A. (2015) 'Social accountability: What does the evidence really say?', *World Development*, 72: 346–361. https://doi.org/10.1016/j.worlddev.2015.03.011

Fung, A. and Wright, E.O. (2001) 'Deepening democracy: Innovations in empowered participatory governance', *Politics & Society*, 29(1): 5–41. https://doi.org/10.1177%2F0032329201029001002

Gaventa, J. and Barrett, G. (2012) 'Mapping the outcomes of citizen engagement', *World Development*, 40(12): 239–241. https://doi.org/10.1016/j.worlddev.2012.05.014

Helmke, G. and Levitsky, S. (2006) *Informal institutions and democracy: Lessons from Latin America*, Baltimore: Johns Hopkins University Press.

Hernández, A., Ruano, A.L., Hurtig, A.K., Goicolea, I., San Sebastián, M. and Flores, W. (2019) 'Pathways to accountability in rural Guatemala: A qualitative comparative analysis of citizen-led initiatives for the right to health of indigenous populations', *World Development*, 113: 392–401. https://doi.org/10.1016/j.worlddev.2018.09.020

Hossain, N. (2009) 'Rude accountability in the unreformed state: Informal pressures on frontline bureaucrats in Bangladesh', *IDS Working Paper 319*, Sussex: Institute of Development Studies.

International IDEA (2014) *Democratic accountability in service delivery: A practical guide to identify improvements through assessments*, Stockholm: International IDEA.

Islam, M.M., Sallu, S., Hubacek, K. and Paavola, J. (2014) 'Vulnerability of fishery-based livelihoods to the impacts of climate variability and change: Insights from coastal Bangladesh', *Regional Environmental Change*, 14(1): 281–294. https://doi.org/10.1007/s10113-013-0487-6

Joshi, A. (2013) 'Do they work? Assessing the impact of transparency and accountability initiatives in service delivery', *Development Policy Review*, 31(S1): 29–48. https://doi.org/10.1111/dpr.12018

Keefer, P. (2007) 'Seeing and believing: Political obstacles to better service delivery', in Devarajan, S. and Widlund, I. (eds.), *The politics of service delivery in democracies. better access for the poor*, Stockholm: Expert Group on Development Issues, pp. 42–55.

Keefer, P. and Khemani, S. (2005) 'Democracy, public expenditures, and the poor: Understanding political incentives for providing public services', *The World Bank Research Observer*, 20(1): 1–27.

Loevinsohn, B. (2000) *Contracting for the delivery of primary health care in Cambodia: Design and initial experience of a large pilot test*, Washington, DC: World Bank. Available at: https://openknowledge.worldbank.org/handle/10986/20097

Lynch, U., McGrellis, S., Dutschke, M., Anderson, M., Arnsberger, P. and Macdonald, G. (2013) 'What is the evidence that the establishment or use of community accountability mechanisms and processes improves inclusive service delivery by governments, donors and NGOs to communities?', *EPPI Report 2107*, London: EPPI Centre.

Mansuri, G. and Rao, V. (2013) *Localizing development: Does participation work?* Washington, DC: World Bank.

Mcloughlin, C. and Batley, R. (2012) *The politics of what works in service delivery: An evidence-based review*, Manchester: Effective States and Inclusive Development Research Centre.

Midorikawa, Y., Midorikawa, K., Sangsomsack, B., Phoutavan, T., Chomlasak, K., Watanabe, T., Vannavong, N., Habe, S., Nakatsu, M., Kosaka, Y., Akkhavong, K., Boupha, B., Strobel, M. and Nakamura, S. (2010) 'Water, livelihood and health in Attapeu province in Lao PDR', *Japanese Journal of Southeast Asian Studies*, 47(4): 478–498. https://doi.org/10.20495/tak.47.4_478

Misra, V., Ramasankar, P., Durga, L., Murty, JVR., Agarwal, S. and Shah, P. (2007) 'Social accountability series case study 1: Andhra Pradesh, India: Improving health services through community score cards', *Social accountability series note no. 1*, Washington, DC: World Bank. https://openknowledge.worldbank.org/handle/10986/25021

Nelson, J.M. (2007) 'Elections, democracy, and social services', *Studies in Comparative International Development*, 41(4): 79–97. https://doi.org/10.1007/BF02800472

O'Donnell, G.A. (2003) 'Horizontal accountability: The legal institutionalization of mistrust', in Mainwaring, S. and Welna, C. (eds.), *Democratic accountability in Latin America*, Oxford: Oxford University Press, pp. 34–55.

Palmer, N. (2000) 'The use of private sector contracts for primary health care: Theory, evidence, and lessons for low income and middle income countries', *Bulletin of the World Health Organization*, 78(6): 821–829. https://apps.who.int/iris/handle/10665/268154

Pande, R. (2003) 'Can mandated political representation increase policy influence for disadvantaged minorities? Theory and evidence from India', *American Economic Review*, 93(4): 1132–1151. https://doi.org/10.1257/000282803769206232

Przeworski, A., Stokes, S.C. and Manin, B. (1999) *Democracy, accountability, and representation*, Cambridge: Cambridge University Press.

Ramesh, M. (2004) *Social policy in East and Southeast Asia: Education, health, housing and income maintenance*, London: Routledge.

Ringold, D., Holla, A., Koziol, M. and Srinivasan, S. (2012) *Citizens and service delivery: Assessing the use of social accountability approaches in the human development sectors*, Washington, DC: World Bank.

Schedler, A. (2013) *The politics of uncertainty: Sustaining and subverting electoral authoritarianism*, Oxford: Oxford University Press.

Serrat, O. (2017) *Knowledge solutions: Tools, methods, and approaches to drive organizational performance*, Singapore: Springer.

Singh, R. and Vutukuru, V. (2010) 'Enhancing accountability in public service delivery through social audits: A case study of Andhra Pradesh, India', in *Accountability initiative*, New Delhi: Accountability Initiative.

Sirker, K. and Cosic, S. (2007) *Empowering the marginalized: Case studies of social accountability initiatives in Asia*, Washington, DC: World Bank.

Smulovitz, C. and Peruzzotti, E. (2000) 'Societal accountability in Latin America', *Journal of Democracy*, 11(4): 147–158. http://doi.org/10.1353/jod.2000.0087

Sugiyama, N. (2008) 'Ideology and networks: The politics of social policy diffusion in Brazil', *Latin American Research Review*, 43(3): 82–108. www.jstor.org/stable/20488151

Svensson, J. and Bjorkman, M. (2007) *Power to the people: Evidence from a randomized field experiment of a community-based monitoring project in Uganda*, Washington, DC: World Bank.

UNDP (United Nations Development Programme) (2006) *Incentive systems: Incentives, motivation, and development performance*, New York: UNDP.

Unsworth, S. (2010) *An upside down view of governance*, Brighton: Institute of Development Studies.

Wild, L., Chambers, V., King, M. and Harri, D. (2012) 'Common constraints and incentive problems in service delivery', *Working Paper 251*, London: Overseas Development Institute.

World Bank (2021) *World development indicators*. Available at: http://data.worldbank.org/data-catalog/world-development-indicators (Accessed 20 January 2021).

19
ADVOCATING FOR LIVELIHOODS THROUGH SOCIAL MOVEMENTS

Stefan Rzedzian, Margherita Scazza and Elodie Santos Vera

Introduction

In this chapter we take a social movements' perspective to understanding how livelihoods are promoted and defended in the Global South. A social movements' perspective provides a valuable way for looking at livelihoods due to the particular role of social movements in shaping social, cultural and political relations both from the grassroots level but also across scale. Geographically, we centre this chapter in Latin America, and thematically we centre this chapter on issues pertaining to rights. For decades, social movements in Latin America have advocated for the preservation, mobilisation and recovery of livelihoods through grassroots struggles for rights. We focus our discussion on three key areas of rights struggles: territorial rights, environmental rights and cultural rights. Importantly, we stress the interconnected nature of these different struggles and the multidimensionality of social movements, demonstrating this throughout the chapter. Conceptually, we place these discussions within a framework of coloniality, understood as a matrix of power that exerts control over social relations and political structures that is increasingly resisted and contested by social movements. We ground the chapter in a series of case studies from across Latin America, primarily drawing on the experiences of various Indigenous and *campesino* social movements in their efforts to preserve, mobilise and recover livelihoods.

We begin with a discussion of territorial rights, considering how social movements in Latin America have engaged over time with place-based forms of mobilisation in their engagements with livelihoods. Here we draw attention to the ways in which socio-territorial movements have evolved as a result of ever-changing multi-scalar political landscapes. Next, we turn our attention to environmental rights, discussing how social movements have mobilised against extractivism, and in favour of issues such as water rights in order to preserve and defend their livelihoods. Finally, we discuss the promotion and defence of cultural rights, specifically in Indigenous politics, and consider how the social construction of Indigeneity has played a central role in shaping and re-shaping social movements' claims to cultural rights, and their intersection with Indigenous livelihoods. Underpinning our approach is an understanding of livelihoods as enabled not only by access to material resources, such as land and water, but also by cultural self-determination and collective identity preservation.

Advocating for territorial rights

Most social movements engage with space and develop across it through their mobilisations and claims. However, for many social movements in Latin America, space (particularly land and territory) constitutes a central dimension of struggle. The history of the continent is one marked by colonial dispossession of land and plundering of resources. Territorial struggles are as rooted in Latin American history as much as coloniality – as the set of 'long-standing patterns of power that emerged as a result of colonialism' – is (Maldonado-Torres, 2007: 243). The peoples inhabiting the region that is today referred to as America have been resisting the theft of their lands and their epistemicide from the beginning of the Conquest (Santos, 2016). Indigenous and peasant communities' exclusion and exploitation, which underpin the colonial system, continue to be reflected in the high degree of land concentration and in the highest land Gini coefficient in the world (Oxfam International, 2016).

Over the course of the 20th century, social movements advocating for land rights and redistribution were formed by peasant communities demanding agrarian reforms across the region. Thanks to their pressure, many countries experienced important social transformations and implemented a range of agrarian reforms (Teubal, 2009). Nevertheless, conditions of inequality and social injustice persisted, leading to a strategic shift in focus from land claims to territorial claims. Mobilising the notion of territory as both 'an analytical and empirical unit', social movements started emphasising the multiplicity of territorialities existing within national states (Haesbaert, 2013; Pahnke et al., 2015: 1075). Territory is thereby reinvented and redefined as a socially constructed object, beyond colonial and modernist notions cantered around the state's sovereignty (Porto-Gonçalves and Leff, 2015). A conceptual object which peasant and Indigenous movements have employed 'to merge the importance of control over land and resources with questions of racism and dispossession' (Bryan, 2012: 216). Demands for the recognition of land rights are articulated in terms of claims over property, which leave the state's socio-spatial order unquestioned. Conversely, within the so-called 'territorial turn', Latin American social movements have been advocating for a change in the way territory is understood and organised, proposing alternative forms of territorialisation, while asserting their identity and their rights to difference (Offen, 2003). Encompassing land and a variety of other resources, as well as human and non-human subjects – or 'earth-beings' – the ideas of territory that these alternatives seek to enact are founded on radically distinct ontologies (de la Cadena and Blaser, 2018). Concrete examples are, among others, the Embera people's 'vertical territoriality', described by Ulloa (2015), or the model of 'integral territories' advocated by the Awajún and Wampis peoples (Garra and Gala, 2014).

The image of a mosaic, evoked by Ronaldo Munck (2020), perfectly reflects the array of labour, peasant, Indigenous, place-based, feminist and environmental groups currently involved in struggles for territory in the region. Building on Fernandes' notion of socio-territorial movements, we provide here an overview of the most emblematic movements and their struggles (Fernandes, 2005). Halvorsen et al. (2019) have described socio-territorial movements as those for whom territory is a central objective and strategy, and whose mobilisation for the production of territory contributes to the creation of collective identities, new values and new political subjectivities. Three main moments can be identified when considering the recent history of socio-territorial movements in Latin America: agrarian reforms, the territorial turn and the most recent eco-territorial turn.

Agrarian reforms

Since the beginning of the 20th century, peasants and rural movements have been calling for a redistribution of land, highly concentrated in the hands of a few big landholders and at the root of widespread social inequalities. The rise of strong and extensive social movements in the second half of the century, like the Zapatista Army of National Liberation (EZLN) in Mexico and the Landless Rural Workers' Movement (MST) in Brazil, pushed governments to pass substantial agrarian reforms, whose success and implementation varied greatly from country to country (Teubal, 2009). These reforms were gained through a diverse range of strategies of resistance, including uprisings and armed rebellions, as well as land occupations. These land struggles have generally been analysed through structuralist lenses and considered as class struggles, rather than through a focus on the identity of their members (Petras and Veltmeyer, 2001). New agrarian movements which emerged in the 1990s, however, face new antagonising forces and actors, set in motion by globalisation and neoliberal processes affecting agricultural production. Their land struggles, as illustrated by Bebbington et al. (2008), have turned into struggles for a 'territorial question', whereby communities mobilise on the basis of their identity to oppose the neoliberal agenda and regain legal access to their lands, while simultaneously resisting cultural repression and racial discrimination (Veltmeyer, 2019).

The territorial turn

Identity-based socio-territorial movements, such as the Indigenous and Afro-descendant ones, have been central actors within the phenomenon that Offen (2003) has termed the 'territorial turn'. Representing an unprecedented trend in land titling, the large-scale recognition of Indigenous and Afro-descendants' collective land rights which characterised the 1990s in Latin America, is recognised as the product of a unique historical conjuncture (Offen, 2003). Forces traditionally antagonistic, such as new identity and place-based social movements, development and non-governmental organisations (NGOs) and international financial institutions converged upon the legal demarcation of 200 million hectares of communities' territories (Bryan, 2012; Offen, 2003). In a context of neoliberal expansion and rise of multiculturalism, the promotion of equality was operated through the recognition of ethnic and cultural rights, permitted as long as contained within the boundaries of the neoliberal political project (Di Giminiani, 2018; Hale, 2005). This trend was reflected in a number of constitutional reforms, e.g. in Colombia (1991) and Ecuador (1998), which, along with the protection of Indigenous peoples and Afro-descendant communities' rights to territory and to consultation, emphasised the *plurinational* and *multicultural* nature of their societies (Correia, 2019). The World Bank's ethnodevelopment approach, which promoted development through an apparent celebration of difference and the advancement of communities' land security, also figures among the forces underpinning the territorial turn (Bryan, 2012). Despite providing a site for communities to challenge conventional territorialities and carve out autonomous spaces, as occurred with the *Proceso de Comunidades Negras* (PCN) documented by Escobar (2010a) in Colombia, the recognition of their territorial claims is still deeply subordinated to modernist colonial logics (Halvorsen, 2019). For this reason, critics of the territorial turn, like Wainwright (2008) and Bryan (2012: 221), call for 'an entirely new spatial ontology' based on the deconstruction of the concept of territory and its political implications. Territorial claims and movements for the defence of alternative territorialities, however, remain central to current struggles, for the first time bringing together a wide variety of activist groups.

The eco-territorial turn

Since collective land titles did not guarantee the application of appropriate mechanisms for Free Prior and Informed Consent (FPIC), nor greater control over natural resources and autonomy, territorial and environmental conflicts have not stopped proliferating. Although natural resource extraction is not a new phenomenon in the region, over the past two decades it has greatly intensified, assuming a different form. As a development model founded on 'the appropriation of natural resources in large volumes and/or high intensity, where half or more are exported as raw materials, without industrial processing or with limited processing', extractivism has engendered new conflicts for the production of territory (Bebbington, 2007; Gudynas, 2018: 62). These circumstances have led to an unprecedented convergence of ideological matrices and resistances opposing destructive extractive projects. The eco-territorial turn, as Maristella Svampa (2019) called it, sees the formation of alliances between different movements concerned with the defence of territory, the commons and the lifeworlds associated to them. If the territorial turn encompassed ethnic movements, the eco-territorial narrative also mobilises environmentalist and feminist movements, enabling the creation of 'a new common language of valuation on territoriality' as well as an epistemological dialogue through which movements can identify shared challenges and diversify their strategies (Svampa, 2019: 40). Indeed, never before have socio-territorial movements been as interconnected as they are now.

Advocating for environmental rights

Social movements have a long history of advocating for rights around environmental issues, including issues of water security (Boelens et al., 2010), pollution (Bridge, 2004), and, as just discussed, land and territorial rights. To speak of 'environmental rights' is to speak of a broad range of issues which cannot be boiled down to one single category or used as an individual analytical tool. For example, social movement struggles over the environment in the context of livelihoods are most often grounded in struggles over place (Escobar, 2010b), that can materialise across highly diverse circumstances, ranging from issues such as lack of political representation to struggles for preservation of heritage and everyday ways of being and knowing (Martinez-Alier, 2014). Broadly framed as 'environmental conflicts' (Guha and Alier, 2013), struggles over environmental rights reflect complex social and political relations across time, place, and scale. As Alier (2014) argues, these conflicts are embedded within exploitative international power relations, dictated by global capitalism's necessity for the commodification of non-human nature at the expense of local populations' health, wellbeing and agency. Furthermore, the resource and capital flows that contextualise these environmental conflicts in the Global South manifest as examples of the continual effects that colonialism has on the world in which we live, where economically poorer populations are exploited at the behest of private interests and dominant global powers. Therefore, and as scholars such as Williams and Mawdsley (2006) and Doyle and Chaturvedi (2010) have discussed extensively, social movements in the Global South that mobilise for the promotion and defence of environmental rights are often also engaged in decolonial and anti-colonial struggles that reach far beyond environmental or resource-based issues.

Extractivism

As touched on in the previous section on territorial rights, a key factor in many social movement mobilisations in Latin America (and indeed the Global South more broadly) pertain to the extraction of natural resources, particularly resources such as petroleum and minerals. Anti-oil and anti-mining agendas have often typified social movement mobilisations in countries such as Ecuador, Peru and Bolivia, due to these states relying heavily on income generated by extractive industries to fund large-scale national development and social welfare programs (Arsel et al., 2016). Extractive activities have resulted in the violation of human rights, manifesting through environmental issues such as the poisoning of populations through polluted waterways, and the harassment and intimidation of local populations by extractive corporations (Davidov, 2013).

In Ecuador in particular, this conflict between extractivism, environmental rights, and development is exceptionally visible in the context of the government's developmental philosophy of *Buen Vivir* (Living Well), as articulated within the country's 2008 constitution. While one of the integral facets which is intended to underpin the concept of Buen Vivir is living 'in harmony with nature' (manifesting through the simultaneous constitutionalisation of 'rights of nature' in the country), the Ecuadorian state has struggled to reconcile this with its pursuit of economic development, fuelled by oil drilling and increasingly mining. A predominantly student-led social movement known as the *Yasunídos* arose in response to the radically contradictory standpoint held by the government in its decision to drill for oil in the Yasuní National Park, one of the most biodiverse places in the world, in the name of pursuing its development framework of Buen Vivir. Focusing on issues of biodiversity (and by extension, the newly constitutionalised rights of nature) as well as Indigenous rights, relating to those communities who live in the Yasuní National Park, the Yasunídos mobilise a conceptualisation of environmental rights that binds together a fundamentally anti-extractivist ethos with matters of territory, identity, and biodiversity, capable of being communicated across a variety of scales, places, and cultures (Apostolopoulou and Cortes-Vazquez, 2018). The Yasunídos conduct activism both on the streets, in political institutions, and online, however the vast majority of their activism is based on social media. For this reason, their primary method of resistance has been critiqued as having limited reach, given that access to the internet and social media is by no means universal in Ecuador (Coryat, 2015: 3751). However, given their social media power, their movement spread beyond Ecuador, with Yasunídos Facebook pages emerging in places such as The Netherlands, the UK, and the United States. In this sense, the power of social media enabled the Yasunídos to scale-up their movement, and bring their issues to a global audience, reminiscent of Dwivedi's 'globalisation of environmental protest' (Dwivedi, 2001: 21).

As extractivism continues to shape the socio-environmental spaces of livelihoods across the world, it is important that we continue to analyse the ways in which environmental rights are impacted, and the political, social, and cultural dynamics of resistance and advocacy manifested by social movements in response.

Water

The politics of water offers another useful locus at which many issues around environmental rights can be synthesised in the context of livelihoods. As Boelens et al. (2010) note, water is deeply embedded within processes of life, culture and power, across both material and symbolic social relations. Social movements in Latin America have been highly influential in navigating, negotiating and shaping the politics of water in many of their countries (Terhorst et al., 2013). Water-related issues include matters such as pollution, access, privatisation and territory, and may often also pertain to particular spiritual or cosmological politics, especially in the cases of many Indigenous groups and communities (Zimmerer, 2000).

As we have already discussed matters of pollution and territory in the *Extractivism* and *Territorial Rights* sections of this chapter, we take this opportunity now to consider how privatisation has influenced livelihoods through shaping and denying access to water, and how this has been resisted by social movements. Water politics in the Global South is intricately intertwined with histories and present continuations of (neo)colonialism, neoliberalism, development and globalisation. Across much of Latin America, water privatisation is synonymous with the late 1990s and early 2000s wave of neoliberalism, fuelled by the hyper-commodification of non-human nature, rollback of the state, and *laissez-faire* economic rationality that deified the 'invisible hand of the market' and demonised public subsidies of even the most basic of goods and services (Harris and Roa-García, 2013).

Social movements across Latin America have mobilised against water privatisation in a variety of ways, including forms of direct action such as marches and protest, legal challenges, as well as the monitoring of private firms' activities on the ground (Borgias, 2018; Olivera et al., 2004). These mobilisations can demonstrate challenges to the state, corporations, or international institutions. Consequently, scholars have often analysed social movements' resistance in the context of scale, particularly in the sense of how resistance is mobilised across scale from the local, to the state, and the global through a series of 'advocacy networks' (Dupuits, 2019). In this sense, social movement mobilisations against water privatisation have frequently been interconnected with transnational alter-globalisation networks of resistance that have drawn on the lived-experience of those whose livelihoods have been dismantled or eroded as a result of water privatisation, driven by global market demands (Laurie and Crespo, 2007) (see Box 19.1).

Box 19.1 Contesting neoliberalism in Bolivia: the Cochabamba water wars

Arguably the most famous of water struggles in Latin America to date were the Cochabamba Water Wars in Bolivia. Between the years 1999 and 2000, Indigenous and campesino social movements mobilised against the privatisation of the municipal water supply in Cochabamba. Bolivia had been undergoing a long series of structural adjustment policies which had seen the country's resources become increasingly privatised and sold off to multi-national corporations and placed further and further out of financial reach for vast swathes of the country's population. When the government enacted new laws to privatise all water, even rainwater, mass protests erupted in the city of Cochabamba. Social movements rejected the new laws, citing water as a 'social and ecological good that guarantees the wellbeing of the family, the collective, and social and economic development' (Assies, 2003). Consequently, the protection of water was directly linked to the protection of livelihoods through a lens of human rights and social justice, predicated on a communitarianism that rejected the rampant individualism espoused by neoliberal frameworks of social organisation.

Water politics continue to be fertile ground for social movement mobilisations in the Global South, as livelihoods remain under threat from processes of privatisation, pollution and increasingly desertification as a result of climate change.

Advocating for cultural rights

Indigenous social movements and their allies have forged strong pathways toward the advocation of cultural rights in their efforts to defend and recover livelihoods. Cultural rights include broad issues related to identity, ways of knowing and ways of being, such as language rights, the right to self-government, cultural autonomy and various other social practices that constitute particular forms of cultural distinctiveness from settler colonial societies (de la Cadena and Starn, 2007). In recent years, Indigenous social movements have mobilised across geographical scales and through transnational networks in order to shape understandings of matters pertaining to Indigenous cultural rights across the world. In doing this, they have continued to engage in discussions that explore, recontextualise and redefine what it means to be 'Indigenous' (Davis, 2008). This dialogue produces and constructs meaningful resources and pathways through which the defence and recovery of livelihoods can occur. As we explore in this section, the construction and mobilisation of Indigeneity as identity and cultural political artefact has had significant impacts on the ways in which Indigenous peoples have engaged with, and been engaged by, settler colonial societies and institutions.

In order to understand how social movements have advocated for Indigenous cultural rights in Latin America over the years, we must first understand the context within which Indigeneity as a form of identity and as a concept has shifted and changed over time. The notion of what it means to 'be Indigenous' has a problematic and deeply political history across much of the Global South; always diametrically opposed to that which is 'not Indigenous' (i.e. settler colonial societies), Indigeneity as a concept and signifier of identity has always existed as a comparison to colonisers through a binary of difference and opposition. Consequently, the notion of being Indigenous is always regarded in relative terms, and functions often in relation to that which it is *not* (de la Cadena, 2010).

It is crucial to remember that classifications as to who is and who is not Indigenous have often been implemented through colonial systems and institutions in a top-down and culturally oppressive fashion (Postero, 2013). More recently, it is through a recovering of this process of identification that some Indigenous social movements strive to establish a core platform through which other claims pertaining to livelihoods can be made (Canessa, 2007). This is why social movement influence in spaces such as the United Nations has been significant in promoting systems of self-identification as to whether one is Indigenous or not (Escárcega, 2010).

From mestizaje to multiculturalism to plurinationality

In Latin America particularly, cultural rights frameworks have been defined by the political projects of *mestizaje*, *neoliberal multiculturalism* and finally *plurinationality*. Mestizaje (mixing) saw Indigenous populations and settler-colonial societies engaging in intercultural and/or interracial mixing, which, for their descendants, led to the boundaries between who is Indigenous and who is not to become blurred. The underlying ethos behind mestizaje was to produce a singular, homogenous society – a cultural melting pot with just one ethnicity. However, the process of mestizaje has been critiqued for eroding away Indigenous culture, in favour of the settler-colonial ways of being and ways of knowing which flourished under modernisation (Chaves Chamorro and Zambrano Escobar, 2006). Consequently, during the 1990s and early 2000s, many countries sought to embrace the notion of neoliberal multiculturalism, espousing instead a philosophy of one society with different cultures (Indigenous and non-Indigenous). However, neoliberal multiculturalism was predicated on the homogenisation of Indigenous cultures into one monolithic experience, history and character, and while opening up new spaces of political participation for Indigenous populations, it also functioned as a way to govern and discipline them (Hale, 2002).

Along with the broader waves of anti-neoliberal resistance across Latin America, some countries, particularly Ecuador and Bolivia, saw issues of Indigenous cultural rights be placed at the forefront of wider social movement mobilisations. Consequently, both Ecuador and Bolivia ushered in a new form of political project known as plurinationality, enshrined in each country's respective constitutions. In contrast to the projects of mestizaje and neoliberal multiculturalism before it, plurinationality is predicated instead on issues of *autonomy* rather than *recognition*, and on *rights* rather than *governance* (Albro, 2010). Therefore, frameworks of plurinationality instead endeavour to embolden Indigenous populations through greater levels of agency and self-determination, a central tenet of Indigenous social movements' efforts to advocate for cultural rights. This has come to manifest around issues such as the greater inclusion of Indigenous languages in state education (Howard, 2009), as well as the recognition of rights-based frameworks of protection for Indigenous languages (Haboud et al., 2016). However, plurinational projects in the Andes have been critiqued for struggling to represent *truly plural* legal, political and cultural realisations of Indigeneity and Indigenous livelihoods, instead favouring certain communities, cultures and populations over others (Albro, 2010; Becker, 2012). Consequently, many social movements recognise that significant effort still remains in the defence and promotion of Indigenous cultural rights, even in countries where much progress has been made.

Conclusion

In this chapter we have discussed how social movements across Latin America have mobilised for the preservation and recovery of livelihoods through grassroots struggles for rights. Grounding our analysis in a broad understanding of livelihoods that goes beyond material conditions, we have examined how social movements' strategies for territorial rights have evolved and adapted over time in order to reject colonial constraints. Similarly, environmental movements have been contesting extractivism and the commodification of nature as threats to livelihoods driven by global capitalism. Finally, we have presented how movements struggling for peoples' self-determination and right to difference have engaged strategically with notions of Indigeneity and plurinationality to protect and pursue autonomous livelihoods.

Key Points

- Territorial rights, environmental rights, and cultural rights are often deeply intertwined with one another.
- Social movement mobilisations are often deeply embedded within interconnected processes of colonialism, neoliberalism, and development, both in historical and contemporary senses.
- Rights-based issues form a crucial platform upon which social movements can advocate for the defence and preservation of livelihoods.

Further reading

Albro, R. (2010) 'Confounding cultural citizenship and constitutional reform in Bolivia', *Latin American Perspectives*, 37(3): 71–90. https://doi.org/10.1177/0094582X10364034

Boelens, R., Getches, D. and Guevara-Gil, A. (2010) *Out of the mainstream: Water rights, politics and identity*, New York: Routledge.

Munck, R. (2020) 'Social movements in Latin America: Paradigms, people, and politics', *Latin American Perspectives*, 47(4): 20–39. https://doi.org/10.1177/0094582X20927007

References

Albro, R. (2010) 'Confounding cultural citizenship and constitutional reform in Bolivia', *Latin American Perspectives*, 37(3): 71–90. https://doi.org/10.1177/0094582X10364034

Apostolopoulou, E. and Cortes-Vazquez, J.A. (2018) *The right to nature: Social movements, environmental justice and neoliberal natures*, New York: Routledge.

Arsel, M., Hogenboom, B. and Pellegrini, L. (2016) 'The extractive imperative and the boom in environmental conflicts at the end of the progressive cycle in Latin America', *The Extractive Industries and Society*, 3(4): 877–879. https://doi.org/10.1016/j.exis.2016.10.013

Assies, W. (2003) 'David versus goliath in Cochabamba: Water rights, neoliberalism, and the revival of social protest in Bolivia', *Latin American Perspectives*, 30(3): 14–36. https://doi.org/10.1177/0094582X03030003003

Bebbington, A. (2007) 'Elementos para una ecología política de los movimientos sociales y el desarrollo territorial en zonas mineras', in Bebbington, A. (ed.), *Minera, movimientos sociales y respuestas campesinas*, Lima: IEP, CEPES, pp. 23–46.

Bebbington, A., Abramovay, R. and Chiriboga, M. (2008) 'Social movements and the dynamics of rural territorial development in Latin America', *World Development*, 36(12): 2874–2887. https://doi.org/10.1016/j.worlddev.2007.11.017

Becker, M. (2012) 'Building a plurinational Ecuador: Complications and contradictions', *Socialism and Democracy*, 26(3): 72–92. https://doi.org/10.1080/08854300.2012.710000

Boelens, R., Getches, D. and Guevara-Gil, A. (2010) *Out of the mainstream: Water rights, politics and identity*, New York: Routledge.

Borgias, S.L. (2018) '"Subsidizing the state": The political ecology and legal geography of social movements in Chilean water governance', *Geoforum*, 95: 87–101. https://doi.org/10.1016/j.geoforum.2018.06.017

Bridge, G. (2004) 'Contested terrain: Mining and the environment', *Annual Review of Environment and Resources*, 29: 205–259. https://doi.org/10.1146/annurev.energy.28.011503.163434

Bryan, J. (2012) 'Rethinking territory: Social justice and neoliberalism in Latin America's Territorial Turn', *Geography Compass*, 6(4): 215–226. https://doi.org/10.1111/j.1749-8198.2012.00480.x

Canessa, A. (2007) 'Who is indigenous? Self-identification, indigeneity, and claims to justice in contemporary Bolivia', *Urban Anthropology and Studies of Cultural Systems and World Economic Development*, 36(3): 195–237. www.jstor.org/stable/40553604

Chaves Chamorro, M. and Zambrano Escobar, M. (2006) 'From blanqueamiento to reindigenización: Paradoxes of mestizaje and multiculturalism in contemporary Colombia', *European Review of Latin American and Caribbean Studies*, 80: 5–23. https://doi.org/10.18352/erlacs.9652

Correia, J.E. (2019) 'Unsettling territory: Indigenous mobilizations, the territorial turn, and the limits of land rights in the Paraguay-Brazil Borderlands', *Journal of Latin American Geography*, 18(1): 11–37. https://doi.org/10.1353/lag.2019.0001

Coryat, D. (2015) 'Extractive politics, media power, and new waves of resistance against oil drilling in the Ecuadorian Amazon: The case of Yasunidos', *International Journal of Communication*, 9: 3741–3760.

Davidov, V. (2013) 'Mining versus oil extraction: Divergent and differentiated environmental subjectivities in "post-neoliberal" Ecuador: Mining versus oil extraction', *Journal of Latin American and Caribbean Anthropology*, 18(3): 485–504. https://doi.org/10.1111/jlca.12043

Davis, M. (2008) 'Indigenous struggles in standard-setting: The United Nations declaration on the rights of indigenous peoples commentary', *Melbourne Journal of International Law*, 9: 439–471.

de la Cadena, M. (2010) 'Indigenous cosmopolitics in the Andes: Conceptual reflections beyond "politics"', *Cultural Anthropology*, 25(2): 334–370. https://doi.org/10.1111/j.1548-1360.2010.01061.x

de la Cadena, M. and Blaser, M. (2018) *A world of many worlds*, Durham: Duke University Press.

de la Cadena, M. and Starn, O. (2007) *Indigenous experience today*, New York: Routledge.

De Sousa Santos, B. (2016) *Epistemologies of the South – justice against epistemicide*, New York: Routledge. https://doi.org/10.1111/j.1468-5906.2008.00423.x

Di Giminiani, P. (2018) 'Indigenous activism in Latin America', in Di Giminiani, P. (ed.), *Routledge handbook of Latin American development*, New York: Routledge, pp. 225–235. https://doi.org/10.4324/9781315162935-20

Doyle, T. and Chaturvedi, S. (2010) 'Climate territories: A global soul for the Global South?', *Geopolitics*, 15(3): 516–535. https://doi.org/10.1080/14650040903501054

Dupuits, E. (2019) 'Water community networks and the appropriation of neoliberal practices: Social technology, depoliticization, and resistance', *Ecology and Society*, 24(2): 20. https://doi.org/10.5751/ES-10857-240220

Dwivedi, R. (2001) 'Environmental movements in the Global South: Issues of livelihood and beyond', *International Sociology*, 16(1): 11–31. https://doi.org/10.1177/0268580901016001003

Escárcega, S. (2010) 'Authenticating strategic essentialisms: The politics of indigenousness at the United Nations', *Cultural Dynamics*, 22(1): 3–28. https://doi.org/10.1177/0921374010366780

Escobar, A. (2010a) *Territorios de diferencia: Lugar, movimientos, vida, redes*, Popayán, Colombia: Universidad del Cauca. https://doi.org/10.2307/j.ctvpv504m

Escobar, A. (2010b) *Una minga para el postdesarrollo: lugar, medio ambiente y movimientos sociales en las transformaciones globales*, Del. Coyoacán, Ciudad de México: Universidad Nacional Mayor de San Marcos, Programa Democracia y Transformación Global. http://bdjc.iia.unam.mx/items/show/46

Fernandes, B. (2005) 'Movimentos socioterritoriais e movimentos socioespaciais: Contribuição teórica para uma leitura geográfica dos movimentos sociais', *Revista Nera*, 6(8): 14–34. https://doi.org/10.47946/rnera.v0i6.1460

Garra, S. and Gala, R. (2014) 'Por el curso de las quebradas hacia el "territorio integral indígena": Autonomía, frontera y alianza entre los awajún y wampis', *Anthropologica*, 32(32): 41–70. https://revistas.pucp.edu.pe/index.php/anthropologica/article/view/9443

Gudynas, E. (2018) 'Extractivisms: Tendencies and consequences', in Munck, R. and Wise, R.D. (eds.), *Reframing Latin American development*, London: Routledge, pp. 61–76.

Guha, R. and Alier, J.M. (2013) *Varieties of environmentalism: Essays north and south*, New York: Routledge.

Haboud, M., Howard, R., Cru, J. and Freeland, J. (2016) 'Linguistic human rights and language revitalization in Latin America and the Caribbean', in Coronel-Molina, S. and McCarty, T. (eds.), *Indigenous language revitalization in the Americas*, New York: Routledge.

Haesbaert, R. (2013) 'A global sense of place and multi-territoriality', in Featherstone, D. and Painter, J. (eds.), *Spatial politics: Essays for Doreen Massey*, Oxford: John Wiley & Sons.

Hale, C. (2005) 'Neoliberal multiculturalism: The remaking of cultural rights and racial dominance in Central America', *Political and Legal Anthropology*, 28(1): 10–28. www.jstor.org/stable/24497680

Hale, C.R. (2002) 'Does multiculturalism menace? Governance, cultural rights and the politics of identity in Guatemala', *Journal of Latin American Studies*, 34(3): 485–524. https://doi.org/10.1017/S0022216X02006521

Halvorsen, S. (2019) 'Decolonising territory: Dialogues with Latin American knowledges and grassroots', *Progress in Human Geography*, 43(5): 790–814. https://doi.org/10.1177/0309132518777623

Halvorsen, S., Fernandes, B.M. and Torres, F.V. (2019) 'Mobilizing territory: Socioterritorial movements in comparative perspective', *Annals of the American Association of Geographers*, 109(5): 1454–1470. https://doi.org/10.1080/24694452.2018.1549973

Harris, L.M. and Roa-García, M.C. (2013) 'Recent waves of water governance: Constitutional reform and resistance to neoliberalization in Latin America (1990–2012)', *Geoforum*, 50: 20–30. https://doi.org/10.1016/j.geoforum.2013.07.009

Howard, R. (2009) 'Education reform, indigenous politics, and decolonisation in the Bolivia of Evo Morales', *International Journal of Education Development*, 29(6): 583–593. https://doi.org/10.1016/j.ijedudev.2008.11.003

Laurie, N. and Crespo, C. (2007) 'Deconstructing the best case scenario: Lessons from water politics in La Paz – El Alto, Bolivia. Geoforum, Pro-Poor Water?', *The Privatisation and Global Poverty Debate*, 38(5): 841–854. https://doi.org/10.1016/j.geoforum.2006.08.008

Maldonado-Torres, N. (2007) 'On the coloniality of being: Contributions to the development of a concept', *Cultural Studies*, 21(2–3): 240–270. https://doi.org/10.1080/09502380601162548

Martinez-Alier, J. (2014) 'The environmentalism of the poor', *Geoforum*, 54: 239–241. https://doi.org/10.1016/j.geoforum.2013.04.019

Munck, R. (2020) 'Social movements in Latin America: Paradigms, people, and politics', *Latin American Perspectives*, 47(4): 20–39. https://doi.org/10.1177/0094582X20927007

Offen, K.H. (2003) 'The territorial turn: Making black territories in Pacific Colombia', *Journal of Latin American Geography*, 2(1): 43–73. https://doi.org/10.1353/lag.2004.0010

Olivera, O., Lewis, T. and Chiva, V. (2004) *¡Cochabamba! Water war in bolivia*, Boston: South End Press.

Oxfam International (2016) *Unearthed: Inequality, land, power in Latin America*, Oxford: Oxfam International.

Pahnke, A., Tarlau, R. and Wolford, W. (2015) 'Understanding rural resistance: Contemporary mobilization in the Brazilian countryside', *Journal of Peasant Studies*, 42(6): 1069–1085. https://doi.org/10.1080/03066150.2015.1046447

Petras, J. and Veltmeyer, H. (2001) 'Are Latin American peasant movements still a force for change? Some new paradigms revisited', *Journal of Peasant Studies*, 28(2): 83–113. https://doi.org/10.1080/03066150108438767

Porto-Gonçalves, C.W. and Leff, E. (2015) 'Political ecology in Latin America: The social re-appropriation of nature, the reinvention of territories and the construction of an environmental rationality', *Desenvolvimento E Meio Ambiente*, 35: 65–88. https://doi.org/10.5380/dma.v35i0.43543

Postero, N. (2013) 'Introduction: Negotiating indigeneity', *Latin American Caribbean Ethnic Studies*, 8(2): 107–121. https://doi.org/10.1080/17442222.2013.810013

Svampa, M. (2019) *Neo-extractivism in Latin America socio-environmental conflicts, the territorial turn, and new political narratives*, Cambridge: Cambridge University Press. https://doi.org/10.1017/9781108752589

Terhorst, P., Olivera, M. and Dwinell, A. (2013) 'Social movements, left governments, and the limits of water sector reform in Latin America's left turn', *Latin American Perspectives*, 40(4): 55–69. https://doi.org/10.1177/0094582X13484294

Teubal, M. (2009) 'Agrarian reform and social movements in the age of globalization: Latin America at the dawn of the twenty-first century', *Latin American Perspectives*, 36(4): 9–20. https://doi.org/10.1177/0094582X09338607

Ulloa, A. (2015) 'Environment and development: Reflections from Latin America', in Perreault, T., Bridge, G. and McCarthy, J. (eds.), *Routledge handbook of political ecology*, Abingdon, UK: Routledge, pp. 320–331.

Veltmeyer, H. (2019) 'Resistance, class struggle and social movements in Latin America: Contemporary dynamics', *Journal of Peasant Studies*, 46(6): 1264–1285. https://doi.org/10.1080/03066150.2018.1493458

Wainwright, J. (2008) *Decolonizing development: Colonial power and the Maya, Antipode*, New York: Wiley.

Williams, G. and Mawdsley, E. (2006) 'Postcolonial environmental justice: Government and governance in India', *Geoforum*, 37(5): 660–670. https://doi.org/10.1016/j.geoforum.2005.08.003

Zimmerer, K.S. (2000) 'Rescaling irrigation in Latin America: The cultural images and political ecology of water resources', *Ecumene*, 7(2): 150–175. https://doi.org/10.1177/096746080000700202

20
EDUCATION AND LIVELIHOODS IN THE GLOBAL SOUTH

Vikas Maniar and Meera Chandran

Introduction

Preparing for future livelihoods is one of the important goals of education. This chapter discusses how education policies and practices in the Global South address this aim. Most education policies are influenced by Human Capital Theory (HCT) that sees education as an investment in future returns for the individual through improved earnings, as well as for the society in terms of greater economic growth (Psacharopoulos, 1981). This focus manifests itself through various strategies adopted by the state to enable economic growth, including a focus on basic literacy, numeracy and life skills in primary schools, vocational and technical education in secondary schools and beyond, and academic higher education for administrative jobs in public and private sectors. Often, however, the aspirations of students and their families foreground the social mobility promised through education. This is particularly true for students from poor and socially marginalised social groups. The realisation of this goal is uneven, disadvantaging students from poor and marginalised backgrounds. Specifically, children from marginalised communities attend poorer-quality schools and disproportionately drop out, making it difficult to participate in formal employment (Bhatty and Namala, 2014).

This chapter begins by introducing HCT and its critique in the context of livelihoods in the Global South. Next, it discusses the educational aspirations and experiences of individuals and communities during their education in the Global South. After this, it discusses alternative conceptions of the education-livelihoods relationship that address the challenges identified earlier in the chapter. Lastly, the conclusion outlines the key learnings from these discussions.

Human capital theory

Until the late 1960s, economists treated spending on education as consumption. The pioneering work of Becker (1962), Schultz (1959) and Mincer (1958) established education as an investment that people and societies make to reap future earnings. These returns accrue as the skills acquired through education allow for better productivity in the workplace. At the personal level, an investment in education is expected to result in higher future earnings, whereas at the societal level, it is expected to accelerate economic growth. Subsequently, private and social rates of returns from education were empirically established (Psacharopualos and Hinchliffe, 1973).

Even when intangible (non-economic) benefits of education were discounted from such calculations, these returns were significant enough to motivate policy interest in education. As such, HCT is now a dominant framework for policymaking for countries in the Global South and has highlighted education as a key mechanism for the economic growth of societies.

The education-livelihoods relationship in HCT has been conceptualised in two different ways. In a more general conception, education is believed to enable the development of skills and dispositions that allow students to be productive in industrial and service sector jobs. This includes the acquisition of foundational skills like literacy and numeracy, and the shaping of dispositions to work in 'factory-like' environments. More recently, the discourse surrounding 21st century skills illustrates the need to train the future workforce to adapt to rapidly emerging information and communication technologies (Dede, 2010). In a stricter conception, the so-called skills discourse suggests that vocational, technical and professional education imparts skills that are directly useful for employment in select industries and, hence, useful for acquiring these jobs.

Close to a decade after the initial euphoria about HCT, Blaug (1976) suggested that the human-capital research programme was in 'crisis'. Many of Blaug's criticisms of HCT are particularly salient in postcolonial societies. 'According to human-capital theory, the labor market is capable of continually absorbing workers with ever higher levels of education, provided that education-specific earnings are flexible downwards' (Blaug, 1976: 845). This assumption of 'unlimited absorption of labour' is untenable in postcolonial economies that already have high rates of unemployment or underemployment due to the structure of their economies. In such a context, education predominantly acts as a signalling mechanism and a focus on education results in credentialisation of jobs (Dore, 1976; Stiglitz, 1975) rather than improving the material well-being of the educated. Reports of thousands of overqualified people applying for a handful of jobs in countries such as India are commonplace. The typical response to this by the proponents of HCT is to attribute this unemployment to a 'skills mismatch' between the demand and supply of labour. However, Steven J. Klees (quoted in Vally and Motala, 2014: vii) asserts,

> unemployment is not a worker supply problem, but a structural problem of capitalism. There are two or more billion un- or under-employed people on this planet, not because they don't have the right skills, but because full employment is neither a feature nor a goal of capitalism.

School completion certificates and higher education qualifications are often a prerequisite to be considered for formal employment. Thus, the poor completion rates prevent a large proportion of youth from accessing these job markets. Even for those who manage to access and complete higher education, claims of unemployability of college graduates because of 'skill deficits' in higher education are common in popular and academic reports. For example, an annual report on skills and employability in India claimed that only about 46.21% of students studying in higher education institutions in 2020 were employable (CII, 2021). Similarly, in South Africa, a Department of Higher Education and Training (DHET) report asserts that of those employed, 28% were underqualified for their job and 32% were mismatched by their field of study (DHET, 2019). As these reports usually take the point of view of the employers seeking candidates for open job positions, they hide the structural constraint on employment where the labour market is unable to absorb educated workers. While post-school education improved the prospect of employment in South Africa, in India, the unemployment rate for college-educated youth was roughly three times the national average as a result of minimal formal job opportunities (SSA, 2019; Basole, 2018). In early 2017, the West Bengal (a state in India) government held an examination for 6000 jobs in the Class IV or Group D category, the lowest category of permanent employment in government service. Two and a half million people took the exam, many of them holders of graduate and postgraduate degrees (Basole, 2018). In contrast, a college degree may improve job prospects in South Africa, although the completion rates for higher education were very low. In both cases, many youths remained unemployed or underemployed.

School systems in the Global South

Impressive gains were made by countries over the last few decades in achieving near-universal enrolment in schools. However, the challenge of achieving the Sustainable Development Goal 4 (SDG4) 'to ensure all girls and boys, complete free equitable and quality, primary and secondary education for all' is far from over. In countries of the Global North, the risk from educational choices typically manifests in the post-schooling phase, as schooling is universal and largely uniform. In the Global South, unequal schooling experiences are the norm because school choices are made in a setting of school diversity, resource constraints and information scarcity (Verger et al., 2016).

In the final decade for achieving the SDG4 targets, there are as many as 64 million students of primary school age, or 9% of children, reported to be out of school globally. Low-income countries have a much higher percentage of children out of school at the primary levels (Sub-Saharan Africa – 21%) and secondary levels (Central and South Asia – 17%) (UNESCO, 2018). Further, there are issues with the enumeration of data on children out of school. For instance, the figures for out of school children in India is estimated to be anywhere from 6 to 20 million according to one pilot study (CPR, 2016). As the study notes, the lack of a uniform definition, diversity of school administration systems and mechanisms of data collection, lead to problems in estimation. In addition, children out of school are also most likely to belong to highly vulnerable groups such as street children, child labourers and children of parents in stigmatised occupations such as waste pickers and sex workers. Children out of school in the 6 to 13 years age group is the highest for those with disabilities (34%), followed by Muslims (7.67%), Dalits (5.96%), Adivasis (5.6%) and girls (4.71%) (Bhatty and Namala, 2014).

However, ensuring that more children enter school is not the end of the road. Retaining them in school once enrolled, ensuring their transition from one stage to the next, and enabling successful school completion is equally challenging. In India between 2015 and 2016, 27.1 million children were enrolled in grade one, while the enrolment in grade 12, the final year of schooling, had dropped to 7 million[1] (Ministry of Education, 2021). The equivalent numbers for South Africa from 2016 were 1.2 million in grade one and 704,500 in grade 12 (Basic Education, 2016). Dropping out is a structural phenomenon given its higher incidence among rural, lower caste and female children in India (UNICEF, 2014). A large proportion of the increase in primary school enrolments is comprised of first-generation school attendees of historically marginalised social groups susceptible to poor school participation, absences and eventual dropout. Cultural factors such as early marriage and school-related factors such as harassment and bullying disproportionately affect school dropout rates among female students of marginalised communities (Prakash et al., 2017) while systemic push factors operate within school systems leading to early dropout of specific groups of children (Govinda and Bandyopadhyay, 2010). Studies point towards the importance of understanding parental notions of learning and value on early employment as a strategic life choice in high-risk contexts (Balagopalan, 2008).

Access to higher education is even lower in the Global South because of low matriculation pass rates and students' financial constraints. Secondary school completion rates for 25 to 35 years olds in India is 22% and the higher education completion rate is 14%. The same figures for South Africa are 77% and 6% respectively (OECD, 2020). The effects of poor school quality are reflected in dismal rates of acquisition of literacy and numeracy for elementary school students. In 2016, one in three children in grade eight was unable to read at a first grade level, and 43.3% were unable to solve a simple division problem (ASER, 2017). Similarly, 2014 Annual National Assessment reports in South Africa indicate that only 63% of grade 6 learners demonstrated grade-level learning outcomes for their first language. This number for mathematics was 43% (Basic Education, 2016).

Following the Jomtien world declaration on Education for All (EFA) in 1990, as the state set out to achieve UEE, there was a parallel expansion of private provisioning in school education. Stratification of schooling systems in the Indian context, although partly a colonial legacy, may be attributed to the privatisation of the sector. Education stratification mirrors the social stratification with students from economically weaker sections attending poorly resourced government schools and the privileged classes attending elite private schools (Vasavi, 2019). A study on public financing of education in South Asia shows that although state provisioning for public education increased as a proportion of the budget, it is significantly low from a spending-per-pupil point of view. The authors note that nearly one-third of children in primary or secondary school in the region attend a private school (Dundar et al., 2014a). This trend is reflected in the receding enrolment in government schools in India (Kingdon, 2019) leading to several state governments moving to rationalise provisioning and resources by closing down schools with low enrolments (Rao et al., 2017). This is a particularly worrying trend from a social equity perspective because private schools are unaffordable for a majority of poor and low caste communities and religious minorities (Härmä, 2009).

According to a recent review of the Indian school system, only 11% of students in the lowest wealth quintile are likely to even reach the secondary level. One reason for this is because 15% of secondary schools currently enrolling students are in the private sector and out of reach for the poor (Mehendale and Mukhopadhyay, 2020). Equity concerns in education are not limited to increased privatisation of schooling but understood as a phenomenon of the commodification of education through low-cost private schooling for the poor to tap into the education market across the Global South (Nambissan, 2014; Verger et al., 2016). A global education reform movement constituted by principles of large-scale assessment, tests-based accountability, standards and decentralisation are promoted as cheaper policy alternatives to 'profound structural equity reforms' (Verger et al., 2019: 24). Studies of school systems in South Asia focus on how the underperformance of students limits their chances for higher productivity and economic growth, and make recommendations about improving learning outcomes, teacher effectiveness and fiscal accountability (Dundar et al., 2014b; Pritchett and Aiyar, 2014). The state's failure to address structural barriers to educational equity gets compounded by excessive focus on outcomes, efficiency and fiscal accountability.

The structural issue in education today has been termed a crisis of confidence in the purpose of schooling (Dow et al., 1999) owing to the social stratification of access, differentiated experience of students and the stratified nature of social mobility. A crisis of schooling has far reaching implications for the democratic and social justice ideals of education (Ellis et al., 2019), responses to which must be founded on a thorough understanding of the micro-processes at the school level along with corresponding policy attention to address these factors. Changes in social structures and relations must be understood to execute appropriate curricular changes responsive to students at risk (Morgan, 2014). The 'capacity to aspire' as Appadurai (2004) puts it, operates at the cultural rather than individual level, is constrained by material conditions and diminished by limitations in opportunities to navigate to a better future. It follows that the systemic lack of educational opportunities for the poor and marginalised constrain their very capacity to aspire for social mobility, let alone equitable participation in the economy and polity.

The cases that follow (Boxes 20.1 and 20.2) illustrate the interplay of economic contexts and the state of school systems in shaping the livelihood opportunities for youth in the Global South.

Box 20.1 Livelihood strategies in rural India

I studied in the village Lower Primary School (grades 1 to 5) for grades 1 and 2 but did not do well because I was distracted. I started again in grade 1 in an Ashram Shala (a residential school) that was far away from my village. I studied there 'til grade 7, after which I joined a residential school run by the Tribal Development Department in Chhota Udepur for my 8th grade. For my secondary education (from grade 9), I joined SF High School (a popular day school in Chhota Udepur) while I lived in a student hostel because that was the only place where I could pursue the science stream. For higher secondary schooling, I attended the Don Bosco residential school in Chhota Udepur. There, I failed my 12th exam on the first attempt but passed it in a supplementary exam. While I was pursuing my secondary education, I started developing my English language skills through private tuition and interactions with my teachers who were fluent in the language. After 12th grade, I joined the local college in Chhota Udepur and majored in Chemistry and Maths, but after I failed Maths in my first year, I switched to Chemistry and Biology. After obtaining my bachelor's degree, I joined a master's degree college in Ahmedabad but had to drop out as it was financially unviable. Then I joined the local teacher's training college for my B.Ed. (Bachelor of Education), as it had better job prospects at the time. I passed the degree with good grades, but by this time another competitive test (Teacher Eligibility Test, TET) was introduced for recruitment for a teacher's job. I took the test but failed on my first attempt.

Now with a degree but without a job, I tried my hand at multiple things. I tried joining a master's course again but that did not work out. Then I tried my hand at a job with an emergency medical response unit but soon got overwhelmed with the stress of the job. After this, I got a job as a contract teacher in a model school run by the Tribal Development Department. While I was at this job, I cleared the TET in my second attempt but now the recruitment for government teachers had stopped. I also got a government job with the state police, which I quit on the very first day as the job would be very boring. My family and community ridiculed me for abandoning the much-coveted government job. Finally, in the next round of recruitment of teachers in government higher primary schools, I managed to get a regular government teacher's job. I now want to complete my master's from an open university, take an eligibility test for the headteacher position (HTAT), and eventually become a headteacher in a primary school.

Source: adapted from Maniar (2019)

Box 20.2 Livelihood strategies in rural South Africa

When I was a child, my father worked in the mines in Joburg. Later when my father was unwell, my mother joined him there. In primary school, my sister and I were in the same class. My sister was a smart student, but I struggled with academics. My sister and I stayed with a lady who was nice to us, so I could bunk school whenever I wanted. I failed grade four while my sister passed.

In primary school, I was inspired by a radio programme to become a pilot but by the time I was in grade 11, I realised that it would not work out for me. I was struggling with academics and had to change subjects. I lost another year because of this. In the meantime, my sister passed matric before I did. By the time I was in matric, my sister had left for college in Port Elisabeth. I managed my home with my younger brother at that time as my mother was away for initiation school for traditional healers (sangoma). I had to cook and take care of cattle because I was the only adult in the house and could not do well in exams. I passed matric in 2007 but failed to get a bursary. By this time my father had retired, so he asked me to work to support my sister financially for her higher education. We expected her to support me financially for tertiary education when she started earning. Unfortunately, I lost my sister to an unknown medical condition. Her son now stays with us. My younger brother is now working as a driver in PE (Port Elizabeth). He dropped out of school in grade 11 (standard 9).

I found a paid apprenticeship at a hotel to train as a waiter but could not get employment. Then I went to Joburg to work on construction sites. First, I worked as a manual labourer but later obtained a contract job as a machine operator. When my contract expired, I went to Cape Town to look for better opportunities. I struggled for two years there and could not find work. I came back in 2011 and bought the bush cutter machine. Now I run a vegetation management business. I also plough a small portion of our garden, but cannot plough the whole land because of the drought. We grow cabbage, spinach, beetroot, cauliflower, broccoli but the produce is just enough for our consumption. Other youths are not interested in going back to agriculture. I do not want to go back to the city to find work based on my two-year experience in Cape Town, although I keep applying for jobs from here. I have now registered my business. Last year I earned a three-month tender for lawn care from my church. I have now started a computer/business management course. I have started to make a business plan. I am inspired by President Zuma who never went to school but is now our president.

Source: adapted from Maniar (2019)

Alternative conceptions of education – livelihoods relationship

Alternate models of schooling that respond to the economic, cultural, social and political contexts of communities have been employed with various degrees of success. Nationalist leaders such as Gandhi, Nyerere and Senghor, proposed national education systems that were sensitive to the contexts of their respective countries after their independence from colonial rule. These were often based on pre-colonial indigenous systems of education, reformed to respond to the present contexts of their societies (see for example Kumar, 1993). Gandhi's educational vision was shaped by his imagination of a good society where village republics had a prominent place. He imagined a village that was largely political and economically autonomous, but interdependent with other villages when necessary (Gandhi, 1989). In his conception, each child would receive seven years of free, compulsory basic education in their mother tongue. This would be centred on some form of manual and productive work, most likely a craft related to meeting the basic needs of the people. The vision for New Education, similar to efforts by visionary nationalist leaders in other countries (Nyerere's Education for Self-Reliance, for example; see Nyerere, 1967), was based on the rejection of received wisdom from the west and a re-imagining of a good society that was sensitive to local contexts. In particular, it recognised the primacy of serving the needs of rural communities through education. These experiments have now succumbed to the hegemonic models of Western schooling.

In recent years, the critique of HCT prompted alternative ways of conceptualising the education-development interlinkage through the Human Rights and Human Development approach. These approaches view the role of education as being about more than the instrumentality of labour participation and acknowledge the constitutive role of education in preparing people for a dignified and good life. The Education for All (EFA) declaration in Jomtien, Thailand, demanded that basic learning needs of all the children, youth and adults be met.

> These needs comprise both essential learning tools (such as literacy, oral expression, numeracy, and problem-solving) and the basic learning content (such as knowledge, skills, values, and attitudes) required by human beings to be able to survive, to develop their full capacities, to live and work in dignity, to participate fully in the development, to improve the quality of their lives, to make informed decisions, and to continue learning.
>
> *(Inter-Agency Commission, 1990 [Article 1.1])*

The Dakar Framework for Action 2000 spelt out the strategies for meeting these EFA goals. It called for a free and compulsory primary education of good quality for all children by 2015. The EFA declaration provides the normative force that drives the universalisation of primary education across the world. The central role of education in enhancing opportunities for a dignified life has resulted in the inclusion of educational achievements in development indicators such as the Human Development Index (HDI), Millennium Development Goals (MDG) and Sustainable Development Goals (SDG). The justiciable obligation on states to enable these universal rights facilitate conditions for a dignified life for a majority of the population and enable access to dignified livelihoods. However, in the absence of adequate resource availability and the executive capacity of the state, proclamations for these rights become rhetorical. This is particularly the case for states in the Global South that are resource-strapped and often lack the capacities to uniformly enforce these rights for all their citizens.

The Capabilities Approach (Sen, 1999; Nussbaum, 2011) provides another useful lens to understand the education-livelihoods relationship. It focuses on expanding the substantive opportunities for people to lead a life that they value, which includes creating opportunities for decent livelihoods through education. Basic education is identified as one of the basic capabilities in Sen's theorisation of the capabilities approach. As argued by Nussbaum (2011), education has a central role in forming 'internal capabilities' that lead to enhanced employment opportunities, political participation, and an ability to interact productively with others in society. The emphasis on the complementarity of individual agency with societal arrangements in the Capabilities Approach (Sen, 1999: xii) suggests that education should respond to the societal context to expand substantive opportunities of material wellbeing. This need for such contextualisation is particularly salient in the Global South.

Conclusion

Education as a means for livelihood translates into education policies influenced to a great extent by the rationale of human capital theory. In this conception, successful education is an investment for gainful individual employment and future economic growth. The limited capacity of the system for absorbing an ever-increasing workforce calls into question the fundamental goal of the capitalist system, particularly in the post-colonial Global South. The structure of labour markets and high rates of unemployment in the economies of the Global South reveal the struggle for livelihood among the youth, most of whom fall back on their traditional livelihoods or remain in the uncertain informal sector. Understanding the intersectionality of education and livelihoods prompts examination of structural inequities evident from data on educational access and quality. Rates of children kept out of school, dropping out of school or failure to successfully complete schooling are disproportionately high among marginalised groups in the Global South. Their capacity to aspire for education as a pathway for social mobility and livelihood options are severely curtailed by systemic factors and educational processes. Education tends to reproduce inequalities among marginalised groups and an absence of social capital curtails aspirations. This calls for alternative ways of conceptualising the education-livelihoods relationship in the Global South. The efforts by the nationalist leaders in the wake of decolonisation have been subdued by the dominant modes of schooling. The focus on the rights-based approach and the capabilities approach to education in recent years is promising but needs to be developed further to address the contexts of the Global South.

Key points

The following key points emerge from this discussion on education and livelihoods in Global South.

- Human Capital Theory (HCT), a dominant conceptualisation of education and livelihoods interlinkages, is premised on assumptions that do not necessarily hold true in the Global South. For example, instead of unlimited absorption of educated workforce into the formal economy as assumed in HCT, these economies are characterised with a high rate of unemployment of educated youth.
- At the same time, access to livelihoods through education is a key aspiration in the Global South. However, educational systems in these societies fall short of preparing youth to realise this aspiration, particularly for marginalised groups.
- This highlights the need for alternate ways of conceptualising the education-livelihoods interlinkages. There are some promising developments working towards this goal, but significant work still remains to be done.

Further reading

Basole, A. (2018) *State of working India 2018*, Bangalore: Azim Premji University.
Nelson Mandela Foundation (2005) *Emerging voices: A report on education in South African rural communities*, Cape Town: HSRC Press.
Robeyns, I. (2006) 'Three models of education: Rights, capabilities and human capital', *Theory and Research in Education*, 4(1): 69–84. https://doi.org/10.1177/1477878506060683

Notes

1 While accurate estimations of school dropout rates should be based on panel data, yearly snapshots provide a good indication of what these rates might be.

References

Appadurai, A. (2004) 'The capacity to aspire: Culture and the terms of recognition', in Rao, V. and Walton, M. (eds.), *Culture and public action: A cross-disciplinary dialogue on development policy*, Stanford: Stanford University Press.
ASER Center (2017) *Annual status of education report 2016 – Rural*, New Delhi: ASER Center. http://img.asercentre.org/docs/Publications/ASER%20Reports/ASER%202016/aser_2016.pdf
Balagopalan, S. (2008) 'Memories of tomorrow: Children, labour, and the panacea of formal schooling', *Journal of the History of Childhood and Youth*, 1(2): 267–285. https://doi.org/10.1353/hcy.0.0005
Basic Education, Department of (2016) *Report on progress in the schooling sector against key learner performance and attainment indicators*, Pretoria: Republic of South Africa.

Basole, A. (2018) *State of working India 2018*, Bangalore: Centre for Sustainable Employment, Azim Premji University. https://cse.azimpremjiuniversity.edu.in/state-of-working-india/swi-2018/

Becker, G.S. (1962) 'Investment in human capital: A theoretical analysis', *Journal of Political Economy*, 70(5, Part 2): 9–49. https://doi.org/10.1086/258724

Bhatty, K. and Namala, A. (2014) *India exclusion report 2013–14*, Bangalore: Books for Change. www.indianet.nl/pdf/IndiaExclusionReport2013-2014.pdf

Blaug, M. (1976) 'The empirical status of Human Capital Theory: A slightly jaundiced survey', *Journal of Economic Literature*, 14(3): 827–855. www.jstor.org/stable/2722630

Centre for Policy Research (CPR) (2016) *A pilot study of estimating out-of-school children in India*, New Delhi: Centre for Policy Research. http://uis.unesco.org/sites/default/files/documents/a-pilot-study-of-estimating-out-of-school-children-in-india-2016-en.pdf

Confederation of Indian Industry (CII) (2021) *India skill report 2021*, New Delhi: Confederation of Indian Industry. www.cii.in/PublicationDetail.aspx?enc=tqg4pk6AY1Uj+SwRfF02NhuILvTKQ6CEkGTvjhN8hYA=

Dede, C. (2010) 'Comparing frameworks for 21st century skills', in Bellanca, J. and Brandt, R. (eds.), *21st century skills: Rethinking how students learn*, Bloomington: Solution Tree Press.

Dore, R. (1976) *The diploma disease: Education, qualification and development*, Berkeley: University of California Press.

Dow, A., Hattam, R., Reid, A., Shacklock, G. and Smyth, J. (1999) *Teachers' work in a globalizing economy*, 1st ed., Abingdon: Routledge. https://doi.org/10.4324/9780203979693

Dundar, H., Béteille, T., Riboud, M. and Deolalikar, A. (2014a) 'Financing for quality education', in Beteille, T., Deolalikar, A.B., Dundar, H. and Riboud, M. (eds.), *Student learning in South Asia: Challenges, opportunities, and policy priorities*, Washington, DC: World Bank, pp. 261–293. https://doi.org/10.1596/978-1-4648-0160-0_ch7

Dundar, H., Béteille, T., Riboud, M. and Deolalikar, A. (2014b) 'What and how much are students learning?', in Beteille, T., Deolalikar, A.B., Dundar, H. and Riboud, M. (eds.), *Student learning in South Asia: Challenges, opportunities, and policy priorities*, Washington, DC: World Bank, pp. 85–128. https://doi.org/10.1596/978-1-4648-0160-0_ch2

Ellis, V., Souto-Manning, M. and Turvey, K. (2019) 'Innovation in teacher education: Towards a critical re-examination', *Journal of Education for Teaching*, 45(1): 2–14. https://doi.org/10.1080/02607476.2019.1550602

Gandhi, M.K. (1989) *Hind Swaraj: Or, Indian home rule*, Ahmedabad: Navajivan Publishing House.

Govinda, R. and Bandyopadhyay, M. (2010) 'Social exclusion and school participation in India: Expanding access with equity', *Prospects*, 40(3): 337–354. https://doi.org/10.1007/s11125-010-9160-8

Härmä, J. (2009) 'Can choice promote education for all?: Evidence from growth in private primary schooling in India', *Compare*, 39(2): 151–165. https://doi.org/10.1080/03057920902750400

Higher Education and Training, Department of (DHET) (2019) *Skills supply and demand in South Africa*, Pretoria: Department of Higher Education and Training. www.dhet.gov.za/SiteAssets/Report%20on%20Skills%20Supply%20and%20Demand%20in%20South%20Africa_%20March%202019.pdf

Inter-Agency Commission (1990) *World declaration on education for all and framework for action to meet basic learning needs*, Jomtien: Inter-Agency Commission (UNDP, UNESCO, UNICEF, World Bank).

Kingdon, G.G. (2019) 'Trends in private and public schooling', in Mamgain, R.P. (ed.), *Growth, disparities and inclusive development in India: Perspectives from the Indian state of Uttar Pradesh*, Singapore: Springer. https://doi.org/10.1007/978-981-13-6443-3_15

Kumar, K. (1993) 'Mohandas Karamchand Gandhi', *Prospects*, 23(3–4): 507–517. https://doi.org/10.1007/BF02195132

Maniar, V. (2019) *Education and wellbeing: A case study in postcolonial contexts*, Doctoral dissertation, Mumbai: Tata Institute of Social Sciences.

Mehendale, A. and Mukhopadhyay, R. (2020) 'School system and education policy in India', in Sarangapani, P. and Pappu, R. (eds.), *Handbook of educational systems in South Asia*, Singapore: Springer, pp. 1–35. https://doi.org/10.1007/978-981-13-3309-5_13-1

Mincer, J. (1958) 'Investment in human capital and personal income distribution', *Journal of Political Economy*, 66(4): 281–302. https://doi.org/10.1086/258055

Ministry of Education (2021) *Enrolment by social category*, New Delhi: Government of India. Available at: http://dashboard.seshagun.gov.in/#/reportDashboard/sReport (Accessed 13 June 2021).

Morgan, J. (2014) 'Michael Young and the politics of the school curriculum', *British Journal of Educational Studies*, 63(1): 5–22. https://doi.org/10.1080/00071005.2014.983044

Nambissan, G.B. (2014) 'Poverty, markets and elementary education in India', *TRG Poverty and Education Working Paper Series 3*, Bonn: Max Weber Stiftung. https://cprindia.org/sites/default/files/events/Geetha%20Nambissan%20paper.pdf

Nussbaum, M.C. (2011) *Creating capabilities: The human development approach*, Cambridge: Belknap Press of Harvard University Press.

Nyerere, J.K. (1967) 'Education for self-reliance', *The Ecumenical Review*, 19(4): 382–403. https://doi.org/10.1111/j.1758-6623.1967.tb02171.x

Organization for Economic Cooperation and Development (OECD) (2020) *Education at a glance 2020*, Paris: OECD Publishing. https://doi.org/10.1787/69096873-en

Prakash, R., Beattie, T., Javalkar, P., Bhattacharjee, P., Ramanaik, S., Thalinja, R., Murthy, S., Davey, C., Blanchard, J., Watts, C., Collumbien, M., Moses, S., Hesie, L. and Isac, S. (2017) 'Correlates of school dropout and absenteeism among adolescent girls from marginalized community in North Karnataka, South India', *Journal of Adolescence*, 61: 64–76. https://doi.org/10.1016/j.adolescence.2017.09.007

Pritchett, L. and Aiyar, Y. (2014) 'Value subtraction in public sector production: Accounting versus economic cost of primary schooling in India', *Working Paper No. 391*, Washington, DC: Center for Global Development. https://dx.doi.org/10.2139/ssrn.2623073

Psacharopoulos, G. (1981) 'Returns to education: An updated international comparison', *Comparative Education*, 17(3): 321–341. https://doi.org/10.1080/0305006810170308

Psacharopoulos, G. and Hinchliffe, K. (1973) *Returns to education: An international comparison*, Amsterdam: Elsevier Scientific Publishing Company.

Rao, S., Ganguly, S., Singh, J. and Ranu Dash, R. (2017) *School closures and mergers: A multi-state study of policy and its impact on public education system – Telangana, Odisha and Rajasthan*, New Delhi: Save the Children.

Schultz, T.W. (1959) 'Investment in man: An economist's view', *Social Service Review*, 33(2): 109–117. https://doi.org/10.1086/640656

Sen, A. (1999) *Development as freedom*, 1st ed., New York: Knopf.

Statistics South Africa (SSA) (2019) *Labour market dynamics in South Africa, 2019*, Pretoria: Republic of South Africa.

Stiglitz, J.E. (1975) 'The theory of screening, education, and the distribution of income', *The American Economic Review*, 65(3): 283–300. https://doi.org/10.7916/D8PG22PM

United Nations Educational, Scientific and Cultural Organization (UNESCO) (2018) *Global education monitoring report 2019: Migration, displacement and education: Building bridges, not walls*, 2nd ed., Paris: UNESCO Publishing. https://doi.org/10.18356/22b0ce76-en

United Nations International Children's Emergency Fund (UNICEF) (2014) *Global initiative on out-of-school children: A situational study of India*, New Delhi: UNICEF. http://uis.unesco.org/sites/default/files/documents/out-of-school-children-south-asia-country-study-education-2014-en.pdf

Vally, S. and Motala, E. (eds.) (2014) *Education, economy & society*, Pretoria: Unisa Press.

Vasavi, A.R. (2019) 'School differentiation in India reinforces social inequalities', *The India Forum*. Available at: www.theindiaforum.in/article/school-differentiation-india-reinforcing-inequalities (Accessed 29 May 2021).

Verger, A., Lubienski, C. and Steiner-Khamsi, G. (2016) 'The emergence and structuring of the global education industry: Towards an analytical framework', in Verger, A., Lubienski, C. and Steiner-Khamsi, G. (eds.), *World yearbook of education 2016: The global education industry*, New York: Routledge. https://doi.org/10.4324/9781315720357

Verger, A., Parcerisa, L. and Fontdevila, C. (2019) 'The growth and spread of large-scale assessments and test-based accountabilities: A political sociology of global education reforms', *Educational Review*, 71(1), 5–30. https://doi.org/10.1080/00131911.2019.1522045

21
YOUTH LIVELIHOODS
Negotiating intergenerationality and responsibility

Sukanya Krishnamurthy

Youth: between ages and generations

Children and young people living in informal areas of the Global South play multiple roles in the household. Their various roles, such as contributing to the household (money, time, etc.) and caring responsibilities, are influenced by an array of cultural, economic and social factors. Across the world, 'youth' is conceptualised as the period of transition between childhood dependence and adult independence. This chronological conceptualisation of age has been problematised and discussed within childhood and youth studies for a few decades now. In particular, authors such as Laz (1998: 90) have questioned the sociology of age:

> how age as a concept and institution is created, maintained, challenged, and transformed; how assumptions and beliefs about age in general and about particular age categories inform and are reinforced by social statuses, norms, roles, institutions, and social structures; and how age patterns individual lives and experiences even as individuals accomplish age.

Other examples include the work of Scott (1998), who questioned chronological age as one of state simplification, Twum-Danso Imoh and Ansell (2013), who posed questions around rights, and Hopkins and Hill (2010), who worked on positioning migrant status. What several of these authors and others have highlighted is that the chronological conceptualisation of age has been cemented by the work of several state actors, international agencies (see Table 21.1) and non-governmental organisations. In response to this, scholars such as Clark-Kazak (2009) and Huijsmans (2013) used the term 'social age', defined as 'the socially constructed meanings applied to physical development and roles attributed to infants, children, young people, adults and elders, as well as their intra- and inter-generational relationships' (Clark-Kazak, 2009: 1310). Throughout this chapter, the term 'social age' is used because its definition aligns well with the multiple roles that youth perform in and outside of the household, as well as their intergenerational position within the household.

Another term that needs further reflection and has gained significant attention in the last decade is 'generation'. Though age and generation are sometimes used interchangeably, generation is useful to understand how societies are structured on the basis of age-based groupings (Thorne, 2004). Age is understood as chronological (based on calendar years), social (roles

DOI: 10.4324/9781003014041-24

Table 21.1 Summary of different definitions of what age ranges constitute 'youth' based on various UN agencies and other organisations

Entity/instrument/organisation	Age	Reference
UN Secretariat/UNESCO/ILO	Youth: 15–24	UN Instruments
UN Habitat (Youth Fund)	Youth 15–32	Agenda 21
UNICEF/WHO/UNFPA	Adolescent: 10–19, Young People: 10–24, Youth: 15–24	UNFPA
UNICEF /The Convention on Rights of the Child	Child until 18	UNICEF
The African Youth Charter	Youth: 15–35	African Union

Source: Author

attributed to people in a household) and relative (where one is older or younger). Hopkins and Pain (2007: 291) for example call for 'more relational geographies of age'. Authors such as Punch (2015), Evans (2015), and Xu (2015) identified that intrahousehold relations need to be understood through existing generational dimensions and the concept of 'inter-generational contracts'. Inter-generational contracts can be understood as the flexible practices, knowledge and resources shared between generations (parent-child relations) that can take on various forms across the world (see work of Huijsmans, 2016; McGregor et al., 2000). For this chapter, it is important to position the wide range of roles and responsibilities that children and youth take on. I do this by focusing on their position in the household and their contribution to livelihoods through the notion of inter-generational contracts rather than age-based roles.

Framed from a Global South perspective, the chapter explores the multiple ways through which the discourse around youth livelihoods transitions between age, rights, responsibilities and roles in their communities. It begins with an examination of the roles and responsibilities that young people and children undertake both in and around the household. Next, the discussion focuses on young people's mobility and how this shapes their access to livelihoods. Lastly, the chapter ends with an outline of examples of the strategies used to address the challenges associated with young people's livelihoods and suggestions for avenues of research that should be the focus of future research.

Roles, responsibilities and intergenerationality

Within the literature, young people from a very young age are treated as active, competent individuals who are expected to shoulder several roles and responsibilities in order to maintain the functioning of a household (Chatterjee, 2012; Sultana, 2019). Household composition, birth order and gender all have an influence over the kind of responsibilities that children and young people undertake. Beall (2002) stressed the importance of understanding the complex web of social relations within the household and the community when analysing urban livelihoods for young people and children. The role of intra-household dynamics, intergenerational relationships, gender, age and family size all impact on the lives of those living within a household, thus they all require consideration and understanding while exploring livelihoods.

In a study looking at children living in the slums of Delhi, children were given diaries to record their daily routines (Chatterjee, 2012). In one example, a child picked up milk in the morning to make tea for her family, went to school, then came home to watch over her younger siblings, cook and grocery shop. Sometimes when the child's mother went out to work as a domestic helper, she also had to take care of the family shop and/or her younger siblings to relieve her mother's burden. Similarly, in a study on poverty and vulnerability in Dhaka slums, it was found that in households with working children, children contributed about 34% of the household income, while among female-headed households, the contribution was much more, ranging from 30 to 50% (Pryer, 2003).

Children's responsibilities also shift as they grow older. At a young age, they may be helping out with household work or caretaking responsibilities, but as they grow older and become physically stronger, gender-based variances in responsibilities emerge. For example, Putnick and Bornstein (2016) found that when children are 5 to 14 years old in the Global South, boys tended to be involved in work outside of the home, such as driving a rickshaw, brickmaking or contract labour, while girls tended to be engaged in household chores, such as cooking, cleaning and caring for younger siblings. These findings support Heissler's (2012) study where they found that girls took up increasingly skilled, physically challenging, and time-consuming work in and around the household. At times, girls also accompany their mothers to work as domestic servants or maids (Sultana, 2019).

In some cases, children and young people may migrate to other cities (permanently or seasonally), or from the village to the city to work at factories or as a domestic helper (Heissler, 2012). In Bangladesh and India research showed that children migrated to cities or towns to work, or parents 'rented' them out to relatives to reduce household costs (Heissler, 2012; Sayibu, 2016; Putnick and Bornstein, 2016). When a divorce, sudden death or illness of one parent occurred, older children were expected to fill in this position and, sometimes, they found themselves taking up a parental role. For example, in single-female parent households in Sri Lanka, mothers spoke to their children like equals, sons were treated like fathers, and daughters were treated like mothers (Ruwanpura, 2004). While the roles that children and young people take on in a household are multiple and varied, the range of responsibilities they take on to support household livelihoods is well documented.

Quite often when the mother was working away from the home, older siblings, often girls, were used as what Porter et al. (2010: 797) call 'domestic anchors' to take care of infants and younger siblings. Mothers would leave some money for the children to buy food and groceries for themselves and the infant/younger sibling (Kabir and Maitrot, 2017). Parents argued that early involvement of daughters in household duties and caretaking tasks was a strategy to socialise them to become good mothers and wives in the future (Dyson, 2014; Froerer, 2011; Husain, 2005; Porter et al., 2010).

In some cultures, ideas of reciprocity and debt governed relations between household members. Parents have a moral obligation to take care of their children, while children are said to be 'indebted' to their parents and are therefore expected to 'repay' this gratitude by taking care of them when they age (author quotes). In Southeast Asia, this aligns with the Filipino concept of 'utung na loob', which means 'internal debt' (Kaut, 1961). In Sayibu's (2016) study for example on young children accompanying adult beggars in Ghana, he pointed out the contradiction where international norms which prohibit children from begging simultaneously coexist with local understandings of being a 'good child' that justifies children's involvement in begging. When a child worked and contributed to the household, whether economically or through household work, they were seen as active contributors to the household, thereby being a good child (Belay, 2016; Sayibu, 2016; Wongboonsin and Tan, 2018).

Analysing intergenerational strategies within a household, Beall (2002) highlighted how long-term security investment in household capacity was accomplished through schooling for young children. Children's help with domestic chores was highly valued, and parents considered it their children's social responsibility to contribute to the wellbeing of their families (Verhoef, 2005). Beall (2002) also pointed out that girls tended to drop out of schooling early because they had to assist in household work, especially in patriarchal cultures, and it was considered unnecessary and wasteful to invest in a daughter's education. Similarly, Dyson (2014) provided a detailed account of the everyday lives of children in the rural village of Bemni, situated in the Indian Himalayas. The work also discussed children's relationship with the environment, friendships and identities, while balancing tasks associated with the household and school. Research over the last decade has shown the intersection between youth contributions to family's livelihoods and empowerment (Abebe, 2013; Ansell, 2005; Ennew et al., 2005). Interestingly, this research showed how children and youth tracked their growth, competence and maturity based on the interrelationship between schooling and various household responsibilities. These various responsibilities helped strengthen and develop one's social networks to advance their individual interests while providing them with social status and a peer support group.

It is also worth noting that several studies undertaken both in and outside of India argued for the intersection between child labour and schooling, where children use the economic resources they gain from work in order to fund their education (Husain, 2005; Swanson, 2009). This contributes to broader debates around complexities of child labour in the Global South, negating assumptions that child labour necessarily hinders schooling (Bourdillon, 2019). Punch (2002) went on to outline the role that work plays in culturally bounded notions of responsibility, where these choices impact future opportunities.

Aside from literature on schooling and work, there have also been an array of reports assessing the impacts of vocational interventions on the livelihoods of adolescent girls living in slums (e.g. Mensch et al., 2004). This literature found that though girls had greater awareness of safe spaces in the area, better social skills and better knowledge of sexual and reproductive health, vocational interventions did not have significant influence over attitudes towards marriage or their livelihoods in general. Authors of these reports attributed this to lack of involvement and communication with the adolescents' family members and the community. In other cases, vocational training (not specific to girls) and literary skills are seen as forms of support to enhance employability, particularly in Latin America (UNESCO, 2012).

Youth, environment and everyday mobility

Youth and adolescence are often conceptualised as a period of *becoming*, a transitional period between the arbitrary categories of childhood and adulthood. It is also enmeshed in unequal power relationships related to age and gender (Hansen, 2014) and a period where young people begin to negotiate their independence, maturity and competence while dealing with societal responsibilities that are thrust upon them. One of the many reasons why children and young people step in to help with household responsibilities is due to infrastructural woes, traffic problems and lack of transportation faced by the parents (Porter et al., 2010). With parents working long hours and/or are stuck in traffic, children step in to run errands, take care of younger siblings and do household tasks.

For many young caregivers, walking is a common way of getting around to perform various tasks either because other forms of transportation are unavailable or are too expensive (Gough, 2008). However, this form of local mobility brings with it a wide range of problems. Porter et al. (2010) found that walking (their case focused on poorer urban areas in Ghana, Malawi and South Africa) can present challenges, such as fear of local thugs, dead bodies along the road, smokers, shooting at night, roaming dogs and lizards, witchcraft and harassment (see Box 21.1 for example from India). This is further evidenced by exclusions children and young people face from school as they are often sprayed by dust and dirt along the road. From the young people's point of view, these restrictions on their physical mobility placed a lot of limitations on their social and economic mobility. Adolescent girls who reached puberty were suddenly removed from spaces where they socialised with friends (Nallari, 2014). While 'roaming around' may be viewed negatively by the family and the community, urban youth in Ghana, Lusaka and Lima saw physical and spatial mobility as essential in building their social networks and finding jobs (ASOARTE et al., 2002; Gough, 2008; Porter et al., 2010). In Ghana and Lusaka, young people spent their time walking to search for whatever work opportunities they come across. These are young residents who not only lived in the city centre, but also in suburban towns and nearby rural villages, where they made frequent travels to the city to find work or education opportunities (Gough, 2008).

Another mobility constraint that young people face relates to when they wish to socialise and meet with their peers. They tend to socialise and meet up with each other on the streets. However, the streets are often associated with gangs and crime, and thus, they feel they have no public/recreational space that they can use to hang out and meet others (ASOARTE et al., 2002). Young people who 'roam' around on the streets can be viewed negatively under the assumption that they are 'up to no good', or for girls, that they are putting themselves at risk of illicit affairs. This pressure is added to the conceptualisation of a 'good child' as one who doesn't move far away, roam around, or stay out late at night (Heissler, 2012). Parents therefore often seek ways to restrict their children's mobility, for example by giving them additional household work, beating them, denying food, or threatening to send them back to the village (Porter et al., 2010). The ability for movement can also be gendered. For example, Nallari (2014) found that parents restricted their daughters' mobility rather than their sons, as adolescent boys engaged in income-earning work to support the family, which tended to be viewed with respect and pride. Parents often found themselves facing a dilemma since the ability of young people to be mobile is valued for livelihoods, but at the same time, if their children are 'too mobile' the chances of bringing loss and shame to the family increases.

Children and young people in Lusaka (Zambia) for example, enjoyed the long 6 km walk to school because they felt it was a time where they were free from adult control and surveillance, and social bonds between peers could be strengthened. Daily exchanges for basic needs such as running errands, loaning money, food etc. were all essential for peer interaction (ASOARTE et al., 2002). Urban youth in Ghana felt the ability to become somebody was associated with being able to move around (Langevang and Gough, 2009). When they were stuck and physically restricted due to parental control or limited transportation networks, as was the case in Lusaka, young people commented that they felt this restricted their opportunities and growth. This also relates to the politics of time-pass and waiting (see work of Jeffrey et al., 2008), where 'sitting around' or 'doing nothing' symbolises 'young people's socio-spatial feelings of being stuck and getting nowhere, their social and physical immobility' (Langevang and Gough, 2009: 749). Faced with the desire to become independent and to be socially acknowledged, boys in particular spent time at their friends' houses as a way to reduce reliance on their parents (Banks, 2016). Work from Hansen (2014) also found that young people engage in a wide range of interpersonal relationships (social, economic, cultural) that are important to their everyday existence and their future. Importantly Hansen's work asked researchers to reflect on politico-economic changes that are impacting young people today.

Box 21.1 Youth livelihoods and future aspirations

As part of an ongoing research project titled 'Shaping Youth Futures' (2020–2022, British Academy 102666), two partners on the project, Youth for Unity and Voluntary Action (YUVA) and Fields of View (FoV) carried out a two-day workshop in Mumbai (India) to introduce a group of young people to the concept of livelihoods and related public policies. The research focused on identifying pathways of youth engagement in policy and practice with regards to access to livelihood options.

Through a series of exercises, participants focused on discussions around policies that impact access to livelihood options, possible roles that youth can play in monitoring policies, feedback on existing policies, and ways to engage with existing and developing new priorities for policies. Over the two days, it became evident that the capacity of the young people to identify themes and concepts within policies related to livelihoods that impact them in the short, medium and long term was high driven by personal interest. The participants were keen to differentiate between livelihoods, jobs and careers, and felt that they would ideally see a convergence between aspirations, jobs and goals.

Apart from working with the policy landscape, participants also identified how their gender, caste and religion shaped their livelihood options and future opportunities. The women in particular highlighted the harassment they faced on an almost daily basis, impacting their access to jobs and opportunities. Through vision boards they were asked to picture the future of youth livelihoods in India and their responses included: rights-based discussions and universal access, women empowerment, addressing the role of the caste system and access to opportunities.

Box 21.2 Youth protagonism

Lloyd (2005) in the Growing Up Global study argues that youth, in the transition between childhood and adulthood, are most susceptible to poverty in vulnerable contexts. Arnot and Swartz (2012) writing about civic competence of youth emphasises the knowledge, capability and awareness that youth are expected to have as active stakeholders in society. As highlighted by the World Bank, 'youth citizenship is crucial for development outcomes. The youth experience of citizenship is formative and has lasting effects on the extent and kind of political participation throughout life. Citizenship affects development outcomes through three channels: by enhancing the human and social capital of individuals, by promoting government accountability for basic service delivery, and by enhancing the overall climate for investment and private decision-making' (World Bank, 2006: 161).

Finding ways to support young people's voice can build their agency, while exercising their rights to citizenship (Arnot and Swartz, 2012). Protagonism, going beyond participation, contains advocacy as well. Protagonism includes the agency of children and youth, and the roles that they play within existing social structures (Liebel, 2007). Grown from studies in children's movement in Latin America, Liebel (2007: 62) highlighted that 'protagonism increases awareness of young people's capabilities and demands their independent and influential role in society'. Nuggehalli (2014: 14) furthered this by stating that 'protagonists view participation as a political intervention and as a right to intervene in changing their environment as active subjects'.

The work of Youth for Unity and Voluntary Action (YUVA) falls into the category of youth protagonism as they identify the agency of youth in determining their choices through various means- gender, religion, caste, class etc. Working toward community-level transformations, but enabling youth to actively understand their situation and environment, YUVA builds capacity through what they call 'City Caravans'. Each City Caravan, akin to capacity building, focuses on supporting community members to identify constraints and opportunities to express their agency. By embracing protagonism as a key tenet, YUVA showcases how youth can lead change on issues that matter to them. More information on their work: https://yuvaindia.org.

Conclusions

Sixteen percent of the global population can be classified as youth aged between 15–24 years (United Nations, 2019), and therefore, building a better understanding of livelihood options for this population is key to the future of planning and policy. Various authors have drawn attention to the crisis of youth employment and the challenges they face (see Banks, 2015, 2016). They also highlighted how closely intertwined social and economic realities are within young people's lives (Hansen, 2014). As the chapter highlights, various factors shape youth livelihoods, such as gender, family, caste, education, social pressures and intergenerational responsibilities (Box 21.1).

Rather than a universal understanding of youth as an age of transition, what we find here is that there are various models for what youth entails. Rapid urbanisation is shaping young people's access to opportunities and future research needs to address how young people's experiences have shifted in the last decade. Addressing changes within the markets, youth protagonism (Box 21.2), the emphasis on the entrepreneurship, and household arrangements, will further inform studies around youth livelihoods.

Key points

- 'Youth' is conceptualised as the period of transition between childhood dependence and adult independence. To understand youth livelihoods, intrahousehold relations need to be understood through existing generational dimensions and the concept of inter-generational contracts. Where knowledge and resources shared between generations can take various forms and shape, from contributing to the household's upkeep to caring responsibilities.
- The role that the surrounding environment plays is key to how youth avail themselves of various opportunities. There are intersectional considerations that need to be accounted for here as well such as gender, age, religion, caste etc. that shape their interaction with the environment.
- The social, political and economic realities shaping young people's lives are complex and understanding how this is impacting the youth of today will be key to determining future directions of research and policies that shape their lives.
- Further research is needed to understand how rapid urbanisation and youth livelihoods are impacted and have shifted/shifting.

Further reading

Huijsmans, R., Ansell, N. and Froerer, P. (2021) 'Introduction: Development, young people, and the social production of aspirations', *The European Journal of Development Research*, 33: 1–15. https://doi.org/10.1057/s41287-020-00337-1

Mabala, R. (2011) 'Youth and "the hood" – livelihoods and neighbourhoods', *Environment and Urbanization*, 23(1): 157–181. https://doi.org/10.1177/0956247810396986

Nuggehalli, R.K. (2021) 'Youth protagonism in Urban India', in Swartz, A., Cooper, A., Batan, C.M. and Kropff Causa, L. (eds.), *The Oxford handbook of Global South youth studies*, Oxford: Oxford University Press.

References

Abebe, T. (2013) 'Interdependent rights and agency: The role of children in collective livelihood strategies in rural Ethiopia', in Hanson, K. and Nieuwenhuys, O. (eds.), *Reconceptualizing children's rights in international development: Living rights, social justice, translations*, Cambridge: Cambridge University Press.

Ansell, N. (2005) *Children, youth, and development*, Oxon: Routledge.

Arnot, M. and Swartz, S. (2012) 'Youth citizenship and the politics of belonging: Introducing contexts, voices, imaginaries', *Comparative Education*, 48(1): 1–10. https://doi.org/10.1080/03050068.2011.637759

ASOARTE, AMM, GEAC and MAFUM (2002) 'Exploring youth and community relations in Cali, Colombia', *Environment and Urbanization*, 14(2): 149–156. https://doi.org/10.1177/095624780201400212

Banks, N. (2015) 'Understanding youth: Towards a psychology of youth poverty and development in sub-Saharan African cities', *Brooks World Poverty Institute Working Paper Series No 216*, Manchester: University of Manchester.

Banks, N. (2016) 'Youth poverty, employment and livelihoods: Social and economic implications of living with insecurity in Arusha, Tanzania', *Environment and Urbanization*, 28(2): 437–454. https://doi.org/10.1177/0956247816651201

Beall, J. (2002) 'Living in the present, investing in the future: Household security among the urban poor', in Rakodi, C. and Lloyd-Jones, T. (ed.), *Urban livelihoods*, London: Earthscan Publications Ltd., pp. 71–87.

Belay, D.G. (2016) "Being small is good': A relational understanding of dignity and vulnerability among young male shoe-shiners and lottery vendors on the streets of Addis Ababa, Ethiopia', in Huijsmans, R. (ed.), *Generationing development. Palgrave studies on children and development*, London: Palgrave Macmillan, pp. 151–174.

Bourdillon, M. (2019) "Child labour' and children's lives', in Twum-Danso Imoh, A., Bourdillon, M. and Meichsner, S. (ed.), *Global childhoods beyond the north-south divide*, London: Palgrave Macmillan, pp. 35–55.

Chatterjee, S. (2012) 'Children growing up in Indian slums: Challenges and opportunities for new urban imaginations', *Early Childhood Matters*, June: 17–23.

Clark-Kazak, C.R. (2009) 'Towards a working definition and application of social age in international development studies', *Journal of International Development*, 45(8): 1307–1324. https://doi.org/10.1080/00220380902862952

Dyson, J. (2014) *Working childhoods: Youth, agency and the environment in India*, Cambridge: Cambridge University Press.

Ennew, J., Myers, W.E. and Plateau, D.P. (2005) 'Defining child labor as if human rights really matter', in Weston, B.H. (ed.), *Child labor and human rights: Making children matter*, London: Lynne Rienner Publishers.

Evans, R. (2015) 'Negotiating intergenerational relations and care in diverse African context', in Vanderbeck, R. and Worth, N. (ed.), *Intergenerational space*, London/New York: Routledge, pp. 199–213.

Froerer, P. (2011) '"Learning, livelihoods, and social mobility: Valuing girls" education in central India', *Anthropology and Education Quarterly*, 43(4): 344–357. https://doi.org/10.1111/j.1548-1492.2012.01189.x

Gough, K.V. (2008) '"Moving around": The social and spatial mobility of youth in Lusaka', *Geografiska Annaler. Series B, Human Geography*, 90(3): 243–255. www.jstor.org/stable/40205050

Hansen, K.T. (2014) 'Cities of youth: Post-millennial cases of mobility and sociality', *Working Paper, No. 2014/001*, Helsinki: The United Nations University World Institute for Development Economics Research.

Heissler, K. (2012) 'Children's migration for work in Bangladesh: The policy implications of intra-household relations', *Development in Practice*, 22(4): 498–509. https://doi.org/10.1080/09614524.2012.673555

Hopkins, P. and Hill, M. (2010) 'Contested bodies of asylum-seeking children', in Hörschelmann, K. and Colls, R. (eds.), *Contested bodies of childhood and youth*, Basingstoke/ New York: Palgrave Macmillan, pp. 136–147.

Hopkins, P. and Pain, R. (2007) 'Geographies of age: Thinking relationally', *Area*, 39(3): 287–294. https://doi.org/10.1111/j.1475-4762.2007.00750.x

Huijsmans, R. (2013) 'Doing gendered age: Older mothers and migrant daughters negotiating care work in rural Lao PDR and Thailand', *Third World Quarterly*, 34(10): 1896–1910. https://doi.org/10.1080/01436597.2013.851952

Huijsmans, R. (ed.) (2016) *Generationing development: A relational approach to children, youth and development*, London: Palgrave Macmillan.

Husain, Z. (2005) 'Analysing demand for primary education: Muslim slum dwellers of Kolkata', *Economic and Political Weekly*, 40(2): 137–147. www.jstor.org/stable/4416043

Jeffrey, C., Jeffery, R. and Jeffery, P. (2008) *Degrees without Freedom? Education, masculinities, and unemployment in north India*, Stanford, California: Stanford University Press.

Kabir, A. and Maitrot, M.R. (2017) 'Factors influencing feeding practices of extreme poor infants and young children in families of working mothers in Dhaka slums: A qualitative study', *PLoS One*, 12(2): e0172119. https://doi.org/10.1371/journal.pone.0172119

Kaut, C. (1961) 'Utang Na Loob: A system of contractual obligation among tagalogs', *Southwestern Journal of Anthropology*, 17(3): 256–272. https://doi.org/10.1086/soutjanth.17.3.3629045

Langevang, T. and Gough, K.V. (2009) 'Surviving through movement: The mobility of urban youth in Ghana', *Social and Cultural Geography*, 10(7): 741–756. https://doi.org/10.1080/14649360903205116

Laz, C. (1998) 'Act your age', *Sociological Forum*, 13(1): 85–113. https://doi.org/10.1023/A:1022160015408

Liebel, M. (2007) 'Paternalism, participation and children's protagonism', *Children, Youth and Environments*, 17: 56–73. www.jstor.org/stable/10.7721/chilyoutenvi.17.2.0056

Lloyd, C.B. (2005) *Growing up global: The changing transitions to adulthood in developing countries*, Washington, DC: Institute of Medicine.

McGregor, J.A., Copestake, J.G. and Wood, G.D. (2000) 'The inter-generational bargain: An introduction', *Journal of International Development*, 12(4): 447–451. https://doi.org/10.1002/1099-1328(200005)12:4%3C447::AID-JID682%3E3.0.CO;2-O

Mensch, B., Grant, M.J., Sebastian, M.P., Hewett, P.C. and Huntington, D. (2004) 'The effect of a livelihoods intervention in an urban slum in India: Do vocational counselling and training alter the attitudes and behavior of adolescent girls?', *Policy Research Division Working Paper No. 194*, New York: Population Council.

Nallari, A. (2014) *The meaning, experience, and value of 'common space' for women and children in urban poor settlements in India*, PhD Thesis, New York: City University of New York.

Nuggehalli, R.K. (2014) 'Children and young people as protagonists and adults as partners', in Westwood, J., Larkins, C., Moxon, D., Perry, Y. and Thomas, N. (eds.), *Participation, citizenship and intergenerational relations in children and young people's lives: Children and adults in conversation*, London: Palgrave Pivot.

Porter, G., Hampshire, K., Abane, A., Robson, E., Munthali, A., Mashiri, M. and Tanle, A. (2010) 'Moving young lives: Mobility, immobility and inter-generational tensions in urban Africa', *Geoforum*, 41(5): 796–804. https://doi.org/10.1016/j.geoforum.2010.05.001

Pryer, J. (2003) *Poverty and vulnerability in Dhaka slums: The urban livelihoods study*, London: Routledge.

Punch, S. (2002) 'Youth transitions and interdependent adult-child relations in rural Bolivia', *Journal of Rural Studies*, 18(2): 123–133. https://doi.org/10.1016/S0743-0167(01)00034-1

Punch, S. (2015) 'Youth transitions and migration: Negotiated and constrained interdependencies within and across generations', *Journal of Youth Studies*, 18(2): 262–276. https://doi.org/10.1080/13676261.2014.944118

Putnick, D.L. and Bornstein, M.C. (2016) 'Girls' and boys' labor and household chores in low- and middle-income countries', *Monographs of the Society for Research in Child Development*, 81(1): 104–122. https://doi.org/10.1111/mono.12228

Ruwanpura, K.N. (2004) 'Dutiful daughters, sacrificing sons', *Domains*, 1(1): 8–37.

Sayibu, W. (2016) '"We don't even use our older children": Young children accompanying blind adult beggars in Tamale, Ghana', in Huijsmans, R. (ed.), *Generationing development. palgrave studies on children and development*, London: Palgrave Macmillan, pp. 175–198.

Scott, J.C. (1998) *Seeing like a state: How certain schemes to improve the human condition have failed*, New Haven/London: Yale University Press.

Sultana, I. (2019) 'Social factors causing low motivation for primary education among girls in the slums of Karachi', *Bulletin of Education and Research*, 41: 61–72.

Swanson, K. (2009) *Begging as a path to progress: Indigenous women and children and the struggle for Ecuador's urban spaces*, Atlanta: University of Georgia Press.

Thorne, B. (2004) 'Editorial: Theorizing age and other differences', *Childhood*, 11(4): 403–408. https://doi.org/10.1177/0907568204047103

Twum-Danso Imoh, A. and Ansell, N. (eds.) (2013) *Children's lives in the era of children's rights: The progress of the convention on the rights of the child in Africa*, Oxford: Routledge

UNESCO (2012) *Youth and skills: Putting education to work*, Paris: United Nations Educational, Scientific and Cultural Organization.

United Nations, Department of Economic and Social Affairs, Population Division (2019) *World population prospects 2019: Data booklet* (ST/ESA/SER.A/424), New York: UNDESA.

Verhoef, H. (2005) '"A child has many mothers": Views of child fostering in northwestern Cameroon', *Childhood*, 12(3): 369–390. https://doi.org/10.1177/0907568205054926

Wongboonsin, P. and Tan, J. (2018) *Care relations in southeast Asia: The family and beyond*, Leiden: BRILL.

World Bank (2006) *World development report 2007: Development and the next generation*, Washington, DC: World Bank.

Xu, Q. (2015) 'One roof, different dreams', in Vanderbeck, R.M. and Worth, N. (eds.), *Intergenerational space*, London/New York: Routledge.

22
THE GOVERNANCE AND REGULATION OF THE INFORMAL ECONOMY

Implications for livelihoods and decent work

Julian Walker, Andrea Rigon and Braima Koroma

Introduction

This chapter examines the informal economy as a key site of livelihoods activities, particularly in the Global South, in the context of the deregulation of labour governance in the globalised economy, increasingly also in high-income countries. Most definitions of the informal economy characterise it in terms of the lack of state regulation. This implies, amongst other things, that those deriving their livelihoods from the informal economy do not benefit from state regulation designed to ensure their access to labour protection and decent work. In this vein, it is often assumed that formalisation, through the extension of state regulation into the informal economy, supports women and men engaged in informal livelihoods by improving their rights and protections as workers. However, a body of literature also questions the role of state regulation in ensuring decent working conditions for informal workers (Chen, 2005), and highlights the ways in which state interventions can, instead, displace livelihoods in the informal economy (Omoegun et al., 2019). In addition, research also points to the parallel importance of social regulation as a means of protecting workers in the informal economy (Song, 2016). Based on these discussions, this chapter explores the scope for synergies between state and social regulation in protecting those deriving their livelihoods from the informal economy.

Livelihoods and decent work in the informal economy

Official data suggests that 81% of enterprises globally are informal (OECD/ILO, 2019), and over 60% of the world's workers are in informal employment, representing 70% of all employment in developing and emerging countries, with significant regional disparity, ranging from 86% in Africa to 25% in Europe and Central Asia (ILO, 2018). While this demonstrates that the informal economy is far more prevalent in emerging economies, a body of research also suggests that there is increasing informality of working conditions in high income countries, resulting from policy processes promoting deregulation and labour market flexibility to increase competitivity in global markets (Standing, 1997) as well as phenomena such as the emerging 'gig economy', which eat away at employer responsibilities for labour rights (De Stefano, 2015).

As discussed below, the informal economy is primarily defined in terms of the absence of state governance. This means that women and men deriving their livelihoods from the informal economy are outside state regulations and so more likely to lack basic labour protections and access to decent work. According to the International Labour Organization, decent work comprises of four pillars: international labour standards and fundamental principles and rights at work; employment creation; social protection; and, social dialogue and tripartism (ILO, 2013b). State actors, such as ministries of labour and employment, should take a leading role in coordinating the delivery of these pillars.

However, the ILO argues that as a result of inadequate state regulation, informal workers are often unable to realise these four pillars of decent work, stating that 'the 2 billion women and men who make their living in the informal economy are deprived of decent working conditions' (ILO, 2018: v). Research has revealed how those securing their livelihoods from the informal economy are often unable to secure key labour rights, such as a minimum wage, the right to rest, the right to employment benefits such as pensions, sick pay and maternity cover, or protection from child labour (ILO, 2018; Lund, 2009; Schlyter, 2002). Examples of industries in which informality is pervasive and where workers are routinely denied key labour rights include: the garment industry, where formal transnational companies often outsource along supply chains to informal enterprises (Merck, 2014); domestic work, where employers are private households and so often inaccessible to labour inspection (Oelz, 2014); or, agriculture, where many workers cannot access labour rights because they are either seasonal or migrant workers (Fudge and Olsson, 2014) or (unpaid) contributing family workers (ILO, 2018). In view of the inability of many informal workers to realise decent work, there has therefore been a global policy focus on strategies of 'formalisation' such as the ILO Recommendation 204 on 'Transition from the Informal to the Formal Economy' (ILO, 2015).

Defining the informal economy

The nature of specific strategies of formalisation to extend decent work to informal economy workers depend on the understanding of what 'informality' constitutes. There is ongoing debate around how to define the informal economy, with two competing schools of thought: one, which understands the informal economy in terms of the lack of state regulation, and a second, which defines it in terms of the organisation of enterprises and their processes of production (ILO, 2013a). However, of the two, the former, with its focus on lack of state regulation, is the most commonly used, both in academic literature and in terms of the working definitions used in national and international policy. In a review of more than 200 pieces of literature on formal – informal economy linkages, Meagher found that 'the prevailing definition accepted across disciplinary and ideological boundaries is that the informal economy refers to income generating activities that operate *outside the regulatory framework of the state*' (Meagher, 2013: 2, authors' italics).

Following on from such definitions, the treatment of informality in policy frequently characterises it as a problem, linked, at least in part, to the lack of state registration/regulation, for example with the policy treatment of street traders in some states in Nigeria (Godswill et al., 2016). This implies a strategy of formalisation through extending state centric systems of regulation.

However, the idea of a clear formal/informal economy dichotomy bounded by activities and spaces in which the state is, or is not, present does not respond well to empirical scrutiny. It is now generally accepted that, rather than being one half of a formal/informal binary, the informal sector is part of a continuum, ranging through economic activities with more or less regulatory inputs, across a variety of arrangements with different levels and types of state, and social, regulation (Boananda-Fuchs and Boananda-Fuchs, 2018; Bunnell and Harris, 2012). Economic sectors, sites of economic activity and enterprises tend to be managed by state actors to a greater or lesser extent, and for a variety of different purposes.

Another blurring of the boundaries between the formal and the informal is in the institutional and spatial 'sites' of informal economic activities. Institutionally, much informal employment now takes place in formal enterprises (Williams and Lansky, 2013), while, spatially, informal economic activities can be pervasive in 'formal' areas of the city (Rigon et al., 2020). Furthermore, in many cities in the Global South, while informality is often the norm, value chains and services contain both formal (state regulated) and informal elements that are interdependent (Myers, 2010). At the same time, the informal sector, rather than being characterised by the absence of the state, is frequently in practice an 'assemblage' of state and non-state actors and their associated processes (Dovey, 2012).

In view of such research, which problematises the boundaries of informality, institutions working to support those who depend on informal livelihoods, such as the NGO WIEGO (Women in Informal Employment: Globalizing and Organizing) highlight the multiple forms of regulation that 'formality' can encompass, and the different outcomes for livelihoods in relation to workers' wellbeing and rights (Chen, 2012). This implies that (a) increasing state regulation of the informal economy does not necessarily result in better labour protection for informal sector workers, and (b) that state regulation is not the only means of extending decent work arrangements for workers in the informal economy.

Informality and the role of state regulation of livelihoods

Even if the definitional boundaries of informality are understood as blurred, a strong normative analysis of this continuum persists in many policy circles – that formalisation (i.e. moving towards formality understood as increased regulation by the state) is the preferred trajectory. In this view, the problems of informal economic activities and employment are defined by their lack of regulation and social protection, and the implied solution is the extension of state regulation, and policy-makers therefore envisage a leading role for state bodies in strategies of formalisation (Chen, 2005; ILO, 2015).

However, at the same time, a range of authors have questioned the scope of the state for ensuring that those carrying out their livelihoods in the informal sector have access to decent work in many contexts. Firstly, an immediate pragmatic concern relates to the ability of poorly resourced states to deliver formality. Even where economic activities or urban spaces are officially regulated by the state, this may not be applied in practice, meaning that there are often overlaps between formal regulation and de facto informality. As Meagher notes in her analysis of informality in Africa, 'even states have become informalised as public officials govern in ways that contravene formal relations, and downsizing public sectors concede an increasing range of governance activities to community organizations' (Meagher, 2007: 406).

Secondly, even where the state does have the capacity to extend its regulation of the informal economy, this begs the question of what exactly the state constitutes, and the capacity of its various branches to regulate economic processes and support livelihoods. The idea of the state as monolithic and consensual is problematic and empirical scrutiny of the state as a regulatory actor reveals complexity and contradictions (Corbridge et al., 2005). If, instead, the state is understood as a 'collection of heterogeneous administrative and bureaucratic fields, together with governmental and non-governmental institutions within which social actors struggle over authority, rules, legislation and discourses' (Bourdieu and Wacquant, 1992: 111), heterogeneity and conflict within the state can be recognised. In this vein, state actors involved in the regulation of livelihoods range from the police, labour inspectors, environmental inspectors, to courts at different levels and with different purposes. Each of these have different priorities for what to regulate, how and at what scale, and some of these may be in contradiction with each other. As Chen points out (2005: 26), 'in the past, the management or regulation of informal activities has often been relegated to social policy departments or, in urban areas, to those departments (such as the police or traffic) that deal with law and order issues', thereby treating the informal economy as a social concern or a law and order issue, rather than a focus for economic policy-makers or ministries of labour concerned with the governance of livelihoods.

In addition to the question of 'what state' should be involved in formalisation is the question of 'what regulations' are prioritised. Economic activities may be regulated in some ways but not in others (Benjamin et al., 2014) meaning, for example, that an enterprise might be regulated vis-à-vis taxation or quality control of output, but not in terms of minimum wage legislation or social protection of workers. Advocacy organisations such as WIEGO have therefore highlighted the need to de-bundle the diversity of forms of regulation by the state and their different purposes (e.g. tax collection, the protection of private property and intellectual property, or the promotion of decent work) with reference to their impact on workers' wellbeing. In this vein, Chen argues that 'it is important to ensure that formalisation offers the benefits and protections that come with being formal and does not simply impose the cost of being formal' (Chen, 2012: 15).

A final question concerns whether it should be assumed that, even if it is feasible, the extension of the power of the state over the informal economy is necessarily desirable. A substantial body of literature argues that states and their regulatory practices typically serve the interests of economic and social elites and have historically not represented the interests of other groups, such as women, ethnic minorities or the poor (Dagnino, 2007; Lister, 2007). Others highlight that the growing purview of the state can imply the growing reach of systems of coercion and exploitation (Ferguson, 1994) rather than the protection of the social contract. In this view, informality, rather than a condition of those who cannot reach the state, can be viewed as a deliberate withdrawal from the state, as implied in Holston's work on insurgent citizenship (Holston, 1999, 2009) or the work on the everyday encroachment by slum-dwellers (Bayat, 2000). In terms of livelihoods, there are a number of authors who see state regulatory intervention as inimical to the livelihoods of the poor. On the one hand, authors such as de Soto argue that many entrepreneurs choose to remain informal to avoid the burdens of state bureaucracy (de Soto, 2000). On the other hand, it has been argued that many state interventions actively displace informal livelihoods. For example, state-led evictions to support urban planning or environmental protection result in the large-scale destruction of informal livelihoods through the eviction of informal markets or street traders (Brown et al., 2015; Omoegun et al., 2019). Furthermore, the stated public interest rationale for such interventions (e.g. environmental protection or improving the city aesthetic) are often a pretext for an underlying interest in releasing land values and supporting commercial interests (Walker et al., 2020; Penz et al., 2011; Bhan, 2016).

Social governance of informal livelihoods

The other side of the coin from questioning the central role of the state is considering the role of other actors or relations in the governance of livelihoods. If 'governance is ultimately concerned with creating the conditions for ordered rule and collective action' (Stoker, 1998: 17) then it is about the negotiation of collective norms that guide group interactions, and thus the rights and duties of citizens. In practice, the actions of the state describe only part of these processes. Governance can be undertaken *by* and *with* a number of actors. This has been recognised by a range of authors, in particular those who have emphasised the importance of societal or community centred forms of governance. This has been key in developing an understanding of how common property regimes work (Ostrom, 2010), as well as the regulation of private property through 'informal' land markets (Hornby et al., 2017). The concept of governmentality, which constitutes the 'organised practices through which we are governed and through which we (consciously and unconsciously) govern ourselves' (Cleaver, 2007: 228), also highlights the ways in which governance can be structured through a range of institutional forms, including through internalised social norms. As pointed out by McFarlane and Waibel (2012: 2),

> informal institutions can replace . . . formal ones in contexts where the state is unable or unwilling to implement its formal rules. In this sense, informality contributes to formal institutions by organising social interaction in the absence of the state, for example, during periods of rapidly changing socio-economic contexts, rapid urban development.

However, as a caution, other authors note that social and insurgent processes of governance are not necessarily benign (see for example Meth (2010), or Monson (2015)).

Broadening the understanding of informality to validate social governance of livelihoods therefore implies that it should not be assumed that the extension of state governance is the only means of regulating informal livelihoods to ensure workers' access to income and the protection of their labour. However, it should be emphasised that social governance, like state governance, is not intrinsically benign, and therefore both social and state modes of regulation should be interrogated vis-à-vis their normative purpose, and whether this is worker wellbeing or other purposes (such as extraction of rent).

The collective management of the informal livelihoods in Freetown (see Box 22.1) highlights the role that locally organised self-regulation can play in governing livelihood practices, and their social and environmental outcomes. At the same time, however, they show that state regulation is nonetheless needed to address some aspects of labour protection, and reveal the potential for cooperative regulation between state and non-state actors.

Box 22.1 Social governance of informal livelihoods in Freetown, Sierra Leone

The majority of women and men in Sierra Leone derive their livelihoods from the informal economy and the capacity for state regulation of livelihoods in the country is limited. According to the 2015 census, 92.9% of the economically active population in Sierra Leone were in informal employment, as opposed to 7.1% in paid employment (Statistics Sierra Leone, 2015) and, at the same time, state capacity to govern the economy is low, characterised by what the African Development Bank refer to as 'persistent challenges in the governance environment' (AfDB, 2020: 24).

In the context of post–civil war Sierra Leone, with the state slowly developing its capacity, different types of collective action emerged as autonomous processes of self-governance, filling state governance gaps. Research in four informal settlements in Freetown looking at livelihood practices in typical sectors in which low-income women and men work (fishing, quarrying, sand-mining or trading) revealed that state interventions affecting livelihoods practices in the communities were very limited. The only direct involvement of state officials in livelihoods activities discussed by research respondents was the role of the police (in dealing with disputes around land ownership and other disputes related to the work in the sectors) and the practice of the National Protected Area Authority (NPAA) of fining people engaged in sand mining in the protected area around tidal mangrove swamps next to one of the settlements. It should be further noted that sand miners argue that the purpose of the NPAA officers is extraction of rent through fining, rather than the stated purpose of environmental protection, as if they pay the fine, the NPAA officers allow them to continue sand mining.

In contrast to the lack of state regulation, all of the settlements had highly complex self-organised social regulation processes, which were used to structure behaviour and relations of those involved in the livelihood sectors. These include occupational associations (e.g. for fishers, or porters in the stone quarrying sector), nominally registered with Freetown City Council but in practice self-organised. Research respondents explained to us that these associations play a role in managing disputes across the sector, drawing on locally developed informal community 'bylaws' that regulate work in the sector and disputes over payment, and appropriate behaviour. Such norms are crucial in managing property relations and economic transactions which are central to people's livelihoods. Penalties for breaking such bylaws are fines, which are used by the community to fund infrastructure projects such as road maintenance which are crucial for the operation of the livelihood sectors. Local associations also play a social protection function, through practices of advance payment, and loans, amongst different actors across sectoral value chains and also through norms which ensure open access to livelihood resources (e.g. stone and sand quarries, or cockle picking sites) for poor residents who lack any other form of basic income.

While these practices of social regulation have more impact on the governance of the livelihood sectors than state regulation, they have their own weaknesses in terms of labour protection (for example, in all of the sectors examined, the pragmatic acceptance of child labour). In addition, it is noteworthy that such social regulation does not always operate in isolation from state regulation. One example of co-production of regulation between state bodies and local social actors was a 2008 policy from the Ministry of Fisheries and Marine Resources banning fishing of immature fish. There was no state capacity for enforcement of this ban, but through their associations, fishermen had adopted the requirement to stop using fine nets which catch immature fish ('fingerlings'), despite the costs implied in replacing nets. Furthermore, this had become normalised across the sector so that, for example, net menders refuse to mend fine grade nets.

Source: This case study draws on the findings of the project 'Urban Livelihoods in Freetown's Informal Settlements' funded by Comic Relief and its outputs (Rigon et al., 2020; Walker et al., 2021).

Governance of informal livelihoods through co-production

Building on the example of co-production from Freetown, one fruitful avenue to extend protection to workers who derive their livelihoods form the informal sector is the co-production of livelihoods governance between formal and informal/social actors (Lindell, 2019; Song, 2016), rather than pursuing strategies of formalisation in which the extension of state regulation displaces the existing social regulation of livelihoods. Such co-production can result in what Song (2016) refers to as 'positive hybridity' between formal and informal processes of governance.

For example, the ILO's campaign to promote the labour rights of domestic workers (a notoriously informal and hard to govern area of employment) in line with ILO convention 189, has, in addition to promoting regulation and policy development by the state, also promoted non-state regulatory arrangements. It has done this by influencing social norms around the employment of domestic workers and changing relations between employers and employees through means such as 'the development of model contracts, assistance to domestic workers in understanding their terms and conditions and, more generally, information and outreach activities to inform workers and employers of applicable laws' (Oelz, 2014: 164–165).

In this vein, despite widespread discourses that characterise informal economies as unregulated and ungoverned, they are often highly regulated through complex and hybrid governance systems which play a fundamental social and economic function, particular where the state has limited capacity to intervene. While there tends to be a state-centric logic of many processes of 'formalisation', non-state governance arrangements can also provide social protection, employment and livelihoods which are critical for the wellbeing of women and men deriving their livelihoods from the informal economy.

Key points

This chapter has examined the literature defining the informal economy, to question the relevance of a formal/informal dichotomy, and argue for the need to interrogate the role of the state and other actors in collective governance, unpacking their normative aims and what they are able to achieve in practice. This has been argued along the following lines:

- The informal economy is largely defined in the literature as the parts of the economy which operate outside the regulatory purview of the state.
- A large proportion of workers in the Global South (70%) derive their livelihoods from the informal economy, and policies promoting economic competition through deregulation mean that work in high-income economies is also increasingly informal.
- Women and men deriving their livelihoods from the informal economy are more likely to experience unprotected and exploitative labour conditions. Policy approaches to the informal economy such as the ILO Recommendation 204 therefore recommend strategies of formalisation.
- State actors play a crucial role in formalisation to extend protection to workers in the informal economy. However, the state may lack the capacity to regulate the informal economy in some contexts, and some state actions may have a negative impact on the livelihoods of informal sector workers.
- Social regulation can also have an important role to play in protecting the livelihoods and labour rights of informal sector workers.
- It is therefore argued that state and social governance intending to promote the livelihoods of informal economy workers should be co-produced, rather than state regulation displacing social regulation.

Recommended reading

Brown, A. and Roever, S. (2016) *Enhancing productivity in the urban informal economy*, Nairobi: United Nations Human Settlements Programme (UN-Habitat).
Chen, M.A. (2012) 'The informal economy: Definitions, theories and policies', *WIEGO Working Paper, No. 1*, Manchester: WIEGO.
Guha-Khasnobis, B., Kanbur, S.M.R. and Ostrom, E. (2006) 'Beyond formality and informality', in Guha-Khasnobis, B., Kanbur, S.M.R. and Ostrom, E. (eds.), *Linking the formal and informal economy: Concepts and policies*, Oxford: Oxford University Press, pp. 1–18.References
African Development Bank (AfDB) (2020) *Sierra Leone country diagnostic note*, Freetown: AfDB Country Economic Department, Regional Directorate General, West Africa, Sierra Leone Country Office.

Bayat, A. (2000) 'From "dangerous classes" to "quiet rebels" politics of the urban subaltern in the Global South', *International Sociology*, 15(3): 533–557. https://doi.org/10.1177%2F026858000015003005

Benjamin, N., with Beegle, K., Recanatini, F. and Santini, M. (2014) 'Informal economy and the World Bank', *Policy Research Working Paper 6888*, Washington, DC: World Bank

Bhan, G. (2016) *In the public's interest: Evictions, citizenship, and inequality in contemporary Delhi*, Athens: University of Georgia Press. www.jstor.org/stable/j.ctt19x3jp8

Boananda-Fuchs, A. and Boananda Fuchs, V. (2018) 'Towards a taxonomic understanding of informality', *International Development Planning Review*, 40(4): 397–420. https://doi.org/10.3828/idpr.2018.23

Bourdieu, P. and Wacquant, L. (1992) *An invitation to reflexive sociology*, Chicago: University of Chicago Press.

Brown, A., Msoka, C. and Dankoco, I. (2015) 'A refugee in my own country: Evictions or property rights in the urban informal economy?', *Urban Studies*, 52(12): 2234–2249. https://doi.org/10.1177%2F0042098014544758

Bunnell, T. and Harris, A. (2012) 'Re-viewing informality: Perspectives from urban Asia', *International Development Planning Review*, 34(4): 339–348. https://doi.org/10.3828/idpr.2012.21

Chen, M. (2005) 'Rethinking the informal economy: linkages with the formal economy and the formal regulatory environment', *WIDER Research Paper, No. 2005/10*, Helsinki: The United Nations University World Institute for Development Economics Research (UNU-WIDER).

Chen, M.A. (2012) 'The informal economy: Definitions, theories and policies', *WIEGO Working Paper No. 1*, Manchester: WIEGO

Cleaver, F. (2007) 'Understanding agency in collective action', *Journal of Human Development and Capabilities*, 8(2): 223–244. https://doi.org/10.1080/14649880701371067

Corbridge, S., Williams, G., Srivastava, M. and Véron, R. (2005) *Seeing the state: Governance and governmentality in India*, Cambridge: Cambridge University Press.

Dagnino, E. (2007) 'Citizenship: A perverse confluence', *Development in Practice*, 17(4–5): 549–556. https://doi.org/10.1080/09614520701469534

De Soto, H. (2000) *The mystery of capital: Why capitalism triumphs in the west and fails everywhere else*, London: Bantam Press

De Stefano, V. (2015) 'The rise of the just-in-time workforce: On-demand work, crowd work, and labor protection in the gig-economy', *Comparative Labour, Law & Policy Journal*, 37(3): 471–504.

Dovey, K. (2012) 'Informal urbanism and complex adaptive assemblage', *International Development Planning Review*, 34(4): 349–368. http://doi.org/10.3828/idpr.2012.23

Ferguson, J. (1994) *Anti-politics machine: Development, depoliticisation, and bureaucratic power in Lesotho*, Minnesota: University of Minnesota Press.

Fudge, J. and Olsson, P.H. (2014) 'The EU Seasonal Workers Directive: When immigration controls meet labour rights', *European Journal of Migration and Law*, 16(4): 439–466. https://doi.org/10.1163/15718166-12342065

Godswill, O.C., Chinweoke, N., Ugonma, O.V. and Ijeoma, E.E. (2016) 'The resilience of street vendors and urban public space management in Aba, Nigeria', *Developing Country Studies*, 6(11): 83–93.

Holston, J. (1999) *Cities and citizenship*, Durham: Duke University Press.

Holston, J. (2009) 'Insurgent citizenship in an era of global urban peripheries', *City & Society*, 21(2): 245–267. https://doi.org/10.1111/j.1548-744X.2009.01024.x

Hornby, D., Kingwill, R., Royston, L. and Cousins, B. (2017) *Untitled: Securing land tenure in urban and rural South Africa*, Pietermaritzburg: University of KwaZulu-Natal Press http://doi.org/10.17159/1727-3781/2018/v21i0a3406

International Labour Office (2013a) *Measuring informality: A statistical manual on the informal sector and informal employment*, Geneva: International Labour Office (ILO).

International Labour Office (2013b) *Decent work indicators: Guidelines for producers and users of statistical and legal framework indicators*, Geneva: ILO.

International Labour Office (2015) *Transition from the Informal to the formal economy, recommendation No. 204*, Geneva: ILO.

International Labour Office (2018) *Women and men in the informal economy: A statistical picture*, Geneva: ILO

Lindell, I. (2019) 'Introduction: Re-spatialising urban informality: Reconsidering the spatial politics of street work in the Global South', *International Development Planning Review*, 41(1): 3–21. https://doi.org/10.3828/idpr.2019.2

Lister, R. (2007) 'Inclusive citizenship: Realising the potential', *Citizenship Studies*, 11(1): 49–61. https://doi.org/10.1080/13621020601099856

Lund, F. (2009) 'Social protection, citizenship and the employment relationship', *WIEGO Working Paper (Social Protection) No. 10*, Women in Informal Employment: Globalizing and Organizing (WIEGO), Manchester: WIEGO.

McFarlane, C. and Waibel, M. (2012) 'Introduction: The informal-formal divide in context', in McFarlane, C. and Waibel, M. (eds.), *Urban informalities: Reflections on the formal and informal*, Farnham: Ashgate, pp. 1–12.

Meagher, K. (2007) 'Introduction: Special issue on "informal institutions and development in Africa"', *Africa Spectrum*, 42(3): 405–418. www.jstor.org/stable/40175202

Meagher, K. (2013) 'Unlocking the informal economy: A literature review on linkages between formal and informal economies in LMICs', *WIEGO Working Paper No. 27, Women in Informal Employment Globalizing and Organizing*, Manchester: WIEGO.

Merk, J. (2014) 'The rise of tier 1 firms in the global garment industry: Challenges for labour rights advocates', *Oxford Development Studies*, 42(2): 259–277. https://doi.org/10.1080/13600818.2014.908177

Meth, P. (2010) 'Unsettling insurgency: Reflections on women's insurgent practices in South Africa', *Planning Theory & Practice*, 11(2): 241–263. https://doi.org/10.1080/14649351003759714

Monson, T. (2015) 'Everyday politics and collective mobilisation against foreigners in a South African shack settlement' *Africa: The Journal of the International African Institute*, 85(1): 131–152. www.jstor.org/stable/24525608

Myers, G.A. (2010) *Seven themes in African urban dynamics*, Uppsala: Nordiska Afrikainstitutet.

OECD/ILO (2019) *Tackling vulnerability in the informal economy*, Paris: Development Centre Studies, OECD Publishing. https://doi.org/10.1787/939b7bcd-en

Oelz, M. (2014) 'The ILO's domestic workers convention and recommendation: A window of opportunity for social justice', *International Labour Review*, 153(1): 143–172. https://doi.org/10.1111/j.1564-913X.2014.00200.x

Omoegun, A.O., Mackie, P. and Brown, A. (2019) 'The aftermath of eviction in the Nigerian informal economy', *International Development Planning Review*, 41(1): 107–129. https://doi.org/10.3828/idpr.2018.30

Ostrom, E. (2010) 'The challenge of self-governance in complex contemporary environments', *The Journal of Speculative Philosophy*, 24(4): 316–332. https://doi.org/10.5325/jspecphil.24.4.0316

Penz, G.P., Drydyk, J. and Bose, P.S. (2011) *Displacement by development: Ethics, rights and responsibilities*, Cambridge: Cambridge University Press

Rigon, A., Walker, J. and Koroma, B. (2020) 'Beyond formal and informal: Understanding urban informalities from Freetown', *Cities*, 105: 1028–1048. https://doi.org/10.1016/j.cities.2020.102848

Schlyter, C. (2002) 'International labour standards and the informal sector: Developments and dilemmas', *Working Paper on the Informal Economy*, Geneva: ILO.

Song, L.K. (2016) 'Planning with urban informality: A case for inclusion, co-production and reiteration', *International Development Planning Review*, 38(4): 359–381. https://doi.org/10.3828/idpr.2016.21

Standing, G. (1997) 'Globalization, labour flexibility and insecurity: The era of market regulation', *European Journal of Industrial Relations*, 3(1): 7–37. https://doi.org/10.1177%2F095968019731002

Statistics Sierra Leone (2017) *Sierra Leone 2015 population and housing census. Thematic report on economic characteristics*, Freetown, Sierra Leone: Statistics Sierra Leone, pp. 4–5.

Stoker, G. (1998) 'Governance as theory: Five propositions', *International Social Science Journal*, 50(155): 17–28. https://doi.org/10.1111/issj.12189

Walker, J., Koroma, B., Sellu, S.A. and Rigon, A. (2021) 'The social regulation of livelihoods in unplanned settlements in Freetown: Implications for strategies of formalisation', *International Development Planning Review*, online first. https://doi.org/10.3828/idpr.2021.3

Walker, J.H., Lipietz, B., Ohaeri, V., Onyebueke, V. and Ujah, O. (2020) 'Displacement and the public interest in Nigeria: Contesting developmental rationales for displacement', *Development in Practice*, 30(3): 332–344. https://doi.org/10.1080/09614524.2019.1694642

Williams, C.C. and Lansky, M.A. (2013) 'Informal employment in developed and developing economies: Perspectives and policy responses', *International Labour Review*, 152(3–4): 355–380. https://doi.org/10.1111/j.1564-913X.2013.00196.x

23
DISABILITY AND SUSTAINABLE LIVELIHOODS
Towards inclusive community-based development

David Cobley

Introduction

With the adoption of the United Nations Convention on the Rights of Persons with Disabilities (CRPD) in 2006, since ratified by the vast majority of nation states, and the unveiling of a disability-inclusive Agenda for Sustainable Development in 2015, disability is rapidly emerging as a priority development issue. It is now widely recognised that disabled people, thought to comprise around 15% of the world's population, are disproportionately vulnerable to poverty and subject to multiple violations of their basic human rights (WHO and World Bank, 2011).

Traditionally, rehabilitation services for disabled people have been rooted in medicalised perceptions of disability, thus focusing mainly on medical interventions designed to improve body function and to enable them to live more independently within an environment that is often inhospitable to their needs (Oliver, 1996). In line with the gradual emergence of a rights-based approach to disability, however, the concept of rehabilitation has now broadened to encompass more holistic approaches, focusing on the limitations of society as well as the impairment-based limitations of disabled people (Cobley, 2018). Such approaches tend to go beyond addressing healthcare and physical rehabilitation needs, crossing into sectors such as employment, education and social welfare (WHO and World Bank, 2011). Furthermore, they are often now immersed within local communities, rather than being confined to institutional settings, in line with the CRPD's call for health and rehabilitation services to be delivered as close as possible to where disabled people live. These broader, community-based approaches can be collectively referred to as community-based rehabilitation (CBR), an umbrella term for rehabilitation strategies aimed at drawing on local community resources to support the empowerment and inclusion of disabled people (ILO et al., 2004).

The CBR model is consistent with the sustainable livelihoods approach (SLA), and this chapter begins with a discussion around three core principles that are common to both: empowerment, holism and sustainability. It goes on to consider how changing perceptions of disability have prompted a shift towards a rights-based view of disability, before examining the close relationship between disability and poverty. The CBR model is then presented as a tool for reducing poverty among disabled people and supporting them to achieve their livelihood goals.

Adopting a livelihoods perspective allows for a broader, more complete analysis of how people survive and flourish than would be possible if focusing purely on employment or income generation (Coleridge, 2016). For example, the livelihoods approach allows for consideration of the instrumental role that families and communities can play in supporting and empowering disabled people, particularly in Southern contexts where individual lives are often characterised by strong familial and communal ties (Singal, 2012). The latter part of this chapter highlights the importance of this role, arguing that community-based approaches need to foster strong and mutually supportive linkages between disabled people, their families, their representative organisations and stakeholders within the wider community. This argument is developed through secondary use of a survey of livelihood project coordinators working in nine countries across East Africa and South Asia, conducted by the author in 2012 on behalf of the UK-based international disability organisation Leonard Cheshire (LC). Evidence is also drawn from field research visits to disability-focused community livelihood projects in Kenya (2010), India (2011) and Sierra Leone (2012).

The sustainable livelihoods approach (SLA): three underlying principles

This section discusses three of the core principles of the SLA (see Chapter 2 for a deeper discussion around the SLA), particularly in terms of how these principles relate to disability.

Empowering approach

The SLA emphasises the need for people-centred approaches, aimed at enabling poor people to build on their own strengths in order to reach their potential (DFID, 1999). This can only be achieved by adopting an empowering approach, in which poor people are the key decision-makers right from the outset. Disabled people often lack self-esteem and confidence, particularly when they come from a background of poverty and have faced discrimination and exclusion all their lives (Martinelli and Mersland, 2010). As a result, they may be perceived as lacking motivation, or not wanting to be empowered. This can change very quickly, however, within an enabling environment, in which participants feel that their views are important and that they can take control of their own lives. In South Korea, for example, where disabled people are typically viewed as unproductive and lacking in interpersonal skills, various Government initiatives designed to foster the growth of micro-enterprises run by disabled entrepreneurs have created 'an opportunity for disabled people's lives to be fashioned by what they can do instead of what they cannot do' (Hwang and Roulstone, 2015: 126).

The concept of empowerment aligns closely with the principle of 'nothing about us, without us' (Charlton, 1998), a slogan that is well known as the mantra of the international disability rights movement. As emphasised in the CRPD, development policymakers and practitioners need to put this principle into practice by working in close partnership with disabled people's organisations (DPOs) – representative organisations that are run by and for disabled people – in all matters concerning disability. The pivotal role of DPOs within community development is examined later in this chapter.

Holistic approach

Poor people typically experience a wide range of deprivations that go far beyond a lack of income and basic material necessities to encompass 'a cruel mix of human deprivation in knowledge, health, dignity and rights, obstacles to participation and lack of voice' (UNDP,

2013: 1). Adopting a holistic approach implies taking account of all dimensions of poverty, as well as considering the broader context within which a livelihoods project operates. In the Ghanaian context, for example, Grischow (2015) describes how a young man named Kwesi, newly disabled due to a motorcycle accident, engaged in a long and often painful struggle to achieve his livelihood objectives, largely due to discriminatory attitudes and practices within various community institutions, including the police force, the justice system and the health system. This chapter considers how development interventions can take account of the local context, particularly in terms of influencing community stakeholders and encouraging them to act as facilitators, rather than presenting barriers to the inclusion and participation of disabled people.

Sustainable approach

The SLA highlights the need for poor people to be supported to build their livelihood assets in a sustainable way (DFID, 1999). Working collaboratively with community stakeholders can increase the sustainability of a livelihoods intervention, since many of these stakeholders are likely to be present in the community long after the development agencies have left. Raising community awareness of disability and challenging negative stereotypes, in order to promote more inclusive and accessible societies, is also vital to ensuring the long-term sustainability of disabled people's livelihoods. This was illustrated by a study conducted in rural Zimbabwe (Munsaka and Charnley, 2013), which revealed that traditional beliefs and practices had directly led to the exclusion of disabled participants from various aspects of community life, from schooling to community projects and public meetings. The authors argue that, in such a context, a critical re-examination of cultural understandings of disability is essential to bringing about the long-term inclusion of disabled people in development processes.

Changing perceptions of disability

Historically, perceptions of disability have reflected an 'individual model', underpinned by 'personal tragedy theory', in which the 'problem' of disability is located within the impaired individual body (Oliver, 1983). This model prioritises professional expertise over the experiences and opinions of disabled people themselves (Iriarte, 2016), with resulting practices tending to focus on treatment or rehabilitation aimed at enabling disabled people to conform to societal norms (Oliver, 1996).

Dissatisfaction with individual model perspectives, particularly among disabled people themselves, has led to the rise of the social model of disability (Oliver, 1983), in which disability is created and perpetuated through the limitations of society. As Oliver himself acknowledges, this model emerged from ideas that were expressed in a seminal document, entitled 'The Fundamental Principles of Disability', published in 1976 by a group of disabled activists in the UK who had come together to form the Union of Physically Impaired Against Segregation (UPIAS) – an early example of a DPO. The document establishes a clear distinction between disability and impairment, as reflected in the following extract:

> In our view, it is society which disables physically impaired people. Disability is something imposed on top of our impairments by the way we are unnecessarily isolated and excluded from participation in society. Disabled people are therefore an oppressed group in society
>
> (UPIAS, 1976: 3–4)

The social model has been subject to much debate, particularly in terms of its applicability to Southern contexts. For example, Singal (2012: 422) argues that the model 'should not be over-emphasised in contexts where disability is most likely to be the result of disease, malnutrition or other treatable or preventable factors'. However, the social model has provided a clear political focus for the international disability movement over the past half-century, as well as gaining increasing acceptance within the mainstream international development arena, with disability increasingly now framed as a human rights issue rather than as a welfare issue. This influence is reflected in the CRPD, which views disability in the following way:

> *Disability results from the interaction between persons with impairments and attitudinal and environmental barriers that hinders their full and effective participation in society on an equal basis with others*
>
> *(UN, 2006: Preamble (e)).*

As this statement suggests, it is necessary to see beyond impairments or health conditions and to take into account the impact of societal features, in order to gain a full understanding of disability. In particular, as the CRPD repeatedly reiterates, there is a need to identify and remove societal barriers that limit the participation of disabled people (Kanter, 2014). The CRPD also highlights the close relationship between disability and poverty, recognising 'the critical need to address the negative impact of poverty on persons with disabilities' (UN, 2006: Preamble (t)). This relationship is explored in the next section.

The relationship between disability and poverty

There is little doubt that disability and poverty are closely associated. A systematic review of 150 studies, all conducted within low and middle-income countries (LMICs), found that 122 of the studies reported a statistically significant positive relationship between disability and economic poverty (Banks et al., 2017). Furthermore, it is widely recognised that disabled people are more likely to experience multiple deprivations than the general population, thus reinforcing the SLA's multidimensional view of poverty. Evidence of this can be found in the first ever 'World Report on Disability' (WHO and World Bank, 2011), which draws on numerous studies from around the world to conclude that disabled people are significantly disadvantaged, particularly in terms of access to education, healthcare and employment.

The relationship between disability and poverty has long been characterised as a 'vicious cycle' (DFID, 2000), in which disability and poverty are mutually reinforcing. However, a growing body of evidence around the links between disability and poverty has provided an increasingly nuanced picture of the relationship. Some studies (see, for example, Trani and Loeb's (2010) study on Afghanistan and Zambia) even suggest that within extremely poor communities there is little difference between the assets and living conditions of households with and without disabled family members. However, a disturbing trend – sometimes referred to as the 'disability and development gap' (Groce and Kett, 2013) – is that as countries rise out of poverty there is a tendency for the poverty gap between disabled people and the general population to widen. This was illustrated through an analysis of World Health Survey data from 15 LMICs across Asia, Africa, Latin America and the Caribbean, revealing an 'adverse relation between economic development and the disability/poverty association' (Mitra et al., 2013: 11). While much progress has been made in terms of reducing poverty across the globe, with the number of people living in extreme poverty declining by more than one half between 1990 and 2015 (UN, 2015), it seems that disabled people are being left behind.

Community-based rehabilitation (CBR)

This section introduces the CBR model, which has gradually evolved from a narrow approach to delivering primary health care and basic medical rehabilitation services to disabled people, particularly those living in low-income countries (WHO, 2010), to the much broader strategy encapsulated in the definition below.

> CBR is a strategy within general community development for rehabilitation, equalization of opportunities and social inclusion of all people with disabilities. CBR is implemented through the combined efforts of people with disabilities themselves, their families, organisations and communities
>
> (ILO et al., 2004: 2).

The CBR model has been adopted in more than 90 countries around the world and has increasingly come to be regarded as an appropriate mechanism for promoting rights and opportunities for disabled people (WHO and World Bank, 2011). Extending the reach of CBR is among the objectives identified in the WHO Global Disability Action Plan 2014–2021, which recognises the potential of CBR to empower disabled people and their families living in countries with limited resources (WHO, 2015). CBR can also be viewed as strategy for reducing poverty among disabled people (WHO, 2010), thus helping to address the 'disability and development gap'.

Detailed guidance on CBR is contained within the WHO's (2010) CBR Guidelines. The guidelines present CBR as a multi-sectoral approach, as captured by the CBR matrix in Figure 23.1. This matrix highlights five major components of CBR: health; education; livelihoods; social and empowerment. As Coleridge (2016) observes, the various elements within these five components link closely to the assets portfolio that lies at the core of the SLA, thus implying an assets-based approach to community development. The guidelines also highlight increased awareness of the need to involve DPOs, as lessons have been learnt from practice. DPOs are vital to the success of CBR initiatives, as they are usually in a strong position to gain the trust of disabled people and their families, and often have long-established linkages with other community stakeholders (Young et al., 2016). Hence, they have an important role to play from the initiation stage through to implementation and evaluation (ILO et al., 2004).

The CBR model aligns closely with the three underlying principles of the SLA, outlined earlier in this chapter. It emphasises the involvement of disabled people themselves in identifying their own rehabilitation priorities and aims to empower them to become 'active contributors to the community and society at large' (ILO et al., 2004: 3). It recognises the multi-dimensional nature of poverty and aims to support disabled people to build their livelihood assets in a holistic way. In terms of sustainability, there is a strong focus on working collaboratively with community stakeholders to promote social inclusion. Underpinned by a rights-based approach to disability and consistent with the SLA, the CBR model thus serves as a useful guiding framework for community-based development interventions designed to support and empower disabled people.

Working with families and communities to build livelihoods

This section draws on the findings of the LC Livelihoods Survey, as well as research conducted in India, Kenya and Sierra Leone, to consider how community-based approaches to livelihoods development can engage effectively with families and communities. There is a particular focus on the critical role that DPOs can play within community development, as highlighted in the previous section, given the opportunity and appropriate support.

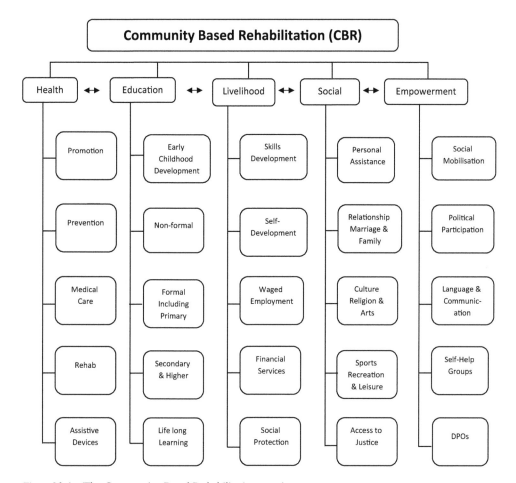

Figure 23.1 The Community-Based Rehabilitation matrix
Source: Adapted from World Health Organization (2010: 25)

Engaging with families

Family members often make great sacrifices to ensure that disabled people's care needs are met, as well as playing a crucial role in supporting them to achieve their livelihood ambitions. They can be a vital source of emotional support and often provide practical support, such as transport to work or filling in gaps in the supply chain for disabled people that are running their own business enterprises from home (Moodie, 2010). Community-based livelihood interventions can support families in this role, as revealed through a research visit to Jan Madhyam, a special education centre in New Delhi. The centre had initiated various economic empowerment initiatives including a home-based production scheme, which involved training school leavers together with their parents and then supporting them to set up home-based enterprises. The project counsellor explained how the process worked for a young woman with an intellectual impairment, whose family had used a business loan to purchase a spice-grinding machine.

> *We helped the girl and her mother to set up a work corner at home, so that the business would be respected within the family, and provided a cabinet for storing raw materials and products. We also helped them to design a work timetable, which would not interfere with home chores. Eventually, the brother and father also got involved, helping to deliver products to customers. Now the family has a small shop.*
>
> (Research Interview, 2011).

As the counsellor went on to explain, working closely with families can be mutually beneficial. Families can benefit through simply discovering that they are not alone in supporting and caring for their disabled family members. The project itself can benefit through access to the wealth of valuable knowledge that a family will normally have, and through the likelihood of vastly improved outcomes if families understand what the project is trying to achieve and are willing to give their support.

It is almost inevitable that the empowerment process will, on occasions, challenge or conflict with the opinions of family members. Ransom (2010) observes that younger disabled people, in particular, are often unable to develop skills at home because parents want to protect them from harm, sometimes believing that they are unable to learn or to make any useful contribution to their household or local community. One respondent to the LC Livelihoods Survey, based in Tanzania, reported that some project participants were aware that their income was being controlled by family members, but felt unable to challenge this due to dependence on them for support in other areas. He went on to explain that these issues need to be handled with great sensitivity, in order to ensure that families remain supportive, or at least accepting, of the project, rather than becoming obstructive.

Engaging with communities

Local communities often contain a wealth of useful resources, which can potentially support the participation and empowerment of disabled people (IDDC, 2006). The success and sustainability of a community-based livelihoods intervention thus depends largely on the extent to which these community resources can be harnessed. Working in partnership with community stakeholders can help to raise disability awareness, to promote mainstream inclusion and to ensure that project resources are not wasted in providing services that other agencies are better placed to provide. Research conducted within a CBR project in rural Kenya (Cobley, 2012) revealed a strong focus on institutional awareness raising within the local community. Project staff worked with government agencies, hospitals, training centres, employers, media outlets and even the local police to raise awareness of disability rights and to support them in making their services more inclusive. As the project manager explained, this was an important aspect of the scheme because disabled people are far more likely to achieve sustainable livelihood outcomes within an inclusive society.

In order to engage effectively with local communities, a livelihoods project may need to face up to the challenge of combating negative attitudes and stigmatising beliefs within these communities. Virtually all of the respondents to the LC Survey identified discriminatory attitudes and beliefs as among the main barriers to livelihood development, with one respondent from Uganda observing that it was not uncommon for the public to 'boycott products made by disabled people, especially foodstuffs'. Other respondents noted that such attitudes tend to be internalised by disabled people themselves, damaging their confidence and will to succeed. Disabling community attitudes and beliefs are also widely reported in the literature. In Cambodia, for example, 'social attitudes associate disability with incapacity, unworthiness . . . thus precluding their access to basic education, economic opportunities and other livelihood opportunities' (Gartrell and Hoban, 2013: 200). Attitudes can change, however, as disability awareness grows – a view that emerged strongly from the LC Survey:

> *We have rebuilt the profile of disability. Especially the belief that disabled persons also are capable of making success rather than useless and worthless has been emerging among the project intervened communities. Plus, the advocacy activities such as materials distribution, talents show, public education and project stakeholder seminars have contributed evidently to removing prejudice and discrimination'*
>
> *(Livelihoods Project Coordinator in China).*

Disabled people themselves can have a powerful influence on community attitudes, as well as becoming positive role models for other disabled people. Lorenzo and Coleridge (2019) observe how the achievements of disabled people can energise the disability sector and transform its image within the wider community. This was recognised by a survey respondent from Sri Lanka, who revealed that wherever possible she tries to involve successful disabled participants in community awareness-raising activities, because 'seeing is believing'.

The role of Disabled People's Organisations (DPOs)

DPOs fulfil a variety of roles, but one of their key functions is to provide a space where disabled people can share experiences and exchange information with their peers. In Ethiopia, for example, Katsui and Mojtahedi (2015) observe the powerful impact of peer support activities within a women's self-help group, particularly in terms of building self-esteem and creating a general sense of belonging. Group membership can also transform perceptions of disabled people within the wider community. This was illustrated through research conducted with self-help groups in South India (Cobley, 2013: 450), revealing how group membership can enhance social status. As one group member commented, 'we now have respect in the community. People fear us because we are a group and we know our rights'.

Virtually all of the respondents to the LC Livelihoods Survey identified DPOs as among the most important strategic partners in terms of developing livelihood programmes within their local communities. Most also recognised the need to strengthen these organisations, particularly in terms of building their capacity to deliver livelihood services for their members. This finding is supported by a systematic review of 11 DPO studies conducted in LMICs, with most of these studies referring to a 'lack of financial and human resources as negatively impacting on the functional capacity of DPOs' (Young et al., 2016: 65). The potential value of capacity-building support was revealed through a project evaluation in the Kono district of Sierra Leone, conducted by the author on behalf of LC, which included an observation of a meeting of the Kamadu Community Group – a DPO with 25 members, many of whom became disabled as a result of Sierra Leone's long-running and brutal civil war. LC had supported this organisation to draft a constitution and provided training in various aspects of group management, including democratic processes and record keeping. They had also provided seed money to set up a Village Loans and Savings Association. Members were able to buy shares in the savings scheme and to receive business loans, enabling them to set up small business enterprises. At the meeting, one group member described how group members had been empowered to work together effectively and democratically, concluding that 'we will not put money to waste. We will show that we are responsible traders. This group is a stepping stone to our development'.

Discussion

CBR approaches have been criticised in some quarters, particularly around insufficient government funding, which can create an over-dependence on donor agencies (see, for example, Bongo et al.'s (2018) study on rural Zimbabwe), and the lack of effective impact evaluations (Weber et al., 2016). The success of a CBR programme is also dependent on the support of the local community, which cannot be guaranteed. A study of CBR programmes in Ghana, Guyana and Nepal showed that community commitment was often limited, particularly when it came to the allocation of resources, thus limiting the effectiveness of the programmes (WHO and SHIA, 2002). It is vital therefore to strengthen the linkages between disabled people, their families and the wider community, as suggested by the research findings discussed in the previous section. Stronger linkages can facilitate the accumulation of social capital, as they enable disabled people to build relationships of trust and reciprocity within their communities, while also making it easier for them to access community institutions that can support their livelihood ambitions. Awareness-raising activities can support this process by encouraging community stakeholders to recognise the participation rights of disabled people and to view them as potential contributors, rather than as a burden on society.

The SLA and the CBR model both emphasise the need for holistic, multi-sectoral strategies that recognise the multidimensional nature of poverty and the multiple factors that influence poor people's livelihoods. The SLA differs from CBR in that it is not a disability-specific approach. However, given the close relationship between disability and poverty, with evidence of the relationship growing stronger as countries move out of poverty (Mitra et al., 2013), it has particular relevance to disability. The need for a holistic approach to livelihood development links closely to the concept of disability mainstreaming, which is about trying to ensure that all mainstream community activities and services are disability-inclusive (WHO and World Bank, 2011). A mainstreaming approach may not be sufficient on its own, however, because disabled people often lack the confidence, knowledge and skills necessary to take advantage of the opportunities that arise from mainstreaming. Adopting a holistic approach thus requires a 'twin-track approach' (DFID, 2000; WHO, 2010), in which both disability-specific and mainstreaming activities are implemented. Disability-specific activities involve empowering disabled people, either individually or within groups, to build on their own resources and to take advantage of the livelihood opportunities available to them. The twin-track approach, combining targeted support for disabled people and DPOs with broader initiatives designed to promote mainstream inclusion, is now viewed as a key strategy within the CBR model (WHO, 2010). It is also perhaps the most realistic approach to meeting the challenge of dismantling disabling barriers and supporting disabled people to achieve their full potential in life.

Key points

- Livelihoods development is essential to breaking the vicious cycle of poverty and disability, and to ensuring that disabled people are not left behind as countries move out of poverty.
- The CBR model links closely to the livelihood assets at the core of the SLA, providing a holistic, multi-sectoral framework to guide community-based approaches to building the livelihoods of disabled people.
- Development agencies can empower DPOs through capacity-building support, while drawing on their expertise and knowledge to enhance the reach, effectiveness and sustainability of community-based approaches.
- In line with the SLA, inclusive community-based development initiatives should be empowering, holistic and sustainable. This can best be achieved through adopting a twin-track approach to community development.

Further reading

Cobley, D. (2018) *Disability and international development: A guide for students and practitioners*, London: Routledge.

Coleridge, P. (2016) 'Access to livelihoods', in Iriarte, E., McConkey, R. and Gilligan, R. (eds.), *Disability and human rights: Global perspectives*, London: Palgrave, pp. 189–204.

References

Banks, L.M., Morgon, H. and Pollack, S. (2017) 'Poverty and disability in low- and middle-income countries: A systematic review', *PLoS One*, 13(9): 1–19. https://doi.org/10.1371/journal.pone.0189996

Bongo, P., Dziruni, G. and Muzenda-Mudavanhu, C. (2018) 'The effectiveness of community based rehabilitation as a strategy for improving quality of life and disaster resilience for children with disability in rural Zimbabwe', *Jamba: Journal of Disaster Risk Studies*, 10(1): 442. https://doi.org/10.4102/jamba.v10i1.442

Charlton, J. (1998) *Nothing about US without Us*, Berkeley: University of California Press.

Cobley, D. (2018) *Disability and international development: A guide for students and practitioners*, London: Routledge.

Cobley, D.S. (2012) 'Towards economic empowerment: Segregation versus inclusion in the Kenyan context', *Disability and Society*, 27(3): 371–384. https://doi.org/10.1080/09687599.2012.654988

Cobley, D.S. (2013) 'Towards economic participation: Examining the impact of the convention on the rights of persons with disabilities in India', *Disability and Society*, 28(4): 441–455. https://doi.org/10.1080/09687599.2012.717877.

Coleridge, P. (2016) 'Access to livelihoods', in Iriarte, E., McConkey, R. and Gilligan, R. (eds.), *Disability and human rights: Global perspectives*, London: Palgrave, pp. 189–204.

Department for International Development (DFID) (1999) *Sustainable livelihoods guidance sheets*, London: Department for International Development.

Department for International Development (DFID) (2000) *Disability, poverty and development*, London: Department for International Development.

Gartrell, A. and Hoban, E. (2013) 'Structural vulnerability, disability and access to non-governmental organisation services in rural Cambodia', *Journal of Social Work in Disability & Rehabilitation*, 12(3): 194–212. https://doi.org/10.1080/1536710X.2013.810100

Grischow, J. (2015) '"I nearly lost my work": Chance encounters, legal empowerment and the struggle for disability rights in Ghana', *Disability and Society*, 30(1): 101–113. https://doi.org/10.1080/09687599.2014.982786

Groce, N. and Kett, M. (2013) 'The disability and development gap', *Working Paper Series No. 21*, London: Leonard Cheshire Disability and Inclusive Development Centre.

Hwang, K. and Roulstone, A. (2015) 'Enterprising? Disabled? The status and potential for disabled people's micro-enterprise in South Korea', *Disability and Society*, 30(1): 114–129. https://doi.org/10.1080/09687599.2014.993750

International Disability and Development Consortium (IDDC) (2006) 'Inclusive development and the UN convention', *IDDC Reflection Paper*, Brussels: IDDC.

International Labour Organization (ILO), United Nations Educational, Scientific and Cultural Organization and World Health Organization (2004) 'CBR a strategy for rehabilitation, equalization of opportunities, poverty reduction and social inclusion of people with disabilities', *Joint Position Paper*, Geneva: United Nations.

Iriarte, E. (2016) 'Models of disability', in Iriarte, E., McConkey, R. and Gilligan, R. (eds.), *Disability and human rights: Global perspectives*, London: Palgrave, pp. 10–32.

Kanter, A. (2014) *The development of disability rights under international law: From charity to human rights*, Oxford: Routledge.

Katsui, H. and Mojtahedi, M. (2015) 'Intersection of disability and gender: Multi-layered experiences of Ethiopian women with disabilities', *Development in Practice*, 25(4): 563–573. https://doi.org/10.1080/09614524.2015.1031085.

Lorenzo, T. and Coleridge, P. (2019) 'Working together: Making inclusive development a reality disabilities', in Watermeyer, B, McKenzie, J. and Swartz, L. (eds.), *The Palgrave handbook of disability and citizenship in the Global South*, London: Palgrave MacMillan, pp. 233–247.

Martinelli, E. and Mersland, D. (2010) 'Microfinance for people with disabilities', in Barron, T. and Ncube, J. (eds.), *Disability and poverty*, London: Leonard Cheshire Disability, pp. 215–260.

Mitra, S., Posarac, A. and Vick, B. (2013) 'Disability and poverty in developing countries', *World Development*, 41: 1–18. https://doi.org/10.1016/j.worlddev.2012.05.024

Moodie, B. (2010) 'Self-employment for people with disabilities', in Barron, T. and Ncube, J. (eds.), *Poverty and disability*, London: Leonard Cheshire Disability, pp. 261–285.

Munsaka, E. and Charnley, H. (2013) 'We do not have chiefs who are disabled: Disability, development and culture in a continuing complex emergency', *Disability and Society*, 28(6): 756–769. https://doi.org/10.1080/09687599.2013.802221.

Oliver, M. (1983) *Social work with disabled people*, Basingstoke: MacMillan

Oliver, M. (1996) *Understanding disability: From theory to practice*, Basingstoke: MacMillan

Ransom, B. (2010) 'Lifelong learning in education, Training and skills development', in Barron, T. and Ncube, J. (eds.), *Disability and poverty*, London: Leonard Cheshire Disability, pp. 145–176.

Singal, N. (2012) 'Doing disability research in a southern context: Challenges and possibilities', *Disability and Society*, 25(4): 415–426. https://doi.org/10.1080/09687591003755807

Trani, J. and Loeb, M. (2010) 'Poverty and disability: A vicious circle? Evidence from Afghanistan and Zambia', *Journal of International Development*, 24: S19–S52. https://doi.org/10.1002/jid.1709

Union of Physically Impaired Against Segregation (UPIAS) (1976) *Fundamental principles of disability*, London: UPIAS.

United Nations (UN) (2006) *Convention on the rights of persons with disabilities and optional protocol*, Washington, DC: United Nations.

United Nations (UN) (2015) *The millennium development goals report*, New York: United Nations.

United Nations Development Programme (UNDP) (2013) *Poverty reduction and UNDP*, New York: United Nations Development Programme.

Weber, J., Grech, S. and Polack, S. (2016) 'Towards a "mind map" for evaluative thinking in Community Based Rehabilitation: Reflections and learning', *Disability and the Global South*, 3(2): 951–979.

World Health Organization (WHO) (2010) *CBR guidelines*, Geneva: WHO.

World Health Organization (WHO) (2015) *WHO global disability action plan 2014–2021: Better health for all people with disability*, Geneva: WHO.

World Health Organization and Swedish Organizations of Disabled Persons International Aid Association (SHIA) (2002) *Part one. Community-based rehabilitation as we experienced it . . . voices of persons with disabilities*, Geneva: WHO.

World Health Organization and World Bank (2011) *World report on disability*, Geneva: WHO.

Young, R., Reeve, M. and Grills, N. (2016) 'The functions of disabled people's organisations (DPOs) in LMICs: A literature review', *Disability, CBR and Inclusive Development*, 27(3): 45–71. https://doi.org/10.5463/dcid.v27i3.539

PART III

Generating livelihoods

24
ENVIRONMENTAL INCOME AND RURAL LIVELIHOODS

Carsten Smith-Hall, Xi Jiao and Solomon Zena Walelign

Introduction

Rural households in the Global South pursue diverse livelihood activities (e.g. Davis et al., 2010; Ellis, 2000) such as growing crops, raising livestock and working off-farm (e.g. Nielsen et al., 2013). These activities result in subsistence income, e.g. through own-consumption of firewood, and/or cash income, e.g. from selling medicinal plants to a trader. Research has shown that environmental income is an important source of household income in the Global South. Such income is derived from fuel, food, fodder, medicine, construction materials and a string of other products harvested across a range of non-cultivated habitats including forests, meadows, mangroves and rivers; it also includes wages from natural resource-based activities and transfer payments for environmental services (PEN, 2007). A comparative study of almost 8000 households in 24 tropical and sub-tropical countries in Africa, Asia and Latin America found environmental income to constitute an average of 27.5% of total household income, or 508±693 USD (in 2011 purchasing power parity per adult equivalent), comparable to the average agricultural income share of 28.7% (Angelsen et al., 2014). These results are arguably representative of the millions of households in smallholder-dominated rural areas in the tropics and sub-tropics, with moderate-to-good access to forest resources, except where population density is very high (Wunder et al., 2014a).

Environmental income is realised by rural households throughout the Global South, generated by a myriad of products that have tended to go unnoticed and unrecorded. Whether calculated per household or per unit area, this income is mainly derived from accessible state forests (Jagger et al., 2014), often de facto open access (Robinson, 2016), even if there are large geographical variations, e.g. Pouliot and Treue (2013) found that environmental income in Burkina Faso and Ghana is mainly derived from non-forest environments. A substantial but unspecified part of environmental product harvesting takes place outside formal legal frameworks ('t Sas-Rolfes et al., 2019). In general, both men and women fairly equitably harvest products to generate environmental income, with some variation in the roles associated to forest product collection across regions, such as women collecting more subsistence products in Africa while men are more engaged in wood and bushmeat harvesting (Sunderland et al., 2014). Men are the main contributors in generating income from forest products (especially processed forest products), though the pattern can be very different in some locations, e.g. the collection and sale of shea nuts in Burkina Faso has traditionally been a source of cash income for female household members (Pouliot, 2012).

Recent decades have seen a wealth of studies documenting the economic importance of environmental income in rural livelihoods in the Global South. The pioneering work on developing methods to estimate environmental incomes was done by Cavendish (2000, 2002) and further standardised and rolled out by the Poverty Environment Network (Angelsen et al., 2011) and the Food and Agriculture Organization of the United Nations (FAO, 2016). These advances allow estimation of the vast majority of subsistence and commercial environmental products though challenges persist regarding valuation of cultural products and water (Wunder et al., 2011). The household-level economic importance of environmental income can be viewed through different lenses, such as the sustainable livelihoods framework or ecosystem services (e.g. Nunan, 2015). In this chapter, we use the three forest functions (Vedeld et al., 2007) to provide an overview of the economic importance of environmental income to rural households in the Global South distinguishing contributions to: (i) current consumption, including patterns of absolute and relative income, (ii) gap filling and safety nets, and (iii) poverty reduction. The chapter ends by specifying the research frontier, including the need to better understand intra-household factors and employ new data generating methods.

Current consumption

There is consensus in the rural livelihood literature that environmental income is important in supporting current consumption (e.g. Angelsen et al., 2011, 2014; Miller and Hajjar, 2020; Rasmussen et al., 2017; Robinson, 2016), including through subsistence consumption, cash incomes and providing agricultural inputs. Environmental income matters and is important in preventing (further) poverty. This agreement is supported by a range of non-timber forest product and ethnobotanical studies documenting the ubiquitous use of environmental products in rural settings throughout the Global South (e.g. Balick and Cox, 2020; Belcher et al., 2005; Shackleton et al., 2011; Sheppard et al., 2020). In aggregated environmental income terms (exemplified in Figure 24.1), documented patterns are: (1A) higher environmental income reliance among poorer households (the lower the total household income, the higher the share of environmental income), and higher absolute environmental income among more well-off households, and (1B) constant relative importance but increasing absolute environmental income with rising total household income. While the first pattern is much more common, uncovering the prevalent pattern in particular localities is important to develop appropriate interventions. And it should be noted that *all* households in rural settings have environmental incomes; they are not only important in indigenous populations living inside forests. While environmental income may commonly be relatively more important to poorer households, absolute environmental income is on average five-fold higher among better-off households (the highest income quintile vs the two lowest quintiles, Angelsen et al., 2014) that are thus likely to extract larger amounts of environmental products. Blaming environmental reliant poorer households for habitat destruction and degradation may thus be misleading (Angelsen et al., 2014).

Recent studies on environmental income in the Global South have documented a very diverse range of environmental product harvesting. For instance, 120 environmental products were recorded in the PEN study in Cambodia (Ra et al., 2011). While many products are harvested infrequently and in low volume, their consumption may be widespread, e.g. Hickey et al. (2016) found that 77% of households in the global PEN sample were involved in wild food collection even if the wild food income only averaged 4% of total household income. Detailed investigations of households' own-reported product value estimates (Rayamajhi and Olsen, 2008; Uberhuaga and Olsen, 2008; Wunder et al., 2011) found unbiased own-reported values with satisfactory properties that can be aggregated into product-level

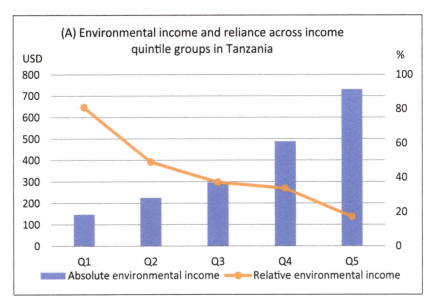
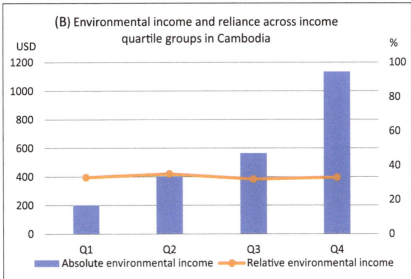

Figure 24.1 Variations in absolute and relative environmental household-level income patterns: (A) the common type: decreasing reliance and increasing absolute income with rising total income, here in the Tanzanian part of the Serengeti-Mara Ecosystem; and (B) the uncommon type: constant reliance and increasing absolute income, here in Cambodia

Source: derived from Jiao et al. (2019) and Jiao et al. (2015).

price estimates. There are usually price variations as: (i) these products are not homogeneous, displaying quality and size differences, e.g. firewood can be made up of different species and hunted mammals can have different sizes even for the same species, (ii) values vary across the year, e.g. before and after harvest, and (iii) values may vary across sites, e.g. due to differences in market access.

Most environmental income is subsistence income, generated by products harvested for own consumption; this is true for both poorer and more well-off households, with the most common pattern of the latter generating higher absolute levels of both subsistence and cash environmental incomes (and the poorer households displaying higher relative income levels) (Angelsen et al., 2014). However, subsistence products are more important to poorer households and cash products more important in better-off households. Both cash and subsistence incomes are realised through harvests of a wide range of products, for energy, medicine, food, livestock feed, farm inputs, etc.

Gap filling and safety nets

In addition to supporting current consumption and thus preventing poverty, environmental income has two other potential roles, in response to expected challenges (gap filling) and unexpected shocks (safety nets). Households know what assets they have (e.g. ½ ha of irrigated land, 1 ha of rain-fed land, two cows, five goats, 12 chickens, two adults (with primary school education), three children, a small house, 20 USD in cash, one set of gold earrings and a bicycle) and that the income generated using these assets will feed the household, say for nine months in a year. Environmental income can serve to fill the known income gap (shortfall) for the remaining three months of the year – this is known as gap filling in response to ex ante risks (seasonality). Figure 24.2 presents data from an empirical study in Tanzania, showing the seasonal variation in income composition and inter alia indicating that the share of environmental income increases during quarter 3 (Q3) when there was limited crop income (Jiao et al., 2019). Households are, however, also subject to unexpected events such as the theft of goats, loss of crops due to drought or the death of an income earning adult member. In such cases, household members can generate environmental income to make up for the unexpected income loss – this is known as the safety net function in response to ex post shocks. Shocks can be covariate, such as a drought affecting all households in an area, or

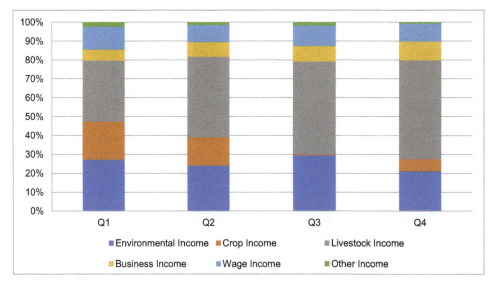

Figure 24.2 Variations in income composition (in %) across four survey quarters (Q1–Q4) in the Greater Serengeti-Mara Ecosystem in Tanzania, reflecting the seasonality and gap-filling effect of environmental income

Source: derived from Jiao et al. (2019)

idiosyncratic, such as a theft only affecting one particular household. In the same case study in Tanzania (Jiao et al., 2019), households were faced with drought during the survey-year and the environmental reliance of the poorest households (the 20% of the sampled households with the lowest income) averaged over 80% due to negative crop income (see Figure 24.1A), illustrating the critical role environmental income can have during hard times with unexpected income loss.

While there is agreement in the literature on the importance of environmental income's gap-filling role, the safety net role is debated in the empirical literature. Examining the PEN global dataset, Wunder et al. (2014b) found a limited safety net response effect, with increased environmental income more likely in connection to covariate shocks but remaining the primary response choice for only 8.0% of households suffering a shock. Later studies confirm these findings and point to the higher importance of spending savings, reducing consumption, selling assets, obtaining loans through social networks and engaging in wage labour (e.g. Møller et al., 2019). It should be noted, however, that there are locations where the safety net function may be important, e.g. a case study from the uplands of Vietnam showed that households experiencing shocks, whether covariate or idiosyncratic, extracted more resources from forests (Völker and Waibel, 2010).

Poverty reduction

The two above functions are mainly associated with poverty *prevention* – support to current consumption, gap filling and the safety net functions help prevent households from getting worse off and falling (deeper) into poverty. Environmental income may also contribute to poverty *reduction*, allowing households to climb above the poverty line. Inclusion of environmental income in estimating total household income allows more accurate calculation of income (Vedeld et al., 2007) and can assist policy makers in obtaining a more complete picture of poverty incidence and dynamics (Walelign et al., 2016). In itself, including previously unrecorded environmental income in household income estimates reduces the number of households living beneath the poverty line (e.g. Cavendish, 2000; Robinson, 2016) – without, obviously, making any difference on the ground. For environmental income to reduce household-level experienced poverty, it must contribute to allowing households to move along pathways that will take them above the poverty line, e.g. by facilitating cash income generation through value-added processing and marketing of products that are in demand (Belcher, 2005).

There is emerging evidence indicating the nation-level economic importance of environmental incomes in some countries, such as for shea nuts in Burkina Faso (Pouliot, 2012; Rousseau et al., 2015), timber in Mexico (Bray et al., 2006; Samii et al., 2014) and medicinal plants in Nepal (Olsen and Helles, 2009). Thus, while environmental incomes do not appear to generally play a common role in providing pathways out of poverty for rural households in the Global South (Walelign et al., 2019), the issue of scale is important: environmental incomes vary across space and time (Angelsen et al., 2014). As argued by Pullanikkatil and Shackleton (2019), there may be much scope in investigating and learning from local examples of successful poverty reduction through environmental incomes; rather than pursuing economy-wide solutions to combat poverty, substantial progress may be realised for households by learning what has worked locally and scaling that up, e.g. at district levels. While there is a lack of panel data on household-level environmental income, such a three-wave dataset exists for three sites in Nepal (Larsen et al., 2014). Lessons learned from this dataset are (i) time-series analysis including environmental income improves household-level poverty assessments and dynamics (Walelign et al., 2016), (ii) combining income and asset data provides a better understanding of livelihood strategies and household movements between strategies over time than using only income or asset data (Walelign et al., 2017), (iii) livelihood strategies are dynamic, with most households changing strategies between observation periods (Walelign et al., 2017), (iv) a common pathway out of poverty included accumulating assets through farming, petty trading, and migratory work, with no major role played by environmental income (Walelign et al., 2017), and (v) while total environmental income does not help households accumulate assets over time, the non-forest component of environmental income allows some household asset accumulation (Walelign et al., 2019, 2021). However, variations across short distances should be kept in mind; in another nearby location, Timmermann and Smith-Hall (2019) used household-level panel data and found that environmental income (in the form of high altitude commercial medicinal plants) was the major route out of poverty for rural households.

While much work has been done on poverty traps (Haider et al., 2018), there is very limited work on the theory of environmental reliance traps, i.e. mechanisms that keep households poor and reliant on environmental income at the same time. It has been speculated that households with high reliance on environmental income are more likely to remain in poverty as they are trapped in low return activities with no other options, e.g. due to limited skills, assets and capital (Angelsen and Wunder, 2003). However, using an environmentally augmented three-wave panel household income and asset dataset from Nepal, Walelign et al. (2020) found no evidence of an environmental reliance trap. The adverse consequences of the existence of environmental reliance traps – likely to lock already marginalised people in poverty – warrant further theoretical and empirical work. Figure 24.3 illustrates households' possible movement patterns into and out of poverty over time: they can stay where they are (remain well-off or poor), drop into poverty, or move out of poverty.

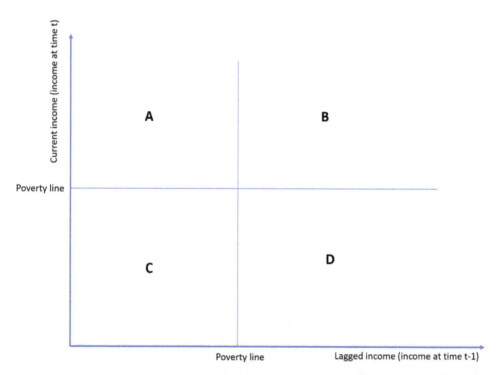

Figure 24.3 Graphic illustrations of the four household-level possible movements into and out of poverty over time (adapted from Charlery and Walelign, 2015): (A) Transient poor: moving out, (B) Non-poor: staying out, (C) Poor: staying in, and (D) Transient poor: moving in. Notes: the vertical and the horizontal lines represent poverty lines in time 1 and 2, respectively. The desired movement is toward B.

We need to better understand structural supply rigidities (Wunder et al., 2014a) and other production network constraints to identify under what conditions and at what scales environmental income can contribute to move rural households out of poverty. This includes the structure and function of domestic, regional and international markets, including the nature and characteristics of demand drivers and supply side issues that move beyond the simplistic Homma (1996) model (emphasising increasing resource scarcity as the product-level driver of price changes, leading to a shift from wild harvesting to cultivation, or overexploitation and economic extinction) and the associated classic extractive boom-and-bust cycles to better understand the dynamics of cultivation processes. It is also important to pay attention to production networks/value chains: existing environmental income studies are primarily focused on estimating incomes in harvesting households, with limited attention to wider factors and opportunities, including in upgrading and processing. There are few studies documenting incomes created in global production networks and there may be much unrealised potential in paying more attention to industrial subsectors. For instance, in a study of commercial medicinal plant processing in Nepal, Caporale et al. (2020) argued that removal of domestic barriers to change could facilitate a move from supplying raw materials to neighbouring countries and producing lower-value consumer products for the domestic market to becoming an exporter of sustainably sourced higher-value products.

Moving forward

Moving beyond the search for an economy-wide silver bullet that will provide a generic approach to enhancing the economic importance of environmental products to rural livelihoods in the Global South, the potential appears to lie in uncovering characteristics of products and contexts that are already making a difference, such as the Chinese caterpillar fungus *Ophiocordyceps sinensis* that is currently transforming rural livelihoods in Tibet and throughout high altitude communities in the Himalayan range states (e.g. Pouliot et al., 2018; Winkler, 2008). There is also scope for deepening our understanding by establishing productive connections across the natural and social sciences, such as integrating data from ecology, economics and political science and using that for scaling up results and identifying recommendations for interventions. Recent inspiring work in this direction include data integration and upscaling for African pangolins (Ingram et al., 2018) and Mexican orchids (Ticktin et al., 2020). In the longer term, work could move towards designing new research that strives for consilience (Wilson, 1998), e.g. by integrating environmental product demand studies with biological psychology.

There is also straightforward scope for starting to consider environmental incomes in the Global South as part of the rapidly expanding bioeconomic scholarship. This is presently dominated by biotechnological approaches, with most published studies focused on northern and western EU (Bugge et al., 2016; Staffas et al., 2013). While other bioeconomic approaches exist (Bugge et al., 2016), their assumptions, causal linkages, and lines of reasoning have not been characterised in any detail, hindering systematic analysis and identification of green (bio-based) transition pathways (Dietz et al., 2018), including those connected to renewable natural resources. The dominant biotechnological approach is decoupled from sustainability issues (Bugge et al., 2016) and its emphasis on natural science and patents makes it less relevant to low-income countries in the Global South. Likewise, proposed ways to measure the bioeconomy (Wesseler and Braun, 2017) are inappropriate for low-income countries, e.g. national accounts cannot be used to estimate the economic contributions of renewable natural resource sectors (e.g. Angelsen et al., 2014; FAO, 2016). While bioeconomic attention is lacking, low-income countries continue The Great Convergence through their escape from poverty (e.g. Pinker, 2018; Radelet, 2015) utilising their key natural resource assets (Collier, 2010) thus adding atmospheric carbon and increasing biodiversity conservation concerns (e.g. Maxwell et al., 2016). However, economic growth need not be synonymous with burning carbon and trashing sustainability. In low-income countries in the Global South, the bioeconomy potentially offers an opportunity to limit emissions while reducing poverty and contributing to conservation. While a few low-income countries have started to pursue elements of bioeconomic strategies, these are heavily oriented towards biotechnological interventions (BioStep, 2020; Dietz et al., 2018) with limited attention to issues of combining economic growth and sustainable natural resource management (Georgeson et al., 2017). This may reflect transference-of-thinking, replicating the EU focus on biotechnology in other countries, as well as difficulties in identifying positive growth-sustainability linkages.

At the level of individual households, there is ample scope to further increase our understanding of the economic importance of environmental income to current consumption – inside the household, who generates environmental income? Who controls environmental income expenditures? While some attention has been paid to within-household inequalities (e.g. Agarwal, 2018; Pouliot, 2012), such knowledge remain scant in connection to environmental products, with almost all studies focused on differences between groups and/or households. The advent of cheap technology also allows methodological progress – what methods will allow continued measurement of environmental income in real time? The use of smartphones, apps and tablets provide a possible avenue to move beyond the problems of recall and expensive single-year recording of environmental product harvesting.

Key points

- Understanding environmental income is critical to obtain accurate understanding of rural livelihoods in the Global South.
- Environmental income contributes to poverty prevention, through support to current consumption and sometimes through gap filling and functioning as a safety net.
- Environmental income does not appear to provide general pathways out of poverty but can do so in certain locations. Lessons from local examples of successful poverty reduction needs to be reviewed.
- Further research on environmental incomes should include work on environmental reliance traps, intra-household issues and method development to gain from new cheap technology.

Further reading

Issue 64 (supplement 1) of the World Development journal contains a string of papers on environmental incomes and rural livelihoods in the Global South.

The Center for International Forestry Research (CIFOR) website includes a page with an overview of all publications coming out of the Poverty Environment Network (https://www2.cifor.org/pen/pen-publications, maintained up to April 2017).

References

Agarwal, B. (2018) 'Gender equality, food security and the sustainable development goals', *Current Opinion in Environmental Sustainability*, 34: 26–32. https://doi.org/10.1016/j.cosust.2018.07.002

Angelsen, A., Jagger, P., Babigumira, R., Belcher, B., Hogarth, N., Bauch, S., Börner, B., Smith-Hall, C. and Wunder, S. (2014) 'Environmental income and rural livelihoods: A global-comparative analysis', *World Development*, 64: S12–S28. https://doi.org/10.1016/j.worlddev.2014.03.006

Angelsen, A., Larsen, H.O., Lund, J.F., Smith-Hall, C. and Wunder, S. (eds.) (2011) *Measuring livelihoods and environmental dependence – methods for research and fieldwork*, London: Earthscan. https://doi.org/10.17528/cifor/003341

Angelsen, A. and Wunder, S. (2003) *Exploring the forest-poverty link: Key concepts, issues and research implications*, Bogor: Center for International Forestry Research.

Balick, M.J. and Cox, P.A. (2020) *Plants, people, and culture – the science of ethnobotany*, New York: CRC Press. https://doi.org/10.1201/9781003049074

Belcher, B.M. (2005) 'Forest product markets, forests and poverty reduction', *International Forestry Review*, 7(2): 82–89. https://doi.org/10.1505/ifor.2005.7.2.82

Belcher, B., Ruiz-Pérez, M. and Achdiawan, R. (2005) 'Global patterns and trends in the use and management of commercial NTFPs: Implications for livelihoods and conservation', *World Development*, 33(9): 1435–1452. https://doi.org/10.1016/j.worlddev.2004.10.007

BioStep (2020) *Overview of political bioeconomy strategies. Promoting stakeholder engagement and public awareness for a participative governance of the European bioeconomy*. Available at: www.bio-step.eu/background/bioeconomy-strategies.html (Accessed 31 May 2020).

Bray, D.B., Antinori, C. and Torres-Rojo, J.M. (2006) 'The Mexican model of community forest management: The role of agrarian policy, forest policy and entrepreneurial organization', *Forest Policy and Economics*, 8(4): 470–484. https://doi.org/10.1016/j.forpol.2005.08.002

Bugge, M.M., Hansen, T. and Klitkou, A. (2016) 'What is the bioeconomy? A review of the literature', *Sustainability*, 8: 691. https://doi.org/10.3390/su8070691

Caporale, F., Mateo-Martin, J., Usman, M.F. and Smith-Hall, C. (2020) 'Plant-based sustainable development – the expansion and anatomy of the medicinal plant secondary processing sector in Nepal', *Sustainability*, 12: 5575. https://doi.org/10.3390/su12145575

Cavendish, W. (2000) 'Empirical regularities in the poverty – environment relationship of rural households: Evidence from Zimbabwe', *World Development*, 28(11): 1979–2003. https://doi.org/10.1016/S0305-750X(00)00066-8

Cavendish, W. (2002) 'Quantitative methods for estimating the economic value of resource use to rural households', in Campbell, B.M. and Luckert, M. (eds.), *Uncovering the hidden harvest: Valuation methods for woodland and forest resources*, London: Earthscan, pp. 17–65.

Charley, L. and Walelign, S.Z. (2015) 'Assessing environmental dependence using asset and income measures: Evidence from Nepal', *Ecological Economics*, 118: 40–48. https://doi.org/10.1016/j.ecolecon.2015.07.004

Collier, P. (2010) *The plundered planet*, London: Penguin Books.

Davis, B., Winters, P., Carletto, G., Covarrubias, K., Quiñones, E.J., Zezza, A., Stamoulis, K., Azzarri, C. and DiGiuseppe, S. (2010) 'A cross-country comparison of rural income generating activities', *World Development*, 38(1): 48–63. https://doi.org/10.1016/j.worlddev.2009.01.003

Dietz, T., Börner, J., Förster, J.J. and von Braun, J. (2018) 'Governance of the bioeconomy: A global comparative study of national bioeconomy strategies', *Sustainability*, 10: 3190. https://doi.org/10.3390/su10093190

Ellis, F. (2000) *Rural livelihoods and diversity in developing countries*, Oxford: Oxford University Press.

Food and Agriculture Organization (FAO) (2016) 'National socioeconomic surveys in forestry', *FAO Forestry Paper 179*, Rome: FAO.

Georgeson, L., Maslin, M. and Poessinouw, M. (2017) 'The global green economy: A review of concepts, definitions, measurement methodologies and their interactions', *Geo: Geography and Environment*, 4(1): e00036. https://doi.org/10.1002/geo2.36

Haider, L.J., Boonstra, W.J., Peterson, G.D. and Schlüter, M. (2018) 'Traps and sustainable development in rural areas: A review', *World Development*, 101: 311–321. https://doi.org/10.1016/j.worlddev.2017.05.038

Hickey, G.M., Pouliot, M., Smith-Hall, C., Wunder, S. and Nielsen, M.R. (2016) 'Quantifying the economic contribution of wild food harvests to rural livelihoods: A global-comparative analysis', *Food Policy*, 62: 122–132. https://doi.org/10.1016/j.foodpol.2016.06.001

Homma, A.K.O. (1996) 'Modernisation and technological dualism in the extractive economy in Amazonia', in Ruiz-Perez, M. and Arnold, J.E.M. (eds.), *Current issues in non-timber forest products research*, Bogor: Center for International Forestry Research, pp. 59–82.

Ingram, D.J., Coad, L., Abernethy, K.A., Maisels, F., Stokes, E.J., Bobo, K.S., Breuer, T., Gandiwa, E., Ghiurghi, A., Greengrass, E., Holmern, T., Kamgaing, T.O.W., Obiang, A.-M.N., Poulsen, J.R., Schleicher, J., Nielsen, M.R., Solly, H., Vath, C.L., Waltert, M., Whitham, C.E.L., Wilkie, D.S. and

Scharlemann, J.P.W. (2018) 'Assessing Africa-wide pangolin exploitation by scaling local data', *Conservation Letters*, 11(2): 1–9. https://doi.org/10.1111/conl.12389

Jagger, P., Luckert, M.K., Duchelle, A.E., Lund, J.F. and Sunderlin, W.D. (2014) 'Tenure and forest income: Observations from a global study on forests and poverty', *World Development*, 64: S43–S55. https://doi.org/10.1016/j.worlddev.2014.03.004

Jiao, X., Smith-Hall, C. and Theilade, I. (2015) 'Rural household incomes and land grabbing in Cambodia', *Land Use Policy*, 48: 317–328. https://doi.org/10.1016/j.landusepol.2015.06.008

Jiao, X., Walelign, S.Z., Nielsen, M.R. and Smith-Hall, C. (2019) 'Protected areas, household environmental incomes and well-being in the Greater Serengeti-Mara Ecosystem', *Forest Policy and Economics*, 106: 101948. https://doi.org/10.1016/j.forpol.2019.101948

Larsen, H.O., Rayamajhi, S., Chhetri, B.B.K., Charlery, L.C., Gautam, N., Nirajan, K., Puri, L., Rutt, R.L., Shivakoti, T., Thorsen, R.S. and Walelign, S.Z. (2014) 'The role of environmental incomes in rural Nepalese livelihoods 2005–2012: contextual information', *IFRO Documentation #4*, Frederiksberg: University of Copenhagen, Department of Food and Resource Economics.

Maxwell, S.L., Fuller, R.A., Brooks, T.M. and Watson, J.E.M. (2016) 'Biodiversity: The ravages of guns, nets and bulldozers', *Nature*, 536: 143–145. https://doi.org/10.1038/536143a

Miller, D.C. and Hajjar, R. (2020) 'Forests as pathways to prosperity: Empirical insights and conceptual advances', *World Development*, 125: 104647. https://doi.org/10.1016/j.worlddev.2019.104647

Møller, L.R., Smith-Hall, C., Meilby, H., Rayamajhi, S., Herslund, L.B., Larsen, H.O., Nielsen, Ø.J. and Byg, A. (2019) 'Empirically based analysis of households coping with unexpected shocks in the central Himalayas', *Climate and Development*, 11(7): 597–606. https://doi.org/10.1080/17565529.2018.1518812

Nielsen, Ø.J., Rayamajhi, S., Uberhuaga, P., Meilby, H. and Smith-Hall, C. (2013) 'Quantifying rural livelihood strategies in developing countries using an activity choice approach', *Agricultural Economics*, 44: 57–71. https://doi.org/10.1111/j.1574-0862.2012.00632.x

Nunan, F. (2015) *Understanding poverty and the environment*, London: Routledge.

Olsen, C.S. and Helles, F. (2009) 'Market efficiency and benefit distribution in medicinal plant markets: Empirical evidence from South Asia', *International Journal of Biodiversity Science and Management*, 5(2): 53–62. https://doi.org/10.1080/17451590903063129

Pinker, S. (2018) *Enlightenment now – the case for reason, science, humanism, and progress*, New York: Viking.

Pouliot, M. (2012) 'Contribution of "women's gold" to West African livelihoods: The case of shea (*Vitellaria paradoxa*) in Burkina Faso', *Economic Botany*, 66(3): 237–248. https://doi.org/10.1007/s12231-012-9203-6

Pouliot, M., Pyakurel, D. and Smith-Hall, C. (2018) 'High altitude organic gold: The production network for *Ophiocordyceps sinensis* from far-western Nepal', *Journal of Ethnopharmacology*, 218: 59–68. https://doi.org/10.1016/j.jep.2018.02.028

Pouliot, M. and Treue, T. (2013) 'Rural people's reliance on forests and the non-forest environment in West Africa: Evidence from Ghana and Burkina Faso', *World Development*, 43: 180–193. https://doi.org/10.1016/j.worlddev.2012.09.010

Poverty Environment Network (PEN) (2007) *PEN technical guidelines version 4*, Poverty Environment Network, Bogor: CIFOR.

Pullanikkatil, D. and Shackleton, C.M. (2019) 'Poverty reduction strategies and non-timber forest products', in Pullanikkatil, D. and Shackleton, C.M. (eds.), *Poverty reduction through non-timber forest products*, Cham: Springer, pp. 3–13.

Radelet, S. (2015) *The great surge: The ascent of the developing world*, New York: Simon & Schuster.

Ra, K., Pichdara, L., Dararath, Y., Jiao, X. and Smith-Hall, C. (2011) 'Towards understanding household-level forest reliance in Cambodia: Study sites, methods, and preliminary findings', *Forest & Landscape Working Paper No. 60–2011*, Copenhagen: University of Copenhagen.

Rasmussen, L.V., Watkins, C. and Agrawal, A. (2017) 'Forest contributions to livelihoods in changing agriculture-forest landscapes', *Forest Policy and Economics*, 84: 1–8. https://doi.org/10.1016/j.forpol.2017.04.010

Rayamajhi, S. and Olsen, C.S. (2008) 'Estimating forest product values in Central Himalaya – methodological experiences', *Scandinavian Forest Economics*, 42: 468–488. https://doi.org/10.22004/ag.econ.198840

Robinson, E.J.Z. (2016) 'Resource-dependent livelihoods and the natural resource base', *Annual Reviews of Resource Economics*, 8: 281–301. https://doi.org/10.1146/annurev-resource-100815-095521

Rousseau, K., Gautier, D. and Wardell, A. (2015) 'Coping with the upheavals of globalization in the shea value chain: The maintenance and relevance of upstream shea nut supply chain organization in Western Burkina Faso', *World Development*, 66: 413–427. https://doi.org/10.1016/j.worlddev.2014.09.004

Samii, C., Lisiecki, M., Kulkarni, P., Paler, L., Chavis, L., Snilstveit, B., Vojtkova, M. and Gallagher, E. (2014) 'Effects of Payment for Environmental Services (PES) on deforestation and poverty in low and middle income countries: A systematic review', *Campbell Systematic Reviews*, 10(1): 1–95. https://doi.org/10.4073/csr.2014.11

Shackleton, S., Shackleton, C. and Shanley, P. (eds.) (2011) *Non-timber forest products in the global context*, Berlin: Springer.

Sheppard, J.P., Chamberlain, J., Agúndez, D., Bhattacharya, P., Chirwa, P.W., Gontcharov, A., Sagona, W.C.J., Shen, H., Tadesse, W. and Mutke, S. (2020) 'Sustainable forest management beyond the timber-oriented status quo: Transitioning to co-production of timber and non-wood forest products – A global perspective', *Current Forestry Reports*, 6: 26–40. https://doi.org/10.1007/s40725-019-00107-1

Staffas, L., Gustavsson, M. and McCormick, K. (2013) 'Strategies and policies for the bioeconomy and bio-based economy: An analysis of official national approaches', *Sustainability*, 5: 2751–2769. https://doi.org/10.3390/su5062751

Sunderland, T., Achdiawan, R., Angelsen, A., Babigumira, R., Ickowitz, A., Paumgarten, F., Reyes-García, V. and Shively, G. (2014) 'Challenging perceptions about men, women and forest product use: A global comparative study', *World Development*, 64: S56–S66. https://doi.org/10.1016/j.worlddev.2014.03.003

Ticktin, T., Mondragón, D., Lopez-Toledo, L., Dutra-Elliott, D., Aguirre-León, E. and Hernández-Apolinar, M. (2020) 'Synthesis of wild orchid trade and demography provides new insight on conservation strategies', *Conservation Letters*, 13(2): e12697. https://doi.org/10.1111/conl.12697.

Timmermann, L. and Smith-Hall, C. (2019) 'Commercial medicinal plant collection is transforming high-altitude livelihoods in the Himalayas', *Mountain Research and Development*, 39(3): R13–R21. https://doi.org/10.1659/MRD-JOURNAL-D-18-00103.1

't Sas-Rolfes, M., Challender, D.W.S., Hinsley, A., Veríssimo, D. and Milner-Gulland, M.J. (2019) 'Illegal wildlife trade: Scale, processes, and governance', *Annual Review of Environment and Resources*, 44: 201–228. https://doi.org/10.1146/annurev-environ-101718-033253

Uberhuaga, P. and Olsen, C.S. (2008) 'Can we trust the data? Methodological experiences with forest product valuation in lowland Bolivia', *Scandinavian Forest Economics*, 42: 508–524.

Vedeld, P., Angelsen, A., Bojö, J., Sjaastad, E. and Berg, G.K. (2007) 'Forest environmental incomes and the rural poor', *Forest Policy and Economics*, 9(7): 869–879. https://doi.org/10.1016/j.forpol.2006.05.008

Völker, M. and Waibel, H. (2010) 'Do rural households extract more forest products in times of crisis? Evidence from the mountainous uplands of Vietnam', *Forest Policy and Economics*, 12(6): 407–414. https://doi.org/10.1016/j.forpol.2010.03.001

Walelign, S.Z., Charlery, L., Smith-Hall, C., Chhetri, B.B.K. and Larsen, H.O. (2016) 'Environmental income improves household-level poverty assessments and dynamics', *Forest Policy and Economics*, 71: 23–35. https://doi.org/10.1016/j.forpol.2016.07.001

Walelign, S.Z., Charlery, L.C. and Pouliot, M. (2021) 'Poverty trap or means to escape poverty? Empirical evidence on the role of environmental income in rural Nepal', *The Journal of Development Studies*, online first: 1–27. https://doi.org/10.1080/00220388.2021.1873282

Walelign, S.Z., Jiao, X. and Smith-Hall, C. (2020) 'Environmental reliance traps and pathways – theory and analysis of empirical data from rural Nepal', *Frontiers in Forests and Global Change*, 3: 571414. https://doi.org/10.3389/ffgc.2020.571414

Walelign, S.Z., Nielsen, M.R. and Larsen, H.O. (2019) 'Environmental income as a pathway out of poverty? Empirical evidence on asset accumulation in Nepal', *Journal of Development Studies*, 55: 1507–1526. https://doi.org/10.1080/00220388.2017.1408796

Walelign, S.Z., Pouliot, M., Larsen, H.O. and Smith-Hall, C. (2017) 'Combining household income and asset data to identify livelihood strategies and their dynamics', *Journal of Development Studies*, 53(6): 769–787. https://doi.org/10.1080/00220388.2016.1199856

Wesseler, J. and von Braun, J. (2017) 'Measuring the bioeconomy: Economics and policies', *Annual Review of Resource Economics*, 9: 275–298. https://doi.org/10.1146/annurev-resource-100516-053701

Wilson, E.O. (1998) *Consilience*, London: Abacus.

Winkler, D. (2008) 'Yartsa gunbu (*Cordyceps sinensis*) and the fungal commodification of Tibet's rural economy', *Economic Botany*, 62(3): 291–305. https://doi.org/10.1007/s12231-008-9038-3

Wunder, S., Angelsen, A. and Belcher, B. (2014a) 'Forests, livelihoods, and conservation: Broadening the empirical base', *World Development*, 64: S1–S11. https://doi.org/10.1016/j.worlddev.2014.03.007

Wunder, S., Börner, J., Shively, G. and Wyman, M. (2014b) 'Safety nets, gap filling and forests: A global-comparative perspective', *World Development*, 64: S29–S42. https://doi.org/10.1016/j.worlddev.2014.03.005

Wunder, S., Luckert, M. and Smith-Hall, C. (2011) 'Valuing the priceless: What are non-marketed products worth?', in Angelsen, A., Larsen, H.O., Lund, J.F., Smith-Hall, C. and Wunder, S. (eds.), *Measuring livelihoods and environmental dependence*, London: Earthscan, pp. 127–145.

25
FORESTS AND LIVELIHOODS

Clare Barnes

Introduction

For many people living in rural areas of the Global South, forests are intricately woven into their social fabrics, their livelihoods and their wellbeing. They are places to make a living, a part of peoples' identities and histories, sources of food and medicines, and sites of conflict and contestation. Estimates of the number of people with forest-related livelihoods vary, however the Food and Agriculture Organization (FAO) of the UN report that the number of people relying on forests and forest products for livelihoods stands at around 1.3 billion (FAO, 2014), whereas the World Bank puts this figure at 1.6 billion (World Bank, 2001) which includes 350 million people depending on dense forests for their daily needs (Miller and Hajjar, 2020).

There is a wealth of evidence of the environmental services and livelihoods provided by forests for millions of people around the world (Oldekop et al., 2020). Environmental services include biodiversity provision, regulation of the carbon and water cycles, soil formation and watershed protection (FAO, 2020; Nesha et al., 2021; Rist et al., 2012) which in turn support livelihoods for those living both close to, and far from, forests. The direct benefits of forests for livelihoods of those living in or near forests in the Global South have been widely reported. They include being places for grazing animals, a source of timber and a wide range of non-timber forest products (NTFPs) such as food, fibre, medicines (Angelsen et al., 2014: FAO, 2020; Rist et al., 2012; Thanichanon et al., 2013; Vedeld et al., 2007), as well as providing aesthetic, spiritual, educational, and recreational benefits (FAO, 2020). Therefore, in this chapter I employ a broad conceptualisation of livelihoods, recognising that forests provide much more than just a means to make a living. What forests mean to those living in or near them will vary over time, and place to place, household to household and person to person.

I firstly outline some key processes through which forests and livelihoods are interwoven, paying attention to forest-poverty linkages, heterogeneity of forest use within communities, and cultural aspects of forests for livelihoods. I then move on to discuss what forest policies mean for livelihoods, recognising that the way a forest is defined has important consequences for the livelihood practices permitted or acknowledged in forests.

Forest-livelihood interlinkages

Can forests help to reduce poverty?

The ways in which forests help to reduce the poverty of people living in or near them continues to be widely debated in academic literature (see Porro et al., 2015, and Chapter 24 in this volume). This line of research focusses on the material aspects of livelihoods and often analyses how forests can meet people's day to day needs for their own survival as well as providing some surplus timber or NTFPs to sell to others. This is captured in Thanichanon et al.'s (2013: 1288) definition of forest dependency as 'the level to which residents rely on forest products and functions to fulfil their subsistence and commercial needs'. There has been much discussion as to whether forest dependency can result in forest loss, creating a downward spiral known as a poverty trap (Sunderlin et al., 2005), or whether income inequality can be reduced through use of forest products (Cavendish, 2000). Given the large number of people relying on forests, we may expect attention in development, or pro-poor policy circles for supporting forest-poverty linkages. However, this is limited (Miller and Hajjar, 2020; Wunder et al., 2014). Even in the Sustainable Development Goals, Miller and Hajjar (2020) argue that forests are limited to one goal ('Life on Land') and one additional target (within 'Clean Water and Sanitation'). The largest aid donor in the forest sector, the World Bank, allocates less than 1% of its funding to forest-related programmes or projects (ibid). Miller and Hajjar (2020) go on to argue that this is partly due to the complexities of forest-livelihood linkages meaning quantifying and monetising goods and services from forests is difficult. Similarly, Wahlén (2017) argues that national governments undervalue the importance of forest goods and services for livelihoods and economic development, as they are often not traded in formal markets. Wunder et al. (2014) suggest that the expensive per capita cost of providing services in forest areas, and the weak presence of state institutions in forested landscapes also play a role in explaining the lack of livelihood programmes in forest areas.

In contrast to the lack of attention in development policy arenas, academic literature on forest – poverty linkages has steadily grown since the 2000s (see Angelsen and Wunder, 2003) with the aim of improving our understanding of the contribution of forests to subsistence and commercial livelihoods. The contribution of forest products to forest dwellers' subsistence livelihoods can be significant (Rasmussen et al., 2017); food, shelter, fuel or medicines are at least partially derived from forest resources (Miller and Hajjar, 2020). That said, a complex picture of factors affecting household forest use decisions emerges, making it difficult to generalise forest-poverty linkages across the Global South. For example, with their focus on how tenure regimes (state owned vs privately owned vs community forests) affect the forest incomes of villages located in/near forests across the Global South, Jagger et al. (2014) found that higher cash incomes are generated in state-owned forests compared to privately owned or community forests. Overall, both commercial and subsistence income from community forests is lower than the other tenure regimes (ibid), which Ribot (2004) argues is due to states often devolving only degraded or low-value forests for community use. Other research investigates why households clear forests for livelihoods: in their global study, Babigumira et al. (2014) found that households with adult males available for clearing land, and households with more land and livestock tend to clear larger forest areas. They argue that their research shows that it is not households with extreme asset poverty that drive forest clearance for engaging in agriculture, as such 'local forest clearing is not a simple needs-based story' (Babigumira et al., 2014: S77). Other research reveals the interrelated role of market integration and social structures, such as Porro et al. (2015) in their study in Ucayali, Peru. The authors found that mestizo households (of mixed indigenous and European heritage) were able to take advantage of existing legal and illegal trade mechanisms to turn timber availability into financial capital. In remote indigenous communities, market integration has not supported local economies as the market is unregulated with little benefit remaining in the communities. Their research points to the need to pay attention to specific ethnicity/location configurations and the wider political economy in understanding forest-poverty linkages.

Researchers should also look beyond simplistic binaries of subsistence vs. commercial livelihoods as the picture across the Global South is more nuanced and complex in many ways (Rasmussen et al., 2017). For example, NTFPs are collected for both subsistence and commercial purposes making their disaggregation challenging (Wahlén, 2017), NTFPs may become commercialised for some forest users, and not for others (Sunderland et al., 2014), NTFP commercialisation can be short-term, insecure and reliant on the interests of support agencies or NGOs, creating fluctuations between these categories (Barnes, 2019), and forest use is not usefully separated from other aspects of livelihoods (Cleaver, 2012) (see Box 25.1, Chapter 26 of this volume and Rasmussen et al. (2017) for a discussion of agriculture-forest livelihoods).

Box 25.1 Forest livelihoods understood as interwoven into other aspects of livelihoods

Thanichanon et al.'s (2013) study in Luang Prabang, northern Laos in an area of mountainous secondary forest, shows how road accessibility affects livelihoods. This research helps to expose the threads weaving through forest use and other aspects of livelihoods, challenging any attempt to neatly separate forest livelihoods from an analysis of other aspects of forest dwellers' lives.

In the research district, NTFPs from secondary forests comprise 44% of household subsistence (firewood, construction materials and food) and 55% of household income (mostly from paper mulberry and broom grass) (Thanichanon et al., 2013) and villagers also engage in agriculture, selling rice and other cash crops, engage in shifting cultivation and raise livestock. Thanichanon et al. (2013) studied three villages along a gradient of accessibility, and found that road accessibility affects the influence of government, markets and public services on livelihoods, which had interwoven positive and negative effects on livelihoods. In the most accessible village, in-migration (encouraged by government policy, infrastructure and market access) has led to declining availability of NTFPs and land shortage, but at the same time increased access to markets, education, healthcare and extension workers has increased income from cash crop cultivation, tree plantations and NTFP trading (ibid). Government land allocation and village relocation policies are more likely in accessible locations and Ducourtieux et al. (2005, in Thanichanon et al. (2013)) found that the impact of such policies in Laos differs depending on the wealth of households, and especially plantations in more accessible villages can increase social inequalities through forest clearance and removing access to forests.

In summary, Thanichanon et al.'s (2013: 1296) identifies the interwoven processes linking village accessibility to forest livelihoods as:

'(1) the influence of new and powerful actors determining the use of natural resources (traders and immigrants),
 (2) the neglect of weaker and marginalised smallholders not able to benefit from improved accessibility, leading to the overexploitation of forests and agricultural land, and
 (3) the lack of policy implementation in terms of extension and support to farmers but also in terms of institutional capacity to protect forest and land resources'.

Within this literature, there has been much discussion as to whether forests may help support livelihoods during times of shocks. Angelsen and Wunder (2003) argue that they may provide a form of insurance, especially to those with fewer assets (physical, human and social) who have limited alternative options for subsistence and for earning cash. This can come in the form of land for growing food, collecting NTFPs or harvesting timber for a cash income (Wunder et al., 2014). In their analysis of the Centre for International Forestry Research's Poverty and Environment Network (PEN) global dataset which includes 24 countries across tropical forested regions in Africa, Asia and Latin America, Wunder et al. (2014) found great variation in the use and importance of forest-based strategies between households, dependent on the type and combination of shocks (e.g. illness, market price for a good fluctuating, climatic event) and region of the world. Their analysis also revealed that forests generally do not serve as seasonal gap fillers for other income sources as the income to be gained from forests varies throughout the year, with its peak correlating with when non-forest household income options are most prevalent (ibid). However, the authors again highlight that specific locations may not align with this broad analysis, encouraging researchers and practitioners to move beyond generalised statements about the role of forests for poverty reduction in the Global South.

Such studies which focus on resource use decisions at community or household level have also been critiqued for not paying sufficient attention to the wider socio-political processes which have allowed some communities to remain vulnerable to shocks, as well as the micro-politics which explain why some individuals within communities are left marginalised and others favoured (Faye and Ribot, 2017). In their study in Tambacounda region, Senegal, Faye and Ribot (2017) show how forest-villagers engaged in selling charcoal are marginalised through being denied access to forest resources, urban markets and influence over the government by the urban merchant class and government officials. Therefore, instead of a focus on building capacities of forest villagers to respond to shocks through technical training programmes, Faye and Ribot (2017: 16) argue that transformation of their livelihoods would require 'a focus on patronage and corruption, impunity and illegality, movement toward substantive representation in which people could hold their leaders to account and in which leaders – at least the local ones – could exercise the powers given to them by law'. In line with this approach which centres power dynamics and social relations in understanding livelihoods, it is essential to interrogate what is meant by a forest *community* and the heterogeneity the term can mask.

Heterogeneity of forest use within communities

Research on forest-livelihood linkages challenges the myth of community (Hall et al., 2014) i.e. a romanticised view of a static, homogenous forest community harmoniously making decisions on sustainable forest use (Blaikie, 2006). Rather, who and what belongs in a community is subject to continual negotiation (Nightingale, 2019) and research has shown that there is a large degree of heterogeneity within communities in terms of how the forest is used for livelihoods (Agrawal and Gibson, 1999; Ostrom, 2005). Social institutions within communities such as gender, age, caste, ethnicity, marital status and wealth affect forest use. Gendered forest use has been particularly widely researched (see Mai et al., 2011; Sunderland et al., 2014) with women often seen as being more likely than men to engage in subsistence forest use for food security (e.g. Cavendish, 2000). However, Gugurani (2002) warns of a romanticised view of women which claims biological and cultural essentialism in assertions about women's interactions with nature (also see Chapter 16 of this volume). Sunderland et al. (2014) analyse the PEN global dataset (see above for description of PEN). In terms of *processed* forest products, they found that men's share of income generated is 61%, compared to 25% for women, across all three regions. In contrast, there is great regional variance in the share value from *unprocessed* forest products between men and women: in Latin America men bring in seven times more income than women, whereas in Africa women's contribution is higher than that of men (ibid). Across their categories of forest products (firewood; charcoal; food; structural and fibre; medicine, resins and dyes; food; fodder and manure; other) they found that men collect a higher diversity than women, but with great regional variation. For example, wild plant food collection was mostly carried out by women in Asia and Africa, but not in Latin America, which the authors argue is likely due to cultural differences across regions (ibid).

Other researchers have taken an intersectional approach which moves beyond simple men vs. women binaries. Indeed, Mai et al. (2011) warn of taking a narrow interpretation of gender to mean a focus only on women, or on collecting sex-disaggregated data without attention for the drivers of gendered relationships. Colfer et al. (2018: 1) argue that 'simplistic and stereotypical narratives, particularly those that dichotomise mainstream men versus mainstream women, may serve to render the variety of differences invisible and instead force complex social realities into a binary gendered model'. Yet researching and writing about such complex social realities without focussing on one aspect of identity is difficult (Nightingale, 2011). This approach means foregrounding overlapping identities and interconnections between gender, caste, wealth etc. in forest use rather than seeing each as categories of social difference which can be usefully studied separately. For example, in ethnographic work in Nepal, Staddon et al. (2014) show how knowledge of the forest can differ between women due to marital status, as it is the custom that newly married women move village to live with their husband's family. Involvement of women in community institutions making decisions about forest use can also differ due to age, wealth, caste and ethnicity (Mwangi et al., 2011).

What do forests mean to people?

So far I have discussed forest livelihoods as ways to make a living. However Miller and Hajjar (2020) encourage a shift in research to broader discussions of forests and *prosperity*. This means moving beyond material concerns towards attention for equity, health, education, safety, subjective wellbeing, culture, freedom of choice (see Miller and Hajjar (2020) for their full list of indicators of prosperity). In short, a wider understanding of livelihoods (see Chapter 2 of this volume). Such a sentiment was incorporated into the Millennium Ecosystem Assessment (MEA) through the language of ecosystems services literature. In the MEA (2005: 40), cultural services are defined as 'the nonmaterial benefits people obtain from ecosystems through spiritual enrichment, cognitive development, reflection, recreation, and aesthetic experiences' which includes cultural diversity, spiritual and religious values, knowledge systems, educational values, inspiration, aesthetic values, social relations, sense of place, cultural heritage values, recreation and ecotourism. In contrast to this rather apolitical notion of culture, Gugurani (2002: 231) argues that 'forests are more than a source of fuel or a 'natural resource': they shape myriad social relations that are locally specific and respond to the ecological and geographical contexts of a place'. As such, forests are situated in particular cultural and historical contexts and power relations can maintain, undermine or alter certain cultural relations with forests.

The following research in Vietnam, Cameroon and Nepal offers examples of why culture needs to be foregrounded in forest livelihood studies. Ngo et al. (2021) found in their study in northern Vietnam that forests are central to key moments for the Tay and Dao ethnic groups such as through the 'phạt mộc' ceremony when moving house, during which the ghost of trees is invited back to the forest. Culture is also located in the meaning afforded to practices such as hunters seeing hunting as an act of pleasure, or the sense of peace gained when their interviewees heard singing birds or lapping water. In their research in north-western Cameroon, Cuni-Sanchez et al. (2019) link individual place identity to cultural practices including the sense of 'home' and family history tied up in claimed land. They also found differences between cultural groups in how they viewed wildlife: pastoralists view wildlife as having an 'existence' value, whereas for farmers, wildlife is a food or cultural ornament. Drawing on the concept of socionatures, in which the social and environmental are seen as entangled rather than separate, Nightingale's (2019: 27) research in Nepal details her respondents' 'deep emotional connection to the forest' in which one respondent described the forest as 'our heart'. The notion of socionatures challenges the separation of culture from the 'natural' forest as if they are objective distinct entities, such as seen in the MEA. As an example of how forests/culture/social power relations in communities cannot be neatly separated from each other, Box 25.2 discusses sacred groves and how they are enmeshed in forest communities' cultural and political relations.

Informal forest community norms have also been the subject of forest conservation research with the rather instrumental aim of understanding how some traditional local norms may support conservation efforts. Sinthumule and Mashau (2020) note how taboos have played an important role in the conservation of forest and wildlife globally and provide examples such as taboos around leaving a path/road so as not to disturb ancestral spirits, or refraining from taking fallen wood from sacred forests.

Box 25.2 Sacred groves

Sithole (2004:122) describes sacred groves as 'areas where the community has established a covenant with deities or other sacred entities to refrain from certain uses of the environment' The responsibility for maintaining the covenant lies with the whole community, though often there is a traditional authority (fetish priest, spirit medium, chief or clan head) and a committee/group to set and enforce the rules (ibid). Similarly, Chouin (2002) argues that a sacred grove is a social rather than a natural entity and its consecration can be seen as a spiritual devise. In her work in dambos (wetlands) in Zimbabwe, Sithole (2004) found that rules in place in the sacred groves may be held secret from outsiders and can include norms of access, permissions, acceptable clothing, areas where silence is required and others where clapping is expected (ibid). In seeing the control of groves in Ghana as a reflection of the power of the ruling group, Chouin (2002: 45) reveals how this control allows such groups to 'monopolise the production of truth and sustain their domination'. Similarly, Sithole (2004) argues that the forest groves on dambos in Zimbabwe are political instruments and the status of the ruling elite is tied to their presence.

In their research on devithans (Nepali Hindu community's sacred groves) in the state of Sikkim, northeast India, Acharya and Ormsby (2017) unpack the enmeshed cultural and political nature of sacred groves. In order to appreciate their significance for the Nepali community's livelihoods, sacred groves need to be understood within wider citizenship issues faced by Nepalis in the state. Some Nepalis have struggled to prove their citizenship of the state and devithans are one way of substantiating their claims to autochthony. However, in contrast to Buddhist sacred sites afforded state protection as places of worship, the government does not recognise devithans as being sacred sites. Yet they provide a sense of belonging, cultural identity and express Nepali association with the forest. As well as their role in validating autochthony, power relations around devithans can be seen at the local level, even within households. As devithans are mostly on private land, they are claimed by the owner's caste or clan, or even family, leading to contested claims between households. Such conflicts in themselves indicate that devithans are sources of pride, status and social mobility.

Livelihood implications of forest policies

Why does the definition of forests matter for livelihoods?

So far I have discussed the many aspects of livelihoods in/near forest areas without defining what I mean by forests. But the way forests are defined and interpreted varies, as different people see or value certain characteristics of forests and not others (Côte et al., 2018; Fairhead and Leach, 1998). For example, a forest may be defined by tree height, canopy cover and size (such as the UN's Food and Agriculture Organization's definition of a forest as being an area of more than 0.5 ha and with more than 10% tree cover), by the amount of carbon it can sequester, by its biodiversity, by its location relative to an urban area (e.g. remote, accessible), by a term related to a perceived level of quality (e.g. degraded, rich, diverse, productive) and by views on whether people live, or *should* live, there or make use of it for their livelihoods (e.g. empty, inhabited, wasteland). Viewing the forest as a 'fuzzy and political category' (Côte et al., 2018: 254), recognises that the way forests are defined matters for livelihoods. Some definitions, or ways of knowing, the forest gain traction amongst international organisations, policy makers and scientists, and become institutionalised in forest policies, with consequences for livelihoods. A *desirable* forest will look very different for conservationists, heterogenous forest users, or those involved in different industries in forested areas such as timber or ecotourism (Leach and Fairhead, 2002) and narratives of elite stakeholders can exclude other views of forests (Scheba and Mustalahti, 2015), leading to conflict, which often involves contestations over livelihoods (Borras Jr. et al., 2020; Davis and Robbins, 2018). Box 25.3 provides an example of the discursive power of forest land classification and how associated policies may erode unacknowledged forest livelihoods.

Box 25.3 Livelihood implications of defining forests as wastelands

Baka's (2017, 2019) research reveals the livelihood consequences of the Government of India's classification of marginal or degraded forests as 'wastelands'. This classification was used when forests were evaluated as unproductive and opened possibilities for such forests to be 'improved' so that economic benefits could ensue. By extension, people who used wastelands as a commons for gathering fuelwood and fodder, were labelled 'backward' and incapable of generating revenue from the forest (Baka, 2019). 'Improvement' of the forests was enacted through two policies which both included tree plantations and practices of enclosure. The first policy was the Social Forestry Programme of the 1970s which aimed to improve fuelwood supplies in rural areas through plantations of *Prosopis juliflora* (Prosopis). This tree species became an invasive species throughout the dryland regions of India and therefore Prosopis lands were again classified as wastelands. This led to these areas again being available for 'improvement', this time in the form of the second policy: National Mission on Biodiesel of 2003 (Planning Commission of India, 2003). Prosopis was replaced by *Jatropha curcas* (Jatropha) with Baka (2017) arguing that this shift in species was due to the government's evolving view of what constitutes modern energy consumption. The consequences of this shift in policy in Tamil Nadu, southern India, were to remove livelihood opportunities for landless labourers employed in Prosopis cutting crews, exclusion of landless communities from forests for grazing animals and collecting fuelwood, the rise of land brokers attempting to secure plots for biofuel companies, and ultimately the migration of affected land users to urban areas in search of wage labour Baka (2019).

How can forest policies influence livelihoods?

Policies at work in forested landscapes encompass and perpetuate particular visions of forest-livelihood relationships. As is now clear, forests in the Global South are far from 'empty', therefore policies aimed at managing forests (or land which was or could be forested) can have huge direct impacts on forest dwellers' livelihoods. In this final section I briefly discuss two broad policy approaches and their livelihoods implications: 1) decentralised forest governance to community institutions to promote forest conservation, 2) policies aiming to control forests for conservation purposes.

The 1980s saw a strengthening of discourses in development and natural resource governance scholarships around decentralisation from national to local government, local democracy, pluralism and rights (Agrawal, 2007; Pacheco et al., 2012). In-line with this broad discourse, donors funded, and civil society coalitions advocated for, decentralised forest management programmes across the Global South (Blaikie and Springate-Baginski, 2007; Ribot, 2002). Forest dwellers were increasingly seen as being cheaper (Somanathan et al., 2009) and better guardians of forests (Fortmann and Ballard, 2011; RRI, 2020) than state governments because the forest-livelihoods linkages discussed above provide incentives to sustainably manage resources (Arts and de Koning, 2017). Forest decentralisation is underpinned by a belief that decisions about forest use should be made at a level where social and environmental effects of forest practices are directly felt (Lund et al., 2018).

Forest decentralisation takes many forms, with communities enjoying varying rights and responsibilities (Lund et al., 2018). The Rights and Resources Initiative maintains a forest tenure database to track 58 countries' forest area. Their latest figures from 2017 reveal that Indigenous Peoples, local communities and Afro-Descendants in Africa, Asia and Latin America now have legally recognised rights to 28% of forests in these regions (RRI, 2021) though the types of rights (e.g. access rights, exclusion rights) and duration of rights varies country to country. These varying rights mean that it is not very helpful to generalise about whether forest decentralisation per se has a positive or negative impact on livelihoods. Rather, the literature reveals some key processes mediating and explaining the impact of decentralisation policies on forest communities' livelihoods: whether resulting institutions are upwardly or narrowly accountable (Lund et al., 2018), whether existing local institutions have been ignored in favour of externally designed community institutions (Cleaver and de Koning, 2015; see Chapters 4 and 44 in this volume), the involvement of local knowledges (Fortmann and Ballard, 2011; Sinthumule and Mashau, 2020), the role of local elite and power dynamics within and between communities (Kashwan, 2015; Persha and Andersson, 2014; see Chapter 15 in this volume), and the social relations affecting access to resources (Krul et al., 2020; Myers and Hansen, 2020; Ribot and Peluso, 2003). Therefore, there is a need to analyse on a case-by-case basis to understand how decentralisation policies are implemented and contested in order to reveal whose livelihoods are affected and in what ways.

The second set of policies I discuss contrasts greatly with the ethos (if not implementation) of forest decentralisation policies, as these policies aim to separate people from forests, thereby excluding people from forest livelihoods. Such policies are often based on a premise that forest users' livelihoods threaten forests and 'inviolate' areas should be created to preserve forests (and other landscapes) (Rai et al., 2021). Implementation of these policies is about realising this image and imposing a view of conservation as being something that can only happen when forest users are removed from landscapes. The controlling practice of excluding people from areas as they are deemed to threaten conservation goals is captured by the widely used term 'fortress conservation' (Brockington, 2002) and can take many forms (sometimes including force). The rights and livelihoods implications of this form of conservation have been widely reported in various landscapes (Brockington, 2002; Büscher and Fletcher, 2019; Rai et al., 2021). Tauli-Corpuz et al. (2018: 4) argue that advancing such forest conservation practices is 'creating chronic patterns of abuse and human-rights violations'. The loss of access to forests for livelihoods can also have wider repercussions for forest users' representation and inclusion in governance, as their exclusion from forest livelihoods reinforces pejorative labels used against them (Leach and Fairhead, 2002).

According to Büscher and Fletcher (2019), the neoprotectionists' call to increase protected areas devoid of people is currently experiencing a resurgence. As proposals to tackle climate change through simplistically appealing policies such as tree-planting gain international attention, there is a critical need to analyse their implications for livelihoods. This requires attention for their impacts on social relations, systems of livelihood production, and cultural claims to forests, as well as scrutinising discourses which privilege particular knowledges or visions of forests at the expense of others. Contestations around forest visions can clearly be seen in debates surrounding the Post-2020 Global Biodiversity Framework being drafted in 2021 by the UN's Convention on Biological Diversity. It has met strong resistance from international organisations such as the Forest Peoples Programme (see FPP, 2021), Survival International (see SI, 2021) and many other civil society organisations due to its goal of creating protected areas and other area-based conservation measures across 30% of the planet by 2030. Concerns around human rights abuses and displacement and eviction of forest users lead many civil society organisations to call for greater safeguarding of societies and livelihoods in such conservation policies. Büscher and Fletcher (2019) argue that such a stance fails to challenge the underlying discourse that forests need to be protected *from* people. Instead, and as part of their call for convivial conservation, they propose a transition from protected to promoted areas, conceptualised as 'fundamentally encouraging places where people are considered welcome visitors, dwellers or travellers rather than temporary alien invaders upon a nonhuman landscape' (Büscher and Fletcher, 2019: 4).

Conclusion

In taking a broad approach to livelihoods this chapter has discussed the complex ways in which forests and livelihoods are interwoven for many living in rural areas of the Global South. Forests can generate significant material and non-material livelihood benefits, but such benefits are not experienced equally across and within communities, and do not remain unchallenged. In viewing livelihoods as part of the social processes shaping lives in forest areas, this chapter has drawn attention to the role of power dynamics across scales in influencing whose livelihoods are supported, altered, challenged or ignored. Hegemonic discourses of what a forest should look like deem some livelihoods acceptable and others as insignificant or irrelevant to wider development or conservation goals. The institutionalisation of such discourses through policies and bureaucratic processes has material and nonmaterial implications for millions of people. Therefore, making forest livelihood dynamics visible is essential for analysing and challenging policies at work in forest areas.

Key points

- Forests provide opportunities for subsistence and commercial livelihoods for millions of people across rural areas of the Global South. How people materially benefit from these opportunities is mediated by the intersection of powerful social institutions such as gender, caste, age, wealth, ethnicity etc. operating across scales, including within households.

- Forests in the Global South are often enmeshed in the cultural and social relations of those living in or near them. What forests mean to people is affected by the particular cultural and historical contexts in which forests are situated, and these meanings are woven into community power relations.
- Visions of an ideal forest differ in terms of whether forest livelihoods are seen as promoting, being compatible with, or posing a threat to, forest conservation. This means forest policies are contentious and how they are implemented can have far-reaching livelihood implications.

Further reading

Colfer, C.J.P., Basnett, B.S. and Ihalainen, M. (2018) 'Making sense of "intersectionality": A manual for lovers of people and forests', *CIFOR Occasional Paper 184*, Bogor, Indonesia: CIFOR. https://doi.org/10.17528/cifor/006793

Fairhead, J. and Leach, M. (1998) *Reframing deforestation. global analyses and local realities: Studies in West Africa*, London: Routledge.

Wunder, S., Angelsen, A. and Belcher, B. (2014) 'Forests, livelihoods, and conservation: Broadening the Empirical Base', *World Development*, 64(S1): S1–S11. https://doi.org/10.1016/j.worlddev.2014.03.007

References

Acharya, A. and Ormsby, A. (2017) 'The cultural politics of sacred groves: A case study of devithans in Sikkim, India', *Conservation and Society*, 15(2): 232–242. https://doi.org/10.4103/cs.cs

Agrawal, A. (2007) 'Forests, governance, and sustainability: Common property theory and its contributions', *International Journal of the Commons*, 1(1): 111–136. http://doi.org/10.18352/ijc.10

Agrawal, A. and Gibson, C.C. (1999) 'Enchantment and disenchantment: The role of community in natural resource conservation', *World Development*, 27(4): 629–649. https://doi.org/10.1016/S0305-750X(98)00161-2

Angelsen, A., Jagger, P., Babigumira, R., Belcher, B., Hogarth, N.J., Bauch, S., Börner, J., Smith-Hall, C. and Wunder, S. (2014) 'Environmental income and rural livelihoods: A global-comparative analysis', *World Development*, 64(S1): S12–S28. https://doi.org/10.1016/j.worlddev.2014.03.006

Angelsen, A. and Wunder, S. (2003) 'Exploring the forest – poverty link: Key concepts, issues and research implications', *CIFOR Occasional Paper 40*, Indonesia: Centre for International Forestry Research.

Arts, B. and de Koning, J. (2017) 'Community forest management: An assessment and explanation of its performance through QCA', *World Development*, 96: 315–325. https://doi.org/10.1016/j.worlddev.2017.03.014

Babigumira, R., Angelsen, A., Buis, M., Bauch, S., Sunderland, T. and Wunder, S. (2014) 'Forest clearing in rural livelihoods: Household-level global-comparative evidence', *World Development*, 64(S1): S67–S79. https://doi.org/10.1016/j.worlddev.2014.03.002

Baka, J. (2017) 'Making space for energy: Wasteland development, enclosures, and energy dispossessions', *Antipode*, 49(4): 977–996. https://doi.org/10.1111/anti.12219

Baka, J. (2019) 'Do wastelands exist? Perspectives on "productive" land use in India's rural energyscapes', *RCC Perspectives*, 2: 57–64.

Barnes, C. (2019) 'Community governance of common-pool resources: Exploring institutional interfaces', in Nunan, F. (ed.), *Governing renewable natural resources theories and frameworks*, London: Routledge: pp. 110–128. https://doi.org/10.4324/9780429053009

Blaikie, P. (2006) 'Is small really beautiful? Community-based natural resource management in Malawi and Botswana', *World Development*, 34(11): 1942–1957. https://doi.org/10.1016/j.worlddev.2005.11.023

Blaikie, P. and Springate-Baginski, O. (2007) 'Understanding the policy process', in Springate-Baginski, O. and Blaikie, P. (eds.), *Forests, people and power: The political ecology of reform in South Asia*, Oxon: Earthscan, pp. 61–91.

Borras Jr., S.M., Franco, J.C. and Nam, Z. (2020) 'Climate change and land: Insights from Myanmar', *World Development*, 129: 1–11. https://doi.org/10.1016/j.worlddev.2019.104864

Brockington, D. (2002) *Fortress conservation: The preservation of the Mkomazi game reserve, Tanzania*, Indiana: Indiana University Press.

Büscher, B. and Fletcher, R. (2019) 'Towards convivial conservation', *Conservation & Society*, 17(3): 283–296. https://doi.org/10.4103/cs.cs_19_75

Cavendish, W. (2000) 'Empirical regularities in the poverty-environment relationship of rural households: Evidence from Zimbabwe', *World Development*, 28(11): 1979–2003. https://doi.org/10.1016/S0305-750X(00)00066-8

Chouin, G. (2002) 'Sacred groves in history: Pathways to the social shaping of forest landscapes in coastal Ghana', *IDS Bulletin*, 33(1): 39–46.

Cleaver, F. (2012) *Development through bricolage: Rethinking institutions for natural resource management*, London: Routledge.

Cleaver, F. and de Koning, J. (2015) 'Furthering critical institutionalism', *International Journal of the Commons*, 9(1): 1–18. http://doi.org/10.18352/ijc.605

Colfer, C.J.P., Basnett, B.S. and Ihalainen, M. (2018) 'Making sense of "intersectionality": A manual for lovers of people and forests', *CIFOR Occasional Paper 184*, Bogor, Indonesia: CIFOR. https://doi.org/10.17528/cifor/006793

Côte, M., Wartmann, F. and Purves, R. (2018) 'Introduction: The trouble with forest: Definitions, values and boundaries', *Geographica Helvetica*, 73(4): 253–260. https://doi.org/10.5194/gh-73-253-2018

Cuni-Sanchez, A., Ngute, A.S.K., Sonké, B., Sainge, M.N., Burgess, N.D., Klein, J.A. and Marchant, R. (2019) 'The importance of livelihood strategy and ethnicity in forest ecosystem services' perceptions by local communities in north-western Cameroon', *Ecosystem Services*, 40: 1–13. https://doi.org/10.1016/j.ecoser.2019.101000

Davis, D.K. and Robbins, P. (2018) 'Ecologies of the colonial present: Pathological forestry from the taux de boisement to civilized plantations', *Environment and Planning E: Nature and Space*, 1(4): 447–469. https://doi.org/10.1177/2514848618812029

Ducourtieux, O., Laffort, J.R. and Sacklokham. S. (2005) 'Land policy and farming practice in Laos', *Development and Change*, 36(3): 499–526. https://doi.org/10.1111/j.0012-155X.2005.00421.x

Fairhead, J. and Leach, M. (1998) *Reframing deforestation. Global analyses and local realities: Studies in West Africa*, London: Routledge.

Faye, P. and Ribot, J. (2017) 'Causes for adaptation: Access to forests, markets and representation in Eastern Senegal', *Sustainability*, 9(2): 1–20. https://doi.org/10.3390/su9020311

Food and Agriculture Organization of the United Nations (FAO) (2014) *State of the world's forests 2014: Enhancing the socioeconomic benefits from forests*, Rome: FAO.

Food and Agriculture Organization of the United Nations (FAO) (2020) *Global forest resources assessment 2020 – Key findings*, Rome: FAO. https://doi.org/10.4060/ca8753en

Forest Peoples Programme (FPP) (2021) *Human rights in the post-2020 global biodiversity framework: Options for integrating a human-rights based approach to achieve the objectives of the convention on biological diversity*, Anderen: Forest Peoples Programme. www.forestpeoples.org/en/briefing-paper/2021/human-rights-post-2020-global-biodiversity-framework-options-integrating-human

Fortmann, L. and Ballard, H. (2011) 'Sciences, knowledges, and the practice of forestry', *European Journal of Forest Research*, 130(3): 467–477. https://doi.org/10.1007/s10342-009-0334-y

Gugurani, S. (2002) 'Forests of pleasure and pain: Gendered practices of labor and livelihood in the forests of the Kumaon Himalayas, India', *Gender, Place and Culture*, 9(3): 229–243. https://doi.org/10.1080/0966369022000003842

Hall, K.V., Cleaver, F., Franks, T.R. and Maganga, F. (2014) 'Capturing critical institutionalism: A synthesis of key themes and debates', *European Journal of Development Research*, 26(1): 71–86 https://doi.org/10.1057/ejdr.2013.48

Jagger, P., Luckert, M.M.K., Duchelle, A.E., Lund, J.F. and Sunderlin, W.D. (2014) 'Tenure and forest income: Observations from a global study on forests and poverty', *World Development*, 64(S1): S43–S55. https://doi.org/10.1016/j.worlddev.2014.03.004

Kashwan, P. (2015) 'Forest policy, institutions, and REDD+ in India, Tanzania, and Mexico', *Global Environmental Politics*, 15(2): 95–117. https://doi.org/10.1162/GLEP_a_00313

Krul, K., Ho, P. and Yange, X. (2020) 'Incentivizing household forest management in China's forest reform: Limitations to rights-based approaches in Southwest China', *Forest Policy and Economics*, 111: 1–11. https://doi.org/10.1016/j.forpol.2019.102075

Leach, M. and Fairhead, J. (2002) 'Introduction: Changing perspectives on forests: Science/policy processes in wider society', *IDS Bulletin*, 33(1): 1–12.

Lund, J.F., Rutt, R.L. and Ribot, J. (2018) 'Trends in research on forestry decentralization policies', *Current Opinion in Environmental Sustainability*, 32: 17–22. https://doi.org/10.1016/j.cosust.2018.02.003

Mai, Y.H., Mwangi, E. and Wan, M. (2011) 'Gender analysis in forestry research: Looking back and thinking ahead', *International Forestry Review*, 13(2): 245–258. https://doi.org/10.1505/146554811797406589

Millennium Ecosystem Assessment (MEA) (2005) *Ecosystems and human well-being: Synthesis*, Washington, DC: Island Press.

Miller, D.C. and Hajjar, R. (2020) 'Forests as pathways to prosperity: Empirical insights and conceptual advances', *World Development*, 125: 1–13. https://doi.org/10.1016/j.worlddev.2019.104647

Mwangi, E., Meinzen-Dick, R. and Sun, Y. (2011) 'Gender and sustainable forest management in East Africa and Latin America', *Ecology and Society*, 16(1): 17. https://doi.org/10.5751/ES-03873-160117

Myers, R. and Hansen, C.P. (2020) 'Revisiting a theory of access: A review', *Society & Natural Resources*, 33(2): 146–166. https://doi.org/10.1080/08941920.2018.1560522

Nesha, M.K., Herold, M., De Sy, V., Duchelle, A.E., Martius, C., Branthomme, A., Garzuglia, M., Jonsson, O. and Pekkarinen, A. (2021) 'An assessment of data sources, data quality and changes in national forest monitoring capacities in the Global Forest Resources Assessment 2005–2020', *Environmental Research Letters*, 16(5): 054029. https://doi.org/10.1088/1748-9326/abd81b

Ngo, T.T.H., Nguyen, T.P.M., Duong, T.H. and Ly, T.H. (2021) 'Forest-related culture and contribution to sustainable development in the northern mountain region in Vietnam', *Forest and Society*, 5(1): 32–47. https://doi.org/10.24259/fs.v5i1.9834

Nightingale, A.J. (2011) 'Bounding difference: Intersectionality and the material production of gender, caste, class and environment in Nepal', *Geoforum*, 42(2): 153–162. https://doi.org/10.1016/j.geoforum.2010.03.004

Nightingale, A.J. (2019) 'Commoning for inclusion? commons, exclusion, property and socio-natural becomings', *International Journal of the Commons*, 13(1): 16–35. https://doi.org/10.18352/ijc.927

Oldekop, J.A., Rasmussen, L.V., Agrawal, A., Bebbington, A.J., Meyfroidt, P., Bengston, D.N., Blackman, A., Brooks, S., Davidson-Hunt, I., Davies, P., Dinsi, S.C., Fontana, L.B., Gumucio, T., Kumar, C., Kumar, K., Moran, D., Mwampamba, T.H., Nasi, R., Nilsson, M., Pinedo-Vasquez, M.A., Rhemtulla, J.M., Sutherland, W.J., Watkins, C. and Wilson, S.J. (2020) 'Forest-linked livelihoods in a globalized world', *Nature Plants*, 6(12): 1400–1407. https://doi.org/10.1038/s41477-020-00814-9

Ostrom, E. (2005) *Understanding institutional diversity*, Princeton NJ: Princeton University Press.

Pacheco, P., Barry, D., Cronkleton, P. and Larson, A. (2012) 'The recognition of forest rights in Latin America: Progress and shortcomings of forest tenure reforms', *Society and Natural Resources*, 25(6): 556–571. https://doi.org/10.1080/08941920.2011.574314

Persha, L. and Andersson, K. (2014) 'Elite capture risk and mitigation in decentralized forest governance regimes', *Global Environmental Change*, 24: 265–276. https://doi.org/10.1016/j.gloenvcha.2013.12.005

Planning Commission of India (2003) *Report of the committee on development of biofuel. Planning Commission*, New Delhi: Government of India.

Porro, R., Lopez-Feldman, A. and Vela-Alvarado, J.W. (2015) 'Forest use and agriculture in Ucayali, Peru: Livelihood strategies, poverty and wealth in an Amazon frontier', *Forest Policy and Economics*, 51: 47–56. https://doi.org/10.1016/j.forpol.2014.12.001

Rai, N.D., Devy, M.S., Ganesh, T., Ganesan, R., Setty, S.R., Hiremath, A.J., Khaling, S. and Rajan, P.D. (2021) 'Beyond fortress conservation: The long-term integration of natural and social science research for an inclusive conservation practice in India', *Biological Conservation*, 254: 108888. https://doi.org/10.1016/j.biocon.2020.108888

Rasmussen, L.V., Watkins, C. and Agrawal, A. (2017) 'Forest contributions to livelihoods in changing agriculture-forest landscapes', *Forest Policy and Economics*, 84: 1–8. https://doi.org/10.1016/j.forpol.2017.04.010

Ribot, J. and Peluso, N. (2003) 'A theory of access', *Rural Sociology*, 68(2): 153–181. https://doi.org/10.1111/j.1549-0831.2003.tb00133.x

Ribot, J.C. (2002) *Democratic decentralization of natural resources: Institutionalizing popular participation*, Washington, DC: World Resources Institute.

Ribot, J.C. (2004) *Waiting for democracy: The politics of choice in natural resource decentralization*, Washington, DC: World Resources Institute.

Rights and Resources Initiative (RRI) (2020) *The opportunity framework 2020: Identifying opportunities to invest in securing collective tenure rights in the forest areas of low- and middle-income countries*, Washington, DC: RRI. https://rightsandresources.org/publication/the-opportunity-framework-2020/

Rights and Resources Initiative (RRI) (2021) *Forest and land tenure*. Available at: https://rightsandresources.org/tenure-tracking/forest-and-land-tenure/ (Accessed 7 May 2021).

Rist, L., Shanley, P., Sunderland, T., Sheil, D., Ndoye, O., Liswanti, N. and Tieguhong, J. (2012) 'The impacts of selective logging on non-timber forest products of livelihood importance', *Forest Ecology and Management*, 268: 57–69. https://doi.org/10.1016/j.foreco.2011.04.037

Scheba, A. and Mustalahti, I. (2015) 'Rethinking "expert" knowledge in community forest management in Tanzania', *Forest Policy and Economics*, 60: 7–18. https://doi.org/10.1016/j.forpol.2014.12.007

Sinthumule, N.I. and Mashau, M.L. (2020) 'Traditional ecological knowledge and practices for forest conservation in Thathe Vondo in Limpopo Province, South Africa', *Global Ecology and Conservation*, 22: 1–11. https://doi.org/10.1016/j.gecco.2020.e00910

Sithole, B. (2004) 'Sacred groves on dambos in Zimbabwe: Are these private resources for the rural elites?', *Forests Trees and Livelihoods*, 14(2–4): 121–135. https://doi.org/10.1080/14728028.2004.9752487

Somanathan, E., Prabhakar, R. and Mehta, B.S. (2009) 'Decentralization for cost-effective conservation', *PNAS*, 106(11): 4143–4147. https://doi.org/10.1073/pnas.0810049106

Staddon, S., Nightingale, A. and Shrestha, S.K. (2014) 'The social nature of participatory ecological monitoring', *Society & Natural Resources*, 27(9): 899–914. https://doi.org/10.1080/08941920.2014.905897

Sunderland, T., Achdiawan, R., Angelsen, A., Babigumira, R., Ickowitz, A., Paumgarten, F., Reyes-García, V. and Shively, G. (2014) 'Challenging perceptions about men, women, and forest product use: A global comparative study', *World Development*, 64(S1): S56–S66. https://doi.org/10.1016/j.worlddev.2014.03.003

Sunderlin, W.D., Angelsen, A., Belcher, B., Burgers, P., Nasi, R., Santoso, L. and Wunder, S. (2005) 'Livelihoods, forests, and conservation in developing countries: An overview', *World Development*, 33(9): 1383–1402. https://doi.org/10.1016/j.worlddev.2004.10.004

Survival International (SI) (2021) *Survival international launches campaign to stop '30x30'* – 'the biggest land grab in history'. Available at: www.survivalinternational.org/news/12570 (Accessed 7 May 2021).

Tauli-Corpuz, V., Alcorn, J. and Molnar, A. (2018) *Cornered by protected areas: Replacing 'fortress' conservation with rights-based approaches helps bring justice for indigenous peoples and local communities, reduces conflict, and enables cost-effective conservation and climate action*, Washington, DC: Rights and Resources Initiative. https://rightsandresources.org/wp-content/uploads/2018/06/Cornered-by-PAs-Brief_RRI_June-2018.pdf

Thanichanon, P., Schmidt-Vogt, D., Messerli, P., Heinimann, A. and Epprecht, M. (2013) 'Secondary forests and local livelihoods along a gradient of accessibility: A case study in Northern Laos', *Society and Natural Resources*, 26(11): 1283–1299. https://doi.org/10.1080/08941920.2013.788429

Vedeld, P., Angelsen, A., Bojö, J., Sjaastad, E. and Berg, G.K. (2007) 'Forest environmental incomes and the rural poor', *Forest Policy and Economics*, 9(7): 869–879. https://doi.org/10.1016/j.forpol.2006.05.008

Wahlén, C.B. (2017) 'Opportunities for making the invisible visible: Towards an improved understanding of the economic contributions of NTFPs', *Forest Policy and Economics*, 84: 11–19. https://doi.org/10.1016/j.forpol.2017.04.006

World Bank (2001) *Making sustainable commitments: An environment strategy for the World Bank*, Washington, DC: The World Bank.

Wunder, S., Angelsen, A. and Belcher, B. (2014) 'Forests, livelihoods, and conservation: Broadening the empirical base', *World Development*, 64(S1): S1–S11. https://doi.org/10.1016/j.worlddev.2014.03.007

26
AGRICULTURAL LIVELIHOODS, RURAL DEVELOPMENT POLICY AND POLITICAL ECOLOGIES OF LAND AND WATER

Exploring new agrarian questions

Cristián Alarcón, Johanna Bergman Lodin and Flora Hajdu

Introduction

Using livelihood approaches to analyse the activities of people depending on agricultural production in the Global South implies a close analysis of the resources used for own consumption and/or for exchange at the level of rural communities and households (Bebbington, 1999; Ellis and Freeman, 2004; Li et al., 2021; Scoones, 2009).[1] Livelihood approaches, when applied to individuals, households and communities whose livelihoods depend directly on different forms of agricultural activities for food consumption and exchange of other agricultural products, stress the need for understanding the specific agrarian contexts surrounding these activities. Recognition of the different contexts and the varieties of agriculture in agricultural livelihoods requires an understanding of how key local resources for production and consumption are entangled in the wider political and geographical contexts where agriculture takes place. Thus, a strict focus only on the local context of resource use risks undermining the analytical potential of a livelihood approach to agricultural livelihoods. As Scoones (2016) has argued, livelihood approaches also need to be analytically articulated in relation to wider processes of political economy. In this regard, Scoones highlights that: 'Locality, place, context and the specific, textured understandings of differentiated livelihood strategies and their changes all are vital, but are merely descriptive without a wider appreciation of political economy and structural forces of power and politics that intersect with them' (Scoones, 2016: 228). Additionally, we argue that questions of ecology and how these intersect with social relations and resource use are key for the definition and deeper analysis and explanation of agricultural livelihoods in the Global South. A political ecology approach (see Chapter 16 in this volume) focuses the analysis of agricultural livelihoods on the different political processes produced through different scales and timeframes needed for the use of ecological resources in agriculture (Giraldo, 2019). A political ecology lens also brings our attention to how scientific agricultural knowledge and local knowledges about agriculture are produced in the context of the politicisation of agricultural relations and rurality (Fairhead et al., 2012; Sumberg et al., 2012). Within this context, land and water are two key ecological resources interplaying in the dynamics of agricultural livelihoods (FAO, 2013; IAASTD, 2008; Reij et al., 2013).

From this follows three important analytical starting points for our approach to agricultural livelihoods in this chapter. First, agricultural livelihoods cannot be analytically detached from wider political and economic processes at play in agrarian change. Second, access to and control over land and water interplays with, and at times produces and reproduces, very specific social relations in agriculture. These, in turn, enable or constrain prospects of agricultural livelihoods in the context of rural change, often privileging some groups over others. Third, the current situation with climate change and the global sustainability crisis further highlights the need to pay attention to different forms of knowledge about land and water use in the understanding of agricultural livelihoods and rural development. These analytical starting points are particularly relevant when approaching political decisions concerning new policies for water and land. Especially important in this context is to consider how land and water are today defined as resources for climate action such as carbon forestry and proposals for climate change mitigation through agriculture. This in turn becomes associated with new forms of governance of the resource nexus and local policies for mitigation and adaptation to climate change (Rasul and Sharma, 2016; Taylor, 2014). These processes are characterised by global scientific discussions that often lead to contested knowledges on how to give new meanings to agriculture in relation to sustainability and climate change concerns (McNeill, 2019). Despite the global nature of these discussions, nation states continue playing key roles in the local processes of (i) reframing agricultural development in light of these concerns and (ii) framing the potential of different forms of agriculture to address reduction of emissions, adapt to climate change and meet sustainability goals. In addition, nation states can enable or limit the role of large agricultural and forestry corporations in how land and water is used in specific places. Thus, processes identified as land and water grabbing and their implications for agricultural livelihoods need to be understood through the interplay between nation states and corporations (Dell'Angelo et al., 2017; Franco et al., 2013).

These contemporary concerns for agriculture bring new areas for the analysis of agricultural livelihoods. We specifically consider rural transformations, the development of large-scale schemes for agricultural development and land use, and the labour and gender dimensions of green revolutions and their implications for agricultural livelihoods in African and South American contexts. Of particular importance here is also the growing interest in alternative approaches to land and water use put forward by agroecological movements in the Global South (see Chapter 19 in this volume). Thus, the next section delves into key questions concerning these areas.

Agricultural livelihoods and agroecological alternatives: labour, land and water in the context of sustainability crisis

Livelihood perspectives pay attention to lived everyday realities and local people's own perspectives on their basic needs and everyday concerns surrounding how to make a living. In this regard, the resources implied in agricultural livelihoods, such as land and water, can have multiple and competing functions and uses. The study of agricultural livelihoods can improve our understanding of the following relationships concerning livelihoods and agriculture: 1) relationships between various types of livelihood activities e.g. between agriculture, forestry, off-farm rural work, migratory employment (and sometimes dependency of rural livelihoods on migrant household members) and small rural businesses, 2) relationships between different scales in the use of resources, ranging from individual and household scales through the village/community scale and up to regional, national and global scales, and 3) relationships between effects of temporal changes in livelihoods and longitudinal perspectives on livelihoods (see Chapter 10 in this volume). It is also important to highlight certain critical issues in the study of agricultural livelihoods (Hajdu et al., 2020). First, studies that focus on one type of rural activity risk missing the complex interlinkages between multiple activities and resources used by rural populations. Second, studies that only use a household lens may miss intra-household inequalities of gender, age or generational nature, as well as inter-household relationships where certain parts of a community have more power and influence than others. Third, studies that focus on surveys or interviews performed at single points in time may miss out the continuously changing and dynamic aspects of livelihoods. Also, it is important to consider the effects of spatial variations in local conditions for agricultural livelihoods and avoid using a few 'case study locations' to draw too far-ranging conclusions based on situations that may be specific to certain places. These critical issues in the study of livelihoods are important when addressing key contemporary processes in agrarian change and their implication for agricultural livelihoods. Below, we highlight five relevant processes concerning agrarian change and rural transformations identified in the literature.

First, the *global rise of flex crops* such as sugar cane, maize, oil palm and soybeans. According to Borras et al. (2016), flex crops are crops that can be flexibly interchanged while some consequent supply gaps can be filled by other flex crops. More importantly, since flex crops have multiple uses as for example, food, fuel, fibre or industrial material, their flexibility implies multiple relationships among various crops, their components and how they are used. This means that flex crops, as commodities and as components of commodities, are not in themselves disadvantageous to poor people or to the environment in rural areas. Thus, multiple uses of flex crops for agricultural livelihoods implies asking questions about how flex crops are produced, who controls the wealth produced from these commodities, and for what strategic purposes they are produced (Borras et al., 2016). In this regard, potential problems with flex crops are the concentration of production to certain areas or countries and that agribusinesses can respond to shifting market prices to maximise profits by determining whether they process flex crops into food or other final products as for example biofuels. While in some cases producers can benefit from these market possibilities, these options can also lead to major social and environmental problems such as deforestation, large use of chemical inputs and biodiversity loss, which are often associated with cash crops and monocropping. Thus, flex crops are entangled in politically contested questions about their use and goals (ibid).

Second, *the process of resource extractivism in rural areas*, as argued by Svampa (2012: 45), is part of a type of accumulation based on an 'over-exploitation of – largely non-renewable – natural resources as well as the expansion of frontiers to territories formerly considered "unproductive"'. Extractivism in this sense also includes the expansion of agribusiness and biofuel sectors. While extraction and processing of agricultural resources are argued to benefit national level interests and generate local employment, Svampa argues that extractivism leads to monocultures, destruction of biodiversity, concentration of landownership and a destructive re-configuration of territories, which parallels the critique of flex crops explained above.

Third, and closely connected to the previous points, the *process of de-agrarianisation*. *De-agrarianisation* has been defined as a 'long-term process of: (1) occupational adjustment, (2) income-earning reorientation, (3) social identification, and (4) spatial relocation of rural dwellers away from strictly peasant modes of livelihood' (Bryceson, 1997: 4). Both livelihoods and the social world of rural populations are thus increasingly de-linked from agriculture, due to processes of dispossession or lack of competitiveness of small-scale production in comparison to industrial-scale production. While expanding industrial production may offer employment to these rural populations, this is increasingly not the case due to mechanisation of production and the need for less staff with a more high-skill profile and rural populations today therefore often experience *jobless de-agrarianisation*. Hajdu et al. (2020) have shown how this has created an increasing dependence on state grants and remittances from urban migrants in Eastern Cape, South Africa. De-agrarianisation implies transformations and adaptation of agricultural livelihoods and in some cases the disappearance of agricultural livelihoods as those represented in peasants' production and reproduction.

Fourth, the process of *land inequality in rural areas*. As argued by Wegerif and Guereña (2020), this type of inequality needs to be addressed from different angles, including stark inequalities in the size and/or value of land that people access or hold, including the completely landless. These authors point to three key issues connected to land inequality, namely, the level of security of tenure that people have (including the ability to defend their land when encroached), the actual control over land that people have (which includes their decision-making power over land) and the control of the benefits from the land, i.e. the ability to appropriate value from the land (Wegerif and Guereña, 2020).

Fifth, *changes in rural employment, wage relations and rural markets*. As stressed by Oya and Pontara (2015), in the analysis of rural employment and wage relations it is important to address processes of accumulation in relation to different trajectories of capitalist development. This includes consideration of the expansion of agricultural commodity production, labour migration and emerging industrialisation. For these authors it is crucial to pay attention to the considerable range of interventions by state and other non-state agencies that are central to the formation of rural labour markets and the conditions of labour. For the context of Latin America and the Caribbean, the International Labour Organization (ILO) (2016) has concluded that the marked long-term downward trend in agricultural employment as a share of total rural employment has meant a decrease in rural employment. This can be traced to the fact that the surplus agricultural labour force has not been absorbed by the manufacturing sector (secondary sector) and only partly transferred to the tertiary, or service, sector made up of jobs in administration and education services for instance. This sector, in the estimation of ILO, went from representing less than 30% of total rural employment in 1950 to more than 60% in 2010 (ILO, 2016). While in some cases rural inhabitants leave agriculture for better off farm opportunities, in other cases they are forced to leave because they cannot compete or cannot survive through their agricultural activities. These shifts in rural employment and the processes through which livelihoods become de-linked from farming have important consequences for the future of agricultural livelihoods not only in Latin America, but across the Global South (Rigg, 2006).

From the perspectives of agricultural livelihoods, these five rural processes are connected to local decisions at the level of both means of subsistence and production and at the level of institutions regulating rural transformations and agrarian change. The case of water use illustrates this well. Locally, and sometimes regionally, water stress, droughts and the already noticeable major climate induced changes in the water systems are directly affecting water availability and water management in local areas (Turral et al., 2011). This in turn threatens vital local agricultural practices and requires institutional responses for water and land use. Political decisions at the state level may respond by defining new rules for either the maintenance or transformation of ecological and economic objectives for land and water use. These directly or indirectly connect agricultural livelihoods to the local actions of state agencies, international organisations and corporations responding to and acting on these new rules. Another example of this are large-scale schemes for agricultural development and the consequences of their implementation and failure for agricultural livelihoods and rural labour. As became apparent in the context of the global discussion on land grabbing that began in the late 2000s, the labour-intensive characteristics of family farms means that for local populations and local employment, outgrower schemes including the inclusion of smallholders in properly negotiated schemes would be better options than large scale plantations where wage labour is used without proper inclusion of smallholders' interests (De Schutter, 2009). Yet, research supporting the Special Rapporteur on the Right to Food's assessment of the consequences of large-scale agricultural projects showed that land leases in Africa during the land grabbing wave reported during late 2000s were characterised by government incentives for 'establishing company-managed plantations rather than promoting contract farming approaches' (Cotula et al., 2009: 86). Thus, the establishment of such plantations and the wage labour relations that followed became drivers of rural transformations and changes in agricultural livelihoods. In addition, water requirements for production implied in such land deals are often inextricably linked to politics of water management for small or large agriculture projects (Woodhouse, 2012). Similarly, the example of new forestry and forest conservation schemes proposed to deal with climate change shows the tensions and contradictions that such initiatives bring for agricultural livelihoods as well (Duker et al., 2019; Fischer and Hajdu, 2018). As highlighted in several social and environmental assessments of Payment for Ecosystems Services (PES) and schemes for Reducing Emissions from Deforestation and Degradation (REDD), one of the main changes these projects bring to rural areas concerns labour possibilities or limitations and changes in terms of access to territories where previously resources had been extracted by local communities. On the other hand, the production of agricultural products and the exchange of those products for agricultural livelihoods relies on the capacity of rural inhabitants to use their labour power in the process of production, consumption and exchange based on access to local resources. In this regard, conflicts surrounding REDD and PES projects in the Global South often show the key role of different forms of labour in agricultural livelihoods in connection to forestry and in other rural activities defined by how people access and work the land and make a living from that work (Hoang et al., 2019; To et al., 2017).

Looking at changes in labour conditions in agricultural livelihoods allows the visualisation and analysis of the key role of gender relations in agrarian change. As argued by Razavi (2009: 221), it is precisely around 'the "fragmentation of labour"' that 'some of the more useful exchanges between political economy of agrarian change and gender studies have materialised'. In this regard, an agrarian political economy approach to livelihoods means that the localised nature of livelihoods and the differentiated labour relations thereby need to be addressed in order to understand the drivers of agrarian change that are external and internal to those local livelihoods. Such interplay between internal and external forces transforming livelihoods can be exemplarily observed in the process of agrarian change connected to green revolutions in which labour, gender and ecological dynamics affect rural livelihoods (Patel, 2013) For example, in the context of the 'New Green Revolution for Africa', Andersson et al. (2021) note how scholars have analysed and challenged the gendered and often unequal outcomes of interventions for agricultural mechanisation (Badstue et al., 2020; Daum et al., 2021; Kansanga et al., 2019), new and improved seeds and breeds (Addison and Schnurr, 2016; Bergman Lodin et al., 2012; Wangui, 2008), irrigated agriculture (Nation, 2010), inorganic fertiliser and pesticide use (Christie et al., 2015; Luna, 2020) and market integration (Quisumbing et al., 2015; Tavenner and Crane, 2018). In addition, as observed in other parts of the world where green revolution schemes have been implemented, these were based on large-scale transformations of ecological relations (for example, greater use of pesticides and fertilisers) and new labour and gender relations (for example, in the feminisation of agricultural tasks under wage relations).

As stated earlier, in many rural settings, land use change and agricultural livelihoods are closely linked to dynamics of forest use, deforestation and forestry development. In their study on the relationship between deforestation and forest-dependent livelihoods, Barraclough and Ghimire (1995) discussed a case where rural inhabitants dependent on small-scale agriculture, hunting, fishing and gathering in a part of the Amazon during the 1970s, ended up severely affected by deforestation of the land they depended on. Their responses to this new situation included attempting to regenerate agroforestry practices, but due to the disruption, their search for new livelihoods implied unsustainable use of resources. In understanding how agriculture and forest livelihoods are linked, it is also important to pay attention to how gender mediates access to and use of forest resources (Arora-Jonsson, 2009). In this regard, Chiwona-Karltun et al. (2017) report on how a number of studies over the last two decades have shown that gender inequalities in access to forest resources also contribute to higher levels of poverty and food insecurity, and that women often play an important but largely unrecognised role in managing forest resources (Mwangi et al., 2011; Suna et al., 2011; Agarwal, 1997). A study from West Africa notes how women allocate more time collecting and preserving non-timber forest products (NTFPs) than men. Forest products can make an important contribution to diets of rural communities by providing supplemental calories, enhancing the dietary diversity including the intake of micronutrients and vitamins, and generating income. These contributions are particularly important during the agricultural lean season, especially improving the food security of vulnerable households (Hasalkar and Jadhav, 2004; Ogle, 1996).

Within this context, we observe that the prospects for agricultural livelihoods are often defined by local responses to crises of labour in agriculture and in other agrarian activities, such as forestry. In both forestry and agriculture, mechanisation and automation make important segments of the agricultural labour force redundant (Alarcón, 2021a). At the same time, partial mechanisation may also increase the labour burdens and drudgery for some groups in relation to the remaining non-mechanised activities. One illustrative example is when subsidised tractor hire services for ploughing were availed to farmers in western Uganda without consideration of the prevailing gendered division of labour. Consequently, these services mainly replaced and saved men's labour, while simultaneously increasing women's labour, especially in weeding. This could be traced to the fact that the tractors permitted an expansion of the area under cultivation compared to when land preparation was done manually with a hoe. Yet, no similar services or technologies were availed to save women's labour with regard to weed control (Bergman Lodin et al., 2012). Therefore, this replacement of labour and its implications for agricultural livelihoods is not gender neutral.

Following these insights, one can frame several current processes of agrarian change as threatening agricultural livelihoods in terms of crises of labour for people depending on both agricultural livelihoods and off farm agricultural employment. In addition, as the effects of climate change show, changes in land and water use jeopardise the means of subsistence and the resources necessary for the reproduction of agricultural livelihoods through labour. This creates the conditions for combined crises of labour and sustainability in agricultural livelihoods. These crises are also seen as economic possibilities for rural areas. Particularly relevant in this context is the work of international organisations such as the World Bank (WB) and the Organisation for Economic Co-operation and Development (OECD), which promote economic policies for agriculture and development to incorporate sustainability concerns into a growth paradigm. During recent years, these organisations have intensively articulated efforts to promote agricultural transformations in rural contexts characterised by crises of labour and sustainability. In particular, new prospects for agriculture are framed in the context of new relations for food production as for example when food security concerns are addressed through markets and international trade under the premises of neoliberal globalisation and international trade. This can be observed in the inception of the New Rural Paradigm by the OECD in 2006 and the World Bank Agriculture for Development Report from 2008 (OECD, 2006; World Bank, 2007).

The efforts to give a new turn to agricultural development while operating under the premises of neoliberal globalisation are not the only recent responses to the crises of labour and sustainability in agriculture. Alternatives for agricultural livelihoods are discernible in the widespread development of agroecological alternatives, the reassessment of peasants' role in sustainable agriculture and the conceptualisation of new peasantries. In the terms of authors such as Altieri and Toledo (2011) and van der Ploeg (2008, 2021), agroecology and peasantries show the potential of alternative agrarian pathways enabling the maintenance, development and scaling up of agricultural livelihoods currently threatened by the forces of neoliberal globalisation and the corporate food regime (McMichael, 2013). Here the potential of smallholders, peasants and rural activists to bring together ecological and social concerns into agricultural practices is given by their material capacity to politically articulate alternative approaches for sustainable land and water use. As Altieri et al. (2017) show, agricultural practices and strategies in agroecology include crop rotations, polycultures, agroforestry systems, cover crops and mulching and crop-livestock mixtures. In scaling up these practices agroecological movements aim at changing the terms of states' support to corporate agricultural interests as well. In this regard, we can observe that the crisis of agricultural labour indicated above runs in parallel with a growing tension between the development of capitalist agriculture and forestry and the new possibilities for agricultural livelihoods and practices within an agroecological perspective (Akram-Lodhi, 2021; Selwyn, 2021). Also, as observed in the context of agrarian struggles in Brazil, widening livelihoods strategies have been a way for farmers to reduce vulnerability. This includes diversification through pluriactivity, creating markets for local foods, production oriented to self-provisioning and internalisation of resources, de-commodification, and co-production (Schneider and Niederle, 2010). These types of agricultural alternatives call attention to the new meanings of rurality under discussion today, and how this paves the way for new rural development projects. We turn our attention to this issue in the following section.

Rural development policy and the political ecology of agricultural livelihoods

The local political ecologies of land and water are key in both the maintenance and the prospects of agricultural livelihoods. Diverging political ecology visions contribute to diverging rural development projects as well. Rural development policies can be seen as one of those key overarching political processes through which agricultural livelihoods are signified and re-signified over time. As De Janvry (1981: 224) observed during the context of rethinking the meaning of rural development 40 years ago, rural development projects can be contrasted according to their different 'clientele, goals, and instruments' and one could distinguish between 'political rural development projects' and 'economic rural development projects'. For De Janvry, when both types of rural development projects are combined into a single project, 'we obtain what has been referred to as 'integrated rural-development projects'. Today, and in addition to political and economic meanings of rural development projects, discussions on rural development increasingly include an explicit and politicised concern with sustainability and ecological resources. Here the very meaning of agricultural livelihoods becomes entangled with diverging rural development projects and struggles around ecological sustainability. To think in terms of rural development projects implies resisting generalisation of the idea of rural development and understanding rural development as specific projects that are materialised in different contexts, with different political orientations and often through political conflicts. In this regard, thinking in terms of rural development projects allows a more specific and analytical assessment of the wide array of particular projects for rural areas today and their implications for agricultural livelihoods.

A key characteristic of rural development projects is the effort to define what rural areas are today and what they could be in the future (Pain and Hansen, 2019). This is often associated with defining rural areas vis-à-vis the need to reinforce economic growth, technological development and increased productivity in agriculture. This approach has been particularly powerful in the case of Chile, where a new rural development policy for the country has been defined through a process of political and legislative discussions premised on the need for new rural development for economic growth in the country. This includes the decision to implement a national policy for rural development approved in 2020, which was largely framed in terms of the OECD Rural Policy Review of Chile from 2014 (OECD, 2014). In this case, the very notion of rural areas is being redefined by taking into account the OECD proposed categorisation for rural areas, which in contrast to the previous national definition, dramatically elevates the number of rural inhabitants in the country from 13.1% to 34.6% of the total population. This has important consequences for the policy making process concerning rurality, and this also carries important ideological views on agricultural livelihoods. One example of this is the new rural paradigm put forward by the OECD and used in Chile (see Table 26.1). As shown below, key aspects of this new rural paradigm are neoliberal goals of increasing the competitiveness of rural areas and exploitation of unused resources.

Table 26.1 The OECD's New Rural Paradigm

	Old approach	New approach
Objectives	Equalisation, farm income, farm competitiveness	Competitiveness of rural areas, valorisation of local assets, exploitation of unused resources
Key target sector	Agriculture	Various sectors of rural economies (ex. rural tourism, manufacturing, ICT industry, etc.)
Main tools	Subsidies	Investments
Key actors	National governments, farmers	All levels of government (supra- national, national, regional and local), various local stakeholders (public, private, NGOs)

Source: Reprinted from *The New Rural Paradigm: Policies and governance* (p. 60) by OECD (2006) Paris: Organisation for Economic Co-operation and Development.

The new rural paradigm aims at moving the focus of rural development from the farm level to the level of the whole rural area. While this reflects new insights about rural livelihoods going beyond agriculture, the way the paradigm is implemented in Chile brings new contestations in the process of agrarian change in the country. In this regard, we can observe that efforts to implement this new rural paradigm have taken place in the context of growing politicisation and contestation over control and use of land and water resources in the country. This not only reflects increasing contestation in rural areas but has also led to growing national conflicts around how land and water should be used, and who should control the use of land and water for irrigation in the country (Panez et al., 2020). Thus, access to and use of land in and for agricultural livelihoods are deeply connected to struggles for water in the country. More specifically, these struggles are about the legitimacy and social and environmental consequences of the legal private property rights to use water resources, which were imposed as a regulation during the implementation of neoliberal reforms under the military dictatorship (1973–1990). Today, contestation of the premise of this water property system also means contestation over state action in rural development and its aims of favouring agribusiness and forestry for export (Alarcón, 2021b; Madariaga et al., 2021).

This rural orientation of the state can be seen in the past rural development projects for the country which have aimed to reinforce an export orientation for the rural areas, but at the same time aimed to capitalise on the new global needs for food and raw materials under sustainable development discourses.

This brings to the front of the political discussion the social and ecological dimensions of private property rights of land and water in the export orientation of the country and how this interplay with agricultural livelihoods. The cases of industrial pine and eucalyptus plantations and industrial avocado plantations in rural areas serve to illustrate this point more deeply. While native forests are global biodiversity hotspots (Mittermeier et al., 2004), they are used for timber, firewood and charcoal production. Replacing them with plantation forestry has made Chile the world's fifth-largest wood pulp exporter. Exotic tree species in plantations however bring biodiversity loss, local water scarcity and risks for large forest fires (Durán and Barbosa, 2019). Regions with plantation forestry are also dealing with persistent poverty, inequalities and emigration from rural areas (Andersson et al., 2016; Hofflinger et al., 2021). These factors generate important conflicts between, on the one hand industrial agriculture and forestry, and on the other, small scale and peasant oriented use of trees and farming for local food production (Alarcón, 2019, 2015). Similarly, large scale plantations of avocado trees for export are a more recent case of agrarian change with severe consequences for agricultural livelihoods in the country (Alarcón, 2021b). These plantations require high quantities of water to grow, and the expansion of avocado plantations to, on the one hand, drier areas of the country, and on the other, areas with other crop systems, has led to competition for land and water. This affects directly the livelihoods of rural inhabitants depending on water sources for human consumption and other production systems, and this also diminishes land availability for local food production as land investment for avocado plantations, along with other fruit tree plantations for export, has meant prices of agricultural lands have soared (Medovic and Villagrán, 2021). In addition, this takes place in a context of growing mechanisation and automation of agriculture which in the estimations of the leader of the Chilean Fruit Exporters' Association could mean the elimination of about 300,000 agricultural jobs over the next 5 or 10 years (Bown, 2020).

In comparative terms, we can take here the case of rural development in Ethiopia, which in several ways resembles the interaction between agriculture, forestry and water in the context of current rural development policies in Chile. Ethiopia (Federal Democratic Republic of Ethiopia/FDRE) is one of Africa's fastest-growing economies. The forest sector is touted for its potential contribution to furthering economic growth and realising environmental goals. Such hopes are evident in national plans for the economy (FDRE, 2019) and climate (FDRE, 2017, 2011). However, deforestation has been a persistent trend for decades. In this regard, rural Ethiopia provides a complex agrarian context where there is both socio-economic (poverty, unemployment, food insecurity) and biophysical (land-degradation, cropland expansion, biodiversity loss) tension between social and environmental goals in rural development. It has also been observed that state action concerning land in Ethiopia has implied in some cases dispossession of peasants in order to bring about development goals (Makki, 2014). During the last two decades, water has become central for the national development policies of the country (Beyene, 2018). Today the incorporation of forestry activities as a new source of development is entangled with such historical patterns of land and water control. These processes bring new conflicts, and, in geopolitical terms, agricultural development connects to struggles for water, which is exemplified in the tensions between Ethiopia and Egypt over control of the Nile waters (Conniff et al., 2013; Wheeler et al., 2020). The current rural development policy for the country emphasises the need for coordinated efforts to address the combined challenges of water use and forestry development vis-à-vis the strong role of agriculture in rural development (Government of Ethiopia, 2003). Yet, this takes places in a context of tensions linked to a historical trajectory characterised by the interactions between landlessness, urbanisation and deagrarianisation (Habtu, 1997). The emergence and promotion of export-oriented agriculture of selected agricultural commodities, as the floriculture industry indicates, follows a process of production and market specialisation that brings both possibilities and new constraints and threats for agricultural livelihoods (Lavers, 2012; Oqubay, 2015). This development takes place in a context where despite ambitious pledges by Ethiopia to improve natural resources management, there are still major weaknesses in the generation of data and an evidence-base for policies concerning water and forests (Gebreyohannis Gebrehiwot et al., 2021).

In Table 26.2, key comparative aspects of today's discursive and material transformations of rural areas through rural development projects are presented for the cases of Chile and Ethiopia.

Table 26.2 Context and selected content of rural development projects in Chile and Ethiopia

	Chile	*Ethiopia*
Key agriculture and forestry facts	• Forest area: 15.9 million hectares representing 21 % of total area • Agricultural land: 15.693 million ha • Forestry sector contribution to GDP: 1.5% (an important part of exports)	• Forest area: 21 million hectares representing 18% of total area • Agricultural land: 37.903 million ha • Forestry sector contribution to GDP: 3.8% (not an important part of exports)
OECD report on rural development	Published in 2014: 'OECD Rural Policy Reviews: Chile 2014'	Published in 2020: 'Rural Development Strategy Review of Ethiopia: Reaping the Benefits of Urbanisation, OECD'
National rural development strategies	Chile's National Rural Development Policy from 2020	The Rural Development Policy and Strategy from 2003
Relevant forestry and agriculture issues in the rural development strategies adopted by the two countries	'Promoting the increase in added value, based on the comparative advantages of rural territories, to allow the consolidation of the agri-food, forestry, fishing, mining, tourism, conservation and energy sectors, among others' (p. 8).	'There are a variety of activities subsumed under agriculture. Animal resources development is one such activity, which itself embraces several different products and operations. There is, of course, also crop production as well as fruits, vegetables and spice production, etc., all of which also include different products and operations. The same goes for soil and water conservation and agro-forestry. In short, agriculture is a wide sector comprising many sub-sectors and a wide range of products. Therefore, even though we will promote specialization across regions and agro-ecological zones, we nevertheless expect farmers to engage in mixed farming (diversified production). Thus, our strategy combines both specialization across regions and diversified production on individual farms' (p. 30).

(Continued)

Table 26.2 (Continued)

	Chile	Ethiopia
Relevant water issues in the development strategies adopted by the two countries	'Promoting the integral management of water resources through normative, regulatory, planning instruments (that consider climate projections) and investment that help guide public and private decisions, prioritizing access and use for human consumption and the conservation of aquatic systems' (p. 10).	'Improving water development policies and directives and their proper implementation, educating and training the necessary manpower, engaging farmers in water development and conservation practices, and encouraging private-sector participation in large scale water resource development projects are the main strategies through which we plan to develop our water resources' (pp. 29–30).

Authors' own. *Sources:* FAO (2021), Government of Ethiopia (2003), Government of Chile (2020), OECD (2014, 2020).

The implementation of rural development policies in Chile and Ethiopia shows the political interplay and prominence of new struggles for control over land and water as key discursive and material relations governing rural production and leading to agrarian change. We note that the political ecology of land and water in the two countries shows how ecology is understood in the interplay between political processes underpinning current economic forces and historical trajectories of rural development, and how ecological dynamics impact political decisions not only at the national level but also at the level of agricultural livelihoods. Hence, these cases illustrate some of the dynamics of agrarian change approached in the recent efforts to conceptualise contemporary agrarian questions. Below we close this chapter developing some perspectives from an agrarian question framework to offer theoretical and empirical insights into local agricultural livelihoods in rural contexts of the Global South.

Agricultural livelihoods and new agrarian questions

As the cases of Chile and Ethiopia indicate, ongoing rural development projects, and how they interplay with agricultural livelihoods, prompt new questions about agriculture and forestry development and the key role of water resources. Also, they show how land and water use are at the centre of the prospects for local resource control and agricultural livelihoods. In that regard, these processes of agrarian change and their implication for agricultural livelihoods can be appropriately addressed through an agrarian question framework. As Akram-Lodhi and Kay (2010a: 177) argue, an agrarian question framework is a

> rigorously flexible framework by which to undertake a historically-informed and country-specific analysis of the material conditions governing rural production, reproduction, and the process of agrarian accumulation or its lack thereof, a process that is located within the law of value and market imperatives that operate on a world scale

(see also Akram-Lodhi and Kay, 2010b). In this regard, and for the analysis of the interplay between rural production, rural accumulation and rural politics, Akram-Lodhi and Kay (2010b) identified the following key questions defining the contemporary agrarian questions: questions concerning class forces in agrarian change, path dependency and decoupling between capitalism and agrarian change, role of a global reserve army in agriculture, the role of the corporate food regime and questions of ecology and gender. As we can observe from the examples above, all these questions have profound implications for the analysis of agricultural livelihoods. In relation to food regimes in times of combined global social and ecological crises, McMichael (2013: 65) summarises the contemporary agrarian question as concerning 'who shall farm the land and to what socio-ecological ends?'. As illustrated above, agricultural livelihoods in the context of new agrarian questions in Chile and Ethiopia raise some key political global issues about the future of land and water in agriculture, and the provision of food and other basic needs beyond these two particular contexts. These questions transcend the limits of agriculture and the livelihoods associated with agrarian change and force us to think about the implications for agricultural livelihoods of the process of rapid and uncertain transformations of land and water use in the Global South. Within this context, understanding prospects and barriers for agricultural livelihoods is connected to how the ecological and labour crisis will continue unfolding in the Global South. Especially relevant in this context are struggles around control of land and water use and how this defines those material conditions governing rural production in times of the climate crisis and the politics of climate change.

As stressed by Taylor (2014: 196), the conceptualisation of climatic change in relation to agricultural livelihoods 'need to be fundamentally intertwined with existing divisions of labour, control of productive assets and uneven distributions of consumption power'. Thus, from the perspective of a political ecology of land and water, the relationships between land use, water use and agricultural livelihoods at the local level are inherently outcomes of political processes. Local knowledge production concerning the state of local resources and the institutionalisation of such knowledge in governance structures in rural areas becomes a major issue for rural development projects and agricultural livelihoods. Importantly, issues of agricultural knowledge in connection to the new agrarian questions in the Global South open important venues to rethink data production and assessments of local resources, such as land and water. In this regard, new prospects for the democratisation of science and local rural participation towards transformative local governance in rural contexts become a key aspect of how rural relations can change to address new sustainability and climate change concerns (Alarcón et al., 2021). These rural transformations imply important rearrangements of the institutional conditions for the interlinkages between agrarian change, rural development policies and agricultural livelihoods. Also, and as the trajectories of agricultural development under neoliberal premises indicate, the interplay between states and corporations is linked to the imposition and regulation of market forces with great impacts on agricultural livelihoods worldwide. Within this context, new policies for agriculture are often also manifested in establishing expanding and contested property relations over resources. However, this also produces local political responses to such state and corporate-led rural transformations, including those expressed in the agroecological movements and their struggles at the level of agricultural livelihoods and rights to land and water.

Stressing the importance of a focus on control of land and water for the analysis of agricultural livelihoods is important because this serves the aim of incorporating in the analysis the role that both land and water used for activities other than agriculture plays in the context of rural livelihoods. Within this context, forestry is today, as it has been in the past, a particularly relevant form of land and water control and use interacting directly with agricultural livelihoods in many agrarian settings of the Global South. This adds an important dimension to the often-complicated and conflict-laden relationships between agriculture and other forms of land and water resource use at the local level, which is exacerbated today by climate change. In addition, addressing the role of land and water control and their use in agricultural livelihoods means integrating consideration of how climate change politics at both global and local levels interact with political decisions in the local transformation of the rural settings where agricultural livelihoods are situated.

Conclusion

This selective overview of key links between agricultural livelihoods, rural development policies and agrarian change shows how agricultural livelihoods are today greatly defined by political questions concerning access to land and water in the context of the sustainability crisis, climate change and new rural development projects. Our examples have shown that conflictive relations between control of land and water for agriculture and forestry in rural settings of the Global South are key issues to be taken into account when thinking about the prospects and barriers for agricultural livelihoods today.

Key points

- Worldwide, agricultural livelihoods are at a crossroad. This is a key aspect of the global sustainability and climate change crisis. A focus on how land and water are controlled, accessed and used allows the visualisation of key material conditions governing rural production and leading to agrarian change.
- Complementing a livelihood approach with agrarian political economy and political ecology perspectives allows a deeper analysis and understanding of the external and internal processes defining prospects and barriers for agricultural livelihoods today. As the growing interests in agroecology and peasant farming show, there exist strong alternative agricultures today and these can offer viable options to the propositions of neoliberal agricultural development and the imperatives of economic growth and the problems these one-sided approaches have produced.
- The examination of agricultural livelihoods through the lens of an agrarian question framework and political ecology opens important empirical and theoretical possibilities to explore agricultural livelihoods, both at the local level where they materialise through the use of local resources, and also from a comparative perspective.

Further reading

Akram-Lodhi, A.H. and Kay, C. (2010a) 'Surveying the agrarian question (part 1): Unearthing foundations, exploring diversity', *The Journal of Peasant Studies*, 37(1): 177–202. https://doi.org/10.1080/03066150903498838

Akram-Lodhi, A.H. and Kay, C. (2010b) 'Surveying the agrarian question (part 2): Current debates and beyond', *The Journal of Peasant Studies*, 37(2): 255–284. https://doi.org/10.1080/03066151003594906

McMichael, P. (2013) *Food regimes and agrarian questions*, Nova Scotia: Fernwood Publishing.

Razavi, S. (2009) 'Engendering the political economy of agrarian change', *The Journal of Peasant Studies*, 36(1): 197–226. https://doi.org/10.1080/03066150902820412

Scoones, I. (2009) 'Livelihoods perspectives and rural development', *The Journal of Peasant Studies*, 36(1): 171–196. https://doi.org/10.1080/03066150902820503

Svampa, M. (2012) 'Resource extractivism and alternatives: Latin American perspectives on development', *Journal für Entwicklungspolitik*, 28(3): 117–143. https://doi.org/10.20446/JEP-2414-3197-28-3-43

Notes

1 For more general presentations and critical reviews on livelihood approaches in the context of agriculture see: Li et al. (2021), Oya and Pontara (2015) and Small (2007).

References

Addison, L. and Schnurr, M. (2016) 'Growing burdens? Disease-resistant genetically modified bananas and the potential gendered implications for labor in Uganda', *Agriculture and Human Values*, 33(4): 967–978. https://doi.org/10.1007/s10460-015-9655-2

Agarwal, B. (1997) '"Bargaining" and gender relations: Within and beyond the household', *Feminist Economics*, 3(1): 1–51. https://doi.org/10.1080/135457097338799.

Akram-Lodhi, A.H. (2021) 'The ties that bind? Agroecology and the agrarian question in the twenty-first century', *The Journal of Peasant Studies*, 48(4): 687–714. https://doi.org/10.1080/03066150.2021.1923010.

Akram-Lodhi, A.H. and Kay, C. (2010a) 'Surveying the agrarian question (part 1): Unearthing foundations, exploring diversity', *The Journal of Peasant Studies*, 37(1): 177–202. https://doi.org/10.1080/03066150903498838

Akram-Lodhi, A.H. and Kay, C. (2010b) 'Surveying the agrarian question (part 2): Current debates and beyond', *The Journal of Peasant Studies*, 37(2): 255–284. https://doi.org/10.1080/03066151003594906

Alarcón, C. (2015) 'Forests at the limits. Forestry, land use and climate change from political ecology and environmental communication perspectives: the case of Chile & Sweden', *Doctoral Thesis No. 2015: 5*, Uppsala: Faculty of Natural Resources and Agricultural Sciences, Swedish University of Agricultural Sciences. https://pub.epsilon.slu.se/11926/

Alarcón, C. (2019) 'Transforming wood energy in Sweden and Chile: Climate change, environmental communication and a critical political ecology of international forestry companies', *Critical Perspectives on International Business*, 16(4): 361–377. https://doi.org/10.1108/cpoib-05-2018-0039

Alarcón, C. (2021a) 'Agrarian questions, digitalisation of the countryside, immigrant labour in agriculture and the official discourses on rural development in the Uppsala region, Sweden', *Italian Review of Agricultural Economics*, 76(1): 19–32. https://oajournals.fupress.net/index.php/rea/article/view/12824

Alarcón, C. (2021b) 'Power, conflicts and environmental communication in the struggles for water justice in rural Chile: Insights from the epistemologies of the south and the anthropology of power', in Sjölander Lindqvist, A., Murin, I. and Dove, M.E. (eds.), *Anthropological perspectives on environmental communication*, London: Palgrave Macmillan.

Alarcón, C., Jönsson, M., Gebreyohannis Gebrehiwot, S., Chiwona-Karltun, L., Mark-Herbert, C., Manuschevich, D., Powell, N., Do, T., Bishop, K. and Hilding-Rydevik, T. (2021) 'Citizen science as democratic innovation that renews environmental monitoring and assessment for the sustainable development goals in rural areas', *Sustainability*, 13(5): 2762. https://doi.org/10.3390/su13052762

Altieri, M.A., Nicholls, C.I. and Montalba, R. (2017) 'Technological approaches to sustainable agriculture at a crossroads: An agroecological perspective', *Sustainability*, 9(3): 349. https://doi.org/10.3390/su9030349

Altieri, M.A. and Toledo, V.M. (2011) 'The agroecological revolution in Latin America: Rescuing nature, ensuring food sovereignty and empowering peasants', *The Journal of Peasant Studies*, 38(3): 587–612. https://doi.org/10.1080/03066150.2011.582947

Andersson, K., Bergman Lodin, J. and Pettersson, K. (2021) 'Gendered inequalities in Rwanda's agriculture policy – discursive practices, gendering effects and alternative problematizations', *Manuscript Under Review*.

Andersson, K., Lawrence, D., Zavaleta, J. and Guariguata, M. (2016) 'More trees, more poverty? The socioeconomic effects of tree plantations in Chile, 2001–2011', *Environmental Management*, 57(1): 123–136. https://doi.org/10.1007/s00267-015-0594-x.

Arora-Jonsson, S. (2009) 'Discordant connections: Discourses on gender and grassroots activism in two forest communities in India and Sweden', *Signs: Journal of Women in Culture and Society*, 35(1): 213–240. https://doi.org/10.1086/599259

Badstue, L., Eerdewijk, A. van, Danielsen, K., Hailemariam, M. and Mukewa, E. (2020) 'How local gender norms and intra-household dynamics shape women's demand for laborsaving technologies: Insights from maize-based livelihoods in Ethiopia and Kenya', *Gender, Technology and Development*, 24(3): 341–361. https://doi.org/10.1080/09718524.2020.1830339

Barraclough, S. and Ghimire, K. (1995) *Forests and livelihoods: The social dynamics of deforestation in developing countries*, London: Springer.

Bebbington, A. (1999) 'Capitals and capabilities: A framework for analyzing peasant viability, rural livelihoods and poverty', *World Development*, 27(12): 2021–2044. https://doi.org/10.1016/S0305-750X(99)00104-7

Bergman Lodin, J., Senaga, M. and Paulson, S. (2012) 'New seeds, gender norms and labor dynamics in Hoima district, Uganda', *Journal of Eastern African Studies*, 6(3): 405–422.

Beyene, A. (2018) 'Large-scale canal irrigation management by smallholder farmers', in Beyene, A. (ed.), *Agricultural transformation in Ethiopia: State policy and smallholder farming*, Uppsala: Zed Books, pp. 63–79.

Borras, S.M., Franco, J.C., Isakson, S.R., Levidow, L. and Vervest, P. (2016) 'The rise of flex crops and commodities: Implications for research', *The Journal of Peasant Studies*, 43(1): 93–115. https://doi.org/10.1080/03066150.2015.1036417

Bown, R. (2020) 'Fruticultura chilena de exportación', in Yuri, J.A. and Meller, P. (eds.), *Cuando la fruta es más que solo fruta: Chile y Perú*, Talca: Universidad de Talca/CIEPLAN.

Bryceson, D.F. (1997) 'De-agrarianisation in sub-Saharan Africa: Acknowledging the inevitable', in Bryceson, D.F. and Jamal, V. (eds.), *Farewell to farms: De-agrarianisation and employment in Africa*, Farnham: Ashgate, pp. 3–20.

Chiwona-Karltun, L., Kimanzu, N., Clendenning, Bergman Lodin, J., Ellingson, C., Lidestav, G., Mkwambisi, D., Mwangi, E., Nhantumbo, I., Ochieng, C., Petrokofsky, G. and Sartas, M. (2017) 'What is the evidence that gender affects access to and use of forest assets for food security? A systematic map protocol', *Environmental Evidence*, 6(1): 1–10. https://doi.org/10.1186/s13750-016-0080-9

Christie, M.E., Van Houweling, E. and Zseleczky, L. (2015) 'Mapping gendered pest management knowledge, practices, and pesticide exposure pathways in Ghana and Mali', *Agriculture and Human Values*, 32(4): 761–775. https://doi.org/10.1007/s10460-015-9590-2

Conniff, K., Molden, D., Peden, D. and Awulachew, S. (2013) 'Nile water and agriculture: Past, present and future', in Awulachew, S.B., Smakhtin, V., Molden, D. and Peden, D. (eds.), *The Nile river basin: Water, agriculture, governance and livelihoods*, Abingdon: Routledge, pp. 27–51.

Cotula, L., Vermeulen, S., Leonard, R. and Keeley, J. (2009) *Land grab or development opportunity? Agricultural investment and international land deals in Africa*, London: International Institute for Environment and Development.

Daum, T., Capezzone, F. and Birner, R. (2021) 'Using smartphone app collected data to explore the link between mechanization and intra-household allocation of time in Zambia', *Agriculture and Human Values*, 38(2): 411–429. https://doi.org/10.1007/s10460-020-10160-3

De Janvry, A. (1981) *The agrarian question and reformism in Latin America*, Baltimore: Johns Hopkins University Press.

De Schutter, O. (2009) *Large-scale land acquisitions and leases: A set of core principles and measures to address the human rights challenge*, Geneva: UN Office of the High Commissioner for Human Rights.

Dell'Angelo, J., D'Odorico, P. and Rulli, M.C. (2017) 'Threats to sustainable development posed by land and water grabbing', *Current Opinion in Environmental Sustainability*, 26: 120–128. https://doi.org/10.1016/j.cosust.2017.07.007

Duker, A.E.C., Tadesse, T.M., Soentoro, T., de Fraiture, C. and Kemerink-Seyoum, J.S. (2019) 'The implications of ignoring smallholder agriculture in climate-financed forestry projects: Empirical evidence from two REDD+ pilot projects', *Climate Policy*, 19(sup1): S36–S46. https://doi.org/10.1080/14693062.2018.1532389

Durán, A.P. and Barbosa, O. (2019) 'Seeing Chile's forest for the tree plantations', *Science*, 365(6460): 1388–1388. https://doi.org/10.1126/science.aaz2170

Ellis, F. and Freeman, H.A. (2004) *Rural livelihoods and poverty reduction policies*, London: Routledge.

Fairhead, J., Leach, M. and Amanor, K. (2012) 'Anthropogenic dark earths and Africa: A political agronomy of research disjunctures', in Sumberg, J. and Thompson, J. (eds.), *Contested agronomy*, London: Routledge. pp. 76–97.

FAO (2013) *The State of the world's land and water resources for food and agriculture: Managing systems at risk*, Abingdon: FAO and Earthscan.

FAO (2021) *FAOSTAT statistical database*. Available at: www.fao.org/faostat/en/#data/RL (Accessed 02 July 2021).

FDRE (2011) *Ethiopia's climate-resilient green economy green economy strategy*, Addis Ababa: Federal Democratic Republic of Ethiopia.

FDRE (2017) *National forest sector development program, Ethiopia, volume I: Situation analysis, volume II: Program pillars, action areas and targets and volume III: Synthesis report*, Addis Ababa: Ministry of Environment, Forestry and Climate Change, Federal Democratic Republic of Ethiopia.

FDRE (2019) *Ethiopia's climate-resilient green economy green economy: National adaptation plan*, Addis Ababa: Federal Democratic Republic of Ethiopia.

Fischer, K. and Hajdu, F. (2018) 'The importance of the will to improve: How "sustainability" sidelined local livelihoods in a carbon-forestry investment in Uganda', *Journal of Environmental Policy & Planning*, 20(3): 328–341. https://doi.org/10.1080/1523908X.2017.1410429

Franco, J., Mehta, L. and Veldwisch, G.J. (2013) 'The global politics of water grabbing', *Third World Quarterly*, 34(9): 1651–1675. https://doi.org/10.1080/01436597.2013.843852

Gebreyohannis Gebrehiwot, S., Woldeamlak, B., Mengistu, T., Nuredin, H., Alarcón Ferrari, C. and Bishop, K. (2021) 'Monitoring and assessment of environmental resources in the changing landscape of Ethiopia: A focus on forests and water', *Environmental Monitoring and Assessment*, 193: 624. https://doi.org/10.1007/s10661-021-09421-3

Giraldo, O.F. (2019) *Political ecology of agriculture: Agroecology and post-development*, Cham: Springer.

Government of Chile (2020) *Chile's national rural development policy*, Sanitago: Government of Chile.

Government of Ethiopia (2003) *Rural development policy and strategy*, Addis Ababa: Government of Ethiopia.

Habtu, Y. (1997) 'Farmers without land: The return of landlessness to rural Ethiopia', in Bryceson, D.F. and Jamal, V. (eds.), *Farewell to farms: De-agrarianisation and employment in Africa*, Farnham: Routledge, pp. 41–60.

Hajdu, F., Neves, D. and Granlund, S. (2020) 'Changing livelihoods in rural Eastern Cape, South Africa (2002–2016): Diminishing employment and expanding social protection', *Journal of Southern African Studies*, 46(4): 743–772. https://doi.org/10.1080/03057070.2020.1773721.

Hasalkar, S. and Jadhav, V. (2004) 'Role of women in the use of non-timber forest produce: A review', *Journal of Social Sciences*, 8(3): 203–206. https://doi.org/10.1080/09718923.2004.11892415

Hoang, C., Satyal, P. and Corbera, E. (2019) '"This is my garden": Justice claims and struggles over forests in Vietnam's REDD+', *Climate Policy*, 19(sup1): S23–S35. https://doi.org/10.1080/14693062.2018.1527202

Hofflinger, A., Nahuelpan, H., Boso, À. and Millalen, P. (2021) 'Do large-scale forestry companies generate prosperity in indigenous communities? The socioeconomic impacts of tree plantations in Southern Chile', *Human Ecology*, online first. https://doi.org/10.1007/s10745-020-00204-x

ILO (2016) *ILO Working in rural areas in the 21st century reality and prospects of rural employment in Latin America and the Caribbean*, Lima: International Labour Organization.

International Assessment of Agricultural Knowledge, Science and Technology for Development (IAASTD) (2008) *Agriculture at a crossroads. A global report*, Washington, DC: Island Press.

Kansanga, M.M., Antabe, R., Sano, Y., Mason-Renton, S. and Luginaah, I. (2019) 'A feminist political ecology of agricultural mechanization and evolving gendered on-farm labor dynamics in Northern Ghana', *Gender, Technology and Development*, 23(3): 207–233. https://doi.org/10.1080/09718524.2019.1687799

Lavers, T. (2012) 'Patterns of agrarian transformation in Ethiopia: State-mediated commercialisation and the "land grab"', *The Journal of Peasant Studies*, 39(3–4): 795–822. https://doi.org/10.1080/03066150.2012.660147

Li, J., Li, S., Daily, G.C. and Feldman, M. (2021) *Rural livelihood and environmental sustainability in China*, Singapore: Palgrave Macmillan.

Luna, J.K. (2020) '"Pesticides are our children now": Cultural change and the technological treadmill in the Burkina Faso cotton sector', *Agriculture and Human Values*, 37(2): 449–462. https://doi.org/10.1007/s10460-019-09999-y

Madariaga, A., Maillet, A. and Rozas, J. (2021) 'Multilevel business power in environmental politics: the avocado boom and water scarcity in Chile', *Environmental Politics*, online first. https://doi.org/10.1080/09644016.2021.1892981

Makki, F. (2014) 'Development by dispossession: Terra nullius and the social-ecology of new enclosures in Ethiopia', *Rural Sociology*, 79(1): 79–103. https://doi.org/10.1111/ruso.12033

McMichael, P. (2013) *Food regimes and agrarian questions*, Nova Scotia: Fernwood Publishing.

McNeill, D. (2019) 'The contested discourse of sustainable agriculture', *Global Policy*, 10(S1): 16–27. https://doi.org/10.1111/1758-5899.12603

Medovic, A. and Villagrán, J. (2021) *Irrupción de fondos de inversión y family offices dispara precios de predios agrícolas*. Available at: www.latercera.com/pulso/noticia/irrupcion-de-fondos-de-inversion-y-family-offices-dispara-precios-de-predios-agricolas/ZDCZGNHMX5EDBGKRIPGVYJ62KY/ (Accessed 04 July 2021).

Mittermeier, R., Robles Gil, P., Hoffmann, M., Pilgrim, J., Brooks, T., Mittermeier, C., Lamoreux, J. and Da Fonseca, G. (2004) *Hotspots revisited. Earth's biologically richest and most endangered terrestrial ecoregions*, Chicago: University of Chicago Press.

Mwangi, E., Meinzen-Dick, R. and Sun, Y. (2011) 'Gender and sustainable forest management in East Africa and Latin America', *Ecology and Society*, 16(1): 17. www.ecologyandsociety.org/vol16/iss1/art17/

Nation, M.L. (2010) 'Understanding women's participation in irrigated agriculture: A case study from Senegal', *Agriculture and Human Values*, 27(2): 163–176. https://doi.org/10.1007/s10460-009-9207-8

OECD (2006) *The new rural paradigm: Policies and governance*, Paris: OECD Publishing. https://doi.org/10.1787/9789264023918-en.

OECD (2014) *OECD Rural policy review of Chile*, Paris: OECD Publishing. https://doi.org/10.1787/9789264222892-en.

OECD (2020) *Rural development strategy review of Ethiopia. Reaping the benefits of urbanisation*, Paris: OECD Publishing. https://doi.org/10.1787/a325a658-en.

Ogle, B. (1996) 'People's dependency on forests for food security', in Ruiz Pérez, M. and Arnold, J.E.M. (eds.), *Current issues in non-timber forest products research*, Bogor: Center for International Forestry Research, pp. 219–241.

Oqubay, A. (2015) *Made in Africa: Industrial policy in Ethiopia*, Oxford: Oxford University Press.

Oya, C. and Pontara, N. (2015) 'Introduction: Understanding rural wage employment in developing countries', in Oya, C. and Pontara, N. (eds.), *Rural wage employment in developing countries*, Abingdon: Routledge, pp. 23–58.

Pain, A. and Hansen, K. (2019) *Rural development*, Abingdon: Routledge.

Panez, A., Roose, I. and Faúndez, R. (2020) 'Agribusiness facing its limits: The re-design of neoliberalization strategies in the exporting agriculture sector in Chile', *Land*, 9(3): 66. https://doi.org/10.3390/land9030066

Patel, R. (2013) 'The long green revolution', *The Journal of Peasant Studies*, 40(1): 1–63. https://doi.org/10.1080/03066150.2012.719224

Quisumbing, A.R., Rubin, D., Manfre, C., Waithanji, E., van den Bold, M., Olney, D.K., Johnson, N.L. and Meinzen-Dick, R.S. (2015) 'Gender, assets, and market-oriented agriculture: Learning from high-value crop and livestock projects in Africa and Asia', *Agriculture and Human Values*, 32(4): 705–725. https://doi.org/10.1007/s10460-015-9587-x

Rasul, G. and Sharma, B. (2016) 'The nexus approach to water – energy – food security: An option for adaptation to climate change', *Climate Policy*, 16(6): 682–702. https://doi.org/10.1080/14693062.2015.1029865

Razavi, S. (2009) 'Engendering the political economy of agrarian change', *The Journal of Peasant Studies*, 36(1): 197–226. https://doi.org/10.1080/03066150902820412

Reij, C., Scoones, I. and Toulmin, C. (2013) *Sustaining the soil: Indigenous soil and water conservation in Africa*, Abingdon: Routledge.

Rigg, J. (2006) 'Land, farming, livelihoods, and poverty: Rethinking the links in the rural South', *World Development*, 34(1): 180–202. https://doi.org/10.1016/j.worlddev.2005.07.015

Schneider, S. and Niederle, P.A. (2010) 'Resistance strategies and diversification of rural livelihoods: The construction of autonomy among Brazilian family farmers', *The Journal of Peasant Studies*, 37(2): 379–405. https://doi.org/10.1080/03066151003595168

Scoones, I. (2009) 'Livelihoods perspectives and rural development', *The Journal of Peasant Studies*, 36(1): 171–196. https://doi.org/10.1080/03066150902820503

Scoones, I. (2016) 'Livelihoods, land and political economy: Reflections on Sam Moyo's research methodology', *Agrarian South: Journal of Political Economy*, 5(2–3): 221–239. https://doi.org/10.1177%2F2277976016683750

Selwyn, B. (2021) 'A green new deal for agriculture: For, within, or against capitalism?', *The Journal of Peasant Studies*, 48(4): 778–806. https://doi.org/10.1080/03066150.2020.1854740.

Small, L.A. (2007) 'The sustainable rural livelihoods approach: A critical review', *Canadian Journal of Development Studies/Revue canadienne d'études du développement*, 28(1): 27–38. https://doi.org/10.1080/02255189.2007.9669186

Sumberg, J., Thompson, J. and Woodhouse, P. (2012) 'Contested agronomy: Agricultural research in a changing world', in Sumberg, J. and Thompson, J. (eds.), *Contested agronomy: Agricultural research in a changing world*, Abingdon: Routledge.

Suna, Y., Mwangi, E. and Meinzen-Dick, R. (2011) 'Is gender an important factor influencing user groups' property rights and forestry governance? Empirical analysis from East Africa and Latin America', *International Forestry Review*, 13(2): 205–219. https://doi.org/10.1505/146554811797406598

Svampa, M. (2012) 'Resource extractivism and alternatives: Latin American perspectives on development', *Journal für Entwicklungspolitik*, 28(3): 117–143. https://doi.org/10.20446/JEP-2414-3197-28-3-43

Tavenner, K. and Crane, T.A. (2018) 'Gender power in Kenyan dairy: Cows, commodities, and commercialization', *Agriculture and Human Values*, 35(3): 701–715. https://doi.org/10.1007/s10460-018-9867-3

Taylor, M. (2014) *The political ecology of climate change adaptation: Livelihoods, agrarian change and the conflicts of development*, Abingdon: Routledge.

To, P., Dressler, W. and Mahanty, S. (2017) 'REDD+ for Red Books? Negotiating rights to land and livelihoods through carbon governance in the Central Highlands of Vietnam', *Geoforum*, 81: 163–173. https://doi.org/10.1016/j.geoforum.2017.03.009

Turral, H., Burke, J. and Faurès, J.-M. (2011) 'Climate change, water and food security', *FAO Water Report 36*, Rome: Food and Agriculture Organization of the United Nations.

Van der Ploeg, J.D. (2008) *The new peasantries: Struggles for autonomy and sustainability in an era of empire and globalization*, Abingdon: Routledge.

Van der Ploeg, J.D. (2021) 'The political economy of agroecology', *The Journal of Peasant Studies*, 48(2): 274–297. https://doi.org/10.1080/03066150.2020.1725489

Wangui, E.E. (2008) 'Development interventions, changing livelihoods, and the making of female Maasai pastoralists', *Agriculture and Human Values*, 25(3): 365–378. https://doi.org/10.1007/s10460-007-9111-z

Wegerif, M.C. and Guereña, A. (2020) 'Land inequality trends and drivers', *Land*, 9(4): 101. https://doi.org/10.3390/land9040101

Wheeler, K.G., Jeuland, M., Hall, J.W., Zagona, E. and Whittington, D. (2020) 'Understanding and managing new risks on the Nile with the Grand Ethiopian Renaissance Dam', *Nature Communications*, 11(1): 5222. https://doi.org/10.1038/s41467-020-19089-x

Woodhouse, P. (2012) 'New investment, old challenges. Land deals and the water constraint in African agriculture', *The Journal of Peasant Studies*, 39(3–4): 777–794. https://doi.org/10.1080/03066150.2012.660481

World Bank (2007) *World development report 2008: Agriculture for development*, Washington, DC: The World Bank.

27
PASTORALISM AND LIVELIHOODS IN THE GLOBAL SOUTH

Lenyeletse Vincent Basupi

Introduction

Drylands cover about a third of the world's land mass and host nearly one third of its human population (Dong, 2016). They are water limited, non-equilibrium systems, characterised by high temporal and spatial variability and low levels of human development (Nyberg et al., 2015). They experience short nutrient storage cycles and their soils generally have low water and nutrient holding capacity (Costantini et al., 2016). Important land uses that support dryland livelihoods include pastoralism, agriculture, tourism (often relating to wildlife) and mining, each of which is typically managed through their own sectoral policies (Basupi et al., 2019a).

Extensive pastoralism occurs on about a quarter of the Earth's land area, mostly in the Global South, from the drylands of Africa and the Arabian Peninsula to the highlands of Asia and Latin America where intensive crop cultivation is not always possible (Dong et al., 2016). Examples of pastoral societies include but are not limited to: Bedouin of West Africa and the Arabian Peninsula, Afar of the Horn of Africa, Maasai, Pokot, Samburu and Turkana of East Africa, Fula people of Sahelian West Africa, Toubou of Niger and Chad, Kuchis of Afghanistan, Tuvans of Mongolia, Ranghar of North India and Pakistan, Zulu of South Africa and Herero/Mbanderu of Namibia and Botswana. From the production viewpoint, pastoralism is animal husbandry, the branch of agriculture concerned with the care, tending and use of livestock in rangeland areas (Leal Filho et al., 2020). Agro-pastoralism is a set of practices that combine crop production with pastoral livelihoods (Basupi, 2018). Pastoralism and agro-pastoralism are extremely important as sources of livelihoods and the most prevalent land-use in arid and semi-arid environments of the world (Leal Filho et al., 2020).

Pastoralism and pastoral livelihoods in drylands

From the perspective of livelihoods, the means of securing the necessities of life, pastoralism is a subsistence living pattern of tending herds of large animals on marginal drylands through livestock herding (Leal Filho et al., 2020). Pastoralists herd animals such as cattle, sheep, goats and camels. Owning livestock, and successfully managing a herd, carries 'social capital' in pastoralist societies (Basupi et al., 2019b). Livestock provide a measure of wealth and social status among pastoralists. They also directly contribute to household and community consumption through meat and milk produce (Mutu, 2017). Traditional mechanisms for redistribution and resource (food) sharing exist among the local populations in specific pastoral regions (Birhanu et al., 2017). One such option is gifting and receiving animals. These exchanges can take the form of loaning milking animals to reduce temporal food shortage in times of need. In exchange for herding, the borrower is entitled to the milk and to use the cattle as draft power, as well as keeping some of the offspring of the herded cattle (Basupi et al., 2017a). At the core of such exchanges is the social mechanism for sharing livestock-based resources and helping households in the temporary or medium term to meet their own needs. Pastoralists make choices that safeguard their standing within the social group and hence their ability to call on that group in possible future times of need (Glowacki, 2020). Apart from keeping livestock, sedentary pastoralists also carry out opportunistic crop production (growing maize, millet, vegetables and other crops) though the success rate has been very much limited due to climatic conditions and other man-made vagaries (Basupi et al., 2019b).

In pastoral areas, many climate risks and hazards (e.g. droughts) often lead to losses of livestock (Mutu, 2017). This, in turn, results in significant damage to household, social and economic structures, worsening already poor living conditions and leading to higher levels of poverty (Leal Filho et al., 2020). Furthermore, climate change in pastoral areas is leading to an increase in frequency of flooding and the spread of human and livestock diseases that thrive during the wet season (Kimaro et al., 2018). A diversity of animals (grazers: sheep, cattle, donkeys, horse; browsers: goats) reduces risk from disease, droughts and parasites. Risk is further controlled by redistributing assets through mutual support, including splitting herds between pastures (Basupi et al., 2017a). Mitigating risk from drought may involve diversification into distant labour or trading markets, as well as expanding trade in wild products (Basupi et al., 2019b).

Pastoral communities hold their land under customary tenure, based on customary laws. Customary laws include a body of extremely diverse rules and regulations (usually unwritten) founding their legitimacy in 'tradition' (Chauveau and Colin, 2007). This management regime is critical because it creates shared communal rights of access, providing an ideal framework for communities to exploit scarce resources across various agro-ecological conditions, which in turn reduces the level of vulnerability (Agrawal, 2001). Large and diverse ranges comprising wet, dry and drought-time grazing areas are managed as common property resources under common property regimes (Basupi et al., 2017a). This adaptive mobility and pastoralist flexibility is crucial to sustainable land and livestock management in these environments because it helps make use of highly variable rangeland resources (Turner, 2011). Water rights are crucial to the sustainable management of land resources. Water management is tightly controlled, and rights are negotiated, along with range management, and the availability of water often gives livestock access to valuable pastures (Basupi et al., 2017a). In Botswana and Kenya for example, the decision by governments to allow enclosure of natural water ponds in private ranches has weakened traditional rangeland management systems, deprived pastoralists of valuable assets and fostered conflicts over the remaining water resources (Basupi et al., 2017b).

Pastoral communities have remained stable for centuries, particularly through flexible responses to short-term variations in climatic conditions (Dong et al., 2016). They evolved a system employing migratory, semi-sedentary and deferred grazing practices to optimise the use of rangeland resources (Basupi et al., 2017a). Nomadic pastoralism, the movement of livestock and herders according to seasonal and ecological conditions, is practised in some African drylands such as Kenya's Maasailand and the Mongolian steppe (Dong, 2016). This customary activity continues to provide a livelihood to a significant population in these drylands (Liao et al., 2020).

Livestock mobility, over space and time, optimises use of the range where rainfall is spatially and temporally varied (Vetter, 2005).

The spatial knowledge systems held by herders help them determine what the temporal and spatial distribution of resources might be in any given year and are central to sustainable pastoral herd mobility (Oba, 2013). Knowledge of when wild species, particularly trees, yield food helps to supplement reduced milk yields during dry times (Basupi et al., 2017a). Tree conservation is vital for conserving fodder, providing shade, fruits, medicinal products and other benefits, some of which can be sold. Historically pastoralists have been able to follow rainfall or specific pasture resources through space and time in order to meet the needs of their animals and prevent rangeland degradation caused by the concentration of animals in smaller territories (Adriansen and Nielsen, 2002). It is this flexibility that provides a measure of security in times of drought or other ecological disasters by creating reciprocal expectations of resource sharing between groups.

The rationale for traditional pastoralism of herd mobility and flexibility has been reinforced by the recognition that drylands systems are non-equilibria in nature and that resource sustainability is largely a function of spatial and temporal variability in rainfall and/or fire regimes (Dougill et al., 2016). The survival of herds in drylands depends on the pastoralists' ability to respond to variability or uncertainty and hence move to better areas with available fodder (Vetter, 2005). Therefore, extensive spatial scales of exploitation become a prerequisite for a successful pastoral production system (Moritz et al., 2013).

Pastoralism functions within a complex system balancing physical and anthropogenic change with environmental and socio-economic limitations (Basupi, 2018). Mobility, which is the primary means of survival for pastoralists, is constrained by population growth, communal land subdivision and privatisation of land for commercial farms and ranches, expansion of national wildlife parks and conflicts over the scarce resources (Basupi et al., 2017a). Sustainable land use, the need for regular income, expropriation of land, modernisation, loss of livestock mobility and climate variability are all pastoral concerns in an era of social transition.

Sedentarisation of pastoralists in the drylands: land tenure transformation

Customary pastoralism in the Global South involves flexible institutions for securing access to seasonally variable common pool water and pasture resources, and coordination across levels of social organisation in times of environmental stress (Mwangi and Ostrom, 2009). Access to land is widely considered as a precondition for access to other livelihood opportunities (Toulmin, 2009). Providing security of tenure is often seen as a precondition for better natural resource management and sustainable rural livelihoods in pastoral societies. These flexible institutions have been modified by several state-led interventions such as land tenure transformation beginning in the colonial era and the 1970s World Bank intervention programmes (Basupi, 2018).

Numerous political, economic and social changes have influenced customary institutions and interacted with herders' norms of livestock mobility. Rangelands, being a common pool resource, were thought to be overused by pastoral communities, which keep large herds of livestock, resulting in rangeland degradation (Rohde et al., 2006). Inspired by Hardin's Tragedy of the Commons (Hardin, 1968) and market liberalisation (Simbizi et al., 2014) theories, many drylands countries (some through World Bank initiatives) revisited their customary tenure arrangements in pastoralists' areas, reforming institutions for the administration of land rights and liberalising tenure arrangements by embarking on individualisation, rangeland enclosures, commercialisation and privatisation of communal lands (Adams, 2013). Indigenous land tenure practices were often blamed for discouraging private incentives to manage the communal pastures and encouraging higher stocking rates which exceeded the ecological carrying capacity of the land; livestock herders were seen to overexploit an area and move on (Rohde et al., 2006). This view portrays pastoralism as a destructive, backward and maladaptive system, which needs to be changed. De Soto's support has been particularly singled out, with his theoretical argument stating that the conditions and terms of negotiation under which land is held under customary tenure only encourage low rates of productivity-enhancing investments (De Soto, 2000). De Soto refers to land held under customary tenure as 'dead capital' because it cannot be used as collateral in a formal banking system. Enclosures, subdivisions and individualisation of communal grazing lands are the policy prescription that emerges from this analysis, since only private individuals or the state were or are seen by governments as capable of managing resources sustainably where the incentives to do so under communal system are said to be weak or absent (Peters, 1994; Rohde et al., 2006).

Mobility and flexibility have diminished as land ownership has become more rigid and fixed, with different land uses separated by fences and other administrative barriers (Letai and Lind, 2013). In sub-Saharan Africa, for example, the impact of government enclosure policy includes privatisation of the best land whilst also leading to overgrazing, violent conflicts and increased wealth inequalities (Galaty, 2013; Lesorogol, 2008). Governments have been very slow to recognise the rationale of pastoral mobility and its implication – that over-grazing is caused by impeding movements with barriers, boundaries and regulation, rather than by allowing moving herds to redistribute grazing pressure and thereby better conserve ecosystems (Liao et al., 2020). In Niger and northern Nigeria, for example, early studies of Fulani groups showed how social or kinship relations were reflected in their distribution in space during the rainy season, and seasonal variations in pasture condition determined their transhumance (Majekodunmi et al., 2017).

Land use policies that focus on sedentarisation of pastoral communities continue to cause accelerated pressure on natural resources leading to rangeland resource degradation, wildlife declines and pastoralist vulnerability (Western et al., 2009). Most rangeland privatisation policies have not yet yielded the intended benefits (Homewood, 2004). Where land degradation existed, it has not been halted (Dougill et al., 2016) and traditional livestock management institutions have been disoriented, undermining traditional livelihoods and rangeland management systems (Boles et al., 2019). Ecological studies conducted in Inner Mongolia (Hilker et al., 2014), Botswana and South Africa (Rohde et al., 2006), among others, revealed that rangeland overstocking, overgrazing and degradation remains a challenge in the limited grazing areas.

As the per capita availability of livestock and land continues to reduce due to increasing subdivision of communal land and constrained mobility of livestock, more pastoralists are likely to experience severe economic hardship. Only a small minority of pastoral elites have been able to take advantage of government incentives that have facilitated private commercial ranching (Basupi et al., 2017b, 2019a; Galaty, 2013). As such, the failure of rangeland privatisation programs is almost universal (German et al., 2013; Homewood, 2004). Low levels of economic development and deficiencies in markets also make it almost impossible to achieve environmental sustainability objectives (Thomas, 2008). Further evidence from the literature suggests that the perceived benefits of tenure transformation have acted as a justification for the concentration of land in the hands of a few, especially politically connected individuals, exacerbating insecurity of land tenure for the rural poor (Boone, 2014).

Pastoral livelihoods and conservation efforts

Support for communal land subdivision and privatisation has been accompanied by expropriation of land for the expansion of conservation areas and the construction of fences for disease control purposes (Galvin et al., 2008; McGahey, 2011). The significance of conservation and related dislocation of indigenous people has been to create wilderness or nature spaces separate from communities, livestock and related livelihood activities (McGahey, 2011). Most protected areas in the tropics, including private game reserves, were created on traditionally inhabited lands, used by indigenous communities for pastoralism (Basupi, 2018). Drylands likely constitute the main remaining land area available for conversion to conservation areas, but at great social and economic cost to their pastoralist inhabitants (Kieti et al., 2020). The unavailability of alternative resources and livelihood options for pastoral communities sometimes makes them hostile towards conservation efforts and creates challenges for sustainable wildlife management and peaceful coexistence of livestock and wildlife (Basupi, 2018).

Pastoralist livelihoods are squeezed between competing land uses, driven by global capital, for large scale private game farms, and national and international conservation interest (Basupi, 2018). In sub-Saharan Africa, for example, the loss of pastoral lands to biodiversity conservation and alternative land uses through privatisation of rangelands is the most commonly reported manifestation of land grabbing (Galaty, 2013). Elsewhere in drylands, conflicts pitting pastoralists and conservation authorities have been on the rise (Basupi et al., 2017a). Persistent human-wildlife conflict experiences around conservation areas is at the centre of political-economic dynamics, where resource inequalities and contesting desires for land use have resulted in a complex group of wildlife tolerant and wildlife intolerant factions of stakeholders (Kieti et al., 2020). Although the goal of local herding communities is not necessarily wildlife conservation, their grazing practices have ensured species diversity in the process of meeting utilitarian needs, mainly through the maintenance of mobility (Basupi, 2018). As it stands, current policies related to grasslands do not provide incentives for conservation, thus reduced mobility resulting from prevailing land use policies is contributing to lower species diversity.

There has been a growing appreciation of wildlife conservation as a social and political process with special emphasis on the need to incorporate local communities in the sustainable use of natural resources (Kieti et al., 2020). Wildlife tourism is portrayed as a means for pastoral groups to diversify, generate revenues and improve wellbeing (Basupi et al., 2017a). In principle, these competing forces are meant to be integrated at the local level through community-based natural resource management (CBNRM), allowing communities and households to develop socially, economically and environmentally sustainable compromises. CBNRM further seeks to give natural resources a meaningful use-value to rural communities who bear the cost of human-wildlife conflict and habitat conservation (Kieti et al., 2020). Challenges experienced, however, include CBNRM's failure to be inclusive of stakeholders, especially communities, in decision-making processes, domination by more powerful individuals and political elites, elite land capture and safari operators (Kieti et al., 2020). Linking conservation efforts with human development offers the most promising course of action for long term sustainability of pastoral livelihoods. If the problems in pastoral areas such as access to grazing lands, livestock disease and human wildlife conflicts are addressed and the community envisages benefits from a combined conservation/pastoralism development scheme, then cooperation and long-term sustainability is possible.

Pastoralists adaptations to environmental and policy change

Pastoralists survive in fragile ecosystems that are adversely affected by drought and are frequently threatened by climatic related challenges; deterioration of rangeland, scarcity of water, prevalence of livestock disease, and low livestock market value (Ahmad and Afzal, 2021). Therefore, pastoralists in the Global South are among the poorest populations, they lack the most important infrastructure such as accessible roads, electricity and communications, making them increasingly isolated. However, pastoralism is seen as a resilient production system and a viable livelihood strategy for millions of people in the drylands (Leal Filho et al., 2020). Context-specific factors mediate complex human-environment interactions as they shape livelihoods and the ability to cope with and adapt to variables such as climatic conditions and other man-made vagaries (Basupi et al., 2019b).

The ability to cope with and adapt to environmental and policy conditions is increasingly recognised as structured by the interaction of institutional, environmental and household factors at multiple scales, leading to a complex biophysical and social basis of vulnerability with differential abilities to respond in different contexts (Scoones, 2009). To mitigate their increased vulnerability, some households have taken to harvesting natural products, adding value through manufacturing simple objects, labouring and otherwise exploiting the multiple – though poorly rewarded – work opportunities offered locally, as well as migrating further afield (Basupi et al., 2019b; Opiyo et al., 2015). Whether such strategies are sustainable depends to a large extent on whether the strategy can improve the capacities of the communities in terms of livelihoods sustenance and environmental sustainability (Valdivia and Barbieri, 2014). These invariably depend on the availability of effective administrative tools within a community (Scoones, 2009).

Pastoralists' adaptive capacity is challenged by multiple factors; infrastructural services, access to markets, insecure property rights and lack of political will to address pastoralist problems in marginal areas (Basupi et al., 2019b). Understanding pastoralists' adaptive capacity and factors which influence their adaptation is a key element in adaptation planning. Thus, policy frameworks with clear direction for action towards building adaptive capacity and socio-ecological sustainability are required. Strategies can improve livelihoods, or the landscape, provided the underlying barriers to adaptation are overcome (Kihila, 2018). Availability of infrastructure for example, especially access to water sources, accessible road networks, pasture land and access to markets, could enhance livestock productivity and transform pastoralists' livelihoods by creating different economic opportunities (Ambelu et al., 2017).

Conclusion

Pastoralists continue to face constraints in accessing the available rangelands and water resources to pursue livelihoods options. In many rangelands, restrictions to pastoral mobility and the undermining of indigenous natural resource governance have led to land degradation and increasing pastoralist vulnerability. Land use competition is a problem in that where market forces are left to prevail, essential land uses, perhaps with lower economic rents like pastoralism, run the risk of being out-competed and relegated to less suitable areas. Greater recognition and support are needed for sustainable pastoral and agro-pastoral systems in view of their contributions to sustaining the livelihoods of millions of people residing in the Global South. While communal and customary land tenure systems of land and resource ownership are commonly targeted for tenure reform, pastoral communities still rely on these institutions as functional strategies for secure livelihoods and sustainable management of resources. Pastoral households are adopting mixed strategies, keeping one foot in pastoralism while exploiting other avenues of livelihoods. However, there are many constraints affecting pastoralists' adaptation in pastoral areas such as remoteness, livestock diseases, land use conflicts, human-wildlife conflicts, tenure insecurity, poor infrastructural services and policies/institutional frameworks that undermine or even fail to understand innovative pastoral responses to change.

Policies and land management strategies are needed to tackle the interconnected dryland challenges of land and water resource degradation, human wildlife conflicts, biodiversity loss, climate change, population growth and pastoralists' adaptations. The challenge for governments and development agencies is to develop and implement policies that protect the various aspects of existing traditional tenure systems and pastoralists' access to key water and grazing resources, and support pastoralists' adaptations and paradigms that promote sustainable livelihoods. Ongoing socioeconomic, environmental and land tenure transformations in drylands make it imperative to examine the trade-offs and co-benefits between the goals of food security for pastoralists in drylands and the ecosystem health of communal lands. Drylands not only support the livelihoods of millions of pastoralists but also host diverse flora and fauna. Targeted support by governments, civil society organisations, development agencies, pastoral networks, development practitioners and researchers are needed to harness opportunities in pastoral regions.

Key points

- Rangelands are important for pastoral communities in the Global South, as remote drylands provide valuable grazing lands for livestock and native wildlife.
- The lack of land rights is a serious challenge for pastoralists, posing threats to pastoral livelihood sustainability and viability.
- Climate change has been threatening pastoral communities in the Global South as their economies are mainly based on rain-fed agriculture.
- There is a need to address pastoralist productivity, address economic and environmental issues and develop sustainable livelihoods for millions of rural communities in the Global South.

Further readings

African Union (2010) *Policy framework for pastoralism in Africa: Securing, protecting and improving the lives, livelihoods and rights of pastoralist communities*, Addis Ababa: African Union.
Ali, A. and Hobson, M. (2009) *Social protection in pastoral areas*, London: Humanitarian Policy Group, Overseas Development Institute.
Catley, A., Lind, J. and Scoones, I. (eds.) (2013) *Pastoralism and development in Africa: Dynamic change at the margins*, Abingdon: Routledge.
Niamir-Fuller, M. and Huber-Sannwald, E. (2020) 'Pastoralism and achievement of the 2030 agenda for sustainable development: A missing piece of the Global Puzzle', in Lucatello, S., Huber-Sannwald, E., Espejel, I. and Martínez-Tagüeña, N. (eds.), *Stewardship of future drylands and climate change in the Global South*, Cham: Springer, pp. 41–55. https://doi.org/10.1007/978-3-030-22464-6_3

References

Adams, M. (2013) 'Reforming communal rangeland policy in southern Africa: Challenges, dilemmas and opportunities', *African Journal of Range & Forage Science*, 30(1–2): 91–97. https://doi.org/10.2989/10220119.2013.819527
Adriansen, H.K. and Nielsen, T.T. (2002) 'Going where the grass is greener: On the study of pastoral mobility in Ferlo, Senegal', *Human Ecology*, 30(2): 215–226. https://doi.org/10.1023/A:1015692730088
Agrawal, A. (2001) 'Common property institutions and sustainable governance of resources', *World Development*, 29(10): 1649–1672. https://doi.org/10.1016/S0305-750X(01)00063-8
Ahmad, D. and Afzal, M. (2021) 'Impact of climate change on pastoralists' resilience and sustainable mitigation in Punjab, Pakistan', *Environment, Development and Sustainability*, 23: 11406–11426. https://doi.org/10.1007/s10668-020-01119-9
Ambelu, A., Birhanu, Z., Tesfaye, A., Berhanu, N., Muhumuza, C., Kassahun, W., Daba, T. and Woldemichael, K. (2017) 'Intervention pathways towards improving the resilience of pastoralists: A study from Borana communities, Southern Ethiopia', *Weather and Climate Extremes*, 17: 7–16. https://doi.org/10.1016/j.wace.2017.06.001
Basupi, L.V. (2018) *Pastoralism and land tenure transformation: Policy implications and livelihoods adaptations in Botswana*, PhD Thesis, Leeds: University of Leeds.

Basupi, L.V., Dougill, A.J. and Quinn, C.H. (2019a) 'Institutional challenges in pastoral landscape management: Towards sustainable land management in Ngamiland, Botswana', *Land Degradation & Development*, 30(7): 839–851. https://doi.org/10.1002/ldr.3271

Basupi, L.V., Quinn, C.H. and Dougill, A.J. (2017a) 'Using participatory mapping and a participatory geographic information system in pastoral land use investigation: Impacts of rangeland policy in Botswana', *Land Use Policy*, 64: 363–373. https://doi.org/10.1016/j.landusepol.2017.03.007

Basupi, L.V., Quinn, C.H. and Dougill, A.J. (2017b) 'Pastoralism and land tenure transformation in Sub-Saharan Africa: Conflicting policies and priorities in Ngamiland, Botswana', *Land*, 6(4): 89. https://doi.org/10.3390/land6040089

Basupi, L.V., Quinn, C.H. and Dougill, A.J. (2019b) 'Adaptation strategies to environmental and policy change in semi-arid pastoral landscapes: Evidence from Ngamiland, Botswana', *Journal of Arid Environments*, 166: 17–27. https://doi.org/10.1016/j.jaridenv.2019.01.011

Birhanu, Z., Ambelu, A., Berhanu, N., Tesfaye, A. and Woldemichael, K. (2017) 'Understanding resilience dimensions and adaptive strategies to the impact of recurrent droughts in Borana Zone, Oromia Region, Ethiopia: A grounded theory approach', *International Journal of Environmental Research and Public Health*, 14(2): 118. https://doi.org/10.3390/ijerph14020118

Boles, O.J., Shoemaker, A., Mustaphi, C.J.C., Petek, N., Ekblom, A. and Lane, P.J. (2019) 'Historical ecologies of pastoralist overgrazing in Kenya: Long-term perspectives on cause and effect', *Human Ecology*, 47(3): 419–434. https://doi.org/10.1007/s10745-019-0072-9

Boone, C. (2014) *Property and political order in Africa: Land rights and the structure of politics*, Cambridge: Cambridge University Press.

Chauveau, J.P. and Colin, J.P. (2007) 'Changes in land transfer mechanisms: Evidence from West Africa', in Cotula, L. (ed.), *Changes in "customary" land tenure systems in Africa*, London: International Institute for Environment and Development, pp. 65–79.

Costantini, E.A., Branquinho, C., Nunes, A., Schwilch, G., Stavi, I., Valdecantos, A. and Zucca, C. (2016) 'Soil indicators to assess the effectiveness of restoration strategies in dryland ecosystems', *Solid Earth*, 7(2): 397–414. https://doi.org/10.5194/se-7-397-2016

De Soto, H. (2000) *The mystery of capital: Why capitalism triumphs in the west and fails everywhere else*, New York: Civitas Books.

Dong, S. (2016) 'Overview: Pastoralism in the world', in Dong, S., Kassam, K-A.S., Tourrand, J.F. and Boone, R.B. (eds.), *Building resilience of human-natural systems of pastoralism in the developing world*, Cham: Springer, pp. 1–37.

Dong, S., Liu, S. and Wen, L. (2016) 'Vulnerability and resilience of human-natural systems of pastoralism worldwide', in Dong, S., Kassam, K-A.S., Tourrand, J.F. and Boone, R.B. (eds.), *Building resilience of human-natural systems of pastoralism in the developing world*, Cham: Springer. pp. 39–92.

Dougill, A.J., Akanyang, L., Perkins, J.S., Eckardt, F.D., Stringer, L.C., Favretto, N. and Mulale, K. (2016) 'Land use, rangeland degradation and ecological changes in the southern Kalahari, Botswana', *African Journal of Ecology*, 54(1): 59–67. https://doi.org/10.1111/aje.12265

Galaty, J.G. (2013) 'Land grabbing in the Eastern African rangelands', in Catley, A., Lind, J. and Scoones, I. (eds.), *Pastoralism and development in Africa: Dynamic change at the margins: Pathways to sustainability*, Abingdon: Routledge, pp. 143–153.

Galvin, K.A., Reid, R.S., Behnke, R.H. and Hobbs, N.T. (eds.) (2008) *Fragmentation in semi-arid and arid landscapes. Consequences for human and natural systems*, Dordrecht: Springer.

German, L., Schoneveld, G. and Mwangi, E. (2013) 'Contemporary processes of large-scale land acquisition in Sub-Saharan Africa: Legal deficiency or elite capture of the rule of law?', *World Development*, 48: 1–18. https://doi.org/10.1016/j.worlddev.2013.03.006

Glowacki, L. (2020) 'The emergence of locally adaptive institutions: Insights from traditional social structures of East African pastoralists', *Biosystems*, 198: 104257. https://doi.org/10.1016/j.biosystems.2020.104257

Hardin, G. (1968) 'The tragedy of the commons', *Science*, 162(3859): 1243–1248. https://doi.org/10.1126/science.162.3859.1243

Hilker, T., Natsagdorj, E., Waring, R.H., Lyapustin, A. and Wang, Y. (2014) 'Satellite observed widespread decline in Mongolian grasslands largely due to overgrazing', *Global Change Biology*, 20(2): 418–428. https://doi.org/10.1111/gcb.12365

Homewood, K.M. (2004) 'Policy, environment and development in African rangelands', *Environmental Science & Policy*, 7(3): 125–143. https://doi.org/10.1016/j.envsci.2003.12.006

Kieti, D., Nthiga, R., Plimo, J., Sambajee, P., Ndiuini, A., Kiage, E., Mutinda, P. and Baum, T. (2020) 'An African dilemma: Pastoralists, conservationists and tourists – reconciling conflicting issues in Kenya', *Development Southern Africa*, 37(5): 758–772. https://doi.org/10.1080/0376835X.2020.1747988

Kihila, J.M. (2018) 'Indigenous coping and adaptation strategies to climate change of local communities in Tanzania: A review', *Climate and Development*, 10(5): 406–416. https://doi.org/10.1080/17565529.2017.1318739

Kimaro, E.G., Mor, S.M. and Toribio, J.A.L. (2018) 'Climate change perception and impacts on cattle production in pastoral communities of northern Tanzania', *Pastoralism*, 8(1): 1–16. https://doi.org/10.1186/s13570-018-0125-5

Leal Filho, W., Taddese, H., Balehegn, M., Nzengya, D., Debela, N., Abayineh, A., Mworozi, E., Osei, S., Ayal, D.Y., Nagy, G.J., Yannick, N., Kimu, S., Balogun, A-L., Alemu, E.A., Li, C., Sidsaph, H. and Wolf, F. (2020) 'Introducing experiences from African pastoralist communities to cope with climate change risks, hazards and extremes: Fostering poverty reduction', *International Journal of Disaster Risk Reduction*, 50: 101738. https://doi.org/10.1016/j.ijdrr.2020.101738

Lesorogol, C.K. (2008) 'Land privatization and pastoralist well-being in Kenya', *Development and Change*, 39(2): 309–331. https://doi.org/10.1111/j.1467-7660.2007.00481.x

Letai, J. and Lind, J. (2013) 'Squeezed from all sides: Changing resource tenure and pastoralist innovation on the Laikipia Plateau, Kenya', in Catley, A., Lind, J. and Scoones, I. (eds.), *Pastoralism and development in Africa: Dynamic change at the margins*, New York: Routledge, pp. 164–176.

Liao, C., Agrawal, A., Clark, P.E., Levin, S.A. and Rubenstein, D.I. (2020) 'Landscape sustainability science in the drylands: Mobility, rangelands and livelihoods', *Landscape Ecology*, 1–15. https://doi.org/10.1007/s10980-020-01068-8

Majekodunmi, A.O., Dongkum, C., Langs, T., Shaw, A.P. and Welburn, S.C. (2017) 'Shifting livelihood strategies in northern Nigeria-extensified production and livelihood diversification amongst Fulani pastoralists', *Pastoralism*, 7(1): 1–13. https://doi.org/10.1186/s13570-017-0091-3

McGahey, D.J. (2011) 'Livestock mobility and animal health policy in southern Africa: The impact of veterinary cordon fences on pastoralists', *Pastoralism: Research, Policy and Practice*, 1(14): 1–29. https://doi.org/10.1186/2041-7136-1-14

Moritz, M., Scholte, P., Hamilton, I.M. and Kari, S. (2013) 'Open access, open systems: Pastoral management of common-pool resources in the Chad Basin', *Human Ecology*, 41(3): 351–365. https://doi.org/10.1007/s10745-012-9550-z

Mutu, P.L. (2017) 'Drought coping mechanisms among the Turkana nomadic pastoral community of Ilemi triangle region in Northern Kenya', *Research in Health Sciences*, 2(2): 104–146. https://doi.org/10.22158/rhs.v2n2p104

Mwangi, E. and Ostrom, E. (2009) 'A century of institutions and ecology in East Africa's rangelands: Linking institutional robustness with the ecological resilience of Kenya's Maasailand', in Beckmann, V. and Padmanabhan, M. (eds.), *Institutions and sustainability*, Dordrecht: Springer, pp. 195–222.

Nyberg, G., Knutsson, P., Ostwald, M., Öborn, I., Wredle, E., Otieno, D.J., Mureithi, S., Mwangi, P., Said, M.Y., Jirström, M., Grönvall, A., Wernersson, J., Svanlund, S., Saxer, L., Geutjes, L., Karmebäck, V., Wairore, J.N., Wambui, R., De Leeuw, J. and Malmer, A. (2015) 'Enclosures in West Pokot, Kenya: Transforming land, livestock and livelihoods in drylands', *Pastoralism*, 5(1): 1–12. https://doi.org/10.1186/s13570-015-0044-7

Oba, G. (2013) 'The sustainability of pastoral production in Africa', in Catley, A., Lind, J. and Scoones, I. (eds.), *Pastoralism and development in Africa: Dynamic changes at the margins*, London: Routledge, pp. 29–36.

Opiyo, F., Wasonga, O., Nyangito, M., Schilling, J. and Munang, R. (2015) 'Drought adaptation and coping strategies among the Turkana pastoralists of northern Kenya', *International Journal of Disaster Risk Science*, 6(3): 295–309. https://doi.org/10.1007/s13753-015-0063-4

Peters, P.E. (1994) *Dividing the commons: Politics, policy, and culture in Botswana*, Charlottesville: University Press of Virginia.

Rohde, R.F., Moleele, N.M., Mphale, M., Allsopp, N., Chanda, R., Hoffman, M.T., Magole, L. and Young, E. (2006) 'Dynamics of grazing policy and practice: Environmental and social impacts in three communal areas of southern Africa', *Environmental Science & Policy*, 9(3): 302–316. https://doi.org/10.1016/j.envsci.2005.11.009

Scoones, I. (2009) 'Livelihoods perspectives and rural development', *The Journal of Peasant Studies*, 36(1): 171–196. https://doi.org/10.1080/03066150902820503

Simbizi, M.C.D., Bennett, R.M. and Zevenbergen, J. (2014) 'Land tenure security: Revisiting and refining the concept for sub-Saharan Africa's rural poor', *Land Use Policy*, 36: 231–238. https://doi.org/10.1016/j.landusepol.2013.08.006

Thomas, R.J. (2008) 'Opportunities to reduce the vulnerability of dryland farmers in Central and West Asia and North Africa to climate change', *Agriculture, Ecosystems & Environment*, 126: 36–45. https://doi.org/10.1016/j.agee.2008.01.011

Toulmin, C. (2009) 'Securing land and property rights in sub-Saharan Africa: The role of local institutions', *Land Use Policy*, 26(1): 10–19. https://doi.org/10.1016/j.landusepol.2008.07.006

Turner, M.D. (2011) 'The new pastoral development paradigm: Engaging the realities of property institutions and livestock mobility in dryland Africa', *Society and Natural Resources*, 24(5): 469–484. https://doi.org/10.1080/08941920903236291

Valdivia, C. and Barbieri, C. (2014) 'Agritourism as a sustainable adaptation strategy to climate change in the Andean Altiplano', *Tourism Management Perspectives*, 11: 18–25. https://doi.org/10.1016/j.tmp.2014.02.004

Vetter, S. (2005) 'Rangelands at equilibrium and non-equilibrium: Recent developments in the debate', *Journal of Arid Environments*, 62(2): 321–341. https://doi.org/10.1016/j.jaridenv.2004.11.015

Western, D., Groom, R. and Worden, J. (2009) 'The impact of subdivision and sedentarization of pastoral lands on wildlife in an African savanna ecosystem', *Biological Conservation*, 142(11): 2538–2546. https://doi.org/10.1016/j.biocon.2009.05.025

28
FISHERIES LIVELIHOODS

Deo Namwira and Fiona Nunan

Introduction

Fisheries are of crucial importance globally, supplying a significant proportion of the world's food and livelihoods. The fisheries sector is often categorised into industrial, small-scale/artisanal and recreational activities, depending on the scale and purpose of operation, fishing area, gear/fleet used, and target species. In terms of fishing areas, fisheries are comprised of marine fisheries, inland fisheries and aquaculture, also known as fish farming. The Food and Agricultural Organization of the United Nations (FAO) (2020) estimated that 59.5 million people were employed worldwide in fishing in 2018, of which 39 million people were employed in fisheries and 20.5 million people in aquaculture, with a majority being in the Global South and mostly in small-scale fisheries. Notably, of all of the people engaged in small-scale fisheries and aquaculture, the highest numbers are in Asia (85%) and in Africa (9%) (FAO, 2020). Out of the estimated 59.5 million people, women are reported as accounting for 14% of the total, with 19% in aquaculture and 12% in fisheries (FAO, 2020). According to FAO (2020), in most regions, women are less involved in offshore and long-distance capture fishing, reflecting a clear gendered division of labour that characterises small-scale fisheries and fishing communities in the Global South.

As well as providing employment and income-generation, fisheries also support livelihoods through food and nutrition. Available statistics show that fisheries provide around 3.2 billion people with almost 20% of their average animal protein intake, with an even higher proportion in the Global South (FAO, 2018). Fish are also critical for nutrition security through providing multiple micronutrients and fatty acids, with the potential to address undernourishment and nutritional deficiencies (Bennett et al., 2018).

A global trend of increasing overfishing has led to widespread concern that fish stocks are decreasing throughout much of the world (FAO, 2020; Hilborn et al., 2020). The latest FAO State of World Fisheries and Aquaculture report indicates that global fisheries production in 2018 had increased by 5.4% from the average of the previous three years, reaching a record of 96.4 million tonnes (FAO, 2020). The FAO has estimated that the percentage of fish stocks fished within biologically sustainable levels decreased from 90% in 1974 to 65.8% in 2017, reflecting increasing overexploitation of fish stocks (FAO, 2020). Data collection and management efforts often focus on large-scale commercial fisheries, paying much less attention to small-scale fishing activities (Harper et al., 2020), which suggests that overall fisheries production could be even higher than data collected indicates.

This chapter introduces and reviews literature on fisheries livelihoods in the Global South, focusing on small-scale fisheries. The chapter begins by reviewing literature on the nature and role of small-scale fisheries, providing insights into their characteristics. It then identifies who generates livelihoods within fisheries, taking a value chain approach and gender lens to identify the main types of actors and how a gendered division of labour is often found. The chapter goes on to review how poverty has been understood within fisheries and key sources of vulnerability for fisherfolk. Finally, the chapter reviews how fisheries management regimes affect access to fisheries and livelihood benefits. The chapter concludes with a summary of key points and recommended readings.

Small-scale fisheries

An overwhelming majority of the world's fisheries are small-scale (Nayak et al., 2014). In the 2012 World Bank report 'Hidden Harvest: the Global Contribution of Capture Fisheries', it was estimated that 90% of the employment in the marine fisheries sector was small-scale (World Bank, 2012). The term 'small-scale fisheries' (SSF) is inclusive of fish harvesting activities generally performed by self-employed fisherfolk, using diverse low-technology types of fishing methods, gears and boats, and targeting a wide range of fish (de la Torre-Castro et al., 2014; Salas et al., 2007). Other characteristics of SSF identified in the literature include that they are: i) mainly artisanal, ii) often decentralised in their operations, iii) low in investment, iv) conducted in short fishing trips and near-shore, and v) intended primarily for subsistence (International Finance Corporation, 2015; Tietze, 2016). Additionally, SSF are mainly located in the Global South (de la Torre-Castro et al., 2014; Salas et al., 2007), and known to use relatively small amounts of mechanical energy (Tietze, 2016). Kalikoski et al. (2019) observe that small-scale fisheries constitute a truly global sector that displays enormous diversity, including in terms of various organisations of fishers and fish workers, such as forms of cooperatives that help them to sustain their livelihoods and enhance their fisheries resources collectively.

The focus on defining small-scale fisheries in terms of the size of boats and types of fishing gear has, however, been critiqued, both in terms of the variation masked within such defining characteristics and in terms of focus, which Smith and Basurto (2019: 15) suggest, 'places undue emphasis on male-dominated harvesting at the expense of a more expansive view of the social and ecological relations along the SSF value chain that span land and sea'. In recognising the diversity of experience and nature of SSF, the FAO (2015) Voluntary Guidelines for Securing Sustainable Small-Scale Fisheries do not prescribe a definition and leaves the application of the guidelines to national actors, including the delineation of what small-scale fisheries are.

Small-scale fisheries have also been characterised as experiencing significant constraints including high post-harvest losses due to limited infrastructure (Salas et al., 2007), such as processing equipment and cold storage facilities. With underdeveloped infrastructure, post-harvest fish losses are estimated at 20 million tonnes per year, accompanied by financial loss since the spoiled or poorly processed fish are discarded or sold at a low price (Ahmed, 2008). In the case of the Lake Victoria sardine (*dagaa*) value chain, physical losses are particularly high during the rainy season when moisture spoils drying fish (Funge-Smith and Bennett, 2019). The fish losses, both in quantity and quality, are driven by inefficiencies related to accessing electricity, potable water, roads, ice, cold storage and refrigerated transport (FAO, 2020).

Whose livelihoods?

Those involved in fisheries livelihoods go beyond the act of fishing itself, with a wide range of other fishing-related activities taking place, including fish portering, processing, trading and shipping, as well as repairing gears and vessels. Even the act of fishing can be complex, in terms of defining who is a fisher, who fishes and what counts as fishing. There are several ways of investigating these questions, including undertaking a value chain analysis to identify actors involved and analyse the share of value along the chain, and taking a gendered lens to analyse who is involved in different small-scale fisheries activities and why.

A value chain refers to 'the full range of activities which are required to bring a product or service from conception, through the different phases of production (involving a combination of physical transformation and the input of various producer services), delivery to final consumers, and final disposal after use' (Kaplinsky and Morris, 2001: 4). This definition describes a 'simple value chain'; as Kaplinsky and Morris (2001) note, however, value chains can be a lot more complicated, with additional links and interaction within the chain; one value chain may be connected to other value chains, complicating analysis and transparency of the flow of information, value and responsibility; and, value chains can be global, bringing further challenges to transparency, equity and governance. In relation to understanding fisheries livelihoods, value chain analysis can be used to identify what types of occupations, activity and relations are found within fish value chains.

From an analysis of eight seafood value chains in the Philippines, Rosales et al. (2017: 14) report that actors include 'boat owners, small-scale fishers, commercial fishers, traders, processors, wholesalers, market vendors and consumers', with each actor providing different services to add value. There are often multiple traders involved as fish is traded along the chain, from primary buyer to wholesaler, and traders will buy from many fishers. Critical to the value chain are relationships between actors, facilitating the flow of fish and credit along the chain, relying often on informal relationships (i.e. with no contract), with trust and reciprocal relations being essential.

In some fisheries, these reciprocal relations reflect 'patron-client relations', where unequal relations of power, status and wealth facilitate reciprocal relations, though with generally greater advantages for the patrons. As part of the exchange of goods and services, social obligations accrue, particularly offering credit and support during ill-health (Powell, 1970). Patron-client relations therefore have developed to secure the flow of goods and services, minimising risks that may jeopardise that flow, reducing fluctuations in the flow of goods and services, but also in the availability of labour and credit (Platteau, 1995). These patron-client relations may be found along a fish value chain. Boat crew depend upon boat owners not only for employment and their salaries, but for loans when income from fisheries is reduced or challenges arise, such as ill-health; boat owners depend on boat crew for their skills, trustworthiness in fishing operations, including looking after fishing vessels and gears, and boat engines, and on traders for credit when catches are low, or there is need for repairs; and, traders rely on fishers for reliable supply of fish and on each other for credit (Johnson, 2010; Nunan et al., 2018; Nunan et al., 2020). So whilst these relationships can be exploitative, they are reciprocal and serve an essential function in situations with limited access to alternative income sources and credit.

These patron-client relations are reflected in the unequal share of value along the chain reported in analyses of fish value chains, with middlemen and exporters receiving much of the value, particularly in situations where there is limited competition at these levels (Purcell et al., 2018). Measures to increase the share of value to the producers include the formation of cooperatives and the implementation of certification schemes, such as the Marine Stewardship Council (MSC) Scheme (Purcell et al., 2018). Ponte (2012) reported in a review of the MSC scheme that few certified fisheries were found in countries of the Global South. Since that publication, the MSC has developed a Global Accessibility Program and, since 2015, a Global Accessibility Fund, in recognition of the limited number of certified fisheries in the Global South (MSC, 2021).

In identifying different occupations and actors benefiting from fisheries, it should also be acknowledged that not everyone involved in fisheries does so on a full-time or permanent basis. In particular, fishers may also be farmers for part of the year, or farmers may be fishers, with engagement in both activities reducing the risks associated with each. Geheb and Binns (1997) found that many fishers on the shores of Lake Victoria in Kenya were also engaged in farming and Yuerlita et al. (2013) categorised fishing households on Singkarak Lake, West Sumatra, Indonesia, as 'fishing farmers', 'farming fishers' and poorly-diverse fishers. Diversity of livelihoods is essential for many fisheries households in coping with variations in fish availability, prices and income (Allison and Ellis, 2001).

Taking a gendered lens to analyse who is involved in different small-scale fisheries activities responds the long perception and portrayal of fisheries being dominated by men, reflecting the focus on the act of fishing rather than how the fishing enterprise was funded, who buys the fish and what happens further along the value chain (Harper et al., 2017; Kleiber et al., 2015; Weeratunge et al., 2010). There is wide recognition that roles within fisheries are often 'gendered', with men mainly associated with fishing and owning boats and women being more associated with processing and trading fish (Kleiber et al., 2015). However, women do engage in fishing, particularly gathering or gleaning shellfish in the intertidal zones, with gender norms and relations influencing the roles of men and women within fisheries and the type and location of fishing generally undertaken by women (Kleiber et al., 2015; Weeratunge et al., 2010). Weeratunge et al. (2010: 406) argue that 'fisheries employment itself begins to look like a female sphere if you account for the roles of gleaning, trading, processing and fish farming'. However, it is also observed that the income that women gain from fisheries is generally substantially less than men receive, explained by the cultural constraints on their roles and mobility, limited access to credit and markets, and the extent of their household responsibilities (Bradford and Katikiro, 2019).

Fisheries and poverty

While small-scale fisheries are a significant source of wealth, their participants often do not reap the benefits of the contributions they make, being under constant stress coming from poverty, powerlessness and marginalisation (Jentoft et al., 2018). In the fisheries literature, there is a perception that SSF in the Global South and poverty are closely related (Béné et al., 2003b). Small-scale fishers in the Global South are often thought to include the most impoverished people in communities (Béné and Friend, 2011), with fisheries offering an occupation of 'last resort' (Masae and McGregor, 1996). The most visible, and possibly the first sign of this state of fishers' extreme destitution, has been cited as food insecurity linked to low income (Béné et al., 2003b). Other indicators include landlessness, poor access to quality health services, lack of education and political marginalisation (Béné and Friend, 2009). However, for Béné et al. (2003b), there is no systematic, or even clear, relationship between fishing and poverty, they argue that fishing households are not necessarily the poorest of the poor. Béné and Friend (2009) contend that the complex relationship between fisheries and poverty is not well-understood. Using a wealth ranking exercise on the Lake Chad Basin, in Chari delta and Yaéré communities along the western shores, Béné et al. (2003b) identified wealthier and poorest households within the two fishing communities, characterised by whether or not they had fishing as their primary source of income and the differing extent of access to the water. Thus, given the wealth differentiation within fisheries, Béné et al. (2003b) demonstrated how the issue of marginalisation mainly concerns the poorest and most deprived fishing households.

Attribution of the origin of poverty amongst small-scale fisherfolk to the fluctuating nature of fisheries resources stresses the significance of external factors, including climate change, conflict, seasonality and natural disasters (Ahmed et al., 2013; Berkes, 2015). These external factors create conditions such as decreasing availability of fish, low fish prices, inadequate provision of public services at landing sites and low status of fisherfolk in society (Allison and Ellis, 2001; Ellis and Freeman, 2004; Sarch and Allison, 2000). On Chilika Lake in eastern India, for example, increasing poverty has been explained as being associated with changes in the lake's ecology (including habitat destruction, water quality deterioration and species composition degradation), which gradually affected social conditions through loss of fish productivity, fisher livelihoods and outmigration (Nayak et al., 2014).

Institutional factors have also been identified as being associated with fisherfolk poverty, in the sense that fisheries livelihoods are facilitated by institutions through fisheries management mechanisms and arrangements (Béné, 2004; Ellis and Freeman, 2004; Lewins, 2004). This institutional-focused view considers the role of institutions in influencing how and to what extent people are able to access available natural resources (Allison and Horemans, 2006; Nunan et al., 2015). According to this view, poverty is more related to institutional factors than to fluctuating physical and ecological conditions (Béné, 2004; Ellis and Freeman, 2004; Lewins, 2004). Thus, well-intended policies for the benefit of fisheries ecosystems can be at the same time stressors to fisherfolk (Brown, 2016), setting adverse institutional contexts that can make and keep people poor (Allison et al., 2006; Béné, 2003). In addition, the poor provision of public services in fisheries communities is seen to contribute to a situation of poverty, with inadequate access to health, education and reliable roads (Béné, 2003).

Small-scale fisheries and aquaculture have been recognised as important opportunities for enhancing household food security (Kawarazuka and Béné, 2010). The role of small-scale fisheries as a livelihood support is critical for many fisherfolk households, especially in areas with limited alternative employment and scarce or non-existent public services (Béné et al., 2007). While the large majority of fisherfolk households in the Global South do not make high economic gains, they are enabled to sustain their livelihoods through fishing-related activities. For instance, in Africa, in parts of Gabon, the Democratic Republic of Congo and Equatorial Guinea, populations rely on fish for their food and livelihood despite their low income approaching the poverty line (Belhabib et al., 2015).

With a role of strengthening the livelihoods of people through both food security and cash-income generation, small-scale fisheries can play a fundamental role in local economies, especially in remote rural areas (Béné et al., 2009). Within this perspective, fisheries can serve as a labour buffer or safety valve, which refers to its capacity to absorb surplus labour (Béné et al., 2010). Labour surplus in a rural context has been defined as 'the total number of rural employed persons not involved in any of these three activities: agricultural employment (AG), rural nonagricultural employment (RNAG), and rural urban employment (RUE)' (Yinhua and Peng, 2012: 40–42).

The safety valve or safety-net function of fisheries provides temporary assistance to people's livelihoods when faced with shocks such as a major flood, drought or family illness (Coomes et al., 2010). Fisheries can also serve a safety net function in situations of conflict, where conditions are said to 'compel affected households to turn to fisheries as alternative employment' (Béné et al., 2010: 338). Thus, when in pressing need for fast access to food or income due to shocks, fishing may be preferred over forest extraction or agriculture (Coomes et al., 2010; Martin et al., 2013).

In having the capacity to absorb unskilled labour and serve as a safety net or valve, fisheries are said to offer a 'welfare function' to local societies and economies (Béné, 2006; Béné et al., 2010; Nunan, 2014). Due to the often open-access nature of fisheries, a large number of people may be attracted to benefits from fisheries, especially in areas where alternatives for other economic activities are limited or impeded (Béné et al., 2003b).

Vulnerability: sources and coping responses

Many sources of vulnerability are associated with fisheries, from the weather affecting migration of fish and fishing conditions, to daily income and attitudes to risk, being in turn associated with risky sexual behaviour and relatively high prevalence of HIV/AIDS among fishing communities in Africa and Asia (Smolak, 2014). Bad weather conditions are a key source of vulnerability to drowning, with Whitworth et al. (2019) and Kobusingye et al. (2017) reporting substantially higher incidences of drowning on the shores of Tanzania and Uganda shores of Lake Victoria than the rest of the two countries. High numbers of drowning are also attributed to fishers often not being able to swim and not using a lifejacket (Whitworth et al., 2019).

To cope with the potential dangers and precarious nature of fishing, it has been reported that fishermen often turn to alcohol, with alcohol consumption found to be a usual practice (Ribeiro et al., 2017). As well as reported effects on health (Ribeiro et al., 2017), high levels of alcohol consumption in fishing communities are associated with high levels of casual sex, and in turn high prevalence of HIV/AIDS (Sileo et al., 2016), and with domestic violence (Coulthard et al., 2020).

In coping with seasonal variations in fish stocks and fluctuating prices of fish, fishers often resort to diversifying their income sources (Allison and Ellis, 2001), or they may move from one fishing area to another, staying at other locations for months at a time. Fishers move in search of better catches and higher prices and hence income (Marquette et al., 2002; Nunan, 2010; Randall, 2005; Wanyonyi et al., 2016), though over time, migration may become a way of life, with social networks facilitating movement (Wanyonyi et al., 2016). Migration then can reduce vulnerability associated with income, but it can also create new vulnerabilities, associated with moving to new locations and the type of behaviour and norms referred to above, in terms of alcohol consumption, being away from the permanent partner and engaging in new sexual relationships (Nunan, 2010). In some areas, this has been associated with the high prevalence of HIV/AIDS found in fishing communities (Kissling et al., 2005).

A further source of vulnerability for fishers is associated with working conditions. In marine industrial fisheries, these can be particularly alarming, with forced labour and unsafe working conditions. In Thailand, marginalised groups, mainly migrant fishers, have been reported to be subject to abuse through forced labour (Chantavanich et al., 2016). Forced labour entails several coercive practices, including abuse of vulnerability and abusive working and living conditions, occurring along the labour chain from recruitment to exit (Chantavanich et al., 2016). Thus, while migrant fishers' vulnerability is exploited to trap them into enduring various forms of abuse, it keeps them in conditions that, in turn, continue their vulnerable situations. In the context of Cambodia and the Philippines, this vulnerability for migrant fishers is associated with structures and conditions that produce precarity in their work, with documented occurrences of human trafficking in the fisheries settings (Yea, 2019).

Fisheries management and livelihoods

Fisheries are subject to management regimes due to concerns about depletion of fish stocks, with associated impacts on biodiversity, ecosystems and sustainability. These management regimes, in turn, have implications for livelihoods as they aim to control, at least to some extent, access to fisheries and how much can be taken, including which species, from which areas. The measures adopted within management regimes include regulations to control the types of gears and fishing vessels that can be used, with restrictions on gears that enable immature fish to be caught or gears banned that may be easily lost and remain harmfully fishing for decades to come.

Challenges in many fisheries stem from their open-access nature which creates irrational incentives, including overuse, inefficient use and a race for resources (FAO, 2008; Mansfield, 2004). Open access regimes give free access to anyone, with no exclusions, with no owner of the resource, and is recognised as a key causal factor behind fish stock depletion (Servos et al., 2013). The impact on livelihoods of fish stock depletion is challenging throughout the Global South, where many fisheries are inland, de facto open access and generally in situations of poor management (de la Torre-Castro et al., 2014), with high levels of dependence on fisheries resources (Allison and Ellis, 2001).

Given recognition of fish stock depletion, regulations to control and limit access to fisheries have been developed by many governments (Allison and Ellis, 2001; Basurto et al., 2012). The state property regime, a top-down state resources management system, includes measures such as requiring boat or operator licenses, restrictions on vessel capacity, closed seasons and closed fishing zones, as well as technical measures such as the imposition of minimum mesh size for nets and prohibition of certain types of gears (Allison and Ellis, 2001; Basurto et al., 2012). However, state management regimes have also largely failed to limit fishing effort due to factors including the lack of capacity and lack of inclination to enforce (Cepić and Nunan, 2017), as well as the lack of adaptability to both unpredictable variability in fish stocks and the fisher-fish relationship (Allison and Ellis, 2001). The fisher-fish relationship means that resource scarcity does not stop fishers from fishing but can encourage changes in how fishing is conducted. For examples, fishery fluctuations in African inland lakes affect fishers' mobility, leading them to extend their geographic areas through migrant fishing (Allison and Ellis, 2001). Another response by fishers to regulations is to increase the number of gears they use or to fish illegally, using prohibited gears, which may have too small a mesh size or the hook size may be smaller than permitted (Njiru et al., 2007).

Besides state-led management access control mechanisms, many fisheries are managed and governed through community-based or collaborative management approaches, with fishing communities working with government in managing fisheries. Such community-based approaches may be based on common property regimes and traditional practices, where a collective of users has rights over the resource and can exclude people not in their collective (Agrawal, 2001). Restriction mechanisms can define exactly who has access and under which conditions, including payment of access rights, fees, licences, etc. (Béné et al., 2003a). Studies have shown that the use of property rights in the commons offers protection against overfishing if they reflect the joint interests of a community of users rather than of isolated individual actors (Ostrom, 1977). Community-based and collaborative approaches to management often seek outcomes associated with improved livelihoods as well as greater sustainability of natural resources, often realised through empowerment and greater access to decision-making, rather than necessarily improvements in income, as described in Chapter 42 of this volume on collective organisations.

Conclusion

Fisheries provide a livelihood for millions of people in the Global South – a livelihood in the sense of income, food and nutrition and identity associated with being involved in fisheries. Income generated through fisheries enables households to access healthcare and send children to school. Such livelihoods go beyond the catching of fish to include the investment behind the boat and gears, the processing and trading of fish and the provision of food and accommodation to migrant fishers. From the literature, strong gender relations are apparent within fisheries though the situation can be more complex than a gendered division of labour, with gender relations having implications for livelihoods in terms of access to fish and income from fisheries.

Multiple sources of vulnerability that fisherfolk face will be exacerbated by climate change, as increasing temperatures affect the location and breeding of fish, with implications for fish stocks and catches. The management of fisheries in this context will be even more critical in this context, to maintain sustainable fish stocks, ecosystems and livelihoods.

Key points

- Fisheries livelihoods in the Global South are strongly associated with small-scale fisheries, often characterised by artisanal boats, low levels of capital investment and nearshore fishing. However, this characterisation has been critiqued for masking diversity of activity and scale.
- Fisheries occupations and livelihoods are often distinguished in terms of the roles played along the value chain, with inequalities along the chain in terms of power relations and share of economic value.
- Small-scale fisheries are generally characterised by gendered labour roles, with men associated with fishing and owning boats and women with processing and trading fish. Though these are generalisations, gendered relations strongly influence fisheries livelihoods.
- Fisheries have at times been characterised as the 'occupation of last resort', offering a labour buffer and safety net role. However, the situation may be more complicated, with many fishers having regular access to cash income but with fishing communities experiencing poverty through poor access to public services, including healthcare, education and good roads.
- Fluctuations in fish catches and prices, bad weather and dangerous working conditions contribute sources of vulnerability to fisheries livelihoods, with fisherfolk responding through coping mechanisms that include migration and illegal fishing, which themselves can create further vulnerabilities.
- Fisheries are largely managed through state, community-based or collaborate regimes, though many fisheries operate as open access regimes with limited restrictions on access and extraction.

Further reading

Allison, E.H. and Ellis, F. (2001) 'The livelihoods approach and management of small-scale fisheries', *Marine Policy*, 25: 377–388. https://doi.org/10.1016/S0308-597X(01)00023-9

Béné, C., Macfadyen, G. and Allison, E.H. (2007) 'Increasing the contribution of small-scale fisheries to poverty alleviation and food security', *FAO Fisheries Technical Paper No. 481*, Rome: Food and Agriculture Organization.

Berkes, F. (2015) *Coasts for people: Interdisciplinary approaches to coastal and marine resource management*, New York: Routledge.

Neiland, A. and Béné, C. (2004) *Poverty and Small-scale fisheries in West Africa*, Amsterdam: Kluwer Academic Publishers.

References

Agrawal, A. (2001) 'Common property institutions and sustainable governance of resources', *World Development*, 29(10): 1649–1672. https://doi.org/10.1016/S0305-750X(01)00063-8

Ahmed, A.A. (2008) 'Post-harvest losses of fish in developing countries', *Nutrition and Health*, 19(4): 273–287. https://doi.org/10.1177/026010600801900403

Ahmed, N., Rahman, S. and Bunting, S.W. (2013) 'An ecosystem approach to analyse the livelihood of fishers of the Old Brahmaputra River in Mymensingh region, Bangladesh', *Local Environment*, 18(1): 36–52. https://doi.org/10.1080/13549839.2012.716407

Allison, E.H. and Ellis, F. (2001) 'The livelihoods approach and management of small-scale fisheries', *Marine Policy*, 25: 377–388. https://doi.org/10.1016/S0308-597X(01)00023-9

Allison, E.H., Horemans, B. and Béné, C. (2006) 'Vulnerability reduction and social inclusion: strategies for reducing poverty among small-scale fisherfolk', Paper presented at the *Wetlands, Water and Livelihoods Workshops*, Wetland International. 30 January–2 February, St. Lucia, South Africa.

Allison, E.H. and Horemans, B. (2006) 'Putting the principles of the sustainable livelihoods approach into fisheries development policy and practice', *Marine Policy*, 30: 757–766. https://doi.org/10.1016/j.marpol.2006.02.001

Basurto, X., Cinti, A., Bourillón, L., Rojo, M., Torre, J. and Weaver, A.H. (2012) 'The emergence of access controls in small-scale fishing commons: A comparative analysis of individual licenses and common property-rights in two Mexican communities', *Human Ecology*, 40(4): 597–609. https://doi.org/10.1007/s10745-012-9508-1

Belhabib, D., Sumaila, U.R. and Pauly, D. (2015) 'Feeding the poor: Contribution of West African fisheries to employment and food security', *Ocean & Coastal Management*, 111: 72–81. https://doi.org/10.1016/j.ocecoaman.2015.04.010

Béné, C. (2003) 'When fishery rhymes with poverty: A first step beyond the old paradigm on poverty in small-scale fisheries', *World Development*, 31(6): 949–975. https://doi.org/10.1016/S0305-750X(03)00045-7

Béné, C. (2004) 'Poverty in small-scale fisheries: A review and some further thoughts', in Neiland, A.E. and Béné, C. (eds.), *Poverty and small-scale fisheries in West Africa*, Dordrecht: Springer, pp. 61–82.

Béné, C. (2006) 'Small-scale fisheries: Assessing their contribution to rural livelihoods in developing countries', *FAO Fisheries Circular 1008*, Rome: Food and Agricultural Organization.

Béné, C. and Friend, R.M. (2009) 'Water, poverty and inland fisheries: Lessons from Africa and Asia', *Water International*, 34(1): 47–61. https://doi.org/10.1080/02508060802677838

Béné, C. and Friend, R.M. (2011) 'Poverty in small-scale fisheries: Old issue, new analysis', *Progress in Development Studies*, 11(2): 119–144. https://doi.org/10.1177/146499341001100203

Béné, C., Hersoug, B. and Allison, E.H. (2010) 'Not by rent alone: Analysing the pro-poor functions of small-scale fisheries in developing countries', *Development Policy Review*, 28(3): 325–358. https://doi.org/10.1111/j.1467-7679.2010.00486.x

Béné, C., Macfadyen, G. and Allison, E.H. (2007) 'Increasing the contribution of small-scale fisheries to poverty alleviation and food security', *FAO Fisheries Technical Paper No. 481*, Rome: Food and Agriculture Organization.

Béné, C., Neiland, A., Jolley, T., Ladu, B., Ovie, S., Sule, O., Baba, M., Belal, E., Mindjimba, K., Tiotsop, F., Dara, L., Zakara, A. and Quensiere, J. (2003a) 'Natural-resource institutions and property rights in inland African fisheries: The case of the Lake Chad Basin region', *International Journal of Social Economics*, 30(3): 275–301. https://doi.org/10.1108/03068290310460161

Béné, C., Neiland, A., Jolley, T., Ovie, S., Sule, O., Ladu, B., Mindjimba, K., Belal, E., Tiotsop, F., Baba, M., Dara, L., Zakara, A. and Quensiere, J. (2003b) 'Inland fisheries, poverty, and rural livelihoods in the Lake Chad Basin', *Journal of Asian and African Studies*, 38(1): 17–51. https://doi.org/10.1177/002190960303800102

Béné, C., Steel, E., Luadia, B.K. and Gordon, A. (2009) 'Fish as the "bank in the water": Evidence from chronic-poor communities in Congo', *Food Policy*, 34(1): 108–118. https://doi.org/10.1016/j.foodpol.2008.07.001

Bennett, A., Patil, P., Kleisner, K., Rader, D., Virdin, J. and Basurto, X. (2018) 'Contribution of fisheries to food and nutrition security: Current knowledge, policy, and research', *NI Report 18–02*, Durham, NC: Duke University.

Berkes, F. (2015) *Coasts for people: Interdisciplinary approaches to coastal and marine resource management*, New York: Routledge.

Bradford, K. and Katikiro, R.E. (2019) 'Fighting the tides: A review of gender and fisheries in Tanzania', *Fisheries Research*, 216: 79–88. https://doi.org/10.1016/j.fishres.2019.04.003

Brown, K. (2016) *Resilience, development and global change*, London: Routledge.

Cepić, D. and Nunan, F. (2017) 'Justifying non-compliance: The morality of illegalities in small scale fisheries of Lake Victoria, East Africa', *Marine Policy*, 86: 104–110. https://doi.org/10.1016/j.marpol.2017.09.018

Chantavanich, S., Laodumrongchai, S. and Stringer, C. (2016) 'Under the shadow: Forced labour among sea fishers in Thailand', *Marine Policy*, 68: 1–7. https://doi.org/10.1016/j.marpol.2015.12.015

Coomes, O.T., Takasaki, Y., Abizaid, C. and Barham, B.L. (2010) 'Floodplain fisheries as natural insurance for the rural poor in tropical forest environments: Evidence from Amazonia', *Fisheries Management and Ecology*, 17(6): 513–521. https://doi.org/10.1111/j.1365-2400.2010.00750.x

Coulthard, S., White, C., Paranamana, N., Sandaruwan, K.P.G.L., Manimohan, R. and Maya, R. (2020) 'Tackling alcoholism and domestic violence in fisheries: A new opportunity to improve well-being for the most vulnerable people in global fisheries', *Fish and Fisheries*, 21: 223–236. https://doi.org/10.1111/faf.12426

de la Torre-Castro, M., Di Carlo, G. and Jiddawi, N.S. (2014) 'Seagrass importance for a small-scale fishery in the tropics: The need for seascape management', *Marine Pollution Bulletin*, 83(2): 398–407. https://doi.org/10.1016/j.marpolbul.2014.03.034

Ellis, F. and Freeman, H.A. (eds.) (2004) *Rural livelihoods and poverty reduction policies*, Abingdon: Routledge.

FAO (2015) *Voluntary guidelines for securing sustainable small-scale fisheries in the context of food security and poverty eradication*, Rome: Food and Agricultural Organization.

FAO (2018) *The state of world fisheries and aquaculture 2018*, Rome: Food and Agricultural Organization.

FAO (2020) *The state of world fisheries and aquaculture 2020*, Rome: Food and Agricultural Organization. https://doi.org/10.4060/ca9229en

Food and Agriculture Organization (FAO) (2008) *The state of world fisheries and aquaculture 2008*, Rome: Food and Agricultural Organization.

Funge-Smith, S. and Bennett, A. (2019) 'A fresh look at inland fisheries and their role in food security and livelihoods', *Fish and Fisheries*, 20(6): 1176–1195. https://doi.org/10.1111/faf.12403

Geheb, K. and Binns, T. (1997) '"Fishing farmers" or "farming fishermen"? The quest for household income and nutritional security on the Kenyan shores of Lake Victoria', *African Affairs*, 96(382): 73–93. https://doi.org/10.1093/oxfordjournals.afraf.a007822

Harper, S., Adshade, M., Lam, V.W.Y., Pauly, D. and Sumaila, U.R. (2020) 'Valuing invisible catches: Estimating the global contribution by women to small-scale marine capture fisheries production', *PLoS One*, 15(3): 0228912. https://doi.org/10.1371/journal.pone.0228912

Harper, S., Grubb, C., Stiles, M. and Sumaila, U.R. (2017) 'Contributions by women to fisheries economies: Insights from five maritime countries', *Coastal Management*, 45(2): 91–106. https://doi.org/10.1080/08920753.2017.1278143

Hilborn, R., Amoroso, R.O., Anderson, C.M., Baum, J.K., Branch, T.A., Costello, C., de Moor, C.L., Faraj, A., Hively, D., Jensen, O.P., Kurota, H., Little, L.R., Mace, P., McClanahan, T., Melnychuk, M.C., Minto, C., Osio, G.C., Parma, A.M., Pons, M., Segurado, S., Szwalski, C.S., Wilson, J.R. and Ye, Y. (2020) 'Effective fisheries management instrumental in improving fish stock status', *Proceedings of the National Academy of Sciences*, 117(4): 2218–2224. https://doi.org/10.1073/pnas.1909726116

International Finance Corporation (2015) *Addressing project impacts on fishing-based livelihoods. A good practice handbook: Baseline assessment and development of a fisheries livelihood restoration plan*, Washington, DC: International Finance Corporation.

Jentoft, S., Bavinck, M., Alonso-Población, E., Child, A., Diegues, A., Kalikoski, D., Kurien, J., McConney, P., Onyango, P., Siar, S. and Rivera, V.S. (2018) 'Working together in small-scale fisheries: Harnessing collective action for poverty eradication', *Maritime Studies*, 17(1): 1–12. https://doi.org/10.1007/s40152-018-0094-8

Johnson, D.S. (2010) 'Institutional adaptation as a governability problem in fisheries: Patron-client relations in the Junagadh fishery, India', *Fish and Fisheries*, 11: 264–277. https://doi.org/10.1111/j.1467-2979.2010.00376.x

Kalikoski, D.C., Jentoft, S., McConney, P. and Siar, S. (2019) 'Empowering small-scale fishers to eradicate rural poverty', *Maritime Studies*, 18(2): 121–125. https://doi.org/10.1007/s40152-018-0112-x

Kaplinsky, R. and Morris, M. (2001) *A handbook for value chain research*, Ottawa: International Development Research.

Kawarazuka, N. and Béné, C. (2010) 'Linking small-scale fisheries and aquaculture to household nutritional security: An overview', *Food Security*, 2(4): 343–357. https://doi.org/10.1007/s12571-010-0079-y

Kissling, E., Allison, E.H., Seeley, J.A., Russell, S., Bachmann, M., Musgrave, S.D. and Heck, S. (2005) 'Fisherfolk are among groups most at risk of HIV: Cross-country analysis of prevalence and numbers infected', *AIDS*, 19: 1939–1946. https://doi.org/10.1097/01.aids.0000191925.54679.94

Kleiber, D., Harris, L.M. and Vincent, A.C.J. (2015) 'Gender and small-scale fisheries: A case for counting women and beyond', *Fish and Fisheries*, 16(4): 547–562. https://doi.org/10.1111/faf.12075

Kobusingye, O., Tumwesigye, N.M., Magoola, J., Atuyambe, L. and Olange, O. (2017) 'Drowning among the lakeside fishing communities in Uganda: Results of a community survey', *International Journal of Injury Control and Safety Promotion*, 24(3): 363–370. https://doi.org/10.1080/17457300.2016.1200629

Lewins, R. (2004) 'The sustainable livelihoods approach: The importance of policies, institutions and process', in Neiland, A.E. and Béné, C. (eds.), *Poverty and small-scale fisheries in West Africa*, Dordrecht: Springer, pp. 37–45.

Mansfield, B. (2004) 'Neoliberalism in the oceans: "rationalization," property rights, and the commons question', *Geoforum*, 35(3): 313–326. https://doi.org/10.1016/j.geoforum.2003.05.002

Marine Stewardship Council (MSC) (2021) *Our global accessibility program*. Available at: www.msc.org/for-business/fisheries/developing-world-and-small-scale-fisheries/global-accessibility-program (Accessed 28 May 2021).

Marquette, C.M., Koranteng, K.A., Overå, R. and Aryeetey, E.B.D. (2002) 'Small-scale fisheries, population dynamics, and resource use in Africa: The case of Moree, Ghana', *AMBIO: A Journal of the Human Environment*, 31(4): 324–337. https://doi.org/10.1579/0044-7447-31.4.324.

Martin, S.M., Lorenzen, K. and Bunnefeld, N. (2013) 'Fishing farmers: Fishing, livelihood diversification and poverty in rural Laos', *Human Ecology*, 41(5): 737–747. https://doi.org/10.1007/s10745-013-9567-y

Masae, A. and McGregor, J.A. (1996) Sustainability under conditions of rapid development: A freshwater fishery in Southern Thailand, *Occasional Paper 01/96*, Bath: Centre for Development Studies, University of Bath.

Nayak, P.K., Oliveira, L.E. and Berkes, F. (2014) 'Resource degradation, marginalization, and poverty in small-scale fisheries: Threats to social-ecological resilience in India and Brazil', *Ecology and Society*, 19(2): 73. http://dx.doi.org/10.5751/ES-06656-190273

Njiru, M., Nzungi, P., Getabu, A., Wakwabi, E., Othina, A., Jembe, T. and Wekesa, S. (2007) 'Are fisheries management, measures in Lake Victoria successful? The case of Nile perch and Nile tilapia fishery', *African Journal of Ecology*, 45: 315–323. https://doi.org/10.1111/j.1365-2028.2006.00712.x

Nunan, F. (2010) 'Mobility and fisherfolk livelihoods on Lake Victoria: Implications for vulnerability and risk', *Geoforum*, 41: 776–785. doi.org/10.1016/j.geoforum.2010.04.009

Nunan, F. (2014) 'Wealth and welfare? Can fisheries management succeed in achieving multiple objectives? A case study of Lake Victoria, East Africa', *Fish and Fisheries*, 15(1): 134–150. https://doi.org/10.1111/faf.12012

Nunan, F., Cepić, D., Mbilingi, B., Odongkara, K., Yongo, E., Owili, M., Salehe, M., Mlahagwa, E. and Onyango, P. (2018) 'Community cohesion: Social and economic ties in the personal networks of fisherfolk', *Society and Natural Resources*, 31(3): 306–319. https://doi.org/10.1080/08941920.2017.1383547

Nunan, F., Cepić, D., Onyango, P., Salehe, M., Yongo, E., Mbilingi, B., Odongkara, K., Mlahagwa, E. and Owili, M. (2020) 'Big fish, small fries? The fluidity of power in patron-client relations of Lake Victoria fisheries', *Journal of Rural Studies*, 79: 246–253. https://doi.org/10.1016/j.jrurstud.2020.08.021

Nunan, F., Hara, M. and Onyango, P. (2015) 'Institutions and co-management in East African inland and Malawi fisheries: A critical perspective', *World Development*, 70: 203–214. https://doi.org/10.1016/j.worlddev.2015.01.009

Ostrom, E. (1977) 'Collective action and the tragedy of the commons', in Hardin, G. and Bader, J. (eds.), *Managing the commons*, W.H. Freeman: San Francisco, pp. 173–181.

Platteau, J-P. (1995) 'A framework for the analysis of evolving patron-client ties in Agrarian economies', *World Development*, 23(5): 767–786. https://doi.org/10.1016/0305-750X(95)00011-Z

Ponte, S. (2012) 'The Marine Stewardship Council (MSC) and the making of a market for "sustainable fish"', *Journal of Agrarian Change*, 12(2&3): 300–315. https://onlinelibrary.wiley.com/doi/abs/10.1111/j.1471-0366.2011.00345.x

Powell, J.D. (1970) 'Peasant society and clientelist politics', *American Political Science Review*, 64(2): 411–429. https://doi.org/10.2307/1953841

Purcell, S.W., Crona, B.I., Lalavanua, W. and Eriksson, H. (2018) 'Distribution of economic returns in small-scale fisheries for international markets: A value-chain analysis', *Marine Policy*, 86: 9–16. https://doi.org/10.1016/j.marpol.2017.09.001

Randall, S. (2005) *Review of literature on fishing migrations in West Africa – from a demographic perspective. sustainable fisheries livelihoods programme*, Rome: FAO/DFID.

Ribeiro, C.R.B., Saboia, V.M. and Pereira, C.M. (2017) 'Alcohol consumption among fishermen: An integrative review', *Rev Fund Care Online*, 9(2): 575–582. http://dx.doi.org/10.9789/2175-5361.2017.v9i2.575-582

Rosales, R.M., Pomeroy, R., Calabio, I.J., Batong, M., Cedo, K., Escara, N., Facunla, V., Anecita Gulayan, A., Narvadez, M., Sarahadil, M. and Sobrevega, M.A. (2017) 'Value chain analysis and small-scale fisheries management', *Marine Policy*, 83: 11–21. https://doi.org/10.1016/j.marpol.2017.05.023

Salas, S., Chuenpagdee, R., Seijo, J.C. and Charles, A. (2007) 'Challenges in the assessment and management of small-scale fisheries in Latin America and the Caribbean', *Fisheries Research*, 87(1): 5–16. https://doi.org/10.1016/j.fishres.2007.06.015

Sarch, M-T. and Allison, E.H. (2000) 'Fluctuating fisheries in Africa's inland waters: Well adapted livelihoods, maladapted management', *Proceedings of the 10th International Conference of the Institute of Fisheries Economics and Trade, Corvallis, 9–14 July 2000*, Corvallis: International Institute of Fisheries Economics and Trade (IIFET).

Servos, M.R., Munkittrick, K.R., Constantin, G., Mngodo, R., Aladin, N., Choowaew, S., Hap, N., Kidd, K.A., Odada, E., Parra, O., Phillips, G., Ryanzhin, S. and Urrutia, R. (2013) 'Science and management of transboundary Lakes: Lessons learned from the global environment facility program', *Environmental Development*, 7: 17–31. https://doi.org/10.1016/j.envdev.2013.04.005

Sileo, K.M., Kintu, M., Chanes-Mora, P. and Kiene, S.M. (2016) '"Such behaviors are not in my home village, I got them here": A qualitative study of the influence of contextual factors on alcohol and HIV risk behaviors in a fishing community on Lake Victoria, Uganda', *AIDS Behavior*, 20(3): 537–547. https://doi.org/10.1007/s10461-015-1077-z

Smith, H. and Basurto, X. (2019) 'Defining small-scale fisheries and examining the role of science in shaping perceptions of who and what counts: A systematic review', *Frontiers in Marine Science*, 6: 236. https://doi.org/10.3389/fmars.2019.00236

Smolak, A. (2014) 'A meta-analysis and systematic review of HIV risk behavior among fishermen', *AIDS Care*, 26(3): 282–291. https://doi.org/10.1080/09540121.2013.824541

Tietze, U. (2016) 'Technical and socio-economic characteristics of small-scale coastal fishing communities, and opportunities for poverty alleviation and empowerment', *FAO Fisheries and Aquaculture Circular No. 1111*, Rome: Food and Agricultural Organization.

Wanyonyi, I.N., Wamukota, A., Tuda, P., Mwakha, V.A. and Nguti, L.M. (2016) 'Migrant fishers of Pemba: Drivers, impacts and mediating factors', *Marine Policy*, 71: 242–255. https://doi.org/10.1016/j.marpol.2016.06.009

Weeratunge, N., Snyder, K.A. and Sze, C.P. (2010) 'Gleaner, fisher, trader, processor: Understanding gendered employment in fisheries and aquaculture', *Fish and Fisheries*, 11(4): 405–420. https://doi.org/10.1111/j.1467-2979.2010.00368.x

Whitworth, H.S., Pando, J., Hansen, C., Howard, N., Moshi, A., Rocky, O., Mahanga, H., Jabbar, M., Ayieko, P., Kapiga, S., Grosskurth, H. and Watson-Jones, D. (2019) 'Drowning among fishing communities on the Tanzanian shore of Lake Victoria: A mixed-methods study to examine incidence, risk factors and socioeconomic impact', *BMJ Open*, 9: e032428. https://doi.org/10.1136/bmjopen-2019-032428

World Bank (2012) *Hidden harvest: The global contribution of capture fisheries*, Washington, DC: World Bank Group.

Yea, S. (2019) 'Secondary precarity in Asia: Family vulnerability in an age of unfree labour', *Journal of Contemporary Asia*, 49(4): 552–567. https://doi.org/10.1080/00472336.2019.1572772

Yinhua, M. and Peng, X. (2012) 'Estimating rural labour surplus in China: A dynamic general equilibrium analysis', *The Chinese Economy*, 45(6): 38–59. https://doi.org/10.2753/CES1097-1475450603

Yuerlita, Perret, S.R. and Shivakoti, G.P. (2013) 'Fishing farmers or farming fishers? Fishing typology of inland small-scale fishing households and fisheries management in Singkarak Lake, West Sumatra, Indonesia', *Environmental Management*, 52: 85–98. https://doi.org/10.1007/s00267-013-0050-8

ns of work in this sector (Birkbeck, 1978;
29
COMPLEXITY AND HETEROGENEITY IN THE INFORMAL ECONOMY OF WASTE

Problems and prospects for organising and formalising

Aman Luthra

Introduction

The waste collection and recycling sector around the world is comprised of a complex network of workers who derive their livelihood from collecting, sorting and recycling materials. In doing so, they provide crucial economic, municipal and environmental services (Dias, 2016; Scheinberg et al., 2010; UN-HABITAT, 2010). Workers in this sector can be informal or formal or quasi-formal, waged or self-employed (Dias, 2011). Yet, these categories may have limited analytical utility in understanding the nature and relations of work in this sector (Birkbeck, 1978; Dinler, 2016; Millar, 2014). This chapter provides an overview of the structure and heterogeneity of the informal waste sector, and their implications for organising and formalising workers. A series of boxes provide a deeper understanding of some of the issues dealt with in this chapter through an examination of workers in the Indian context.

In cities of the Global South, informal workers in this sector provide 50% to 100% waste collection coverage, thus heavily subsidising municipal services often without being paid for service provision (Scheinberg et al., 2010). Through waste diversion alone, these informal workers save cities between 15 and 20% of their waste management budgets (UN-HABITAT, 2010). Although estimates of their contribution to the recycling economy vary by city, a survey of six cities shows that the informal sector recycles as much as 66% of solid waste in one city and more than 20% on average (Scheinberg et al., 2010). Benefits of recycling aside, their work diverts recyclable materials from the waste stream and therefore reduces the environmental burden associated with disposal. In addition, solid waste collection and recycling is a crucial environmental service reducing greenhouse gas emissions and the urban environmental footprint (Chintan, 2009; Vergara et al., 2016).

Despite their crucial economic, municipal and environmental contributions, workers in this sector worldwide are facing challenges, particularly from formal privatisation of waste management services that threaten to displace and dispossess them from their means of subsistence (Samson, 2009; Sandhu et al., 2017). These challenges stem from four broad interrelated pressures on municipal governments. First, in many low- and middle-income countries, quantities of waste (both in per capita and absolute terms) are increasing with urbanisation and economic growth (Hoornweg and Bhada-Tata, 2012). Second, increasingly stringent environmental regulations governing the management of waste are putting pressures on municipal governments to increase service coverage while reducing costs (McBride, 2009). Third, rapidly growing cities are facing pressures, particularly from middle class residents to improve waste management service delivery to enhance urban quality of life in terms of environmental and public health indicators (Luthra, 2019). Fourth, faced with fiscal constraints, municipal governments have been cutting public sector spending and privatising waste management services over the past three decades, often with technical support from international development institutions such as the World Bank (Spronk, 2010). Different cities are facing combinations of these pressures to different extents but many have sought to formally outsource waste management service provision to domestic and foreign firms for whom this serves as a lucrative business opportunity (McBride, 2009; Schoenberger, 2003). In response to these pressures, informal workers in cities around the world have organised and formed networks at the local, national and international scales to articulate their demands for safe, secure and stable livelihoods, with varying degrees of success between and within countries (Dias, 2016; Samson, 2009).

Estimates of the number of workers involved in this sector vary widely depending on who is counting, who is being counted and the purpose of counting. A review of 43 data sets on informal workers in waste management suggests that 0.6% of the urban population is engaged in such work but estimates vary widely between cities and countries from a low of 0.02% to a high of 2.56% (Linzner and Lange, 2013). Most existing data and estimates are of the population of waste pickers, a term that replaced derogatory terms such as 'scavengers' and 'ragpickers' at the First International Conference of Waste Pickers in Bogota, Colombia, organised by Women in Informal Employment Globalizing and Organizing (WIEGO) in 2008 (ILO, 2013). Waste pickers are also only one, albeit foundational, part of the wider recycling economy, thus even these statistical snapshots, as variable as they are, may leave out those who do not count as waste pickers per se, yet provide crucial labour that sustains the global recycling system. For instance, ILO (2013) applies the term waste picker to those who collect and sort recyclable materials, as well as those who are involved in the processing of recyclable waste. But, the extent to which the latter are included in various estimates is unclear (Linzner and Lange, 2013).

Understanding exactly who is being counted matters for at least two interrelated reasons. First, policy models and prescriptions often homogenise the functional and demographic diversity of actors within this sector, generating new kinds of exclusions, even though 'inclusion' is the demand they are purportedly responding to and nominally satisfying. Second, and more importantly, understanding the functional and demographic complexity of the sector is important for collective organising projects so that policy exclusions may be combated more effectively and all workers' livelihoods secured. WIEGO (2020) has proposed two typologies of informal waste workers: a functional one that categorises workers according to the work that they perform (dump/landfill waste pickers, street waste pickers, doorstep waste pickers, on route/truck waste pickers, itinerant buyers, sorters and handlers of organic waste), and another one that classifies workers in terms of their relationship to organisations (unorganised or autonomous workers, organised workers, and workers with a contract). Box 29.1 applies WIEGO's (2020) functional typology of work and workers in this sector to the Indian context in order to expose its complex and heterogeneous nature.

Box 29.1 Structure of the informal economy of waste in India

In the Indian context, the informal recycling system is often functionally represented as a pyramid (see Figure 29.1): at the base are waste pickers who collect recyclable materials and sell them to intermediate dealers who in turn sell them to reprocessors who turn those materials into new commodities (Agarwal et al., 2005; Chintan, 2003). This pyramidal representation is useful for conceptualising not only the increasing specialisation and decreasing number of workers as one moves upward in the pyramid (Agarwal et al., 2005), it is also helpful in understanding the movement and concentration of surplus value into fewer hands as waste is turned into value through the labours of numerous waged and unwaged workers along the process. Much scholarly work and policy interventions has focused on workers at the bottom of this pyramid even though the entire sector is largely informal and so are those who work in it: such as those who work for intermediary dealers manually segregating recyclables into ever finer and more precise categories, and those who work for reprocessors transforming those hyper-segregated materials into new commodities, turning polyethylene terephthalate (PET) plastics into pellets and finally into fleece yarn for example.

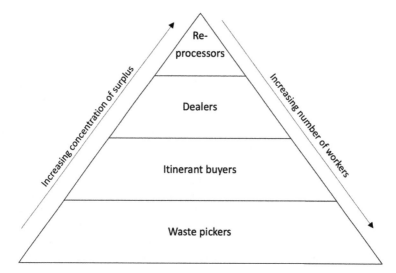

Figure 29.1 Informal waste sector in urban India
Source: Author

The following sections show how different types of waste work are circumscribed by hierarchies of class, race, caste, ethnicity and gender. Formalisation programmes that have benefited some waste pickers, have nonetheless reinscribed those hierarchies (Samson, 2010).

Complexity and heterogeneity of work and workers

There is little doubt that the work of waste (both formally and informally) is performed by those who are socially, politically and economically marginal globally. Forms of marginality are expressed through a myriad of social, political and economic relations that include mutually constitutive class, race, caste and gender categories (Samson, 2010). Governmental and non-governmental organisations highlight certain markers of vulnerability in their generalised representations of workers even though the composition of the informal economy along these categories is highly variable between and within cities and countries. In doing so, dominant representations imagine and design programmes to benefit a homogeneous population of workers that can have contradictory effects for the intended and unintended objects of their interventions. Box 29.2 elaborates how hierarchies of gender and caste structure the work of waste in contemporary urban India.

Gender: Globally, informal workers in the waste sector are often generically described as 'highly vulnerable, particularly women and children' (Medina, 2011: 9). Yet, there are significant variations in the ways waste work is gendered globally (ILO, 2013). While men were found to dominate work in this sector in Brazil and five of the seven West African cities surveyed, women's involvement in this work was greater in urban India and the other two of the seven West African cities (ILO, 2013). These numerical differences obscure functional differences however; gender not only plays a crucial role in structuring access to particular kinds of work, structural changes within the sector also have gendered effects (Samson, 2003).

Race, caste and ethnicity: In addition to being gendered, the work of waste is stratified by race, caste and ethnicity. Critical scholars have long pointed to capitalism's production of superfluous waste matter and superfluous humans; the latter survives (barely) by putting the former to use (Bauman, 2004; Millington and Lawhon, 2018). Race and caste often connect the two. For example, a Brazilian study of waste pickers at the largest open dump in Latin America revealed that most waste pickers were either Black, brown or of mixed race (Cruvinel et al., 2019). In Cairo, the work of waste collection is performed primarily by the *zabaleen*, a Coptic Christian community (Fahmi and Sutton, 2006). In caste-based societies such as India, notions of purity and pollution demarcate clean and unclean work; indeed, caste privilege manifests in the ability to distance oneself from dirt and pollution (Butt, 2019; Kornberg, 2019; Prashad, 2000).

Class: The category of waste picker as a catch-all phrase obscures the class stratifications among workers at the base of the pyramid which includes doorstep collectors and street and dump/landfill waste pickers. While some scholars have suggested that dualistic distinctions of waged versus unwaged and self-employed versus employed for describing waste pickers are of little analytical value (Birkbeck, 1978; Dinler, 2016), these distinctions might be useful for helping us think through one crucial difference between the two: that of ownership of their means of subsistence or recyclable materials. While self-employment may indeed be an illusion, because workers are in essence selling their labour power to dealers and ultimately to reprocessors who often have control over prices of recyclable materials that they purchase from waste pickers (Birkbeck, 1978; Dinler, 2016), pickers' control over the materials they pick offers them the potential for upward mobility: to 'make gains from spatial and temporal arbitrage, and accumulate' (Chaturvedi and Gidwani, 2011: 132).

Box 29.2 Dynamics of gender, class and caste in informal waste work in India

Descriptions of Indian waste pickers as being primarily women and children from the 'most vulnerable', 'weakest', 'most deprived' and 'least organised' populations are reiterated in many governmental and non-governmental documents (MoUD, 2014: 346; Planning Commission, 1995: 13; World Bank, 2008: 134). In India, there are considerable differences between cities in the division of labour in this sector along gender, caste and class lines.

- *Gender*: While there are more women waste pickers than men in Pune and Ahmedabad, in Delhi the situation is reversed (Chikarmane and Narayan, 2000; Dave et al., 2009; Hayami et al., 2006). In Pune, for instance, while women dominate waste picking, itinerant buyers are primarily men; gendered functional differences also connote class differences (Chikarmane and Narayan, 2000). In Delhi on the other hand, women's labour is more invisible since they likely do the work of sorting recycling materials at home (ILO, 2004).
- *Caste*: Caste distinctions are most noticeable between those who 'scavenge' for low-value recyclable materials (waste pickers) and those who 'trade' in high-value recyclable materials (itinerant buyers) (Gill, 2010). In some cities such as Delhi, although Bengali-speaking Muslim migrants from the eastern states of India, are now increasingly involved in waste collection, this has not meant the demise of caste distinctions (Kornberg, 2019). Instead, this new group of workers is subjected to the power of caste hierarchies: Bengali Muslim waste collectors have been 'casteified' (Kornberg, 2019).
- *Class*: Commonly lumped into a broad category of 'waste picker' at the base of the pyramid (see Box 29.1), the category obscures class differences between groups of works: doorstep waste collectors typically earn almost twice as much as street and dump/landfill pickers (Hayami et al., 2006). Another class and caste-based difference is between those who buy recyclable materials from waste generators (itinerant buyers) and waste pickers (Gill, 2010). While waste pickers collect and extract low value recyclables from waste, itinerant buyers purchase high value recyclables from waste generators (Luthra, 2020b).

Representing the informal economy of waste in particular ways betrays more about the political interests of those engaging in such representational practices than about the objects of those representations. Casting workers as those who are marginalised and deprived due to their social positions provides justification for particular kinds of interventions. Hence, examining the representations and their political intent shows us that particular representational practices serve to discursively reduce workers from being economic subjects to objects of social intervention (see Box 29.3). One specific technology of 'emancipation' that waste picker advocacy organisations have embraced is 'formalisation', a process that relies on and reproduces gender, class and caste hierarchies even as it secures livelihoods for some in the informal economy of waste.

Box 29.3 The politics of representation of waste pickers in India

The following two examples serve to illustrate the politics behind certain representational practices. The first relates to the casting of waste pickers as 'illegal' immigrants from Bangladesh, a representational practice deployed by right-wing Hindu nationalist political parties that is also often used by the police to stigmatise and extract bribes from Bengali Muslim waste pickers in Delhi and other cities. This issue has politically intensified during the time leading up to and since the passing of the Citizenship Amendment Act in 2019. The second involves calls by several actors to ban waste picking altogether, calls that have been realised at least once in Delhi's history when for a short period waste picking was banned during the 1994 plague outbreak (Chaturvedi and Gidwani, 2011). The discourses enrolled in the legitimation of bans purport to do so in the interest of workers themselves. Chaturvedi et al. (2008: 79) recount an interview with a government official, who asked, 'aren't they all children? Should we not be banning them?'. Calls for banning are accompanied by calls for formal privatisation that would concentrate market power in the hands of corporate firms by displacing informal service providers. In framing their interventions however, these actors, often representing firms in the waste management industry, articulate their roles as the providers of good jobs for the very same people they would be displacing, indeed even claiming a moral responsibility for doing so: 'Corporations, as well as municipalities, do have a responsibility to take care of them. What we propose is rehabilitation of these workers, not a continuation of the existing system' (Rao, 2013). Disguised as moral imperative, representational practices serve to simultaneously delegitimise workers and legitimise the processes of dispossession associated with the formal privatisation of waste management services.

Formalisation and its pitfalls

Definitions of formalisation vary widely depending on who defines, what is being defined and for what purpose. Even as the necessity and importance of the informal sector in underwriting and subsidising the formal sector is widely recognised, it nonetheless continues to be associated with illegitimacy, disorder and illegality, functioning as the 'unplannable' object in the epistemology of urban planning in developing countries (Porter, 2011; Roy, 2005). It is these representations and practices of classification of the informal as illegal, disorderly and illegitimate that have animated calls for formalisation from state planners and policymakers 'above', and grassroots and advocacy organisations 'below'. These calls from opposing directions have found common expression in projects that formalise waste pickers in cities all over the world (Chikarmane and Narayan, 2000; Dias, 2016; O'Hare, 2020; Rosaldo, 2019; Samson, 2009).

Although there are marked differences among these formalisation projects, some features are common to them. First, there is some form of recognition or acknowledgement of the labours of workers through policy at the local (city level) and/or larger scales (regional or national level) (Dias, 2016). Second, such recognition involves the formation of entities such as cooperatives that the state recognises as formal organisations that it can negotiate agreements or contracts with (Gutberlet, 2012; Medina, 2000). Third, some form of identification such as ID cards allows workers to be rendered nominally legible and legitimate (Agarwala, 2016; Millar, 2014). Fourth, such programmes try to 'professionalise' workers and make them more visible through requiring the use of uniforms for instance (Machado-Borges, 2015; Millar, 2018; O'Hare, 2019). Yet, even as formalisation programmes might make livelihoods safer and more secure for some, they have also had contradictory effects.

In describing such a project that formalised waste pickers in Uruguay, O'Hare (2020) uses the term 'quasi-formalisation' to alert us to the programme's neglect of para-formal dimensions of work that underpinned and subsidised the transition to formalisation. In recounting the experiences of formalising waste pickers in Brazil, Millar (2014: 35) shows us how formalised waste pickers returned to the dump as informal workers rather than continue to work in the formal programme in order to maintain their 'relational autonomy': workers returned to the dump because 'relative degree of control over work activities and time enables catadores [waste pickers] to sustain relationships, fulfil social obligations, and pursue life projects in an uncertain everyday', as they simultaneously upheld an affective attachment to the Fordist promises of formal, stable employment. Rosaldo (2019: 6–7) points to the tensions 'between the imperatives of combatting exploitation and dispossession' that emerge in such formalisation programmes: 'policy schemes that seek to reduce the exclusionary traits of informal work through regulation, mechanisation or social enclosure may also undermine its inclusive traits by reducing the quantity and accessibility of available jobs'. Indeed, formalisation of some has been accompanied by dispossession of others (O'Hare, 2019), a process that Tucker (2017) has described as 'dispossession through formalisation' in the context of informal street vendors in Paraguay.

Feminist scholars such as Nightingale (2006) have urged that we pay attention to gender as a process, that is, gender not only defines divisions of labour but is constituted through work. Their insights have been useful for thinking through the mutually constitutive relations of social categories and reconfigurations of informal work through formalisation programmes (Samson, 2010). Formalised doorstep waste collection programmes in urban India operate by legitimising existing doorstep waste collectors as formal service providers. In Delhi, most doorstep collectors are men while women perform the more invisible work of waste segregation in the private spaces of the home or warehouses of waste dealers. In Pune, on the other hand, most waste collectors are women. If in the former case, the formalisation programme masculinises doorstep waste collection work, in the latter case, it is feminised. The extent to which workers reproduce these gender categories in relation to their work is an empirical question warranting further examination, although Samson (2010) provides evidence in support of this claim in the South African context. Class divisions similarly might both produce and be reproduced through formalisation programmes. Formalisation programmes, at least in the Indian context, focus on doorstep waste collectors who are already in a slightly better socio-economic position than waste pickers (Hayami et al., 2006). As a result, formalisation creates new class cleavages such as between those without IDs and those with IDs, uniformed and un-uniformed, as O'Hare (2019) has demonstrated in the case of Uruguayan waste collectors. However, workers themselves uphold these subtle class distinctions to distinguish themselves as legitimate, honest and law-abiding (Millar, 2018).

Conclusion

Although informal workers in the waste sector provide crucial environmental and public health services in cities of the Global South, privatisation of municipal services threatens their livelihoods. Organising informal workers is going to continue to be crucial in their struggles against this persistent threat of displacement and dispossession. The classical conception of the working class as well as the standard model of worker organising such as trade unionism needs to be rethought as informality and precarity for workers is increasing worldwide both in the Global South as well as the North (Breman and van der Linden, 2014). While formalisation programmes discussed above have rightly been understood as efforts to 'increase the number of entitled beneficiaries [of state welfare programs] relative to earlier formal workers' programs' (Agarwala, 2016: 120), the new kinds of exclusions that these programmes have generated – exclusions that are rooted in an imagined homogeneity of workers in the sector – need to be scrutinised further (see Box 29.4 for a comparison of worker organising in two Indian cities). Recognising and understanding the heterogeneity and complexity of work and workers needs to be starting point for developing appropriate organising strategies if 'formalisation' is indeed able to secure livelihoods for all workers in this sector.

Box 29.4 The implications of heterogeneity for worker organising in Pune and Delhi

In the Indian context, the contrasting experiences of waste pickers in Delhi and Pune offer some insight into the implications of heterogeneity for worker organising. In Pune, the largely female *dalit* population of waste pickers is more homogenous in a socio-economic sense than those in Delhi where ethno-religious and linguistic divisions are much more pronounced. Workers' organisations in Pune have also been much more successful at securing livelihoods for workers through a formal contract with the municipality to provide waste collection services than in Delhi where workers have merely been able to secure 'permission' from the municipality (as opposed to a formal contract) to provide those services in scattered parts of the city (Luthra, 2020a). Workers in Pune have also been able to secure state welfare entitlements more so than workers in Delhi, especially those whose socio-economic positions (Bengali Muslims) are more likely to serve as justification for their treatment as non-citizens and thus ineligible for receiving those state entitlements. Thus, models and strategies of worker organising that work in Pune have not been able to work in cities such as Delhi.

Key points

- In cities of the Global South, workers in the informal economy provide crucial waste collection and recycling services. In doing so, they not only provide their own means of subsistence, they also provide crucial environmental and public health services to cities, often without being paid by governments.
- Workers in this sector often belong to socially and economically marginalised groups; relations of gender, race, class and caste constitute and are constituted through work in this sector.
- In response to displacement and dispossession from their means of livelihood, workers across the world have organised, and have made significant political gains in securing access to their livelihoods. Formalisation of informal workers as service providers through formal entities such as cooperatives has been a common strategy across countries.
- While formalisation has improved outcomes for workers in some cases, it has also generated new gender, class, and caste distinctions in others. Organising strategies need to be cognizant of the heterogenous and complex nature of work and workers in this sector in order to be more effective in securing livelihoods.

Recommended reading

Dias, S. (2016) 'Waste pickers and cities', *Environment and Urbanization*, 28(2): 375–390. https://doi.org/10.1177/0956247816657302

Samson, M. (2010) 'Producing privatization: Re-articulating race, gender, class and space', *Antipode*, 42(2): 404–432. https://doi.org/10.1111/j.1467-8330.2009.00752.x.

References

Agarwal, A., Singhmar, A., Kulshrestha, M. and Mittal, A.K. (2005) 'Municipal solid waste recycling and associated markets in Delhi, India', *Resources, Conservation & Recycling*, 44(1): 73–90. https://doi.org/10.1016/j.resconrec.2004.09.007

Agarwala, R. (2016) 'Redefining exploitation: Self-employed workers' movements in India's garments and trash collection industries', *International Labor and Working-Class History*, 89: 107–130. https://doi.org/10.1017/S0147547915000344

Bauman, Z. (2004) *Wasted lives: Modernity and its outcasts*, Cambridge: Polity Press.

Birkbeck, C. (1978) 'Self-employed proletarians in an informal factory: The case of Cali's garbage dump', *World Development*, 6(9/10): 1173–1185. https://doi.org/10.1016/0305-750X(78)90071-2

Breman, J. and van der Linden, M. (2014) 'Informalizing the economy: The return of the social question at a global level', *Development and Change*, 45(5): 920–940. https://doi.org/10.1111/dech.12115

Butt, W. (2019) 'Beyond the abject: Caste and the organization of work in Pakistan's waste economy', *International Labour and Working Class History*, 95: 18–33. https://doi.org/10.1017/S0147547919000061

Chaturvedi, B. and Gidwani, V. (2011) 'The right to waste: Informal sector recyclers and struggles for social justice in post-reform urban India', in Ahmed, W., Kundu, A. and Peet, R. (eds.), *India's new economic policy: A critical analysis*, New Delhi: Routledge, pp. 125–153.

Chaturvedi, B., Chikarname, P. and Narayan, L. (2008) *Recycling livelihoods: Integration of the informal recycling sector in solid waste management in India*, Eschborn, Germany: Deutsche Gesellschaft für Internationale Zusammenarbeit (GIZ) GmbH.

Chikarmane, P. and Narayan, L. (2000) 'Formalising livelihood: Case of wastepickers in Pune', *Economic and Political Weekly*, 35(41): 3639–3642. www.jstor.org/stable/4409829

Chintan (2003) *Space for waste: Planning for the informal recycling sector*, New Delhi: Chintan Environmental Research and Action Group.

Chintan (2009) *Cooling agents an analysis of greenhouse gas mitigation by the informal recycling sector in India*, New Delhi: Chintan Environmental Research and Action Group.

Cruvinel, V., Marques, C.P., Cardoso, V., Carvalho, M.R., Novaes, G., Araujo, W.N., Angulo-Tuesto, A., Escalda, P.M.F., Galato, D., Brito, P. and da Silva, E.N. (2019) 'Health conditions and occupational risks in a novel group: Waste pickers in the largest open garbage dump in Latin America', *BMC Public Health*, 19: 581. https://doi.org/10.1186/s12889-019-6879-x

Dave, J., Shah, M. and Parikh, Y. (2009) 'The self-employed women's association (SEWA) – organising through union and co-operative in India', in Samson, M. (ed.), *Refusing to be cast aside: Waste pickers organising around the world*, Cambridge, MA: Women in Informal Employment: Globalizing and Organizing (WIEGO), pp. 27–32.

Dias, S. (2011) 'Statistics on waste pickers in Brazil', *WIEGO Statistical Brief No. 2*, Cambridge, MA: WIEGO.

Dias, S. (2016) 'Waste pickers and cities', *Environment and Urbanization*, 28(2): 375–390. https://doi.org/10.1177/0956247816657302

Dinler, D.Ş. (2016) 'New forms of wage labour and struggle in the informal sector: The case of waste pickers in Turkey', *Third World Quarterly*, 37(10): 1834–1854. https://doi.org/10.1080/01436597.2016.1175934

Fahmi, W. and Sutton, K. (2006) 'Cairo's Zabaleen garbage recyclers: Multi-nationals' takeover and state relocation plans', *Habitat International*, 30(4): 809–837. https://doi.org/10.1016/j.habitatint.2005.09.006.

Gill, K. (2010) *Of poverty and plastic: Scavenging and scrap trading entrepreneurs in India's urban informal economy*, New Delhi: Oxford University Press.

Gutberlet, J. (2012) 'Informal and cooperative recycling as a poverty eradication strategy: Informal and cooperative recycling', *Geography Compass*, 6(1): 19–34. https://doi.org/10.1111/j.1749-8198.2011.00468.x

Hayami, Y., Dikshit, A.K. and Mishra, S.N. (2006) 'Waste pickers and collectors in Delhi: Poverty and environment in an urban informal sector', *Journal of Development Studies*, 42(1): 41–69. https://doi.org/10.1080/00220380500356662

Hoornweg, D. and Bhada-Tata, P. (2012) 'What a waste: A global review of solid waste management', *Urban Development Series Knowledge Papers No. 15*, Washington, DC: World Bank.

ILO (2004) *Addressing the exploitation of children in scavenging (waste picking): A thematic evaluation on action on child labour*, Geneva: International Labour Organization.

ILO (2013) *Women and men in the informal economy: A statistical picture*, Geneva: International Labour Organization.

Kornberg, D. (2019) 'From Balmikis to Bengalis', *Economic and Political Weekly*, 54(47): 48–54.

Linzner, R. and Lange, U. (2013) 'Role and size of informal sector in waste management: A review', *Proceedings of the Institution of Civil Engineers – Waste and Resource Management*, 166(2): 69–83. https://doi.org/10.1680/warm.12.00012

Luthra, A. (2019) 'Municipalization for privatization's sake: Municipal solid waste collection services in urban India', *Society and Business Review*, 14(2): 135–154. https://doi.org/10.1108/SBR-11-2017-0102

Luthra, A. (2020a) 'Efficiency in waste collection markets: Changing relationships between firms, informal workers, and the state in urban India', *Environment and Planning A: Economy and Space*, 52(7): 1375–1394. https://doi.org/10.1177/0308518X20913011

Luthra, A. (2020b) 'Housewives and maids: The labor of household recycling in urban India', *Environment and Planning E: Nature and Space*, 4(2): 475–498. https://doi.org/10.1177/2514848620914219

Machado-Borges, T. (2015) '"Have you ever seen a thief wearing a uniform?!" Struggling for dignity in urban southeastern Brazil', *Ethnography*, 16(2): 207–222. https://doi.org/10.1177/1466138114547623

McBride, E. (2009) *Talking rubbish a special report on waste*, London: The Economist.

Medina, M. (2000) 'Scavenger cooperatives in Asia and Latin America', *Resources, Conservation and Recycling*, 31(1): 51–69. https://doi.org/10.1016/S0921-3449(00)00071-9

Medina, M. (2011) 'The informal waste sector', in Gunsilius, E., Spies, S. and Garcia-Cortes, S. (eds.), *Recovering resources, creating opportunities: Integrating the informal sector into solid waste management*, Eschborn, Germany: Deutsche Gesellschaft für Internationale Zusammenarbeit (GIZ) GmbH, pp. 8–9.

Millar, K. (2014) 'The precarious present: Wageless labor and disrupted life in Rio de Janeiro, Brazil', *Cultural Anthropology*, 29(1): 32–53. https://doi.org/10.14506/ca29.1.04

Millar, K. (2018) *Reclaiming the discarded: Life and labour on Rio's garbage dump*, Durham: Duke University Press.

Millington, N. and Lawhon, M. (2018) 'Geographies of waste: Conceptual vectors from the Global South', *Progress in Human Geography*, 43(6): 1044–1063. https://doi.org/10.1177/0309132518799911

MoUD (2014) *Municipal solid waste management manual*, New Delhi: Ministry of Urban Development, Government of India.

Nightingale, A. (2006) 'The nature of gender: Work, gender, and environment', *Environment and Planning D*, 24(2): 165–185. https://doi.org/10.1068/d01k

O' Hare, P. (2019) '"The landfill has always borne fruit": Precarity, formalisation and dispossession among Uruguay's waste' pickers', *Dialectical Anthropology*, 43(1): 31–44. https://doi.org/10.1007/s10624-018-9533-6

O'Hare, P. (2020) '"We looked after people better when we were informal": The "Quasi-formalisation" of Montevideo's waste-pickers', *Bulletin of Latin American Research*, 39(1): 53–68. https://doi.org/10.1111/blar.12957

Planning Commission (1995) *Urban solid waste management in India: Report of the high power committee*, New Delhi: Planning Commission, Government of India.

Porter, L. (2011) 'Informality, the commons and the paradoxes for planning: Concepts and debates for informality and planning', *Planning Theory & Practice*, 12(1): 115–153. https://doi.org/10.1080/14649357.2011.545626

Prashad, V. (2000) *Untouchable freedom: A social history of a Dalit community*, Oxford: Oxford University Press.

Rao, R. (2013) 'Making sense of Delhi's waste wars', *Business Standard*, 27 May. Available at: www.business-standard.com/article/economy-policy/making-sense-of-delhi-s-waste-wars-113052700034_1.html (Accessed 14 December 2020).

Rosaldo, M. (2019) 'The Antinomies of Successful Mobilization: Colombian Recyclers Manoeuvre between Dispossession and Exploitation', *Development and Change*, Online First. https://doi.org/10.1111/dech.12536

Roy, A. (2005) 'Urban informality: Toward an epistemology of planning', *Journal of the American Planning Association*, 71(2): 147–158. https://doi.org/10.1080/01944360508976689

Samson, M. (2003) *Dumping on women: Gender and the privatisation of waste management*, Woodstock, South Africa: South Africa Municipal Workers Union.

Samson, M. (2009) *Refusing to be cast aside waste pickers organising around the world*, Cambridge, MA: WIEGO.

Samson, M. (2010) 'Producing privatization: Re-articulating race, gender, class and space', *Antipode*, 42(2): 404–432. https://doi.org/10.1111/j.1467-8330.2009.00752.x

Sandhu, K., Burton, P. and Dedekorkut-Howes, A. (2017) 'Between hype and veracity; privatization of municipal solid waste management and its impacts on the informal waste sector', *Waste Management*, 59: 545–556. https://doi.org/10.1016/j.wasman.2016.10.012

Scheinberg, A., Simpson, M., Gupt, Y., Anschütz, J., Haenen, I., Tasheva, E., Hecke, J., Soos, R., Chaturvedi, B., Garcia-Cortes, S. and Gunsilius, E. (2010) *Economic aspects of the informal sector in solid waste management*, Eschborn, Germany: GTZ.

Schoenberger, E. (2003) 'The globalization of environmental management: International investment in the water, wastewater and solid waste industries', in Peck, J. and Yeung, H. (eds.), *Remaking the global economy: economic-geographic perspectives*, London: Sage, pp. 83–98.

Spronk, S. (2010) 'Water and sanitation utilities in the Global South: Re-centering the debate on "efficiency"', *Review of Radical Political Economics*, 42(2): 156–174. https://doi.org/10.1177/0486613410368389

Tucker, J. (2017) 'City-stories: Narrative as diagnostic and strategic resource in Ciudad del Este, Paraguay', *Planning Theory*, 16(1): 74–98. https://doi.org/10.1177/1473095215598176

UN-HABITAT (2010) *Solid waste management in the world's cities*, Washington, DC: UN-HABITAT/Earthscan.

Vergara, S.E., Damgaard, A. and Gomez, D. (2016) 'The efficiency of informality: Quantifying greenhouse gas reductions from informal recycling in Bogotá, Colombia: Quantifying GHG reductions from informal recycling in Colombia', *Journal of Industrial Ecology*, 20(1): 107–119. https://doi.org/10.1111/jiec.12257

WIEGO (2020) *Basic Categories of waste pickers*. Available at: www.wiego.org/basic-categories-waste-pickers (Accessed 8 December 2020).

World Bank (2008) *Improving solid waste management in India: A sourcebook for policy makers and practitioners*, Washington, DC: The World Bank.

30
PLANNING FOR SUSTAINABLE URBAN LIVELIHOODS IN AFRICA

Lauren Andres, Stuart Paul Denoon-Stevens, John R. Bryson, Hakeem Bakare and Lorena Melgaço

Introduction

This chapter explores the role, successes and failures of spatial planning in shaping African cities and its influence on livelihoods, particularly lower-income livelihoods. Spatial planning is understood in this context not as a regulatory mechanism but a means to facilitate, ideally, the delivery of better places for the future and particularly a more holistic, more strategic, inclusive, integrative and attuned approach to spatial, urban and sustainable development (Haughton et al., 2009). We build upon Chambers' and Conway's (1992) definition to understand livelihoods as capabilities, material and social assets, and activities that support everyday living. Such livelihoods are sustainable if they can be maintained, recover from shocks, and allow for the survival and wellbeing of individuals and households, in various and diverse urban contexts. This is particularly critical for the poorest and more vulnerable communities whose daily survival often relies on informal activities and arrangements (Wilkinson et al., 2020).

Unpacking the challenges behind the process of planning for sustainable livelihoods is especially important in overcoming poverty, inclusion and health inequalities, future pandemics and climate change. Urban planning plays an important role in developing sustainable livelihoods; given this, there is a need to better understand the ways in which planning visions, mechanisms, and challenges impact them. To date, planning in Africa has struggled to meet the needs of the poor, especially concerning livelihoods. Drawing upon these considerations, we structure our argument as follows. First, we explore how the legacy of colonial planning has impacted the segregation of spaces and hence of livelihoods, particularly those of the poorer communities. We then discuss the challenges faced by planning in response to the informal nature of the livelihoods of lower-income communities. Finally, we sketch out how planning for sustainable livelihoods can be tackled in Africa in the future.

The development of African cities: colonisation, race and segregation of the poor

The way cities and neighbourhoods have been shaped in Africa needs to be understood as a legacy of colonial planning. This has created path-dependency that has played an important role in creating differences in the livelihoods of the poor from those in other cohorts.

The legacy of colonisation: shaping African cities

Patterns of spatial ordering can be observed in indigenous populations before modern European colonisation (Silva, 2016). One common feature was the tendency to prioritise collective action rather than segregation, with the structuring of public spaces including market squares, farms, and playgrounds (Amankwah-Ayeh, 1996). Here, the so-called planned way of using and managing spaces was linked to how spaces were used to support everyday living. With colonisation and the transfer of European planning systems and models in the African continent at the end of the 19th century, space became managed according to the interests of colonial elites (Parnell, 2002), thereby contributing to inequalities and having long-term consequences for contemporary cities.

Planning colonial cities responded to the need for effective occupation of colonial territories and conditions for white colonisation of rural areas (Silva, 2015). This meant developing effective strategies to enable the plundering of resources from colonial territories (Ross and Telkamp, 1985), the possibility for extraction of agricultural surplus and provide services, and political control (King, 1985). Planning was also influenced by eugenic theories (Munanga, 2016), which underpinned spatial segregation based on race. In many colonies, such segregation was justified by a discourse on health, as 'sanitation has correctly been identified as the metaphor which colonialists first invoked to justify the establishment of segregated locations that facilitated the control of urbanised African workers' (Parnell, 1993: 488). As planning relies on the physical organisation of space and its diverse infrastructures (transport, water, electricity, housing etc.) such models of development have a long-term legacy, particularly in dividing and segregating spaces.

Later planning was one of the active forces in shaping cities further. From 1948, it played a critical role during the apartheid period in South Africa (Harrison et al., 2007). These types of planning interventions dictated the racial development of spaces, and hence of livelihoods, with white livelihoods being far more sustainable and resilient than any others. This cohort was provided with better quality housing, more reliable infrastructures (water, energy) and occupied better locations (e.g. mostly away from high-risk flooding areas). In contrast, the more vulnerable (black livelihoods) were forced to live in more marginal areas, including the edges of cities, with limited access to reliable basic services and very little acknowledgement of their everyday needs.

Colonial planning and its long-lasting impact on the livelihoods of the poor

Colonial planning generated uneven development between regions and countries, with clear patterns of rapid urbanisation, spatialised racial segregation and social inequalities that are still visible today. Poor settlements – known as informal dwellings, slums and townships – are still characterised by significant burdens which fail to be addressed by planning and other connected policies. There are concerned with poor health outcomes (Marais and Cloete, 2014; Satterthwaite et al., 2019) including exposure to indoor and outdoor air pollution and high TB levels. Residents are at risk of greater exposure to harm due to hazards including fire, floods, and other environmental related risks (ibid). Communities face entrenched poverty linked to increased difficulties in practising urban agriculture and home-based livelihoods due to the lack of space and services (Crush et al., 2011).

Contemporary African cities are thus still designed to have formal, well-provisioned areas for the middle-/high-income communities, with less consideration given to low-income areas. The provision of and access to key services is a major issue. There has been a move in some African countries to start providing supermarkets in low-income areas (Battersby, 2017). This has had mixed effects. It has eased the travel burden for poor households who typically use a range of formal and informal food providers to access food. However, it also resulted in a decline in the number of spazas (small, home-based shops) following the opening of supermarkets; hence livelihoods were affected. Worryingly, planning was identified as one of the key obstacles to the operation of spazas owing to the lengthy and costly process of obtaining planning approval. There are also examples of municipalities in South Africa using planning law to push out informal traders in favour of supermarkets and formal stores, with this partly being motivated by the fact that the latter pay municipal taxes, whereas the former do not (Competition Commission, 2019). This observation reveals the main tension between planning and providing for sustainable livelihoods, addressing informality.

Planning, informality and the livelihoods of lower-income communities

Planning to date has failed to account for sustainable livelihoods, particularly those of the poorer communities. This is linked to three main challenges: land management, the design of lower-income neighbourhoods and zoning and its account of commercial uses which all struggle to account for the importance of informal uses and activities.

Land management

While informality is a key component of African cities, it is not or is barely accounted for by existing planning regulations and policies (Pieterse, 2014). This choice is conditioned by the lack of other economic alternatives and by the inability of the system to recognise an informal place of living and its diversity. For entrepreneurs to access formal institutions, from banks to government, proof of address or proof of ownership is often required. This has an impact on livelihoods as only an estimated 10% of land in Sub-Saharan Africa is formally registered as private property (Bah et al., 2018). This lack of formal ownership has multiple implications. For one, properties without formal ownership cannot be used for loan guarantees. Furthermore, the various approvals necessary for businesses to become formally registered often require the consent of the property owner to make the application. If 'formal' ownership is unclear, these applications cannot be made. This results in a situation of 'enforced informality' where formality becomes an impossibility for entrepreneurs (Charman et al., 2013); by essence this constraint the development of more profitable businesses and economic activities hence affecting livelihoods.

Land use management often aggravates this situation as, even if property ownership can be proven, most town planning schemes are based on modernist ideas that perpetuate segregation, considering a house only as a place to live, not a place for business. This forces households to make an application for planning rights when they want to formalise their home-based businesses. However, this process is costly and takes considerable time. Low-income households typically cannot afford, nor can they wait for, approval before operating, as often the home-based business is their primary means of survival (Charman et al., 2017). Again, this has clear implications on the livelihoods of the most vulnerable left with little options to secure regular and stable incomes.

Poor design of lower-income neighbourhoods

Beyond issues of planning rights, the lack of consideration of the multi-dwelling, multi-use nature of poor households leads to the inadequate provision of services and poorly planned and designed housing units. An example of this from South African cities are backyard dwellings, which are either rented out to generate a livelihood, or used to accommodate wider family networks; they constitute the second most common form of housing in the country. However, settlements are rarely designed with backyard shacks in mind nor with the idea of being used for various purposes (living/working/providing additional incomes); thus, the infrastructure often cannot handle the additional demand on utilities. This results in unsafe illegal electrical connections, which pose a fire risk, and increased demand on infrastructure (sanitation, power supply, etc.) often resulting in utility failures (Lemanski, 2009).

The presence of backyard shacks often leads to the sharing of sanitation facilities which were only designed to accommodate a single household. This results in poor hygiene and breakages, which in turn generates a high incidence of diarrhoea and other water-related ailments (Govender et al., 2011). Similar issues have been noted for home-based enterprises, as the provision of state-subsidised housing does not usually provide a separate space for livelihood activities, leading to an overlap of business and home activities, which can hinder business operations (Gough et al., 2003) thus impacting the profitability of the activities with wider implications on the households' incomes. This testifies from the disconnection between how the most vulnerable live, work and survive and how space, buildings and infrastructures are designed and planned.

Zoning and commercial uses

Often the zoning of low-income areas allows no, or limited, provision for commercial land. This makes it difficult for larger businesses to exist in low-income areas, as there may be no sites large enough to accommodate their needs. Even if the sites are large enough, they are forced to go through a cumbersome approval process, and the servicing of the site may require significant upgrades to accommodate higher-order land uses (Parnell and Pieterse, 2010).

Informal traders who operate in formal business areas are rarely accepted by authorities and face regular evictions. For example, Zimbabwe in 2005 launched Operation Murambatsvina/ Restore Order, which in three months almost decimated the whole informal economy of Zimbabwe. The motivations behind this campaign, which are echoed regularly in similar exercises that occur throughout Africa, was to 'restore order'; informal traders were accused of hiding criminals, practising illegal activities, and not practising hygiene, thereby 'spoiling' the image of town and country. The very name of the operation, Murambatsvina, conveyed this message, as it translates to 'drive out the rubbish' (Rogerson, 2016). While planners were not directly involved in this operation, planning was complicit in that many of the modernist planning arguments were used by the authorities to justify and rationalise this operation (Kamete, 2009).

It is thus apparent that Africa has struggled to understand, account for and support sustainable livelihoods for the lower-income communities. Failure of the planning system and profession has been a key factor in that respect. We now turn to which shifts are needed to allow planning to support the development of sustainable livelihoods.

How to overcome the challenges to planning for sustainable livelihoods

The failure to plan for sustainable livelihoods needs to be connected back to four issues: planning resource and scarcity, inclusion, the need for people-centric and localised approaches, and political commitment.

Tackling planning resource and scarcity

The lack of planning resources and the scarcity of the profession is a significant burden in Africa, with huge discrepancies from one country to another. Such skill shortages and a limited number of planners to deliver planning (UN-Habitat and APA, 2013) are not only associated with financial resources, lack of investment in skill development, access to education (Mateus et al., 2014) but also with path-dependent socio-economic and political factors. For too long, African cities have been ignored and developing the planning profession was not considered to be a priority. Capital and major cities have attracted the most skilled practitioners whereas smaller cities and rural territories struggle. Such polarisation leads to unequal distribution of planning skills and capabilities across countries; this is reflected in the capacity and resources of local governments to develop and implement planning interventions to enhance livelihoods (Watson and Odendaal, 2013).

As a result, African urban planning tends to be outsourced to multinational consultancy firms (Watson, 2014), who are often based in the Global North. Unrealisable visionary planning futures are created, ignoring the political subtleties and everyday needs of local communities, particularly of the most vulnerable. This 'planning as best practice' approach focuses on creating neo-liberal models of developments, influenced by models applied in cities of the Global North, is underpinned by governments rejecting any 'improper' solutions, defined according to Northern standards. Often the outcome is the criminalisation of the livelihoods and shelter strategies of the poor and most vulnerable (Charlton, 2018). This reinforces the segregation of spaces and the lack of acknowledgement of informal needs and practices. Providing more resources for the planning profession, while accounting for the needs of the many and not the few, is essential if planning is going to contribute to shaping sustainable livelihoods. This rests upon an inclusive and socially just approach towards urban planning.

Addressing inclusivity

To date, planning in Africa is not inclusive enough. It does not account for the diversity of needs and practices of individuals and communities, particularly those with low incomes (Andres et al., 2020a). A socially just form of African planning should mandate the inclusion of informal traders, microenterprises and public transport facilities in all-new shopping centres, thereby fostering integration, not exclusion (Denoon-Stevens, 2016). This can be delivered in practice. The City of Johannesburg (South Africa) is currently implementing this through a zoning scheme requirement which states that all retail areas larger than 5000 m² must make provision for public transport facilities and informal trading facilities and ablutions (City of Johannesburg, 2019).

To be more inclusive, planning can encourage specific forms of development, for example, by proactively changing the zoning rights of properties; commercial land uses can also be encouraged in areas that work best for the city as a whole (Denoon-Stevens and Nel, 2020). An example of this is Eveline Street in Windhoek, Namibia. Property owners along this street were informed that their properties were to be rezoned to permit business rights. This contributed to the doubling and diversification of microenterprises along this street between 2008 and 2016. This was also supported by the rectangular plot sizes and wide pavements, which created spatial conditions that were conducive to a mixed-use environment (Tonkin et al., 2018).

Other inclusive actions that planners can take include making provision for home-based livelihoods in state-subsidised low-income housing developments. In Mathare 4A, Kenya, live-work units were provided to housing recipients who had previously run a home-based business, which provided separate space for living and business activities in one building. These were clustered on main roads, providing better access to potential markets, and the increased space allowed businesses to grow, while the separation of work and living spaces led to a higher quality of life for business operators (Kigochie, 2001).

Promoting a people-centric and localised approach to planning

To tackle the lack of planners and planning resources, a new form of people-centric alternative approaches to planning is needed, recognising the importance of informality in Africa. This will enable individuals and communities to shape their livelihoods with a focus on more responsible and realistic place-based outcomes (Andres et al., 2021). Such approaches are very limited to date. Such a shift rests upon empowering local communities, with assistance from social welfare departments and humanitarian agencies, to acquire and apply place-making skills. This must build on existing capabilities and activities, resulting in a process of localised inclusive place-making which would transform liveability and livelihoods. At the centre of this responsible inclusive planning agenda is the creation of a new place-based partnership between people and planning that will alter futures through releasing the transformative power of citizen-centric innovation, largely based on survival mechanisms (Andres et al., 2020a) and the collective power of small actions (Dittmar, 2020).

Accounting for locality complements empowering residents to co-create living spaces with urban planners through developing workable and pragmatic practices, which feeds into how new forms of spatial planning have been identified (Haughton et al., 2009). Localised approaches can be delivered through land use plans (e.g. subdivision plans) which favour ownership and influences urban forms (Lai and Davies, 2020); they can also be implemented through micro-scale approaches allowing flexibility in the use of space, specifically towards informal practices.

People-centric approaches to planning sustainable livelihoods require solutions to contextualised needs; this is essential to ensure preparedness and resilience to crisis. The recent African epidemics (Ebola), the COVID-19 global pandemic, and the ongoing challenge of TB have highlighted the need for planning to focus on planning out opportunities for disease transmission. This is important in the African context where WASH facilities – in other words, the provision of water, sanitation, health care waste management, hygiene and environmental cleaning infrastructure, and services – are often unavailable.

There is a real risk that attempts to sustain livelihoods will rely on non-locally tailored approaches, shaped for Northern-type cities, and sold under a credo of being 'pandemic-resilient' (Andres et al., 2020a). Planning for sustainable and healthy livelihoods requires the development of local solutions based on contextualised spatial planning; this is not about land use, but rather about a vision of development that is both sustainable and resilient. Local knowledge is key to ensuring that interventions and their enforcement contribute to sustainable livelihoods rather than creating perverse outcomes. It is estimated that the standardised and even stricter lockdown mechanisms that were applied in South Africa during the 2020 pandemic led to an increase in mortality and morbidity greater than that caused by COVID-19 (Denoon-Stevens and du Toit, 2021). These interventions ignored the livelihoods and food strategies of the poor. The strategies intended to limit the impacts of COVID-19 needed to include a focus on minimising related impacts on the livelihoods of the most vulnerable (ibid). Such a localised and inclusive approach to sustainable livelihoods relies on considering spatial planning as a flexible process that can respond and adapt to predictable and unpredictable events.

Political commitment

Finally, whatever shifts occur, the reality is that planning for sustainable livelihoods is by essence highly political and requires political commitment. Planning operates through a variety of institutional arrangements, which in the African context and in many other policy fields are affected by patronage and clientelism. This contributes to concerns related to shadow governance coupled with extortion, corruption, and patronage across Africa (Olver, 2017). Urban planners must navigate these power dynamics, and situations differ from one country to another. South African planners have managed to secure considerable power to reshape the built environment, particularly in major urban centres (Andres et al., 2020b). In contrast, the situation in Ghana is very different (Cobbinah and Darkwah, 2017), as urban planning outcomes are dominated by political elites with little understanding of the role planning can play in supporting the development of sustainable livelihoods.

Community needs, and hence any attempts to provide for sustainable livelihoods, are not accounted for and decisions are driven by political strategies. Planners, in many African countries, still have very little influence, or power, in shaping the built environment; influence is constrained by a lack of knowledge, mapping tools (for example regarding land uses), data, and political support including implementation (ibid). This is a significant constraint with very few solutions.

Conclusion

Planning for sustainable livelihoods across Africa must consider the distinction between universal or more generic approaches to planning and the experience of particular places and people, including informal needs and practices and lower-income livelihoods. It must distinguish between living and the material structures that support life and different forms of life. It must also engage with the question posed by Fassin (2018) regarding forms of life, namely, 'are forms of life shared by the whole human species or is it inscribed in a given space and time?' (2018: 20). The concept of 'forms of life' highlights the distinction between the universal and the particular' and the 'tension between the biological and the biographical' (Fassin, 2018: 41–42).

Across Africa, livelihoods are constrained with extant forms of life forcing people to focus on survival rather than providing opportunities to shape biographies that also reflect sustainable and resilient livelihoods and lifestyles. Thus, the form of life of African urban residents is founded upon the constraints and possibilities provided by their external environment. Planning has been part of the problem and it is time for an alternative approach to emerge that support forms of life and encourage the formation of inclusive sustainable livelihoods. Planning must create the conditions that enable smaller-scale and highly contextualised initiatives to flourish (Dittmar, 2020). This is about balancing the tensions between structures imposed by planning and enabling opportunities for alternative substitute place-making or localised improvisation to occur. The focus must be on planning as a policy tool to support the formation of sustainable urban livelihoods across Africa.

Key points

- Planning can deliver both positive and negative impacts and must be crafted very carefully to develop sustainable and resilient livelihoods, especially when dealing with informality.
- Localised and contextualised approaches are critical, and planning must be clearly delivered on behalf of or for all people rather than reflecting the interests of advantaged cohorts above the most vulnerable.
- The issue of resource scarcity – financial and human capacity – needs to be addressed, as it hinders the development and implementation of an integrated and inclusive approach towards facilitating sustainable urban livelihoods.
- Planning must recognise the diversity of forms of life (livelihoods and liveability), land use, and management systems; urban livelihoods differ significantly from rural livelihoods and lower-income livelihoods are distinct from any other forms of life.

Further reading

Bhan, G., Srinivas, S. and Watson, V. (eds.) (2017) *The Routledge companion to planning in the Global South*, Oxford and New York: Routledge.
Charman, A., Petersen, L. and Govender, T. (2020) *Township economy: People, spaces and practices*, Cape Town: HSRC Press.
Pieterse, E. and Parnell, S. (eds.) (2014) *Africa's urban revolution*, London and New York: Zed Books.

References

Amankwah-Ayeh, K. (1996) 'Traditional planning elements of pre-colonial African towns', *New Contree*, 39: 60–76.
Andres, L., Bakare, H., Bryson, J.R., Khaemba, W., Melgaço, L. and Mwaniki, G.R. (2021) 'Planning, temporary urbanism and citizen-led alternative-substitute place-making in the Global South', *Regional Studies*, 55(1): 29–39. https://doi.org/10.1080/00343404.2019.1665645
Andres, L., Bryson, J.R., Denoon-Stevens, S., Bakare, H., du Toit, K. and Melgaço, L. (2020a) 'Calling for responsible inclusive planning and healthy cities in Africa', *Town Planning Review*, 92(2): 195–201. https://doi.org/10.3828/tpr.2020.49
Andres, L., Jones, P., Denoon Stevens., S. and Melgaço, L. (2020b) 'Negotiating polyvocal strategies: Re-reading de Certeau through the lens of urban planning in South Africa', *Urban Studies*, 57(12): 2440–2455. https://doi.org/10.1177/0042098019875423
Bah, E.M., Faye, I. and Geh, Z.F. (2018) *Housing market dynamics in Africa*, London: Palgrave Macmillan. https://doi.org/10.1057/978-1-137-59792-2
Battersby, J. (2017) 'Food system transformation in the absence of food system planning: The case of supermarket and shopping mall retail expansion in Cape Town, South Africa', *Built Environment*, 43(3): 417–430. https://doi.org/10.2148/benv.43.3.417
Chambers, R. and Conway, G. (1992) 'Sustainable rural livelihoods: Practical concepts for the 21st century', *IDS Discussion Paper 296*, Brighton: Institute of Development Studies.
Charlton, S. (2018) 'Confounded but complacent: Accounting for how the state sees responses to its housing intervention in Johannesburg', *The Journal of Development Studies*, 54: 2168–2185. https://doi.org/10.1080/00220388.2018.1460465
Charman, A., Tonkin, C., Denoon-Stevens, S.P. and Demeestere, R. (2017) *Post-apartheid spatial inequality: Obstacles of land use management on townships micro-enterprise formalisation*, Cape Town: Sustainable Livelihoods Foundation.
Charman, A.J., Petersen, L.M. and Piper, L. (2013) 'Enforced informalisation: The case of liquor retailers in South Africa', *Development Southern Africa*, 30(4–05): 580–595. https://doi.org/10.1080/0376835x.2013.817306
City of Johannesburg (2019) 'Land use scheme 2018', *Provincial Gazette*, 25(1), Pretoria: The Province of Galteng. www.gpwonline.co.za/Gazettes/Gazettes/1_02-01-2019_Gauteng.pdf
Cobbinah, P.B. and Darkwah, R.M. (2017) 'Urban planning and politics in Ghana', *GeoJournal*, 82: 1229–1245. https://doi.org/10.1007/s10708-016-9750-y
Competition Commission (South Africa) (2019) *The grocery retail market inquiry: Final report (non-confidential)*. Available at: www.compcom.co.za/retail-market-inquiry/ (Accessed 24 March 2021).
Crush, J., Hovorka, A. and Tevera, D. (2011) 'Food security in Southern African cities: The place of urban agriculture', *Progress in Development Studies*, 11(4): 285–305. https://doi.org/10.1177/146499341001100402
Denoon-Stevens, S.P. (2016) 'Developing an appropriate land use methodology to promote spatially just, formal retail areas in developing countries: The case of the city of Cape Town, South Africa', *Land Use Policy*, 54: 18–28. https://doi.org/10.1016/j.landusepol.2016.01.010
Denoon-Stevens, S.P. and du Toit, K. (2021) 'The job-food-health nexus in South African townships and the impact of COVID-19', in Bryson, J.R., Andres, L., Ersoy, A. and Reardon, L. (eds.), *Living with pandemics: Places, people, policy and rapid mitigation and adaptation to covid-19*, Cheltenham: Edward Elgar.
Denoon-Stevens, S.P. and Nel, V. (2020) 'Towards an understanding of proactive upzoning globally and in South Africa', *Land Use Policy*, 97: 104708. https://doi.org/10.1016/j.landusepol.2020.104708
Dittmar, H. (2020) *DIY city: The collective power of small actions*, Washington: Island Press.
Fassin, D. (2018) *Life: A critical user's manual*, Cambridge: Polity Press.

Gough, K.V., Tipple, A.G. and Napier, M. (2003) 'Making a living in African cities: The role of home-based enterprises in Accra and Pretoria', *International Planning Studies*, 8(4): 253–277. https://doi.org/10.1080/1356347032000153115

Govender, T., Barnes, J.M. and Pieper, C.H. (2011) 'The impact of densification by means of informal shacks in the backyards of low-cost houses on the environment and service delivery in Cape Town, South Africa', *Environmental Health Insights*, 5: 23–52. https://doi.org/10.4137/ehi.s7112

Harrison, P., Todes, A. and Watson, V. (2007) *Planning and transformation: Learning from the post-apartheid experience*, London and New York: Routledge. https://doi.org/10.4324/9780203007983

Haughton, G., Allmendinger, P., Counsell, D. and Vigar, G. (2009) *The new spatial planning: Territorial management with soft spaces and fuzzy boundaries*, London: Routledge. https://doi.org/10.1080/14649357.2010.525378

Kamete, A.Y. (2009) 'In the service of tyranny: Debating the role of planning in Zimbabwe's urban "clean-up" operation', *Urban Studies*, 46(4): 897–922. https://doi.org/10.1177/0042098009102134

Kigochie, P.W. (2001) 'Squatter rehabilitation projects that support home-based enterprises create jobs and housing: The case of Mathare 4A, Nairobi', *Cities*, 18(4): 223–233. https://doi.org/10.1016/s0264-2751(01)00015-4

King, A.D. (1985) 'Colonial cities: Global pivots of change', in Ross, R. and Telkamp, G.J. (eds.), *Colonial cities: Essays on urbanism in a colonial context*, Dordrecht: Martinus Nijhoff Publishers, pp. 7–32. https://doi.org/10.1007/978-94-009-6119-7_2

Lai, L.W. and Davies, S.N. (2020) '"Surveying was a kind of writing on the land": The economics of land division as town planning', *Planning Theory*, 19(4): 421–444. https://doi.org/10.1177/1473095220912791

Lemanski, C. (2009) 'Augmented informality: South Africa's backyard dwellings as a by-product of formal housing policies', *Habitat International*, 33(4): 472–484. https://doi.org/10.1016/j.habitatint.2009.03.002

Marais, L. and Cloete, J. (2014) '"Dying to get a house?" The health outcomes of the South African low-income housing programme', *Habitat International*, 43: 48–60. https://doi.org/10.1016/j.habitatint.2014.01.015

Mateus, A.D., Allen-Ile, C. and Iwu, C.G. (2014) 'Skills shortage in South Africa: Interrogating the repertoire of discussions', *Mediterranean Journal of Social Sciences*, 5(6): 63–73. https://doi.org/10.5901/mjss.2014.v5n6p63

Munanga, K. (2016) 'Desenvolvimento, construção da democracia e da nacionalidade nos países africanos: Desafio para o milênio', *Cadernos CERU*, 27(2): 45–56.

Olver, C. (2017) *How to steal a city*, Cape Town: Jonathan Ball.

Parnell, S. (1993) 'Creating racial privilege: The origins of South African public health and town planning legislation', *Journal of Southern African Studies*, 19(3): 471–488. https://doi.org/10.1080/03057079308708370

Parnell, S. (2002) 'Winning the battles but losing the war: The racial segregation of Johannesburg under the natives (urban areas) act of 1923', *Journal of Historical Geography*, 28(2): 258–281. https://doi.org/10.1006/jhge.2001.0401

Parnell, S. and Pieterse, E. (2010) 'The "right to the city": Institutional imperatives of a developmental state', *International Journal of Urban and Regional Research*, 34(1): 146–162. https://doi.org/10.1111/j.1468-2427.2010.00954.x

Pieterse, E. (2014) 'Filling the void: An agenda for tackling African urbanisation', in Pieterse, E. and Parnell, S. (eds.), *Africa's urban revolution*, London and New York: Zed books, pp. 200–220. https://doi.org/10.5040/9781350218246.ch-011

Rogerson, C.M. (2016) 'Responding to informality in urban Africa: Street trading in Harare, Zimbabwe', *Urban Forum*, 27(2): 229–251. https://doi.org/10.1007/s12132-016-9273-0

Ross, R. and Telkamp, G.J. (1985) 'Introduction', in Telkamp, G.J. and Ross, R. (eds.) *Colonial cities: Essays on urbanism in a colonial context*, Dordrecht: Martinus Nijhoff Publishers, pp. 1–6. https://doi.org/10.1007/978-94-009-6119-7_1

Satterthwaite, D., Sverdlik, A. and Brown, D. (2019) 'Revealing and responding to multiple health risks in informal settlements in sub-Saharan African cities', *Journal of Urban Health*, 96(1):112–122. https://doi.org/10.1007/s11524-018-0264-4

Silva, C.N. (ed.) (2015) *Urban planning in sub-Saharan Africa: Colonial and post-colonial planning cultures*, London and New York: Routledge. https://doi.org/10.4324/9781315797311-9

Silva, C.N. (2016) *Urban planning in North Africa*, London and New York: Routledge. https://doi.org/10.4324/9781315548753

Tonkin, C., Charman, A. and Govender, T. (2018) 'Development-oriented township land use management: Learning from Eveline Street, Katutura, Windhoek', in Bickford, G. (ed.), *The urban land paper series: Volume 2*, Braamfontein: South African Cities Network, pp. 54–64.

UN-Habitat and APA (African Planning Association) (2013) *The state of planning in Africa*, Nairobi: UN-Habitat.

Watson, V. (2014) 'African urban fantasies: Dreams or nightmares? *Environment and Urbanisation*, 26(1): 215–231. https://doi.org/10.1177/0956247813513705

Watson, V. and Odendaal, N. (2013) 'Changing planning education in Africa: The role of the association of African planning schools', *Journal of Planning Education and Research*, 33(1): 96–107. https://doi.org/10.1177/0739456x12452308

Wilkinson, A. and contributors (2020) 'Local response in health emergencies: Key considerations for addressing the COVID-19 pandemic in informal urban settlements', *Environment and Urbanisation*, 32(2): 503–522. https://doi.org/10.1177/0956247820922843

31
ARTISANAL MINING AND LIVELIHOODS IN THE GLOBAL SOUTH

Roy Maconachie

Introduction

Over the past three decades, rural livelihoods in resource-rich countries across the Global South have been rapidly transformed by the growth of the artisanal and small-scale mining (ASM) sector – low-tech, labour-intensive, mineral extraction and processing. However, surprisingly, up until the 1990s, ASM was barely acknowledged in academic literature or policy discussions on poverty alleviation and economic development in poor countries (Hilson and Gatsinzi, 2014). In contrast to early narratives, which portrayed ASM as an activity pursued by rogue operators and associated with a host of environmental, health and safety, and social concerns, a growing body of research has more recently demonstrated that ASM is a vital livelihood activity. This work, in fact, suggests that the revenue generated by artisanal mining is relied upon heavily at the local level, providing a 'life-line' for poor households and making it possible for them to pay for household expenses, including school fees, medical expenditures and funeral costs.

In many countries, ASM has become a 'magnet' for young, single, unemployed, unskilled labourers in search of job opportunities in employment constrained economies (see Figure 31.1). In doing so, it has become an 'engine of employment' and a driver of economic development in rural areas, with some observers going as far as to suggest that it is now the most important non-farm rural livelihood activity in the developing world. Moreover, if the conditions are right, evidence suggests that the ASM sector can help to foster a platform for wealth creation, generating vital 'start up' capital for upstream and downstream activities, and providing opportunities to build sustainable and innovative economies. This could, in turn, serve as a foundation for spawning entrepreneurship and catalysing sustainable economic linkages locally, both of which are important processes for attaining the UN Sustainable Development Goals (SDGs) (Hilson and Maconachie, 2020; Maconachie and Conteh, 2021; Sturman et al., 2020).

This chapter provides an introduction and critical overview of ASM and its contribution to livelihoods in countries of the Global South. Drawing upon case study examples predominantly from sub-Saharan Africa, a range of different minerals that are extracted artisanally are explored. Following this introduction, the chapter first sketches out the contours of the ASM sector, detailing the scope and scale of artisanal mining activities across the Global South. It then goes on to review some of the key literature which explores the potential of artisanal mining as a vital livelihood activity in resource-rich developing countries. In examining both the challenges and

Figure 31.1 Artisanal gold miners in Kono District, Sierra Leone
Source: Photo credit: Roy Maconachie

benefits that artisanal operators face, it is argued that the ASM sector continues to be defined by its intractable informality. The chapter concludes by exploring the merits of a formalised ASM sector as a possible way of safeguarding livelihoods and ensuring that more benefits accrue to host communities where extraction takes place.

Scope, scale and emerging trends in ASM

Numerous minerals and resources are now mined artisanally in the Global South, including precious metals and minerals (e.g. diamonds, gemstones, gold), industrial minerals (bauxite, coltan, cobalt, coal), and even construction materials (e.g. sand, gravel, dimension stones). Recognising the increasing prevalence of artisanal operators who extract so-called 'development minerals' – materials that are mined, processed, manufactured and used domestically in industries such as construction, manufacturing and agriculture (Franks et al., 2016) – in 2016, the UNDP and EU launched the African, Caribbean Pacific (ACP)-EU Development Mineral Programme to improve the management and profile of this important sector. Although often referred to as 'low value minerals', because they have a low price in relation to their weight, development minerals are not only crucial for generating employment, but also for stimulating in-country economic linkages and national development. Such minerals are vital elements in a wide range of manufactured products, including toothpaste, paints, glass, lightbulbs, ceramic tiles, cutlery, soap and plasterboard (UNDP, 2016).

In donor and policy circles, however, artisanally mined 'development minerals' have, until recently, largely been neglected. Rather, there has been a fixation on the industrial and precious minerals

Table 31.1 A snapshot of global trends in the ASM sector

Key trend	Estimate
Number of people working directly in ASM in 2017	40 million
Number of dependents who rely on ASM across 80 countries in the Global South	150 million
Percentage of the global gold supply produced by the ASM sector	20%
Percentage of the global diamond supply produced by the ASM sector	20%
Percentage of the global sapphire supply produced by the ASM sector	80%
Percentage of the global tantalum supply produced by the ASM sector	26%
Percentage of the global tin supply produced by the ASM sector	25%
Percentage of the ASM workforce in Africa who are women	40–50%
Percentage of small-scale miners who are informal	70–80%

Source: Adapted from IGF (2017)

segments of the ASM sector, in large part due to their integration in global supply chains (UNDP, 2016). Over the past two decades, much of the interest in supporting the ASM sector has focused on those engaged in the extraction of precious metals and stones, with gold being one of the most popular minerals being mined (Hilson and McQuilken, 2014). Estimates suggest that ASM operators extract 10.6 million ounces of gold annually across 70 countries (Sturman et al., 2020). According to some sources, as much as 10% of the world's gold supply, 20% of the global diamond supply, and 80% of the global sapphire supply originate from the ASM sector (see UNECA, 2011; African Union, 2009; IGF, 2017). Table 31.1 summarises some of the key global figures and trends that characterise the ASM sector.

Recent literature also now suggests that in opposition to the widespread belief that ASM is motivated by opportunism and a desire to 'get rich quickly' (Barry, 1996; USAID, 2005; Alpan, 1986; Noetstaller, 1987), it is more often a livelihood activity that is driven by poverty (e.g. see Maconachie and Hilson, 2016; Spiegel, 2009; Banchirigah, 2006; Yelpaala and Ali, 2005). Beginning in the late 1970s and early 1980s, the implementation of 'conditionalities' associated with World Bank structural adjustment programs had dramatic impacts on poor people in the Global South. Widespread neo-liberal economic reforms in many debt-ridden developing countries led to a significant downsizing of industries, made life increasingly challenging for smallholder farmers, and was responsible for widespread redundancies in the public sector. Building on the experience of structural adjustment, Poverty Reduction Strategy Plans (PRSPs) continued to focus on liberalisation and opening up developing economies to foreign direct investment. As global mineral prices have soared and heightened resource demands have emerged from the world's emerging economies, investments in large-scale mining operations have often driven spectacular economic growth at the national level. However, much of this resource extraction has been concentrated in exclusionary spatial 'enclaves', and there has been very little 'trickle down' to local populations. The ASM sector has offered one of the few employment opportunities available to rural dwellers, many of whom have found themselves in precarious financial situations. Much evidence suggests that the ASM economy and the downstream industries it spawns have absorbed millions of people who were adversely affected by these changes (Hilson and McQuilken, 2014).

Although large-scale, capital-intensive mining projects have long captured the imaginations of governments, policy makers and development professionals (Luning, 2014), artisanal mining remains a much more significant employer at the point of extraction. To date, no comprehensive census has been carried out on the extent of ASM across the Global South, but current estimates indicate that ASM provides a direct livelihood for as many as 40 million people globally, compared to estimates of 7 million people working in industrial mining (IGF, 2017). However, the World Bank maintains that for every individual directly employed in ASM, at least six additional jobs are created in ancillary industries. These estimates suggest that the ASM sector has played a critical role in helping to meet the livelihood needs of a wide range of marginalised social groups, including unemployed youth (Howard, 2014), migrants (Gratz, 2009), women (Jenkins, 2014; Lahiri-Dutt and Macintyre, 2006) and children (Maconachie and Hilson, 2016). Drawing upon both academic and policy literature, Table 31.2 summarises ASM employment estimates and key minerals mined for selected developing countries.

Table 31.2 Employment estimates for ASM, and key minerals extracted, in selected developing countries

Country	Directly working in ASM	Estimated number of dependents	Key minerals mined on a small and artisanal scale
Angola	150,000	900,000	Diamonds
Argentina	5,800	34,800	Gold
Bolivia	72,000	432,000	Gold
Brazil	250,000	1,500,000	Gold, diamonds, gemstones
Burkina Faso	200,000	1,200,000	Gold
CAR	400,000	2,400,000	Gold, diamonds
Chad	100,000	600,000	Gold
Chile	12,000	72,000	Gold, copper, silver
China	15,000,000	90,000,000	Gold, coal, construction materials
Colombia	200,000	1,200,000	Gold, gemstones
Cote D'Ivoire	100,000	600,000	Gold, diamonds
DRC	200,000	1,200,000	Diamonds, gold, coltan
Ecuador	92,000	552,000	Gold
Eritrea	400,000	2,400,000	Gold
Ethiopia	500,000	3,000,000	Gold
French Guiana	10,000	60,000	Gold
Ghana	1,100,000	6,600,000	Gold, diamonds, sand
Guinea	300,000	1,500,000	Gold, diamonds
Guyana	20,000	120,000	Gold, diamonds
India	500,000	3,000,000	Gold, tin, coal, gemstones
Indonesia	109,000	654,000	Gold
Liberia	100,000	600,000	Gold, diamonds
Madagascar	500,000	3,000,000	Coloured gemstones, gold
Malawi	40,000	240,000	Coloured gemstones, gold
Mali	400,000	2,400,000	Gold
Mongolia	120,000	720,000	Gold, fluorspar, coal
Mozambique	100,000	600,000	Coloured gemstones, gold
Myanmar	14,000	84,000	Gold, tin, jade, coloured gemstones
Niger	450,000	2,700,000	Gold
Nigeria	500,000	2,500,000	Gold
PNG	60,000	360,000	Gold
Peru	30,000	180,000	Gold
Philippines	300,000	1,800,000	Gold

Country	Directly working in ASM	Estimated number of dependents	Key minerals mined on a small and artisanal scale
South Africa	20,000	120,000	Gold
Sierra Leone	300,000	1,800,000	Gold, diamonds, coltan
South Sudan	200,000	1,200,000	Gold
Sri Lanka	165,000	990,000	Gold, coloured gemstones
Suriname	20,000	120,000	Gold
Tanzania	1,500,000	9,000,000	Coloured gemstones, gold, diamonds
Uganda	150,000	900,000	Gold
Venezuela	40,000	240,000	Gold, coltan
Zimbabwe	500,000	3,000,000	Gold, diamonds, coloured gemstones

Source: Adapted from Hilson and Maconachie (2017)

Data extracted from ILO (1999), Dreschler (2001), Mutemeri and Petterson (2002) and Hinton (2005)

ASM as a driver of sustainable livelihoods

In the development studies literature, there is now an established body of scholarship that has explored livelihood diversification as a crucial survival strategy for rural households in poor, developing countries (Barrett et al., 2001; Bryceson, 1996, 1999; Ellis, 2000; Reardon, 1997). This work has convincingly demonstrated that although still of central importance, subsistence farming on its own has increasingly become unable to provide a sufficient means of survival in rural areas (Jayne et al., 2010; Meagher and Mustapha, 1997). Notably, early work by Bryceson and Jamal (1997) argued that as African rural populations have moved away from a reliance on smallholder agriculture, 'de-agrarianisation' through urban migration and the expansion of non-agricultural activities in rural areas has provided new income sources, livelihood possibilities and social identities for rural dwellers.

Building on this research, 'sustainable livelihood' approaches became popular with donors and governments at the beginning of the new millennium, offering a new framework for analysing livelihoods and identifying entry points for development interventions. According to this model, poor households possess different combinations of 'capital assets' in their livelihood portfolios and at any given time and depending on the circumstances, they may convert one category of assets into another (Stocking and Murnaghan, 2001). In improving and safeguarding their livelihood bases, rural households do not merely rely on agriculture alone, but rather they draw on a wide range of other resources, including non-farm activities and migration. Following on from this work, over the past two decades, a rich body of literature has emerged that has convincingly demonstrated how for tens of millions of individuals across the Global South, ASM has become an integral component of their dynamic livelihood portfolios. This has not only allowed them to make ends meet, but also to strengthen resilience by galvanising linkages across sectors in response to changing circumstances (Maconachie and Binns, 2007; Maconachie and Hilson, 2018).

In evaluating its wider linkages to the rural economy, there is perhaps no better example of ASM's positive 'domino effect' on livelihoods than its symbiotic connection to the smallholder agriculture sector. Over the past two decades, a sizeable body of analysis has emerged which explores the connections between the two activities, confirming the important income supplement that ASM plays for agricultural households. To provide one example, research carried out in Sierra Leone explored the inter-locking nature of rice farming and artisanal diamond mining (Maconachie, 2011; Maconachie and Binns, 2007). Here, the seasonality of farming and mining allows individuals to 'straddle' both activities, undertaking mining in the dry season when rivers are low, and farming during the rainy season when planting conditions are optimal. This symbiotic relationship between livelihood activities allows miners to reinvest their mining income into farming activities during the rainy season and farmers to sell surplus crops to mining areas for higher prices in the dry season. Such a tendency to diversify portfolios is widespread in the Eastern Province of Sierra Leone, where actors pursue complex and dynamic livelihood strategies that involve the intertwining of the farming and mining economies. The resilience of this dual economy, and the varying contributions generated by farming – mining linkages are an essential strategy towards maintaining the sustainable livelihoods of poor households.

Much evidence also suggests that the farming-mining linkages noted above have a strong gender dimension in many countries of the Global South. In sub-Saharan Africa, for example, recent scholarship suggests that women produce 60–70% of food for home consumption through smallholder cultivation, but they are constrained by access to land, capital, agricultural inputs and in some cases poor market linkages (SDGC, 2017). However, other work has suggested that women's simultaneous engagement in ASM activities has helped them overcome some of these challenges (Hilson and Maconachie, 2020). Although women in Africa are constrained by what has been described as a 'triple burden' of labour, on productive, reproductive and social fronts (e.g. see Barreto et al., 2018), the income derived from ASM has played a role in empowering women and helping them address the agricultural challenges they face. Returning to Sierra Leone once again, research suggests that women managed to gain control of the artisanal gold panning sector in many parts of the country, into which they moved very easily as men were more preoccupied with mining diamonds (Maconachie and Hilson, 2011). In this case, women have used their earnings from artisanal gold mining to finance agricultural 'businesses' and trading ventures, and to purchase a wide range of fresh foodstuffs from nearby villages, including cassava, oranges, limes and mangoes, which they transport to the country capital of Freetown to sell for much higher prices at the Calabar Town market (Maconachie and Hilson, 2011). They have also used this income to establish informal savings groups and revolving credit schemes, which has played an important role in giving women the ability to have more control over their finances and meet their livelihood needs.

In Sierra Leone and elsewhere in the Global South, women have creatively used the ability to move freely within the informal artisanal mining 'space' to support their households and re-invest mining income into other economic activities. It therefore remains critical that an appreciation of this achievement is reflected in gender-sensitive policies and programmes that support the ASM sector. While the gendered aspects of artisanal mining have been explored, to a limited degree, in recent literature (e.g. see, Bashwira et al., 2014; Buss et al., 2019; Fisher, 2007; Hilson et al., 2018), new policy frameworks designed to support ASM must take on board women's empowerment and gender equity as a primary concern.

Safeguarding livelihoods through formalisation

Despite the undisputed livelihood importance of the ASM sector, these benefits have often come at a cost. As noted earlier, most ASM activities in the Global South take place in the informal sphere, with mining operations occurring in remote areas where governance is poor, regulatory enforcement is virtually non-existent and elite capture is widespread. As the demand for key industrial minerals – such as the '3 Ts' (tin, tungsten and tantalum) and cobalt – has soared in recent years, artisanal mineral supply chains that feed production in the major electronic companies have also come under increasing scrutiny. The COVID-19 pandemic further

Figure 31.2 Women gold miners in Kono District, Sierra Leone
Source: Photo credit: Roy Maconachie

exposed the complexity and challenges of the mineral supply chains that power the global economy.

Research exploring the 'darker side' of ASM has examined its link to human rights abuses, shadow state economies, money laundering and criminal and terrorist networks (FATF, 2013). Illicit transactions within the informal ASM economy pose a major challenge to governments, marking a significant loss of revenue that could be used to promote economic growth and development in the countries where minerals are extracted. Alongside this research, other scholars have focused on the sector's environmental impacts, including widespread land degradation (Laing and Moonsammy, 2021; Odumo et al., 2018) and mercury contamination associated with artisanal gold mining (Clifford, 2014; Tschakert and Singha, 2007; Veiga and Hinton, 2002). Still other scholars have centred their attention on the numerous 'social ills' that regularly accompany ASM activities, including the use of child labour (Maconachie and Hilson, 2016), hazardous labour practices (Smith et al., 2016), the spread of prostitution (Bryceson et al., 2013) and the abuse of drugs and alcohol in mining camps (Hinton, 2005; Donoghue, 2004). Coverage of the sector's negative attributes has portrayed the unchecked growth of ASM as a problem, which has created widespread international public concern.

In light of growing awareness about the social and environmental impacts of ASM that take place in informal 'spaces', a number of recent initiatives have emerged to address these challenges. In particular, the so-called 'traceability agenda', has gained considerable traction over the past decade, as consumers demand more transparency in the mineral supply chains that feed the manufactured products they consume. While initiatives such as the US conceived Dodd-Frank Act have reduced the opportunities for armed groups and criminal networks that benefit from the informal ASM economy, they have had a much wider impact on the livelihoods of miners. Section 1502 of the Act requires publicly traded companies to disclose to the US Securities and Exchange Commission whether or not they are using certain minerals in their products that originate from the Democratic Republic of the Congo (DRC) or surrounding Great Lakes Region countries. This has, in effect, created a ban on Congolese gold and mineral exports, having a dramatic impact on the lives of miners and their families at the bottom of the supply chain. According to Geenen (2012), the most important non-farm income generating activity has been removed from the rural economy, having devastating knock-on impacts on all those living in and around mining areas. Some estimates suggest that between 1 and 2 million Congolese artisanal miners have been put out of work by the ban (Seay, 2012). This has had a significant ripple effect in rural areas, reducing 'start up' capital for upstream and downstream activities, and placing stress on farming-mining linkages.

For many governments and policy makers in resource-rich countries of the Global South, the ASM sector has overwhelmingly been defined by its illegality and informality, flourishing outside formal state regulation, with most activities unlicensed and operating with little or no formal support (International Labour Organization, 1999; Siegel and Veiga, 2009). The unregulated nature of ASM reinforces its informality, as transactions among elite actors are strengthened by informal practices within the confines of informal 'spaces' (Hilson et al., 2017), leading to loss of revenues for governments and low returns for artisanal miners at the bottom of the value chain.

Unsurprisingly, the many missed opportunities for mineral-rich countries to effectively harness ASM's potential have led to persistent calls for formalisation, as a potential panacea capable of permanently solving the problems of artisanal operators (Maconachie and Hilson, 2011; Siwale and Siwale, 2017). The case for the formalisation of ASM has been made by academics, practitioners and international development agencies, such that it has now almost become irrational for governments not to pursue it, given the many potential benefits (Siegel and Veiga, 2009). For example, in the context of Zambia, Siwale and Siwale (2017) argued that proponents of formalisation envisage that the process will lead to operators having secured titles, invariably leading to the consolidation of property laws and their enforcement by states. Formalisation can also promote miners' visibility and provide a framework through which governments can facilitate technical support and lines of credit for miners. It is thus argued that formalisation will invariably result in governments being able not only to tax miners, but to effectively provide a framework to 'better govern and consequently manage the social and environmental impacts of mining' (Salo et al., 2016: 1058–1059). So if the ASM sector has the potential to reduce unemployment, support livelihoods and alleviate poverty in developing countries, while simultaneously boosting government revenues, why has reform continued to be undermined? Indeed, in many resource-rich countries of the Global South, it would appear that in contrast to the increasing recognition of the importance of ASM by developing countries' governments and donors (Hilson et al., 2017; Maconachie and Hilson, 2011), the sector's informality has become entrenched and intractable (Jónsson and Fold, 2014; Schoneveld et al., 2018).

Conclusion

Across the Global South, the failure to recognise the growing economic and livelihood importance of ASM has played a role in retarding the sector's development, and stifling its potential. In most countries, ASM remains on the margins of poverty reduction strategies, despite much evidence now suggesting that it should be at the centre. Consequently, in recent years, many donors and policy makers have become increasingly committed to formalising ASM activities.

A formalised ASM sector, it is argued, could strengthen the delivery of much-needed livelihood benefits, while at the same time addressing a series of pressing development challenges. However, in order for such developmental benefits to accrue and reach those most in need, a more dynamic policy approach is urgently required. Not only are new incentives needed to bring illicit miners into a formalised, legal domain, but this must be complemented by a more nuanced understanding of the day-to-day realities of mining communities. For example, a formalised ASM 'space' should enable operators to move flexibly between different economic sectors, allowing for a symbiotic circulation of labour between dovetailing economies, as is the case with Sierra Leone's farming-mining linkages discussed earlier in the chapter.

The World Bank and various departments of the United Nations are now putting pressure on many governments in the Global South to make ASM the centrepiece of their rural development strategies and provide support to the millions of workers who operate within the sector. This stems largely from the growing recognition of ASM's economic and livelihood impact, the need to gain greater control over its sprawling activities, and the importance of putting regulators in an improved position to tackle the host of environmental problems and social 'ills' mentioned in the previous section of the paper (also see, Hentschel et al., 2002; UNEP, 2012). Although to date there has been little effort made to develop the intervention models required to address these problems, it has been argued by many that the formalisation of ASM is the key to empowering unregistered miners, safeguarding their livelihoods and making the sector more sustainable.

Key points

- Across the Global South, ASM has experienced meteoric growth over the past three decades, largely due to the rising value of mineral prices, policy reforms in the mining sector and the increasing difficulty of earning a living in rural, agrarian settings.
- Estimates suggest that ASM provides a direct livelihood to as many as 40 million people across 80 countries globally, with an additional 150 million people benefiting indirectly from the upstream and downstream industries it spawns.
- Agriculture and ASM should be viewed as complementary livelihood activities; there are many important links between the two sectors which allow rural households to supplement their farming earnings with mining income, and invest in farming and farm inputs.
- Most ASM activities in the Global South take place in the informal sphere, where governance and regulatory enforcement are often poor.
- Informality can foster negative socioeconomic, health and environmental impacts, all of which can trap miners and communities in cycles of poverty and exclude them from formal support and protection.
- A formalised ASM sector could strengthen the delivery of much-needed livelihood benefits, while simultaneously addressing a series of pressing development challenges.

Further reading

Geenen, S. (2012) 'A dangerous bet: The challenges of formalizing artisanal mining in the Democratic Republic of Congo', *Resources Policy*, 37(3): 322–330. https://doi.org/10.1016/j.resourpol.2012.02.004

Maconachie, R. and Binns, T. (2007) '"Farming miners" or "mining farmers"? Diamond mining and rural development in post-conflict Sierra Leone', *Journal of Rural Studies*, 23(3): 367–380. https://doi.org/10.1016/j.jrurstud.2007.01.003

Siegel, S. and Veiga, M.M. (2009) 'Artisanal and small-scale mining as an extralegal economy: De Soto and the redefinition Of "formalization"', *Resources Policy*, 34: 51–56. https://doi.org/10.1016/j.resourpol.2008.02.001

Siwale, A. and Siwale, T. (2017) 'Has the promise of formalizing artisanal and small-scale mining (ASM) failed? The case of Zambia', *The Extractive Industries and Society*, 4: 191–201. https://doi.org/10.1016/j.exis.2016.12.008

References

African Union (2009) *Africa mining vision: Boosting artisanal and small-scale mining*, Addis Ababa: African Union.
Alpan, S. (1986) 'The role of government in promoting small-scale mining', *Natural Resources Forum*, 10(1): 95–97. https://doi.org/10.1111/j.1477-8947.1986.tb00783.x
Banchirigah, S.M. (2006) 'How have reforms fuelled the expansion of artisanal mining? Evidence from sub-Saharan Africa', *Resources Policy*, 31(3): 165–171. https://doi.org/10.1016/j.resourpol.2006.12.001
Barreto, L., Schein, P., Hinton, J. and Hruschka, F. (2018) *The impact of small-scale mining operations on economies and livelihoods in low-to middle-income countries*, Westcombe: Pact Global UK, Envigado: Alliance for Responsible Mining.
Barrett, C.B., Reardon, T. and Webb, P. (2001) 'Nonfarm income diversification and household livelihood strategies in rural Africa: Concepts, dynamics, and policy implications', *Food Policy*, 26(4): 315–331. https://doi.org/10.1016/S0306-9192(01)00014-8
Barry, M. (ed.) (1996) 'Regularizing informal mining: A summary of the proceedings of the international roundtable on artisanal mining', *Industry and Energy Department Occasional Paper No. 6*, Washington, DC: World Bank.
Bashwira, M.R., Cuvelier, J., Hilhorst, D. and van der Haar, G. (2014) 'Not only a man's world: Women's involvement in artisanal mining in eastern DRC', *Resources Policy*, 40: 109–116. https://doi.org/10.1016/j.resourpol.2013.11.002
Bryceson, D. (1996) 'Deagrarianization and rural employment in sub-Saharan Africa: A sectoral perspective', *World Development*, 24(1): 97–111. https://doi.org/10.1016/0305-750X(95)00119-W
Bryceson, D. (1999) 'African rural labour, income diversification and livelihood approaches: A long-term development perspective', *Review of African Political Economy*, 26(80): 171–189. https://doi.org/10.1080/03056249908704377
Bryceson, D.F. and Jamal, V. (eds.) (1997) *Farewell to farms: De-agrarianisation and employment in Africa*, Aldershot: Ashgate.
Bryceson, D.F., Jønsson, J. and Verbrugge, H. (2013) 'Prostitution or partnership? Wifestyles in Tanzanian artisanal gold-mining settlements', *The Journal of Modern African Studies*, 51(1): 33–56. http://dx.doi.org/10.1017/S0022278X12000547
Buss, D., Rutherford, B., Stewart, J., Cote, G.A., Sebina-Zziwa, A., Kibomo, R., Hinton, J. and Lebert, J. (2019) 'Gender and artisanal and small-scale mining: Implications for formalization', *The Extractive Industries and Society*, 6(4): 1101–1112. https://doi.org/10.1016/j.exis.2019.10.010
Clifford, M. (2014) 'Future strategies for tackling mercury pollution in the artisanal gold mining sector: Making the Minamata convention work', *Futures*, 62(Part A): 106–112. https://doi.org/10.1016/j.futures.2014.05.001
Donoghue, A. (2004) 'Occupational health hazards in mining: An overview', *Occupational Medicine*, 54(5): 283–289. https://doi.org/10.1093/occmed/kqh072
Dreschler, B. (2001) *Small-scale mining and sustainable development within the SADC region, Mining, Minerals and Sustainable Development (MMSD)*, London: International Institute for Environment and Development and World Business Council for Sustainable Development.
Ellis, F. (2000) *Rural livelihoods and diversity in developing countries*, Oxford: Oxford University Press.
Financial Action Task Force (FATF) (2013) *Money laundering and terrorist financing through trade in diamonds*, Paris: Financial Action Task Force/OECD.
Fisher, E. (2007) 'Occupying the margins: Labour integration and social exclusion in artisanal mining in Tanzania', *Development and Change*, 38(4): 735–760. https://doi.org/10.1111/j.1467-7660.2007.00431.x
Franks, D.M., Pakoun, L. and Ngonze, C. (2016) *Development minerals: Transforming a neglected sector in Africa, the Caribbean and the Pacific*, Belgium: United Nations Development Programme.
Geenen, S. (2012) 'A dangerous bet: The challenges of formalizing artisanal mining in the Democratic Republic of Congo', *Resources Policy*, 37(3): 322–330. https://doi.org/10.1016/j.resourpol.2012.02.004
Gratz, T. (2009) 'Moralities, risk and rules in West African artisanal gold mining communities: A case study of Northern Benin', *Resources Policy*, 34(1–2): 12–17. https://doi.org/10.1016/j.resourpol.2008.11.002

Hentschel, T., Hruschka, F. and Priester, M. (2002) 'Global report on artisanal and small-scale mining', *Minerals Mining and Sustainable Development (MMSD) Project*, London: International Institute for Environment and Development.

Hilson, G. and Gatsinzi, A. (2014) 'A rocky road ahead? Critical reflections on the futures of small-scale mining in sub-Saharan Africa', *Futures*, 62: 1–9. https://doi.org/10.1016/j.futures.2014.05.006

Hilson, G., Hilson, A., Maconachie, R., McQuilken, J. and Goumandakoye, H. (2017) 'Artisanal and small-scale mining (ASM) in sub-Saharan Africa: Re-conceptualizing formalization and "illegal" activity', *Geoforum*, 83: 80–90. https://doi.org/10.1016/j.geoforum.2017.05.004

Hilson, G., Hilson, A., Siwale, A. and Maconachie, R. (2018) 'Female faces in informal "spaces": Women and artisanal and small-scale mining in sub-Saharan Africa', *Africa Journal of Management*, 4(3): 306–346. https://doi.org/10.1080/23322373.2018.1516940

Hilson, G. and Maconachie, R. (2020) 'Artisanal and small-scale mining and the sustainable development goals: Opportunities and new directions for sub-Saharan Africa', *Geoforum*, 111: 125–141. https://doi.org/10.1016/j.geoforum.2019.09.006

Hilson, G. and McQuilken, J. (2014) 'Four decades of support for artisanal and small-scale mining in sub-Saharan Africa: A critical review', *The Extractive Industries and Society*, 1(1): 104–118. https://doi.org/10.1016/j.exis.2014.01.002

Hinton, J. (2005) *Communities and small-scale mining: An integrated review for development planning*, Washington, DC: World Bank.

Howard, N. (2014) 'Teenage labor migration and antitrafficking policy in West Africa', *American Academy of Political and Social Science*, 653(1): 124–140. https://doi.org/10.1177%2F0002716213519242

Intergovernmental Forum on Mining, Minerals, Metals and Sustainable Development (IGF) (2017) *Global trends in artisanal and small-scale mining (ASM): A review of key numbers and issues*, Winnipeg: International Institute for Sustainable Development.

International Labour Organization (1999) *Social and labour issues in small-scale mines: Report for discussion at the tripartite meeting on social and labour issues in small-scale mines*, Geneva: International Labour Organization.

Jayne, T.S, Mather, D. and Mghenyi, E. (2010) 'Principal challenges confronting smallholder agriculture in sub-Saharan Africa', *World Development*, 38(10): 1384–1398. https://doi.org/10.1016/j.worlddev.2010.06.002

Jenkins, K. (2014) 'Women, mining and development: An emerging research agenda', *The Extractive Industries and Society*, 1(2): 329–339. https://doi.org/10.1016/j.exis.2014.08.004

Jónsson, J.B. and Fold, N. (2014) 'Dealing with ambiguity: Policy and practice among artisanal gold miners', in Bryceson, D.F., Fisher, E., Jónsson, J.B. and Mwaipopo, R. (eds.), *Mining and social transformation in Africa: Mineralizing and democratizing trends in artisanal production*, New York: Routledge, pp. 113–129.

Lahiri-Dutt, K. and Macintyre, M. (2006) *Women miners in developing countries: Pit women and others*, Aldershot: Ashgate Publishing Ltd.

Laing, T. and Moonsammy, S. (2021) 'Evaluating the impact of small-scale mining on the achievement of the sustainable development goals in Guyana', *Environmental Science & Policy*, 116: 147–159. https://doi.org/10.1016/j.envsci.2020.11.010

Luning, S. (2014) 'The future of artisanal miners from a large-scale perspective: From valued pathfinders to disposable illegals?' *Futures*, 62: 67–74. https://doi.org/10.1016/j.futures.2014.01.014

Maconachie, R. (2011) 'Re-agrarianizing livelihoods in post-conflict Sierra Leone? Mineral wealth and rural change in artisanal and small-scale mining communities', *Journal of International Development*, 23(8): 1054–1067. https://doi.org/10.1002/jid.1831

Maconachie, R. and Binns, T. (2007) '"Farming miners" or "mining farmers"? Diamond mining and rural development in post-conflict Sierra Leone', *Journal of Rural Studies*, 23(3): 367–380. https://doi.org/10.1016/j.jrurstud.2007.01.003

Maconachie, R. and Conteh, F.M. (2021) 'Artisanal mining policy reforms, informality and challenges to the sustainable development goals in Sierra Leone', *Environmental Science and Policy*, 116: 38–46. https://doi.org/10.1016/j.envsci.2020.10.011

Maconachie, R. and Hilson, G. (2011) 'Artisanal gold mining: A new frontier in post-conflict Sierra Leone?', *Journal of Development Studies*, 47(4): 595–616. https://doi.org/10.1080/00220381003599402

Maconachie, R. and Hilson, G. (2016) 'Re-thinking the child labor "problem" in rural sub-Saharan Africa: The case of Sierra Leone's "half shovels"', *World Development*, 78: 136–147. https://doi.org/10.1016/j.worlddev.2015.10.012

Maconachie, R. and Hilson, G. (2018) 'The war whose bullets you don't see': Diamond digging, resilience and Ebola in Sierra Leone', *Journal of Rural Studies*, 61: 110–122. https://doi.org/10.1016/j.jrurstud.2018.03.009

Meagher, K. and Mustapha, A.R. (1997) 'Not by farming alone', in Bryceson, D.F. and Jamal, V. (eds.), *Farewell to farms: De-agrarianisation and employment in Africa*, Africa Studies Centre Leiden, Research Series 1997/10, Ashgate, Aldershot, pp. 63–84.

Mutemeri, N. and Petterson, F.W. (2002) 'Small-scale mining in South Africa: Past, present and future', *Natural Resources Forum*, 26(4): 286–292. https://doi.org/10.1111/1477-8947.t01-1-00031

Noetstaller, R. (1987) *Small-scale mining: A review of the issues*, Washington, DC: World Bank.

Odumo, B., Nanos, N., Carbonell, G., Torrijos, M., Patel, J.P. and Rodriguez Martin, J.A. (2018) 'Artisanal gold mining in a rural environment: Land degradation in Kenya', *Land Degradation and Development*, 29(10): 3285–3293. https://doi.org/10.1002/ldr.3078

Reardon, T. (1997) 'Using evidence of household income diversification to inform study of the non-farm labor market in Africa', *World Development*, 25(5): 735–747. https://doi.org/10.1016/S0305-750X(96)00137-4

Salo, M., Hiedanpää, J., Karlsson, T., Ávila, L.C., Kotilainen, J., Jounela, P. and García, R.R. (2016) 'Local perspectives on the formalization of artisanal and small-scale mining in the Madre de Dios Gold Fields, Peru', *The Extractive Industries and Society*, 3: 1058–1066. https://doi.org/10.1016/j.exis.2016.10.001

Schoneveld, G., Chacha, M., Njau, M., Jónsson, J.B., Cerutti, P.O. and Weng, X. (2018) *The new face of informality in the Tanzanian mineral economy: Transforming artisanal mining through foreign investment*, London: IIED.

Seay, L.E. (2012) 'What's wrong with Dodd-Frank 1502? Conflict minerals, civilian livelihoods, and the unintended consequences of western advocacy', *Working Paper 284*, Washington, DC: Centre for Global Development.

Siegel, S. and Veiga, M.M. (2009) 'Artisanal and small-scale mining as an extralegal economy: De Soto and the Redefinition of "Formalization"', *Resources Policy*, 34: 51–56. https://doi.org/10.1016/j.resourpol.2008.02.001

Siwale, A. and Siwale, T. (2017) 'Has the promise of formalizing artisanal and small-scale mining (ASM) failed? The case of Zambia', *The Extractive Industries and Society*, 4: 191–201. https://doi.org/10.1016/j.exis.2016.12.008

Smith, N., Ali, S., Bofinger, C. and Collins, N. (2016) 'Human health and safety in artisanal and small-scale mining: An integrated approach to risk mitigation', *Journal of Cleaner Production*, 129: 43–52. https://doi.org/10.1016/j.jclepro.2016.04.124

Spiegel, S.J. (2009) 'Resource policies and small-scale gold mining in Zimbabwe', *Resources Policy*, 34: 39–44. https://doi.org/10.1016/j.resourpol.2008.05.004

Stocking, M. and Murnaghan, N. (2001) *Handbook for the field assessment of land degradation*, London: Earthscan Publications.

Sturman, K., Toledano, P., Akayuli, C.F.A. and Gondwe, M. (2020) 'African mining and the SDGs: From vision to reality', in Ramutsindela, M. and Mickler, D. (eds.), *Africa and the sustainable development goals*, Sustainable Development Goals Series, Cham: Springer, pp. 59–69.

Sustainable Development Goals Centre for Africa (SDGC) (2017) *Africa 2030. How Africa can achieve the sustainable development goals*, Kigali: Sustainable Development Goals Centre for Africa and Sustainable Development Solutions Network.

Tschakert, P. and Singha, K. (2007) 'Contaminated identities: Mercury and marginalization in Ghana's artisanal mining sector', *Geoforum*, 38(6): 1304–1321. https://doi.org/10.1016/j.geoforum.2007.05.002

United Nations Development Programme (UNDP) (2016) *Development Minerals in Africa the Caribbean and the Pacific: Background Study 2016*, ACP-EU Development Minerals Programme, Brussels: UNDP.

United Nations Economic Commission for Africa (UNECA) (2011) *Minerals and Africa's development: The international study group report on Africa's mineral regimes*, Addis Ababa: United Nations Economic Commission for Africa and African Union.

United Nations Environment Programme (UNEP) (2012) *Analysis of formalization approaches in the artisanal and small-scale gold mining sector based on experiences in Ecuador, Mongolia, Peru, Tanzania and Uganda*, Geneva: UNEP, Division of Technology, Industry and Economics (DTIE), Chemical Branch.

USAID (2005) *Minerals and conflict: A toolkit for intervention. Office of Conflict Management and Mitigation*, Washington, DC: USAID.

Veiga, M.M. and Hinton, J.J. (2002) 'Abandoned artisanal gold mines in the Brazilian Amazon: A legacy of mercury pollution', *Natural Resources Forum*, 26: 15–26. https://doi.org/10.1111/1477-8947.00003

Yelpaala, K. and Ali, S.H. (2005) 'Multiple scales of diamond mining in Akwatia, Ghana: Addressing environmental and human development impact', *Resources Policy*, 30(3): 145–155. https://doi.org/10.1016/j.resourpol.2005.08.001

PART IV

Enabling livelihoods

32
CONCEPTUALISING MIGRATION AND LIVELIHOODS
Perspectives from the Global South

Mariama Zaami

Introduction

The significance of migration has continued growing in livelihood strategies, and its potential for the structural transformation of societies is increasingly acknowledged (Skeldon, 2012). In particular, the labour migration within the Global South has had an impact on livelihoods of many households (van der Geest, 2011). Migration has shaped the economic and social livelihoods of the Global South. In 2019, the number of migrants across the world reached 272 million, which is 3.5% of the world's population, an increase of 51 million from 221 million in 2010 (UNDESA, 2019). South-South migration amounted to about 60 million migrants between 1990 to 2005, which increased to 105 million in 2019 (UNDESA, 2019). With these changes in South-South migration, it is imperative to review the literature to focus on these trends and identify who is migrating, what the migration trajectories are, and what the diverse migrant livelihoods entail. It is estimated that about two-fifths of all international migrants have moved from one country of the Global South to another (UNDESA, 2019). In 2019, about 39% of all international migrants who were born in a low-income country resided in another low-income country (UNDESA, 2019). Within this context, South-South migration flows, often overlooked in the migration literature, but which account for up to an estimated two thirds of migration in the Global South, play a key role as a medium to long-term livelihood strategy to diversify income sources and mitigate risks faced by households (Bakewell, 2009).

This chapter focuses on South-South migration and how livelihoods are being transformed by giving examples of different livelihoods in the Global South. The chapter is organised into three sections: the first section focuses on conceptualising migration by defining migration and reviewing theories and migration typologies. The second section concentrates on migrant livelihood diversification, specifically addressing migrant work in the informal sector. The third section discusses the linkages between migration and livelihoods and the importance of migration for individuals, households, and communities in the Global South.

Conceptualising migration in the Global South

Migration scholars have put forward definitions to explain the types of movement taking place (see Box 32.1). These categorisations are usually not distinct. For instance, if an individual enters a country legally and does not leave after their visa and work permit have expired, then the person has moved from being a regular to an irregular migrant (see Hagen-Zanker, 2008).

Box 32.1 Types of migrants in South-South migration (SSM)

Regular vs. irregular migrants: Regular migrants are people who enter a country with legal documents. Irregular migrants enter a country either without documents or with forged documents or enter legally and stay until their visa or work permit has expired.

Internal vs. international migrants: An internal migrant is someone who migrates within their country. International migration is the crossing of borders from one country to another.

Temporary vs. permanent migration: A temporary migrant migrates to a new destination and stays for a short period. A permanent migrant is someone who migrates and stays at destination for at least a year.

Voluntary vs. forced migration: Voluntary migration occurs because of economic reasons, sometimes referred to as labour migrants. Forced migration is the movement of people who are forced to leave their home country or are displaced due to conflict, persecution or environmental reasons (e.g. drought or famine).

Economic migrants migrate because of job opportunities in the destination country.

Refugees have been forced to leave their countries due to conflicts, political instabilities, environment and climatic factors.

Chain migration is explained as one member of a household or community migrating. With time, more individuals from the same community tend to join the migration process to the same destination and sometimes engage in the same occupations.

Circular migration is when a person migrates back and forth between countries of origin and destination. This form of migration can be temporary or permanent.

Environmental migrants are people, whose life situations are impacted negatively, due to reasons beyond their control, and as a result of abrupt or gradual changes in the environment. These people are compelled or opt to leave their usual place of residence on a temporary or permanent basis. This occurs 'in-country' or 'within countries'.

Transit migrants are people who use other countries to transit to migrate to Europe (for example, sub-Saharan migrants using Morocco as transit to Europe).

Seasonal migrants migrate due to weather patterns.

Sources: Castles et al. (2014), Massey et al. (1993), IOM (2008)

Migration behaviours of individuals have been attributed to the following: 1) individuals' and households' attempts to diversify their livelihoods, 2) family unification, 3) political instabilities/conflicts, and 4) environmental factors (see Castles et al., 2014; De Haas and Fokkema, 2010). An individual might migrate in search of alternative or additional sources of income and/or to join their family in the receiving country or region. Others might engage in the migration process because of the various social networks and ties within their migration trajectories.

Major theories of migration

This section focuses on theorising migration and livelihoods as separate concepts and proceeds to show how these different theories together provide a holistic understanding of the migration process. The drivers of migration in the Global South have been attributed to economic opportunities, social networks, and access to information. People move because of better conditions and differences in wages between sending and receiving countries or regions (Awumbila, 2017; Castles et al., 2014; Skeldon, 2012). Migration theories have been grouped into two main paradigms: the functionalist and historical-structural theories (Skeldon, 2012; Lee, 1966). *Functionalist theory* posits that society is comprised of individuals and actors geared towards achieving equilibrium. The theory highlights migration as a positive phenomenon that aims to serve most people's interests and contributes to the greater good within and between sending and receiving countries (Castles et al., 2014). Lee (1966) argues that people migrate by weighing the positive and negative factors in both origin and destination areas. The migrant needs to weigh the intervening obstacles such as distance, physical barriers, immigration laws, and personal factors. These are termed the push and pull factors. The push and pull concept argues that unfavourable economic, demographic, and environmental factors tend to push people out of their places of origin and pull them to destination places (Lee, 1966; Sjaastad, 1962). Pull factors include a demand for labour, land availability, job opportunities, and freedom from political oppression that exist in destination countries and regions (Lee, 1966).

Historical-structural theorists argue that migration is more concerned with accessing cheap labour from less developed countries, which works against the sending-countries because it deprives them of their labour and skills (Skeldon, 2012). Under this theoretical umbrella, it is believed that economic and political power is unevenly distributed among and between more developed and less developed countries and regions. Migration has introduced multinational companies into these periphery regions – some of the rural areas are used to supply cheap labour in multinational companies, depriving rural development but enabling urban growth. The two paradigms were critiqued for assuming that: 1) people are rational actors who tend to maximise income based on costs and benefits between sending and receiving countries (De Haas, 2011; Massey, 1990), 2) migrants have knowledge about the wage differentials in destination countries, and 3) labour markets are in perfect condition and accessible to everyone, specifically the poor (De Haas, 2011; Massey et al., 1993). These assumptions are unrealistic in real life because conditions such as poverty and other intervening constraints (e.g. visa acquisition procedures) prevail in sending and receiving countries, potentially limiting an individual's migration options for destinations and routes. Again, it can be observed that the push-pull and neoclassical theories have not recognised human agency, the individual's ability to make personal choices and changes to the structural conditions. Ironically, constraints such as limited access to money, social networks, and information, not only commonly affect individuals when migrating to a new destination but can also act as catalyst for migration (Massey et al., 1993; Stark, 1991).

New economics and household approaches

The 'new economics' and 'household' approaches were introduced to address gaps in earlier migration theories (Massey et al., 1993; Stark, 1991). Stark (1991) explains that migration is not merely an individual decision and that migration decisions are sometimes made by families and households. New Economics Labor Migration (NELM) sees migration not as an individual decision to maximise profit but as a form of risk-sharing by families or households (Stark and Bloom, 1985). In view of this, household approaches are useful in explaining migration patterns in less developed countries where people tend to experience social security and income risk issues during droughts and other environmental conditions. While NELM addressed a gap in earlier theories, it failed to recognise intra-household dynamics such as inequalities associated with gender, age, and generation (De Haas and Fokkema, 2010). Intra-household dynamics, for instance, also determine who the family chooses as the person who should migrate. These decisions are influenced by a myriad of factors shaped by the conditions in the country of origin. Thus, migration is not tied to a rational individual but is a household decision that responds to diversifying household livelihoods (see Castles et al., 2014; De Haas and Fokkema, 2010; Stark, 1991).

Further crucial elements in the theorising of migration are the ties, networks, and identities formed between the sending and receiving countries through flows of information, ideas, money, goods and in-kind services. These are termed the network, transnationalism and migration theories (Castles et al., 2014). There is a form of unification between these theories as they give priority to the individual, who has the agency to create social, economic and cultural structures and act as an agent for the continuation of the migration process. It is argued that when migrants combine these structures, they can collectively negotiate and fight structural constraints such as poverty, social exclusion and other governmental restrictions. As such, migration network theory emerged to explain how migrants create and maintain social ties with other migrants, friends, and family in the sending and receiving countries (Castles et al., 2014; Vertovec, 2009). These ties usually lead to the creation of social networks. These social networks tend to act as channels for migration (Vertovec, 2009). Research has shown that migrants use social networks developed before and during their migration in their journey, and several studies have shown that the more networks a migrant has before migrating, the more it helps find jobs, housing, and other necessities (Vertovec, 2013; Banerjee, 1983).

Gender, migration and livelihoods: who is migrating and where?

In the Global South, the migratory pattern is incredibly heterogenous. This section focuses on gender, migration, and the diversification of livelihoods.

With regards to the decision to migrate, De Haas and Fokkema's (2010) study on migration decision-making and intra-household conflicts found that migration and return migration is not an individual decision but a household decision, and, within the household, there are power dynamics at play to decide who migrates and who stays at home. Using Morocco as an example of a patriarchal society, women and children do not have an equal opportunity in the migration decision making (De Haas and Fokkema, 2010). As such, women and children in the Moroccan patriarchal family system do not have any agency to migrate from the household. However, some studies have also found that women tend to dominate trends in labour migration. Compared to the preceding situation, there is a dominance of male migration in the Global South to work in plantations. This does not only diversify the household livelihoods, but it also serves as a form of economic empowerment (Awumbila et al., 2019; Awumbila and Ardayfio-Schandorf, 2008; Deshingkar, 2019). Women labour migrants can contribute to household incomes, thereby increasing their negotiating power, access to employment opportunities and education, and can occupy respected positions in society (see De Haas and Fokkema, 2010). Awumbila and Ardayfio-Schandorf's (2008) study of women migrants from Northern Ghana found that the household decision-making process regarding migration was dominated by men. During the pre-colonial and colonial era, migration was dominated by men. Women and children only migrated to join the women's husbands. However, since the 2000s, there has been a change in the migration trajectories, as now girls migrate independently as young as age 8 to work in the informal economy in urban markets within the capital city (Accra) (Awumbila and Ardayfio-Schandorf, 2008) (see Box 32.2).

Migrant livelihoods, work and vulnerabilities

Labour migration is the predominant form of migration in the Global South pursued by migrants to diversify their livelihoods (Awumbila et al., 2019; Deshingkar et al., 2019; Nawyn et al., 2016; Picherit, 2019). Migration is not always a smooth transition from countries of origin to destination areas. Migration scholars have explored the vulnerabilities, migration brokerage, and the precarious nature of migrants and undocumented migrants' jobs in the Global South (Buckley, 2014; Deshingkar et al., 2019). Deshingkar et al. (2019) explored the brokering practices and agency of migrants focusing on Bangladeshi migrants, showing how brokers and other actors in the migration process organised migrants by assigning them fake identities to enable them to circumvent border controls. This process enables these migrants to access jobs through the falsification of legal documents concerning their skills and other qualities. Deshingkar et al. (2019) explained that due to these migrants' documentation, they end up in precarious and exploitative conditions in Qatar. In Ghana, Awumbila et al. (2019) identified urban brokers and how negotiations are made in engaging rural girls and women in domestic work. The authors identified that brokers assist these young women in gaining access to domestic jobs. Hence, the brokers can negotiate with potential employers on wages and working conditions. Based on the above examples in Ghana and Bangladesh, the bargaining powers and subjectivities are in the hands of the brokers who exploit these migrants for their personal gain.

On a similar note, Arab States host 19% of the world's domestic workers and these are largely women domestic workers (ILO, 2015). The nature of jobs executed by domestic workers are precarious, and includes cleaning homes, caring for family members and elderly populations in private households (Awumbila et al., 2019; Deshingkar et al., 2019; ILO, 2015). Most of these caregivers are temporary workers and are exploited by being paid low wages, being made to pay back illegal recruitment fees, can often have unpaid wages, made to work long hours, and their employer may interfere their privacy (see Awumbila et al., 2019; Deshingkar et al., 2019). Domestic workers in the Arab States are regulated by a sponsorship system known as *kafala*. The *kafala* system ties migrant immigration and legal system to the employer (*kafeel*) until the contract ends. Within the period of the contract, the migrant worker is not able to leave the country, resign from the job, ask for a transfer to another job, and, to some extent, cannot leave the country without obtaining consent from the employer who signed the contract with the migrant (Kagan, 2017).

Regarding South-South migration, there has been an influx of Chinese migrants in Africa (Lu and Van Staden, 2013; Marfaing and Thiel, 2013). Chinese migrants tend to engage in the wholesale and retail of Chinese goods, and open restaurants and clinics. For example, in Cape Verde, Chinese migrants' main labour activity is not to join any established wage-earning labour, but rather to set up their businesses such as the *baihuo shops*, meaning general merchandise (Haugen and Carling, 2005). The Chinese worker, in applying for a visa to Cape Verde, would need a contract with details of monthly salary, accommodation and healthcare. Findings from the study indicate that the written contract used in applying for the visa is a formality. Oral agreements are also made before the young adult embarks on their journey to Cape Verde. While in Cape Verde, the working conditions sometimes change during the employment period. These young adults must work to repay the cost of their tickets. Haugen and Carling (2005) also observed that the workers are recruited among relatives or from strong social ties. Although these Chinese businesses are well organised, these young adults experience forms of abuse and vulnerabilities (Haugen and Carling, 2005).

The Chinese businesses mostly employ Chinese migrants. However, they also employ Cape Verdeans, who are tasked with menial jobs, such as assisting customers and acting as security guards to prevent shoplifting. They are exploited by working more hours for less pay than their Chinese counterparts. Due to trust issues with the Cape Verdeans, the Chinese tend to grow their businesses by employing family members and people with close family ties from China resulting in the growth of Chinese businesses in Cape Verde and across Africa. Considering the above, chain migration is ongoing in Cape Verde.

Box 32.2 illustrates why young women migrate from the northern part of Ghana and the vulnerabilities they encounter in the destination areas.

Box 32.2 Vulnerability, gender and livelihoods

Awumbila and Ardayfio-Schandorf's (2008) study on gender and vulnerabilities examined females and their work trajectories in urban markets in Accra, Ghana. They identified that female migrants from Northern Ghana were engaged in migration as a livelihood strategy to alleviate poverty. The authors identified some intersectionality between gender, poverty and vulnerabilities. They highlighted the coping strategies these female migrants used to navigate spaces and places they now call home. The migration trends were seasonal. Most of the women migrated during the prolonged dry season and returned during the rainy season to engage in farming activities. In this case, migration was used as an alternative (additional) source of livelihood for households during the dry season. Migration has become a vital livelihood strategy and an engine for poverty reduction and growth in many developing countries.

Their study also highlighted the vulnerabilities of female migrants. Female migrants were exposed to sexually transmitted infections (STIs), rape and other health risks. Despite the vulnerabilities encountered by the female migrants, migration is a response to poverty that can improve women's position in patriarchal societies.

Source: Awumbila and Ardayfio-Schandorf (2008)

Diversifying livelihoods through remittances

In this final section, I discuss the nature of remittances sent by migrants to households with the intention of diversifying incomes and addressing other associated risks. In migration, remittances involve person-to-person flows of finances and other resources from migrants to their family and friends. These remittances include money, goods in-kind, and foodstuffs sent through money transfers or hand-carried to migrants' households (Ratha and Shaw, 2007). Remittances are sent to households to diversify their livelihoods and/or collective remittances are sent to the community through hometown associations for developmental projects (see Ratha and Shaw, 2007).

Globally, and for most countries of the Global South, migrants' remittances constitute an essential income source (Plaza et al., 2011). The total remittance flows to developing countries is three times more than Official Development Assistance (ODA) (Ratha, 2016). In 2018, the global remittance flow totalled about $689 billion, of which $529 billion (77%) went to low- and middle-income countries (LMICs) (ibid). Remittances sent by migrants to help in poverty reduction in LMICs reached $551 billion in 2019 and were projected to reach $597 billion by 2021, of which $54 billion was expected to be sent to sub-Saharan Africa (ibid).

Most of the countries of the Global South manage out-migration as a strategy for economic development to alleviate poverty in migrant households and secure foreign direct investments through remittances (Adamson and Tsourapas, 2019). Remittances sent from South-South migration are not as substantial as the remittances received from South-North migration. However, Ratha and Shaw (2007) argue that even the small increases in migrant incomes are significant enough to improve poor households' livelihoods. Migrants send remittances home to take care of the family's basic needs, paying for medical bills and school fees and investing in migrants' communities (Awumbilla, 2017). Remittances serve as safety nets for poor households. Some scholars have argued that remittances only reach the selective few in each society (Gindling,

2009; Lee, 2010). Therefore, those who do not have international or internal migrants in their households are cut off from the remittances (see Lee, 2010).

Although remittances are important sources of livelihood diversification, Lee's (2010) ethnographic study of Nicaraguan migration to Costa Rica identifies migrants' subjective realities in sending remittances. It reveals that not all migrants become successful and can send remittances. Within the context of migration from the Nicaragua countryside, families struggled to feed their households as they await remittances from the migrant. This observation contrasted with the migration literature that identified migrants sending remittances to families left behind. Lee's (2010) study reflects the characteristics of NELM, where the decision to migrate is viewed as being collectively taken at the household level. As such, the migrant is expected to be successful in destination countries to enable him/her to send remittances back home. As shown in the Nicaraguan case, rural people do not always benefit from the migration process nor does it mitigate the economic risks of individual households. Ratha and Shaw's (2007) study identified some form of bureaucracy involved in sending money from one country of the Global South to another. In countries such as Bangladesh, India, Lesotho and Morocco, before a migrant can remit through formal channels, authorisation is needed from the central bank. These restrictions have discouraged South-South migrants from remitting and so they have resorted to in-person remittances, which are not recorded, thereby affecting information on South-South remittance flows.

Conclusion

This chapter outlined the significance of migration as a livelihood strategy in the Global South, specifically, South-South Migration. It highlighted migrant work in the informal sector and the precarious nature of the work. It is observed that migrants engage in different jobs to diversify their livelihoods in the sending regions while engaging in development in the destination countries. The type of jobs migrants engaged in play a key role in livelihood diversification through the remittances sent to poor households. Migration has become a livelihood strategy, and remittances sent are safety nets for the migrant households and community.

Key Points

- South-South migration accounts for nearly half of all international migration; 74 million migrants from one country of the Global South reside in other countries of the Global South.
- Migrants are not passive actors but rather active beings who make decisions with their families or households.
- Reasons for migrating are complex, multifaceted, and intersectional. Individuals and households migrate to diversify their livelihoods, for family unification and in response to political instabilities/conflicts, environmental factors (flooding and droughts).
- There is a dynamic and complex relationship between migration and livelihoods, extending to migrants' vulnerabilities in destination countries.
- South-South migration has the potential to reduce poverty and inequality and create opportunities for decent work.

Recommended reading

Bryceson, D.F. (1999) 'African rural labour, income diversification and livelihood approaches: a long-term development perspective', *Review of African Political Economy*, 26(80): 171–189. https://doi.org/10.1080/03056249908704377

De Haas, H. and Fokkema, T. (2010) 'Intra-household conflicts in migration decision making: Return and pendulum migration in Morocco', *Population and Development Review*, 36(3): 541–561. https://doi.org/10.1111/j.1728-4457.2010.00345.x

Ellis, F. (1998) 'Household strategies and rural livelihood diversification', *Journal of Development Studies*, 35(1): 1–38. https://doi.org/10.1080/00220389808422553

Haugen, H.Ø. and Carling, J. (2005) 'On the edge of the Chinese diaspora: The surge of Baihuo business in an African city', *Ethnic and Racial Studies*, 28(4): 639–662. https://doi.org/10.1080/01419870500092597

Massey, D.S., Arango, J., Hugo, G., Kouaouci, A., Pellegrino, A. and Taylor, J.E. (1993) 'Theories of international migration: A review and appraisal', *Population and Development Review*, 19(3): 431–466. https://doi.org/10.2307/2938462

Ratha, D. and Shaw, W. (2007) 'South-South migration and remittances', *World Bank Working Paper No. 102*, Washington, DC: World Bank.

References

Adamson, F.B. and Tsourapas, G. (2019) 'Migration diplomacy in world politics', *International Studies Perspectives*, 20(2): 113–128. https://doi.org/10.1093/isp/eky015.

Awumbila, M. (2017) *Drivers of migration and urbanization in Africa: Key trends and issues*, Background paper prepared for the United Nations expert group meeting on sustainable cities, human mobility and international migration, 7–8 September 2017, New York: Population Division, Department of Economic and Social Affairs, United Nations Secretariat.

Awumbila, M. and Ardayfio-Schandorf, E. (2008) 'Gendered poverty, migration and livelihood strategies of female porters in Accra, Ghana', *Norsk Geografisk Tidsskrift-Norwegian Journal of Geography*, 62(3): 171–179. https://doi.org/10.1080/00291950802335772

Awumbila, M., Deshingkar, P., Kandilige, L., Teye, J.K. and Setrana, M. (2019) 'Please, thank you and sorry: Brokering migration and constructing identities for domestic work in Ghana', *Journal of Ethnic and Migration Studies*, 45(14): 2655–2671. https://doi.org/10.1080/1369183X.2018.1528097

Bakewell, O. (2009) 'South-south migration and human development', *Working Paper 15*, Oxford: International Migration Institute, University of Oxford.

Banerjee, B. (1983) 'The role of the informal sector in the migration process: A test of probabilistic migration models and labour market segmentation for India', *Oxford Economic Papers*, 35(3): 399–422. https://doi.org/10.1093/oxfordjournals.oep.a041604

Buckley, M. (2014) 'On the work of urbanization: Migration, construction labor, and the commodity moment', *Annals of the Association of American Geographers*, 104(2): 338–347. https://doi.org/10.1080/00045608.2013.858572

Castles, S., De Haas, H. and Miller, M.J. (2014) *The age of migration*, 5th ed., New York: Guilford Press.

De Haas, H. (2011) 'The determinants of international migration: Conceptualising policy, origin and destination effects', *Working Paper 32*, Oxford: International Migration Institute, University of Oxford.

De Haas, H. and Fokkema, T. (2010) 'Intra-household conflicts in migration decisionmaking: Return and pendulum migration in Morocco', *Population and Development Review*, 36(3): 541–561. https://doi.org/10.1111/j.1728-4457.2010.00345.x

Deshingkar, P. (2019) 'The making and unmaking of precarious, ideal subjects: Migration brokerage in the Global South', *Journal of Ethnic and Migration Studies*, 14(1): 2638–2654. https://doi.org/10.1080/1369183X.2018.1528094

Deshingkar, P., Abrar, C.R., Sultana, M.T., Haque, K.N.H. and Reza, M.S. (2019) 'Producing ideal Bangladeshi migrants for precarious construction work in Qatar', *Journal of Ethnic and Migration Studies*, 45(14): 2723–2738. https://doi.org/10.1080/1369183X.2018.1528088

Gindling, T.H. (2009) 'South – South migration: The impact of Nicaraguan immigrants on earnings, inequality and poverty in Costa Rica', *World Development*, 37(1): 116–126. https://doi.org/10.1016/j.worlddev.2008.01.013

Hagen-Zanker, J. (2008) 'Why do people migrate? A review of the theoretical literature', *Maastricht Graduate School of Governance Working Paper No. 2008/WP002*, Maastricht: Maastricht University. https://doi.org/10.2139/ssrn.1105657

International Labour Organization (ILO) (2015) *Employers' perspectives towards domestic workers in Kuwait: A qualitative study on attitudes, working conditions and the employment relationship*, Regional Office for Arab States, Beirut: International Labour Organization.

International Organization for Migration (IOM) (2008) *Discussion note: Migration and the environment (MC/INF/288–1 November 2007–ninety fourth session)*, Geneva: International Organization for Migration.

Kagan, S. (2017) *Domestic workers and employers in the Arab states: Promising practices and innovative models for a productive working relationship*, ILO White Paper, Beirut, Lebanon: ILO Regional Office for Arab States.

Lee, E.S. (1966) 'A theory of migration', *Demography*, 3(1): 47–57.

Lee, S.E. (2010) 'Development or despair? The intentions and realities of South-South migration', *Encuentro* (87): 6–25. https://doi.org/10.5377/encuentro.v0i87.245.

Lu, J. and Van Staden, C. (2013) 'Lonely nights online: How does social networking channel Chinese migration and business to Africa?', *African East-Asian Affairs*, 1: 94–116. https://doi.org/10.7552/0-1-5

Marfaing, L. and Thiel, A. (2013) 'The impact of Chinese business on market entry in Ghana and Senegal', *Africa*, 83(4): 646–669.

Massey, D.S. (1990) 'The social and economic origins of immigration', *The Annals of the American Academy of Political and Social Science*, 510(1): 60–72. https://doi.org/10.1177/0002716290510001005

Massey, D.S., Arango, J., Hugo, G., Kouaouci, A., Pellegrino, A. and Taylor, J.E. (1993) 'Theories of international migration: A review and appraisal', *Population and Development Review*, 19(3): 431–466. https://doi.org/10.2307/2938462

Nawyn, S.J., Kavakli, N.B., Demirci-Yılmaz, T. and Pantic Oflazoğlu, V. (2016) 'Human trafficking and migration management in the Global South', *International Journal of Sociology*, 46(3): 189–204. https://doi.org/10.1080/00207659.2016.1197724

Picherit, D. (2019) 'Labour migration brokerage and Dalit politics in Andhra Pradesh: A Dalit fabric of labour circulation', *Journal of Ethnic and Migration Studies*, 45(14): 2706–2722. https://doi.org/10.1080/1369183X.2018.1528101

Plaza, S., Navarrete, M. and Ratha, D. (2011) *Migration and remittances household surveys in sub-Saharan Africa: Methodological aspects and main findings*, Washington, DC: World Bank.

Ratha, D. (2016) *Migration and remittances Factbook 2016*, Washington, DC: World Bank.

Ratha, D. and Shaw, W. (2007) 'South-South migration and remittances', *World Bank Working Paper No. 102*, Washington, DC: World Bank.

Sjaastad, L.A. (1962) 'The costs and returns of human migration', *Journal of Political Economy*, 70(5): 80–93. https://doi.org/10.1086/258726.

Skeldon, R. (2012) 'Going round in circles: Circular migration, poverty alleviation and marginality', *International Migration*, 50(3): 43–60. https://doi.org/ 10.1111/j.1468–2435.2012.00751.x

Stark, O. (1991) *The migration of labour*, Cambridge: Harvard University Press.

Stark, O. and Bloom, D.E. (1985) 'The new economics of labor migration', *The American Economic Review*, 75(2): 173–178.

United Nations Department of Economic and Social Affairs (UNDESA) (2019) *International migration 2019 report*, New York: The United Nations.

Van der Geest, K. (2011) 'North-South migration in Ghana: What role for the environment', *International Migration*, 49: e69–e94. https://doi.org/ 10.1111/j.1468–2435.2010.00645.x

Vertovec, S. (2009) *Transnationalism*, Milton Park: Routledge.

Vertovec, S. (2013) 'Circular migration', in Ness, E. (ed.), *The Encyclopedia of Global Human Migration*, Hoboken, NJ: Wiley-Blackwell.

33
INTERNATIONAL MIGRATION AND EXPERIENCES OF INDIAN WOMEN MIGRANTS

A critical analysis of the Kafala system

Jyoti Bania

Introduction

Human migration is not a recent phenomenon, but one that stretches back to the earliest periods of human history (Amin, 1998; Boghean, 2016). International migration, both voluntary and forced, has been a continuous and prominent characteristic of humanity and demographic change throughout the ages (Boghean, 2016; Chamie, 2020) and has created ample opportunities for many people, although it also comes with many challenges (Boghean, 2016). In a globalised world, emigration and immigration continue to provide states, societies and migrants with many opportunities. At the same time, migration has emerged in the last few years as a critical political and policy challenge in matters such as integration, displacement, safe migration and border management.

Millions of men, women and children have crossed international borders to settle in another country, and many millions more would migrate if they could (Esipova et al., 2018; Chamie, 2020). In the recent past, significant changes in the magnitude, sources, causes and consequences of international migration has transformed it into a major global issue, with some considering it to be the most defining issue of the 21st century (Betts, 2015; Chamie, 2020).

With the unprecedented rapid growth of the world's population following the Second World War, the number of international migrants grew rapidly during the second half of the 20th century, from 77 million (2.1% of the world's population) in 1960 to 174 million (2.8% of the world's population) by the close of the 20th century (Chamie, 2020). According to the UNDESA (2019) estimate, there were around 272 million international migrants in the world in 2019, which is 3.5% of the world's population (IOM, 2019). If the current proportion of international migrants were to remain constant at its current level of 3.5%, the projected number of international migrants would reach 343 million by the mid-21st century (Chamie, 2020).

More than half of all international migrants (141 million) live in Europe and North America. The United States of America (USA) has continued to be the top destination country with 50.7 million international migrants in 2019 (IOM, 2019). Germany and Saudi Arabia have the second and third largest number of international migrants worldwide at around 13 million each, followed by the Russian Federation with 12 million, and the United Kingdom with 10 million in 2019 (Chamie, 2020).

More than 40% of all international migrants worldwide in 2019 (112 million) hail from countries in Asia, primarily India, China, Bangladesh, and Pakistan. In particular, India continues to be the largest country of origin of international migrants with 17.5 million people living abroad, followed by Mexico with 11.8 million, and China with 10.70 million (IOM, 2019).

This chapter focuses on Indian workers who migrated to the Gulf states and account for over 9 million out of the total international migrant workers worldwide (Wadhawan, 2018). The central arguments of the chapter are that international migration is becoming increasingly feminised because more women are leaving their countries of origin and that Indian women migrants, especially unskilled workers in search of livelihoods, are at a greater risk of exploitation, discrimination and abuse than the other migrants and skilled workers in general under the Kafala migrant labour management system in the Gulf countries. This chapter argues that policy-makers, both in origin and receiving countries, should develop policies which centre round addressing human rights violations and providing adequate safeguards and protection for the most vulnerable sections of immigrants.

This chapter is divided into three sections. The first section, 'Migration, rights and livelihoods', connects the notion of migration with human rights and livelihoods. It stresses the lack of rights of women migrants and how that affects their livelihoods. Rights are the most violated when it comes to unskilled Indian women migrants from poor socio-economic backgrounds in search of livelihoods. The second section is the 'Analysis of migrant rights under the Kafala system in the context of legal framework of human rights'. It discusses the major migrant rights conventions and laws and their effectiveness in the context of international migration. It further discusses the migrant workers' rights under the Kafala system while providing an overview of the Kafala system and its reforms. The third section highlights the 'Experiences of abuse and slavery like situations' faced by migrant workers in the Gulf countries, while outlining the debates surrounding migration and trafficking. It discusses the concept of 'trafficking' and its definition within the larger legal framework of international law and how it can be applied to the Kafala system. The conclusion argues that the Kafala system cannot be completely dismantled because of socio-political and economic interests of the Gulf countries. It also argues that the reformation of the Kafala system needs to be treated as a global problem dealt equally with other immigration policies of developed nations such as the USA, UK and Europe. Lastly, it suggests strategies to be taken by governments of both migrant sending and receiving countries to ensure the rights and safety of the migrant workers.

Migration, rights and livelihoods

The effects of migration directly impact the policy decisions of both migrant origin and destination countries. Therefore, exploring the effects of migration on livelihoods is key to understanding and proposing policy debates surrounding migration because it helps to decide the extent to which migration can occur and to generate positive economic outcomes.

Migration has been analysed and approached both in relation to livelihoods and human rights. The approaches of livelihoods and human rights are not mutually exclusive but rather are mutually beneficial. The human rights approach treats migrants as individuals, having rights, including rights to livelihood, which uplift in them a sense of belonging, citizenship, and active participation in society. It emphasises the interests of the migrants rather than the interests of the host nations.

Migration as a phenomenon is an important aspect of livelihood strategies because it positively impacts the labour market of any country by reducing poverty and unemployment rates. Migration affects more people economically than those who migrate, such as the migrants' families and their home countries. The remittances sent by the migrant workers to their home country play an important role in the migrant families' welfare and sustainable livelihood (Koser, 2016). Migrants are sending back home more money than ever before, and the United Nations acknowledged that the remittances sent back home by migrants are one of the major contributors to increased development and reduced poverty of migrant origin countries (IOM, 2017). In many low- and middle-income countries, such as India, Philippines, Nepal and Sri Lanka, this remittance is a higher source of income than the official aid provided by richer countries (IOM, 2017).

Feminisation of migration

In India, the first wave of migration occurred in the early 1950s and continued into the mid-1970s. During this phase, the majority of migration was to developed countries like the United Kingdom, United States, Canada, Australia and many of the western European countries (Naidu, 1991). The second wave began in the early 1980s when there was large-scale migration to Middle East countries driven by the hike in oil prices (Naidu, 1991). It is in this phase of international migration, that the 'feminisation of migration' became a dominant aspect.

The 'feminisation of migration' started in the 1980s and accelerated during the 1990s. Today, women are an important component of international migration, as they account for nearly half of all international migrants (Dhar, 2012). Traditionally, the male bias in research and policy is undoubtedly based on the assumption that most women migrate only for reasons like family reunification (Gülçür and Ilkkaracan, 2002; Kofman et al., 2000). This is epitomised in the International Labour Organization's (ILO) Migrant Workers (Supplementary Provisions) Recommendation, 1949 (No. 86), where it refers to the migrant worker's family as being 'his wife and minor children' (Dhar, 2012). The result of this assumption is that women are not considered as autonomous agents capable of making independent decisions based on their own volition to migrate in search of better economic and social opportunities. Feminisation of the labour market, demand for and supply of women's labour, and changing views on women's mobility are some of the factors that motivated women to seek employment overseas. Therefore, following the feminisation of migration, women's visibility has become a common feature in international migration (Kofman et al., 2000; Gülçür and Ilkkaracan, 2002) and at present, every second international migrant is a woman (Zlotnik, 1990).

Several factors influence the feminisation of migration, including: a) change in gender roles in both countries of origin and destination, b) a growing demand for women in the destination countries in the informal service sectors e.g. domestic and sex related work, c) dire economic needs that largely impact women, d) freedom from oppressive social environments, and e) transcending traditional gender roles in search for creating better lives (Gülçür and Ilkkaracan, 2002; Kofman et al., 2000; Phizacklea, 1998).

Despite feminisation of migration, women's emigration is restricted from many South Asian countries, such as India, Bangladesh, Pakistan and Nepal, in the name of 'women's safety'. Conversely, in these countries, men's labour migration is highly encouraged by governments as an economic development strategy (Bélanger and Rahman, 2013). When states officially construct women's migration as being negative, harmful or dangerous, the patriarchal ideology is reproduced and creates significant hurdles for aspiring women migrants and returnees (Bélanger and Rahman, 2013). In such circumstances, women migrants face multiple structural, cultural, religious and political obstructions.

To safeguard and regulate the emigration of unskilled Indian women migrants to the Gulf, in 2016 the Government of India introduced the minimum age criterion of 30 years for all migrant women (except nurses) workers on Emigration Check Required (ECR) passports to ECR countries (MEA, n.d.; Wadhawan, 2018). The ECR process is regulated through an 'e-migrate system' for the protection of less educated, 'blue-collar' workers. The 'e-migrate system' is integrated within the Passport Seva Project of the Ministry of External Affairs and the Indian missions in the 18 ECR countries along with the foreign employers and the registered recruiting agents. Thus, all stakeholders, employees, employers and the missions of both the sending and receiving countries are brought together on a single platform called the 'e-migrate' system to simplify the emigration process (MEA, n.d).

Due to the restriction on women's mobility and a lack of effective policies and laws for women migration, women must seek other (illegal) means to migrate to other countries. Therefore, illegal migration is common amongst unskilled women migrants. Overall, attaining legal labour migration is particularly difficult for unskilled women workers from South Asian countries. Therefore, the movement of women from South Asian countries like India, Bangladesh and Nepal to more economically rich nations in search of livelihoods, such as better employment opportunities, increased pay and better lifestyles, leads to significant human rights violations.

An overview of rights: migrants, citizenship and immigration policy

To understand the significance of the rights of migrants, we have to define who can be considered a migrant legally (refer to Box 33.1) and what rights they are entitled to.

Box 33.1 Meaning of the term 'migrant'

Although not defined by international law, a 'migrant' in layman's terms is typically 'a person who moves away from his or her place of usual residence. . . . The term includes a number of well-defined legal categories of people, such as migrant workers; persons whose particular types of movements are legally defined, such as smuggled migrants; as well as those whose status or means of movement are not specifically defined under international law, such as international students' (IOM, 2019: 132).

The legal rights that migrant workers enjoy are significantly influenced by their impacts on the national interests of the host country. Conceptualising migrant rights as a subset of citizenship rights would help to link the interests of nation-states to migrants' rights (Ruhs, 2012). Citizenship rights include economic, cultural and gender rights (Castles and Davidson, 2000; Ruhs, 2012). Unlike human rights, the main characteristic of citizenship rights is that they derive from a relationship with a particular state rather than from universal notions of 'personhood' or 'human dignity' (Ruhs, 2012). Without 'rights', and more specifically 'human rights', the migrant loses access to any kind of legal protection or safeguard from exploitation, torture, and abuse. Although there are international laws that treat 'migrants' rights on par with 'citizen rights', many countries discriminate between 'migrant rights' and the rights of their own citizens (Ruhs, 2012).

In practice, nation-states use their immigration, integration and naturalisation policies to limit and strictly regulate migrants' access to citizenship status and specific citizenship rights. The legal rights of the migrants depend on their immigration and residence status in the host country. The immigration policies of most high-income countries create a number of different types of residence status for migrants, each of which is associated with various rights and restrictions (Ruhs, 2012). The skill level of the migrants plays an important role in determining what rights and restrictions a migrant may or may not have. The immigration policy allows for nation-states to make three fundamental decisions: 1) how to regulate the number of migrants to be admitted, 2) the criteria for the selection of migrants, and 3) what rights should be granted to migrants upon admission. Although some rights generate net costs for the receiving state or country, it does not imply moral justification for condoning or even advocating restrictions on these rights (Ruhs, 2012).

From the skills perspective, over 90% of Indian women migrant workers in the Gulf region are employed in the service or informal sectors and are regarded as semi-skilled or low-skilled workers (Wadhawan, 2018). These sectors are not significantly covered under basic human and labour rights laws, making these unskilled migrant workers highly vulnerable to exploitation and abuse in the Gulf countries.

Analysis of migrant rights under the Kafala system in the context of the legal framework of human rights

Human rights and international migration

Irregular immigration poses a major global challenge as millions of men, women and children have fewer chances of immigrating legally and therefore, risk their lives by migrating illegally (refer to Box 33.2 for definition). Irregular migration is often closely associated with human trafficking and smuggling (Chamie, 2015 cited in Chamie, 2020: 234). Across all the regions of the world recruitment through fraudulent means and criminal agencies are increasingly engaged in smuggling and human trafficking due to the demands for cheap labour force and sexual exploitation, as well as the process having relatively low risks and large profits (Chamie, 2015 cited in Chamie, 2020: 234).

Box 33.2 Definition of asylum seeker, irregular migrant and refugee

Asylum seeker: 'A person who seeks safety from persecution or serious harm in a country other than his or her own and awaits a decision on the application for refugee status under relevant international and national instruments. In the case of a negative decision, the person must leave the country and may be expelled, as may any non-national in an irregular or unlawful situation, unless permission to stay is provided on humanitarian or other related grounds' (IOM, 2011: 12).

Irregular migrant: 'A person who, owing to unauthorised entry, breach of a condition of entry, or the expiry of his or her visa, lacks legal status in a transit or host country' (IOM, 2011: 54).

Refugee: 'A person who, owing to a well-founded fear of persecution for reasons of race, religion, nationality, membership of a particular social group, or political opinions, is outside the country of their nationality and is unable or, owing to such fear, is unwilling to avail themselves of the protection of that country (Art. 1(A) (2), Convention relating to the Status of Refugees, 80 International Migration Law Art. 1A(2), 1951 as modified by the 1967 Protocol)' (IOM, 2011: 79).

Governments across the world have adopted different policies to limit international migration, minimising refugee flows, and rejecting asylum seekers (refer to Box 33.2 for definitions), repatriating unlawful residents and redefining or denying citizenship to certain groups (Chamie, 2020). At the international level, two global compacts such as Global Compact for Safe, Orderly and Regular Migration concerning international migration and the Global Compact on Refugees to deal with refugee crisis were endorsed by a large majority of the UN Member States in the year 2018 (Chamie, 2020).

International migrant rights conventions

Ruhs (2012) maintains that 'Migrant rights are human rights', which is a common argument made by migrant rights advocates around the world. In 1990, the United Nations (UN) adopted a new human rights treaty that specifically deals with the rights of migrant workers and their families. However, it is worrisome to see that most countries do not consider migrant rights as human rights that should be codified and ratified in domestic law (Ruhs, 2012).

The key features and principles of human rights include universality, i.e. human rights apply everywhere and to everyone (including migrants); indivisibility, i.e. there is no hierarchy of rights and certain types of rights cannot be separated from others; inalienability, i.e. human rights cannot be denied to any human being, nor can they be given up voluntarily; and equality and non-discrimination, i.e. all individuals are equal as human beings. Human rights derive from a 'common humanity' and the 'inherent dignity of each human person' rather than from citizenship of a particular country (Ruhs, 2012).

The three most significant international legal instruments that address the rights of migrant workers are:

1. UN's International Convention on the Protection of the Rights of All Migrant Workers and Members of Their Families (CMW), 1990
2. ILO's Migration of Employment Convention, (No.97), 1949
3. ILO's The Migrant Workers (Supplementary Provisions) Convention (No. 143), 1975

The CMW has become a cornerstone of the rights-based approach to migration advocated by many international organisations and non-governmental organisations concerned with the protection of migrant workers (Ruhs, 2012). Based on more than a decade of negotiations, the CMW incorporates and builds on ILO Conventions 97 and 143. It has a very broad set of rights for migrants, including those living and/or working abroad illegally. The CMW includes 93 articles (compared to 23 articles of ILO Convention 97 and 24 articles of ILO Convention 143) and extends fundamental human rights to all migrant workers, both regular and irregular, with additional rights recognised for regular migrant workers and members of their families. Importantly, the CMW is based on the principle of equal treatment of migrants and nationals rather than on 'minimum standards' characterising many other international legal instruments (Lönnroth, 1991). Thus, the CMW provides a more accurate interpretation of the human rights of migrant workers as it applies to the entire migration continuum, including the recruitment process and migrant rights once they are admitted (De Guchteneire and Pécoud, 2009).

However, the ratifications (less than 50) of the CMW and ILO conventions on migrant workers by state parties have been very disappointing, in both absolute terms (i.e. considering the total number of UN and ILO member states) and in relative terms (i.e. compared to the ratifications of other human rights treaties and ILO conventions) (Ruhs, 2012). The few countries that have ratified the CMW are predominantly countries of net emigration rather than net immigration. These countries with three quarters having low or medium human development index are all classified as 'low-income' or 'middle-income' by the World Bank (Ruhs, 2012; UNDP, 2009). These figures suggest that despite accepting the idea of human rights, the world's high-income countries – where migrants are most heavily concentrated – clearly do not accept that these rights should also apply to migrants living in their territories (Ruhs, 2012).

Hune and Niessen (1994) argue that the low level of ratification of the CMW has been partly attributed to various contextual factors such as an adverse economic and social climate. The high unemployment and welfare dependency make it difficult for governments in high-income countries to promote migrant workers' rights. The rights granted to migrant workers relate to the national interests and critically depend on the perceived and/or real benefits and costs, be it socio-economic, political or cultural (Ruhs, 2012). Concerns about costs placed on host countries to grant rights to irregular migrant workers was another major obstacle to the ratification of the CMW (Hillman and Koppenfels, 2009; Ruhs, 2012). Clearly, national interests and impacts – and not concerns about the protection of the human rights of migrants – were at the forefront of the politics of the drafting process of the CMW (Ruhs, 2012), which meant that the low level ratification of the CMW directly impacted the livelihood strategies and opportunities for migrants.

Given the low levels of ratification in its first 20 years, it is hard to deny that the CMW has been largely unsuccessful in achieving its main stated aim of providing an effective framework for protecting the rights of migrant workers in the global economy (Ruhs, 2012). Even though the Convention is not ratified widely, it plays an important political and strategic role, and has become an instrument of reference for states' parties and non-ratifying countries, including those that have stated explicitly that they do not wish to ratify it (Ruhs, 2012).

Migrant workers' rights under the Kafala system

Amin (1998) maintains that there were no laws to deal with the migrant workers during the first waves of migration in the Gulf countries. However, the governments of Gulf countries were gradually forced to copy some of the international conventions to regulate and protect the rights of migrant workers, although these laws remain superficial in almost all Gulf countries (Amin, 1998).

Purpose, reform and analysis of the Kafala system

The Kafala system (refer to Box 33.3) is the predominant mode of labour management system for millions of migrants. It began in the 1950s and is based on the Bedouin principles of hospitality, which prescribes a set of obligations for the treatment and protection of foreign guests. As migration increased, this traditional system of hospitality became a major mode of securing manpower in the region. Hence, it is the central institution of identities, rights and obligations for employers and migrant workers (Khan and Harroff-Tavel, 2011; Malaeb, 2015).

Box 33.3 Origin and meaning of 'Kafala'

'Historically, the kafala described a relationship that functioned as a mechanism for hosting foreigners in the close and genealogically framed societies typical of the Arabian Peninsula. Through the kafala, members of these societies vouched for the foreign visitor, essentially assuming responsibility for that foreigner's presence and behavior during his stay. In addition to temporarily placing that individual in the local social matrix, the kafala relationship also implied the sponsor's responsibility for the safety and protection of that individual. As migration to the region increased, however, this traditional mechanism for temporarily incorporating foreigners into local society was extrapolated to a comprehensive (if piecemeal) legal and state-based system for governing foreign labor in the Gulf'.

Source: Gardner (2011: 8)

The Kafala is typically the most strictly enforced system for all migrant workers irrespective of their occupational categories into the Gulf states (Khan and Harroff-Tavel, 2011). Labour migration to the Gulf states and other middle eastern countries has long been regulated by the Kafala sponsorship system (Qadri, 2020) and has become the only legal basis for residency and employment of migrant workers (Khan and Harroff-Tavel, 2011). The central purpose of the system is to meet labour demands through migrant workers while at the same time maintaining only the temporary resident status of the migrants. The laws and procedures of the Kafala system vary from state to state (Qadri, 2020).

The system is typically constituted by the employers, or the 'Kafeels', who determine their demand for labour and meet the demand through intermediaries, such as private employment agencies. Depending on the identification of jobs, authorisation is obtained for select migrant workers to enter the country. The employment permits are issued by the Ministry of Labour through an administrative process that also conducts health screening of the migrants. On their arrival at the destination country, the migrants come under the direct control and responsibility of the Kafeel, and their residency status is dependent on their continued employment by the Kafeel (Khan and Harroff-Tavel, 2011). The Kafeel is in control of all mobility and movement of the migrant workers. The migrant workers cannot enter the country, transfer employment nor leave the country without an explicit written permission from the employer or the Kafeel (Migration Forum in Asia, n.d.). The Kafeel further exerts control by confiscating the passport and travel documents of the migrant worker, despite some countries declaring confiscation of the passport as illegal (Migration Forum in Asia, n.d.). Amin (1998) points out that irrespective of the number of years that migrant workers spend in the Gulf states, they have no right to free movement and mobility without the consent of the employers and that they cannot become citizens. Furthermore, the migrant workers have no right to belong to any trade unions nor organise any other associations which can defend their rights or take up their causes in all Gulf countries except for Kuwait. Hence, the migrant workers find themselves under the mercy of an oppressive employer (Amin, 1998).

Reformation of the Kafala system began in 2009. Bahrain and Kuwait have made the greatest attempt to reform the Kafala system, while other Gulf countries attempted limited reforms to the system and some simply refused any attempt to reform the system, instead choosing to maintain the status quo (Migration Forum in Asia, n.d.; Khan and Harroff-Tavel, 2011). One of the main obstructions to a complete dismantling or reformation of the Kafala system, and the hesitation by most Gulf countries, is directly related to the geopolitics of the region. As correctly pointed out by Qadri (2020), the Kafala system remains a critical component in the state's role in ensuring political control in a situation where only a minority of its population has citizenship rights and the legal and social power rests in the hands of a small group of royal families. Further, Qadri (2020) makes an important point on the role of the Kafala in creating a social contract between the state and the citizens, which promises the citizens ready sources of revenue and significant control over migrant workers in exchange for reduced social and political freedoms. Also, he adds that the UAE authorities have long viewed the unequal legal status of migrants as a matter concerning national security. This allowed for the state to preserve its national identity, provide significant economic and social advantages to individuals with citizenship, and be able to maintain a safe and stable society rooted in the rule of law of the region. All of this comes at the cost of reduced social and political freedoms for its citizens in terms of disallowing them from forming or joining a union, participating in collective bargaining, and exercising the rights of a peaceful assembly, association or protest (Qadri, 2020).

Given the hesitancy and unwillingness to dismantle or completely reform the Kafala system by the Gulf countries, the system continues to be exploitative by its very underlying nature.

This makes the migrant workers, especially unskilled women, more vulnerable to abuse and exploitation by situating them into slavery-like situations.

Experiences of abuse and slavery like situations

Horrifying stories of abuse, assault and exploitation are very common among Indian migrant domestic workers who go to the Gulf countries. Women workers' anecdotes include hopes for improved living conditions and reduction of financial burdens and sharing similar stories, like local agents luring them with claims of a comfortable job with good wages, abuse at the hands of the employers and the failure to receive timely help from local or Indian authorities (Mantri, 2017). A Human Rights Watch report (2008) also documented a wide range of abuses against migrant domestic workers in Saudi Arabia, including deception during recruitment, violations of freedom of movement, physical and sexual abuse, labour exploitation and double victimisation under the criminal justice system.

Trafficking, forced labour and slavery

Migrant domestic workers in Gulf countries are the victims of trafficking, forced labour, and enslavement (refer to Box 33.4 for definitions). Illegal confinement of domestic workers within their workplace, withholding their passports, placing them at the risk of arrest and subsequent punishment if they escape are kinds of forced labour cases that have been reported. Thus, migrant domestic workers in Gulf countries migrate voluntarily with full or partial information but end up in situations of forced labour (Human Rights Watch, 2008).

Box 33.4   Definition of forced or compulsory labour, slavery and trafficking in persons

Forced or Compulsory Labour: 'All work or service which is exacted from any person under the menace of any penalty and for which the said person has not offered himself/herself voluntarily' (IOM, 2019: 75).

Slavery: 'The status or condition of a person over whom any or all the powers attaching to the right of ownership are exercised (Art. 1, Slavery Convention, 1926 as amended by 1953 Protocol). Slavery is identified by an element of ownership or control over another's life, coercion and the restriction of movement and by the fact that someone is not free to leave or to change employer' (IOM, 2011: 91).

Trafficking in persons: 'The recruitment, transportation, transfer, harbouring or receipt of persons, by means of the threat or use of force or other forms of coercion, of abduction, of fraud, of deception, of the abuse of power or of a position of vulnerability or of the giving or receiving of payments or benefits to achieve the consent of a person having control over another person, for the purpose of exploitation (Art. 3(a), UN Protocol to Prevent, Suppress and Punish Trafficking in Persons, Especially Women and Children, Supplementing the UN Convention against Transnational Organized Crime, 2000)' (IOM, 2011: 99–100).

A Human Rights Watch report (2018) argues that migration and trafficking are interconnected, because traffickers often exploit the processes by which individuals migrate. In addition, the employers may deceive migrant domestic workers about their actual working conditions, and trafficking victims may be found in situations of forced domestic labour, sexual, and marital arrangements.

The debates surrounding migration and trafficking

Jureidini (2010) questions how the concept of trafficking, as defined by the 'Trafficking in Persons Protocol', may be applied to migrant domestic workers with an emphasis on those under the Kafala sponsorship system in the Middle East who are obligated to use a local sponsor. The Migrant workers who are trafficked into forced labour are typically innocent victims of recruitment agents who charge them a very high recruitment fee in the promise of securing work permits and a job in the Gulf States (Qadri, 2020). Victims are often ignorant of the terms of the contract that they sign with the recruitment agents. This leads to all kinds of abuse and exploitation at various levels such as confiscation of passports, long hours of working, low wages, withholding wages and poor working conditions. While 'trafficking' as an act of coercion and exploitation is part of migration, legally and technically 'migration' and 'trafficking' represent two different concepts, and it is important to make the distinction. Jureidini (2010) points out the challenges to defining 'trafficking in persons' and its related terms, such as 'exploitation', and how these classifications affect our understanding and acknowledgement of what constitutes trafficking activities. Simply bringing a migrant into a foreign country on the promise of getting a job or securing a visa by misleading the victim on the terms or conditions of the contract would not constitute 'trafficking' or 'an activity of trafficking' (Jureidini, 2010).

The UN Special Rapporteur to the UAE and Gulf states in 2013 stated that 'trafficking' as a problem specifically concerns women and children targeted for commercial sex exploitation. This is because the country's policies separate or de-link 'trafficking' and 'labour migration' and the governments refuse to intervene in situations concerning violations due to labour exploitation (Qadri, 2020). According to the US State Report 2019 on 'Trafficking in Persons', what constitutes as indicators of trafficking activity is passport confiscation, delayed or non-payment of wages, and contract switching (Qadri, 2020). So, the Kafala sponsorship system that provides the employers with substantial power and control over the migrants, while providing the migrant workers with unequal legal status and no protection or safeguard against exploitation, can allegedly be considered a form of modern-day slavery. On the other hand, the citizens of the GCC countries who support the Kafala system strongly disagree with the term 'slavery' used by the international community to describe the plight of migrant workers (Maleab, 2015).

Psychological, physical and sexual abuse

In many cases, domestic workers experience a combination of psychological, physical or sexual abuse. Most domestic workers report some form of psychological or verbal abuse, including shouting, insults, belittlement, threats, and humiliation. Domestic workers' isolation in private homes and power relations between employers and workers heighten the risk of such abuse (Human Rights Watch, 2008).

Sexual violence is a major complaint among women seeking help. Typically, male employers or relatives, including teenage or adult sons, were the perpetrators of such (sexual) abuse, involving a range of actions, from inappropriate touching, hugging and kissing, to repeated rape. In some cases, employers harassed women by offering money for sex or threatening to withhold their salaries unless they submitted to rape (Human Rights Watch, 2008).

Labour abuses and exploitation

Domestic workers may experience several types of labour rights violations, including unpaid wages, excessively long working hours and lack of rest periods, rest days, worker's compensation, and other benefits. Many domestic workers continue to face highly exploitative working conditions due to the absence of any legal regulation of minimum standards, punishment for abuse, or ways to ease the forced isolation of domestic workers within homes (Human Rights Watch, 2008). Apart from these common known violations, other examples of abuse include refusing payment of the promised wage on time, unstructured nature of work, long work hours without any regular time off and deducting the costs of clothes, food, or lodging from wages (Human Rights Watch, 2007; Khan and Harroff-Tavel, 2011; Pande, 2013).

These violations and abuses of migrant domestic workers directly affect their livelihood opportunities. The common perception that migrating to countries like those in the Gulf would significantly improve their livelihood and their families back home is completely shattered by the fact that these domestic workers end up in conditions where they feel cheated, trapped in their host countries, and without legal protection. In some cases, the migrants had to return home empty handed while in other cases their families had to send money to the Kafeel (employer) for their return home. In extreme conditions, some domestic workers are also trapped for life due to the failure of receiving timely support from both the local and Indian authorities, in which cases they may face deportation or imprisonment.

In cases where the migrant women do return, their social status is further reduced to what it was prior to their migration. They have to negotiate with their family members, reinforcing the belief that women's migration is still stigmatised in a conservative, patriarchal society like India. This points to the fact that migration has contradictory or negative impacts on women migrant workers and their livelihoods, particularly those women who are unskilled workers from poor socio-economic backgrounds. Hence, women's migration intensifies their vulnerability through stigmatisation and exclusion.

Conclusion

In a highly interconnected globalised world, with half of the total international migrants being women and a significant portion of them coming from low-income countries, the issues faced by migrant women, particularly in the domestic or informal sector, requires special emphasis. Women migrant workers from countries of the Global South face multiple challenges at various levels including emotional, psychological, physical, sexual and economic abuse both from their home countries and the destination countries. This is closely connected to the patriarchal societies that they come from, where these attitudes are reinforced time and again without any change in the actual status or position of these women. Despite having the largest number of overseas migrant populations in the world, India does not have a well-structured policy framework on international migration, and, in terms of encouraging women migration, it is almost negligible.

The Kafala system, which has been the prime mode of operation for private employment in the Gulf states, is in desperate need of reform as recommended by ILO and other UN conventions. The GCC states have not ratified important UN conventions like CMW and ILO Conventions 97 and 143 on migrant workers' rights because it contradicts their customary Islamic law and their socio-political and economic interests. Hence, a complete annihilation of the Kafala system remains a utopian dream that is not practically viable without wider socio-political and civil reforms of the Gulf States. Therefore, this Kafala system should be considered as a global problem and to be dealt with in line with other migration policies undertaken by countries like the USA and UK, and other European countries.

In the meantime, the Gulf States should be aware that annihilation of the Kafala system is not a solution for ensuring the protection of the rights of migrant workers because of the socio-political, economic and demographic complexities of this region. Given the complexity of the issue, a great deal of effort and willingness is required to incorporate the legal and humanistic approach that focuses on the rights of workers central to promoting and strengthening of their livelihoods and economic development in the states.

Key points

Listed below are some measures to be taken by the Governments of both the origin and the destination countries regarding protecting migrant workers' rights:

- Ratifications of the CMW and ILO laws by all countries from the Global North as well as the Gulf States.
- Implementation of laws and policies that would encourage women migration, especially unskilled migrant workers by providing education, awareness, and reducing restriction of their movement and ensuring their rights, safety and security.
- As many forms of abuses and violations of human rights are committed by the private sponsors (Kafeel) rather than the government or semi-governmental bodies, it is important for the Gulf States to replace the individual sponsors with the state sponsors.
- A uniform agreement on minimum legal rights, wages and social security of the migrants to be set by all the migrant sending countries through open dialogues and engagements.
- Hold the employers or sponsors accountable and punish them for committing violation of rights, abuse and exploitation of migrant workers.
- Establishment of more human rights organisations and proper implementation of anti-discrimination laws in the Gulf States for defending migrant worker's rights and to prevent any form of discrimination based on their nationality, religion, race, caste, gender, and colour.

Further reading

Castles, S. (2011) 'Bringing human rights into the migration and development debate', *Global Public Policy Journal*, 2(3): 248–258.
Guild, E., Grant, S. and Groenendijk, C.A. (eds) (2018) *Human rights of migrants in the 21st century*, London: Routledge.
Kerwin, D. (2020) 'International migration and work: Charting an ethical approach to the future', *Journal on Migration and Human Security*, 8(2): 111–133. https://doi.org/10.1177/2331502420913228
Piper, N. (ed.) (2008) *New perspectives on gender and migration: Livelihood, rights and entitlements*, New York: Routledge.
Ruhs, M. (2012) 'The human rights of migrant workers', *American Behavioral Scientist*, 56(9): 1277–1293.

Acknowledgement

I am thankful to Narmada P. (Ph.D. Philosophy) from the University of Hyderabad, India, for supporting and helping me through the process of developing my ideas in the chapter.

References

Amin, R.A. (1998) 'Migrant workers in the Gulf', *International Centre for Trade Union Rights*, 5(4): 12–13. www.jstor.org/stable/41935711
Bélanger, D. and Rahman, M. (2013) 'Migrating against all the odds: International labour migration of Bangladeshi women', *Current Sociology*, 61(3): 356–373. https://doi.org/10.1177/0011392113484453
Betts, A. (2015) 'Human migration will be a defining issue of this century. How best to cope?', *The Guardian*. Available at: www.theguardian.com/commentisfree/2015/sep/20/migrants-refugees-asylum-seekers-21st-century-trend (Accessed 6 June 2021).

Boghean, C. (2016) 'The Phenomenon of migration. Opportunities and challenges', *The USV Annals of Economics and Public Administration*, 16(Special): 14–20. Available at: https://ideas.repec.org/a/scm/usvaep/v16y2016i1(23)p14-20.html (Accessed 6 June 2021).

Castles, S. and Davidson, A. (2000) *Citizenship and migration: Globalization and the politics of belonging*, 1st ed., London: Palgrave.

Chamie, J. (2020) 'International migration amid a world in crisis', *Journal on Migration and Human Security*, 8(3): 230–245. https://doi.org/10.1177/2331502420948796

De Guchteneire, P. and Pécoud, A. (2009) 'Introduction: The UN convention on migrant workers' rights', in Cholewinski, R., De Guchteneire, P. and Pécoud, A. (eds.), *Migration and human rights*, Cambridge: Cambridge University Press, pp. 1–46.

Dhar, R. (2012) 'Women and international migration: A cross-cultural analysis', *Social Change*, 42(1): 93–102. https://doi.org/10.1177/004908571104200106

Esipova, N., Pugliese, A. and Ray, J. (2018) 'More than 750 million worldwide would migrate if they could', *Gallup*. Available at: https://news.gallup.com/poll/245255/750-million-worldwide-migrate.aspx (Accessed 6 June 2021).

Gardner, A.M. (2011) 'Gulf migration and the family', *Journal of Arabian Studies*, 1(1): 3–25. https://doi.org/10.1080/21534764.2011.576043

Gülçür, L. and İlkkaracan, P. (2002) 'The "Natasha" experience: Migrant sex workers from the former Soviet Union and Eastern Europe in Turkey', *Women's Studies International Forum*, 25(4): 411–421. https://doi.org/10.1016/S0277-5395(02)00278-9

Hillmann, F. and Koppenfels, A.K.V. (2009) 'Migration and human rights in Germany', in Cholewinski, R., De Guchteneire, P. and Pécoud, A. (eds.), *Migration and human rights*, Cambridge: Cambridge University Press, pp. 322–342.

Human Rights Watch (2007) *Exported and exposed: Abuses against Sri Lankan domestic workers in Saudi Arabia, Kuwait, Lebanon, and the United Arab Emirates*, New York: Human Rights Watch. Available at: www.hrw.org/reports/2007/srilanka1107/ (Accessed 6 June 2021).

Human Rights Watch (2008) *As if I am not human: Abuses against Asian domestic workers in Saudi Arabia*, New York: Human Rights Watch. Available at: www.hrw.org/reports/2008/saudiarabia0708/ (Accessed 6 June 2021).

Hune, S. and Niessen, J. (1994) 'Ratifying the UN migrant workers convention: Current difficulties and prospects', *Netherlands Quarterly of Human Rights*, 12(4): 393–404. https://doi.org/10.1177%2F016934419401200403

International Organization for Migration (IOM) (2011) *Glossary on migration*, International migration law, Geneva: International Organization for Migration. https://publications.iom.int/system/files/pdf/iml25_1.pdf

International Organization for Migration (IOM) (2017) *World migration report 2018*, Geneva: International Organization for Migration.

International Organization for Migration (IOM) (2019) *Glossary on migration*, International migration law, Geneva: International Organization for Migration. https://publications.iom.int/system/files/pdf/iml_34_glossary.pdf

International Organization for Migration (IOM) (2019) *World migration report 2020*, Geneva: International Organization for Migration.

Jureidini, R. (2010) 'Trafficking and contract migrant workers in the Middle East', *International Migration*, 48(4): 142–163. https://doi.org/10.1111/j.1468-2435.2010.00614.x

Khan, A. and Harroff-Tavel, H. (2011) 'Reforming the Kafala: Challenges and opportunities in moving forward', *Asian and Pacific Migration Journal*, 20(3–4): 293–313. https://doi.org/10.1177/011719168-1102000303

Kofman, E., Phizacklea, A., Raghuram, P. and Sales, R. (2000) *Gender and international migration in Europe: Employment, welfare and politics*, 1st ed., London: Routledge.

Koser, K. (2016) *International migration: A very short introduction*, 2nd ed., New York: Oxford University Press.

Lönnroth, J. (1991) 'The international convention on the rights of all migrant workers and members of their families in the context of international migration policies: An analysis of ten years of negotiation', *International Migration Review*, 25(4): 710–736. https://doi.org/10.1177%2F019791839102500404

Malaeb, H.N. (2015) 'The "Kafala" system and human rights: Time for a decision', *Arab Law Quarterly*, 29(4): 307–342. https://doi.org/10.1163/15730255-12341307

Mantri, G. (2017) 'Modern day slaves: How Indian migrant workers' rights are violated in the Gulf', *The News Minute*. Available at: www.thenewsminute.com/article/modern-day-slaves-how-indian-migrant-workers-rights-are-violated-gulf-72179 (Accessed 6 June 2021).

MEA, n.d. 'About eMigrate project', *eMigrate*. Available at: https://emigrate.gov.in/ext/about.action (Accessed 6 June 2021).

Migrant Forum in Asia (n.d.) 'Reform of the Kafala (Sponsorship) System', *Policy Brief No. 2*, Quezon City: Migrant Forum in Asia. http://mfasia.org/migrantforumasia/wp-content/uploads/2012/07/reformingkafala_final.pdf

Naidu, K.L. (1991) 'Indian labour migration to Gulf countries', *Economic and Political Weekly*, 26(7): 349–350.

Pande, A. (2013) 'The paper that you have in your hand is my freedom: Migrant domestic work and the sponsorship (Kafala) system in Lebanon', *International Migration Review*, 47(2): 414–441. https://doi.org/10.1111/imre.12025

Phizacklea, A. (1998) 'Migration and globalization: A feminist perspective', in Koser, K. and Lutz, H. (eds.), *The new migration in Europe: Social constructions and social realities*, London: Palgrave Macmillan, pp. 21–38. www.palgrave.com/gp/book/9780312210052

Qadri, M. (2020) 'The Kafala system: Harmless or human trafficking?' in Page, M. and Vittori, J. (eds.), *Dubai's role in facilitating corruption and global illicit financial flows*, Washington, DC: Carnegie Endowment for International Peace, pp. 79–84.

Ruhs, M. (2012) 'The human rights of migrant workers', *American Behavioral Scientist*, 56(9): 1277–1293. https://doi.org/10.1177/0002764212443815

United Nations Department of Economic and Social Affairs (UNDESA) (2019) *International Migration 2019*, New York: United Nations.

United Nations Development Programme (UNDP) (2009) *Human development report 2009: Overcoming barriers: Human mobility and development*, New York: UNDP. Available at: http://hdr.undp.org/en/content/human-development-report-2009 (Accessed 6 June 2021).

Wadhawan, N. (2018) *India labour migration update 2018*, New Delhi: International Labour Organization. www.ilo.org/wcmsp5/groups/public/ – asia/ – ro-bangkok/ – sro-new_delhi/documents/publication/wcms_631532.pdf

Zlotnik, H. (1990) 'International migration policies and the status of female migrants', *International Migration Review*, 24(2): 372–381. https://doi.org/10.2307/2546556

34
REMITTANCES AND ECONOMIC DEVELOPMENT IN THE GLOBAL SOUTH

Haruna Issahaku, Anthony Chiaraah and George Kwame Honya

Introduction

The main objective of this chapter is to discuss the impact of remittances on economic development in the Global South. Specifically, this chapter reviews trends in worldwide remittances inflows and the impact of remittances on entrepreneurship and job creation, poverty and inequality, and health and human capital development. In addition, the chapter discusses the role of remittances in building resistance against eventualities.

Trends in remittances around the world

Globally, remittance has become a major source of development finance. The role of remittance in promoting development is even more apparent in low- and middle-income countries (LMICs). Table 34.1 reports remittances sent and those projected to be sent to various parts of the world, as reported by the World Bank (2020). Worldwide, the total remittance receipts rose from $437 billion in 2009 to $714 billion in 2019. In the case of LMICs, remittances receipts rose from $307 billion in 2009 to $554 billion in 2019. The $554 billion remittances received by LMICs in 2019 constitutes 78% of global remittance flows. Thus, the bulk of remittance goes to the developing world where it is most needed.

Global remittance inflows were estimated to decline slightly to $572 billion in 2020 on the back of the COVID-19 pandemic, but projected to rise to $602 billion in 2021, according to the World Bank (2020). This represents a decline in the growth rate of remittances by 19.9% in 2020 and an estimated rise of 5.2% in 2021. Similarly, in LMICs, remittances flows were estimated to fall from $554 billion in 2019 to $445 billion in 2020, and thereafter pick up marginally in 2021 to settle at $470 billion. In percentage terms, this embodies a drop in the growth rate of remittances by 19.7% in 2020 and a projected increase of 5.6% in 2021.

A close examination of the growth rate of remittances in dollar terms reveals that remittances are quite sensitive to global economic crises, as well as economic crises in source countries. In 2009, global remittances declined by 5.1% in the wake of the 2007–2009 financial crises and by a colossal 19.9% following the COVID-19 engineered global economic crises according to the World Bank (2020). No doubt, COVID-19 has dug a deeper hole in the purse of economies than other economic crises and pandemics in recent history. For instance, the Spanish Flu pandemic of 1918–1920,

Table 34.1 Trends in remittance flows: 2009–2021

Region of the world	Year						
	2009	2016	2017	2018	2019e	2020f	2021g
	($ billions)						
Low- and middle-income	307	446	487	531	554	445	470
East Asia and Pacific	80	128	134	143	147	128	138
Europe and Central Asia	36	46	55	61	65	47	49
Latin America and Caribbean	55	73	81	89	96	77	82
Middle East and North Africa	33	51	57	58	59	47	48
South Asia	75	111	118	132	140	109	115
Sub-Saharan Africa	29	39	42	48	48	37	38
World	437	597	643	694	714	572	602
	(Growth rate, Percent)						
Low- and middle-income	−5.0	−1.5	9.1	9.0	4.4	−19.7	5.6
Sub-Saharan Africa	−0.2	−8.3	9.3	13.7	−0.5	−23.1	4.0
World	**−5.1**	**−0.9**	**7.7**	**8.0**	**2.8**	**−19.9**	**5.2**

Adapted from World Bank (2020)

which infected about 500 million people globally, led to a fall in GDP of the worst affected countries by an average of 0.8%. The SARS epidemic, which infected about 8,000 people from 2002 to 2004, instigated a fall in GDP in the worst affected economies by an average 3.5%. The Ebola outbreak of 2014–16, which attacked 28,700 people mostly in Guinea, Liberia, and Sierra Leone, led to an average decline in the GDP growth of these countries by 8.6%. Lastly, the global financial crises of 2007–2009 led to a fall in economic growth in the worst affected high-income countries and LMICs respectively by 3.7% and 3.3%. Comparatively, the COVID-19 (2019–2020) pandemic resulted in a decline in the growth of the worst affected economies by 8.7%.

Based on data from the International Organization for Migration (2020), the top 5 remittance recipients in absolute dollar terms in 2018 were India ($78.61 billion), China ($67.41 billion), Mexico ($35.66 billion) and the Philippines ($33.83 billion). Apart from France and Germany, all top 10 remittance receiving countries in 2018 were in the Global South, underscoring the role of remittances in financial development in this region. From being the third largest remittance recipient in 2005, India overtook China to become the largest remittance recipient over the last decade (World Bank, 2006).

The top 10 remittance sending countries are mostly higher-income countries (HICs) (International Organization for Migration, 2020). The top 5 remittance sending countries in 2017 were United States ($67.96 billion), United Arab Emirates ($44.37 billion), Saudi Arabia ($36.12 billion), Switzerland ($26.60 billion) and Germany ($22.09 billion). China, which is the second largest recipient of remittances is also a significant sending country, though remittances *from* China are on the decline.

Economic development implications of remittances

This section discusses the economic development implications of remittances in respect of entrepreneurship and job creation, poverty and inequality, education and human capital and resilience. The evidence is based on research conducted mainly in countries of the Global South.

Entrepreneurship and employment generation implications of remittances

The importance of credit for micro- and small enterprise formation has been emphasised in the literature (Aryeetey et al., 1994; Filli et al., 2015). Obtaining such credit is difficult due to numerous constraints including the high risks involved, small loan size and collateral problems. All these heighten the transaction costs of obtaining funding for micro- and small enterprises (Aryeetey et al., 1994; Awotide et al., 2015; Mtanthiko, 2012). Hence, remittance money has been identified as key for engaging in micro- and small enterprises for development (Taylor, 1999).

The effect of remittance on entrepreneurship is not straightforward. In some cases, the results are positive, while they are more mixed in other cases (see Acosta, 2007; Naudé et al., 2017; Yang, 2006). On the one hand, remittances can be channelled into self-employment or microenterprise activities to create employment opportunities in limited labour and credit markets. International remittances can contribute to alleviating credit constraints blocking investment in developing countries, which helps remittance recipients to invest more in their own businesses and become increasingly engaged in self-employment activities, as observed by Yang (2006) among Philippine migrants. On the other hand, migration can mean entrepreneurially minded individuals leave their homes because they can more easily migrate, leaving those less entrepreneurially minded at home, as found by Naudé et al. (2017) in a review of global evidence. This is particularly relevant in rural areas with undeveloped infrastructure and where markets are in distant places. Under such circumstances, it is unrealistic to expect migrant families to channel their resources towards risky productive activities.

In addition, remittances could reduce the labour supply if migrant families substitute more leisure for labour. This results in a situation where labour force participation falls in some remittance receiving households leading to low productivity (see Acosta, 2007; Chalise, 2014). Remittances also influence the composition of employment between farm and non-farm employment and between self-employment and salaried employment. Evidence from El Salvador reveals that male remittance recipients are more likely to engage in self-employment related to agriculture, whereas business management in nonfarm enterprises are higher among female remittance recipients (Acosta, 2007). Remittances increase involvement in self-employment while reducing engagement in salaried employment in Mexico and the Philippines (see Shapiro and Mandelman, 2014; Yang, 2006) by providing start-up and business expansion capital for entrepreneurs. As a result, during economic downturns, the recovery of the salaried sector is very sluggish in receiving economies. Remittances also promote investment in capital intensive enterprises (Yang, 2006).

Surprisingly, while remittance increases self-employment among receiving households, returned migrants are generally not interested in investing in their own ventures. Returnees are primarily interested in finding a job rather than becoming entrepreneurs, with entrepreneurship seen as a necessity rather than an opportunity to make use of skills acquired overseas (World Bank, 2013). Returnees wishing to start businesses face limited resources, difficulties in managing businesses due to inadequate skills, lack of business knowledge, and so on. For most returning migrants, investing in either existing businesses and/or starting a new business would require additional sources of business management advice and investment capital.

Remittances, poverty and inequality

Research on the effect of remittance on poverty provides a favourable picture of the relationship between remittance and poverty alleviation (Adams, 1991; Adams and Page, 2003; Abdih et al., 2012; Adenutsi, 2011). In Egypt, Adams (1991) found that international remittance accounts for

12.5% of gross income of surveyed households and 30.4% of gross income of migrant households. In evaluating the impact of international remittance on poverty and income distribution, the study used a predicted income equation to estimate changes in welfare for two scenarios: (1) excluded remittances from household income for households who still have migrants abroad, and (2) included remittances from household income for those households who have migrants abroad. The results indicated that international remittances have a small but positive effect on poverty reduction, with the number of households living in poverty declining by 9.8% in the excluded case and 12 % in the included case.

Adams and Page (2003) examined the relationship between international migration, remittances, inequality, and poverty from 71 developing countries. The results showed that both international migration and remittances significantly reduced the level, depth and severity of poverty in the developing world. The level of poverty measures the proportion of the population living below the poverty line and is often measured using the head count index. The depth of poverty measures the amount by which the mean expenditure of the poor falls below the poverty line. The depth of poverty can be measured using the poverty gap index. The severity of poverty measures how poor a poor person is relative to another poor person. It is usually measured using the squared poverty gap. Adams and Page (2003) found that, on average, a 10% increase in the proportion of international migrants in a country's population will lead to a 2.1% decline in the proportion of people living on less than $1.00 per person per day. A similar result for international remittances showed that a 10% increase in per-capita official international remittances will lead to a 3.5% decline in the number of people living in poverty. Likewise, Adams (2004) used a large, nationally representative household survey to analyse the impact of both international remittances from the USA and internal remittances on poverty in Guatemala. The study revealed that both internal and international remittances reduced the level, depth and severity of poverty in Guatemala. However, remittances have a greater impact on reducing the severity of poverty as opposed to the level of poverty.

Ratha (2007) asserts that remittances directly augment the income of recipient households. They also positively affect poverty and welfare through indirect multiplier and macroeconomic effects. A cross-country regression analysis showed that a 10% increase in per-capita official remittances may lead to a 3.5% decline in the number of poor people (Ratha, 2007). Abdih et al. (2012) also report of strong evidence of the welfare-enhancing benefits of remittances for recipients. For example, remittances are credited with reducing poverty and minimising recipients' consumption volatility. Adenutsi (2011) examined the impact of international remittances on poverty and income inequality in 34 sub-Saharan Africa (SSA) countries between 1980 and 2009 using the system Generalised Methods of Moments to examine a set of dynamic panel data models. The study revealed that remittances have significant poverty-alleviating effects, with the poorest of the poor benefiting the least.

Zhu and Luo (2008) assessed rural-urban migration impacts on rural poverty and inequality for Hubei Province in China using household survey data compiled in 2002. The authors found that by providing alternatives to households with lower marginal labour productivity in agriculture, migration increases rural income. Internal and international migration have been compared by the World Bank (2006), including their relative prospects in affecting poverty in developing countries. The findings indicated that internal migration may play a larger role in reducing poverty than international migration. While the expected wage gain is lower in internal migration, poor workers may have a better chance of finding work domestically than in economies with higher wages, where workers' lack of skills and language ability restrict job opportunities. Moreover, international migration tends to be too costly for the poorest workers. In a related study on

74 developing countries, Adams and Page (2005) showed that internal remittances reduced the level of poverty by 14%, while international remittances reduced the level of poverty by only 5%.

With respect to income inequality, research findings are mixed. For example, Adenutsi (2011) found that international remittances have income equalisation effects in countries with a relatively narrower income gap but intensify income inequality in countries with a relatively wider income gap. Zhu and Luo (2008) offered support for the hypothesis that migration tends to have egalitarian effects on rural income. The reasons given include: (i) migration is rational self-selection – farmers with higher agricultural productivities choose to remain in local agricultural production while those with higher expected returns in urban non-farm sectors migrate, (ii) poorer households facing binding constraints of land shortages are more likely to migrate, and (iii) the poorest of the poor benefit disproportionately from remittances.

Whether international remittances reduce or increase inequality depends on the size of the migrant sock abroad for a given number of households or region. According to Taylor (2006), international migrant remittances increased inequality in regions where only a small percentage of households have migrants abroad, but remittances reduced inequality in the highest-migration region. Remittances from international migrants have little effect on poverty in regions where only a few households have migrants because most of the 'pioneer migrant' households are not poor. However, in high migration regions, increases in international remittances reduce poverty significantly. It appears that even poor households gain access to foreign migration opportunities in regions where international migration has really taken off.

Similarly, Adams (1991) who carried out a study in rural Egypt discovered a negative relationship between international remittances and income distribution. Inequality increases when remittances are included in the per capita household income. One reason is that at the time of the survey, most of the migrants who were still abroad were from the upper income groups within the population.

Internal remittances may tend to have a more favourable effect on inequality than international remittances as evidenced in Adams (1996). Adams (1996) employed a three-year panel data set to compare the effects of remittances from internal and international migration on asset accumulation and income distribution in certain poor rural districts of Pakistan. Remittances from internal migration are found to have an equalising effect on rural income distribution (i.e. inequality mitigating effect). However, international remittances have a negative income equalising effect on rural income distribution (i.e. inequality enhancing effect). The reason for this may be that while internal migrants often come from lower-income households, the richer and better educated people in the community are more likely to migrate internationally because they can afford the high entry costs. The study also found that external remittances have a significant positive impact on the accumulation of income-generating assets, particularly on land and livestock holdings, while internal remittances have no such effect. This is because the richer income groups, who tend to be the main recipients of external remittances, possess complementary assets like farm machinery, and tend to have a higher propensity to invest than do lower income groups. The evidence from Adams (1996) suggests divergent policies for growth-promotion and equity enhancement: promotion of rural asset accumulation as a policy calls for encouraging international migration. However, promotion of income distribution will call for encouraging internal migration.

Altogether, the results of these studies suggest that in places where poor households are the ones to receive remittance, more commonly, the result will have an egalitarian effect on poverty. It will simply raise their income level towards or above the poverty line. In poor nations where the educational system is pro-poor, this egalitarian effect will be more enhanced because migrants among them will be skilled, and given that they come from poor households, they are likely to be moved by altruistic motives to remit large amounts. However, where it is the rich who mainly migrate, their remittances will of course enhance inequality. The migrants may not even be motivated by altruism to remit much. Thus, in poor countries where the educational and skill formation system discriminates against the poor, this unequal situation is likely to be perpetuated.

Remittances, human capital development and health

Remittances can increase investment in human capital (Azizi, 2018). This impact is especially important since human capital is a crucial factor in the promotion of long-term growth in developing countries. Remittances have positive effects on the welfare of children by improving infant mortality and child health by raising household incomes and increasing the health-knowledge of mothers. For example, Terrelonge (2014) examined how remittances and government health spending improve child mortality in developing countries and concluded that remittances reduce mortality through improved living standards from the relaxation of households' budget constraints. Similarly, Chauvet et al. (2013) considered the respective impact of aid, remittances, and medical brain-drain on child mortality rates. Brain-drain refers to the emigration of skilled workers in search of greener pastures abroad, leading to a depletion of human capital in the home country. Their results showed that remittances reduce child mortality while medical brain drain increases it. Health aid also significantly reduces child mortality, but its impact is less robust than the impact of remittances (Chauvet et al., 2009).

There are two theories developed based on empirical studies about the impact of migration and remittances on education. The first was developed by Bouoiyour and Miftah (2016) and states that parents who invest in the education of their children believe it generates a higher rate of return than do the savings incurred if they did not invest in education. Some parents in low-income households invest too little in the education of their children because they cannot afford to finance educational investments regardless of such returns. Hence, it is expected that the reduction of these liquidity constraints will make education more accessible to children from poor families. Therefore, financial transfers from family members can lift the budget constraints and allow parents to invest in the education of their children to the extent that is optimal for them. The second hypothesis was developed by Amuedo-Dorantes and Pozo (2010), who argued that the presence of family members abroad may induce changes in school attendance of children in migrants' households for a variety of reasons. Children may have less time to devote to schooling because they engage in market activities to earn income to defray migration-related expenses of households. Alternatively, children may leave school to perform necessary household chores that the absent migrant no longer attends to. Finally, if children are encouraged to migrate in the future, they may drop out of school if the origin country's education is not generally well-recognised in the destination.

Remittances and agricultural development

In theory, the impact of migration and remittances on the agricultural sector can occur in two ways. For one, the reduced labour due to migration could impact agricultural output negatively and thereby affect productivity (labour effect). Second, an increase in capital investment and the improvement in technology fuelled by remittances could increase productivity and impact agricultural output positively (capital and technology effects). In the end, the net effect of remittances on agriculture will depend on which of the effects dominates: it will be negative if the labour effect dominates and positive if the capital and technology effects dominate.

The determinants of agricultural productivity in a Neoclassical framework are land, labour, and capital as the factors of production (Howard, 1983). Land is fixed, labour is mobile – hence migrates away, and capital is limited. Access to financial resources is a key constraint to investment, explaining why there is such low investment in human and physical capital in developing countries. While our focus is on the individual agricultural productivity, one might expect to observe a decrease in total output due to migration and reduction in labour. This occurs when individual workers leave, leaving fewer people to work on a farm, resulting in a labour shortage. Such a shortage reduces output in the neoclassical model. However, as a migrant starts sending money back home, the family is able to invest in both human and physical capital as well as technology. Hence, remittances relax the financial constraint to investment and, as a result, increase agricultural productivity and output.

Remittances can be used to support improved farming equipment and techniques, which can increase agriculture productivity, depending on how much investment goes into the agriculture sector. For example, Khanal et al. (2015) reported a study of village level data from Tanahu in Nepal that found only 5% of the remittance being used for agriculture purposes – purchase of equipment, inputs, animals, labour, land improvements, etc. Instead, the majority of remittance is used for food and clothing, repayment of loans, health and education.

Remittances provide the investment for capital accumulation, such as a tractor instead of traditional tools like cutlasses and hoes. Also, remittances enable households to buy improved seeds and fertiliser that will contribute to productivity. Additionally, migration and remittances allow households to exploit tools and machineries in a more effective way through the accumulation of productive assets (Chiodi et al., 2012). We call this production effectiveness technology and it comes from an individual's experience in a foreign country, either by the investment in human capital that comes through education, training and the skills acquired through learning by-doing or an improved connectivity (e.g. telephone, internet, social media, etc.).

The relationship between remittance and agriculture development has direct and indirect components as outlined by Bhandari (2019) and Maharjan and Knerr (2019). In most rural and local areas, remittance supplements the financial needs of farmers as they only rely on local credit cooperatives and microfinance for agricultural development and investment. Those who receive remittances are likely to receive loans from these local and informal institutions. In addition, remittances can also impact the farmer's access to the market through information gathered and shared by the use of technology such as the cell phone and internet. This information means farmers may produce more if the market conditions are favourable.

Remittances and socio-economic resilience

Remittance plays a vital role in the wellbeing of the migrants and their families. But there is also growing evidence suggesting that when adults migrate to support their families via remittances, family members left behind often experience poorer physical and mental health (Adhikari et al., 2011; Graham et al., 2015; Ivlevs et al., 2019; Nguyen, 2016). This phenomenon suggests the possibility of an unexpected negative consequence of migration: the financial benefits of receiving remittances from family members who have migrated may be offset by the negative health effects for those who remain in the place of origin.

Remittances should be able to offset any negative effects left behind by the migrant because if that is not done, it would have been better for that person to remain in his or her place of origin. Once this is done, we can say remittance promotes socio-economical resilience. Socio-economic resilience is defined as the systemic analysis of socioeconomic indicators that reproduce the robustness of the development trajectories, guaranteeing progress even in crisis contexts. For remittance to be socio-economically resilient, both the economy and individual levels must ensure the following (Bettin and Zazzaro, 2018):

1. Macroeconomic stability that entails the avoidance of exchange rate restrictions, black market premiums and high inflation.
2. Household income and consumption smoothing, which is looking at the economic opportunities for migrants and migrant destination diversity, as well as the postponement decision to send money.
3. Increased investments for future preparedness in terms of both financial and material wealth. In other words, financial sector development and lowering remittance costs.

Studies have shown that remittances rise when there are disasters such as floods, drought, earthquakes, fires and other crises and calamities in the home country in a bid to ameliorate the suffering of those who remain (Bettin and Zazzaro, 2018; Mohapatra et al., 2009). In addition, remittance receiving households are said to have a good cash buffer to mitigate the effects of crises. As a result, remittance receiving households are less likely to sell assets. Research in Ethiopia revealed that households that received remittances relied more on cash reserves to cope with drought rather than selling livestock (Mohapatra et al., 2009). In Burkina Faso and Ghana, remittance receiving households are reported to have better housing and access to communication equipment and are therefore better positioned to deal with shocks than their non-remittance receiving counterparts (Mohapatra et al., 2009). Moreover, migrant households use remittances as a diversification strategy against income losses, unemployment and other socioeconomic risks. The countercyclical properties of remittances make them an effective tool for economic resilience and in doing so, remittances cushion the home economy in periods of shocks and crises.

Conclusion

This chapter surveyed the impact of remittances on development in the Global South. Based on economic theory, the impact of remittances on inclusive development indicators such as poverty, entrepreneurship, human capital development, health, and financial development is strongly positive. However, evidence has pointed to a lack of unanimity in the impact of remittances on poverty, inequality, entrepreneurship, education and health. However, the evidence tilts in favour of positive effects. Moreover, the empirical literature revealed that remittances make receiving households more resilient against shocks and disasters. The findings point to the need for countries in the Global South to use policies to support receiving households to use remittances to foster economic development. Such policy incentives could include creating conducive business environment, making credit accessible, and providing business management training to migrant households.

Given that remittances reduce poverty while promoting entrepreneurship, human capital development, health and financial development means that remittance is crucial in achieving the Sustainable Development Goals (SDGs). In this regard, economies in the Global South should remove impediments such as the high cost of remitting and capital flow restrictions to allow for the flow of more remittances.

The ability of remittances to build resilience of recipient households against eventualities is a key factor that separates remittances from other capital flows such as foreign direct investment, aid, and private portfolio investments. In turbulent times, portfolio investments disappear, FDI flows decline, but remittance inflows increase. This means countries in the Global South ought to prioritise remittances over other capital flows and fully maximise the stabilising benefits from remittances. Overall, countries in the Global South can do more with remittances. With the elevated role remittances are playing now in the Global South, innovative financing mechanisms involving securitisation of remittances receivable should be utilised to reduce the cost of borrowing.

Key points

- The impact of remittances on poverty, entrepreneurship, human capital development, health and financial development is strongly positive, theoretically.
- In practice, the impact of remittances on poverty, inequality, entrepreneurship, education and health is mixed though the evidence tilts in favour of positive impact.
- Remittances increase the resilience of receiving households against shocks and disasters.
- Remittances can potentially be leveraged to support achievement of the SDGs in the Global South.

Further reading

Acosta, P. (2007) 'Entrepreneurship, labor markets and international remittances', in Ozden, C. and Schiff, M. (eds.), *International migration, economic development and policy*, London: World Bank and Palgrave Macmillan Publishers.

Adams, R.H. Jr. (2011) 'Evaluating the economic impact of international remittances on developing countries using household surveys: A literature review', *Journal of Development Studies*, 47(6): 809–828. https://doi.org/10.1080/00220388.2011.563299

Amidu, M., Abor, J.Y. and Issahaku, H. (2019) 'Left behind, but included: The case of migrant remittances and financial inclusion in Ghana', *The African Finance Journal*, 21(2): 36–62. https://hdl.handle.net/10520/EJC-1b9cd8809c

Issahaku, H. (2019) 'Harnessing international remittances for financial development: The role of monetary policy', *Ghana Journal of Development Studies*, 16(2): 113–137. https://mpra.ub.uni-muenchen.de/id/eprint/97004

References

Abdih, Y., Dagher, J., Chami, R. and Montiel, P. (2012) 'Remittances and institutions: Are remittances a curse?', *World Development*, 40(4): 657–666.

Acosta, P. (2007) 'Entrepreneurship, labor markets and international remittances', in Ozden, C. and Schiff, M. (eds.), *International migration, economic development and policy*, London: World Bank and Palgrave Macmillan Publishers.

Adams, R.H., Jr. (1991) 'The effects of international remittances on poverty, inequality and development in rural Egypt', *Research Report 86*, Washington, DC: International Food Policy Research Institute (IFPRI).

Adams, R.H., Jr. (1996) *Remittances, inequality and asset accumulation: The case of rural Pakistan*, Washington, DC: International Food Policy Research Institute. Reprinted from: O'Connor and Farsakh (eds.), *Development strategy, employment, and migration: Country experiences*: OECD.

Adams, R.H., Jr. (2004) 'Remittances and poverty in Guatemala', *World Bank Policy Research Working Paper 3418*, Washington, DC: World Bank.

Adams, R.H. Jr. and Page, J. (2003) 'International migration, remittances and poverty in developing countries', *World Bank Policy Research Working Paper 3179*, London: World Bank.

Adams, R.H., Jr. and Page, J. (2005) 'Do international migration and remittances reduce poverty in developing countries?', *World Development*, 33(10): 1645–1669. https://doi.org/10.1016/j.worlddev.2005.05.004

Adenutsi, D.E. (2011) 'Do remittances alleviate poverty and income inequality in poor countries? Empirical evidence from sub-Saharan Africa', *MPRA Paper No. 37130*. https://mpra.ub.uni-muenchen.de/37130/

Adhikari, R., Jampaklay, A. and Chamratrithirong, A. (2011) 'Impact of children's migration on health and health care-seeking behavior of elderly left behind', *BMC Public Health*, 11: 143. https://doi.org/10.1186/1471-2458-11-143

Amuedo-Dorantes, C. and Pozo, S. (2010) 'Accounting for remittance and migration effects on children's schooling', *World Development*, 38(12): 1747–1759. https://doi.org/10.1016/j.worlddev.2010.05.008

Aryeetey, E., Baah-Nuakoh, A., Duggleby, T., Hettige, H. and Steel, W.F. (1994) 'Supply and demand for finance of small enterprises in Ghana', *World Bank Discussion Papers*, Washington, DC: Africa Technical Department Series, World Bank.

Awotide, B.A., Abdoulaye, T., Alene, A. and Manyong, V.M. (2015) 'Impact of access to credit on agricultural productivity: Evidence from smallholder cassava farmers in Nigeria', A Contributed paper prepared for the *International Conference of Agricultural Economists* (ICAE), Milan, Italy 9–14 August 2015.

Azizi, S. (2018) 'The impacts of workers' remittances on human capital and labor supply in developing countries', *Economic Modelling*, 75: 377–396. https://doi.org/10.1016/j.econmod.2018.07.011

Bettin, G. and Zazzaro, A. (2018) 'The impact of natural disasters on remittances to low- and middle-income countries', *Journal of Development Studies*, 54(3): 481–500. https://doi.org/10.1080/00220388.2017.1303672

Bhandari, P. (2019) 'Rural households' allocation of remittance income in agriculture in Nepal', *Global Journal of Agricultural and Allied Sciences*, 1(1): 1–10. https://doi.org/10.35251/gjaas.2019.001

Bouoiyour, J. and Miftah, A. (2016) 'Education, male gender preference and migrants' remittances: Interactions in rural Morocco', *Economic Modelling*, 57: 324–331. https://doi.org/10.1016/j.econmod.2015.10.026

Chalise, B. (2014) 'Remittance and its effect on entrepreneurial activities: A case study from Kandebas village development committee, Nepal', *Conference Paper*, Izmir University International Conference in Social Science, Izmir, Turkey, Volume 2. www.researchgate.net/publication/267686732.

Chauvet, L., Gubert, F. and Mesplé-Somps, S. (2009) 'Are remittances more effective than aid to reduce child mortality? An empirical assessment using inter and intra-country data', *G-MonD Working Paper Number 10*, Paris: Paris School of Economics.

Chauvet, L., Gubert, F. and Mesple-Somps, S. (2013) 'Aid, remittances, medical brain drain and child mortality: Evidence using inter and intra-country data', *Journal of Development Studies*, 49(6): 801–818. https://doi.org/10.1080/00220388.2012.742508

Chiodi, V., Jaimovich, E. and Montes-Rojas, G. (2012) 'Migration, remittances and capital accumulation: Evidence from rural Mexico', *Journal of Development Studies*, 48(8): 1139–1155. https://doi.org/10.1080/00220388.2012.688817

Filli, F.B., Onu, J.I., Adebayo, E.F. and Tizhe, I. (2015) 'Factors influencing credits access among small scale fish farmers in Adamawa State, Nigeria', *Journal of Agricultural Economics, Environment and Social Sciences*, 1(1): 46–55.

Graham, E., Jordan, L.P. and Yeoh, B.S. (2015) 'Parental migration and the mental health of those who stay behind to care for children in South-East Asia', *Social Science and Medicine*, 132: 225–235. https://doi.org/10.1016/j.socscimed.2014.10.060.

Howard, M. (1983) 'The neoclassical theory of capital productivity: Profit determined by the supply of and the demand for capital', in *Profits in Economic Theory. Radical Economics*, London: Palgrave, pp. 103–110. https://doi.org/10.1007/978-1-349-17142-2_12.

International Organization for Migration (2020) *World migration report 2020*, Geneva: International Organization for Migration.

Ivlevs, A., Nikolova, M. and Graham, C. (2019) 'Emigration, remittances, and the subjective well-being of those staying behind', *Journal of Population Economics*, 32: 113–151. https://doi.org/10.1007/s00148-018-0718-8

Khanal, U., Alam, K., Khanal, R. and Regmi, P. (2015) 'Implications of out-migration in rural agriculture: A case study of Manapang village, Tanahun Nepal', *The Journal of Developing Areas*, 49(1): 331–352.

Maharjan, A. and Knerr, B. (2019) 'Impact of migration and remittances on agriculture: A micro-macro-analysis', in Thapa, G., Kumar, A. and Joshi, P. (eds.), *Agricultural transformation in Nepal*, Singapore: Springer. https://doi.org/10.1007/978-981-32-9648-0_14.

Mohapatra, S., Joseph, G. and Ratha, G. (2009) *Remittances and natural disasters: Ex-post response and contribution to ex-ante preparedness*, Washington, DC: World Bank. https://doi.org/10.1596/1813-9450-4972

Mtanthiko, D.N. (2012) 'The effective financing of the agriculture sector in Malawi', *The 5th African rural and Agricultural Credit Association (AFRACA) Agribusiness Forum*, Serena Hotel, Kigali, 24–25 July, 2012.

Naudé, W., Siegel, M. and Marchands, K. (2017) 'Migration, entrepreneurship and development: Critical questions', *Journal of Migration*, 6: 5. https://doi.org/10.1186/s40176-016-0077-8

Nguyen, C.V. (2016) 'Does parental migration really benefit left-behind children? Comparative evidence from Ethiopia, India, Peru and Vietnam', *Social Science and Medicine*, 153: 230–239. https://doi.org/10.1016/j.socscimed.2016.02.021.

Ratha, D. (2007) *Leveraging remittances for development*. Available at: www.migrationpolicy.org/research/leveraging-remittances-development (Accessed 10 June 2020).

Shapiro, A.F. and Mandelman, F.S. (2014) 'Remittances, entrepreneurship, and employment dynamics over the business cycle', *Federal Reserve Bank of Atlanta, Working Paper 2014–19*, November 2014.

Taylor, E.J. (1999) 'The new economics of labour migration and the role of remittances in the migration process', *International Migration*, 37(1): 63–88. https://doi.org/10.1111/1468-2435.00066

Taylor, E.J. (2006) 'International migration and economic development', *International Symposium on International Migration and Development*, Population Division, Department of Economic and Social Affairs, United Nations Secretariat, Turin, Italy, 28–30 June 2006.

Terrelonge, S.C. (2014) 'For health, strength, and daily food: The dual impact of remittances and public health expenditure on household health spending and child health outcomes', *Journal of Developing Studies*, 50(10): 1397–1410. https://doi.org/10.1080/00220388.2014.940911

World Bank (2006) *Global economic prospects: Economic implication of remittances and migration*, Washington, DC: World Bank.

World Bank (2013) *Migration and entrepreneurship in Nepal with a focus on youth: An initial analysis*, Nepal: The World Bank Group.

World Bank (2020) *COVID-19 crisis through a migration lens, Migration and development Brief 32, April 2020*, Washington, DC: World Bank.

Yang, D. (2006) 'International migration, remittances and household investment: Evidence from Philippine migrants' exchange rate shocks', *National Bureau of Economic Research Working Paper 12325*. www.nber.org/papers/w12325

Zhu, N. and Luo, X. (2008) 'The impact of remittances on rural poverty and inequality in China', *Policy Research Working Paper 4637*, East Asia and Pacific Region: World Bank.

35
MOBILE MONEY, FINANCIAL INCLUSION AND LIVELIHOODS IN THE GLOBAL SOUTH

Stanley Kojo Dary, Abdulai Adams and Shamsia Abdul-Wahab

Introduction and definitions

Mobile Money (MM) is a relatively new financial innovation in the Global South. Since its introduction by telecommunication networks, there has been an upsurge in the usage of MM services by the poor, unbanked and underbanked as well as the banked in the Global South. The price and non-price barriers to accessing formal financial services, leading to financial exclusion of certain groups, fuelled the MM revolution. MM is a form of electronic money where transactions are made using a mobile phone which is connected to a mobile number as a device. Relative to traditional financial services, MM enables cheap, reliable, convenient and faster money transfers between people as well as businesses. Aron (2017: 4) conceived MM as 'financial transaction services potentially available to anyone owning a mobile phone, including the underbanked and unbanked global poor who are not a profitable target for commercial banks'.

Access and usage of formal financial services remain limited in the Global South (Ky et al., 2021). For instance, women particularly face greater social, economic, cultural, educational, technological, and legal and regulatory barriers in accessing formal financial services (Holloway et al., 2017; Pitcaithly et al., 2016; World Bank, 2017). Also, smallholder farmers have little or no access to formal credit (International Finance Corporation, 2014) as formal financial services are limited in rural areas (International Labour Organization [ILO], 2019). MM has become a pathway to financial inclusion (Groupe Spécial Mobile Association [GSMA], 2020), being 'a process that ensures the ease of access, availability and usage of the formal financial system for all members of an economy' (Sarma, 2008: 3). Financial inclusion has been a goal pursued in many countries due to its role in economic empowerment and poverty and inequality reduction (Munyegera and Matsumoto, 2016; Bongomin et al., 2018). It enables access to broad financial services, particularly for the poor and disadvantaged groups in society, by removing price and non-price barriers (Demirguc-Kunt and Klapper, 2012). Financial inclusion covers a broad range of financial services including credit, savings, insurance, payments and remittances (Dev, 2006). MM facilitates financial inclusion, which tends to enhance the performance of livelihood activities and hence improve livelihood outcomes (Conrad and Meike, 2016; Sekabira and Qaim, 2017), and is also a livelihood activity in itself for merchants and agents in the MM value chain.

The rest of the chapter is organised as follows. In the following section, the traditional financial system is examined in relation to the types of financial institutions and constraints relating to access. The evolution of MM, its interaction with the existing financial system and its impact on financial inclusion follows. The impact of MM on livelihood activities and outcomes are then presented, followed by a conclusion.

Traditional financial institutions in the Global South

The type and nature of traditional financial institutions (FIs) varies from one country to another. However, they are broadly classified as formal, semi-formal and informal FIs (Aryeetey, 2008). The formal FIs are those that are licensed by central banks and incorporated under most countries' laws. The formal FIs consist of commercial and development banks and non-bank financial institutions (NBFIs) such as rural banks, post banks, savings and loan companies, and microfinance banks (Steel, 2006). These institutions are mostly endowed with infrastructure and systems, funds and prospects for portfolio diversification which enables them to offer a variety of services to their clients (Tweneboah Senzu, 2017). Their services are often inaccessible to the rural and urban poor but are available to the urban middle-income and high-income groups (Aryeetey, 2008; Steel and Andah, 2003). In Ghana for instance, Mensah (2004) notes that the low use of formal finance is a result of the undeveloped financial sector with low levels of intermediation, lack of institutional and legal structures that can ease the management of small enterprise lending risk, high cost of borrowing and inflexibility in interest rates.

Semi-formal finance is the second category of traditional finance, and refers to legally registered institutions which are not normally licensed as FIs by central banks. Savings and loan companies, credit unions and other microfinance institutions operate under the semi-formal financial system (Steel, 2006). These institutions are noted for their very active roles in financing small businesses, albeit their services are restricted (Osei-Assibey et al., 2012); for instance, they are barred from services such as current accounts or demand deposits (Tweneboah Senzu, 2017). Notwithstanding their restricted services, Osei-Assibey et al. (2012) observed that these institutions often receive support from development agencies and government because their services are targeted at the poor, women and vulnerable groups. They also give collateral-free small loans to small businesses. Despite this support, the services of semi-formal institutions have not been able to adequately reach the target beneficiaries due to high default rates, over-reliance on government and donors for funding, limited coverage and pervasive political patronage (Mensah, 2004; Schicks, 2013).

The informal FIs are those that are not regulated and have no legal backing within the laws of the countries in which they operate. As such, they are often convenient and less bureaucratic for the rural and urban poor. The informal FIs includes Susu, moneylenders, trade creditors, self-help groups, government credit programmes and personal loans from friends and relatives. The Susu system, which is often referred to as Rotating Savings and Credit Associations (ROSCAs) (see Chapter 36 in this volume), is the most widespread informal financial institution in sub-Saharan Africa (Etang et al., 2011). Although the Susu system has a credit component, its major feature is how it enables clients to build up their own savings plan within a period of one month to two years (Steel and Andah, 2003). Despite the credit and savings advantage offered by the Susu system, it has several challenges including high risk of loss of monies through robberies of Susu collectors, Susu collectors absconding with people's monies and the inability of Susu collectors to lend more to their clients than the amount they have saved (Osei, 2007).

A greater proportion of the population, notably the poor, women and other marginalised groups are financially excluded from the formal financial system and to some extent the semi-formal financial system due to several constraints to access such as costly services, lack of geographical reach, collateral requirements, less flexibility and bureaucratic procedures. This explains why the informal financial system has thrived over the years in the Global South.

Mobile money and financial inclusion in the Global South

The evolution of MM and its role in financial inclusion

MM has been acclaimed as the most successful innovation in the financial system of the Global South (Suri, 2017), where financial exclusion is very high (Ky et al., 2021), particularly among women and persons operating in the informal economy (GSMA, 2020). Thus, MM is more prominent in cash-based countries; it is not prominent in the Global North as it faces competition from entrenched financial products (Aron, 2017). The proliferation of mobile phones has aided the adoption of the MM technology (Lashitew et al., 2019). The MM system has evolved over the years allowing for the performance of more services including (i) storage of money (savings), (ii) bill and utility payments, (iii) person-to-person transfers, (iv) payment for goods and services, (v) borrowing money, (vi) international transfers (remittances), (vii) purchasing airtime and data, (viii) receipt/payment of salaries, wages, and social security/social transfers, (ix) payment of insurance, and (x) payment of dues and fees.

Although the MM technology was first deployed in 2000, 2007 has been generally taken as the effective year for the commencement of the MM industry (Aron, 2017; GSMA, 2017, 2020; Rea and Nelms, 2017) with the launch of M-Pesa in Kenya. The success story of M-Pesa, which has dominated the MM literature, has spearheaded the adoption and growth of MM worldwide. The adoption of MM has been fast-paced and widespread in the Global South, notwithstanding that adoption and success rates differ significantly across and within countries (Lashitew et al., 2019). At the close of 2019, the MM technology was in service in 95 countries, with 1.04 billion MM accounts and $1.9 billion being processed daily (GSMA, 2020). Sub-Saharan Africa is leading in all dimensions of MM – registered accounts, active accounts, transaction volume and transaction value. Within sub-Saharan Africa, East Africa is leading in all dimensions by far (GSMA, 2020). Aron (2017) states that MM has 'leapfrogged' the provision of formal financial services in the Global South. In some countries, there are more MM accounts than bank accounts and more MM agents than bank branches and ATMs (GSMA, 2020; Suri, 2017). MM agents refer to persons or businesses contracted to facilitate MM transactions such as cash-in and cash-out services, new customer registration and customer education (GSMA, 2011).

Anyone with a mobile handset and a mobile number in jurisdictions where MM is operating can register for MM. The registration is done by an MM agent and the only document required is a valid national ID card. In some cases (e.g. Vodafone in Ghana), customers can self-register through the menu on the sim card. The registration process is completed within minutes and there are no registration fees. The MM account is activated by depositing money through any MM agent and thereafter transactions can commence immediately. At the MM agent's kiosk, a customer receives electronic money (e-money) by depositing cash and gives out e-money during cash withdrawal. While there is no minimum MM account balance requirement, there is a limit to how much money a MM account can hold and to transaction sizes, but this varies from country to country. Suri (2017) for instance reports a maximum account balance of $1,000 and $1,200 in Kenya and Uganda, respectively. Deposits into an MM account do not attract transaction fees, only monies leaving the account, and fees charged are usually conditioned on the transaction amount. Aside from a deposit, all MM transactions require a pin code. The Mobile Money Operators (MNOs), i.e. the telecommunication network operators, pay interest on deposits/savings (this is not applicable in all MM jurisdictions). Also, borrowing services are available from some MM operators and the amount available for borrowing is graduated with each successful repayment. Some MNOs have developed MM applications, but this requires customers to have a smartphone.

MM emerged to overcome the barriers of access to formal financial services especially for the poor and uneducated (Aron, 2017; Ky et al., 2021). The barriers underscored in the literature include insufficient funds, distance, weak financial institutions, high cost of running accounts and documentation requirements (Aron, 2017; Demirgüç-Kunt and Klapper, 2012; Ky et al., 2021). These barriers and the unattractiveness of the poor to banks (Aron, 2017) have led to financial exclusion of some populations. MM has thus become the path to financial inclusion in most low-income countries (GSMA, 2020). MM is increasing access of the poor and unbanked/underbanked to financial services (Lashitew et al., 2019; Rea and Nelms, 2017). Access to formal financial services through MM is another pathway to financial inclusion (Aron, 2017; Ky et al., 2021). With MM, more women are using financial services, low-income households are accessing essential utility services and smallholder farmers are getting paid more quickly and conveniently (GSMA, 2020). MM has brought a transformative impact on financial inclusion in rural areas where it is not commercially viable for formal banks to operate (GSMA, 2020). The MM systems have the potential to drive the formalisation and modernisation of the characteristically large informal sectors of most countries in the Global South (Foster and Heeks, 2013).

Unique advantages and challenges of mobile money systems

The MM system has several unique advantages relative to formal financial systems: (a) because of its additional social function, MM has the advantage of increasing the access of the poor, unbanked/underbanked and persons remotely located to financial services, (b) MM operations require less investment in direct and complementary infrastructure such as buildings, electricity and internet (Rea and Nelms, 2017), (c) reduced transaction costs for MNOs, MM agents and MM customers (Aron, 2017; Suri, 2017), (d) MM provides easy, convenient, faster, cheaper and secure means of financial transactions. Most MM transactions can be done at home or with the nearest MM agent, (e) There is no minimum balance requirement and account maintenance or servicing fees, making it attractive to both the banked and unbanked, especially for poor people (Aron, 2017), (f) MM has geographical reach advantage and is found in places where banks cannot thrive (Munyegera and Matsumoto, 2016). Aron (2017) states that the cost of creating bank branches in rural unbanked locations is high, especially when weighed against the returns, and (g) MM system has technological and information advantages over banks. Both MM agents and customers need only a mobile phone to undertake a financial transaction, which most people already have, including poor and illiterate populations.

Notwithstanding these unique advantages of MM systems, a number of challenges exists including poor network coverage and breakdowns, malfunctioning of MM systems leading to service delays or aborted services, activities of fraudsters, liquidity problems on the part of MM agents, robberies of MM agents, limits on MM deposits and transactions, electricity supply in remote areas, and challenges relating to government regulation (Davidson and McCarty, 2011; Donovan, 2012; Rea and Nelms, 2017). In the presence of liquidity challenges, there are issues of favouritism or marginalisation of certain customers on the part of MM agents (Rea and Nelms, 2017).

The interaction between mobile money and existing financial services

MM may serve as a substitute or complement to existing financial services (GSMA, 2020; Ky et al., 2021; Ramada-Sarasola, 2012). Ky et al. (2021) note that MM may act as a substitute for formal financial services particularly for poor people or for those living in unbanked locations. MM is moving people from informal means of storing money to mobile storage via mobile phone. MM can also be a complement to informal financial services to cater for specific financial needs (ibid). For those with access to formal financial services, MM acts as a complement, enhancing the efficiency of existing financial services. Aron (2017) states that MM payments linked to users' bank accounts are common in countries with larger financial sectors relative to the size of their economies. There is evidence that people with bank accounts are also operating MM accounts and patronising MM services (Demirgüç-Kunt and Klapper, 2012; GSMA, 2020). The interlinkages between MM and formal financial services are growing. At the institutional level, a significant development in the finance industry is the increasing integration and interoperability between MM providers and banks. Increasingly, MM providers (about 70%) are partnering with banks to offer credit (GSMA, 2020) and these partnerships are essential for efficient MM services delivery (Lashitew et al., 2019). MM is thus complementing the formal banking sector while also meeting the needs of entirely new customer segments, including traditionally underserved and cash-reliant customers (GSMA, 2020).

At the individual level, the effect of MM on existing financial services can be transformative, where the financially excluded gain access to formal financial services, or additive, where MM further increases the level of financial inclusiveness of persons already with access to formal financial services (Mas and Porteous, 2015; Ky et al., 2021). Hence MM can motivate people to patronise bank services. For bank account holders, MM accounts' subscription and their integration with their bank accounts may facilitate increased savings in the latter (Munyegera and Matsumoto, 2016). GSMA (2020) reports that in 48 of the 95 countries where the MM technology is deployed, customers are able to transfer money between accounts held with different MM providers and banks. The empirical literature exploring MM impact on formal and informal financial services is still at the developmental stage. Notwithstanding, initial research has demonstrated a significant impact of MM on other forms of financial services. Ky et al.'s (2021) study in Burkina Faso showed MM has a positive impact on the patronage of formal channels for savings by individuals using informal channels, individuals with irregular income, the less educated and women. In Kenya, using MM raises and lowers the probability of using formal and informal savings channels, respectively.

Impact of mobile money on livelihood activities and outcomes

Mobile money and employment creation

High unemployment, especially among the teeming youth, is a global concern. This was exacerbated by the COVID-19 pandemic which saw rising levels of unemployment in both the Global South and North. In Africa, MM serves as a vehicle for job creation, with MTN, a leading telecommunication company, claiming that they alone have created some 400,000 jobs (MTN, 2019). The new ecosystem created by MM impacts directly on job creation both at the agent and merchant levels. Firstly, with a small amount of cash and a mobile phone, people are able to register and operate as MM agents. Secondly, innovative entrepreneurs are able to leverage on MM and collect payments digitally or operate as merchants. As at December 2018, there were 120,000 MM merchants working at MTN in Africa (MTN, 2019). MM is therefore central to empowering entrepreneurs who not only grow their business and create more jobs along the chain but also drive social and economic growth. Job creation which is directly linked to the financial inclusion drive hinges on infrastructure provision especially mobile phone subscriptions. Rural and urban unemployment ratios are significant factors that decrease financial inclusiveness (Ngo, 2019). However, issues such as poor network signals and power supply pose major hindrances to the job creation potential of MM especially in remote villages without electricity supply (Adaba et al., 2019).

MM and women empowerment

MM expands people's capabilities and empowers marginalised groups particularly women to effectively participate in financial markets (Adaba et al., 2019). This capability enhances household decision-making processes with better outcomes for health, education and financial wellbeing (Adaba et al., 2019). With MM, women in particular are able to address their urgent needs through remittances received from family members and social networks. However, this feature of MM could diminish people's capability and create long-term dependency especially among vulnerable groups if not checked (Adaba et al., 2019). There is a need to capture MM trends and the wellbeing of women on a broader spectrum. Furthermore, financial services enhance food security especially for women in three ways: (1) social protection payments received by women increases their food expenditures, (2) increased bargaining power with men on household income spent on food, and (3) increased access to quality farm inputs due to separate savings accounts held (Kumaraswamy and Bin-Human, 2019). Evidence of crop sales payments made directly into male and female tobacco farmers' savings accounts in Malawi showed that women have increased access to agricultural inputs, seen by an increase in inputs expenditure of about 13% (Brune et al., 2015).

Mobile money and food security and nutrition

A number of recent studies in Africa have analysed the impact of MM on the welfare of people through interventions in agricultural-related activities and reported positive impacts on income, consumption expenditure and commercialisation in production due to access to inputs (Conrad and Meike, 2016; Kikulwe et al., 2014; Kirui et al., 2013; Peprah et al., 2020). MM services which are now popular among households have helped in addressing credit market failures especially in rural areas where formal banking rarely exists. In agriculture, remittances made through MM help in reducing risks, minimising liquidity constraints and enhancing timely access to farm inputs. The use of MM and the volume transacted is directly linked to food security improvement (Conrad and Meike, 2016).

The link between financial access and food security can be viewed in six dimensions: income for family needs (money), transparent market pricing (gains in households purchasing power for food items), adequate storage infrastructure (minimisation of food losses), resource availability (sustainable supply of food in the value chain) and investments in production (access to quality inputs and equipment, increased yields) (Gencer, 2011). These dimensions of food security are underpinned by access to finance of which MM plays a key role in enabling financial access, especially to those at the base of the pyramid in the Global South. For instance, lack of finance, which is a persistent feature of smallholder agriculture, results in low crop yields leading to low income and poor health of farm families. Also, in the absence of viable crop insurance schemes or credit, farmers are exposed to high production risks (bushfires, drought, floods etc.), reducing potential levels of agriculture investments and contributing to malnutrition and deterioration of the health status of families. It is thus evident that access to financial services impacts positively on all dimensions of food security.

Market participation by smallholder farmers in a more transparent manner through the use of mobile phones is well documented (e.g. Gencer, 2011; Munyegera and Matsumoto, 2016) as it enables farmers to find buyers remotely and receive MM payments easily, thus reducing transaction costs. MM services also impact positively on the ownership of savings accounts by farmers and reduce malnutrition and hunger (Kumaraswamy and Bin-Human, 2019). Other impact pathways such as agricultural marketing and off-farm economic activities that benefit from MM, contribute to high incomes and consumption (Sekabira and Qaim, 2017). For instance, in Uganda, MM users engaged in coffee marketing are reported to earn higher prices (7% more) and sell larger volumes of shelled coffee beans to buyers in high-value markets than non-users (Sekabira and Qaim, 2017). MM thus impacts on agriculture in diverse ways with largely positive outcomes on households and agribusinesses.

MM and social inclusion

The role of social networks in enabling rural financial inclusion is well documented in the literature (e.g. Bongomin et al., 2018; Peprah et al., 2020). The use of MM does not simply promote financial inclusion but also has a direct significant effect on social networks (Bongomin et al., 2018). Social activities such as funerals, weddings and naming ceremonies are better planned and executed as families are able to save in their MM wallets. MM has therefore proven to contribute to smoothing out the income over time for the poor, enabling better liquidity in food trade and creating more efficiency in markets especially for the poor.

MM and health improvement

Access to basic health services is a growing concern although mobile health (mheath) systems' development is on a growth trajectory. Access to mhealth platforms remains limited and their integration with existing healthcare systems poor (GSMA, 2013). The introduction of mobile financial services (MFS) holds the promise of reducing costs and enhancing efficiency in healthcare services delivery (affordability and inclusiveness) systems for the poor. A large proportion of the over 900 mhealth products and services that exist globally are in Africa and this is expected to result in reductions in healthcare costs, expansion in health service coverage, and stimulate value addition (GSMA, 2013). The use of MM by healthcare providers could bring more transparency in salary payments and financial flows. The current cash payment system in place in most countries in the Global South is not only insecure and risky but also cumbersome with cost implications. Salary payments through MM accounts could reduce transaction costs and increase the impact of health services' delivery to patients. For instance, in Pakistan, the use of MM to make payments to tuberculosis (TB) screening health workers in communities led to increased reported cases (GSMA, 2013). This suggests that MM can serve as an incentive tool to reward healthcare performance with dramatic effects on health outcomes. Besides, by providing a more secure easy payment option for households, MM can also spur access to healthcare services, and enable prompt funds mobilisation to cover healthcare costs during emergencies (Obadha et al., 2018).

Conclusion

This chapter examined traditional financial services and the challenges or constraints associated with access. It argued that per the structure and operation of the formal financial systems, certain groups, especially the poor and unbanked/underbanked groups are financially excluded in the Global South. The MM innovation developed in response to the challenges with access to formal financial services. The MM system is in pursuance of not only commercial goals but social goals as well. The evidence that MM is enhancing financial inclusion is overwhelming. There are also interactions between the MM system and the formal financial system. The formal financial intermediaries are adapting to the MM innovation as well. Financial inclusion is an intermediate outcome of MM. Financial inclusion through MM has had a positive impact on livelihoods and poverty in the Global South. Government regulation should take into account the social goals of MM so that evolving regulation will strengthen the social goals as opposed to their abandonment.

Key Points

- MM is a recent financial innovation, predominantly in the Global South where the financial system is undeveloped and financial exclusion is high.
- MM emerged to overcome both the demand-side and supply-side barriers of access to formal financial services.
- MM is fostering financial inclusion by increasing access of the poor, uneducated, unbanked/underbanked and persons in remote locations to financial services.
- MM-led financial inclusion is impacting positively on various livelihood activities and outcomes such as employment, food and nutrition security, health and women empowerment in the Global South.

Further readings

Della Peruta, M. (2018) 'Adoption of mobile money and financial inclusion: A macroeconomic approach through cluster analysis', *Economics of Innovation and New Technology*, 27(2): 154–173. https://doi.org/10.1080/10438599.2017.1322234.

Maurer, B. (2012) 'Mobile money: Communication, consumption and change in the payments Space', *Journal of Development Studies*, 48(5): 589–604. https://doi.org/10.1080/00220388.2011.621944

Suri, T. and Jack, W. (2016) 'The long run poverty and gender impacts of mobile money', *Science*, 354(6317): 1288–1292. https://doi.org/10.1126/science.aah5309

References

Adaba, G.B., Ayoung, D.A. and Abbott, P. (2019) 'Exploring the contribution of mobile money to well-being from a capability perspective', *The Electronic Journal of Information Systems in Developing Countries*, 85(4): 1–11. https://doi.org/10.1002/isd2.12079.

Aron, J. (2017) '"Leapfrogging": A survey of the nature and economic implications of mobile money', *Working Paper Series 2017–02*, Oxford: Centre for the Study of African Economies, University of Oxford.

Aryeetey, E. (2008) 'From informal finance to formal finance in sub-Saharan Africa: Lessons from linkage efforts', paper presented at the *High-Level Seminar on African Finance for the 21st Century*, organized by the IMF Institute and the Joint Africa Institute, Tunis, Tunisia, March 4–5, 2008. http://citeseerx.ist.psu.edu/viewdoc/download?doi=10.1.1.617.4016&rep=rep1&type=pdf

Bongomin, G.O.C., Ntayi, J.M., Munene, J.C. and Malinga, C.A. (2018) 'Mobile money and financial inclusion in sub-Saharan Africa: The moderating role of social networks', *Journal of African Business*, 19(3): 361–384. https://doi.org/10.1080/15228916.2017.1416214

Brune, L., Gine, X., Goldberg, J. and Yang, D. (2015) 'Facilitating savings groups for agriculture: Field experimental evidence from Malawi', *NBER Working Paper Series 20946*, Cambridge, MA: National Bureau of Economic Research. www.nber.org/papers/w20946

Conrad, M. and Meike, W. (2016) 'Mobile money and household food security in Uganda', *Global Food Discussion Paper No. 76*, Göttingen: Georg-August-Universitat. 10.22004/ag.econ.229805.

Davidson, N. and McCarty, Y. (2011) *Driving customer usage of mobile money for the unbanked*, London: GSMA. www.gsma.com/mobilefordevelopment/wp-content/uploads/2012/03/drivingcustomerusagefinallowres.pdf

Demirgüç-Kunt, A. and Klapper, L. (2012) 'Measuring financial inclusion: The global Findex database', *Policy Research Working Paper 6025*, Washington, DC: The World Bank. https://doi.org/10.1596/1813-9450-6025

Dev, S.M. (2006) 'Financial inclusion: Issues and challenges', *Economic and Political Weekly*, 41(41): 4310–4313. www.jstor.org/stable/4418799

Donovan, K. (2012) 'Mobile money for financial inclusion', in World Bank (ed.), *Information and Communications for Development 2012: Maximizing Mobile*, Washington, DC: World Bank, pp. 61–73. 10.1596/978-0-82138991-1

Etang, A., Fielding, D. and Knowles, S. (2011) 'Trust and ROSCA membership in rural Cameroon', *Journal of International Development*, 23(4): 461–475. https://doi.org/10.1002/jid.1686.

Foster, C. and Heeks, R. (2013) 'Innovation and scaling of ICT for the bottom-of-the-pyramid', *Journal of Information Technology*, 28(4): 296–315. https://doi.org/10.1057/jit.2013.19.

Gencer, M. (2011) 'Mobile money: A foundation for food security', *Innovations: Technology, Governance, Globalization*, 6(4): 73–79. https://doi.org/10.1162/INOV_a_00102.

Groupe Spécial Mobile Association (GSMA) (2020) *State of the industry report on mobile money 2019*, London: GSMA. www.gsma.com/sotir/wp-content/uploads/2020/03/GSMA-State-of-the-Industry-Report-on-Mobile-Money-2019-Full-Report.pdf

GSMA (2011) *Annual report: Mobile money for the unbanked*, London: GSMA. www.gsma.com/mobilefordevelopment/wp-content/uploads/2012/06/mmu_report.pdf

GSMA (2013) *Mobile money: Transforming health care in emerging markets*, London: GSMA. www.gsma.com/mobilefordevelopment/programme/mhealth/mobile-money-transforming-healthcare-in-emerging-markets/

GSMA (2017) *State of the industry report on mobile money*, Decade edition: 2006–2016, London: GSMA. www.gsma.com/mobilefordevelopment/wp-content/uploads/2017/03/GSMA_State-of-the-Industry-Report-on-Mobile-Money_2016.pdf

Holloway, K., Niazi, Z. and Rouse, R. (2017) *Women's economic empowerment through financial inclusion: A review of existing evidence and remaining knowledge gaps*, New Haven, Connecticut: Innovations for Poverty Action. www.poverty-action.org/sites/default/files/publications/Womens-Economic-Empowerment-Through-Financial-Inclusion.pdf

International Finance Corporation (2014) *Access to finance for smallholder farmers: Learning from the experiences of microfinance institutions in Latin America*, Washington, DC: International Finance Corporation (IFC). https://documents1.worldbank.org/curated/en/965771468272366367/pdf/949050WP0Box3800English0Publication.pdf

International Labour Organization (ILO) (2019) *Developing the rural economy through financial inclusion: The role of access to finance*, Geneva: International Labour Organization. www.ilo.org/wcmsp5/groups/public/ – ed_dialogue/ – sector/documents/publication/wcms_437194.pdf

Kikulwe, E.M., Fischer, E. and Qaim, M. (2014) 'Mobile money, smallholder farmers, and household welfare in Kenya', *PLoS One*, 9(10): e109804. https://doi.org/10.1371/journal.pone.0109804

Kirui, O.K., Okello, J.J., Nyikal, R.A. and Njiraini, G.W. (2013) 'Impact of mobile phone- based money transfer services in agriculture: Evidence from Kenya', *Quarterly Journal of International Agriculture*, 52(2): 141–162. https://doi.org/10.22004/ag.econ.173644.

Kumaraswamy, S.K. and Bin-Human, Y. (2019) *Three ways financial inclusion improves women's food security*, CGAP Blog, 28 March 2019. Available at: www.cgap.org/blog/3-ways-financial-inclusion-improves-womens-food-security (Accessed 21August 2020).

Ky, S.S., Rugemintwari, C. and Sauviat, A. (2021) 'Friends or foes? Mobile money interaction with formal and informal finance', *Telecommunications Policy*, 45(1): 102057. https://doi.org/10.1016/j.telpol.2020.102057

Lashitew, A.A., van Tulder, R. and Liasse, Y. (2019) 'Mobile phones for financial inclusion: What explains the diffusion of mobile money innovations?', *Research Policy*, 48(5): 1201–1215. https://doi.org/10.1016/j.respol.2018.12.010.

Mas, I. and Porteous, D. (2015) 'Pathways to smarter digital financial inclusion', *CAPCO Institutes Journal of Financial Transformation*, 42: 1–28. http://dx.doi.org/10.2139/ssrn.1858377.

Mensah, S. (2004) 'A review of SME financing schemes in Ghana', A paper presented at the *UNIDO Regional Workshop of Financing SMEs*, Accra, Ghana. http://semcapitalgh.com/downloads/research/SME_Financing_Schemes_in_Ghana.pdf

MTN (2019) *The ripple effect of mobile money is changing the lives of people in Africa and beyond*. Available at: www.mtn.com/our-story/spotlights/mobile-money-contributes-to-400-000-jobs-in-africa/ (Accessed 15 August 2020).

Munyegera, G.K. and Matsumoto, T. (2016) 'Mobile money, remittances, and household welfare: Panel evidence from rural Uganda', *World Development*, 79: 127–137. https://doi.org/10.1016/j.worlddev.2015.11.006.

Ngo, A.L. (2019) 'Index of Financial Inclusion and the Determinants: An Investigation in Asia', *Asian Economic and Financial Review*, 9(12): 1368–1382. https://doi.org/10.18488/journal.aefr.2019.912.1368.1382

Obadha, M., Seal, A. and Colbourn, T. (2018) *Mobile money increasing healthcare access*. Available at: www.scidev.net/sub-saharan-africa/health/opinion/mobile-money-increasing-healthcare-access.html (Accessed 15 August 2020).

Osei, R.D. (2007) *Linking traditional banking with modern finance: Barclays microbanking Susu collectors initiative*, Growing Inclusive Markets Case Study, New York: UNDP. www.inclusivebusiness.net/node/1540

Osei-Assibey, E., Bokpin, G.A. and Twerefou, D.K. (2012) 'Microenterprise financing preference', *Journal of Economic Studies*, 39: 84–105. https://doi.org/10.1108/01443581211192125.

Peprah, J.A., Oteng, C. and Sebu, J. (2020) 'Mobile money, output and welfare among smallholder farmers in Ghana', *SAGE Open*, 10(2): 1–12. https://doi.org/10.1177/2158244020931114.

Pitcaithly, L.A., Biallas, M.O., Japhta, R. and Murthy, P. (2016) 'Research and literature review of challenges to women accessing digital financial services', *Working Paper No. 117455*, IFC mobile money toolkit, Washington, D.C: World Bank Group.

Ramada-Sarasola, M. (2012) *Can mobile money systems have a measurable impact on local development?* Available at: SSRN 2061526, https://papers.ssrn.com/sol3/papers.cfm?abstract_id=2061526 (Accessed 19 April 2020).

Rea, S.C. and Nelms, T.C. (2017) 'Mobile money: The first decade', *Institute for Money, Technology and Financial Inclusion Working Paper, 2017–1*, Irvine: University of California. www.imtfi.uci.edu/files/docs/2017/Rea_Nelms_Mobile%20Money%20The%20First%20Decade%202017_3.pdf

Sarma, M. (2008) 'Index of financial inclusion', *Working Paper No. 215*, New Delhi: Indian Council for Research on International Economic Relations (ICRIER). www.icrier.org/pdf/Working_Paper_215.pdf

Schicks, J. (2013) 'The sacrifices of micro-borrowers in Ghana: A consumer protection perspective on measuring over-indebtedness', *Journal of Development Studies*, 49(9): 1238–1255. https://doi.org/10.1080/00220388.2013.775421

Sekabira, H. and Qaim, M. (2017) 'Mobile money, agricultural marketing, and off-farm income in Uganda', *Agricultural Economics*, 48(5): 597–611. https://doi.org/10.1111/agec.12360.

Steel, W. and Andah, D. (2003) 'Rural and microfinance regulation in Ghana: Implications for development and performance of the industry', *Africa Region Working Paper, No 49*, Washington D.C: World Bank. https://documents1.worldbank.org/curated/en/402221468749372797/pdf/266520AFR0wp49.pdf

Steel, W.F. (2006) *Extending financial systems to the poor: What strategies for Ghana*. 7th ISSER-Merchant Bank Annual Economic Lectures, University of Ghana, Legon.

Suri, T. (2017) 'Mobile money', *Annual Review of Economics*, 9: 497–520. https://doi.org/10.1146/annurev-economics-063016-103638.

Tweneboah Senzu, E. (2017) 'Examining the economic impact and challenges associated with Savings and Loans companies in Ghana', *MPRA Paper No. 94558*. https://mpra.ub.uni-muenchen.de/94558/

World Bank (2017) *The global Findex database: Financial inclusion data*. Available at: https://globalfindex.worldbank.org/ (Accessed 15 August 2020).

36
THE ROLE OF MICROFINANCE IN MEDIATING LIVELIHOODS

Karabi C. Bezboruah

Introduction

Microfinance is a mechanism to provide very small loans to finance and support small business enterprises. Designed as a tool to promote economic self-sufficiency for financially marginalised households in low-and-middle-income countries (LMICs), microfinance institutions offer both borrowing and savings services. Microfinance, in short, refers to small-scale financial services offered to low-income people who may not have the ability to access services provided by commercial banks. Most low-income households have very limited assets and are not able to provide collateral for loans from commercial banks. In order to meet credit needs and borrow money, they resort to private moneylenders who often charge hefty interest rates on the loans. Microfinance institutions (MFIs) aim to provide an alternative to borrowing from private moneylenders by offering small loans at little to no collateral for establishing small-scale businesses in handicrafts, livestock and similar non-farm enterprises (Cull et al., 2009). MFIs provide both financial and non-financial services to their clients. Financial services include deposits, credits and payment services. Non-financial services include insurance services, financial education and mentoring, micro-entrepreneurial development services, business or technical skills training, health education, literacy and language training, and legal advice (Badullahewage, 2020; Herath, 2018).

By providing microloans and other financial services, MFIs aim to assist economically marginalised individuals to be self-reliant through small-scale entrepreneurial activities. As a development strategy, banking and credit services were favoured poverty alleviation mechanisms for many countries between the 1950s and the 1980s, but these produced dismal results such as very low loan repayment rate, increase in the costs of subsidies, and benefits for the politically powerful instead of the intended low-income recipients (Le, 2017; Von Pischke, 1997). This led to the conclusion that more emphasis should be placed on access to credit for non-farm rural enterprises and voluntary savings mobilisation (Adams et al., 1984; Conning and Udry, 2007). In this scenario, microfinance, when introduced in South Asia in the 1970s, proved to be a unique development approach that brought about positive results by harnessing the power of microloans, group-lending and peer accountability to support non-farm enterprises (Akhter and Cheng, 2020; Morduch, 1999). Over the years, microfinance has become popular across the Global South as a developmental tool with various structures and forms of operations.

This chapter begins with a discussion of the history and evolution of microfinance, followed by a review of the pertinent literature on the successes and criticisms of microfinance. Finally, the chapter draws conclusions and identifies key points.

History and evolution of microfinance

The concept of microfinance has existed in various forms around the world and has evolved over the years. Lending systems with characteristics similar to microcredit were found in 10th century Asia and 16th century Europe. Researchers (Bouman, 1995; Seibel, 2005; Zainuddin and Yasin, 2020) refer to the *hui* in China, the *chit* funds in India, the *arisan* in Indonesia, *kye* in Korea, *samity* in Bangladesh, *kou or tanomoshi-kou* in Japan and the *paluwagan* in the Philippines as widespread practices of moneylending among the economically disadvantaged. These funding systems were the traditional versions of the modern-day Rotating Savings and Credit Associations (ROSCA), where members make regular contributions to a fund which is then distributed in rotation to members (Geertz, 1962; Shoaib and Siddiqui, 2020; Zainuddin and Yasin, 2020). The ROSCAs have been in Africa since the 16th century, known as *esusu* in Nigeria, *chita* by the Indian origin people in South Africa, *ayutta* in Somalia, *chamas* in Kenya and *gameya* in Egypt (Bouman, 1995). ROSCAs are similar to MFIs as these are group-based credit and savings schemes and rely on social capital and peer pressure for repayments (Ghatak and Guinnane, 1999; Zainuddin and Yasin, 2020).

Credit cooperatives are another source of finance for low-income people with the earliest form found in Germany in the mid-1800s. In India, the first known cooperative was established in 1906 by Rabindra Nath Tagore who provided collateral free loans to the local people through his bank (Ganguly, 2014). Similar initiatives were also formed in the 1950s in Bangladesh by Akhtar Hameed Khan through a two-tier cooperative system. However, these credit models were not successful due to inefficient management of the unions and borrowers' inability to repay the loans (Zainuddin and Yasin, 2020). These early versions led to the development of credit unions and cooperatives globally with 86,055 such institutions in 118 countries (World Council of Credit Unions, 2019).

Merchant banking and moneylenders are private traders who charge exorbitant interest on loans and are another source of credit for those that cannot access loans from commercial financial institutions. Moneylenders have origins in pre-historic times and remains a funding source among low-income households in Asia. They offer loans for small business and personal needs at very high interest rates and repayment in kind (Habib, 1964; Ray, 2019; Seibel, 2005). The excessive practices used by moneylenders are often used as examples to legitimise the role of MFIs in developing nations (Zainuddin and Yasin, 2020), and they still exist today and compete with MFIs in rural Asia (Ray, 2019; Schrader, 1997; Seibel, 2005). This is because money lenders are situated in the community, know their clients, understand the risks associated with lending to the locals and have information advantage over formal institutions. The borrowers have easy access and depend more on the local lenders than on banks for small personal loans.

The lending systems that existed since the 10th century have evolved into the modern-day MFIs. In the 1970s, a microcredit project in Bangladesh initiated group-lending to women from low-income households, which was regarded as a success because the women used the money to start small businesses and repaid the loans. This led to more group-lending programs in other countries of the Global South and the eventual growth of MFIs. According to the García-Pérez et al. (2020), MFIs worldwide serve more than 140 million borrowers and have a total loan portfolio of about 124 billion USD. The growth of MFIs came with challenges and criticisms. For example, during the 2010s, there were several instances of mismanagement, such as the excessive profits earned by the Mexico based MFI Compartamos, uncontrolled growth of the MFIs and an increase in debt cycle and over indebtedness of borrowers in the Global South. This led to operational reforms and regulation of the MFI sector. Some noteworthy developments are professionalisation and digitisation of MFIs, diversification of services to include agriculture and housing loans, business and employment training, and impact investments. What started as informal nonprofit non-governmental (NGO) credit granting institutions that relied on donations from public institutions and private individuals metamorphosed through a process of commercialisation and growth to be part of the mainstream financial sector (Brown et al., 2016; Microfinance Barometer, 2019; Schulte and Wrinkler, 2019).

Attributes of microfinance institutions

Microfinance provides access to financial services to low-income people and is considered a key strategy for the economic development of LMICs. In 2019, there were 140 million active MFI borrowers globally, of which 80% were women and 65% were rural borrowers (Microfinance Barometer, 2019). The higher participation of women is because MFIs particularly target women borrowers due to the demonstrated socio-economic impact of the loans made to women on their households and on the communities where they reside (Bezboruah and Pillai, 2017; Mukherjee, 2015). While the proportions of borrower type (women and rural borrowers) have remained stable, the number of borrowers has increased from 98 million in 2009 to 140 million in 2019. South Asia has the largest share with 85.6 million borrowers, followed by East Asia and the Pacific with 20.8 million borrowers. The other regions have about an equal number of borrowers (Microfinance Barometer, 2019).

Ownership and regulation

MFIs have the dual objectives of social and financial performance through outreach to financially excluded people, while also remaining financially sustainable (Hermes and Lensink, 2007). There are three categories of MFIs based on their primary funding source – first, financial intermediaries such as commercial banks with emphasis on microfinance; second, savings-driven financial institutions such as credit cooperatives and self-reliant village banks; and third, donor-driven NGOs with special microlending programs (Wisniwski, 1999). There are various forms of ownership of MFIs such as micro-banks, non-bank financial institutions (NBFI), credit unions and cooperatives, and NGOs. Of these, the credit unions and cooperatives and NGOs are socially responsible and focus on making meaningful social impact by serving and benefitting a large number of financially excluded people. The NBFIs and banks are commercially oriented and focus more on profits and financial returns. The type of ownership of the MFIs determines the differences in regulatory and supervisory mechanisms, governance processes, agency issues, and risk preferences of the institution (Gupta and Mirchandani, 2020). Banks and NBFIs must conform to general as well as specific banking rules, while credit cooperatives and financial NGOs are excused from banking laws but must adhere to general and commercial regulations. ROSCAs, self-help groups and similar informal MFI providers are neither subject to any banking laws nor audited by external regulatory agencies (Bezboruah and Pillai, 2015). For example, socially oriented NGOs and cooperatives are more likely to be nonregulated or semi-regulated, self-governed, and reliant on donations.

In the Global South, a majority of the MFIs are in the nonregulated and self-governed NGO sector that strives to reach as many financially excluded people as possible. This is related to the MFIs' social performance which is measured by its outreach – the average loan size is used

Table 36.1 Types of microfinance institutions

Institutions	Types	Services	Examples
Banks	• Commercial banks with microfinance services • Rural banks • Postal banks • Development banks	• Both savings and deposits • Financial education	• Grameen Bank, Bangladesh • Bancosol, Bolivia • Bandhan Bank, India • Compartamos, Mexico
Credit Unions and Cooperatives	• Savings and credit cooperatives • Financial cooperatives	• Savings and deposits • Financial education	• PAMECAS, Senegal • SACCOs, Nepal
Non-bank financial institutions (NBFIs)	• Credit only MFIs • Microcredit organisations • Nonbanking financial company MFIs • Micro-entrepreneur credit companies	• No deposits • Loans (at below market interest rates) • Financial education	• AMEEN, Lebanon • Sa-Dhan, India
NGOs	• Savings • Self-help groups • Credit only MFIs	• Some have both savings and deposits • Financial education	• SEF, South Africa • Organización Privada de Desarrollo, Honduras

to measure the depth of outreach while the number of customers is used for the breadth of outreach. Awusabo-Asare et al. (2009) found that rural MFIs in Ghana had greater outreach to all categories of clients compared to other formal financial institutions. Table 36.1 summarises the various institutions that offer loans and related services and some country specific examples.

Funding

MFIs differ substantially in their funding strategies. In the 1970s, MFIs were primarily in the NGO sector and funded through donations. This dependence on donors posed significant obstacles to organisational growth and resulted in large operational deficits. MFIs now receive financial support from various international sources such as multilateral banks and development organisations (e.g. World Bank, Asian Development Bank, European Union, United Nations, USAID), Ford Foundation, apex organisations such as ACCION, FINCA, KIVA and international private debt and equity providers through Microfinance Investment Vehicles (MIVs) (Brière and Szafarz, 2015; Symbiotics, 2016). These funds are primarily in the form of grants and highly subsidised loans or soft loans. Despite these funds, MFIs still do not have access to enough capital to reach all of the world's financially excluded people.

Savings and deposits such as demand deposits and passbook savings are common mechanisms for generating funds. However, MFIs in the Global South must meet certain capital requirements which can be achieved only when they are in operation for a while before they can re-deploy their deposits. Dang et al. (2021) summarise from previous studies that the development of microfinance institutions occur through four stages, namely, start-up, self-reliance, financial sustainability and profit-making. In the first stage of the start-up, donor grants, savings and soft loans are the primary vehicles of funding. In the second stage of operational self-sufficiency, MFIs extend to private sources such as commercial loans. The third stage is characterised by financial self-sufficiency through retained earnings when MFIs become part of the formal licensed and regulated financial institutions. Here, MFIs use a mix of donations, soft loans, savings, debt and equity. Finally, as MFIs grow, they generate financial returns similar to commercial banks and are eligible to tap into commercial equity. As the MFIs mature in their operations, they are able to mobilise capital, voluntary savings, loans, and issue bonds and stocks. This ensures financial sustainability and profitability. Thus, there is a great diversity in the organisations offering microfinance services and the ability to reach the financially marginalised populations. Organisations that are driven by the mission of poverty alleviation often are more successful in outreach and addressing income inequalities than those focused on earning revenues.

Microfinance and livelihoods – a critique

Impact on individual borrowers

Muhammad Yunus, who revolutionised microfinance in Bangladesh, believed that microcredit programs will assist low-income people with small entrepreneurial ventures, which would then help the local community prosper through job creation and sourcing of materials from other businesses resulting in poverty alleviation. The focus on microfinance as a mechanism to reduce poverty resulted in a significant amount of research since the late 1980s. These studies found positive impacts of microfinance on low-income households by increasing their incomes and purchasing power, and enhancing women's economic decision-making and empowerment (Armendariz and Labie, 2011; Banerjee et al., 2018; Collins et al., 2009; Hashemi et al., 1996; Morduch, 1999). However, Banerjee et al. (2018) found that while microfinance leads to substantial improvements in business outcomes, these gains were mostly seen in businesses that were established prior to seeking microfinance. Félix and Belo (2019) studied the impact of microcredit on poverty reduction by controlling for income, employment, inflation rate and education in 11 countries in south-east Asia. They found that microcredit along with education and employment reduces poverty. Other studies (Bezboruah and Pillai, 2013; Gupta and Mirchandani, 2020) discussed how MFIs assisted and promoted women's microenterprises and business decisions. They found that socially oriented MFIs had the most outreach especially to women and were able to show positive outcomes in terms of entrepreneurial activities, income generation and loan repayments. However, the average loan sizes were very small limiting the borrowers to microenterprises and trapped in a perpetual cycle of debt. Banerjee et al. (2013, 2018) argued that microcredit does not contribute to improving the living conditions of the borrowers as their overall welfare is stymied from more non-financial costs of business such as long work hours, stress and uncertainty. Microfinance benefits have been overhyped as borrowers are entrenched in the cycle of borrowing (Hassan and Islam, 2019) for very small loans with doubtful empowerment opportunities (Bezboruah and Pillai, 2013, 2015), and inadequate evidence of any sustainable impact (Dahal and Fiala, 2020). Entrepreneurship as a way out of poverty will not be possible if borrowers are confined to very small loans, which perpetuate their dependency on microcredit for micro-businesses with limited scope for expansion to bigger enterprises.

Impact on micro-enterprises

MFIs provide financial and non-financial services to low-income people with the objective of improving their economic and social conditions (Herath, 2018; Otero and Rhyne, 1994). The primary argument for MFIs is that if the low-income and financially marginalised populations have access to credit, they will be able to engage in entrepreneurship activities and break the cycle of poverty (Herath, 2018). Lack of access to credit services is considered the primary cause of poverty. MFIs, through loans, provide the start-up capital for establishing small enterprises. These loans not only increase the capacities of small businesses but also provide financial inclusion to the borrowers. The non-financial services assist with education and training to manage businesses and have some social impact. Badullahewage (2020) found that MFI services such as microcredit, technology training, and entrepreneurship programmes had significant positive impact on small enterprises. Another rationale provided by proponents was that MFIs displaced the private moneylender (Graeber, 2011) and made low-cost loans available to borrowers for establishing job generating enterprises. The hype, however, did not last long. Critics argued that countries in the Global South faced the problem of over-crowding of informal microenterprises and self-employment ventures (Bateman, 2010, 2019; Bauchet et al., 2015). With demand for goods and services being stagnant, most microenterprises struggled to survive, and pulled clients away from other microbusinesses. Thus, instead of reducing poverty through income gains, a proliferation of microenterprises led to displacement of other existing businesses. Additionally, due to unsustainable competition from burgeoning microbusinesses, many were forced to close and exit the market or saw a significant decline in their incomes. Research found that globally (Gomez, 2008; McKenzie and Paffhausen, 2017) about 75% of microenterprises do not last beyond their first two years of operation with the highest number of closures occurring in Africa (Nagler and Naudé, 2017; Page and Söderbom, 2012; Patton, 2016). Studies have also reported income decline from these businesses in Latin America and South Africa as a result of severe competition for clients in an economy with stagnant demand (Bateman et al., 2011; Bateman, 2019; Davis, 2006). Entrepreneurs can manage their enterprises better with a mix of financial capital (micro-credit), human capital (training/skill development) and social capital (network of relations) (Hameed et al., 2020). Further, some regulations to check the uncontrolled growth of microenterprises can provide the external support needed.

Impact on women

Operationalised as a group-lending mechanism primarily targeting women borrowers, MFIs were touted to address income and gender inequalities and promote socio-economic development. Microfinance brought about success in increasing household incomes and consumption (Schroeder, 2020), reducing poverty (Félix and Belo, 2019), closing the income inequality gap (Lacalle-Calderon et al., 2019) and increasing women's participation in institutional governance (Bezboruah and Pillai, 2015). Women have been the key targets of MFIs because of their significant impact on social development from participating in microcredit programs. Additionally, MFIs also target women in countries or cultures where they are most likely to experience financial discrimination or dependence (Bezboruah and Pillai, 2017; Drori et al., 2020). Studies showed that access to microcredit services has heightened women's financial independence, entrepreneurial abilities and their roles in household decision-making in the Global South (Felix and Belo, 2019; Fofana et al., 2015; Haile et al., 2015; Mukherjee, 2015; Porter, 2016). The financial independence gained from participating in MFIs have led women to contribute to the economy and gain self-confidence and social respect. Scholars have attributed this to the empowerment of women (Sanyal, 2009). Kabeer's (2005) study in South Asia, where 89% of the microcredit borrowers are women, found that long-term participation by women in MFIs had far-reaching impact in the society through higher levels of political participation, acquisition of practical skills and knowledge and increased awareness and confidence.

Women borrow more from NGOs (Bezboruah and Pillai, 2013) and since these NGOs are unregulated, they aggressively recruit women borrowers to improve their outreach function and raise more support from donor agencies. Women, however, borrow very small loans (Agier and Szafarz, 2013; Bezboruah and Pillai, 2017; D'Espallier et al., 2013; Fletschner, 2009) due to insufficient property ownership or equity to offer as collateral, being risk averse or simply because they are offered very small loans. Consequently, the businesses established with the small loans are not very profitable and inadequate to generate any jobs. Others (Bezboruah and Pillai, 2017; Mahinja et al., 2020; Postelnicu and Hermes, 2018; Samineni and Ramesh, 2020) found that microcredit alone does not reduce poverty or improve livelihoods or the financial condition of women. They found that other social and non-financial support services such as entrepreneurship training, social networks, alternative sources of income and longer loan repayment periods lead to improved business performance, higher self-esteem, increased savings and participation in decision making.

The impact of MFIs as evidenced over the years varies among regions primarily due to ways in which the microfinance services were provided to improve the economic conditions of the financially marginalised. Banerjee et al. (2015) reviewed experimental studies conducted in six countries of the Global South and concluded that while there was not enough evidence that MFIs reduced poverty or improved living standards, it had an impact on increased profits, business expansion, occupational choice and women's decision-making power. In another study conducted in India over a period of ten years, Banerjee et al. (2020) found positive impact on the wellbeing of the very low-income households specifically on their consumption, food security, income and health. The initial start-up capital helped with diversifying the sources of income, investing in non-farm microenterprises and pursuing alternative employment opportunities. In contrast, a similar randomised controlled study by Blattman et al. (2020) in Uganda found that while after four years of receiving the start-up funds, there were noticeable improvement in earnings, after nine years, however, the gains dissipated. The control group were able to catch up with the treated group due to opportunities for wage employment. This suggests that the effects of microfinance across the Global South on livelihoods are somewhat dependent on how the programme works. When non-financial services are offered in addition to microcredit, there is a higher chance of an increase in positive effects on the households. Other external factors such as government regulations on financial institutions, public programmes on employment opportunities and credit provision also play a critical role in the long-term effects of microfinance.

Key points

- Microfinance is considered an effective tool to financially support low-income people to initiate micro-entrepreneurial ventures and close the income inequality gap. For nations in the Global South, MFIs are a way to increase self-employment, community development, and women's empowerment. It can be a powerful instrument in reducing poverty and increasing financial inclusion.
- There are mixed findings from studies that examine microfinance and its impact on people. While several studies demonstrate the positive effects of MFI on household incomes, poverty alleviation, education and empowerment, other studies have highlighted the negative impacts such as over-indebtedness, small loans and less than expected overall benefits.
- Microfinance can work better when additional non-financial services such as financial, technological and management training are provided to the clients. Most clients have no business management experience or understand financial concepts. Such education and training can provide the skillsets and tools necessary to manage their ventures and remain sustainable.

Further reading

Bateman, M., Blankenburg, S. and Kozul-Wright, R. (2019) *The rise and fall of global microcredit: Development, debt and disillusion*, London: Routledge.

Cautero, R.M. (2021) 'Why microfinance is important to small business', *The Balance*. Available at: www.thebalance.com/what-is-microfinance-and-how-does-it-work-4165939#.

Credit Summit (2021) *What is microfinance? A complete guide*. Available at: www.mycreditsummit.com/what-is-microfinance2.html

References

Adams, D.W., Graham, D.H. and Von Pischke, J.D. (1984) *Undermining rural development with cheap credit*, Boulder: Westview Press.

Agier, I. and Szafarz, A. (2013) 'Microfinance and gender: Is there a glass ceiling on loan size?' *World Development*, 42: 165–181. https://doi.org/10.1016/j.worlddev.2012.06.016

Akhter, J. and Cheng, K. (2020) 'Sustainable empowerment initiatives among rural women through microcredit borrowings in Bangladesh', *Sustainability*, 12(6): 2275. https://doi.org/10.3390/su12062275

Armendariz, B. and Labie, M. (2011) *Handbook of microfinance*, Singapore: World Scientific.

Awusabo-Asare, K., Annim, S.K., Abane, A.M. and Asare-Minta, D. (2009) 'Who is reaching whom? Depth of outreach of rural microfinance institutions in Ghana', *International NGO Journal*, 4(4): 132–141. https://doi.org/10.5897/INGOJ.9000164

Badullahewage, S.U. (2020) 'Reviewing the effect of microfinance programmes for alleviating poverty among microenterprises in Sri Lanka', *International Journal of Management Studies and Social Science Research*, 2(2): 45–64.

Banerjee, A., Duflo, E., Glennerster, R. and Kinnan, C. (2013) 'The miracle of microfinance? Evidence from a randomized evaluation', *NBER Working Paper No. 18950*, Cambridge: National Bureau of Economic Research.

Banerjee, A., Duflo, E. and Hornbeck, R. (2018) 'How much do existing borrowers value microfinance? Evidence from an experiment on bundling microcredit and insurance', *Economica*, 85(340): 671–700. https://doi.org/10.1111/ecca.12271

Banerjee, A., Duflo, E. and Sharma, G. (2020) 'Long-term effects of the targeting the ultra-poor program', *Working Paper 28074*, Cambridge, MA: National Bureau of Economic Research. https://doi.org/www.nber.org/papers/w28074

Banerjee, A., Karlan, D. and Zinman, J. (2015) 'Six randomized evaluations of microcredit: Introduction and further steps', *American Economic Journal: Applied Economics*, 7(1): 1–21. http://dx.doi.org/10.1257/app.20140287

Bateman, M. (2010) *Why doesn't microfinance work? The destructive rise of local neoliberalism*, London: Zed Books.

Bateman, M. (2019) 'Impacts of the microcredit model: Does theory reflect actual practice?', in Bateman, M., Blankenburg, S. and Kozul-Wright, R. (eds.), *The rise and fall of global microcredit: Development, debt and disillusion*, London: Routledge, pp. 42–68.

Bateman, M., Duran Ortiz, J.P. and Sinkovic, D. (2011) 'Microfinance in Latin America: The case of Medellín in Colombia', in Bateman, M. (ed.), *Confronting microfinance: Undermining sustainable development*, Sterling, VA: Kumarian Press.

Bauchet, J., Morduch, J. and Ravi, S. (2015) 'Failure vs. displacement: Why an innovative anti-poverty program showed no net impact in South India', *Journal of Development Economics*, 116: 1–16. https://doi.org/10.1016/j.jdeveco.2015.03.005

Bezboruah, K. and Pillai, V. (2017) 'Microcredit and development: A multi-level examination of women's participation in microfinance institutions', *Development in Practice*, 27(3): 328–339. https://doi.org/10.1080/09614524.2017.1298723

Bezboruah, K.C. and Pillai, V. (2013) 'Assessing the participation of women in microfinance institutions: Evidence from a multinational study', *Journal of Social Service Research*, 39(5): 616–628. https://doi.org/10.1080/01488376.2013.816409

Bezboruah, K.C. and Pillai, V. (2015) 'Exploring the participation of women in financial cooperatives and credit unions in developing countries', *VOLUNTAS: International Journal of Voluntary and Nonprofit Organizations*, 26(3): 913–940. https://doi.org/10.1007/s11266-014-9467-9

Blattman, C., Fiala, N. and Martinez, S. (2020) 'The long-term impacts of Grants on poverty: Nine-year evidence from Uganda's youth opportunities program', *American Economic Review: Insights*, 2(3): 287–304. https://doi.org/10.1257/aeri.20190224

Bouman, F.J.A. (1995) 'ROSCA: On the origin of the species', *Savings and Development*, 19: 117–148. www.jstor.org/stable/25830410

Brière, M. and Szafarz, A. (2015) 'Does commercial microfinance belong to the financial sector? Lessons from the stock market', *World Development*, 67: 110–125. https://doi.org/10.1016/j.worlddev.2014.10.007

Brown, M., Guin, B. and Kirschenmann, K. (2016) 'Microfinance banks and financial inclusion', *Review of Finance*, 20(3): 907–946. https://doi.org/10.1093/rof/rfv026

Christen, B. and Mas, I. (2009) 'It's time to address the microsavings challenge, scalably', *Enterprise Development and Microfinance*, 20(4): 274–285. https://doi.org/10.3362/1755-1986.2009.031

Collins, D., Morduch, J., Rutherford, S. and Ruthven, O. (2009) *Portfolios of the poor: How the world's poor live on $2 a day*, Princeton, NJ: Princeton University Press.

Conning, J. and Udry, C. (2007) 'Rural financial markets in developing countries', in Gardner, B.L., Evenson, R.E., Rausser, G.C. and Pingali, P. (eds.), *Handbook of agricultural economics: Agricultural development: Farmers, farm production and farm markets*, Vol. 3, London: Elsevier.

Cull, R., Demirgüç-Kunt, A. and Morduch, J. (2009) 'Microfinance meets the market', *Journal of Economic Perspectives*, 23(1): 167–192. https://doi.org/10.1257/jep.23.1.167

D'Espallier, B., Guerin, I. and Mersland, R. 2013 'Focus on Women in microfinance institutions', *Journal of Development Studies*, 49(5): 589–608. https://doi.org/10.1080/00220388.2012.720364

Dahal, M., & Fiala, N. (2020) 'What do we know about the impact of microfinance? The problems of statistical power and precision', *World Development*, 128: 104773. https://doi.org/10.1016/j.worlddev.2019.104773

Dang, T.T., Vu, Q.H. and Hau, N.T. (2021) 'Impact of outreach on operational self-sufficiency and profit of microfinance institutions in Vietnam', in Ngoc Thach, N., Kreinovich, V. and Trung, N.D. (eds.), *Data science for financial econometrics. Studies in computational intelligence*, Vol. 898, Cham: Springer, pp. 543–565. https://doi.org/10.1007/978-3-030-48853-6_37

Davis, M. (2006) *Planet of Slums*, London: Verso.

Drori, I., Manos, R., Santacreu-Vasut, E. and Shoham, A. (2020) 'How does the global microfinance industry determine its targeting strategy across cultures with differing gender values?', *Journal of World Business*, 55(5): 100985. https://doi.org/10.1016/j.jwb.2019.02.004

Félix, E.G.S. and Belo, T.F. (2019) 'The impact of microcredit on poverty reduction in eleven developing countries in south-east Asia', *Journal of Multinational Financial Management*, 52: 100590. https://doi.org/10.1016/j.mulfin.2019.07.003

Fletschner, D. (2009) 'Rural women's access to credit: Market imperfections and intrahousehold dynamics', *World Development*, 37(3): 618–631. https://doi.org/10.1016/j.worlddev.2008.08.005

Fofana, N.B., Antonides, G., Niehof, A. and van Ophem, J.A.C. (2015) 'How microfinance empowers women in Côte d'Ivoire', *Review of Economics of the Household*, 13(4): 1023–1041. https://doi.org/10.1007/s11150-015-9280-2

Ganguly, S. (2014) 'A poet and a landlord: Rabindranath Tagore and the Nobel Prize money', *Economic & Political Weekly*, 49: 138–142. www.jstor.org/stable/24480471

García-Pérez, I., Fernández-Izquierdo, M.Á. and Muñoz-Torres, M.J. (2020) 'Microfinance institutions fostering sustainable development by region', *Sustainability*, 12(7): 2682. https://doi.org/10.3390/su12072682

Geertz, C. (1962) 'The rotating credit association: A "Middle Rung" in development', *Economic Development and Cultural Change*, 10: 241–263. www.jstor.org/stable/1151976

Ghatak, M. and Guinnane, T.W. (1999) 'The economics of lending with joint liability: Theory and practice', *Journal of Development Economics*, 60: 195–228. https://doi.org/10.1016/S0304-3878(99)00041-3

Gomez, G.M. (2008) *Do micro-enterprises promote equity or growth?* The Hague: Institute of Social Studies.

Graeber, D. (2011) *Debt: The first 5000 years*, New York: Melville House.

Grameen Bank (2004) *What is microcredit?* Available at: www.grameen-info.org/mcredit/defit.html (Accessed 21 August 2020).

Gupta, N. and Mirchandani, A. (2020) 'Corporate governance and performance of microfinance institutions: Recent global evidences', *Journal of Management and Governance*, 24: 307–326. https://doi.org/10.1007/s10997-018-9446-4

Habib, I. (1964) 'Usury in medieval India', *Comparative Studies in Society and History*, 6: 393. https://doi.org/10.1017/S0010417500002267

Haile, H.B., Osman, I., Shuib, R. and Oon, S.W. (2015) 'Is there a convergence or divergence between feminist empowerment and microfinance institutions' success indicators?', *Journal of International Development*, 27(7): 1042–1057. https://doi.org/10.1002/jid.3041

Hameed, W.U., Mohammad, H.B., & Shahar, H.B.K. (2020) 'Determinants of micro-enterprise success through microfinance institutions: A capital mix and previous work experience', *International Journal of Business and Society*, 21(2): 803–823. https://doi.org/10.33736/ijbs.3295.2020

Hashemi, S., Schuler, S.R. and Riley, A.P. (1996) 'Rural credit programs and women's empowerment in Bangladesh', *World Development*, 24(4): 635–653. https://doi.org/10.1016/0305-750X(95)00159-A

Hassan, S.M. and Islam, M.M. (2019) 'The socio-economic impact of microfinance on the poor family: A study from Bangladesh', *Journal of Asian and African Studies*, 54(1): 3–19. https://doi.org/10.1177/0021909618785399

Herath, H.M.W.A. (2018) 'Microfinance as a strategy of empowering the conflict-affected communities in Sri Lanka', *Journal of Social and Development Sciences*, 9(2): 6–21. https://doi.org/10.22610/jsds.v9i2.2377

Hermes, N. and Lensink, R. (2007) 'The empirics of microfinance: What do we know?', *The Economic Journal*, 117: 1–10. https://doi.org/10.1111/j.1468-0297.2007.02013.x

Kabeer, N. (2005) 'Wider impacts: Social exclusion and citizenship', in Copestake, J., Greeley, M., Johnson, S., Kabeer, N. and Simanowitz, A. (eds.), *Money with a mission*, Vol. 1, London, UK: Practical Action Publishing, pp. 96–125.

Lacalle-Calderon, M., Larrú, J.M., Garrido, S.R. and Perez-Trujillo, M. (2019) 'Microfinance and income inequality: New macrolevel evidence', *Review of Development Economics*, 23(2): 860–876. https://doi.org/10.1111/rode.12573

Le, C.A. (2017) *Microfinance: The impacts of a poverty reduction approach and financial systems approach on poor rural households in Vietnam*, University Honors Theses. Paper 454, Portland: Portland State University. https://doi.org/10.15760/honors.451

Mahinja, F., Kiliza, K.R. and Charles, A.M. (2020) 'Microfinance practices and its impact on women's socio-economic welfare: A case study of Kinondoni Municipality in Dar es Salaam, Tanzania', *International Journal of Arts, Humanities and Social Studies*, 2(4): 40–48.

McKenzie, D. and Paffhausen, A. (2017) 'Small firm death in developing countries', *Policy Research Working Paper No. 8236*, Washington, DC: World Bank.

Microfinance Barometer (2019) *Global microfinance figures: What are the trends?* Paris: Convergences.

Morduch, J. (1999) 'The microfinance promise', *Journal of Economic Literature*, 37(4): 1569–1614. https://doi.org/10.1257/jel.37.4.1569

Mukherjee, A.K. (2015) 'Empowerment through government subsidized microfinance program: Do caste and religion matter?', *International Journal of Social Economics*, 42(1): 2–18. https://doi.org/10.1108/IJSE-02-2013-0036

Nagler, P. and Naudé, W. (2017) 'Non-farm entrepreneurship in rural sub-Saharan Africa: New empirical evidence', *Food Policy*, 67: 175–191. https://doi.org/10.1016/j.foodpol.2016.09.019

Otero, M. and Rhyne, E. (1994) *The new world of microenterprise finance*, West Hartford, CT: Kumarian Press.

Page, J. and Söderbom, M. (2012) 'Small is beautiful? Small enterprise, aid and employment in Africa', *UNU WIDER Working Paper No., 2012/94*, November, Helsinki: WIDER.

Patton, A. (2016) 'Uganda is a land of entrepreneurs: but how many start-ups survive?' *The Guardian*, 16 February. Available at: www.theguardian.com/global-development-professionals-network/2016/feb/16/uganda-is-a-land-of-entrepreneurs-but-how-many-startups-survive (Accessed 07 December 2020).

Porter, M. (2016) 'Effects of microcredit and other loans on female empowerment in Bangladesh: The borrower's gender Influences intra-household resource allocation', *Agricultural Economics*, 47(2): 235–245. https://doi.org/10.1111/agec.12225

Postelnicu, L. and Hermes, N. (2018) 'Microfinance performance and social capital: A cross-country analysis', *Journal of Business Ethics*, 153(2): 427–445. https://doi.org/10.1007/s10551-016-3326-0

Qudrat-I Elahi, K. and Lutfor Rahman, M. (2006) 'Micro-credit and micro-finance: Functional and conceptual differences', *Development in Practice*, 16(5): 476–483. https://doi.org/10.1080/09614520600792481

Ray, S. (2019) 'Challenges and changes in Indian rural credit market: A review', *Agricultural Finance Review*, 79: 338–352. https://doi.org/10.1108/AFR-07-2018-0054

Samineni, S. and Ramesh, K. (2020) 'Measuring the impact of microfinance on economic enhancement of women: Analysis with special reference to India', *Global Business Review*, online first. https://doi.org/10.1177/0972150920923108

Sanyal, P. (2009) 'From credit to collective action: The role of microfinance in promoting women's social capital and normative influence', *American Sociological Review*, 74(4): 529–550. https://doi.org/10.1177/000312240907400402

Schrader, H. (1997) *Changing financial landscapes in India and Indonesia: Sociological aspects of monetization and market integration*, Vol. 2, Münster: LIT Verlag.

Schroeder, E. (2020) 'The impact of microcredit borrowing on household consumption in Bangladesh', *Applied Economics*, 52(43): 4765–4779. https://doi.org/10.1080/00036846.2020.1743815

Schulte, M. and Winkler, A. (2019) 'Drivers of solvency risk: Are microfinance institutions different?', *Journal of Banking & Finance*, 106: 403–426. https://doi.org/10.1016/j.jbankfin.2019.07.009

Seibel, H.D. (2005) 'Does history matter? The old and the new world of microfinance in Europe and Asia', paper presented at *From Moneylenders to Microfinance: Southeast Asia's Credit Revolution in Institutional, Economic and Cultural Perspective Workshop*, Asia Research Institute, Department of Economics and Department of Sociology, National University of Singapore 7–8 October 2005.

Shoaib, A. and Siddiqui, M.A. (2020) 'Why do people participate in ROSCA saving schemes? Findings from a qualitative empirical study', *Decision*, 47(2): 177–189. https://doi.org/10.1007/s40622-020-00244-8

Symbiotics (2016) *Symbiotics microfinance investment vehicles survey: Market data & peer group analysis*, Geneva: Symbiotics Group.

Von Pischke, J.D. (1997) 'Poverty, human development and financial services', *Occasional Paper 25*, New York: Human Development Report. United Nations Development Programme.

Wisniwski, S. (1999) 'Microsavings compared to other sources of funds', *Working Group on Savings Mobilization*, Eschborn, Germany: CGAP.

World Council of Credit Unions (2020) *Statistical report – 2019*, Madison: World Council of Credit Unions.

Yunus, M. (2004) 'Expanding microcredit outreach to reach the millennium development goals – Some issues for attention', Paper presented at an international seminar *Attacking Poverty with Microcredit*, 8–9 January 2003, Dhaka.

Zainuddin, M. and Yasin, I.M. (2020) 'Resurgence of an ancient Idea? A study on the history of microfinance', *FIIB Business Review*, 9(2): 78–84. https://doi.org/10.1177/2319714520925933

APPENDIX A: TERMINOLOGY

Term	Meaning	Example
Microfinance	• Describes a broad spectrum of financial services and tools to assist entrepreneurial ventures of financially marginalised people. • It is a developmental approach that combines financial and social intermediation.	• Microloans, micro-savings accounts, micro-insurance, business training, health services, peer support and networking associated with entrepreneurial activities.
Microcredit	• Describes very small loans to financially deprived entrepreneurs for supporting small scale businesses that generate income. • Microcredit institutions distributes and recovers loans and are based on borrower group formation and compulsory savings.	• Does not require collateral from the borrowers, but requires regular savings.
Micro-savings	• Indicates the small savings or deposits required when involved in group-lending to instil prudent financial behaviour and education among the clients. • Borrowers receive a small interest rate on their micro-savings, which contributes to good loan repayment performance and an increase in productivity by using the savings for business improvement.	• Small savings that are stored as demand deposits and passbook savings.
Group Lending	• Known as joint liability lending where borrowers typically form small groups and guarantee each other's loan repayments. • Helps with effective monitoring of the business activities, exclusion of risky borrowers from participation, increasing loan repayments, and mitigating problems associated with information asymmetry.	• Peer pressure among group members assures timely loan repayments

Sources: Grameen Bank (2004), Christen and Mas (2009), Morduch (1999), Qudrat-I Elahi and Rahman (2006), Yunus (2004).

37
GLOBAL MARKETS AND SOUTHERN LIVELIHOODS
Exploring trans-scalar connections

Thaisa Comelli

Introduction

In times of world crises, such as the recent COVID-19 pandemic, it becomes more evident that global markets (GMs) are ever present around us and can strongly impact local lives. They are everywhere – from ventilators that need to move from China to Brazil, to basic food items, medicines and inputs for vaccines that need to cross multiple borders. GMs depend on consumers, producers and investors, as people depend on GMs to navigate capitalism. In broad terms, global markets are conceptualised in this chapter as interconnected and interdependent networks of trade in goods (material and immaterial) and services that pervade local and national borders.

Although markets directly and indirectly impact all lives and countless livelihood systems, such networks can often be complex and hard to fully comprehend. Moreover, it can be unclear whether their impacts are the same for everyone. It has long been argued that the goods and hazards produced by globalisation are not equally distributed around the world. Markets, in particular, interact across geographies that are in constant dispute with each other. Especially in territories marked by longstanding inequality and vulnerability, the impacts of global markets can be acute on some local livelihood systems. Considering this criticism, the chapter asks: what is the significance of global markets for livelihoods in the Global South? How do livelihoods and global markets interact, and which analytical frameworks help to unpack such complex interactions?

Drawing on the above question, this chapter briefly sketches paths that connect global markets and livelihoods in the Global South. To do so, the chapter follows three main steps: the first step is dedicated to exploring some of the rationales, paradigms and structures that enable the global expansion of markets. Next, the chapter unravels two lines of analysis that connect GMs and livelihoods in the Global South: through increased risk and vulnerability (step two), and through dependency (step three). Overall, these connections provoke reflections on underlying ethical and justice dimensions of global market interactions. They also highlight the complex trans-scalar processes that arise in the context of contemporary capitalism.

Unpacking global markets

The term global markets in economic and social theory, as well as for international organisations and agencies, is usually associated with four main structures: transnational trading systems, multinational chains of production and consumption (often described as global value chains), international and foreign investments in national and local economies, and transnational flows of financial capital and stocks (ILO, 2004). Although complex, these market structures have several features that allow them to be included in the umbrella of global markets and part of a larger global capitalistic dynamic. Overall, the chapter highlights three key features:

1. The existence of global markets requires more flexible (or inexistent) local and national borders. Such borders are both physical as well as administrative, in the form of taxes, regulations and fees. Global markets, therefore, have a borderless character and require a transition from physical-local-territorial spaces to virtual-global spaces.
2. GMs are interconnected. In other words, they connect people, institutions, companies, and territories to each other, as well as one market with another. Thus, GMs are usually composed of smaller markets that exchange and trade with each other across the globe. It is also important to stress, as Aspers and Kohl (2015) do, that although most markets are connected, this does not mean that all markets are global or that there is even a 'global economy'. There are still markets in the world that operate within strict borders and with fewer connections. This means that speaking about a global economy brings with it a theoretical reductionism that helps to understand the magnitude of the phenomenon, although inaccurate from an empirical point of view.
3. GMs are marked by interdependence – and not only size, flexible borders and increased connections. There are entire markets that could hardly survive without the engagement of other countries, cities and territories. This interdependence is also one of the major sources of inequality in terms of who reaps the fruits and absorbs the risks and impacts generated by markets worldwide.

To approach the topic of global markets, it is necessary to understand *the market* as an institution and social system linked to contemporary capitalism. Such an entity, although virtual, carries with it assumptions, paradigms and enabling values, referred to as market-driven rationales. They are linked to the 'spirit of capitalism' primarily studied by Weber ([1905] 2012), but which acquired new economic, cultural and political nuances after the end of the Cold War, in a governance configuration commonly referred to in critical theory as the Neoliberal State or Neoliberalism (Gill, 1995; Harvey, 2007a, 2007b). Many contemporary values and ideas illustrate market-driven rationales and their instrumentalisation by neoliberalism: the notion of private property, which made the commodification and enclosure of public goods possible; freedom, which came to be associated with the deregulation of markets by the State; and productivity and competitiveness for economic growth, which came to be seen as a value for individuals in society. These are examples of values and ideas from modernity that acquired new nuances and a new scale in contemporary times (Arndt, 1989; Harvey, 2006).

It is important to stress that the way in which neoliberalism boosts GMs requires a particular role for local and national states, which incorporate and sanction market-driven rationales through policies or their legal-normative framework (Harvey, 2006). The global housing crisis illustrates this point well. Private property, which is protected as a right in most free market societies, has been a facilitator for the commodification of housing and its insertion into real estate markets. In contemporary neoliberal regimes, such commodification and trade reached a global level, as housing started to be absorbed by financial markets and transformed into a pool of assets (e.g. mortgage pool). Within GMs, such assets (and housing itself) become harder to track, regulate and protect by states. Thus, by engaging with the financialisation of housing, market-driven policies and neoliberal states facilitate the expansion and strengthening of global markets (Rolnik, 2019).

Furthermore, no analysis of global markets could ignore the role of technology for their expansion and consolidation in the contemporary world. Many of the market dynamics described thus far actually started in early modernity. However, technology not only added size and intensity to such processes, but it also created new spaces, structures and actors for market exchange. For example, Bitcoin appeared in the late 2000s as a new currency for trading and speculating in digital market spaces. Due to its ability to avoid states' control, the crypto-coin created an expandable environment for both legal and illegal markets (Castells, 1998, 2010; Kethineni and Cao, 2019). In addition, delivery and transportation platforms like Uber or Deliveroo contribute to new global divisions of labour and a so called 'gig-economy' (Fleming, 2017; Graham et al., 2017). In terms of market spaces, platforms such as Amazon or AliExpress are also good examples of the role of technology in boosting GMs. Although the labour that supplies these market spaces is still geographically bounded, buyers and sellers mostly connect with each other through virtual platforms (Graham et al., 2017). Thus, technology has facilitated connections between actors and markets, in addition to the flexibility of borders, which ultimately strengthens GMs and their interdependence.

In conclusion, global markets are part of a complex mix of rationales, actors and practices that are not new in society, but which, in contemporary times, assume an unprecedented scale and intensity, along with a more fluid and complex landscape of power. This landscape also relates to uneven geographies, where some territories and their livelihoods might have greater decision-making power over markets' conditioning factors or an increased vulnerability to their impacts.

Global markets and livelihoods in the Global South: exploring connections

Connections through risk and vulnerability

One of the main points of criticism of globalisation relates to the uneven distribution of the related risks and hazards. Global risks can be distributed unevenly both geographically and socially, as they might affect more vulnerable regions, individuals and/or groups. In other words, risk is more critical when related to people and territories with less capabilities and means to cope with 'the (social) uncertainties that mark their lives' (Beck, 2009: 178).[1] And as localities, nations and economies become more interconnected and interdependent, instability and risk produced on one side of the world might considerably harm lives and livelihoods on another.

The international financial system is a useful entry point to explore the potential risks of GMs and their impacts. With the consolidation of capitalist governance around the world at the end of the 20th century, financial markets became larger and stronger, but also more complex in terms of instruments, products and transaction processes. In other words, these markets became not only too big to fail (Pontell et al., 2014), but also too complex to track and regulate. The 2008 financial crisis illustrates the size and complexity of the impact of global markets and their inherent risk for everyday life. Led by US markets, a flimsy subprime house of cards had gone unnoticed by governments, regulators, investors and other actors for at least two years (Rose and Spiegel, 2012) before the housing bubble finally burst. The crisis was more acute and evident in the US, affecting several vulnerable lives in the country (Desmond, 2012; Karamessini and Rubery, 2014; Rajmil et al., 2014), but it was also felt globally.

The crisis rippled around the world because the interconnectedness and interdependence of GMs intensifies risk in general, even with improved mitigation mechanisms. However, over the past few decades, many economies of the Global South have undergone market opening processes, either through macro adjustment policies (Stewart, 1991) or through local actions aimed at attracting foreign capital (Bolwig et al., 2010). This does not mean that a crisis in GMs will always or necessarily affect Global South economies with greater intensity, as it depends on how (and to what degree) economies are connected through different market instruments. Notwithstanding, a state of scarce financial reserves, vast supply of natural resources or workforce, few instruments of social protection and fewer guarantees of rights can make countries and livelihoods particularly reliant on global markets and, ultimately, more vulnerable to some of the risks and hazards produced by them.

In Latin America, countries such as Brazil and Mexico, which in the 2000s were experiencing economic and social growth, delved into a severe recession in the last quarter of 2008 and in the first quarter of 2009, which inevitably led to thousands of jobs lost (Ocampo, 2012). In Mexico, a country more attached to US markets and where financial mechanisms such as the mortgage securitisation had been introduced in the early 2000s, the blow was even more violent. While the securitisation of mortgages served to cushion the recession for the housing industry, the low-income population, and particularly borrowers, saw a sharp decrease in real wages, finding themselves unable to acquire affordable housing and secure their livelihoods (Soederberg, 2015). The effects of the crisis had also uneven regional impacts, where northern livelihoods (closer to the US border and more connected to international trade) were more affected than southern ones. Arévalo and Herreros (2015) show, for instance, that workers linked to the manufacturing industry experienced a sudden loss of their livelihoods in the form of formal jobs and labour rights. In other words, livelihoods more dependent on foreign markets delved quickly into informality, while the independent ones became more insulated.

The financial crisis has both directly and indirectly contributed to phenomena linked to land markets in the Global South and the livelihoods that depend on these lands, such as rural workers, small farmers and indigenous populations. Due to the importance of the US dollar in global markets and to speculation in future-related market instruments, the subprime housing crisis led to a spike in commodity prices, particularly food, which contributed to increased poverty, hunger and political unrest in many Global South countries (Clapp, 2009). The global food crisis helped to ramp up the phenomenon of land grabbing globally, with severe implications for livelihoods in the Global South (Borras et al., 2011; Cotula, 2012). This means that large portions of land started to be purchased by private or public actors to secure future agricultural production, which can also be considered a global market by itself or a process of financialisation of land and agriculture (Ouma, 2014).

The process of financialisation of land and agriculture has increased food and tenure insecurity in rural areas (Robertson and Pinstrup-Andersen, 2010), while also affecting the urban environment (Zoomers et al., 2017). Other GMs such as foreign investment in tourism and energy markets are suggested to be some of the drivers of contemporary land grabbing (Zoomers, 2010). For example, the demand for vast amounts of land in Ghana coincided with several institutional reforms aimed at market liberalisation. After the mid-2000s, foreign companies started to acquire large portions of land for agricultural production, including several for biofuel feedstock (Schoneveld et al., 2011). This is a country where most of the land is under customary tenure, which is indigenous and/or communal land managed by traditional authorities. Furthermore, a recent study by Aha and Ayitey (2017) suggests that, in addition to the high degree of uncertainty about the future of livelihoods and tenure security, 93% of the local farmers in the Brong Ahafo and Ashanti regions were not even consulted before being dispossessed.

These cases from Latin America and Africa illustrate the complexity of connections between market structures and more vulnerable livelihoods in the Global South. Even though socio-spatial vulnerability is not an exclusive feature of the South, as Lee et al. (2009) highlight, the legacy of global markets in vulnerable countries may represent a form of 'silent violence' (ibid, p. 742), where exposed supplier and/or labour-intensive livelihoods suffer indirect 'catastrophic consequences' of market-related hazards. Yet, studies focusing on the impacts of GMs on livelihoods in the Global South are still scarce. Future research on the subject must address how precisely (and unevenly) the risks of GMs are distributed across vulnerable territories, social sectors and livelihood systems (Kirby, 2013).

Connections through dependency

Within the field of development and globalisation studies, several works attempt to draw connections between governance structures (driven by markets) and local livelihoods. On a global macro-scale, one of the major contributions in this sense was made by dependency and world-system theorists who dedicated their efforts to analysing and challenging development gaps between northern and southern countries. The key idea in these bodies of literature is that development, wealth and wellbeing could not be understood as isolated within national borders; 'core and periphery' countries (as loci of power) are economically and socially interlaced with each other, producing global dependency with an uneven distribution of goods and hazards (Frank, 1967; Ghosh, 2019; Santos, 2017 [2000]; Wallerstein, 2011 [1974]).

More recently, studies pointed to the tendency of transnational companies, businesses and investments to seek settlement in countries and territories that offer lower production and service costs. The phenomenon is often referred to as a 'race to the bottom', and it has offered useful entry points for the analysis of global dependency through trans-scalar connections between local governments, national states, multi-level corporations, and livelihood systems. According to these critiques, the race to the bottom has been particularly damaging for countries of the Global South due to their vast supply of cheap labour and little to no state protection for livelihoods, workers or the environment, which ultimately can boost a vicious cycle of dependency (Singh and Zammit, 2004; Mosley and Uno, 2007).

Although contested[2] (Hecock and Jepsen, 2013), 'race to the bottom'-related theories indicate not only the wide spectrum of actors and the multidimensional impact of neoliberal governance on vulnerable lives,[3] but also how some persistent processes of exploitation and inequality are essential for the very maintenance and expansion of governance structures through GMs and their dynamics. In other words, labour precariousness in the Global South is not only one of the consequences of pervasive market dynamics, but it is, to some extent, one of the premises for certain markets' structures to function. Paradoxically, while the expansion of GMs to the Global South represents jobs and new opportunities for many, those same markets may increase inequality gaps or even create new ones. This is sometimes the case, for instance, with service-offshoring in fields that require pre-existing hard skills. Service offshoring ends up favouring middle-class workers in low-income countries, while the less skilled remain excluded from the formal market. In other cases, women occupy the most precarious and worst paid jobs offered by foreign companies (Bottini et al., 2008; Singh and Zammit, 2004; Wells, 2009). In 2004, a study by Oxfam that analysed more than 12 developing countries showed how companies at the top of global supply chains systematically pushed costs and risks of business to women workers and their families. In Chile, 75% of women in the agricultural sector worked on a casual basis, despite dedicating more than 60 hours a week to those jobs. In the Guangdong province in China, 90% of women in the garment industry had no access to social insurance, while 60% had no formal contracts (Raworth, 2004).

The garment industry in particular has been part of numerous media scandals denouncing sweatshops with slave or semi-slave labour in Asian countries (Rosen, 2002). Under the contemporary 'fast-fashion' paradigm, the complex chains of design, production, distribution, sale and consumption (commonly called Global Value Chains – GVC) sometimes involve dozens of companies, subcontractors, self-employed professionals and, consequently, different livelihood systems, host cities, countries and ethical guidelines (Gereffi et al., 2005). It is also noteworthy that, in many free-market states, those standards and protection systems have been managed by the companies themselves through actions of Corporate Social Responsibility, common in neo-liberal governance systems (Ireland and Pillay, 2010; Utting, 2008; Wells, 2009).

In sum, although the expansion of global markets to the Global South has represented significant opportunities for countries and social sectors to enter the system competitively and even lead some of those markets (Barrientos, 2019; Gereffi and Fernandez-Stark, 2010), there are still many challenges from an ethical and human rights standpoint (Barrientos et al., 2011; Clarke and Boersma, 2017; Siddiqui, 2017). Beyond ethical procedures within markets, the provocation that the dependency perspective offers is even more structural; after all, could some of these markets even survive without the support of precarious low-wage, labour intensive countries and livelihood systems?

Finally, it should be noted that North-South dichotomies from dependency theory are no longer sufficient to read complex market relationships (Horner and Nadvi, 2018). For instance, there is the significant presence of economies of the Global South in the leadership of global markets (particularly China), which can generate new processes of South-South competition and exploitation not covered by traditional dependency descriptions. The growth of new real estate developments in African countries by large multinationals and foreign Asian contractors is, for example, a phenomenon to be closely observed, with many others arising every day (Agbebi and Virtanen, 2017; Chan, 2003).

The power landscape in contemporary capitalism is constantly changing and this requires a constant effort to untangle the complex webs of actors, scales and territories that connect global markets and local livelihoods. Such relations are never static in time and space. But, while it is vital to recognise the insufficiency of a North-South division and the diversity of economies and livelihood systems of the Global South, it is also important to note that the Global South is still an important analytical lens for identifying contemporary dependencies and injustices.

Conclusion

This chapter navigated through broad and transdisciplinary critical debates that connect global markets and livelihoods in the Global South. Through risk/vulnerability and dependency perspectives, we can grasp connections that are underpinned by notions of justice (Fraser, 2010). That is, connections that emphasise that the risks and impact of GMs are *maldistributed*; that the dependency between certain livelihoods and markets is still little visible and *recognised* worldwide; and that Global South geographies (their governance systems and citizens) still do not *participate with parity* in market-related decisions that may directly or indirectly impact their livelihood systems. The major problem that still lingers with globalisation and global markets is not related to size or strength but is instead the price that some people (and livelihoods) must 'pay' for it. In the current scenario, only few have the decision-making power to shape the paths of globalisation and global markets, and this only seems to be more accentuated with new technologies and financial instruments (Harvey, 2006; Kirby, 2013; Rolnik, 2019). In this sense, just as global markets are currently quite strong, interdependent and interconnected, so too must be our democratic systems of governance and decision-making. If risk/vulnerability and interdependence are often pervasive and reach a global scale, the mechanisms for solving these severe problems need to be equally global and deeply democratic.

Key Points

- Risk/vulnerability and dependency are only possible threads that connect global markets and livelihoods in the Global South, not necessarily the main ones. Such connections are part of a complex web of relations, which are not always straightforward. Contemporary domination patterns generated by or through markets should be understood in relation to specific geographies, cultural contexts, market instruments and socio-economic regulations.
- The risk-vulnerability relational thread is not limited to Global South geographies. However, the examples outlined in this chapter suggest that it is still a relevant one. The recent opening of post-colonial states to GMs, coupled with weak social protection instruments, can create a path for impacts that are rarely overcome by certain livelihoods in the Global South. This is particularly true for livelihoods that contribute to global markets as suppliers or are linked to new market instruments without proper protection (e.g. mortgage markets and indebtedness).
- The dependency thread creates more evident connections between GMs and the livelihoods in the Global South. It shows how precarious work and poverty are still conditions that facilitate the existence of certain markets. However, a development curtain that divides North and South is not appropriate as a simplistic explanation for dependency phenomena. While large southern economies are growing, there are South-South intra-exploitation relationships that need to be further explored.
- Understanding the threads that connect global markets and livelihoods in the Global South is important, but useless if the rationales, power structures and actors that facilitate the expansion of markets are not questioned. Studies underpinned by social justice must look at the roots of processes that create and validate certain unfair market relationships.

Further reading

Agbebi, M. and Virtanen, P. (2017) 'Dependency theory: A conceptual lens to understand China's presence in Africa?', *Forum for Development Studies*, 44(3): 429–451. https://doi.org/10.1080/08039410.2017.1281161

Kirby, P. (2013) 'Vulnerability and globalization: The social impact of globalization', in Turner, B.S. and Holton, R.J. (eds.), *The Routledge international handbook of globalization studies*, London and New York: Routledge, pp. 136–156.

Raworth, K. (2004) *Trading away our rights: Women working in global supply chains*, Oxford: Oxfam International.

Notes

1. The chapter draws on Ulrich Beck's (2009) relational and interconnected concepts of *risk* and *vulnerability*.
2. Although it is vastly agreed that companies and investors seek lower production costs, it is contested that they are necessarily motivated by little state spending on social welfare.
3. Authors such as Nita Rudra (2008) argue that the 'race to the bottom' in the Global South is more related to a relative impoverishment of the middle class than an actual impact on the poorest urban and rural livelihoods.

References

Agbebi, M. and Virtanen, P. (2017) 'Dependency theory: A conceptual lens to understand China's presence in Africa?', *Forum for Development Studies*, 44(3): 429–451. https://doi.org/10.1080/08039410.2017.1281161.

Aha, B. and Ayitey, J.Z. (2017) 'Biofuels and the hazards of land grabbing: Tenure (in)security and indigenous farmers' investment decisions in Ghana', *Land Use Policy*, 60: 48–59. https://doi.org/10.1016/j.landusepol.2016.10.012.

Arévalo, J.L. and Herreros, Ó.P. (2015) 'El desigual impacto de la crisis económica de 2008–2009 en los mercados de trabajo de las regiones de México: La frontera norte frente a la región sur', *Contaduría y Administración*, 60: 195–218. https://doi.org/10.1016/j.cya.2015.05.004.

Arndt, H.W. (1989) *Economic development: The history of an idea*, Chicago: University of Chicago Press.

Aspers, P. and Kohl, S. (2015) 'Economic theories of globalization', in Turner, B.S. and Holton, R.J. (eds.), *The Routledge international handbook of globalization studies*, London: Routledge, pp. 61–79.

Barrientos, S. (2019) *Gender and work in global value chains: Capturing the gains?* Cambridge: Cambridge University Press.

Barrientos, S., Gereffi, G. and Rossi, A. (2011) 'Economic and social upgrading in global production networks: A new paradigm for a changing world', *International Labour Review*, 150(3-4): 319–340. https://doi.org/10.1111/j.1564-913X.2011.00119.x.

Beck, U. (2009) *World at risk*, Cambridge: Polity Press.

Bolwig, S., Ponte, S., Du Toit, A., Riisgaard, L. and Halberg, N. (2010) 'Integrating poverty and environmental concerns into value-chain analysis: A conceptual framework', *Development Policy Review*, 28(2): 173–194. https://doi.org/10.1111/j.1467-7679.2010.00480.x.

Borras Jr, S., Hall, R., Scoones, I., White, B. and Wolford, W. (2011) 'Towards a better understanding of global land grabbing: An editorial introduction', *The Journal of Peasant Studies*, 38(2): 209–216. https://doi.org/10.1080/03066150.2011.559005.

Bottini, N., Ernst, C. and Luebker, M. (2008) *Offshoring and the labour market: What are the issues?* Geneva: International Labour Organization.

Castells, M. (1998) *End of millennium: V.3: The information age: Economy, society and culture*, London: Blackwell.

Castells, M. (2010) 'Globalisation, networking, urbanisation: Reflections on the spatial dynamics of the information age', *Urban Studies*, 47(13): 2737–2745. https://doi.org/10.1177/0042098010377365.

Chan, A. (2003) 'Racing to the bottom: International trade without a social clause', *Third World Quarterly*, 24(6): 1011–1028. https://doi.org/10.1080/01436590310001630044.

Clapp, J. (2009) 'Food price volatility and vulnerability in the Global South: Considering the global economic context', *Third World Quarterly*, 30(6): 1183–1196. https://doi.org/10.1080/01436590903037481.

Clarke, T. and Boersma, M. (2017) 'The governance of global value chains: Unresolved human rights, environmental and ethical dilemmas in the apple supply chain', *Journal of Business Ethics*, 143(1): 111–131. https://doi.org/10.1007/s10551-015-2781-3.

Cotula, L. (2012) 'The international political economy of the global land rush: A critical appraisal of trends, scale, geography and drivers', *The Journal of Peasant Studies*, 39(3–4): 649–680. https://doi.org/10.1080/03066150.2012.674940.

Desmond, M. (2012) 'Eviction and the reproduction of urban poverty', *American Journal of Sociology*, 118(1): 88–133. https://doi.org/10.1086/666082.

Fleming, P. (2017) 'The human capital hoax: Work, debt and insecurity in the era of Uberization', *Organization Studies*, 38(5): 691–709. https://doi.org/10.1177/0170840616686129.

Frank, A.G. (2014 [1967]) 'The development of underdevelopment', in Wheeler, S.M. and Beatley, T. (eds.), *Sustainable urban development reader*, Abingdon: Routledge.

Fraser, N. (2010) *Scales of justice: Reimagining political space in a globalizing world*, New York: Columbia University Press.

Gereffi, G. and Fernandez-Stark, K. (2010) 'The offshore services value chain: Developing countries and the crisis', *Policy Research Working Paper, No. WPS 5262*, Washington, DC: World Bank Group.

Gereffi, G., Humphrey, J. and Sturgeon, T. (2005) 'The governance of global value chains', *Review of International Political Economy*, 12(1): 78–104. https://doi.org/10.1080/09692290500049805.

Ghosh, B.N. (2019) *Dependency theory revisited*, Abingdon: Routledge.

Gill, S. (1995) 'Globalisation, market civilisation, and disciplinary neoliberalism', *Millennium*, 24(3): 399–423. https://doi.org/10.1177/03058298950240030801.

Graham, M., Hjorth, I. and Lehdonvirta, V. (2017) 'Digital labour and development: Impacts of global digital labour platforms and the gig economy on worker livelihoods', *European Review of Labour and Research*, 23(2): 135–162. https://doi.org/10.1177/1024258916687250.

Harvey, D. (2006) *Spaces of global capitalism: Towards a theory of uneven geographical development*, London: Verso.

Harvey, D. (2007a) *A brief history of neoliberalism*, Oxford: Oxford University Press.

Harvey, D. (2007b) 'Neoliberalism as creative destruction', *The Annals of the American Academy of Political and Social Science*, 610(1): 21–44. https://doi.org/10.1177/0002716206296780.

Hecock, R.D. and Jepsen, E.M. (2013) 'Should countries engage in a race to the bottom? The effect of social spending on FDI', *World Development*, 44: 156–164. https://doi.org/10.1016/j.worlddev.2012.10.016.

Horner, R. and Nadvi, K. (2018) 'Global value chains and the rise of the Global South: Unpacking twenty-first century polycentric trade', *Global Networks*, 18(2): 207–237. https://doi.org/10.1111/glob.12180.

International Labour Organization (ILO) – World Commission on the Social Dimension of Globalization (2004) *A fair globalization: Creating opportunities for all*, Geneva: International Labour Organization.

Ireland, P. and Pillay, R.G. (2010) 'Corporate social responsibility in a neoliberal age', in Utting, P. Marques, J.C. (eds.), *Corporate social responsibility and regulatory governance*, International Political Economy Series, London: Palgrave Macmillan. https://doi.org/10.1057/9780230246966_4.

Karamessini, M. and Rubery, J. (2014) *Women and austerity: The economic crisis and the future for gender equality*, New York: Routledge.

Kethineni, S. and Cao, Y. (2019) 'The rise in popularity of cryptocurrency and associated criminal activity', *International Criminal Justice Review*, 30(3): 325–344. https://doi.org/10.1177/1057567719827051.

Kirby, P. (2013) 'Vulnerability and globalization: The social impact of globalization', in Turner, B.S. and Holton, R.J. (eds.), *The Routledge international handbook of globalization studies*, London and New York: Routledge, pp. 136–156.

Lee, R., Clark, G.L., Pollard, J. and Leyshon, A. (2009) 'The remit of financial geography: Before and after the crisis', *Journal of Economic Geography*, 9(5): 723–747. https://doi.org/10.1093/jeg/lbp035.

Mosley, L. and Uno, S. (2007) 'Racing to the bottom or climbing to the top? Economic globalization and collective labour rights', *Comparative Political Studies*, 40(8): 923–948. https://doi.org/10.1177/0010414006293442.

Ocampo, J. (2012) 'How well has Latin America fared during the global financial crisis', in Cohen, M. (ed.), *The global economic crisis in Latin America: Impacts and responses*, Abingdon: Routledge.

Ouma, S. (2014) 'Situating global finance in the land rush debate: A critical review', *Geoforum*, 57: 162–166. https://doi.org/10.1016/j.geoforum.2014.09.006.

Pontell, H.N., Black, W.K. and Geis, G. (2014) 'Too big to fail, too powerful to jail? On the absence of criminal prosecutions after the 2008 financial meltdown', *Crime, Law and Social Change*, 61(1): 1–13. https://doi.org/10.1007/s10611-013-9476-4.

Rajmil, L., De Sanmamed, M-J.F., Choonara, I., Faresjö, T., Hjern, A., Kozyrskyj, A.L., Lucas, P.J., Raat, H., Séguin, L., Spencer, N. and Taylor-Robinson, D. and On Behalf of the International Network for Research in Inequalities in Child Health (2014) 'Impact of the 2008 economic and financial crisis on child health: A systematic review', *International Journal of Environmental Research and Public Health*, 11(6): 6528–6546. https://doi.org/10.3390/ijerph110606528.

Raworth, K. (2004) *Trading away our rights: Women working in global supply chains*, Oxford: Oxfam International.

Robertson, B. and Pinstrup-Andersen, P. (2010) 'Global land acquisition: Neo-colonialism or development opportunity?', *Food Security*, 2(3): 271–283. https://doi.org/10.1007/s12571-010-0068-1.

Rolnik, R. (2019) *Urban warfare: Housing under the empire of finance*, London: Verso.

Rose, A.K. and Spiegel, M.M. (2012) 'Cross-country causes and consequences of the 2008 crisis: Early warning', *Japan and the World Economy*, 24(1): 1–16. https://doi.org/10.1016/j.japwor.2011.11.001.

Rosen, E. (2002) *Making sweatshops: The globalization of the US apparel industry*, Berkeley: University of California Press.

Rudra, N. (2008) *Globalization and the race to the bottom in developing countries*, Cambridge: Cambridge University Press.

Santos, M. (2017 [2000]) *Toward another globalization: From the single thought to universal conscience*, Cham: Springer.

Schoneveld, G.C., German, L.A. and Nutakor, E. (2011) 'Land-based investments for rural development? A grounded analysis of the local impacts of biofuel feedstock plantations in Ghana', *Ecology and Society*, 16(4): 10. https://doi.org/10.5751/ES-04424-160410

Siddiqui, K. (2017) 'Globalization, trade liberalisation and the issues of economic diversification in the developing countries: An overview', *Journal of Economics Library*, 4(4): 514–529. https://doi.org/10.1453/jel.v4i4.1486.

Singh, A. and Zammit, A. (2004) 'Labour standards and the "race to the bottom": Rethinking globalization and workers' rights from developmental and solidaristic perspectives', *Oxford Review of Economic Policy*, 20(1): 85–104. https://doi.org/10.1093/oxrep/20.1.85.

Soederberg, S. (2015) 'Subprime housing goes south: Constructing securitized mortgages for the poor in Mexico', *Antipode*, 47(2): 481–499. https://doi.org/10.1111/anti.12110.

Stewart, F. (1991) 'The many faces of adjustment', *World Development*, 19(12): 1847–1864. https://doi.org/10.1016/0305-750X(91)90029-H.

Utting, P. (2008) 'The struggle for corporate accountability', *Development and Change*, 39(6): 959–975. https://doi.org/10.1111/j.1467-7660.2008.00523.x.

Wallerstein, I. (2011) *The modern world-system I: Capitalist agriculture and the origins of the European world-economy in the sixteenth century*, Berkeley: University of California Press.

Weber, M. (2012 [1905]) *The protestant ethic and the 'spirit' of capitalism and other writings*, New York: Routledge.

Wells, D. (2009) 'Local worker struggles in the Global South: Reconsidering northern impacts on international labour standards', *Third World Quarterly*, 30(3): 567–579. https://doi.org/10.1080/01436590902742339.

Zoomers, A. (2010) 'Globalisation and the foreignisation of space: Seven processes driving the current global land grab', *The Journal of Peasant Studies*, 37(2): 429–447. https://doi.org/10.1080/03066151003595325.

Zoomers, A., Van Noorloos, F., Otsuki, K., Steel, G. and Van Westen, G. (2017) 'The rush for land in an urbanizing world: From land grabbing toward developing safe, resilient, and sustainable cities and landscapes', *World Development*, 92: 242–252. https://doi.org/10.1016/j.worlddev.2016.11.016.

38
CONTEXTUALISING URBAN TRANSPORT SYSTEMS AND LIVELIHOODS IN DEVELOPING COUNTRIES

The case of Bus Rapid Transit project

Michael Poku-Boansi, Michael Osei Asibey and Richard Apatewen Azerigyik

Introduction

Rapid urbanisation without the corresponding provision of adequate transport infrastructure has generated numerous transport challenges especially among countries in the Global South. Some of these challenges include but are not limited to heavy traffic congestion in urban centres, informal nature of the sector, low public transport services, the use of few high-occupancy vehicles for mass transportation, increased road accidents and air pollution (Agyemang, 2009; Oteng-Ababio and Agyemang, 2012). To this end, mobility within urban areas is less enjoyable and, sometimes, extremely difficult. This phenomenon is common among the urban poor and low-income earners who constitute the urban vulnerable and live within the 'transport poverty' trap (UN-Habitat, 2013). In many parts of the Global South, public transport systems are largely informal, where operators in many cities operate outside the officially sanctioned public transport sector and thus, are regarded to be a nuisance requiring public intervention and occasionally, eradication (Cervero and Golub, 2007; Poku-Boansi, 2020; Poku-Boansi and Cobbinah, 2017). The informal transport sector has relied primarily on minibuses, taxis, motorcycles and vans (Del Mistro and Behrens, 2015) with several benefits and challenges. There have consequently been attempts by governments in many sub-Saharan African countries to implement Bus Rapid Transit (BRT) systems to address the numerous challenges and complement the role played by the informal public transport sector.

The BRT system is argued to have originated in Curitiba, Brazil, in the 20th century and has spread to many other cities such as Bogota (Columbia), Mexico City (Mexico), and Quito (Ecuador); Asian countries such as Nagoya (Japan), Taipei and Beijing (China), Bangkok (Thailand), Delhi, Pune, and Hyderabad (India), Dhaka (Bangladesh), Jakarta (Indonesia), and Seoul (Korea); and, African countries such as Lagos (Nigeria), Cape Town (South Africa), Dakar (Senegal) and Dar es Salaam (Tanzania) (Agyemang, 2015; Deng and Nelson, 2011; Institute for Transportation and Development Policy, 2003). In many countries, the BRT denotes a high-quality bus transport system that provides quick, comfortable and comparatively less expensive urban transport on dedicated separate lanes to offer high frequency while providing high-quality customer service (Levinson, 2003). The system can also mean a quick mode of transport that amalgamates the quality of rail transport and bus transport flexibility (Levinson, 2003). As of 2016, BRT systems were operating across 45 countries on a projected 5,542 km of dedicated lanes in use (Global BRT Data, 2016).

The BRT system is one of the effective ways of aiding the urban vulnerable to escape the 'transport poverty' trap (Poku-Boansi and Marsden, 2018). The inability of countries of the Global South to mobilise the needed financial resources to revamp the transport sector resulted in implementing the BRT system as a cost-effective approach for mass transportation (Agyemang, 2015). The BRT system since its inception has improved social equity, environmental and atmospheric conditions, and economic prosperity (Deng and Nelson, 2011; International Energy Agency, 2002; Alpkokin and Ergun, 2012). Lessons learned from countries in the Global North that have successfully implemented the BRT system revealed that the system had created direct and indirect jobs for urban dwellers (Spooner, 2011), similar to sub-Saharan Africa, although the system has faced several agitations from informal transport actors (Ferrante et al., 2020; Hidalgo and Gutiérrez, 2013).

The government of Ghana, for example, through the support of development partners such as the World Bank has initiated steps by undertaking the Ghana Urban Transport Project (GUTP), with an overall objective of promoting the use of large occupancy vehicles and a BRT. The project has had both significant positive and adverse influence on the livelihoods of urban dwellers. It is important to further note that the design and implementation of the BRT system comes with several planning challenges that, if not satisfactorily addressed, will have adverse implications on the livelihoods of individuals engaged in the public transport sector, particularly the informal public transport operators (Poku-Boansi and Marsden, 2018; Spooner, 2018). Taking Ghana's GUTP as a case, this chapter examines the BRT system and its effect on urban livelihoods as well as the challenges associated with its implementation to guide future policies and planning decisions.

Urban transport systems and livelihood impacts in the Global South

An individual's mobility is achieved if s/he can move from one area to different spatial locations to access economic and social services. This is determined by the ability of the transport system to link the user to the various locations with the available modal options. The urban poor of the Global South require improved mobility to access their crucial needs. This mobility improvement usually requires affordable, available and effective transport services for the majority of the urban poor community to escape their poverty (Cervero, 2011, 2013; Poku-Boansi and Marsden, 2018). The urban transport system in many countries of the Global South is largely informal, i.e. where actors operate outside the officially sanctioned public transport sector. Urban transport services are mostly provided by the informal or private sector due to the failures of governments to provide efficient transport for all (Del Mistro and Behrens, 2015; Yeboah and Asibey, 2019). On this, Cervero (2011) indicates that in the case of the poorest countries, governments largely hand over the provision of collective-ride mobility to informal operators. In India, the federal government has partnered the private sector to ensure a holistic approach in service provision by opening the public bus sector to private companies (Jaspal, 2016).

Despite the description given to urban public transport services, the sector is very vital in road transportation in Africa, which remains the major mode of travel. In Africa, road transport accounts for 80% of goods and 90% of passenger traffic (UN-Habitat, 2010). This finding is similar to that of Ghana where 90% of passengers and 95% of freight traffic are carried by road transport (Government of Ghana, 2008). As earlier stated, urban transport systems are dominated by private transport operators resulting in an increase in vehicle ownership in Africa (see Adarkwa and Poku-Boansi, 2011; Chakwizira et al., 2014). This increase in vehicle ownership has however not addressed the mobility needs of Africa's urban residents as over 80% of them do not have access to private means of transportation (UN-Habitat, 2010) and thus, rely on public transportation for their daily activities, emphasising the importance of public transport services in Africa. The high proportion of urban residents that rely on public transportation has translated in high volumes of passenger and freight movements.

It is largely documented that urban transport services have significant impact on the livelihoods of people, specifically those living in low-income areas of cities of the Global South in terms of accessing work, education and petty trading through public transport. First, the contribution of urban transport systems to social development is mostly manifested in providing and improving accessibility to major public services such as education, healthcare, energy and sanitation. Several studies have shown the positive relationship between urban transport infrastructure and access to the above services (e.g. Abekah-Nkrumah et al., 2019; Adom-Asamoah et al., 2020; Gyamera, 2016). Further, people with good access to transport services have higher educational attainment than those with poor access. Moreover, it is argued that 'improved road accessibility significantly influenced households to enrol their children in school' (Adom-Asamoah et al., 2020: 57). This was because lack of access hinders school attendance and flow of educational information to persuade parents to enrol their wards. It can also be said that road access facilitates the access to other educational materials such as books and furniture, especially in remote areas of the Global South.

Further, urban transport promotes access to healthcare delivery and to diverse health facilities. It impacts positively on health in three major areas including promoting the usage of health facilities (as opposed to other life-threatening treatment); increasing the range of services; and presenting patients with options to choose the level of (or particular) service they can access (Adom-Asamoah et al., 2020). However, it must be acknowledged that the above correlation between urban transport systems and health could be simplistic given the fact that there could be adverse ramifications in the areas of motor accidents and air pollution (Laumbach and Kipen, 2012; Zimmerman et al., 2012). These indicate that while transport has the overall positive effect on health, particular attention should be given to addressing the issue of increasing mortalities and morbidities associated with its growth for a more beneficial outcome. In addition to promoting access to educational and health facilities, urban transport systems also increase access to sustainable energy such as Liquefied Petroleum Gas (LPG), regular waste collection and sensitisation programmes related to good sanitary practices. In all, urban transport systems have significant impact on social enhancement and enlightenment, both prerequisites for social modernism.

Microeconomic studies (e.g. Banister and Berechman, 2001; Herranz-Loncán, 2007; Pucher et al., 2007) have also revealed that investment in urban transport generates significant impact on economic growth and poverty reduction. Investment in urban transport results in reduction in production and transaction costs which promotes trade through specialisation and enhances opportunities to propel economic growth. In urban centres, roads are *'the first priority'* as they connect farmers to the markets for transactions. Khandker et al. (2006) support this with their study in Bangladesh where large increases in agricultural production, output prices and wages with reduction in input cost and transport cost were recorded. Dercon et al. (2009) hold a similar view, that if the major economic activity is agriculture, then improvement in roads has the potential to increase productivity through reduction in cost of inputs coupled with higher output prices. The labour impacts of urban transport improvement have been documented by some authors (e.g. Adom-Asamoah et al., 2020; Cervero, 2000, 2011), where they reported enhanced service sector employment and incomes.

Transport is again regarded as one of the factors of production and hence a fall in cost means more will be demanded to increase supply. Road transport has been the nexus of economic development by serving as a boost for production, distribution and consumption. Roads help boost agricultural and industrial production, trade, employment and access to economic amenities. Urban transport has the potency of promoting or impeding development through costs of operation; hence, more remote areas would be the last regions to achieve economic development due to the absence of low-cost transport that will serve as a conduit for the growth process (Cervero, 2013; Ferrante et al., 2020; Hidalgo and Gutiérrez, 2013). In summary and in terms of production, transport serves as a channel of access to raw materials, economic opportunities and infrastructure, credit and investment facilities, technology and market. In countries of the Global South, transport improvements lead to improved access to industrial raw materials, yields, extension services and markets.

Due to the importance of the informal sector in providing relevant public transport services and having significant influence on livelihoods, governments in many sub-Saharan African countries, including Ghana, have initiated reforms within the public transport sector, mainly resulting in the implementation of BRT systems as an alternative to the existing informal urban transport systems to ensure efficiency. The BRT systems have similarly had significant effects on urban livelihoods in the Global South but have been confronted with restrictions and agitations from informal actors (Okoye et al., 2010).

The BRT and its effects on urban livelihoods in the Global South

The BRT system has, in general, brought about an improvement in urban livelihoods. There have generally been shorter commuting times to workplaces and educational institutions (Chengula and Kombe, 2017). In line with this, time savings result in expansion of time for economic activities and reduce stressful situations. In general, an improved public transport system 'enables the multiple social connections and interactions that flourish within an urban space and leads to the different levels of cohesion of the multiple parts' (Bocarejo et al., 2016a: 45). BRT buses provide easy access for vulnerable people (children, elderly, physically impaired people), which contribute to transport infrastructure inclusiveness, as observed in many cities across African countries such as Tanzania, Ghana, Kenya, Rwanda and South Africa (Poku-Boansi and Marsden, 2018; Shen et al., 2020). This further supports the Sustainable Development Goal (SDG) target 11.2 which seeks to improve access for vulnerable people.

In Ghana, the BRT employs over 200 staff with the jobs having far-reaching implications for the livelihoods of their families. The BRT's support for local economic development also lies in the traders' access to the central business district in Accra. The buses cover a 22 kilometre-stretch from Amasaman to Tudu in Accra Central. Patronage, has however, been undermined by the lack of dedicated routes for the buses and the attendant adverse effects on travel time. With regards to capacity building and investments in local communities, the Ghana Urban Transport Project trained over 60 females to be employed as bus drivers and conductors. This is a strategy to create jobs and promote gender equality. The implementation of the BRT system in Ghana, specifically the MMT (for intracity transport services), led to human resources training and development (Yobo, 2013). The Government of Ghana supported the recruitment and training of highly skilled human resources towards successfully implementing and managing public transport services. Artisans and drivers were trained to effectively manage the purchased buses (Yobo, 2013). Some of the artisans trained and employed by the MMT included auto mechanics, auto electricians, auto body repairers, welding and fabrication engineers (Birago et al., 2017; Yobo, 2013).

On employment creation, the MMT created about 4,007 jobs for individuals, comprising drivers, conductors, and customer service ambassadors across the country (Yobo, 2013). The findings are similar to that of Bocarejo et al. (2012) who reported that the TransMilenio BRT project in the City of Bogota, Colombia, created permanent employment for about 2,900 bus drivers. In addition, between 1,400 and 1,800 temporary jobs were created for workers during the construction phase of dedicated routes for the BRT in the City of Bogota, Colombia (Bocarejo et al., 2012). This is widely reported in several countries where there is evidence of changes in access to employment opportunities from BRT systems, in terms of ease of access to jobs (Bocarejo and Oviedo, 2012; Bocarejo et al., 2016b; Yañez-Pagans et al., 2018). Again, in many African cities, the BRT is inducing the springing up of economic opportunities. Specifically, by the staging of bus stops as local hubs, street vendors find various locations to operate their businesses due to the increased number of passengers.

Even though the implementation of the BRT in many African cities has created employment for some individuals, there is evidence that the project displaced a substantial proportion of workers employed in the informal public transport sector. For instance, within the Accra Metropolis alone, there was around 12,000 registered minibuses as of the year 2005 out of which around 6,000 of these minibuses operated daily. By the year 2011, a projected 10,000 trotros (a mini-bus with a carrying capacity between 12 and 15 passengers popularly used for intra-city transport) were functioning in the city of Accra (Finn et al., 2011). Inferring from the work of Finn et al. (2011), it can be concluded that over 10,000 drivers operated as 'trotro' drivers for their daily survival. Considering the highly subsided fares that parade the operations of the BRT, the patronage of 'trotro' services, consequently resulting in the loss of jobs or affected income levels of a significant number of 'trotro' operators (Agyemang, 2015). It is also imperative to note that drivers earn an average monthly income of $157.12 (Yusif, 2018), which is almost the same as the national average of $156.77 (Ghana Statistical Service, 2016). The relatively cheaper fares of the BRT and its corresponding high patronage is likely to affect demand for the services of these informal transport operators (drivers) and consequently, reduction in the monthly income of drivers. The reduction in the monthly income could have ripple effects on some sustainable livelihood indicators such as catering for health and education needs, providing shelter and saving for investment purposes.

Similar to the above-described situation of the 'trotro driver', the bus conductor, locally called 'trotro mate', receives a daily average wage of $5.24 (Yusif, 2018). According to Yusif (2018), the 'trotro mates' are individuals engaged by the drivers to assist them in their operations for daily wages. Like the driver, the mate's wage is influenced by the number of passengers who patronise their services. Hence, the low patronage of informal transport services due to the BRT implementation has adverse effect on their wages and consequently other livelihood indicators.

The implication of the foregoing is that the BRT, despite its numerous advantages as earlier indicated, can result in the low patronage of the informal public transport operator's services; consequently affecting the income levels of primary actors within the informal public transport sector. This corroborates findings of a study by Spooner (2018), which established that the proposed Nairobi BRT project would likely create unemployment for more than 50% of individuals engaged by the sector such as drivers, bus conductors and head porters, among others. A similar finding was reported by Timéra et al. (2020) on the labour impact assessment survey conducted on the proposed Dakar Bus Rapid Transit project. The report concluded that the anticipated project is likely to result in a substantial loss of jobs for people employed by the informal transport sector; around 35,000 jobs in the informal transport service could be lost on a planned 180 km route within the city of Dakar, Senegal.

With the advancement of information technology, there has been a rapid rise of on-demand ride-hailing services such as Uber. Ride-hailing companies continue to infiltrate the market of transportation-for-hire services in major urban and metropolitan cities all across the globe. These new technology-enabled transportation services considerably increase the versatility of travel options and access to transportation services without the fixed high cost of ownership of a vehicle (Shaheen et al., 2016). The implementation of the BRT in the Global South, particularly Africa, has adverse implications on the patronage and livelihood of vehicle investors and suppliers. In the case of vehicle investors and vehicle suppliers, these individuals perceive the informal transport sector as an investment opportunity; hence, purchase vehicles for individuals to drive and provide them with income. Examples include investment in taxis and ride-hailing services such as Uber (Cherry, 2016). Yusif (2018) revealed that vehicle suppliers or investors earn an average daily income of $17.46, all other things being equal; hence, they are likely to earn an average monthly income of $523.73.

The other group of actors whose livelihoods have been and could be adversely affected by the BRT, resulting in resistance, are spare parts dealers. These actors supply spare parts to 'trotro' drivers and other informal transport operators. The lucrative nature of this business is exhibited in the total number of people engaged in it. For instance, there are about 10,000 stalls and shops with about 30,000 spare parts dealers in Abossey Okai in Accra alone (Fako, 2019). The significance of the activities of these actors cannot be underestimated. Obeng-Odoom (2010) argued that overaged or used vehicles dominate the commercial vehicles operating in urban areas. The use of aged vehicles with the corresponding low maintenance of these vehicles results in their easy breakdown. Therefore, these informal transport operators depend on the spare part dealers' services as their primary suppliers of vehicle parts to fix their broken-down vehicles.

Relating the discussion to the adverse impact of the BRT system on livelihoods of informal transport workers within the Accra Metropolis, the low patronage of the services of the informal transport operators implies being pushed out of the transport business. The negative multiplier effect of this phenomenon is that it will affect the gross domestic product of the local economy of African cities and the countries' economy at large in terms of job creation and the payment of import duties (Fako, 2019). The foregoing calls for measures to have an integrated framework which secures livelihoods of service operators and facilitate efficient service delivery which should be done in a participatory and inclusive manner to promote sustainable public transport system (Kashi and Carello, 2020).

Conclusion

Urban transport systems in low-income countries, which are essentially dominated by private groups and individuals, have been largely classified as informal and requiring formalisation. The quest by governments to promote efficient and sustainable public transport systems in cities has led to the formation of two schools of thought. The first school suggests the need for public intervention and, occasionally, calls for their eradication as, in most cases, they operate outside the officially sanctioned domain of the state. The second school supports public reform-driven intervention as it emphasises the critical role these private operators play in meeting the mobility demand of the urban population, providing over 50% of transport services in some cities. Governments' efforts over the last few decades have essentially leaned towards the second thought with the implementation of strategies to reform the existing regimes. With support from the World Bank and other partners, governments have invested in BRT systems as an alternative to the existing public transport systems.

The BRT has contributed positively to infrastructure and service delivery, improvement in working environment and socioeconomic development of cities in the Global South. The BRT, since its implementation, has also contributed significantly to local governance, capacity enhancement and employment creation as well as income generation to many urban dwellers. The introduction of the BRT system has indeed created new jobs and generated more income opportunities. The system has considerably reduced commuting times and increased inclusivity of urban residents, including the disadvantaged. Nonetheless, the BRT has not made the desired impact due to, among others, the lack of dedicated routes for the commuter buses to ply, exposing the schemes to competition from other service providers in cities. In contrast, the BRT system, including in Accra, has not improved commuter time. Most importantly, the BRT has had and could have adverse impact on the livelihoods of actors within the informal public transport sector. BRT has resulted in low patronage of services of informal public transport actors, having adverse implications on their livelihoods. The BRT project has thus faced resistance from informal public transport actors due to mistrust and fear of losing their livelihoods.

The chapter, based on the foregoing, suggests that the public urban transportation sector should be reformed to ensure that the informal public transport services complement the BRT. These informal agents in the urban transport subsector should be restricted to the assigned routes to ensure complementarity. This proposal requires the strengthening of the capacity of appropriate agencies to ensure compliance with local bye-laws. Powers from city authorities, whose jurisdiction the BRT will operate, will have to transfer powers to the relevant bodies to ensure compliance with allocated routes by other commercial vehicles. The effectiveness of the BRT can create and sustain livelihoods. In conclusion, the chapter calls for an integrated framework which secures livelihoods of service operators and facilitates efficient service delivery done in a participatory and inclusive manner to promote sustainable public transport systems.

Key Points

- The liberalised and deregulated market for the transport sector has resulted in informal operators taking control of the provision of public transport services in African cities.
- Even though the informal transport services can be difficult to manage and rationalise, they still provide an important component of the public transport subsector.
- Governments and city managers have sought to promote efficient and sustainable public transport systems in cities of the Global South.
- Development partners and governments of the Global South have invested in Bus Rapid Transit systems as an alternative to the existing public transport systems to ensure efficiency, lower transport-related GHG emissions and make cities competitive.

Suggested readings

Colin, Hagans. (2013) 'Livelihoods, land-use and public transport: Opportunities for poverty reduction and risks of splintering urbanism in Nairobi's Spatial Plans', *DPU Working Paper No. 159*, Development Planning Unit, University College London. London.

Esson, J., Gough, K.V., Simon, D., Amankwaa, E.F., Ninot, O. and Yankson, P.W.K. (2016) 'Livelihoods in motion: Linking transport, mobility and income-generating activities', *Journal of Transport Geography*, 55: 182–188. https://doi.org/10.1016/j.jtrangeo.2016.06.020

Krüger, F., Titz, A., Arndt, R., Groß, F., Mehrbach, F., Pajung, V., Suda, L., Wadenstorfer, M. and Wimmer, L. (2021) 'The Bus Rapid Transit (BRT) in Dar es Salaam: A pilot study on critical infrastructure, sustainable urban development and livelihoods', *Sustainability*, 13: 1058. https://doi.org/10.3390/su13031058

References

Abekah-Nkrumah, G., Asuming, P.O. and Telli, H. (2019) 'The effects of the introduction of a bus rapid transit system on commuter choices in Ghana in brief', *Policy Brief 33400*, Legon: The International Growth Centre, University of Ghana.

Adarkwa, K.K. and Poku-Boansi, M. (2011) 'Rising vehicle ownership, roadway challenges, and traffic congestion in Kumasi', in Adarkwa, K.K. (ed.), *The future of the tree: Managing the growth and development of Kumasi*, Kumasi: University Press, pp. 128–152.

Adom-Asamoah, G., Amoako, C. and Adarkwa, K.K. (2020) 'Gender disparities in rural accessibility and mobility in Ghana', *Case Studies on Transport Policy*, 8(1): 49–58. https://doi.org/10.1016/j.cstp.2019.12.006

Agyemang, E. (2009) *Traffic congestion: The bane of a bus rapid transit system in Accra, Ghana?* Master of Philosophy Thesis in Development Studies (specializing in Geography), Trondheim: Norwegian University of Science and Technology.

Agyemang, E. (2015) 'The bus rapid transit system in the Greater Accra Metropolitan Area, Ghana: Looking back to look forward', *Norsk Geografisk Tidsskrift*, 69(1): 28–37. https://doi.org/10.1080/00291951.2014.992808

Alpkokin, P. and Ergun, M. (2012) 'Istanbul Metrobüs: First intercontinental bus rapid transit', *Journal of Transport Geography*, 24: 58–66. https://doi.org/10.1016/j.jtrangeo.2012.05.009

Banister, D. and Berechman, Y. (2001) 'Transport investment and the promotion of economic growth', *Journal of Transport Geography*, 9(3): 209–218. https://doi.org/10.1016/S0966-6923(01)00013-8

Birago, D., Opoku Mensah, S. and Sharma, S. (2017) 'Level of service delivery of public transport and mode choice in Accra, Ghana', *Transportation Research Part F: Traffic Psychology and Behaviour*, 46: 284–300. https://doi.org/10.1016/j.trf.2016.09.033

Bocarejo, J.P., Escobar, D., Hernandez, D.O. and Galarza, D. (2016a) 'Accessibility analysis of the integrated transit system of Bogotá', *International Journal of Sustainable Transportation*, 10(4): 308–320. https://doi.org/10.1080/15568318.2014.926435

Bocarejo, J.P. and Oviedo, D.R. (2012) 'Transport accessibility and social inequities: A tool for identification of mobility needs and evaluation of transport investments', *Journal of Transport Geography*, 24:142–154. https://doi.org/10.1016/j.jtrangeo.2011.12.004

Bocarejo, J.P., Portilla, I. and Meléndez, D. (2016b) 'Social fragmentation as a consequence of implementing a Bus Rapid Transit system in the city of Bogota', *Urban Studies*, 53(8): 1617–1634. https://doi.org/10.1177/0042098015588739

Bocarejo, J.P., Velasquez, J.M., Díaz, C.A. and Tafur, L.E. (2012) 'Impact of bus rapid transit systems on road safety: Lessons from Bogotá, Colombia', *Transportation Research Record*, 2317(1): 1–7. https://doi.org/10.3141/2317-01

Cervero, R. (2000) *Informal transport in the developing world*, Nairobi: United Nations Commission on Human Settlements.

Cervero, R. (2011) 'State roles in providing affordable mass transport services for low- income residents', *Discussion Paper No. 2011-17*, Paris: International Transportation Forum, Organization for Economic and Cooperative Development.

Cervero, R. (2013) 'Bus rapid transit (BRT): An efficient and competitive mode of public transport', *Working Paper 2013-01*, Berkeley, CA, USA: Institute of Urban and Regional Development University of California.

Cervero, R. and Golub, A. (2007) 'Informal transport: A global perspective', *Transport Policy*, 14(6): 445–457. https://doi.org/10.1016/j.tranpol.2007.04.011

Chakwizira, J., Mudau, P.M. and Radali, A.C.O. (2014) 'Managing traffic congestion in small sized rural towns in South Africa: The case of Vhembe District Municipality', Proceedings of the *33rd Southern African Transport Conference* (SATC), Pretoria. 7–10 July 2014 (pp. 806–822).

Chengula, D. and Kombe, K. (2017) 'Assessment of the effectiveness of Dar Es Salaam Bus Rapid Transit (DBRT) system in Tanzania', *International Journal of Sciences: Basic and Applied Research*, 36(8): 10–30.

Cherry, M.A. (2016) 'Beyond misclassification: The digital transformation of work', *Comparative Labor Law and Policy Journal*, 37(3): 577–602.
Del Mistro, R. and Behrens, R. (2015) 'Integrating the informal with the formal: An estimation of the impacts of a shift from paratransit line-haul to feeder service provision in Cape Town', *Case Studies on Transport Policy*, 3(2): 271–277. https://doi.org/10.1016/j.cstp.2014.10.001
Deng, T. and Nelson, J.D. (2011) 'Recent developments in bus rapid transit: A review of the literature', *Transport Reviews*, 31(1): 69–96. https://doi.org/10.1080/01441647.2010.492455
Dercon, S., Gilligan, D.O., Hoddinott, J. and Woldehanna, T. (2009) 'The impact of agricultural extension and roads on poverty and consumption growth in fifteen Ethiopian Villages', *American Journal of Agricultural Economics*, 91(4): 1007–1021. https://doi.org/10.1111/j.1467-8276.2009.01325.x
Fako, P. (2019) *The impact of import duties on the automobile spare parts industry in Ghana*, MSc Development Finance Thesis, Legon: University of Ghana.
Ferrante, C., Ciampoli, L.B., De Falco, M.C., D'Ascanio, L, Presta, D. and Schiattarella, E. (2020) 'Can a fully integrated approach enclose the drainage system design and the flood risk analysis?', *Transportation Research Procedia*, 45: 811–818. https://doi.org/10.1016/j.trpro.2020.02.089
Finn, B., Kumarage, A. and Gyamera, S. (2011) 'Organisational structure, ownership and dynamics on control in the informal local road passenger transport sector', presented at the *12th Conference on Competition and Ownership in Land Passenger Transport*, South Africa, Conference Proceedings, pp. 133–146.
Ghana Statistical Service (2016) *Labour force report (2015)*, Accra: Ghana Statistical Service.
Global BRT Data (2016) *Database of bus rapid transit systems around the world*. Available at: https://brtdata.org/#/location (accessed 17 November 2020).
Government of Ghana (2008) *National transport policy*, Accra, Ghana: Government of Ghana.
Gyamera, S. (2016) 'Organising and regulating the unorganised public transport system in Accra', Greater Accra Passenger Transport Executive (GAPTE), presentation to *UATP Workshop on Promoting Soot-Free and Sustainable Transport in Africa*, Nairobi.
Herranz-Loncán, A. (2007) 'Infrastructure investment and Spanish economic growth, 1850–1935', *Explorations in Economic History*, 44(3): 452–468. https://EconPapers.repec.org/RePEc:eee:exehis:v:44:y:2007:i:3:p:452-468.
Hidalgo, D. and Gutiérrez, L. (2013) 'BRT and BHLS around the world: Explosive growth, large positive impacts and many issues outstanding', *Research in Transportation Economics*, 39(1): 8–13. https://doi.org/10.1016/j.retrec.2012.05.018
Institute for Transportation and Development Policy (2003 Fall) 'Bus rapid transit spreads to African and Asia. In pedestrianizing Asian cities', *Sustainable Transport*, 15: 4–11. https://itdpdotorg.wpengine.com/wp-content/uploads/2014/07/ST15-2003.pdf
International Energy Agency (2002) *Bus systems for the future: Achieving sustainable transport worldwide*, Paris: IEA Publications.
Jaspal, S. (2016) *City public transportation developments in India*. Available at: www.intelligenttransport.com/transport-articles/21458/city-public-transportation-india/ (Accessed 20 November 2020).
Kashi, B. and Carello, S. (2020) *Urban transport solutions for Accra. In Ghana priorities*, Tewksbury, MA: Copenhagen Consensus Center.
Khandker, Shahidur R., ZaidBakt and Gayatri B.K. (2006) 'The poverty impact of rural roads: The evidence from Bangladesh', *World Bank Policy Research Working Paper 3875*, Washington, DC: World Bank.
Laumbach, R.J. and Kipen, H.M. (2012) 'Respiratory health effects of air pollution: Update on biomass smoke and traffic pollution', *Journal of Allergy and Clinical Immunology*, 129(1): 3–11. https://doi.org/10.1016/j.jaci.2011.11.021
Levinson, H.S. (2003) 'Bus rapid transit on city streets how does it work', paper presented at the *2nd Urban Street Symposium* (Anaheim, California), 28–30 July 2003, pp. 1–25. CA, UTRC Icon Mentor City College, New York.
Obeng-Odoom, F. (2010) 'Drive left, look right: The political economy of urban transport in Ghana', *International Journal of Urban Sustainable Development*, 1(1–2): 33–48. https://doi.org/10.1080/19463130903561475
Okoye, V., Sands, J. and Debrah, A.C. (2010) *The Accra pilot bus-rapid transit project: Transport-land use research study*, Accra: Millennium Cities Initiative and Accra Metropolitan Assembly.
Oteng-Ababio, M. and Agyemang, E. (2012) 'Virtue out of necessity? Urbanisation, urban growth and Okada services in Accra, Ghana', *Journal of Geography and Geology*, 4(1): 148–162. https://doi.org/10.5539/jgg.v4n1p148
Poku-Boansi, M. (2020) 'Path dependency in transport: A historical analysis of transport service delivery in Ghana', *Case Studies on Transport Policy*, 8(4): 1137–1147. https://doi.org/10.1016/j.cstp.2020.07.003

Poku-Boansi, M. and Cobbinah, P.B. (2017) 'Land use and urban travel in Kumasi, Ghana', *GeoJournal*, 83: 563–581. http://dx.doi.org/10.1007/s10708-017-9786-7

Poku-Boansi, M. and Marsden, G. (2018) 'Bus rapid transit systems as a governance reform project', *Journal of Transport Geography*, 70:193–202. https://doi.org/10.1016/j.jtrangeo.2018.06.005

Pucher, J., Peng, Z.R., Mittal, N., Zhu, Y. and Korattyswaroopam, N. (2007) 'Urban transport trends and policies in China and India: Impacts of rapid economic growth', *Transport Reviews*, 27(4): 379–410. https://doi.org/10.1080/01441640601089988

Shaheen, S., Chan, N. and Gaynor, T. (2016) 'Casual carpooling in the San Francisco bay area: Understanding characteristics, behaviours, and motivations', *Transport Policy*, 51: 165–173.

Shen, Y., Bao, Q. and Hermans, E. (2020) 'Applying an alternative approach for assessing sustainable road transport: A benchmarking analysis on EU countries', *Sustainability*, 12(24): 10391. https://doi.org/10.3390/su122410391

Spooner, D. (2011) *Transport workers in the urban informal economy: Livelihood profile.* Available at: http://gli-manchester.net/wp-content/uploads/2019/08/Urban-Informal-Transport-Livelihood-Profile.pdf (Accessed 29 May 2021).

Spooner, D. (2018) *Nairobi bus rapid transit labour impact assessment*, Manchester: Global Labour Institute.

Timéra, M.B., Diongue, M. and Sakho, P. (2020) *Dakar bus rapid transit labour impact assessment research report 2020*, Dakar: Laboratory of Human Geography, Cheikh Anta Diop University; Manchester: Global Labour Institute,

UN-Habitat (2010) *State of the world's cities 2010/2011 – Cities for all: Bridging the urban divide*, London: Earthscan.

UN-Habitat (2013) *Planning and design for sustainable urban mobility: Global report on human settlements*, New York: Routledge.

Yañez-Pagans, P., Martinez, D., Mitnik, O., Scholl, L. and Vázquez, A. (2018) 'Urban transport systems in Latin America and the Caribbean: Challenges and lessons learned', *Inter-American Development Bank Technical Note No. IDB-TN-01518, 2018*, Washington, DC: Inter-American Development Bank. http://doi.org/10.18235/0001346

Yeboah, V. and Asibey, M.O. (2019) 'Transport and historical changes in Kumasi's growth and form', *Case Studies on Transport Policy*, 7(4): 802–813. https://doi.org/10.1016/j.cstp.2019.08.001

Yobo, E. (2013) *The politics of public transportation in Ghana: The case of metro mass transit limited*, Legon: University of Ghana.

Yusif, F. (2018) *Farida writes: A day in the life of a TroTro Mate.* Available at: https://citinewsroom.com/2018/11/farida-writes-a-day-in-the-life-of-a-trotro-mate/ (Accessed 29 May 21).

Zimmerman, K., Mzige, A.A., Kibatala, P.L., Museru, L.M. and Guerrero, A. (2012) 'Road traffic injury incidence and crash characteristics in Dar es Salaam: A population based study', *Accident Analysis and Prevention*, 45(2): 204–210. https://doi.org/10.1016/j.aap.2011.06.018

PART V
Contextualising livelihoods

39
LIVELIHOODS AND SOCIAL PROTECTION

Leo de Haan

Introduction

Social protection is usually defined as public policies and interventions that are meant to prevent people's destitution and to promote their welfare. As a development strategy, social protection only entered the development debate in the late 1990s. The origin of modern social protection rather lies in the rise of the European welfare states at the end of the 19th century. Industrialisation, urbanisation and population growth had substantially weakened traditional forms of social protection offered by family networks, churches and guilds, with massive poverty as a result. Then, various societal forces, ranging from labour unions, charities and political parties to churches and civil servants, combined efforts to enforce all sorts of corrective measures, such as health and accident insurances, old-age pensions and unemployment benefits, to prevent deprivation and to promote welfare.

This chapter starts with an overview of social protection as development strategy. The section discusses the rise of social protection in development, explains the main social protection instruments and highlights some controversies. The following section discusses the ways social protection is expected to enhance the poor's livelihoods. Four successive elements are considered – protection, prevention, promotion and transformation – and special attention is given to the graduation approach. Then follows three case sections, each focussing on a notable social protection scheme, the Mahatma Gandhi National Rural Employment Guarantee Act in India, the Productive Safety Net Program in Ethiopia and Bolsa Familia in Brazil. Their impact on livelihoods is examined bearing the four elements of livelihoods enhancement in mind. The final section of this chapter briefly reflects upon these schemes' results.

Social protection as development strategy

For a long time, social protection took up a marginal position in the development debate. It was associated with the welfare state of industrialised countries and considered too costly to extend to entire populations of developing countries. It never grew beyond insurance and pension schemes for a limited number of formal sector employees. Moreover, even in the welfare states of the developed world social protection went out of favour, due to rising costs, economic crises and the concurrent rise of a neo-liberal ideology. In the 1980s, the World Bank strongly opposed broad social protection schemes in developing countries and only agreed to minimal safety nets for the most vulnerable. But from the late 1990s, a paradigm shift became apparent as Merrien (2013) explained. The devasting social effects of structural adjustment programmes in Africa, the Asian economic crisis, persistent global poverty and mounting global social inequalities prompted even the World Bank to embrace social protection as part of poverty reduction strategies. Social protection received a prominent role in the Millennium Development Goals and the International Labour Organization embarked upon a worldwide campaign to extend social protection to developing countries. Other donors followed suit. Nowadays social protection is no longer perceived as a dedicated policy to mitigate instant economic shocks for the most vulnerable but as an overall development policy. However, consensus is not unanimous and social protection remains politically contested.

In the 1980s, safety net programmes were the first social protection schemes to emerge. Fully in line with the neo-liberal *Zeitgeist* at the time, they started off as the most minimalist approach to social protection, i.e. literally the final resort for those who could not keep pace with the restructuring of the economy by structural adjustment programmes. Exemplary were the food-for-work programmes, meant to ensure a minimal level of food security for mostly rural populations that otherwise would suffer from seasonal hunger after the abolishment of food and input subsidies. These programmes offered employment paid in kind, for example in the construction of roads or anti-erosion structures. A self-targeting mechanism reduced participation and costs to a minimum. The idea was to set the food ration so low that those otherwise earning enough wage or producing enough food to feed themselves, would not be interested in the hard work involved (Ellis et al., 2009). However, practice proved obstinate. Clay (1986) concluded that in rural Maharashtra, India, the wage was set so low that although participants worked on average 160 days in the programme, 90% of them were still living under the local poverty line. Food-for-work schemes in Bangladesh showed similarly poor results. On the other hand, while it was often argued that the weakest segments of the population, i.e. the elderly, the disabled and women caring for small children, were not able to profit from the food-for-work programmes, both schemes did draw in those usually disadvantaged in seeking employment. But Clay (1986) also showed that wherever food-for-work programmes put more emphasis on the building of infrastructure than on the distribution of food, i.e. the safety net function, able-bodied men were preferred over other participants.

Eventually it turned out that structural adjustment did not deliver economic growth and thus the idea that only minimal safety nets were needed to cater for the few that were left behind, became outmoded. Nowadays, social protection has become a strategy of social transfers and investments for many (Merrien, 2013). Social protection is usually divided into three main categories. Firstly, social insurance mechanisms, which safeguard participants against exigencies as a result of sickness, maternity, invalidity, old-age and unemployment. Contributions from workers and their employers typically provide the financial basis for these insurance mechanisms. Secondly, social assistance mechanisms, which provide support in various forms to the poor, for example cash transfers, vouchers and school meals. These mechanisms are typically tax-based. Thirdly, labour market regulations which protect workers by enforcing minimum wages, job security and work standards and ensuring trade union rights (Barrientos and Hulme, 2008) – see also Box 39.1. Occasionally, broader definitions of social protection are used. These include all services that enhance human capital, like education, community development, health and sanitation. Sometimes, social protection is broadened even further to include other public policies, even macroeconomic policies. For example, in the 1960s and 1970s, subsidising retail food prices was a popular policy in developing countries, keeping down food prices for consumers of staples liked rice, maize and wheat. For many reasons, like bad management and corruption, increased burden on the state budget and the favouring of urban consumers over rural producers, food subsidies fell into disfavour. However, they did represent a blanket social protection mechanism. Subsidies on agricultural inputs like fertiliser had the same intention. Note that some countries still provide food vouchers to the poor, enabling them to buy food at low prices in government shops (Ellis et al., 2009). Among these broader definitions, transformative social policy is worth mentioning. Transformative social policy aims to go beyond the idea of social protection reducing vulnerability and poverty. It argues in favour of a broader economic, social and political transformation – development as some would call it – grounded in norms of equality and social solidarity. To achieve this transformation, a broad range of policy instruments is needed ranging from education, health and human settlement and housing, to agrarian reform, labour market, equity-affirmative action, family and child care, old-age care, social insurance and fiscal instruments (Adesina, 2011).

Box 39.1 Controversies of social protection in development

Despite points of agreement, social protection in development also remains a subject of fierce debate. Most fundamental is the controversy between the moral argumentation of equity and social justice and the efficiency argument of unleashing human and productive potential (de Haan, 2017a). In plain words, the former departs from the idea that all members of society perform their tasks as well as they can and are given what they need, while the latter argues that poverty is a waste of human productive potential, weakens social unity, risks social unrest and therefore reduces economic growth. Next is the question of universalism versus particularism, i.e. whether social protection should cover the entire population or just particular groups, for example those in need or those who paid contributions. Proponents of universalising social protection argue that social protection is a claimable entitlement, and that reserving it to particular groups risks fragmentation and inadequate accessibility. That would in the end threaten human security and solidarity and push inequality and poverty (UNRISD, 2010). It also follows from this viewpoint that social protection should be a public affair. That does not mean that private sector initiatives cannot play an important role, on the contrary: it does in many countries and some private sector schemes offer even greater social protection than public schemes. The point of universalists is that private sector initiatives should be organised within an overall social provisioning system, in which the government plays the central role (de Haan, 2017b). However, in practice opposing views easily walk together. Few protagonists of universalism would reject safety net programmes targeting particular groups such as the poorest or women. They may even enthusiastically support them, considering it a first stage towards universalising social protection. Other issues of debate are matters of conditionality versus non-conditionality and in-kind contributions versus cash payments. Apart from practical benefits or disadvantages, the controversy reveals in fact underlying patronising or respecting attitudes towards beneficiaries.

Social protection and livelihoods

Livelihood studies grew out of the frameworks and toolboxes of the 1990s into a critical, holistic perspective on how the poor organise their livelihoods (using as active agents the assets they control, though embedded in wider socio-economic, cultural and political contexts) and how they derive meanings from it, in order to understand and counter their poverty. Recognising social exclusion is essential in understanding livelihoods as well as in designing interventions that go beyond simple symptoms such as lack of income and hunger. Recognising social exclusion means looking in an integrated way at multiple, overlapping disadvantages the poor are confronted with and grasping how institutional rules and relationships regulate their access to livelihoods resources and opportunities (de Haan, 2017c).

The way social protection enhances livelihoods and surmounts social exclusion consists of four elements: protection, prevention, promotion and transformation (Barrientos and Hulme, 2008; Ellis et al., 2009; Sabates-Wheeler and Devereux, 2008) – see also Box 39.2.

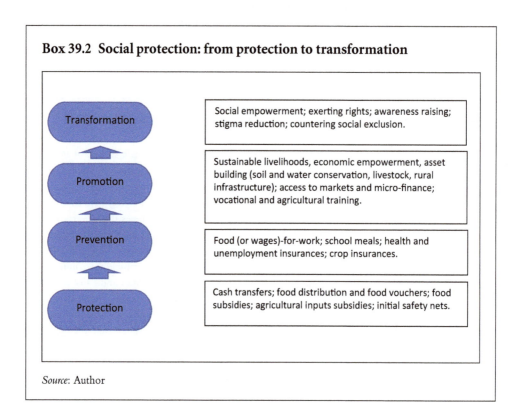

Box 39.2 Social protection: from protection to transformation

Source: Author

Protection is safeguarding minimum levels of consumption of those in poverty or at risk of falling into poverty. Prevention is people, whose livelihoods are suspectable to shocks and stresses, being prevented from having to sell their assets and become even more vulnerable. Promotion is fostering people to become less vulnerable in the future. It is sometimes called graduation out of poverty into sustainable livelihoods (see Box 39.3). If protection, prevention and promotion are considered to be economic components of social protection, then transformation is the social component and the apex, the moral aspiration of social protection. Transformative social protection tackles the root cause of social exclusion. It means strengthening the agency of the poor, building their empowerment and tackling societal structures of domination and marginalisation. Note that in development practice these successive components of livelihoods support may well overlap.

In the following sections, social protection schemes in India, Ethiopia and Brazil and their impact on livelihoods are examined bearing these four elements in mind.

Box 39.3 Graduation out of poverty

The graduation approach is a phased intervention that supports the poorest in an integrated way to achieve sustainable livelihoods and thus move out of poverty, often within a specified period. It integrates for example transfers, training, heath care, savings and credit and community participation. As Sabates-Wheeler et al. (2018) explain, the approach considers asset accumulation as the key to wellbeing and livelihoods security. The extreme poor are unable to use the available resources to pull themselves above the poverty threshold because of lack of (access to) resources, education and health, social exclusion and natural or economic shocks and stresses. Graduation programmes aim to move the poor to a structurally non-poor position, i.e. a sustainable livelihood. Thus, simply exiting a graduation programme at the end of the specified period does not necessarily mean that the household's livelihood has improved, let alone a sustainable livelihood has been achieved. Graduation programmes provide short-term asset, cash or food transfers together with support for asset creation, savings, training and mentoring as a vehicle to economic empowerment. However, it is too simple to expect that a one-size-fits-all approach will do. Household livelihood trajectories are manifold and depend on household characteristics (dependency rate, labour constraints and female headedness), initial resources available (some initial land or small business is usually a better feeding ground to productively absorb cash transfers) and external context of shocks and stresses and exclusion. All these factors determine in a complex combined action whether households will be able to reap quickly the gains from the graduation programme or not. A linear progression out of poverty is therefore not plausible.

Social protection in India: the Mahatma Gandhi National Rural Employment Guarantee Act (MGNREGA)

Starting with the Workmen's Compensation Act as early as 1923, which required employers to compensate workers for accidents at work, social protection in India has a long history, though it only really took off with the country's independence in 1947. Still, Indian social protection has always been a patchwork of schemes, initiated by both central and state governments but poorly coordinated, with limited coverage of the population and unevenly spread benefits among occupations. Most of the schemes only target the extreme poor or are limited (as of old) to workers in the formal sector, while the bulk of the Indian workforce works in the informal sector (Asher, 2017).

Since 2005, India's social protection has gradually moved into the direction of a rights-based approach, instead of looking upon citizens as mere beneficiaries of state provided welfare (Mehrotra et al., 2014) – see also Box 39.4. After 2014, social protection became considered an integral part of India's development strategy rather than an appendage. Resembling the idea of transformative social policy as explained earlier, social protection in India is no longer viewed as only providing monetary payments, but also as delivering public amenities and services to households, in the sense that a household's welfare ultimately depends on consumption of public and private sector goods and services, both monetary and non-monetary. This means that initiatives that reduce household's expenditure, time, energy and effort are contributing to the household's welfare and thus are integral elements of the social protection system. Examples of the latter are schemes for village electrification and for supplying liquid gas to households. By substituting fuelwood and other cooking fuels and reducing indoor pollution, the latter reduces women's time spent on wood collection and cooking and improves the indoor climate. Another scheme insures against crop failure and thus stabilises farmers' incomes (Asher, 2017).

> **Box 39.4 Three rationales for social protection**
>
> Social policy is like a pendulum that swings between three points, each representing a distinct rationale for action – a 'welfarist' notion of benevolence, a 'productivist' public goods instrumentality, or a 'rights/solidarity' based approach. Each rationale embeds a different understanding of the causes of poverty and deprivation, and consequent implications for who should be responsible to address the problem – the individual, the household, the market or the state (Sen and Rajasekhar, 2012: 91).

The Mahatma Gandhi National Rural Employment Guarantee Act (MGNREGA) was established in 2005 following a long history of public work schemes, mainly to relieve famine and drought. The scheme, mainly funded by the central government, grants every household the right of 100 days of paid work per year against a minimum wage. Work is usually provided in local infrastructure projects such as roads, water conservation (like digging ponds) and irrigation channels. If work cannot be provided within two weeks, an unemployment benefit should be paid. MGNREGA is often said to be the world's largest anti-poverty and public works scheme. In 2017–2018 alone it reached 51 million households and generated 2342 million person-days of employment (Government of India, 2020). Studies evaluating MGNREGA show mixed results. On the one hand, it has paid wages to many rural poor who otherwise would have been without income for example during the agricultural slack season. It has also raised rural wages and supported marginal groups, in particular women and scheduled casts and tribes. By doing so, the scheme improved household consumption levels and also reduced distress migration (Parida, 2016). So MGNREGA shows both protective and preventive elements as discussed earlier. But asset building, i.e. livelihoods promotion, has been less successful and mainly limited to water harvesting. For example, Saha (2018) found that roads constructed were of poor quality and even that ponds often dried up in the summer because of their limited depth. Tambe et al. (2019) argue that instead of focussing on rural infrastructure the scheme should pay more attention to building household livelihood assets and improve the productivity of small farms. Focusing on ground water recharge, small irrigation channels, animal sheds, water tanks, horticulture plantations and fodder development would have more effect. In addition, job creation was lowest in states that needed it most due to political unwillingness, indicating that the right to work is not self-evident in India. Corruption gave MGNREGA a bad name (Fraser, 2015). Elite capture, fraud and nepotism siphoned off public funds (Ramya, 2018). MGNREGA's self-targeting mechanisms are imperfect, nevertheless the scheme is found to be much more pro-poor than other big spenders such as food, fertiliser and fuel subsidies which favour the middle-class rather than the poor (Fraser, 2015).

Because MGNREGA includes a number of pro-women features, its results on gender empowerment, i.e. its contribution to livelihoods transformation, are repeatedly debated. MGNREGA stipulates for example a women quota of minimally 33% and work for pregnant and lactating women that requires less effort and is close to home. In addition, provisions for childcare at the work site should be made available if there are five or more children below the age of five present. Chopra (2019) explains that women participation has always been well over the 33% – even reaching 50% after 2012 – but argues that a high participation rate and improved income of women, does not necessarily result in more women's empowerment. She shows that gendered decision-making at the local level and informal gender norms cause women to be allocated jobs with lower wages, with pregnant and lactating women earning even lower wages. In addition, they have difficulties in accessing their earnings because these are paid into individual bank accounts. But opening a bank account, travelling to the post office and withdrawing money are all governed by informal gender norms, making women rely on male relatives or mediators. Again, a new round of contestation may arise about by whom and on what the money is spent.

Social protection in Ethiopia: the Productive Safety Net Program (PSNP)

The Productive Safety Net Program of Ethiopia was launched in 2005 following a number of ad-hoc emergency food aid and food-for-work programmes. Annual numbers of beneficiaries vary between 5 million and 8 million, i.e. 7% and 11% of the population, depending on the incidence of drought (Tadesse, 2018). Though PSNP is heavily donor-financed, it is well embedded in Ethiopia's development ideology, in which the (developmental) state leads the development process. PSNP is very much part of a public goods instrumentality or productivist view on social policy as explained in Box 39.4 Its rationale is not only to protect important political constituencies in Ethiopia, but also to make productive contributions to development by mobilising resources (Lavers, 2019). PSNP aims first to protect food insecure households in food insecure areas by providing resources to secure food consumption in lean periods. Second, the programme prevents asset sales and debts by borrowing. Third, the programme aims to promote livelihoods by building productive assets (Shigute et al., 2017), as illustrated in Box 39.5. Participation in PSNP is the result of a top-down targeting process. The government identifies food insecure districts and subsequently chronically food insecure villages on the basis of past performance. Within the selected villages, households are targeted on the basis of such criteria as having received emergency aid before, assets and income. Local communities themselves can update the list of targeted households annually, according to the actual situation (Shigute et al., 2020). Targeted households without able-bodied members due to age, disability or illness, receive direct support in cash or in kind. Cash transfers to these households are paid through the district financial office or electronically through service providers. The other targeted households only receive payments when participating in public works. Payment is set according to the local purchasing power and their labour contribution. Livelihoods promotion consists of enabling access to credit and technical support, to training for livelihoods diversification and to asset building in agricultural production and non-agricultural activities. Promotion goals are specified in so-called household business plans.

Evaluating an earlier phase of PNSP, Berhane et al. (2014) found that the public works part of the programme had modest effects. Households usually only received tiny sums and PSNP only improved food security by 1.29 months after 5 years participation, i.e. one-third of the hungry season. Together with flanking programmes in food security and asset building, food security rose to 1.5 months, demonstrating that public work together with asset building performs better. Though modest, these are still relevant improvements. But it shows at the same time that protection and prevention at this scale and at this low level of livelihoods security is a matter of endurance.

Box 39.5 PNSP's graduation out of poverty

Despite progression out of poverty as a linear process being contested as explained in Box 39.3, PNSP aims for graduation out of poverty through a number of progressive steps as shown above. Protection is the focus in the first and second step and prevention in the third. Promotion starts from the fourth phase onwards when livelihood development packages are provided to households able and willing to engage in income-generating activities in one of the three designated livelihood pathways,

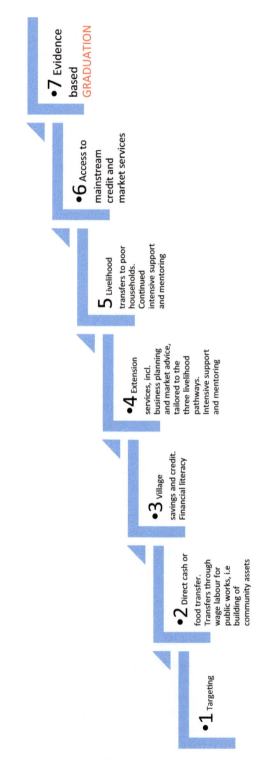

Source: Author, compiled from Sabates-Wheeler and Devereux (2013: 920) and Tadesse (2018: 21)

i.e. on-farm (poultry, dairy, livestock fattening, bee keeping), off-farm (petty trade, handicraft) or employment. Since 2010 only some 315,000 households (some 1.6 million people) entered this step through the credit scheme, but only 14% graduated out of poverty instead of the expected 50%. One reason is that out of the households that developed business plans, 38% did not gain access to credit to implement their plans (FAO, 2020). Moreover, not all poor households are willing to take credit even if it is made accessible to them because of risk-averse behaviour common to the poorest. Therefore, a livelihoods cash transfer pilot to invest in livelihood income activities was established recently, directed towards households too poor to be eligible for credit. They invested most credit in goat and sheep breeding. Low graduation rates can also be due to stronger emphasis on protection rather than on promotion. Moreover, also clientelism between government officials and beneficiaries contributed to the low graduation rate. Transfers have the additional function of mobilising the population for political and social actions. From that perspective graduation means that officials lose their grip on people and so make graduation politically unfeasible for officials (Tadesse, 2018).

Livelihoods promotion within PSNP is also a matter of endurance. Activities take place on the basis of local plans in which all members of local communities are expected to participate (Shigute, 2019). The public works are first of all directed towards achieving food security by building and enhancing communal assets through watershed development. Indeed, 70% of such activities are for water harvesting and soil and water conservation works. Some 250,000 ha have been terraced every year. Afforestation, i.e. plantations on communal land and agroforestry on farmlands, is the second component though it is plagued by low tree survival rates. The rest consists of the construction of feeder roads, schools and health posts, water supplies and small-scale irrigation works and agricultural training. Adimassu and Kessler (2015) pointed out that public works on community assets ignore the necessity of farmers' investments in long-term sustainable land management practices on their own plots. Compared to farmers not participating in public works, farmers participating in public works had less time to invest in sustainable practices on their own lands, such as making soil bunds and stone bunds, and applying inorganic fertiliser, farmyard manure and compost.

Finally, in PNSP too exclusion lurks and not only because the poorest households are not always willing to take credit as explained in Box 39.5. Exclusion is built into PNSP's set-up. Its geographical targeting of food insecure areas raises issues of equity because not targeted (food secure) areas may still have food insecure households. This became apparent in 2015 when large food transfers to non-PNSP areas were necessary due to drought. Worth mentioning is that the social protection effects of public works improve in conjunction with other livelihoods promotion interventions. Shigute et al. (2020) found that individuals participating in PNSP and taking in addition a voluntary health insurance through the Community-Based Health Insurance scheme, yielded substantial additional benefits such as more use of outpatient services, a higher participation in off-farm work, a slight increase in livestock and a decline in debt.

Social protection in Brazil: Bolsa Familia (BF)

Bolsa Familia, said to be the largest conditional cash transfer programme in the world, started in 2003 when the Workers' Party came to power in Brazil. BF emerged from a mix of municipal and federal social protection initiatives aiming at crisis mitigation for poor households, income transfers and educational interventions like Bolsa Escola, a guaranteed income scheme linked to children's schooling.

> **Box 39.6 Bolsa Familia's impact on schooling: a remarkable difference between girls and boys**
>
> Bolsa Familia's impact on schooling is worth discussing more in detail, because impact varies considerably between boys and girls. In their study de Brauw et al. (2015) evaluated BF's effects on schooling outcomes. For both younger and older girls, they found an increase in school participation by 8.2 percentage points, but no impact on boys' participation. They also found a large effect on grade progression for girls living in both rural and urban areas, ranging between 14.6 percentage points and 22.5 percentage points. But they couldn't ascertain any impact on boys' grade progression in either urban or rural areas. For boys, the only statistically significant improvement was a 6.9 percentage point reduction in dropout for young men in urban areas aged 15–17. The authors speculate that girls' grade progression may have to do with a reduction in the amount of time girls spend on domestic work allowing them more time to study. However, especially their overall conclusion is remarkable. They conclude that Bolsa Familia widens a gender gap in schooling, which is already in favour of girls (de Brauw et al., 2015).

Summarising the results of a large number of studies evaluating BF's impact, Barrientos et al. (2016) list an impressive series of achievements: children's weight-for-height and body mass improved; the incidence of immunisations improved; school attendance went up, but mainly for girls (see Box 39.6) and for the North-East; grade repetition reduced; children's entry into the labour market was delayed by a year; pre-natal visits by expectant mothers increased; mothers' influence in decisions over household budgets improved; and contraception-use rose. Significant effects on labour supply were not found, but Santos et al. (2017) analysing data of some 3 million individuals found that beneficiaries of the programme have a greater chance of remaining in their jobs than non-beneficiaries. He estimated the risk of leaving one's job to be 7% to 10% less for beneficiaries, compared to non-beneficiaries. Moreover, BF's inequality reducing power is significant. It explains almost one third of the reduction in household income inequality, mainly because transfers go to households at the bottom of the income pyramid. Interestingly, Chioda et al. (2016) demonstrated that this reduction in inequality was accompanied by reduced crime rates.

While evaluative studies generally confirm BF's successes in prevention and protection, achievements in livelihoods promotion, apart from education, and transformation are more patchily documented. Moreover, the economic crisis in Brazil, which reduced government revenues, as well as the shift towards a new populist administration put BF expenditures under pressure. While the new government first promised payment of an extra 13th month, the admission of new beneficiaries to the program slowed down and the total number of beneficiaries dropped.

Conclusion

With incomes rising worldwide but also poverty persisting and social inequality intensifying, social protection as a development strategy evolved from a marginal role as safety net in the 1990s to counter the most negative effects of economic growth, to a prominent role as promoter of social transfers and investments for many. Departing from the idea that social exclusion is the root cause of poverty, social protection aims to support the livelihoods of the poor through protection, prevention, promotion and transformation. The cases of social protection in India, Ethiopia and Brazil, three countries with large social protection programmes, show that livelihoods protection and prevention are generally accepted components of social protection, which for the most part and leaving room for improvement, attain their goals. However, promotion to sustainable livelihoods has yet to come for large segments of the targeted beneficiaries and without massive successful livelihoods promotion, transformation towards dignified livelihoods as the moral aspiration of social protection seems unlikely. Yet, the cases of India, Ethiopia and Brazil do show some encouraging results in the field of livelihoods transformation. These point at an alternative interpretation. Instead of livelihoods promotion being a necessary condition for livelihoods transformation, livelihoods promotion and transformation go hand-in-and. Improvements in livelihoods promotion trigger progress in transformation and progress in transformation unlocks new opportunities for livelihoods promotion etc. Similar to the (contested) idea of linear progression out of poverty, a strictly linear progression from protection to transformation should be rejected.

Nonetheless, dignified sustainable livelihoods remain a pipe dream for many. That is not surprising for two interrelated reasons. Firstly, despite the fact that universal social protection is affordable even in lower income countries (Ortiz et al., 2017), 4 billion people are still left unprotected (ILO, 2017). This points at a stunning lack of political will of those in power to implement social protection schemes. Secondly, transformation towards dignified sustainable livelihoods entails building the poor's empowerment. Eventually this means tackling societal structures of domination, which in turn explains this lack of political will.

Key points

- Social protection is usually defined as public policies and interventions that are meant to prevent people's destitution and to promote their welfare.
- Social protection as a development strategy evolved from minimal safety nets into a strategy of social transfers and investments to enhance livelihoods.
- Social protection enhances livelihoods through protection, prevention, promotion and transformation.
- Graduation programmes aim to move the poor to a structurally non-poor position, i.e. sustainable livelihoods.
- Promotion to sustainable livelihoods has yet to come for large segments of the targeted beneficiaries; transformation includes social empowerment and social inclusion and is even more difficult to achieve.
- Instead of livelihoods promotion being a necessary condition for livelihoods transformation, promotion and transformation are rather intertwined components of livelihoods enhancement; a strictly linear progression from protection to transformation should be rejected.

Further reading

Ellis, F., Devereux, S. and White, P. (2009) *Social protection in Africa*, Northampton: Edward Elgar. https://doi.org/10.1177%2F146499341001100409

Merrien, F. (2013) 'Social protection as development policy: A new international agenda for action', *International Development Policy*, 5(1): 83–100. https://doi.org/10.4000/poldev.1525

Sabates-Wheeler, R. and Devereux, S. (2013) 'Sustainable graduation from social protection programmes', *Development and Change*, 44(4): 911–938. https://doi.org/10.1111/dech.12047

References

Adesina, J. (2011) 'Beyond the social protection paradigm: Social policy in Africa's development', *Canadian Journal of Development Studies/Revue canadienne d'études du développement*, 32(4): 454–470. https://doi.org/10.1080/02255189.2011.647441

Adimassu, Z. and Kessler, A. (2015) 'Impact of the productive safety net program on farmers' investments in sustainable land management in the Central Rift Valley of Ethiopia', *Environmental Development*, 16: 54–62. https://doi.org/10.1016/j.envdev.2015.06.015

Asher, M. (2017) 'An analysis of post-2014 social protection initiatives in India', *Lee Kuan Yew School of Public Policy Research Paper No. 17–12*, Singapore: National University of Singapore.

Barrientos, A. and Hulme, D. (2008) 'Social protection for the poor and poorest: An introduction', in Barrientos, A. and Hulme, D. (eds.), *Social protection for the poor and poorest. Concepts, policies and politics*, London: Palgrave Macmillan, pp. 3–24.

Barrientos, A., Debowicz, D. and Woolard, I. (2016) 'Heterogeneity in Bolsa Família outcomes', *The Quarterly Review of Economics and Finance*, 62: 33–40. https://doi.org/10.1016/j.qref.2016.07.008

Berhane, G., Gilligan, D., Hoddinott, J., Kumar, N. and Taffesse, A. (2014) 'Can social protection work in Africa? The impact of Ethiopia's productive safety net programme', *Economic Development and Cultural Change*, 63(1): 1–26. https://doi.org/10.1086/677753

Chioda, L., De Mello, J. and Soares, R. (2016) 'Spillovers from conditional cash transfer programs: Bolsa Família and crime in urban Brazil', *Economics of Education Review*, 54: 306–320. https://doi.org/10.1016/j.econedurev.2015.04.005

Chopra, D. (2019) 'Taking care into account: Leveraging India's MGNREGA for Women's empowerment', *Development and Change*, 50(6): 1687–1716. https://doi.org/10.1111/dech.12535

Clay, E. (1986) 'Rural public works and food-for-work; a survey', *World Development*, 14(10/11): 1237–1252. https://doi.org/10.1016/0305-750X(86)90103-8

de Brauw, A., Gilligan, D., Hoddinott, J. and Roy, S. (2015) 'The impact of Bolsa familia on schooling', *World Development*, 70: 303–316. https://doi.org/10.1016/j.worlddev.2015.02.001

de Haan, L. (2017a) 'Rural and urban livelihoods, social exclusion and social protection in sub-Saharan Africa', *Danish Journal of Geography/Geografisk Tidsskrift*, 117(2): 130–141. https://doi.org/10.1080/00167223.2017.1343674

de Haan, L. (2017b) 'Livelihoods in development', *Canadian Journal of Development Studies/Revue canadienne d'études du développement*, 38(1): 22–38. https://doi.org/10.1080/02255189.2016.1171748

de Haan, L. (2017c) 'From poverty to social exclusion', in de Haan, L. (ed.), *Livelihoods and development: New perspectives*, Leiden/Boston: Brill, pp. 1–12.

Ellis, F., Devereux, S. and White, P. (2009) *Social protection in Africa*, Northampton: Edward Elgar. https://doi.org/10.1177%2F1464993341001100409

Food and Agricultural Organization (FAO) (2020) *Strengthening Ethiopia's social protection programme to enhance food and nutrition security*. News Release issued 13 February 2020 by the Food and Agriculture Organization of the United Nations (FAO) Representation in Ethiopia. www.fao.org/ethiopia/news/detail-events/en/c/1261423/ (Accessed 29 June 2020).

Fraser, N. (2015) 'Social Security through Guaranteed Employment', *Social Policy and Administration*, 49(6): 679–694. https://doi.org/10.1111/spol.12164

Government of India (2020) *The Mahatma Gandhi National Rural Employment Guarantee Act. R5.1.1. employment created during the year 2017–2018*, New Delhi: Ministry of Rural Development.

International Labour Organization (ILO) (2017) *World social protection report 2017–19. Universal social protection to achieve the sustainable development goals*, Geneva: International Labour Organization. www.social-protection.org/gimi/gess/ShowWiki.action?id=594

Lavers, T. (2019) 'Social protection in an aspiring "developmental state": The political drivers of Ethiopia's PSNP', *African Affairs*, 184(473): 646–671. https://doi.org/10.1093/afraf/adz010

Mehrotra, S., Kumra, N. and Gandhi, A. (2014) 'India's fragmented social protection system three rights are in place; two are still missing', *Working Paper 2014–18*, Geneva: United Nations Research Institute for Social Development.

Merrien, F. (2013) 'Social protection as development policy: A new international agenda for action', *International Development Policy*, 5(1): 83–100. https://doi.org/10.4000/poldev.1525

Ortiz, I., Durán-Valverde, F., Pal, K., Behrendt, C. and Acuña-Ulate, A. (2017) 'Universal social protection floors: Costing estimates and affordability in 57 lower income countries', *ESS Working Paper No. 58*, Geneva: Social Protection Department, International Labour Organization. www.ilo.org/secsoc/information-resources/publications-and-tools/Workingpapers/WCMS_614407/lang – en/index.htm

Parida, J. (2016) 'MGNREGS, distress migration and livelihood conditions: A study in Odisha', *Journal of Social and Economic Development*, 18(1): 17–39. https://doi.org/10.1007/s40847-016-0021-z

Ramya, T. (2018) 'MGNREGA Vis-à-Vis tribal livelihoods: A study in Kurung Kumey district of Arunachal Pradesh', *Space and Culture, India*, 6(3): 156–169. https://doi.org/10.20896/saci.v6i3.390

Sabates-Wheeler, R. and Devereux, S. (2008) 'Transformative social protection: The currency of social justice', in Barrientos, A. and Hulme, D. (eds.), *Social protection for the poor and poorest. Concepts, policies and politics*, London: Palgrave Macmillan, pp. 64–84.

Sabates-Wheeler, R. and Devereux, S. (2013) 'Sustainable graduation from social protection programmes', *Development and Change*, 44(4): 911–938. https://doi.org/10.1111/dech.12047

Sabates-Wheeler, R., Sabates, R. and Devereux, S. (2018) 'Enabling graduation for whom? Identifying and explaining heterogeneity in livelihood trajectories post-cash transfer exposure', *Journal of International Development*, 30: 1071–1095. https://doi.org/10.1002/jid.3369

Saha, S. (2018) 'A geographical study of the ground realities of rural safety net in India: A village level study of MGNREGS', *Economic Affairs*, 63(3): 779–783. https://doi.org/10.30954/0424-2513.3.2018.26

Santos, D., Leichsenring, A., Filho, N. and Mendes-Da-Silva, W. (2017) 'The impact of the Bolsa Família program on the duration of formal employment for low income individuals', *Brazilian Journal of Public Administration*, 51(5): 708–733. http://doi.org/10.1590/0034-7612171851

Sen, G. and Rajasekhar, D. (2012) 'Social protection policies, experiences and challenges', in Nagaraj, R. (ed.), *Growth, inequality and social development in India. Is inclusive growth possible?* London: Palgrave Macmillan, pp. 91–135.

Shigute, Z. (2019) *Essays on evaluation of social protection programmes in Ethiopia*, The Hague: International Institute of Social Studies of Erasmus University Rotterdam. https://repub.eur.nl/pub/116525

Shigute, Z., Mebratie, A., Sparrow, R., Yilma, Z., Alemu, G. and Bedi, A. (2017) 'Uptake of health insurance and the productive safety net program in rural Ethiopia', *Social Science and Medicine*, 176: 133–141. https://doi.org/10.1016/j.socscimed.2017.01.035

Shigute, Z., Strupat, C., Burchi, F., Alemu, G. and Bedi, A. (2020) 'Linking social protection schemes: The joint effects of a public works and a health insurance programme in Ethiopia', *Journal of Development Studies*, 56(2): 431–448. https://doi.org/10.1080/00220388.2018.1563682

Tadesse, G. (2018) 'Agriculture and social protection: The experience of Ethiopia's productive safety net program', in Wouterse, F. and Tafesse, A. (eds.), *Boosting growth to end hunger by 2025: The role of social protection*, Washington: International Food Policy Research Institute, pp. 16–33.

Tambe, S., Tashi Bhutia, N., Pradhan, S. and Basi, J. (2019) 'Coupling a ladder to the safety net: Reinventing MGNREGA for asset creation', *Development in Practice*, 29(4): 514–524. https://doi.org/10.1080/09614524.2019.1567687

United Nations Research Institute for Social Development (UNRISD) (2010) *Combating poverty and inequality: Structural change, social policy and politics*, Geneva: United Nations Research Institute for Social Development. www.unrisd.org/unrisd/website/document.nsf/(httpPublications)/BBA20D83E347DBAFC125778200440AA7?OpenDocument

40
COLLECTIVE ORGANISATIONS
An introduction to their contributions to livelihoods in the Global South

Molly Atkins

Introduction: collective action through organisations

Collective action broadly refers to voluntary action taken by a group in pursuit of shared interests or common objectives (Meinzen-Dick and Di Gregorio, 2004). The concept encompasses a diversity of social arrangements; from amorphous, sporadic mobilisations that aim to realise a particular objective, to people coming together within highly structured, well-established organisations to achieve a shared goal (Kurien, 2014). Collective action is sometimes a pro-action to new opportunities or, in other instances, a re-action to a shared problem (Jentoft et al., 2018).

This chapter focuses on instances where groups act collectively through an organisation. Kurien (2014: 43) refers to organisations as 'social-cultural structures (tools) that are utilised by a group (members), over time, to achieve collective action objectives'. Collective organisations are formed in a range of circumstances and adapted to address particular grievances and aspirations and hence, their functions and processes can differ enormously, and alter over time (Jentoft et al., 2018). In some cases, these organisations act with encouragement or support from governmental bodies, development organisations or non-governmental organisations (NGOs) (Meinzen-Dick and Di Gregorio, 2004). This chapter introduces three broad types of collective organisation common in the Global South and frequently utilised by external agencies: 1) producer organisations, often favoured for the dissemination of agricultural extension information, 2) savings and credit groups, commonly employed to promote women's economic empowerment, and 3) resource user committees (RUCs), a prominent structure for people-centred community-based natural resource management.

Policy makers, donors and development practitioners routinely work through collective organisations, as participation and community engagement has become increasingly important in development policy and practice (Andersson and Gabrielsson, 2012). Where groups perceived to be appropriate do not already exist, external agencies often drive the process of group formation. Subsequently, collective organisations, in their many forms, are widespread in the Global South.

Theoretical frameworks and allied concepts

Theories of collective action are closely linked with a number of other concepts. These concepts are helpfully brought together in the conceptual framework developed by Ibrahim (2006) and adapted in Figure 40.1. Ibrahim's (2006) framework accounts for the economic, human and

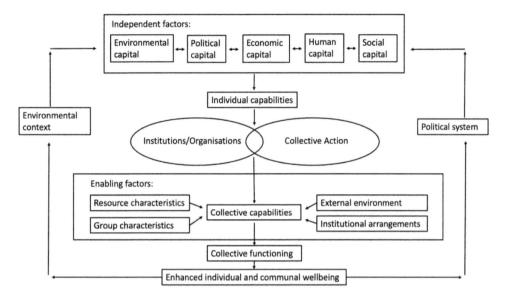

Figure 40.1 Conceptual framework of collective action adapted from Ibrahim's (2006) collective capabilities framework

social factors that affect collective action, presented in Figure 40.1 as capitals (environmental, political, economic, human and social capital). The original framework also incorporates elements from literature on institutions, which Ibrahim (2006) refer to as 'intervening factors' and have been reframed as 'enabling factors' in Figure 40.1 to better align with the collective action literature. The framework illustrates the synergies between concepts prominent in, and most relevant to, collective action literature – collective capabilities and social capital, introduced below. The framework also incorporates four identified enabling factors for collective action which are drawn from institutions literature and discussed later in this chapter.

Collective capabilities

The idea of collective capabilities stems from Amartya Sen's Capability Approach (see Chapter 3 on Capabilities) and are defined by Ibrahim (2006: 404) as 'the newly generated capabilities attained by virtue of their engagement in a collective action or their membership in a social network that helps them achieve the lives they value'. Through this lens, collective action is perceived as both instrumentally valuable (in securing the freedoms that Sen (1999) enumerates: political freedom, economic facilities, social opportunities, transparency and protective security) and intrinsically important (in affecting the formulation of values and beliefs, thus helping to determine which capabilities individuals value) for the expansion of human capabilities (Evans, 2002; Ibrahim, 2006; Stewart, 2005).

Social capital

Interpretations of social capital in collective action studies differ in two distinct ways: 1) from those focused on social capital as socially accessed goods. These studies apply the concept of social capital to analyse the role of networks in providing access to other resources (e.g. natural, physical, human, financial and political capital), and 2) to others focused on the capacity of social capital to facilitate collective action through analyses of how trust or norms affect cooperation. These interpretations are not, however, mutually exclusive (Ramos-Pinto, 2006).

Trust, norms, networks and institutions (described as various forms of social capital) are said to contribute to collective action in 'good' or 'bad' ways (Ostrom and Ahn, 2009; Ramos-Pinto, 2006). Social capital has been observed as both a catalyst for collective action, and an active force that can sustain or limit cooperation among groups (Meinzen-Dick and Di Gregorio, 2004). D'Silva and Pai's (2003) study suggests that collective action is successful among homogenous communities with strong 'bonding social capital' founded on shared cultural values and common identity. However, communities with strong bonding capital, are generally thought to lack 'bridging social capital' gained through social networks with outsiders at micro and macro levels and considered an important source of financial and institutional support for collective efforts (Dahal and Adhikari, 2008; Putnam, 2000). Thus, collective action is regarded most successful 'when communities manage not only to embolden their bonding relations, but expand their linkages so that they can draw benefits from these expanded networks' (Dahal and Adhikari, 2008: 15).

Collective organisations: an introduction to producer organisations, savings and credit groups and natural resource-user committees

Producer organisations

Producer organisations are simply defined as producer-owned and producer-controlled organisations engaged in collective production, marketing and/or processing activities. They can include producers who produce for the market, as well as subsistence producers. Producers organise themselves in various ways, and include cooperatives, farmer associations and collectives (Penrose-Buckley, 2007).

In recent decades, producer organisations have gained popularity in the context of agri-food system transformation (e.g. referring to changes linked to rapid growth in per capita incomes, urbanisation, globalisation, agricultural intensification, climate change, resource use, diets, and food quality standards) owing to their capacity to support small-scale producers to cope with, and adapt to these changes. In rural areas, where many people's livelihoods depend on agricultural production, processing and marketing, producer organisations have been promoted through outside assistance (e.g. governments, NGOs and food companies) (Vicari, 2015).

Services provided through producer organisations vary, but commonly include production services (e.g. input supply, shared labour, access to agricultural equipment, information and technology) and financial services (e.g. access to cash loans and input credit). This coming together to create economies of scale is said to enable small-scale producers to access new and more profitable markets (Kaganzi et al., 2009; Lyon, 2003). Producer groups are reported to support smallholders' participation in organic and Fairtrade schemes which promise farmers better and more stable prices, by reducing the associated high transaction costs and knowledge constraints to certification, traceability and food-safety requirements (Barrett et al., 2001; Nelson et al., 2016).

Additionally, collective marketing can promote important bridges and linkages across the market chain and connect producers, processors and buyers in networks (Markelova et al., 2009). Collaboration between producer organisations, government ministries, NGOs, donors and private companies has improved producers' marketing performance in Tanzania (Barham and Chitemi, 2009), Ethiopia, Kenya and Zambia (Okello et al., 2007). In Malawi, smallholder tea associations have established contractual agreements with commercial tea processors who provide credit for fertilisers and seedlings, market access through direct contracts, tea leaf collection and transportation services, and extension services in farm management to smallholder associations (Chirwa and Kydd, 2009).

Moreover, producer organisations have also been found to facilitate innovation and technological transformations (Devaux et al., 2009), climate adaptations (Andersson and Gabrielsson, 2012; Sudgen et al., 2021) and strategies for improved resource and biodiversity management (Kruijssen et al., 2009) through shared financial risk and knowledge and social learning.

Savings and credit groups

Savings and credit groups, including Rotating Savings and Credit Associations (RoSCAs) and self-help groups (SHGs), are broadly defined as village-based groups designed to provide self-regulating financial aid to members. Members of the group pool their savings through regular 'pay-ins', which can then be used as a source for loans to members (Ban et al., 2015). Peer-to-peer informal lending structures are utilised expansively in South Asia and sub-Saharan Africa to deliver savings and credit facilities to the poor (Alemu et al., 2018). Since the 1980s, development agencies have increasingly delivered development projects through community-based organisations such as SHGs (Gugerty et al., 2019).

Savings and credit associations are used by a wide range of individuals to obtain funds for basic household needs, food, payment of school fees, agricultural inputs, working capital, entrepreneurial investment, and unplanned or emergency expenses (Benda, 2012). These services are particularly important for poor women in rural and urban areas, since they face significant legal, social, cultural and economic restrictions that limit their access to formal financial services (Fletschner, 2009). NGOs, governments and donors similarly encourage youth participation to promote financial inclusion, good savings behaviour, entrepreneurship and asset accumulation, as well as expand and solidify young people's social networks (Flynn and Sumberg, 2018).

According to Ibrahim (2006) participation in SHGs is an investment in an individual's financial, human and social capital. Women have reported improved financial skills alongside increased engagement in, and confidence regarding household money management (Brody et al., 2017). As well as providing vital financial resources, savings and credit groups are said to reproduce and promote democracy, reciprocity and solidarity. In northern Rwanda, participation in RoSCAs was reported to enhance self-esteem, status and confidence, promote sociability, mutual trustworthiness and a sense of belonging among members – all of which contributed to the enhancement of people's wellbeing (Benda, 2012). SHG membership is considered a 'steppingstone' to wider social and political participation (Brody et al., 2017). However, evidence on this is generally mixed and scholars (e.g. Kabeer, 2005) warn against conflating economic benefits with broader improvements to wellbeing.

Resource-user committees

RUCs are groups responsible for the management of natural resources (e.g. water, forests, fisheries) at the user level. These committees have evolved from participatory, decentralised, community-based natural resource management (CBNRM) approaches. Since the 1950s, CBNRM approaches that devolve power and proprietorship to resource users through RUCs diffused rapidly through environmental policy-making (Smith, 2008). In a number of African countries, including Malawi, Kenya and Ethiopia, Water User Associations (WUAs) are legislated power over water provision and maintenance from boreholes and streams for domestic and agricultural (i.e. irrigation) purposes (Aarnoudse et al., 2018; Hegga et al., 2020). In theory, community water systems bestow a sense of ownership and encourage participatory management processes (Terry et al., 2015).

In Nepal, Forest User Groups (FUGs) are also enabled by policy incentives. FUGs often receive training in resource management and work alongside government authorities to manage forest resources, and curb illegal activities. In Nepal, Dev et al. (2003) recorded generally positive impacts of community forestry, including improved tenure rights and benefit flows of forest products, increased household income-generating opportunities and improved social support structures. Beyond forest management costs, FUGs contributed to community development activities (including support to schools, religious institutions, electrification and water supply) which led to improved community infrastructure (Dev et al., 2003). However, the impacts of FUGs in Nepal are diverse within and between groups, as will be discussed below.

In the fisheries sector, CBNRM has generated positive socio-ecological outcomes. Decentralising governance of local fisheries in the Coral Triangle has allowed management policies such as seasonal or spatial fishery closures, to react, respond and adapt to the specific social and economic needs and priorities of communities (Cohen and Steenbergen, 2015). However, managing the multiple objectives (e.g. restoring biodiversity and ecosystems, enhancing livelihoods and asserting access rights) of community-based fisheries management and potential trade-offs is challenging in practice (Cohen et al., 2014).

CBNRM is expected to engender more equitable resource sharing and, as such, there has been much academic and policy interest on the inclusion of women in community forestry programmes (e.g. Nightingale, 2002). Gender quotas formally encourage equitable gender representation in community-based water management committees in Kenya (Hannah et al., 2021) and community-based Beach Management Unites on Lake Victoria (Nunan and Cepić, 2020). However, power relations and social norms continue to constrain the participation by women and other marginalised groups (Chhetri et al., 2013).

Enabling factors for effective collective action

The collective action literature identifies four broad categories of factors important for the effective formation and functioning of groups; resource system characteristics, group characteristics, institutional arrangements and the external environment (Figure 40.1) (Agrawal, 2001; Ostrom, 1990).

Resource system characteristics

The physical characteristics of resources including the mobility, volatility and unpredictability in the flow of benefits from a resource, resource size and resource boundaries are recognised as critical variables for effective collective action (Agrawal, 2001). In producer organisations, resource characteristics can be considered as the types of products and markets and determine the incentives of collective action (Markelova et al., 2009).

Group characteristics

Group characteristics that affect collective action include group size, the homogeneity of identities and interests, power inequalities and social norms. Group size is said to affect trust and cooperation. Social relations are generally considered as less intimate in larger groups and, as a result, the capacity for mutual monitoring and informal sanctioning is reduced and the problem of 'free riding' more likely (Agrawal and Goyal, 2001). Whereas, members often feel stronger identification and sense of ownership with smaller organisations (Jentoft, 1989). Expansive membership can dilute the purpose and impact of an organisation, and some evidence suggests that when organisations grow in scale, internal conflicts may also rise among members (Kurien, 2014).

There is much debate on whether homogeneity or heterogeneity (in terms of wealth, age, ethnicity, religion, gender etc.) results in better group functioning. Collective action is thought to benefit from a sense of shared identity linked to shared world view, common interests, strong social relations (kinship, friendship) and a sense of belonging among membership (Jentoft et al., 2018). Therefore heterogenous groups, for example fisheries organisations which often include members from distinct sub-groups of boat owners, boat crew, fish processors and traders, can lack cohesion and common interest affecting the sustainability of the organisations (Kurien, 2014).

Power structures and social norms present within communities, and in turn groups, influence participation, decision-making and the distribution of benefits. CBNRM models have received notable criticism for their failure to attend to power differences within communities (e.g. Gibbes and Keys, 2010). According to critics, CBNRM interventions imagine and accredit 'the community' as a fair structure to entrust with power despite the highly complex and fluid nature of many communities. Actors within communities can have divergent – sometimes competing or conflicting – interests regarding natural resource management (Agrawal and Gibson, 1999). CBNRM interventions can unwittingly strengthen existing power structures and potentially stimulate local 'elite capture' and reinforce gender inequalities especially when they are based on traditional, usually male-dominated decision-making processes (Westerman and Benbow, 2013). So, although inclusive in theory, CBNRM approaches prove challenging in practice in delivering on inclusivity.

Power relations also affect participation in, and the distribution of, benefits from group-based savings and credit groups and producer organisations. Mayoux (1995) asserts that some group-based savings and credit schemes exclude the poorest women, since they often target existing women entrepreneurs. Groups can be prone to elite capture 'whereby more powerful members dominate them and use them for particular interests' (Godfrey-Wood and Mamani-Vargas, 2017: 76). Mayoux (1995) found that among cooperative groups in Nicaragua and India, women with higher levels of education and fewer or no family commitments dominated group decision-making and management. Moreover, in mixed-gender producer organisations, male dominated power structures may continue to limit women's participation. In Tanzania, comparatively large-scale farmers were observed to have acquired leadership positions and co-opted the group for their own purposes and, due to their greater power associated with the scale of the men's farming operations, women found it difficult to challenge male dominance (Baden, 2013). Clearly, collective action can be gender-blind and thus, gender inequalities may be reproduced within groups.

Institutional arrangements

Institutional arrangements, including access and management rules, enforcement mechanisms, and accountability structures, are also important for the effective formation and functioning of groups. Effective information sharing and simple and understandable rules are said to increase compliance (Kyeyamwa et al., 2008) and provide transparency. Additionally, and possibly in contrast, organisational flexibility is important for members' agency and ownership in collective efforts (Andersson and Gabrielsson, 2012). Kurien (2014) suggests arrangements should be constantly (re)negotiated in order to adapt and respond to new challenges and group interests. In successful cases, individual contributions to, and responsibilities within, groups vary according to the means of individual members and livelihood conditions. Trust, a sense of moral obligation and mutual monitoring among members is crucial to this level of flexibility (Andersson and Gabrielsson, 2012). In sum, there are likely trade-offs between the benefits of clarity in fixed legislation and procedures and the scope for innovation and flexibility.

External environment

Exogenous variables that mediate the outcomes of collective action include support from external agents, policy incentives and multi-scalar networks. Links with state, development and private actors as well as other collective action institutions at regional and global levels, are generally considered to have positive implications for group visibility and voice, market opportunities, financial and institutional support (Devaux et al., 2009). Networked arrangements have become particularly important in the context of globalisation (Kurien, 2014). Innovative institutional arrangements such as vertical or horizontal coordination have helped small-scale livestock producers enter more profitable markets and contribute to increasing farmer's incomes (Kyeyamwa et al., 2008). Heterogeneous groups are generally believed to have greater linking and bridging social capital (Dahal and Adhikari, 2008) and larger groups are also reported to have greater influence in arenas beyond the community (Jentoft et al., 2018).

Conclusion and key points

- Collective organisations are diverse, multi-faceted and dynamic. They include producer organisations, savings and credit groups and natural resource user committees.
- Collective organisations are widely utilised as instrumental platforms for participatory development interventions and the dissemination of state and private socio-economic services.
- In principle, collective organisations provide access to multiple resources (social, economic, environmental, political and human capital) and facilitate intrinsically and instrumentally important cooperation among groups. However, in practice livelihood outcomes are mixed.
- The formation and functioning of collective action are affected by a number of factors that are categorised into four broad categories: resource system characteristics, group characteristics, institutional arrangements and the external environment.
- Power relations and social norms present among groups influence participation and the distribution of benefits. Yet, it is often assumed that problems associated with group characteristics can be addressed with good institutional design and external support.

Key readings

Andersson, E. and Gabrielsson, S. (2012) '"Because of poverty, we had to come together": Collective action for improved food security in rural Kenya and Uganda', *International Journal of Agricultural Sustainability*, 10(3): 245–262. https://doi.org/10.1080/14735903.2012.666029
This article provides a good analysis of collective action as a potential pathway to livelihood and sustainability improvements.

Kurien, J. (2014) 'Collective action and organisations in small-scale fisheries', in Kalikoski, D. and Franz, N. (eds.), *Strengthening organizations and collective action in fisheries – A way forward in implementing the international guidelines for securing sustainable small-scale fisheries. FAO Workshop, 18–20 March 2013, Rome, Italy. FAO Fisheries and Aquaculture Proceedings No. 32*, Rome: FAO, pp. 41–65.
Though specific to fisheries organisations, this paper provides an informative overview of the forms of collective organisations that exist, including hybrid and new networked arrangements.

Ibrahim, S.S. (2006) 'From individual to collective capabilities: The capability approach as a conceptual framework for self-help', *Journal of Human Development*, 7(3): 397–416. https://doi.org/10.1080/14649880600815982

This text provides valuable conceptual insight and a theoretical framework of collective capabilities. The author applies this framework to self-help groups' initiatives in Egypt.

Markelova, H., Meinzen-Dick, R., Hellin, J. and Dohrn, S. (2009) 'Collective action for smallholder market access', *Food Policy*, 34(1): 1–7. https://doi.org/10.1016/j.foodpol.2008.10.001

This text examines the conditions that facilitate effective producer organisations, applying insights from studies of collective action in natural resource management. It also serves as a useful springboard for further reading.

References

Aarnoudse, E., Closas, A. and Lefore, N. (2018) 'Water user associations: A review of approaches and alternative management options for Sub-Saharan Africa', *IWMI Working Paper 180*, Colombo: International Water Management Institute (IWMI). https://doi.org/10.5337/2018.210

Agrawal, A. (2001) 'Common property institutions and sustainable governance of resources', *World Development*, 29(10): 1649–1672. https://doi.org/10.1016/S0305-750X(01)00063-8.

Agrawal, A. and Gibson, C.C. (1999) 'Enchantment and disenchantment: The role of community in natural resource conservation', *World Development*, 27(4): 629–649. https://doi.org/10.1016/S0305-750X(98)00161-2.

Agrawal, A. and Goyal, S. (2001) 'Group size and collective action: Third-party monitoring in common-pool resources', *Comparative Political Studies*, 34(1): 63–93. https://doi.org/10.1177/0010414001034001003.

Alemu, S.H., Van Kempen, L. and Ruben, R. (2018) 'Women empowerment through self-help groups: The bittersweet fruits of collective apple cultivation in highland Ethiopia', *Journal of Human Development and Capabilities*, 19(3): 308–330. https://doi.org/10.1080/19452829.2018.1454407

Baden, S. (2013) 'Women's collective action in African agricultural markets: The limits of current development practice for rural women's empowerment', *Gender and Development*, 21(2): 295–311. https://doi.org/10.1080/13552074.2013.802882

Ban, R., Gilligan, M.J. and Rieger, M. (2015) 'Self-help groups, savings and social capital: Evidence from a field experiment in Cambodia', *Policy Research Working Paper Series: 7382, 2015*, Washington, DC: The World Bank.

Barham, J. and Chitemi, C. (2009) 'Collective action initiatives to improve marketing performance: Lessons from farmer groups in Tanzania', *Food Policy*, 34: 53–59. https://doi.org/10.1016/j.foodpol.2008.10.002

Barrett, H.R., Browne, A.W., Harris, P.J.C. and Cadoret, K. (2001) 'Smallholder farmers and organic certification: Accessing the EU market from the developing world', *Biological Agriculture and Horticulture*, 19(2): 183–199. https://doi.org/10.1080/01448765.2001.9754920

Benda, C. (2012) 'Community rotating savings and credit associations as an agent of well-being : A case study from northern Rwanda', *Community Development Journal*, 48(2): 232–247. https://doi.org/10.1093/cdj/bss039

Brody, C., de Hoop, T., Vojtkova, M., Warnock, R., Dunbar, M., Murthy, P. and Dworkin, S.L. (2017) 'Can self-help group programs improve women's empowerment? A systematic review', *Journal of Development Effectiveness*, 9(1): 15–40. https://doi.org/10.1080/19439342.2016.1206607

Chhetri, B.B.K., Johnsen, F.H., Konoshima, M. and Yoshimoto, A. (2013) 'Community forestry in the hills of Nepal: Determinants of user participation in forest management', *Forest Policy and Economics*, 30: 6–13. https://doi.org/10.1016/j.forpol.2013.01.010

Chirwa, E. and Kydd, J. (2009) 'From statutory to private contracts: Emerging institutional arrangements in the smallholder tea sector in Malawi', in Kirsten, J., Dorward, A., Poulton, C. and Vink, N. (eds.), *Institutional economics perspectives on African agricultural development*, Washington, DC: International Food Policy Research Institute, pp. 213–226.

Cohen, P.J., Jupiter, S., Weeks, R., Tawake, A. and Govan, H. (2014) 'Is community-based fisheries management realising multiple objectives? Examining evidence from the literature', *SPC Traditional Marine Resource Management and Knowledge Information Bulletin*, 34: 3–12.

Cohen, P.J. and Steenbergen, D.J. (2015) 'Social dimensions of local fisheries co-management in the Coral Triangle', *Environmental Conservation*, 42(3): 278–288. https://doi.org/10.1017/S0376892914000423

D'Silva, E. and Pai, S. (2003) 'Social capital and collective action: Development outcomes in forest protection and watershed development', *Economic and Political Weekly*, 38(14): 1404–1415.

Dahal, G.R. and Adhikari, K.P. (2008) 'Bridging, linking, and bonding social capital in collective action', *CAPRi Working Papers*, (79): 1–23. https://doi.org/10.2499/CAPRiWP79

Dev, O., Yadav, N.P., Springate-Baginski, O. and Soussan, J. (2003) 'Impacts of community forestry on livelihoods in the Middle Hills of Nepal', *Journal of Forest and Livelihood*, 3(1): 64–77.

Devaux, A., Horton, D., Velasco, C., Thiele, G., López, G., Bernet, T., Reinoso, I. and Ordinola, M. (2009) 'Collective action for market chain innovation in the Andes', *Food Policy*, 34(1): 31–38. https://doi.org/10.1016/j.foodpol.2008.10.007

Evans, P. (2002) 'Collective capabilities, culture, and Amartya Sen's development as freedom', *Studies in Comparative International Development*, 37(2): 54–60.

Fletschner, D. (2009) 'Rural women's access to credit: Market imperfections and intrahousehold dynamics', *World Development*, 37(3): 618–631. https://doi.org/10.1016/j.worlddev.2008.08.005

Flynn, J. and Sumberg, J. (2018) 'Are savings groups a livelihoods game changer for young people in Africa?', *Development in Practice*, 28(1): 51–64. https://doi.org/10.1080/09614524.2018.1397102

Gibbes, C. and Keys, E. (2010) 'The illusion of equity: An examination of community based natural resource management and inequality in Africa', *Geography Compass*, 4(9): 1324–1338. https://doi.org/10.1111/j.1749-8198.2010.00379.x

Godfrey-Wood, R. and Mamani-Vargas, G. (2017) 'The coercive side of collective capabilities: Evidence from the Bolivian Altiplano', *Journal of Human Development and Capabilities*, 18(1): 75–88. https://doi.org/10.1080/19452829.2016.1199169

Gugerty, M.K., Biscaye, P. and Anderson, L. (2019) 'Delivering development? Evidence on self-help groups as development intermediaries in South Asia and Africa', *Development Policy Review*, 37(1): 129–151. https://doi.org/10.1111/dpr.12381

Hannah, C., Giroux, S., Krell, N., Lopus, S., McCann, L.E., Zimmer, A., Caylor, K. and Evans, T. (2021) 'Has the vision of a gender quota rule been realized for community-based water management committees in Kenya?', *World Development*, 137: 105–154. https://doi.org/10.1016/j.worlddev.2020.105154

Hegga, S., Kunamwene, I. and Ziervogel, G. (2020) 'Local participation in decentralized water governance: Insights from north-central Namibia', *Regional Environmental Change*, 20(3): 105. https://doi.org/10.1007/s10113-020-01674-x

Jentoft, S. (1989) 'Fisheries co-management. Delegating government responsibility to fishermen's organizations', *Marine Policy*, 13(2): 137–154. https://doi.org/10.1016/0308-597X(89)90004-3

Jentoft, S., Bavinck, M., Alonso-Población, E., Child, A., Diegues, A., Kalikoski, D., Kurien, J., McConney, P., Onyango, P. and Siar, S. (2018) 'Working together in small-scale fisheries: Harnessing collective action for poverty eradication', *Maritime Studies*, 17(1): 1–12. https://doi.org/10.1007/s40152-018-0094-8

Kabeer, N. (2005) 'Is microfinance a "Magic Bullet" for women's empowerment? Analysis of findings from South Asia', *Economic and Political Weekly*, 40(October 29): 4709–4718.

Kaganzi, E., Ferris, S., Barham, J., Abenakyo, A., Sanginga, P. and Njuki, J. (2009) 'Sustaining linkages to high value markets through collective action in Uganda', *Food Policy*, 34(1): 23–30. https://doi.org/10.1016/j.foodpol.2008.10.004

Kruijssen, F., Keizer, M. and Giuliani, A. (2009) 'Collective action for small-scale producers of agricultural biodiversity products', *Food Policy*, 34(1): 46–52. https://doi.org/10.1016/j.foodpol.2008.10.008.

Kyeyamwa, H., Speelman, S., Van Huylenbroeck, G., Opuda-Asibo, J. and Verbeke, W. (2008) 'Raising offtake from cattle grazed on natural rangelands in sub-Saharan Africa: A transaction cost economics approach', *Agricultural Economics*, 39(1): 63–72. https://doi.org/10.1111/j.1574-0862.2008.00315.x

Lyon, F. (2003) 'Community groups and livelihoods in remote rural areas of Ghana: How small-scale farmers sustain collective action', *Community Development Journal*, 38(4): 323–331. https://doi.org/10.1093/cdj/38.4.323

Mayoux, L. (1995) 'Beyond naivety: Women, gender inequality and participatory development', *Development and Change*, 26(2): 235–258. https://doi.org/10.1111/j.1467-7660.1995.tb00551.x

Meinzen-Dick, R. and Di Gregorio, M. (eds.) (2004) 'Collective action and property rights for sustainable development', *IFPRI 2020 Vision Focus Briefs 11*. Washington, D.C: International Food Policy Research Institute (IFPRI). http://ebrary.ifpri.org/cdm/ref/collection/p15738coll2/id/129259

Nelson, V., Haggar, J., Martin, A., Donovan, J., Borasino, E., Hasyim, W., Mhando, N., Senga, M., Mgumia, J., Quintanar Guadarrama, E., Kendar, Z., Valdez, J. and Morales, D. (2016) *Fairtrade coffee: A study to assess the impact of Fairtrade for coffee smallholders and producer organisations in Indonesia, Mexico, Peru and Tanzania*, Chatham: Natural Resources Institute, University of Greenwich.

Nightingale, A.J. (2002) 'Participating or just sitting in? The dynamics of gender and caste in community forestry', *Journal of Forest and Livelihood*, 2(1): 17–24.

Nunan, F. and Cepić, D. (2020) 'Women and fisheries co-management: Limits to participation on Lake Victoria', *Fisheries Research*, 224: 105454. https://doi.org/10.1016/j.fishres.2019.10545

Okello, J., Narrod, C.A. and Roy, D. (2007) 'Food safety requirements in African green bean exports and their impact on small farmers', *IFPRI Discussion Paper 00737*, Washington, DC: International Food Policy Research Institute.

Ostrom, E. (1990) *Governing the commons: The evolution of institutions for collective action*, Cambridge: Cambridge University Press.

Ostrom, E. and Ahn, T. (2009) 'The meaning of social capital and its link to collective action', in Svendsen, G.T. and Svendsen, G.L.H. (eds.), *Handbook of social capital: The Troika of sociology, political science and economics*, Cheltenham, UK; Northampton, MA: Edward Elgar, pp. 17–35.

Penrose-Buckley, C. (2007) *Producer organisations: A guide to developing collective rural enterprises*, Oxford: Oxfam GB.

Putnam, R.D. (2000) *Bowling alone*, New York: Free Press.

Ramos-Pinto, P. (2006) 'Social capital as a capacity for collective action', in Edwards, R., Franklin, J. and Holland, J. (eds.), *Assessing social capital: Concept, policy and practice*, Newcastle: Cambridge Scholars Press, pp. 53–69.

Sen, A.K. (1999) *Development as freedom*, Oxford: Oxford University Press.

Smith, J.L. (2008) 'A critical appreciation of the "bottom-up" approach to sustainable water management: Embracing complexity rather than desirability', *Local Environment*, 13(4): 353–366. https://doi.org/10.1080/13549830701803323

Stewart, F. (2005) 'Groups and capabilities', *Journal of Human Development*, 6(2): 185–204. https://doi.org/10.1080/14649880500120517

Sudgen, F., Agarwal, B., Leder, S., Saikia, P., Raut, M., Kumar, A. and Ray, D. (2021) 'Experiments in farmers' collectives in Eastern India and Nepal: Process, benefits, and challenges', *Journal of Agrarian Change*, 21(1): 90–121. https://doi.org/10.1111/joac.12369

Terry, A., McLaughlin, O. and Kazooba, F. (2015) 'Improving the effectiveness of Ugandan water user committees', *Development in Practice*, 25(5): 715–727. https://doi.org/10.1080/09614524.2015.1046421

Vicari, S. (2015) *2014 Annual report on FAO's projects and activities in support of producer organizations and cooperatives*, Rome: Food and Agricultural Organization.

Westerman, K. and Benbow, S. (2013) 'The role of women in community-based small-scale fisheries management: The case of the southern Madagascar octopus fishery', *Western Indian Ocean Journal of Marine Science*, 12(2): 119–132.

41
REBUILDING LIVELIHOODS TO REDUCE DISASTER VULNERABILITIES

Gargi Sen, Vineetha Nalla and Nihal Ranjit

An introduction to disaster risk and vulnerability

Given the increasing frequency of climate-change related extreme events, minimising vulnerabilities of people, systems, and regions is critical for reducing disaster risk. Livelihoods based approaches could enable the reimagining of development pathways aimed at reducing key vulnerabilities and, therefore, disaster risk. This chapter starts by examining key concepts in the discourse surrounding disasters, focusing on vulnerability and disaster risk. Next, it discusses the impacts of disasters in the Global South, particularly on the livelihoods of disaster-affected persons (DAPs). Lastly, it draws out four key learning outcomes from disaster responses in the Global South that focused on rebuilding livelihoods.

The United Nations International Strategy for Disaster Reduction (UNISDR) describes disasters as a 'combination of: the exposure to a hazard; the conditions of vulnerability that are present; and insufficient capacity or measures to reduce or cope with the potential negative consequences' (UNISDR, 2009: 9). While hazards represent climate induced (viz. cyclones, floods, droughts, etc.) as well as human-made (viz. industrial hazards, air pollution, etc.) processes or events that have potential to cause damage, disasters represent the hazard's impacts on society (The World Bank, 2010) that could cause 'damage and disruption that exceeds the affected society's capacity to cope' (Twigg, 2004: 12).

Disasters frequent most parts of the world in varying degrees. However, their impacts, as measured by changes in average percentage of gross domestic product (GDP), are the most substantial in countries of the Global South (CRED and UNISDR, 2018; Twigg, 2004). People in these countries face everyday risks that stem from deeper social predicaments, such as poverty and lack of access to vital resources, making them more vulnerable to the damaging effects of a hazard. These vulnerabilities, which may not be related to the hazard itself (Cannon et al., 2003; Chhotray and Few, 2012; Wisner et al., 2004), can increase people's susceptibility to disastrous effects (Adger, 2006; Krellenberg et al., 2016). For instance, the urban poor are often forced to settle in hazardous locations such as floodplains of rivers, near volcanoes, or on land without secure tenure, due to their adverse economic situations (Wisner et al., 2004). This amplifies disaster risk by increasing their exposure to hazards. Here, exposure is a combination of the physical location and characteristics of the built and natural environment that shapes the external dimension of vulnerability (Ingram et al., 2006).

Considering this, Wisner et al. (2004: 4) argue that studies of disasters 'should not be segregated from everyday living' underlining that 'the risks involved in disasters must be connected with the vulnerability created for many people through their normal existence'. Disaster risk is therefore a composite of external hazards exacerbating existing internal vulnerabilities of elements (people, systems, assets etc.) that influence their propensity to risk (Chhotray and Few, 2012; Few, 2003; Wisner et al., 2004). The risk framework outlined by Davidson (1997) explains that as the capacity to respond to hazards improves, vulnerability reduces and so does the disaster risk (Birkmann, 2013).

A key part of building capacities and reducing vulnerabilities is strong livelihood enhancement programmes. This focus on livelihoods is suggested by the Sendai Framework for Disaster Risk Reduction, which also promotes social safety nets, food security, and access to healthcare and education. Studies have found that a focus on reducing vulnerabilities during the post-disaster reconstruction and recovery period can reduce impacts of future events (Joakim and Wismer, 2015; Pelling, 2003; Wisner et al., 2004).

Disasters and their impacts

It has been established that disasters are first and foremost an impediment to development, particularly for the poor and marginalised (Cannon et al., 2003; Twigg, 2004; Wisner et al., 2004). Their impacts are products of maldevelopment, which in turn cause massive developmental setbacks, and leads to a vicious cycle that undermines economic growth and distorts the developmental goals of a country (FAO, 2015).

In the Global South, CRED and UNISDR (2018) and CRED and UNDRR (2020) report that cyclonic storms and floods are some of the most common hazards recorded. These events are also high intensity and can result in extensive loss of life and assets. The India Meteorological Department (IMD) reports that the number of cyclones originating in both the Arabian Sea and the Bay of Bengal have increased significantly in the last decade, especially in the post-monsoon seasons. One example of how frequent cyclones and floods can adversely impact livelihoods is the 1999 Super Cyclone had long-term impacts on agriculture and livestock rearing in the Indian state of Odisha, where these are primary occupations for large sections of the population (Chhotray and Few, 2012).

Asia records the highest share of flooding events globally, accounting for 41% of all disaster events (CRED and UNDRR, 2020: 17). Between 2000 and 2019, floods affected more than 1.5 billion people in Asia, accounting for 93% of the population affected by floods worldwide (CRED and UNDRR, 2020: 17). Moreover, population growth and changes in land-use patterns have increased human vulnerability to floods (Dewan, 2015), especially in South Asian countries such as Bangladesh and Nepal, where frequent flooding costs lives, as well as damages crops, livestock, and infrastructure. Similarly, in Latin America, close to 74 million people were affected by disasters between 2000 and 2015, predominantly by floods (Nagy et al., 2018). Development in Latin American countries has also been severely impacted by geophysical disasters – such as the 1987 earthquake in Ecuador (Bello et al., 2015).

The 2004 Indian Ocean tsunami is another example of a hydro-meteorological hazard that wreaked widespread damage to the fisheries sector along the Indian subcontinent, including both loss of lives and detrimental impacts on livelihoods. The tsunami caused extensive damage to several thousand boats and fishing equipment, as well as infrastructure such as harbours and fish landing centres (De Silva and Yamao, 2007; United Nations, World Bank and Asian Development Bank, 2006).

Droughts[1] are slow-onset disasters unlike floods, tropical cyclones and tsunamis, and accounted for only 5% of disasters that occurred around the world in the last two decades. However, their impacts are significant considering the long-term indirect or secondary impacts they can have on livelihoods that contribute to mass migration (CRED and UNISDR, 2018: 12). For instance, more than 10 countries in Africa have reported six or more droughts since the 1990s. Recurrent droughts in East African countries result in crop failure and reduction in water levels severely impacting agriculture and allied sectors, such as irrigation and livestock husbandry, all of which are major sources of livelihood in the region. Prolonged periods of droughts can also severely impact food security resulting in famines and starvation, as witnessed during the 2010–2011 drought in Somalia (Haile et al., 2019). A combination of the region's heavy reliance on rain-fed agriculture, frequent epidemics and impairment of key infrastructure sectors (e.g. electricity in Ghana is mainly dependent on hydropower) adversely affected economic activities (Bekoe and Logah, 2013), and therefore, people's capacities to withstand the effects of the drought.

Some hazards are more prominent than others (e.g. coastal inundation, landslides, lightning strikes etc.), but whether rapid or slow onset, they severely impact key economic sectors such as agriculture, livestock and fisheries. Moreover, the impacts on people and their livelihoods are varied as they cope by altering their means of livelihoods, mortgaging or selling assets, and even reducing nutritional intake or deferring medical treatment (Chambers, 1989). Such actions further contribute to the entrenched vulnerabilities of these communities, creating adverse cascading impacts. On this note, Chambers (1989) explains the *ratchet effect* of vulnerability where 'each succeeding event reduces the resources a group or individual has to resist and recover from the next environmental shock or stress' (Pelling, 2003: 16). Moreover, Shughrue et al. (2020: 606) argue that the dominant frameworks used to study impacts of hazards have accounted 'only for direct, local damages for a short period after occurrence . . . neglecting secondary, economic impacts that spread globally among cities from the original site of a disaster'.

Impacts of disasters are not limited only to the location of its occurrence but 'might spread unevenly' from the source location 'guided by societal structures and potentially resulting in inequitable burdens' (Shughrue et al., 2020: 606). The Intergovernmental Panel on Climate Change (IPCC) notes that the risks to livelihoods due to rising temperatures and the consequent increase in frequency and severity of disaster events will almost always affect lower and middle-income groups the most (Hoegh-Guldberg et al., 2018). Furthermore, impacts on key livelihood sectors in developing countries, such as agriculture and fisheries, severely impede the recovery of peoples and economies.

Livelihood and their vulnerabilities

Agriculture and livestock rearing dominate the economy in many Global South countries (FAO, 2015). Optimum soil conditions and water availability are critical for production of food crops, resources that are becoming increasingly scarce due to climate change, and this scarcity is heightened by rapid onset hazards. In particular, the impacts of droughts on agriculture are more pronounced in African countries, especially in terms of declining outputs over years (FAO, 2015) that have cascading impacts on food security, supply chains, output of agro-based industries, and ultimately, the nation's GDP.

Climate change has had negative impacts on small-scale fisheries in developing countries, more so due to the high vulnerability of small-scale fisheries to climate change-induced hazards such as cyclones, floods, droughts, sea-level rise, land erosion and temperature and rainfall fluctuations (IPCC, 2007). Their vulnerability is not only caused by their high exposure to the hazard, but also by their entrenched socio-economic vulnerabilities, limited participatory decision making, low asset ownership (Pomeroy et al., 2006), and lack of adaptive capacity[2] in terms of alternative livelihoods or physical, natural and financial capital (Islam et al., 2014). Marine fisheries-based livelihoods are highly resource dependent and there are often limited alternative livelihoods available due to the isolated locations of these communities (Pomeroy et al., 2006). For instance, in ecologically vulnerable places with low adaptive capacity like Bangladesh, disasters and their repercussions are an extension of the persistent impacts of climate change affecting the local ecology and crippling the fisheries sector.

Thus, the vulnerability of livelihoods is not only attributed to the impact of a hazard but also to geographical, socio-economic and other environmental predispositions. Typical characteristics of developing economies of the Global South, such as densely populated settlements, poverty, mismanagement of environmental resources, loss of valuable biodiversity, degenerating natural resources that are sources of livelihoods for poor and marginalised people, all perpetuate vulnerability (Thomalla et al., 2006).

Responses

Given this wide spectrum of disaster impacts, how then have different nation states responded to and accounted for disaster risk? Most states have used the term 'build back better' for post-disaster work, which is conceptualised as an opportunity for improved redevelopment that reduces vulnerability and disaster risk. However, Joakim and Wismer (2015) argue that rarely do these investments and projects accomplish their intended goals. On examination of past approaches in post-disaster interventions, Pomeroy et al. (2006) conclude that disaster recovery has more commonly focused on physical reconstruction rather than rebuilding livelihoods and communities, which would be more likely to ensure reduced vulnerability and lead to effective longer-term recovery.

Based on approaches to the rehabilitation of flood-affected communities in Mzuzu city in Malawai, Kita (2017) argues that an emphasis on the resettlement of affected communities is obscuring key drivers of vulnerability. He further notes that DRR measures, such as resettlement, should not be standalone interventions but should be accompanied by measures that address the root causes of vulnerability. This would include providing security of tenure, sustainably restoring livelihoods, and ensuring income stability. These goals can be achieved by building capacities, ensuring equity through institutional measures that reduce marginalisation, and promote the protection of natural resources.

Post-disaster measures and development processes in general, which are not attuned to climate and disaster risks and their variabilities, end up amplifying existing vulnerabilities or creating new ones (see Box 41.1). The natural hazards and disaster-focused approach to vulnerability reduction assumes intensity of the hazard and its frequency as key components of vulnerability. While this approach may be beneficial in addressing hazard-specific vulnerabilities, it often overlooks the economic or social conditions that cause existing vulnerabilities (Yamin et al., 2005). Risk reduction strategies for the poor should work toward reducing economic vulnerability and, at the same time, capitalise on (and perhaps nurture) the inherent social and cultural capacities of poor communities (Yodmani, 2001). Given this, since the late 1990s, protecting and rebuilding livelihoods post-disasters as an attempt to reduce vulnerability began to gain recognition (Cannon et al., 2003; Jones et al., 2010). While these approaches are still in the experimental stage, more formalised initiatives have been introduced across the Global South over the last decade that address not only replacement of physical assets related to livelihoods, but also the restoration of crucial social networks, the provision of financial services, and the development of markets (IRP and UNDP-India, 2010).

> **Box 41.1 Post-disaster policies in South Asia**
>
> In South Asian countries, post-disaster measures have often focused on processes such as speed of recovery, funding distribution mechanisms or coordination between governments and humanitarian organisations, which have been 'one dimensional' (Joakim and Wismer, 2015) and 'reactive' (Ingram et al., 2006) as they do not account for reducing vulnerability of impacted communities (Joakim and Wismer, 2015). Housing reconstruction is particularly favoured as it is known to be a means to facilitate economic regeneration (Ahmed, 2017). For example, the Coastal Buffer Zone policy of the Government of Sri Lanka, which was formulated within days of the aftermath of the Indian Ocean Tsunami in 2004, promoted unevenly distributed reconstruction projects that relocated communities without considering the socio-economic impact of the move (Ingram et al., 2006). While the housing measure intended to reduce the vulnerability of DAPs, it spurred gentrification that benefitted high-end tourism businesses and further marginalised the impoverished fishing communities. Moreover, the resettlement that was intended to reduce the vulnerability resulted in exposure to persistent flooding.

Livelihoods approach to reduce vulnerabilities

Livelihoods are formed within social, economic and political contexts and thus, livelihood-based responses to disasters and DRR are highly context-specific (IRP and UNDP-India, 2010). Ideally, effective livelihood-based strategies reinforce vulnerability reduction and vice versa in a cyclical manner. This can improve the overall adaptive capacity of the region and can be introduced at any stage of the disaster cycle (Twigg, 2004). Moreover, the planning and design of livelihood recovery measures in the wake of disasters need to consider that successful initiatives are ones that are highly localised (Joakim and Wismer, 2015; Régnier et al., 2008).

There are several arguments in favour of livelihoods-based approaches to reduce vulnerability to disasters (Ingram et al., 2006; Jain et al., 2016; Pomeroy et al., 2006; Sanderson, 2000). These approaches are primarily people-centric and focus on the capabilities of individuals and households as a means to a livelihood (Chambers and Conway, 1992; Sudmeier-Rieux et al., 2006). For example, as a response to regular flooding in the Karnali River Basin in Western Nepal, the Nepal Flood Resilience Project focused on increasing household income of the communities through skill development training. An evaluative study by Practical Action (2018) found that increased incomes enabled households to invest in education of children, personal health, and hygiene, as well as insurance for family and livestock. Further, the study reported that this also motivated people to invest more in flood-resilient options, such as elevating the grain storage in their houses (ibid). Here, livelihoods-based approaches mainstream disaster recovery into development via poverty-reduction strategies, wherein assets[3] are a means of building capabilities and, therefore, livelihoods (Moser and Dani, 2008). Further, they also provide social protection measures or safety nets that protect communities experiencing sudden loss of resources.

Building on a review of literature and analysis of case studies of post disaster interventions from the Global South, we highlight four key criteria that underlie the effective pathways for implementation of livelihoods-based interventions.

First, livelihoods must be conceptualised to be sustainable. Sustainable livelihoods (SL) have been promoted as a conceptual framework focused on a longer-term goal of poverty reduction (Gyawali et al., 2019). Chambers and Conway (1992) explain that livelihoods are sustainable when they can cope with and recover from shocks and stresses, and build assets and capabilities while not undermining the natural resource base. Twigg (2001: 8) explains that conventional DRR measures take 'disaster/hazard vulnerability as the starting point, viewing livelihoods as an aspect of the question'. Conversely, the SL approach starts from a development standpoint and focuses on livelihoods as the central aspect. The Sustainable Fisheries component of the World Bank-funded Coastal Disaster Risk Reduction Project (CDRRP) in Tamil Nadu and Puducherry in India targeted all aspects of the fisheries supply chain in the region. The project encompassed protection of resources, development of modern infrastructure, upliftment of the marginalised, and livelihood diversification. It addressed multiple facets of fishing livelihoods – infrastructure, resource management, and the safety of fishers. Technology-enabled safety measures, modernised fishing harbours, and introduction of innovative methods through seaweed cultivation, enhanced the fishing output while reducing risks to fishers and the resources they depend on. Additionally, the project formalised community-based management bodies for fishing activities and infrastructure, and further incentivised traditional fishing practices that prevent the exploitation of natural resources (Ramakrishnan, 2020; Santha, 2014).

Second, within the larger ambit of SL, livelihoods must be conceptualised as part of a longer-term resilience-building agenda. Given the importance of mainstreaming DRR into development, relief and rehabilitation measures should contribute to long-term development and reduction of vulnerabilities to future disasters. Strategies should aim to build local capacities and develop infrastructures that will be of value even after the emergency period is over (Gyawali et al., 2019). For instance, after the tsunami in 2004, the NGO Coordination and Resource Center (NCRC) was set up in Nagapattinam, Tamil Nadu, India, which worked on reviving and strengthening livelihoods in the agriculture and fisheries sectors. Having a long-term resilience agenda since its inception, NCRC embedded themselves in the region and continued to work on the recovery of coastal communities beyond the immediate relief and response stage. NCRC was then succeeded by BEDROC (Building and Enabling Disaster Resilience of Coastal Communities) and their mandate changed from providing relief to strengthening existing livelihoods to build the resilience of coastal communities (Nalla et al., 2019).

Third, strategies focused on building assets and capabilities must be participatory in both conceptualisation and implementation. Participation of DAPs in recovery and planning processes has been promoted (Hore et al., 2020; Wisner et al., 2004) to ensure representation of all groups to improve decision making (Hamideh, 2020). In fact, in his review of livelihoods-based frameworks, Twigg (2001: 15) emphasises the need for participation in 'examining livelihoods *and* their context'. For instance, in the case of Koh Klang Island in Thailand's Andaman coast, which was severely affected by flooding from the 2004 tsunami, partnerships were fostered with the communities to manage natural resources, sustain livelihoods, and conserve ecosystems.[4] Amongst the several projects implemented in the region, the EPIC (Ecosystems Protecting Infrastructures and Communities) project restored mangroves as a bio-shield for the village and added silvo-fisheries for local food security. This was implemented by MAP, an NGO, through participatory processes in in two of three villages in the region. But the benefits of participating in decision-making in the two villages garnered the attention of the third village, which had reportedly shown little interest in public affairs before (Lin, 2019). Such projects, by prioritising assets and capacities, empower communities through participation in post-disaster initiatives.

Lastly, livelihood diversification plays an important role in building economic resilience within livelihood-based approaches. Pomeroy et al. (2006) asserts that diversification of livelihoods among coastal communities allows for a wider mix of people to find employment. In Bangladesh, a programme by Solidarités International promoted diversification of crops as a key feature of disaster risk reduction measures. Through grants for sustainable farming practices, the project resulted in the enhanced use of the land yielding sustainable revenues throughout the year. Moreover, the diversification of crops improved food security and provided nutritional benefits to the community contributing to an undisrupted and wider variety of food sources (Maisonnave and Mayans, 2017). Further, Gyawali et al. (2019) argue that in the urgency to provide relief, pre-disaster livelihoods are often restored as they are perceived to be adequate when in fact, they may have been unsustainable even before the disaster. For example, in Nagapattinam in India, an oversupply of boats intended to revive fishing livelihoods after the tsunami instead led to overfishing and exacerbated other pre-existing environmental concerns (Nalla et al., 2019).

While the four criteria discussed above are in no way comprehensive, the central aim is to highlight the importance of livelihood-based strategies in reducing vulnerabilities and therefore, disaster risk. The insights presented here, based on several cases from across the Global South, could contribute to the development of a more comprehensive framework – one that prioritises post-disaster recover measures in development pathways to effectively reduce vulnerability among disaster-affected communities.

Key points

- Disaster risk combines the threat of a hazard with existing vulnerabilities. Disaster impacts are products of maldevelopment and often lead to developmental setbacks. Their impacts on people and livelihoods are varied based on the people's vulnerabilities. Given limited capacities, DAPs cope by altering livelihoods or by compromising on personal wellbeing.
- Development strategies may reduce vulnerabilities by building capacities, leading to effective disaster risk reduction and more resilient communities.
- Post-disaster livelihood interventions should be context specific and ideally aimed towards building adaptive capacities. To promote long-term resilience, the livelihood-based strategies should be designed and implemented in a participatory fashion, focusing on building local capacities and diversifying economic opportunities

Further reading

Twigg, J. (2001) 'Sustainable livelihoods and vulnerability to disasters', *Disaster Management Working Paper*, 2, London: Benfield Greig Hazard Research Centre, University College London.
Minamoto, Y. (2010) 'Social capital and livelihood recovery: Post-tsunami Sri Lanka as a case', *Disaster Prevention and Management: An International Journal*, 19(5): 548–564. https://doi.org/10.1108/09653561011091887
Moench, M. and Dixit, A. (2004) *Adaptive capacity and livelihood resilience: Adaptive strategies for responding to floods and droughts in South Asia*, Kathmandu: Institute for Social and Environmental Transition (ISET).
Wisner, B., Blaikie, P., Cannon, T. and Davies, I. (2004) *At risk: Natural hazards, people's vulnerability, and disasters*, 2nd ed., London: Routledge.

Notes

1 Droughts are characterised as a period of abnormally dry weather long enough to cause serious hydrological imbalances (IPCC, 2018). Droughts are commonly identified using indicators such as precipitation, temperature, evapotranspiration, soil moisture, and groundwater (Haile et al., 2019).
2 Adaptive capacity is defined as the ability of systems, institutions, and individuals to cope with potential damage, to take advantage of opportunities, or to respond to consequences (IPCC, 2018).

3 Assets are a means of production (Gwimbi, 2009: 72) and are 'not only physical (e.g. land), but also social (e.g. good relations with neighbours), human (e.g. good entrepreneurial skills), financial (e.g. savings) and, arguably, political, (e.g. having a say in democratic processes)' (Sanderson, 2000: 97).
4 Community-based natural resource management (CBNRM) was used in post-disaster recovery efforts in Thailand.

References

Adger, W.N. (2006) 'Vulnerability', *Global Environmental Change*, 16(3): 268–281. https://doi.org/10.1016/j.gloenvcha.2006.02.006

Ahmed, I. (2017) 'Resilient housing reconstruction in the developing world', in March, A. and Kornakova, M. (eds.), *Urban planning for disaster recovery*, Oxford: Butterworth-Heinemann, pp. 171–188. http://doi.org/10.1016/B978-0-12-804276-2.00012-8

Bekoe, E.O. and Logah, F.Y. (2013) 'The impact of droughts and climate change on electricity generation in Ghana', *Environmental Sciences*, 1(1): 13–24.

Bello, O., Ortiz, L. and Samaniego, J. (2015) 'Assessment of the effects of disasters in Latin America and the Caribbean, 1972–2010', *ECLAC Environment and Development Series No. 157*, Santiago: United Nations.

Birkmann, J. (2013) *Measuring vulnerability to natural hazards: Towards disaster resilient societies*, 2nd ed., Tokyo: United Nations University Press.

Cannon, T., Twigg, J. and Rowell, J. (2003) *Social vulnerability, sustainable livelihoods and disasters*, report to DFID conflict and humanitarian assistance department (CHAD) and sustainable livelihoods support office, Chatham: Natural Resources Institute, University of Greenwich.

Chambers, R. (1989) 'Vulnerability, coping and policy (editorial introduction)', *IDS Bulletin*, 37(4): 33–40. https://doi.org/10.1111/j.1759-5436.1989.mp20002001.x

Chambers, R. and Conway, G. (1992) 'Sustainable rural livelihoods: Practical concepts for the 21st century', *IDS Discussion Paper, Issue 296*, Brighton: Institute of Development Studies.

Chhotray, V. and Few, R. (2012) 'Post-disaster recovery and ongoing vulnerability: Ten years after the super-cyclone of 1999 in Orissa, India', *Global Environmental Change*, 22(3): 695–702. https://doi.org/10.1016/j.gloenvcha.2012.05.001

CRED and UNDRR (2020) *The human cost of disasters: An overview of the last 20 years (2000–2019)*, Brussels: CRED; Geneva: UNDRR.

CRED and UNISDR (2018) *Economic losses, poverty and disasters: 1998–2017*, Brussels: CRED; Geneva: UNISDR.

Davidson, R. (1997) *An urban earthquake disaster risk index*, Stanford: Stanford University.

De Silva, D. and Yamao, M. (2007) 'Effects of the tsunami on fisheries and coastal livelihood: A case study of tsunami-ravaged southern Sri Lanka', *Disasters*, 31(4): 386–404. https://doi.org/10.1111/j.1467-7717.2007.01015.x.

Dewan, T.H. (2015) 'Societal impacts and vulnerability to floods in Bangladesh and Nepal', *Weather and Climate Extremes*, 7: 36–42. https://doi.org/10.1016/j.wace.2014.11.001

FAO (2015) *The impact of natural hazards and disasters on agriculture, food security and nutrition*, Rome: Food and Agriculture Organization.

Few, R. (2003) 'Flooding, vulnerability and coping strategies: Local responses to a global threat', *Progress in Development Studies*, 3(1): 43–58. https://doi.org/10.1191/1464993403ps049ra

Gwimbi, P. (2009) 'Linking rural community livelihoods to resilience building in flood risk reduction in Zimbabwe', *Journal of Disaster Risk Studies*, 2(1): 71–79. https://doi.org/10.4102/jamba.v2i1.16

Gyawali, S., Tiwari, S.R., Bajracharya, B.S. and Skotte, H. (2019) 'Promoting sustainable livelihoods: An approach to post-disaster reconstruction', *Sustainable Development*, 28(4): 626–633. https://doi.org/10.1002/sd.2013

Haile, G., Tang, Q., Sun, S., Huang, Z., Zhang, X. and Liu, X. (2019) 'Droughts in East Africa: Causes, impacts and resilience', *Earth-Science Reviews*, 193: 146–161. https://doi.org/10.1016/j.earscirev.2019.04.015

Hamideh, S. (2020) 'Opportunities and challenges of public participation in post-disaster recovery planning: Lessons from Galveston, TX', *Natural Hazards Review*, 21(4): 1–16. https://doi.org/10.1061/(ASCE)NH.1527-6996.0000399

Hoegh-Guldberg, O., Jacob, D., Taylor, M., Bindi, M., Brown, S., Camilloni, I., Diedhiou, A., Djalante, R., Ebi, K.L., Engelbrecht, F., Guiot, J., Hijioka, Y., Mehrotra, S., Payne, A., Seneviratne, S.I., Thomas,

A., Warren, R. and Zhou, G. (2018) 'Impacts of 1.5°C global warming on natural and human systems', in Masson-Delmotte, V., Zhai, P., Pörtner, H.-O., Roberts, D., Skea, J., Shukla, P.R., Pirani, A., Moufouma-Okia, W., Péan, C., Pidcock, R., Connors, S., Matthews, J.B.R., Chen, Y., Zhou, X., Gomis, M.I., Lonnoy, E., Maycock, T., Tignor, M. and Waterfield, T. (eds.), *Global warming of 1.5°C. An IPCC special report on the impacts of global warming of 1.5°C above pre-industrial levels and related global greenhouse gas emission pathways, in the context of strengthening the global response to the threat of climate change, sustainable development, and efforts to eradicate poverty*, Geneva: Intergovernmental Panel on Climate Change, pp. 175–311.

Hore, K., Gaillard, J., Davies, T. and Kearns, R. (2020) 'People's participation in disaster-risk reduction: Re-centering power', *Natural Hazards Review*, 21(2): 04020009. https://doi.org/10.1061/(ASCE)NH.1527-6996.0000353

Ingram, J.C., Franco, G., Rumbaitis-del Rio, C. and Khazai, B. (2006) 'Post-disaster recovery dilemmas: Challenges in balancing short-term and long-term needs for vulnerability reduction', *Environmental Science and Policy*, 9(7–8): 607–613. https://doi.org/10.1016/j.envsci.2006.07.006

IPCC (2007) *Climate change 2007: Synthesis report. Contribution of working groups I, II and III to the fourth assessment report of the intergovernmental panel on climate change*, Geneva: IPCC.

IPCC (2018) *Global warming of 1.5°C. An IPCC special report on the impacts of global warming of 1.5°C above pre-industrial levels and related global greenhouse gas emission pathways, in the context of strengthening the global response to the threat of climate change, sustainable development, and efforts to eradicate poverty*, Masson-Delmotte, V., Zhai, P., Pörtner, H.-O., Roberts, D., Skea, J., Shukla, P.R., Pirani, A., Moufouma-Okia, W., Péan, C., Pidcock, R., Connors, S., Matthews, J.B.R., Chen, Y., Zhou, X., Gomis, M.I., Lonnoy, E., Maycock, T., Tignor, M. and Waterfield, T. (eds.), Geneva: Intergovernmental Panel on Climate Change.

IRP and UNDP-India (2010) *Guidance note on livelihood recovery*, Kobe: International Recovery Platform Secretariat.

Islam, M.M., Sallu, S., Hubacek, K. and Paavola, J. (2014) 'Vulnerability of fishery-based livelihoods to the impacts of climate variability and change: Insights from coastal Bangladesh', *Regional Environmental Change*, 14: 281–294. https://doi.org/10.1007/s10113-013-0487-6

Jain, G., Jigyasu, R., Gajjar, S.P. and Malladi, T. (2016) *Cities provide transformational opportunity to reduce risk accumulation*, Bangalore: Indian Institute for Human Settlements.

Joakim, E. and Wismer, S.K. (2015) 'Livelihood recovery after disaster', *Development in Practice*, 25(3): 401–418. https://doi.org/10.1080/09614524.2015.1020764

Jones, L., Jaspars, S., Pavanello, S., Ludi, E., Slater, R., Arnall, A., Grist, N. and Mtisi, S. (2010) 'Responding to a changing climate: Exploring how disaster risk reduction, social protection and livelihoods approaches promote features of adaptive capacity', *Working Paper 319*, London; Overseas Development Institute.

Kita, S.M. (2017) 'Urban vulnerability, disaster risk reduction and resettlement in Mzuzu city, Malawi', *International Journal of Disaster Risk Reduction*, 22: 158–166. https://doi.org/10.1016/j.ijdrr.2017.03.010

Krellenberg, K., Welz, J., Link, F. and Barth, K. (2016) 'Urban vulnerability and the contribution of socio-environmental fragmentation: Theoretical and methodological pathways', *Progress in Human Geography*, 41(4): 408–431. https://doi.org/10.1177/0309132516645959

Lin, P.-S.S. (2019) 'Building resilience through ecosystem restoration and community participation: Post-disaster recovery in coastal island communities', *International Journal of Disaster Risk Reduction*, 39: 1–9. https://doi.org/10.1016/j.ijdrr.2019.101249

Maisonnave, E. and Mayans, J. (2017) *Resilient farming in Sathkhira, Bangladesh*, Clichy: Solidarites International.

Moser, C. and Dani, A.A. (2008) 'Assets and livelihoods: A framework for asset-based social policy', in Moser, C. and Dani, A.A. (eds.), *Assets, livelihoods and social policy*, Washington, DC: World Bank, pp. 43–82. https://doi.org/10.1596/978-0-8213-6995-1

Nagy, G., Leal Filho, W., Azeiterio, U.M., Heimfarth, J., Verocae, J.E. and Li, C. (2018) 'An assessment of the relationships between extreme weather events, vulnerability, and the Impacts on human wellbeing in Latin America', *International Journal of Environmental Research and Public Health*, 15(9): 1802. https://doi.org/10.3390/ijerph15091802

Nalla, V., Udupa, Y. and Dhanapal, G. (2019) 'Recovering from the tsunami: A case of long-term engagement to build resilience in Nagapattinam', *IIHS Case No 1–0036*, Bangalore: Indian Institute for Human Settlements.

Pelling, M. (2003) *The vulnerability of cities: Natural disasters and social resilience*, Abingdon: Earthscan.

Pomeroy, R.S., Ratner, B.D., Hall, S.J., Pimolijinda, J., Vivekanandan, V. (2006) 'Coping with disaster: Rehabilitating coastal livelihoods and communities', *Marine Policy*, 30(6): 786–793. https://doi.org/10.1016/j.marpol.2006.02.003

Practical Action (2018) *Livelihood support: Building flood resilience*, Kathmandu: Practical Action.

Ramakrishnan, D.H. (2020) *Chief minister honours fishing village in Thoothukudi that does not use nets*, Chennai: The Hindu. Available at: www.thehindu.com/news/national/tamil-nadu/chief-minister-honours-fishing-village-in-thoothukudi-that-does-not-use-nets/article30804357.ece (Accessed 10 January 2021).

Régnier, P., Neri, B., Scuteri, S. and Miniati, S. (2008) 'From emergency relief to livelihood recovery', *Disaster Prevention and Management*, 17(3): 410–429. https://doi.org/10.1108/09653560810887329

Sanderson, D. (2000) 'Cities, disasters and livelihoods', *Environment and Urbanization*, 12(2): 93–102. https://doi.org/10.1177/095624780001200208

Santha, S.D. (2014) *Coastal resources, ecosystem services and disaster risk reduction: An analysis of social and environmental vulnerability along the coast of India*, Prepared for the Global Assessment Report on Disaster Risk Reduction 2015, Geneva: UNISDR.

Shughrue, C., Werner, B. and Seto, K.C. (2020) 'Global spread of local cyclone damages through urban trade networks', *Nature Sustainability*, 3: 606–613. https://doi.org/10.1038/s41893-020-0523-8

Sudmeier-Rieux, K., Masundire, H., Rizvi, A. and Rietbergen, S. (2006) 'Ecosystems, livelihoods and disasters', *Ecosystem Management Series No. 4*, Gland: IUCN – World Conservation Union.

Thomalla, F., Downing, T., Spranger-Siegfried, E., Han, G. and Rocktstrom, J. (2006) 'Reducing human vulnerability to climate-related hazards: Towards a common approach between the climate change adaptation and the disaster risk reduction', *Disasters*, 30(1): 39–48. https://doi.org/10.1111/j.1467-9523.2006.00305.x.

Twigg, J. (2001) 'Sustainable livelihoods and vulnerability to disasters', *Disaster Management Working Paper, 2*, London: Benfield Greig Hazard Research Centre, University College London.

Twigg, J. (2004) 'Disaster risk reduction: Mitigation and preparedness in development and emergency programming', *Good Practice Review No. 9*, London: Humanitarian Practice Network, Overseas Development Institute.

UNISDR (2009) *UNISDR terminology on disaster risk reduction*, Geneva: United Nations International Strategy for Disaster Risk Reduction.

United Nations, World Bank, Asian Development Bank (2006) *Tsunami: India: Two years after*, Chennai: United Nations Team for Tsunami Recovery Support; New Delhi: World Bank; New Delhi: Asian Development Bank.

Wisner, B., Blaikie, P., Cannon, T. and Davies, I. (2004) *At risk: Natural hazards, people's vulnerability, and disasters*, 2nd ed., London: Routledge.

The World Bank (2010) *Natural hazards, unnatural disasters: The economics of effective prevention*, Washington, DC: World Bank.

Yamin, F., Rahman, A. and Huq, S. (2005) 'Vulnerability, adaptation and climate disasters: A conceptual overview', *IDS Bulletin*, 36(4): 1–14. https://doi.org/10.1111/j.1759-5436.2005.tb00231.x

Yodmani, S. (2001) *Disaster risk management and vulnerability reduction: Protecting the poor*, Paper presented at The Asia and Pacific Forum on Poverty Organized by the Asian Development Bank, Bangkok: ADB.

42
RELIGION AND LIVELIHOODS STUDIES

Emma Tomalin

Introduction

> For most people of the 'South' spirituality is integral to their understanding of the world and their place in it, and so is central to the decisions they make about their own and their communities' development. Their spirituality affects decisions about who should treat their sick child, when and how they will plant their fields, and whether or not to participate in risky but potentially beneficial social action
>
> *(Ver Beek, 2000: 31)*

In this quotation, Ver Beek is responding to the lack of consideration that development studies, policy and practice have typically paid to the areas of spirituality and religion, despite the significant role that they play in the livelihood strategies of people in the Global South. Since the publication of his landmark paper, *Spirituality: a Development Taboo* (2000), there has been a 'turn to religion' within the aid business, following broader shifts within society reflecting limitations to the secularisation thesis, alongside the emergence of approaches to aid that attempt to be locally grounded (Tomalin, 2013). Livelihoods studies, which 'aim to contribute to the understanding of poor people's lives with the ambition to enhance their livelihoods' (de Haan, 2017: 1), emerged in the 1990s 'as part of a broader stream of people-centred, bottom-up approaches' (de Haan, 2017: 2). In contrast to top-down Global North-led development, livelihoods studies view poor people as having agency to lead solutions to reduce poverty through the creation of sustainable livelihoods where, as Scoones writes:

> A livelihood comprises the capabilities, assets (including both material and social resources) and activities for a means of living. A livelihood is sustainable when it can cope with and recover from stresses and shocks, maintain or enhance its capabilities and assets, while not undermining the natural resource base.
>
> *(2009: 175)*

However, given that access to the assets or resources that people need to maintain sustainable livelihoods is not equally available to everyone within a society, and that such inequality can lead to poverty, livelihoods research also sought to understand what constraints existed, how they were reproduced by local and global power regimes, and how they might be addressed. In addition to material, technological and economic factors, the livelihoods approach also considered the social and cultural factors which underpin norms and values, not least those pertaining to gender, that can shape behaviour and lead to inequality and exclusion (de Haan, 2017: 3).

Despite these shifts to thinking about the role of culture and social exclusion on livelihoods, a focus on religion – as a key influence on social and cultural systems and structures, and therefore on norms and behaviours that limit people's access to assets or resources – has yet to emerge as a serious topic within livelihoods studies, as a subdivision of the broader field of development studies. Moreover, where religion has been taken up in development studies, policy and practice, this remains a niche area. It is one that, more often than not, reflects an instrumentalist approach to harnessing the assets or capitals that religions can offer to development programming rather than a serious engagement with the diversity of ways that religion shapes livelihoods strategies in intersection with other factors, from the perspective of local communities (Jones and Petersen, 2011). An aspect of the instrumentalisation of religion by development actors, has been the tendency to engage with formal international faith-based organisations (IFBOs) and faith actors who are internationally networked, rather than local faith actors and communities who like other local actors 'have not yet acquired the language spoken in the world of development' (Olivier de Sardan, 2005: 183). Moreover, there is a preference for organisations and actors who are not expressive about their faith identity, which can marginalise local faith actors, where those working for faith-based organisations (FBOs) often claim that they 'leave their faith at the door' when they engage with secular development actors (Clarke, 2007: 84). As Karam writes, some have 'voiced their unease that this interest in religion may be another passing fad that seeks to capitalize on their strengths and even attempt to change their way of doing things, almost as if a covert attempt were at hand to secularize the religious' (2011: 10).

In recent years, there has been a new impetus for what has previously been termed 'bottom-up' or 'people-centred' development, the result of a number of global and intersecting public debates and movements, and with a renewed determinism to go further with this than previous efforts. For instance, discourses around the 'localisation' of aid, an outcome of the 2016 World Humanitarian Summit's 'Grand Bargain', have thrust the 'local' to the centre of development discourse. And more recent calls to 'decolonize development' that have rapidly gained momentum as an offshoot of the remarkable impact of the Black Lives Matter (BLM) movement in 2020 and the wider currency that BLM has given to critical race theory, have played a role in forcing development scholars to (again) rethink what development is. What is now up for discussion is not a new topic, but one that is being approached with a renewed vigour and which goes beyond the demand that livelihoods studies engage with religion and faith actors as resources for development projects but that to localise and decolonise development also means being open to the fact that there are 'multiple modernities' that do not match with the neo-liberal secular model (Eisenstadt, 2000; Rosati and Stoeckl, 2012).

In this chapter, I first examine the marginalisation of religion in livelihoods studies and the implications of this for achieving improved livelihoods strategies and outcomes. Next, I look at the relevance of religion in livelihoods studies in two areas: the role that religion can play in shaping the livelihoods that people pursue; and the impact of religion on social inequality and social exclusion, thereby restricting access to assets or resources necessary for sustainable livelihoods. In the final part of the chapter, drawing on the work of Norman Long (2001), I propose an actor-oriented approach to religions and livelihoods in the Global South that moves beyond the instrumentalisation of religion by development actors (Tomalin, 2020).

The marginalisation of religion in livelihoods studies

Given the left wing-secularist underpinnings of development studies since its inception in the post-World War Two period, religion has been marginalised from its theories, concepts and empirical research (Rakodi, 2015). On the one hand, theories about the inevitability of secularisation were dominant within the social and political sciences, and, on the other hand, secularism was seen as progressive for social and political systems. However, rather than the inevitable decline of religion as societies have adopted western models of development, a number of recent studies demonstrate the symbiotic and mutually beneficial relationship between religion and neo-liberalism. For instance, in their study of Islamic gated communities in Istanbul, Pentecostal prayer camps in Lagos, and Pentecostal grassroots movements in the favelas of Rio de Janeiro, Lanz and Oosterbaan demonstrate 'that urban religion should also be regarded as a constitutive force of contemporary capitalism and should therefore be placed at the heart of the neoliberal construction of urban space instead of at its margins' (2016: 487). Other examples of this dynamic can be found in Atia's work on 'pious neoliberalism', where she examines how Islamic development organisations 'promote financial investment, entrepreneurship, and business skills as important components of religiosity' (2011: 808; see also Rudnyckyj, 2015), and Chacko's study on the Hindu Right in India (2019). She writes that 'in the Hindu nationalist conception of market citizenship, the individual is an entrepreneurial consumer who is regulated in his/her behaviour by the cultural framework of Hindu majoritarianism and is driven by a desire to strengthen the Hindu nation' (2019: 380–381).

Since the early 2000s there has been a 'turn to religion' in development studies, which has seen a steady increase in publications on the topic of religion and development, reflecting a trend towards greater engagement with faith actors in development policy and practice (Bompani, 2019; Swart and Nell, 2016; Tomalin, 2013). Scholarship that reflects this religious turn, not only argues that the study of religion is crucial to understanding and therefore addressing the ways in which it can act against human rights and wellbeing, particularly since theories of secularisation have proved to be limited, but also that the sidelining of religion from development studies, policy and practice misses valuable opportunities to engage with faith actors who often share progressive development goals that seek to challenge and alleviate inequalities. Moreover, even when faith actor goals and religiously inflected world views do not align with development goals, the decolonial turn in development studies has taught us that these cannot simply be dismissed as out of date or irrelevant, based on superstition and a lack of rationality. Instead, projects that seek to address gender inequality, for example, and that face opposition from religious norms and structures, can be effective when they work with local community structures and religio-cultural traditions to seek to effect social transformation in ways that do not reproduce oppressive colonial power relations (Deneulin and Bano, 2009; Istratii, 2020).

Such an approach to seeing things from the perspective of people in the Global South is not a new trend in development studies, and is at the core of the emergence of the sub-discipline of livelihoods studies (de Haan, 2017; Scoones, 2009). This shift to a 'bottom up' perspective was induced by a number of pioneers, including Robert Chambers' work on 'participatory rural appraisal' which aimed to 'put the last first' (1983); sociologist Norman Long's work on 'actor-oriented approaches' to development (2001); and critical ethnographies of development by social anthropologists including David Lewis and David Mosse (2006) and Jean Pierre Olivier de Sardan (2005), which challenged the 'top down' approach of the international aid system. This included calling out development actors for adopting the language of participatory and bottom-up development, while continuing to think and act in ways that legitimated their Northern perspective. Bourdieu's (1986) work on social capital *and* power relations has also been highly influential on the livelihoods approach, where Bebbington, for instance, has promoted the view that the assets or capitals that people have access to – 'produced, human, natural, social and cultural capital' (1999: 2022) – are not just material or economic resources 'through which they make a living: they also give meaning to the person's world' (1999: 2022). This move to a more 'holistic approach' towards human development and the view that poverty is therefore multi-dimensional, also influenced by the work on capabilities by Sen (1993) and Nussbaum (2013) (see Chapter 3 of this volume), is a key feature of the livelihoods approach.

Despite the focus in livelihoods studies on participatory and bottom-up approaches, the impact of culture and social norms, and poverty as multidimensional, requiring a holistic approach, the pioneers of the livelihoods approach did not make consideration of religion a key feature of their work. Reflecting the broader neglect of religious issues, concerns and worldviews as well as the role that faith actors play in development, particularly at the local level, serious engagement with religion in the livelihoods approach is unrealised and scattered.

The relevance of religion for livelihoods studies

How religion impacts choice of livelihood

Others are beginning to reflect upon this gap and to direct research to address it, as Ayifah et al. write 'research on the effect of religion on livelihood choice is very sparse . . . In the rural African context, we are not aware of any authors that have specifically investigated this issue' (2021: 2). Their study focused on the effect of religious affiliation on the choice of livelihood activity for women in the Yilo and Lower Manya Krobo Districts of Ghana. Rather than looking at the impact of particular religious teachings, texts or traditions upon livelihood choice, this study focused on the role that *belonging* to a particular religion plays in terms of a shared social identity and the access to social capital that this affords. For instance, in their sample 'Catholic, Pentecostal and Protestant women are more likely to be in farming' where 'the effect of religion on farming for Pentecostals, Catholics and Protestants is due to a very large social capital/social identity effect, and a smaller negative effect of risk aversion, because of belonging to these religious groups' (2021: 9). They argue that this kind of knowledge is essential in considering 'programmes and policy designs aimed at enhancing livelihood activities of women in this area' (2021: 9).

In contrast to this study, Knoop et al. (2020) were interested in the role of religious belief on the hunting habits of the Indigenous Maraguá people in the central Amazon, Brazil. Today the Maraguá practice Catholicism or Seventh-day Adventism, a conservative Evangelical denomination that holds 'meat taboos derived from the Hebrew bible which forbid the consumption of meat other than that of birds and ruminants . . . a novel phenomenon, as Indigenous food taboos, potentially present in the past, had gone extinct before the introduction of Adventism' (2020: 134). While one function of Indigenous taboos is likely to be 'adaptive responses' to conserve local biodiversity, this is not guaranteed with the Adventist taboo systems as they come from outside (2020: 141). Knoop et al. warn that 'the overall effects of conversion to Adventism on wildlife populations are complex and hard to anticipate' (2020: 141) and they 'highlight the importance of considering cultural and religious particularities in research on subsistence hunting and design of management plans for Indigenous lands in Amazonia' (2020: 131).

Studies like this, that focus on the implications for biodiversity conservation of religious taboos and teachings around hunting and other kinds of resource use, underpinned by Indigenous traditions or the newer world religions, are highly relevant to thinking about 'sustainable livelihoods' that became popular following the publication of the Bruntland Commission Report 'Our Common Future' in 1987 (Scoones, 2009). Such studies form part of a broader literature which has emerged from the humanities, as well as the social and physical sciences, mainly since the 1980s, and with various labels including 'eco-theology', 'religious environmentalism' or studies on 'local knowledge' or 'local resource management' (see Tomalin, 2013). This diverse body of literature has done much to raise the significance of paying attention to the ways in which religion could be relevant to thinking about sustainable development through the implementation of specific teachings that prohibit natural resource exploitation or as the foundation for ecological ethics to encourage people to change their behaviour. However, studies in this area often suffer from the adoption of the same kind of instrumentalist approach common in the broader 'turn to religion' within the global development domain (Tomalin, 2013). First, there has been a tendency for work on eco-theology and religious environmentalism to decontextualise religious teachings that appear to promote care for nature and then to offer them back as evidence of an inherent environmental ethic, that communities of believers need to 'rediscover'. Second, conservationists often seek to capture the power of local taboos and traditions in their conservation projects (Peres and Nascimento, 2006). However, as Knoop et al. reflect with respect to their case study with the Maraguá, for 'conservationists, it may be tempting to manipulate taboos according to their expected value for wildlife conservation . . . but such promotion of cultural change risks reinforcing disparate power relationships between scientists and local people' (2020: 142; see also van Vliet, 2018).

It is true that many communities whose livelihoods have a low impact on resource depletion have religious and cultural traditions that are closely connected to animals and the land. However, the symbiotic relationship between livelihoods and religio-cultural traditions have developed and adapted to particular contexts over generations and it cannot be extrapolated from this that they would have a broader significance in other settings or could be 'revived' once they have been superseded by other livelihoods and religio-cultural formations. To avoid essentialising and romanticising the connection between interpretations of religious tradition and support for sustainable development, other studies take a more grounded approach and start from the local level with studies of contemporary organisations and movements that engage with religious teachings and practices in a contextual way alongside livelihoods activities. Recent examples of this approach include Rajkobal's work on the Sarvodaya movement in Sri Lanka which is motivated by Buddhist ethical principles that support the idea of 'holistic development' and 'right livelihood' (artha charya) (2019) or Darlington's work on Buddhist monks' engagement with environmentalism in Northern Thailand (2019).

Religion as a barrier to accessing resources and opportunities to 'make oneself a living' (de Haan, 2017:7)

'Religion' is often included in lists of intersecting factors that can act as a barrier to accessing resources in livelihoods studies, but is rarely unpacked into the variety of different aspects that can influence people's life chances and opportunities, including values, teachings, texts, institutions, leadership, practices or beliefs. Moreover, often the term 'culture' appears instead as a proxy or 'safe word' for religion, where to address 'religion' head on is viewed as too political, disrespectful and potentially risky to partnerships with communities in the Global South. I suggest that this is a legacy of the Christian-influenced western view of the separation of the religious and the secular where, in an attempt to avoid interfering in 'religious matters' and thereby replicating colonialist structures, religion tends to be packaged away as something private, unchanging and ahistorical. However, this view of religion itself serves to reinscribe colonialist structures through imposing a secularist positionality upon communities where instead religion is multi-faceted, public, dynamic and dialectically related to social and political change.

One area where this is particularly marked within development/livelihoods studies is with respect to the relationship between religion and gender, specifically women's rights. There is a broad and diverse literature on religion and gender emerging since the 1970s from within theology and religious studies, but also from within other disciplines, as the hold of theories about secularisation began to weaken. While this literature has exposed the patriarchy within religions traditions that undermines women's rights, as well as how women and men are seeking to reinterpret their religions to promote greater equality for women in their societies (Tomalin, 2011), engagement with religion in Gender and Development (GaD) approaches which emerged in the 1980s has been hesitant and, where it has taken place, is often superficial and instrumentalist. Bucking this trend, the journal *Gender & Development* has published three special issues about the relationship between gender and religion and the implications of this for development policy and practice. In 1999 the first of these thematic issues focused on 'Religion and Spirituality', a second issue, on the theme 'Working with faith-based communities', was published in 2006 and then in 2017 there was an edition on 'Religious Fundamentalisms'. The editorial in 1999 drew attention to how development studies, policy and practice

> has dismissed religion, its rituals and its customs, as at best irrelevant and at worst a barrier to economic, social, and political 'progress'... This marginalisation of a critical area of human activity has had, and continues to have, a dramatic negative impact on economic, social, and political development, and the attainment of equality for women.
>
> *(Sweetman, 1999: 2)*

The chapters included in this issue aimed to highlight the ways in which religion was tied up with women's development and livelihoods, including Catherine Dolan's chapter, 'Conflict and compliance: Christianity and the occult in horticultural exporting'. She describes the ways in which women turned to 'born again' Christianity, and also traditional witchcraft, in order to deal with conflicts over land, labour and income, created via the introduction of cash crops into Meru District, Kenya, in the early 1990s. While Christianity helped them withstand the pressures they were under, they also used traditional witchcraft practices to inflict harm on their husbands in response to the marginalisation from a livelihood activity that was previously part of their domain before it gained its economic value, and as a way of 'regaining control' (1999: 24). Dolan argues that the struggles over land and labour were played out in the spiritual domain, yet the fact that too little attention has been paid by development practitioners to this aspect of people's lives meant that the impact of the development model introduced in Meru has not been fully understood. Similarly, Rebecca Saul's chapter, 'No time to worship the serpent deities: women, economic change, and religion in north-western Nepal', explores how Buddhist rituals were changing in significance and nature in response to the growing tourist trade. In one settlement where tourism had a limited impact, religious traditions had not been affected. However, in a neighbouring settlement that engaged in tourism, women no longer had the time to carry out their traditional religious rituals. Saul argues that assessments of women's work should also consider their spiritual roles and responsibilities. In common with Dolan, Saul raises questions about the ways in which 'spiritual, reproductive, productive, and community roles support (or weaken) each other', and whether or not ideas of economic development and purchasing power ought to be 'the only yardsticks by which household and community well-being should be measured' (1999: 39).

By 2006, the editorial similarly claimed that 'there is little scholarship or capturing of practice that considers the interaction between development practice and faith-based communities from the feminist perspective' (Greany, 2006: 341). Tomalin, in her chapter 'The Thai bhikkhuni movement and women's empowerment' (2006), writes about how sex workers in Thailand refer to Buddhist teachings about *kamma* in describing the reasons they have adopted this livelihood. The money that they can earn through sex work can enable them to give donations to monasteries and to support their families, therefore providing them with an opportunity to earn religious merit in order to improve their *kamma* for a better rebirth in the next life (i.e. as a man) (2006; see also Muecke, 1992; Peach, 2000). This study draws attention to a movement in Thailand that is campaigning for full ordination of women (the *bhikkhuni* ordination) and argues that without women leaders in the Buddhist tradition, negative stereotypes about women are likely go unchecked. Thus, social values that are underpinned by religious norms also play a key role in women's choice of livelihoods and can put up barriers to widening their opportunities. Therefore, development initiatives that seek to address exploitative gendered employment practices, including sex trafficking and other forms of modern slavery that effect women, could benefit from greater knowledge about how religion shapes social norms and values that limit women's livelihood options as well as finding ways to facilitate leadership opportunities for women within their religious traditions.

By the time of the 2017 thematic special issue, through the work of organisations such as AWID (Association of Women in Development), religion had gained a higher profile, but mainly in relation to its role in fundamentalist movements where 'religious fundamentalisms are widely associated with extreme control over gender relations, enforced by violence – including violence against women and girls, but also against dissenting men, and LGBTI people whose existence threatens patriarchal gender relations' (Sweetman, 2017: 5; Imam et al., 2017). While increasing concerns over the impact of fundamentalisms on women's lives have brought religion more strongly to the fore in GaD research, policy and practice, this has not meant a wider attention to how religious dynamics more broadly shape women's choice of and experience within their livelihoods (Khalaf-Elledge, 2021).

An actor-oriented approach to religions and livelihoods in the Global South

From the late 1980s, the sociologist of development Norman Long began to lay out the 'theoretical foundations for an actor-oriented analysis, giving special attention to questions of lived experience, agency, issues of knowledge and power, and to the need for developing 'theory from below' (2001: 4; see also Olivier de Sardan, 2005). He called for

> a systematic ethnographic understanding of the 'social life' of development projects – from conception to realisation – as well as the responses and lived experiences of the variously located and affected social actors . . . Central elements of this ethnographic endeavour focus on the elucidation of internally-generated strategies and processes of change, the links between the 'small' worlds of local actors and the larger-scale 'global' phenomena and actors, and the critical role played by diverse and often conflicting forms of human action and social consciousness in the making of development'.
>
> *(2001: 13)*

While like other scholars at this time he did not reflect upon the role of religion within this approach, elsewhere I have made the case for an actor-oriented approach to religion and development/livelihoods studies (Tomalin, 2020). This would begin with a focus on individual actors and how they negotiate to fulfil their needs alongside other actors, rather than labelling their individual or collective projects as being about 'religion' or 'development'. Whether something is religious or related to development would then emerge from the meaning that individuals assign to their activities rather than the scholar or policy maker categorising them as such from the outset (Spies and Schrode, 2021; Tomalin, 2020).

While it is tempting to go to religious texts, teachings, institutions and leaders to find out how religion might influence people's lives, this approach is limited in that it is divorced from lived experience. A 'lived religion' approach has emerged in religious studies and the sociology of religion over the past couple of decades, that 'enquires into how religion is encountered and experienced – how it comes into play – in different environments: public and private, official and informal, sacred, secular, and religiously "neutral"' (Knibbe and Kupari, 2020: 157; Hall, 1997; McGuire, 2008; Orsi, 2010; Ammerman, 2014). The benefits of this approach are that it allows us to think about the contextualised and embedded nature of religion and to better understand the subtle and not so subtle ways that religion shapes people's ability to secure livelihoods in an intersectional way rather than reifying religion as a *sui generis* phenomenon. From this perspective, religion comes into being through communities of practice who are also shaped by other intersecting factor such as gender, race, ethnicity and class. To use development terminology, looking at religion from the 'bottom up' or through an 'actor oriented' lens (Long, 2001) has the 'added advantage of circumventing the Protestant bias often embedded in conventional academic definitions of religion (e.g. McGuire, 2008: 20–24)' (Knibbe and Kupari, 2020: 159). This includes aspects such as the strict separation of the religious from the secular and the prioritisation of texts and belief over embodied practice. This is not to say that texts, institutions and leaders are unimportant but that if they are the starting point for enquiry then one is not likely to get very far. Instead, these aspects of religion need to be researched in context alongside alternative and informal leaders (e.g. women) and the ways that inherited faith traditions are deeply imbricated with everyday life and survival strategies, using sociological and ethnographic methods.

Conclusion

In this chapter I have demonstrated that despite innovation within the livelihoods approach towards 'bottom-up' and 'holistic' development, the role of religion has been neglected. This is despite the significant role that religion can play in shaping people's choice of livelihood as well as in impacting upon social norms and values that can restrict the access of certain groups to the necessary assets or resources to effectively develop sustainable livelihood strategies. Where livelihoods approaches have engaged with religion, this has been criticised for instrumentalising religious capital and the role of faith actors to achieve Global North goals.

To remedy this situation, I have argued that an actor-oriented approach to religions and livelihoods in the Global South could generate research on religion and livelihoods that is designed to learn what is meaningful for communities as well as the deeper structures that prevent or promote access to resources/assets in ways that do not presuppose definitions of either development or religion.

Key points

- Although 'religion' is significant to understanding the kinds of livelihoods people are able to pursue, it can only exert an influence in intersection with other factors such as gender, race, ethnicity and class and cannot therefore be researched as an isolated phenomenon.

- Although religious texts, teachings, institutions and leaders can be important sources of information about the likely role that religion plays in shaping people's livelihoods, a 'lived religion' approach is likely to offer more reliable and grounded data.
- In attempts to promote sustainable livelihoods it is important not to decontextualise and instrumentalise religious teachings and traditions that appear to promote resource conservation without the participation of local communities to avoid reinforcing damaging power hierarchies.

Further reading

Bompani, B. (2019) 'Religion and development: Tracing the trajectories of an evolving sub-discipline', *Progress in Development Studies*, 19(3): 1–15. https://doi.org/10.1177/1464993419829598, Special issue of the Journal of Contemporary Religion 2020, 35(2) 'Theorizing Lived Religion'.

Tomalin, E. (2020) 'Global aid and faith actors: The case for an actor-orientated approach to the "turn to religion"', *International Affairs*, 96(2): 323–342. https://doi.org/10.1093/ia/iiaa006

References

Ammerman, N.T. (2014) 'Finding religion in everyday life', *Sociology of Religion*, 75(2): 189–207. https://doi.org/10.1093/socrel/sru013

Atia, M. (2011) '"A way to paradise": Pious Neoliberalism, Islam, and faith-based development', *Annals of the Association of American Geographers*, 102(4): 808–827. https://doi.org/10.1080/00045608.2011.627046

Ayifah, E., Romm, A.T. and Kollamparambil, U. (2021) 'The relationship between religion and livelihood activities of women: Empirical evidence from the Yilo and Lower Manya Krobo districts of Eastern Ghana', *Research in International Business and Finance*, 55: 101311, 1–10. https://doi.org/10.1016/j.ribaf.2020.101311

Bebbington, A. (1999) 'Capitals and capabilities: A framework for analyzing peasant viability, rural livelihoods and poverty', *World Development*, 27(12): 2021–2044.

Bompani, B. (2019) 'Religion and development: Tracing the trajectories of an evolving sub-discipline', *Progress in Development Studies*, 19(3): 1–15. https://doi.org/10.1177/1464993419829598

Bourdieu, P. (1986) 'The forms of capital', in Richardson, J.G. (ed.), *Handbook of theory and research for the sociology of education*, New York: Greenwood Press, pp. 241–258.

Chacko, P. (2019) 'Marketizing Hindutva: The state, society, and markets in Hindu nationalism', *Modern Asian Studies*, 53(2): 377–410. https://doi.org/10.1017/S0026749X17000051

Chambers, R. (1983) *Rural development: Putting the last first*, Essex, England: Longmans Scientific and Technical Publishers; New York: John Wiley.

Clarke, G. (2007) 'Agents of transformation? Donors, faith-based organizations and international development', *Third World Quarterly*, 28(1): 77–96. https://doi.org/10.1080/01436590601081880

Darlington, S.M. (2019) 'Buddhist integration of forest and farm in Northern Thailand', *Religions*, 10(9): 521. https://doi.org/10.3390/rel10090521

de Haan, L. (2017) *Livelihoods and development: New perspectives*, Leiden: Brill.

Deneulin, S. with Bano, M. (2009) *Religion in development; rewriting the secular script*, London and New York: Zed.

Dolan, C.S. (1999) 'Conflict and compliance: Christianity and the occult in horticultural exporting', *Gender & Development*, 7(1): 23–30. https://doi.org/10.1080/741922937

Eisenstadt, S.N. (2000) 'The reconstruction of religious Arenas in the framework of "multiple modernities"', *Millennium*, 29(3): 591–611. https://doi.org/10.1177/03058298000290031201

Greany, K. (2006) 'Editorial', *Gender and Development*, 14(3): 341–350. https://doi.org/10.1080/13552070600979593

Hall, D.D. (ed.) (1997) *Lived religion in America: Toward a history of practice*, Princeton, NJ: Princeton University Press.

Imam, A., Gokal, S. and Marler, I. (2017) 'The devil is in the details: A feminist perspective on development, women's rights, and fundamentalisms', *Gender and Development*, 25(1): 15–36. https://doi.org/10.1080/13552074.2017.1286803

Istratii, R. (2020) *Adapting gender and development to local religious contexts: A decolonial approach to domestic violence in Ethiopia*, London and New York: Routledge.

Jones, B. and Petersen, M.J. (2011) 'Instrumental, narrow, normative? Reviewing recent work on religion and development', *Third World Quarterly*, 32(7): 1291–306. https://doi.org/10.1080/01436597.2011.596747

Karam, A. (ed.) (2011) *Religion, development and the United Nations*, New York: Social Science Research Council.

Khalaf-Elledge, N. (2021) *Religion – Gender nexus in development policy and practice considerations*, London and New York: Routledge.

Knibbe, K. and Kupari, H. (2020) 'Theorizing lived religion: Introduction', *Journal of Contemporary Religion*, 35(2): 157–176. https://doi.org/10.1080/13537903.2020.1759897

Knoop, S.B., Morcatty, T.Q., El Bizri, H.R. and Cheyne, S.M. (2020) 'Age, religion, and taboos influence subsistence hunting by indigenous people of the lower Madeira River, Brazilian Amazon', *Journal of Ethnobiology*, 40(2): 131–148. https://doi.org/10.2993/0278-0771-40.2.131

Lanz, S. and Oosterbaan, M. (2016) 'Entrepreneurial religion in the age of neoliberal urbanism', *International Journal of Urban and Regional Research*, 40(3): 487–506. https://doi.org/10.1111/1468-2427.12365

Lewis, D. and Mosse, D. (eds.) (2006) *Development brokers and translators: The ethnography of aid and agencies*, Bloomfield, CT: Kumarian.

Long, N. (2001) *Development sociology: Actor perspectives*, London: Routledge.

McGuire, M.B. (2008) *Lived religion: Faith and practice in everyday life*, Oxford: Oxford University Press.

Muecke, M.A. (1992) 'Mother sold food, daughter sells her body: The cultural continuity of prostitution', *Social Science & Medicine*, 35(7): 891–901. https://doi.org/10.1016/0277-9536(92)90103-W

Nussbaum, M. (2013) *Creating capabilities: The human development approach*, Cambridge, Massachusetts: Harvard University Press.

Olivier de Sardan, J-P. (2005) *Anthropology and development: Understanding contemporary social change*, London and New York: Zed.

Orsi, R.A. (2010) *The Madonna of 115th street: Faith and community in Italian Harlem, 1880–1950*, 3rd ed., New Haven, CT: Yale University Press.

Peach, L. (2000) 'Human rights, religion, and (sexual) slavery', *The Annual of the Society of Christian Ethics*, 20: 65–87. https://doi.org/10.5840/asce2000205

Peres, C.A. and Nascimento, H.S. (2006) 'Impact of game hunting by the Kayapó of South Eastern Amazonia: Implications for wildlife conservation in tropical forest indigenous reserves', *Biodiversity and Conservation*, 15: 2627–2653. https://doi.org/10.1007/978-1-4020-5283-5_16

Rajkobal, P. (2019) *The sarvodaya movement holistic development and risk governance in Sri Lanka*, London and New York: Routledge.

Rakodi, C. (2015) 'Development, religion and modernity', in Tomalin, E. (ed.), *The Routledge handbook of religions and global development*, London and New York: Routledge, pp. 17–35.

Rosati, M. and Stoeckl, K. (eds.) (2012) *Multiple modernities and postsecular societies*, 1st ed., London and New York: Routledge.

Rudnyckyj, D. (2015) 'Religion and economic development', in Tomalin, E. (ed.), *The Routledge handbook on religions and global development*, London: Routledge, pp. 405–417.

Saul, R. (1999) 'No time to worship the serpent deities: Women, economic change, and religion in north-western Nepal', *Gender and Development*, 7(1): 31–39. https://doi.org/10.1080/741922930

Scoones, I. (2009) 'Livelihoods perspectives and rural development', *The Journal of Peasant Studies*, 36(1): 171–196. https://doi.org/10.1080/03066150902820503

Sen, A. (1993) 'Capability and well-being', in Nussbaum, M. and Sen, A. (eds.), *The quality of life*, Oxford: Oxford University Press, pp. 30–53.

Spies, E. and Schrode, P. (2021) 'Religious engineering: Exploring projects of transformation from a relational perspective', *Religion*, 51(1): 1–18. https://doi.org/10.1080/0048721X.2020.1792053

Swart, I. and Nell, E. (2016) 'Religion and development: The rise of a bibliography', *HTS Teologiese Studies/Theological Studies*, 72(4): 1–27. https://doi.org/10.4102/hts.v72i4.3862

Sweetman, C. (1999) 'Editorial', *Gender and Development*, 7(1): 2–6. https://doi.org/10.1080/741922933

Sweetman, C. (2017) 'Introduction: Gender, development and fundamentalisms', *Gender and Development*, 25(1): 1–14. https://doi.org/10.1080/13552074.2017.1304063

Tomalin, E. (2011) *Working in gender and development: Faith and development*, Rugby: Oxfam, Practical Action Publishing.

Tomalin, E. (2013) *Religions and development*, London and New York: Routledge.

Tomalin, E. (2020) 'Global aid and faith actors: The case for an actor-orientated approach to the "turn to religion"', *International Affairs*, 96(2): 323–342. https://doi.org/10.1093/ia/iiaa006

Tomalin, E. (2006) 'The Thai *bhikkhuni* movement and women's empowerment', *Gender and Development*, 14(3): 385–397. https://doi.org/10.1080/13552070600980492

van Vliet, N. (2018) '"Bushmeat crisis" and "cultural imperialism" in wildlife management? Taking value orientations into account for a more sustainable and culturally acceptable wildmeat sector', *Frontiers in Ecology and Evolution*, 6: 112. https://doi.org/10.3389/fevo.2018.00112

Ver Beek, K.A. (2000) 'Spirituality: A development taboo', *Development in Practice*, 10(1): 31–43. https://doi.org/10.1080/09614520052484

43
CLIMATE CHANGE ADAPTATION AND AGRICULTURAL LIVELIHOODS OF SMALLHOLDER FARMERS

Issaka Kanton Osumanu

Introduction

Livelihood sustainability and agricultural adaptation to the effects of climate change in the Global South has progressively become a research imperative (Alhassan et al., 2019; IPCC, 2018; Yaro et al., 2016). Many farmers in the Global South produce food using agricultural systems they have developed through many years of farming under difficult climatic and other adverse environmental conditions (Apuri et al., 2018). However, as the negative impacts of climate change on food production continue to rise, the danger of food supply deficit and food insecurity increase (IPCC, 2018). The absence of governmental support, through effective climate change adaptation policies, that would help smallholder farmers to tackle the future effects of climate change, adds to their vulnerability.

This chapter discusses agricultural livelihoods and climate change adaptation by smallholder farmers. The chapter not only contributes to the sustainable livelihoods discourse by highlighting the effects of climate change on farmers' livelihoods, but it also underscores the adaptive capacity of smallholder farmers for climate change adaptation. Following this introduction, smallholder farming livelihoods are examined focusing on their diversity. This is followed by a discussion of climate change impacts on smallholder agricultural livelihoods and adaptations to climate change, typologies and strategies. The next section discusses adaptive capacity by drawing out differences between smallholder farmers. The chapter then concludes with a list of key points emanating from the discussion.

Smallholder farming livelihoods

Generally, smallholder farmers, who constituted 65% of the entire population in the Global South in 2015 (FAO, 2015), are poor and live below the international poverty line of US $1.90 per day (Tal, 2018). Food crop or mixed crop-livestock production is the primary source of direct employment and income for these famers (IFAD, 2007). Smallholder farmers are identified by the number of hectares of land they cultivate, which is usually 2 ha or less (Calcaterra, 2013). Other classifications include population density as a variable of farm size (World Bank,

2016), in which farmers in areas with high population densities usually cultivate less than 1 ha, increasing up to 10 ha in more sparsely populated arid areas (Rapsomanikis, 2015). In parts of the Global South, particularly in sub-Saharan Africa (SSA), smallholder farmers generally employ traditional methods of farming, which involve the use of rudimentary and obsolete equipment, high seasonal worker oscillations, each family member doing different farm work, and low returns (Tahiru et al., 2019). The World Bank (2016) outlined a wider perspective of 'resource-poor' farmers such as those with low capital (including livestock), fragmented farm holdings and inadequate access to inputs. However, in their research in the Sahel Savannah, Tahiru et al. (2019) found that smallholder farmers vary in individual attributes, farm sizes, resource allocation between different types of crops, livestock ownership, off-farm income earnings, use of non-family inputs and hired labour, the share of food crops traded and household expenditure outlines.

The differential roles in farming by gender is an important dimension that governs smallholder farming households and explains the ways that smallholder households relate to farming and to each other (Rapsomanikis, 2015). At the household level, women, men and children are responsible for different specialised activities in farming. Men often hold tenure to land and have a more proprietary role, while women may perform most of the agricultural tasks (e.g. planting, hoeing/weeding, harvesting, transporting, storing and processing) and have a more productive role. Girls often assist their mothers in farming and tending livestock, and processing or storing food. Boys may also assist in the fields and in tending livestock. There is often a mutual arrangement with men to exchange labour with women, so that women are paid in kind if not in cash (Fan and Rue, 2020; Kydd, 2002; Rapsomanikis, 2015; Tahiru et al., 2019). There are also cultures and circumstances, such as in more egalitarian societies, where men share part of the household tasks, cooking, washing and fetching water and firewood.

Smallholder farming plays a critical role by providing livelihoods for the rural poor who practice small-scale food crop production, depend heavily on family labour and sell part of the farm produce for income (World Bank, 2016). Smallholder farmers are broadly regarded to be the largest producers of food in the Global South with some estimates stating that over 90% of the food produced in this area is derived from smallholders (Rapsomanikis, 2015). Yet, smallholder farming is considered to be the most vulnerable component of the rural agricultural sector because of its dependence on rainfall, cultivation in marginal areas and lack of access to technical or financial support (Alhassan et al., 2019; IPCC, 2014). As a result, smallholder rural households are amongst those most affected by climate change.

Climate change impacts on smallholder agricultural livelihoods

Available evidence has proven that the global climate is changing with noticeable consequences on life-supporting systems including food production (IPCC, 2001). Considerable changes have occurred in precipitation, temperature, sea level and frequency of flood and drought events in both the tropical and subtropical regions (Badege et al., 2013). Climate change has resulted in increased average temperature of around 1 °C and rainfall has generally decreased by 2.5% since 1960, whilst also showing increased inconsistency (IPCC, 2014, 2018). Seasonal variability in rainfall is also apparent and manifest in the Savannah agro-ecological zones.

Climate change impacts on food crop production are varied, with studies reporting both adverse and positive effects. However, a synthesis of outcomes over the past 50 years by IPCC (2018) suggests more negative effects than positive impacts globally. In the tropical sub-region, average temperature increase with a concomitant reduction in mean annual precipitation and increased variability in all the agro-ecological regions over the past four decades, have culminated in a reduction in agricultural production (Rapsomanikis, 2015). There is ample demonstration of farmers in this region being susceptible to crop failure, reduction in crop yields, loss of livestock due to shortage of water and lack of insurance schemes to mitigate the effects (Assan et al., 2009). Susceptibilities to climate change, food production systems and farmers' livelihoods are projected to be severely impacted by climate unpredictability and variation (World Bank, 2013).

Although climate change is a worldwide occurrence, its adverse impacts are reported to be more severe on poor food crop producers in the Global South who rely heavily on climate-sensitive agricultural activities for their livelihoods (Gerber et al., 2013). Yaro et al. (2014) and Alhassan et al. (2019) underscored climate change as a key threat to livelihoods particularly in vulnerable settings like SSA, where there is high dependence on rain-fed agriculture amidst fragile ecosystems. However, the effects of climate related shocks on livelihoods differ based on geographical, economic and social conditions as well as other underlying vulnerabilities. Arid and semi-arid agro-ecological regions, which rely greatly on rain-fed and subsistence farming, have high climate-sensitive farming systems and are more vulnerable to climate change. Farmers in these regions struggle with poverty, food insecurity, poor access to adequate healthcare and other basic human needs (Yaro et al., 2014). There is a generally low economic growth of such regions, which places smallholder farmers at the extreme end of the climate change susceptibility spectrum (Alhassan et al., 2019). In contrast, forest ecosystems are less sensitive to decreases in precipitation than fragile, arid or semi-arid ecosystems due to continual effects of climate change on water availability (Antwi-Agyei et al., 2017).

Disparities in climate change impacts on food production across different agro-ecological zones are due to variations in the levels of exposure, sensitivity and capacity to adapt. Better climatic and other environmental conditions, for instance, make the impacts relatively less severe in the forest zones than the other agro-ecological regions (Green et al., 2010). In the particular case of smallholder farmers, it has been observed that their success or failure is tied to rainfall and temperature and, when there is variability, their livelihood is impeded (Menghistu et al., 2020). It is predicted that crop yield in the Global South is likely to reduce by 10–20% by 2030 or even up to 50% by 2050 as a result of climate change (IPCC, 2014). This is particularly so because agriculture in many parts of the Global South is largely rain-fed (World Bank, 2013) and, therefore, essentially reliant on the whims of the weather. Forecasts of climate change impacts signpost progressively severe adverse effects including prolonged and more severe droughts (Boko et al., 2007; Christensen et al., 2007), rising average yearly temperatures, high evapo-transpiration and decrease in soil moisture content (Boko et al., 2007). According to Sissoko et al. (2011), these distressing projections are expected to reduce net crop returns by 90% by 2100 with rippling consequences that could further worsen current poverty levels and reduce economic opportunities.

Several impacts of climate change on food crop production, which pose serious challenges to farmers' livelihoods in particular and poverty reduction efforts in general, have been reported. Mawere et al. (2013) reported that climate change has caused a reduction in crop yields and agricultural productivity in all its forms including reduction in labour for food crop farming due to health-related risks associated with climate change. For instance, increased temperatures have increased the levels of water-related diseases, including malaria and cholera in the tropics and sub-tropics, and reduced farm labour productivity. Moreover, changing climatic conditions, particularly increasing temperatures, can bring about ozone layer depletion which has the potential to affect food crop production (IPCC, 2014). A shift in the length of the cropping season has been reported in West Africa with a decrease of about 20 days which, according to Abubakari and Abubakari (2015), has led to a decrease in cereal crop yields in arid and semi-arid zones due to a reduction in the growing period and increase in evaporation rates. Changes in precipitation are reported to manifest in a shorter rainy season with severe consequences on crop yields (IPCC, 2018). In other areas, excessive rain leading to prolonged intense erosion carries away the topsoil thereby making farmlands uncultivable by farmers (Osumanu et al., 2016).

Adaptations to climate change: typologies and strategies

Global attempts at finding answers to climate change have recognised the role of both mitigation and adaptation (Badege, 2013; IPCC, 2014). The main aim of adaptation is to abate the negative impacts of unavoidable climate change through activities targeted at vulnerable farming systems and these may include both on-farm and off-farm activities. One of the fundamental ways of reducing vulnerability to climate change is increasing the resilience of farming systems to climatic shocks (Rapsomanikis, 2015). Adaptation could also involve taking measures to seize new opportunities offered by climate change (Acquah and Onumah, 2011), such as building a climate-resilient farming system and greenhouse gas mitigation in agriculture. Adaptation of farmers can be planned or spontaneous (autonomous); public or private; and reactive or proactive (anticipatory) (World Economic Forum, 2019). Whereas proactive adaptation is when a system is amended long before experiencing the impacts of climate change, reactive adaptation is when a system is amended at the onset of experiencing the effects of climate change. Spontaneous adaptation refers to the non-planned efforts of those affected, in comparison to planned adaptation which is the outcome of a thoughtful policy resolution based on changed or anticipated change in conditions which need to be restored to a desirable state (IPCC, 2014). The nature of climate change impact directly and indirectly determines the type of response of stakeholders, especially smallholder farmers who rely on rain-fed farming for their livelihoods.

To deal with the undesirable impacts of global climate change, smallholder farmers have adjusted themselves by developing and adapting livelihoods based on their traditional farming practices (Lennard et al., 2018; Sylla et al., 2016). Apart from farmers obtaining information about climate change from diverse sources, including traditional and new media, their own experiences of the phenomena also play a major role in enhancing their knowledge of the situation (Osumanu et al., 2017). Farmers' decisions to alter their farming practices are not only based on changes in average climatic conditions, but also on several other environmental influences, such as rainfall incidence, timing and intensity, and early or late coming of the dry season, observed using personal experience (Yaro et al., 2016).

Smallholder food crop farmers in the Global South employ a diversity of climate change adaptation strategies, including both modern and traditional methods, to lessen their livelihood vulnerabilities to climate change effects. These coping or reactive mechanisms are the practical responses to climate change in the event of unwelcome situations, and may be considered as short term responses. In other words, they are autonomous or spontaneous adaptations which, according to Osumanu et al. (2017), are regarded to be those that happen invariably in reactive response to climatic stimuli after early impacts are felt by farmers. Widely known climate change adaptation measures among farmers that have been reported are fertiliser use, farming on fallowed land, small-scale irrigation, cultivation of early maturing crops, mixed cropping, crop rotation and changing planting dates (Osumanu et al., 2017). Menghistu et al. (2020) have also reported agriculture adaptation strategies implemented by smallholder farmers to include crop diversification, adopting drought-tolerant crops, tree planting or agroforestry, mulching and construction of fire belts. However, while these measures are well-known to smallholder farmers across agro-ecological zones, mulching is not popular in the arid and semi-arid zones because the amount of precipitation in such areas is quite erratic and does not give farmers enough time to employ soil moisture retention.

Practicing climate change adaptation measures includes employing effective methods of cropping to reduce negative impacts, meaning crop yields are sustained (Begum et al., 2016). Across the Global South, the average crop yield of farmers who practice climate change adaptation interventions tends to be higher (Yaro et al., 2016). Begum et al. (2016) reported that the average food sufficiency months for farmers in arid and semi-arid regions of SSA who practice climate change adaptation are higher and hence they are more food secure. Indicators of improved livelihoods reported also include enriched nutrition due to the availability of different varieties of food (Avea et al., 2016). Different crops react differently to climate change and farmers that cultivate many crops are likely to be less impacted negatively during periods of unfavorable climatic conditions than farmers who cultivate fewer crops (Avea et al., 2016). Thus, agricultural diversification increases smallholder farmers' resilience to climatic stimuli. Osumanu et al. (2016) have indicated that farmers who do not practice adaptation measures are more likely to lose their farmlands due to continual degradation from soil erosion.

Adaptive capacities

Climate change adaptation is a developmental issue. It involves all the aspects that define the susceptibility of farming systems as determined by the climate and other social, economic and political circumstances (Deressa et al., 2009). The difficulty in dealing with climate change is developing the adaptive capacities of those who are most vulnerable, to empower them to reverse its impactsor to adapt to a new state and take advantage of new opportunities. Smallholder farmers, in particular, have been shown to have limited adaptive capacity to mitigate the adverse effects of climate change (Antwi-Agyei et al., 2017), particularly the ability to develop and implement integrated approaches that can enhance their resilience.

Deressa et al. (2009) argued that the choice of particular adaptation techniques depends on several issues which may be environmental, social, economic and institutional factors. Different studies have put forward a wide range of factors found to be influencing the practice of strategies by smallholder farmers to abate the adverse effects of climate change. Factors such as age, gender, level of formal education, the size of the farm, access to agricultural extension services, income, availability of credit and information, and engagement in non-farm economic activities have been reported to influence farmers' adaptive capacities (Bryan et al., 2013; Deressa et al., 2009; Osumanu et al., 2017). Formal education, training, access to extension services, availability of information, farm size and farming experience are very important such that the more these variables increase, the more likely farmers will adapt to climate change (Bryan et al., 2013; Menghistu et al., 2020). These factors determine adaptation strategies such as diversification of livelihoods, appropriate agronomic practices, employing indigenous knowledge, planting drought-tolerant varieties, planting various crops at different times of the year and agro-forestry (Osumanu et al., 2017). In addition, such farmers successfully adapt to climate change because they recognise the need to adapt, have the requisite knowledge of existing opportunities and the relevant capacities and skills to evaluate and execute appropriate opportunities. According to Menghistu et al. (2020), farmers who have adequate access to information on climate change (especially temperature and rainfall), through agricultural extension service providers, are significantly more likely to adapt to climate change.

Income and availability of credit enhances farmers' adaptive capacity by helping them to address challenges such as high cost of farmland, inherited and communal systems of land ownership, and non-availability and high cost of farm labour as well as other inputs (e.g. fertilisers) (Deressa et al., 2009). Gender and age determine the choice of a particular climate adaptation strategy by a farmer (planned or autonomous). In particular, women and young farmers are more inclined towards autonomous adaptation measures (Bryan et al., 2013) due to their lack of access to financial resources, climate and production information and institutional support. In the Savannah region of West Africa, Yaro et al. (2016) found that gender influences the ability to adopt particular fertiliser applications and to shift to non-farm measures. According to their study, men farmers in that region have a higher capacity to adopt to fertiliser application than women due to men's control over production factors, but women have a higher capacity to move to non-farm jobs than men.

Smallholder food crop producers have inherent capacities to adapt to climate change impacts. These capacities are bound up in their ability to act collectively. Decisions on climate change adaptation are made by individual farmers, groups, organisations and governments (Jones and Boyd, 2011). Jones and Boyd (2011) stressed that the ability of individual smallholder farmers to practice climate change adaptation is dependent on the availability of productive assets, their capacity to take collective action (considered as a social capital) amidst the risks posed by climate change, and the effectiveness, efficiency and legitimacy of institutions for resource management. Formal and informal institutions in agricultural-dependent communities, including farmer-based organisations (Wang et al., 2013), shape the livelihood outcomes of climate threats through a variety of essential roles they play including collection and diffusion of information, deployment of productive resources, skills training and capacity building, and provision of leadership and networking with external actors. Generally, communities with strong local institutions and connections with influential governments have better levels of climate change adaptive capacity (Yaro et al., 2016).

Farmers' adaptive capacity at any given point in time determines whether they will act. Farmers' adaptive capacity is also a function of the appreciation for the need to practice particular adaptation measures, trust that adaptation is possible and necessary, willingness to bear the costs of adaptation, availability of resources needed to implement adaptation, ability to use resources in an appropriate manner and the external threats on or impediments to the implementation of adaptation (IPCC, 2018). Available evidence suggests that smallholder farmers' adaptive capacity to climate change is also dependent on national economic development. While farmers in relatively developed countries have sufficient knowledge of climate change, are less vulnerable, and have high adaptive capacity (Codjoe et al., 2013), in many of the less economically endowed countries such as those in SSA, although farmers' knowledge of the causes and impacts of climate change are high (Osumanu et al., 2017), they are highly susceptible to climate change impacts due to the lack of accurate predictions of climatic events, especially reliable information on rainfall and temperature (Asante and Amuakwa-Mensah, 2015; Codjoe et al., 2013). As a result, farmers in SSA often base their adaptation strategies on the conventional approach of 'trial and error' (Apuri et al., 2018). With reference to national economic development, it has been established that high levels of poverty and lack of access to financial support or credit limit farmers' ability and capacity to implement climate change adaptation strategies (IPCC, 2014). This is exacerbated by the inability of mandated state and local institutions to assist smallholder farmers with building capacity for climate change adaptation. Issues such as farmers' lack of skills, capacities, and financial resources have also been found to thwart institutional efforts to assist farmers with appropriately responding to the undesirable impacts of climate change (Wang et al., 2013). These challenges aggravate preexisting social, economic, cultural and political inequalities and vulnerabilities to climate change impacts on agricultural livelihoods.

Conclusion

Climate change poses several threats to the livelihoods of smallholder farmers in the Global South who have the highest poverty levels in the world. In these areas, increasing temperatures, variable rainfall, droughts and floods have resulted in decreased food crop production with serious consequences, including food and livelihood insecurity. Climate change will remain a threat to food crop production, farmers' income, mean net crop revenues and resilience of farmers, but effective adaptation measures can offset such threats. Within these constraints, it is important to enhance farmers' adaptive capacity to enable them to reduce the adverse impacts of climate change. Climate change adaptive measures are already known, therefore fostering adaptive capacity will require building farmers' adaptive capacity to employ these measures, and supporting smallholder farmers' involvement in strengthening existing adaptation strategies using both traditional and modern knowledge.

Key points

- Climate change poses serious challenges to food crop production, livelihoods and poverty reduction efforts.
- Smallholder farmers' livelihood susceptibility to climate change impacts is attributable to environmental, social, economic and political conditions.
- In many regions, gender inequality means that women are more vulnerable to the negative effects of climate change. This is due to gender roles which mean that men have more control over factors of production.
- Livelihoods can be improved by increasing farmers' participation in climate change adaptation interventions which boost their adaptive capacity and overall resilience. The average crop yield of farmers who practice climate change adaptation interventions tends to be higher.
- Farmers' adaptive capacity can be improved if existing adaptation strategies being used by farmers are supported by national policies.

Further reading

Angelsen, A., Jagger, J., Babigumira, R., Belcher, B., Hogarth, N.J., Bauch, S., Börner, J., Smith-Hall, C. and Wunder, S. (2014) 'Environmental income and rural livelihoods: A global-comparative analysis', *World Development*, 64(1): 512–528. https://doi.org/10.1016/jworddev.2014.03.006

Food and Agriculture Organization (FAO) (2016) *Increasing the resilience of agricultural livelihoods*, Rome: FAO.

Steel, G. and van Lindert, P. (2017) *Rural livelihood transformations and local development in Cameroon, Ghana and Tanzania*, London: IIED.

Zimmerer, K.S. (2007) 'Agriculture, livelihoods, and globalization: The analysis of new trajectories (and avoidance of just-so stories) of human-environment change and conservation', *Agriculture and Human Values*, 24: 9–16. https://doi.org/10.1007/s10460–006–9028-y

References

Abubakari, F. and Abubakari, F. (2015) 'Effects of soil conservation on the yield of crops among farmers in upper east region of Ghana', *Academic Research Journal of Agricultural Science and Research*, 3(5): 86–91. https://doi.org/10.14662/ARJASR2015.009

Acquah, F.D. and Onumah, E.E. (2011) 'Farmers perception and adaptation to climate change: An estimation of willingness to pay', *Agris On-Line Papers in Economics and Informatics*, 3(4): 31–39.

Alhassan, H., Kuwornu, J.K.M. and Osei-Asare, Y.B. (2019) 'Gender Dimensions of vulnerability to climate change and variability: Empirical evidence of smallholder farming households in Ghana: A participatory approach', *International Journal of Climate Change Strategies and Management*, 11(2): 195–214. https://doi.org/10.1108/IJCCSM-10-2016-0156

Antwi-Agyei, P., Quinn, C.H., Adiku, S.G.K., Cogjoe, S.N.A., Dougill, A.J., Lamboll, R. and Dovie, D.B.K. (2017) 'Perceived stressors of climate vulnerability across scales in the savannah zone of Ghana: A participatory approach', *Regional Environmental Change*, 7: 213–227. https://doi.org/10.1007/s10113-016-0993-4

Apuri, I., Peprah, K. and Achana, G.T. (2018) 'Climate change adaptation through agroforestry: The Case of Kassena Nankana west district, Ghana', *Environmental Development*, 28: 32–41. https://doi.org/10.1016/j.envdev.2018.09.002

Asante, F.A. and Amuakwa-Mensah, F. (2015) 'Climate change and variability in Ghana: Stocktaking', *Climate*, 3: 78–99. https://doi.org/10.3390/cli3010078

Assan, J.K., Caminade, C. and Obeng, F. (2009) 'Environmental variability and vulnerable livelihoods: Minimising risks and optimising opportunities for poverty alleviation', *Journal of International Development*, 21: 403–418. https://doi.org/10.1002/jid.1563

Avea, A.D., Zhu, J., Tian, X., Baležentis, T., Li, T., Rickaille, M. and Funsani, W. (2016) 'Do NGOs and development agencies contribute to sustainability of smallholder soybean farmers in northern Ghana: A stochastic production frontier approach', *Sustainability*, 8(465): 2–17. https://doi.org/10.3390/su8050465

Badege, B., Neufeldt, H., Mowo, J., Abdelkadir, A.M., Dalle, G. and Assefa, T. (2013) *Farmers' strategies for adapting to and mitigating climate variability and change through agroforestry in Ethiopia and Kenya*, Corvallis, Oregon: Oregon State University.

Begum, A., Roy, D.C., Abedin, J. and Lubaba, N.A. (2016) *Tracing back 2016*, Dhanmondi, Dhaka: Practical Action Bangladesh.

Boko, M., Niang, I., Nyong, A., Vodel, C., Githeko, A., Medany, M., Osman-Elasha, B., Tabo, B. and Yanda, P. (2007) 'Africa: Climate change 2007: Impacts, adaptation and vulnerability', in Parry, M., Canziani, O., Palutikof, J., van der Linden, P. and Hanson, C. (eds.), *Contribution of working group II to the fourth assessment report of the intergovernmental panel on climate change*, Cambridge: Cambridge University Press.

Bryan, E., Ringler, C., Okoba, B., Roncoli, C., Silvestri, S. and Herrero, M. (2013) 'Adapting agriculture to climate change in Kenya: Household strategies and determinants', *Journal of Environmental Management*, 114: 26–35. https://doi.org/10.11016/j/jenvman.2012.10.036

Calcaterra, E. (2013) *Defining smallholders: Suggestion for a RSB smallholder definition*, Lausanne, Switzerland: Aidenvironment and Ecole Polytechnique Federale de Lausanne.

Christensen, J.H., Hewitson, B., Busuioc, A., Chen, A., Gao, X., Held, R., Jones, R., Kolli, R.K., Kwon, W. and Laprise, R. (2007) *Regional climate projections: Climate change (2007): The physical science basis. Contribution of working group I to the fourth assessment report of the IPCC*, Cambridge: University Press.

Codjoe, F.N.Y., Ocansey, C.K., Boateng, D.O. and Ofori, J. (2013) 'Climate change awareness and coping strategies of cocoa farmers in rural Ghana', *Journal of Biology, Agriculture and Healthcare*, 3(11): 19–29.

Deressa, T.T., Hassan, R.M., Ringer, C., Alemu, T. and Yesuf, M. (2009) 'Determinants of farmers' choice of adaptation methods to climate change in the Nile basin of Ethiopia', *Global Environmental Change*, 19(2): 248–255. https://doi.org/10.1016/j.gloenvcha.2009.01.002

Fan, S. and Rue, C. (2020) 'The role of smallholder farms in a changing world', in Gomez, Y., Paloma, S., Riesgo, L. and Louhichi, K. (eds.), *The role of smallholder farms in food and nutrition security*, Geneva, Switzerland: Springer, Cham.

Food and Agriculture Organization (FAO) (2015) *A data portrait of smallholder farmers*, Rome: FAO.

Gerber, P., Steinfeld, H., Henderson, B., Mottet, A., Opio, C., Dijkman, J., Falcucci, A. and Tempio, G. (2013) *Tackling climate change through livestock: A global assessment of emissions and mitigation opportunities*, Rome: FAO.

Green, D., Alexander, L., Mclnnes, K., Church, J., Nicholls, N. and White, N. (2010) 'An assessment of climate change impacts and adaptation for the Torres Strait Islands, Australia', *Climate Change*, 102: 405–433. https://doi.org/10.1007/s10584-009-9756-2

Intergovernmental Panel on Climate Change (IPCC) (2001) *Climate change 2001: The scientific basis*, Cambridge: IPCC.

International Fund for Agricultural Development (IFAD) (2007) *Ghana: Upper east region land conservation and smallholder rehabilitation project*, Rome: IFAD.

IPCC (2014) *Climate change 2014: Impacts, adaptation and vulnerability. Summary for policy makers*, Cambridge: Cambridge University Press.

IPCC (2018) *Global warming of 1.5°C: An IPCC special report on the impacts of global warming of 1.5°C above pre-industrial levels and related global greenhouse gas emission pathways, in the context of strengthening the global response to the threat of climate change, sustainable development, and efforts to eradicate poverty – summary for policymakers*, Geneva, Switzerland: IPCC.

Jones, L. and Boyd, E. (2011) 'Exploring social barriers to adaptation: Insights from western Nepal', *Global Environmental Change*, 21(4): 1262–1274. https://doi.org/10.1016/j.gloenvcha.2011.06.002

Kydd, J. (2002) *Agriculture and rural livelihoods: Is globalization opening or blocking paths out of rural poverty?* London: DFID.

Lennard, C.J., Nikulin, G., Dosio, A. and Moufouma-Okia, W. (2018) 'On the need for regional climate information over Africa under varying levels of global warming', *Environmental Resource Letters*, 13(6): 060401. https://doi.org/10.1088/1748-9326/aab2b4

Mawere, M., Madziwa, B.F. and Mabeza, C.M. (2013) 'Climate change and adaptation in third world Africa: A quest for increase food security in semi-Arid Zimbabwe', *The International Journal of Humanities and Social Studies*, 1(2): 14–22.

Menghistu, T.H., Abraha, A.Z., Tesfay, T. and Mawcha, G.T. (2020) 'Determinant factors of climate change adaptation by pastoral/Agro-pastoral communities and smallholder farmers in sub-Saharan Africa: A systematic review', *International Journal of Climate Change Strategies and Management*, 12(3): 305–321. https://doi.org/10.1108/IJCCSM-07-2019-0049

Osumanu, I.K., Aniah, P. and Yelfaanibe, A. (2017) 'Determinants of adaptive capacity to climate change among smallholder rural households in the Bongo District, Ghana', *Ghana Journal of Development Studies*, 14(2): 142–163. https://doi.org/10.4314/gjds.v1412.8

Osumanu, I.K., Kosoe, E.A. and Nabiebakye, H.N. (2016) 'Land degradation management in the Lawra district of Ghana: Present practices and opportunities for rural farmers in semi-Arid areas', Journal of Natural Resources and Development, 6: 72–80. https://doi.org/10.5027/jnrd.06.v6i0.08

Rapsomanikis, G. (2015) *The economic lives of smallholder farmers: An analysis based on household data from nine countries*, Rome: FAO.

Sissoko, K., van Keulen, H., Verhagen, J., Tekken, V. and Battaglini, A. (2011) 'Agriculture, livelihoods and climate change in the west African Sahel', *Regional Environmental Change*, 11(1): 119–125. https://doi.org/10.1007/s10113-010-0164-y.

Sylla, M.B., Elguindi, N., Giorgi, F. and Wisser, D. (2016) 'Projected robust shift of climate zones over west Africa in response to anthropogenic climate change for the late 21st century', *Climate Change*, 134: 241–253. https://doi.org/10.1007/s10584-015-1522-z

Tahiru, A., Sackey, B., Owusu, G. and Bawakyillenuo, S. (2019) 'Building the adaptive capacity for livelihood improvements of Sahel Savannah farmers through NGO-led adaptation interventions', *Climate Risk Management*, 26: 100197. https://doi.org/10.1016/j.crm.2019.100197

Tal, A. (2018) 'Making conventional agriculture environmentally friendly: Moving beyond the glorification of organic agriculture and the demonization of conventional agriculture', *Sustainability*, 10(1078): 1–17. https://doi.org/10.3390/su10041078

Wang, X.J., Zhang, S., Shahid, E., Amgad, R., He, Z.B. and Mahtab, A. (2013) 'Water resources management strategy for adaptation to droughts in China', *Mitigation and Adaptation Strategies for Global Change*, 17: 923–937. https://doi.org/10.1007/s11027-011-9352-4

World Bank (2013) *Turn down the heat: Climate extremes, regional impacts, and the case for resilience*, Washington, DC: The World Bank.

World Bank (2016) *A year in the lives of smallholder farmers*, Washington, DC: The World Bank.

World Economic Forum (2019) *The global risk report, 14th edition*, Geneva: World Economic Forum.

Yaro, J.A., Teye, J.K. and Bawakyillenuo, S. (2014) 'Local institutions and adaptive capacity to climate change/variability in the northern Savannah of Ghana', *Climate and Development*, 7(3): 235–245. https://doi.org/10.1080/17565529.2014.951018

Yaro, J.A., Teye, J.K. and Bawakyillenuo, S. (2016) 'An assessment of determinants of adaptive capacity to climate change/variability in the rural savannah of Ghana', in Yaro, J.A. and Hesselberg, J. (eds.), *Adaptation to climate change and variability in rural West Africa*, Geneva, Switzerland: Springer. https://doi.org/10.1007/978-3-319-31499-0_5

44
LIVELIHOODS AND DISARMAMENT, DEMOBILISATION AND REINTEGRATION

From security to inclusive development

Henry Staples

Introduction

Since the United Nations Observer Group in Central America mission (1989–1992), the United Nations has supported Disarmament, Demobilisation and Reintegration (DDR) or DDR-related programmes in over 20 countries.[1] Helping ex-combatants to establish new, peaceful livelihoods has always been a core objective of these interventions. Yet the underlying rationale, precise measures, stakeholders, beneficiaries and evidence base have expanded – and evolved – dramatically over the last three decades.

First-generation DDR took place within a *peacekeeping* framework, prioritising state-building and stabilisation. Little could be learned from livelihood studies, which centred on how people organise their lives in poor yet politically stable conditions. Only with the dawn of a people-centred peacebuilding agenda did cross-cutting themes emerge. DDR's second-generation was tasked with contributing to long-term recovery and development, particularly in conflict-affected communities. Suddenly, loss of livelihood was the 'missing link' in explanations of conflict, and livelihood security became central to DDR's success (Cain et al., 2004: 8).

Contemporary peacebuilding involves fragmented armed groups engaging in localised conflict, often with limited chance of a peace deal. In response, a third-generation of DDR has emerged, wherein decisions regarding 'when, how and with whom to engage' are much more complex and nuanced (Muggah and O'Donnell, 2015: 1). This necessitates greater sensitivity to local power dynamics and politics, and an honest assessment of the potentially harmful consequences of interventions. Notwithstanding these additional concerns, analyses of DDR still prioritise economic goals over and above social or political concerns (Shariff, 2018).

In 2020, the United Nations was engaged in five missions, eight special political missions and two additional missions involving DDR; all of which were located in the Global South. While the relationship between poverty, inequality and conflict is notoriously complex, there is evidence that inter-group inequality is a major driver of conflict (Nygård, 2018). Moreover, as

Table 44.1 Key DDR terms

Key terms	Definition
Disarmament	The collection, documentation, control and disposal of small arms, ammunition, explosives and weaponry. Also establishes entry and eligibility criteria for further economic and other support.
Demobilisation	The formal and controlled discharge of active combatants from armed forces or other armed groups. May entail processing of individual combatants or the massing of troops in cantonment sites. Usually includes an assessment of personal and professional interests, skills and education, and medical and other needs.
Reinsertion	The assistance offered to ex-combatants, usually during demobilisation but prior to the longer-term process of reintegration. May include transitional safety allowances, food, clothes, shelter, medical services, short-term education, training, employment and tools.
Reintegration	The process by which ex-combatants acquire civilian status and gain sustainable employment and income.

Source: Adapted from the UN Integrated DDR Standards (United Nations, 2019 – Module 1.20). See also Willibald (2006) for alternative definitions.

inequality between countries falls and within-country inequality rises, poverty itself has become a 'distribution issue'; that is, about ensuring *access* to resources (de Haan, 2017). This means that anyone interested in understanding the interplay between conflict and livelihoods must meaningfully engage with power imbalances, structural constraints and processes of social inclusion and exclusion. If this challenge is met, there is potential for livelihoods research to make a meaningful contribution to the design, implementation and analysis of third-generation DDR.

This chapter charts DDR's evolution, centring on how different understandings of livelihoods have shaped or might better inform interventions. Analysis is based on a review of literature, as well as fieldwork data collected in Colombia in the wake of the 2016 peace agreement; a case firmly rooted in DDR's second-generation but illuminating for much of the discussion throughout.

First-generation DDR

DDR emerged in the wake of the Cold War as a component of multi-lateral peacekeeping missions. Grounded in a state-building rationale, engagement with ex-combatants centred on minimising the threat they posed in peacetime. The youth bulge hypothesis proposed that young men lacking meaningful livelihoods filled the ranks of armed groups, driving violence and conflict (Sommers, 2011). At conflict's end, the micro- and macro-security perspective presupposed that a reduction in income would increase individual propensity to engage in crime, with widespread dissatisfaction potentially leading to collective re-mobilisation (Collier, 1994). Accordingly, the surrender of weapons – which provide personal and economic security – equated to a voluntary social contract between ex-combatants, the government and the international community, based on a credible offer of non-violent livelihood opportunities and security guarantees (Knight and Özerdem, 2004; Phayal et al., 2015). Reinsertion measures would meet immediate needs through cash transfers and other temporary provisions. Economic reintegration, the cornerstone of conventional DDR and distinct from its social and political dimensions, then entailed a transition from the livelihood support mechanism of the armed group to other employment or income-generating activities (Torjesen, 2013).

Early economic reintegration packages were limited in scope, usually consisting of a small pension, with a minority of ex-combatants eligible to work in newly formed security forces (Muggah and O'Donnell, 2015). In rural areas, priority was given to agricultural programmes, based on the assumption that employment opportunities in other sectors were limited (Porto and Parsons, 2003). Such assumptions were often accurate, and research highlights the impact of macro-economic conditions including neoliberal economic reforms enacted as part of broader 'liberal peace' policies (Richmond, 2006). Additionally, economic geographies and spatial variations in labour supply and demand often meant that ex-combatants' skillsets did not match their surroundings, particularly in remote rural areas with sparse opportunities (Peters, 2007).

During early interventions, it also became clear that outcomes were contingent not only on the demand for labour, but on ex-combatants' perceptions and preferences regarding the set of options available to them (Subedi, 2014). To illustrate, in Sierra Leone 70% of households depend on (semi)subsistence agriculture for their livelihoods, but only 15% of ex-combatants chose agricultural training, with most opting for vocational trades on the expectation that this would lead to better paid employment. This expectation was largely unmet and many drifted into unemployment; a failure that revealed the acute dangers of not meeting ex-combatants' expectations, as many also migrated abroad, exacerbating security issues in neighbouring countries (Peters, 2007).

Second-generation DDR: the age of development

DDR's first-generation produced mixed results, logistical failures and negative programme effects (Colletta et al., 1996). It further revealed divergent expectations between the international community, who prioritised 'negative peace' (i.e. the absence of violence), and civil society, who imagined more expansive, positive socio-political change (Alden, 2002). But the 2004 intervention in Haiti signalled that DDR was evolving. Conventional methods were poorly suited to deal with urban armed gangs; instead the UN employed a Community Violence Reduction model aiming to promote local dialogues and create the necessary conditions for peace (Muggah and O'Donnell, 2015). Meanwhile, a 'sustainable livelihoods' approach had become central to development thinking. While still prioritising agency and bottom-up perspectives, this implied closer scrutiny of structural constraints, including the (mal)distribution of land and natural resources, and the impact of combatant and civilian displacement on rural and urban communities – thereby necessitating an 'integrated response on a large scale' (Cain et al., 2004: 72).

The UN's first Integrated DDR Standards, seeking to build on past failures and align with the changing peace-security-development landscape, re-positioned DDR as a comprehensive process contributing to immediate security goals, paving the way for long-term recovery and development (United Nations, 2006). This heralded a paradigm shift from negative to positive peace, with 'subtle, but far-reaching' implications (Muggah and O'Donnell, 2015: 3).

Competing agendas

The 2016 peace agreement between the Revolutionary Armed Forces of Colombia – People's Army (FARC – EP or FARC in Spanish) and the Colombian government arguably represents the pinnacle of DDR's 'positive peace' paradigm shift. The agreement, predicated on the concept of 'territorial peace', promised significant investment in peripheral conflict-affected communities, comprehensive land reform, the promotion of alternative livelihoods in order to overcome dependency on coca cultivation, and protection of the diverse lives and livelihoods of indigenous and Afro-Colombian communities – all with extensive community participation. This appeared to reflect the 'local turn' in peace-building (Mac Ginty and Richmond, 2013), also mirroring livelihood research on the intangible, non-material dimensions of wellbeing and the cultural and symbolic significations imbued in land (de Haan and Zoomers, 2005; Esland, 2017).

The emphasis on community participation – in Colombia and elsewhere – led to attempts to exploit the synergies between economic and social policy measures. Research was already challenging the assumption of passivity of first-generation DDR, instead conceptualising ex-combatants as active, 'moral agents' (McEvoy and Shirlow, 2009). At the same time, notions like repatriation, rehabilitation and reconciliation had entered the policy lexicon, and interventions sought to provide training and employment to ex-combatants in order to promote positive community relations.[2] These initiatives either implicitly or explicitly drew on the concept of social capital, which was central to the Sustainable Livelihoods Framework (SLF), and to the World Bank's Community-Driven Development model of post-conflict recovery (World Bank, 2006). They also drew on evidence that ex-combatants are highly reliant on community support networks for their economic security (Tajima, 2010; Kilroy and Basini, 2018).

Yet, just as critiques of the Sustainable Livelihoods Approach point out its failure to account for power dynamics (Serrat, 2017), research has also questioned the assumption of synergy between economic and social reintegration, instead pointing to directly contradictory effects (Humphreys and Weinstein, 2007). To illustrate, economic support provided to ex-combatants can reduce the incentive to reach out to communities or create resentment among them (Kostner, 2001). Given the numerous everyday problems of post-conflict settings, establishing positive social relations is not always high among community priorities, and their decision to collectively resist, tolerate or accept ex-combatants is a major factor shaping outcomes (Clubb and Tapley, 2018; Eastmond, 2010). Relations can be particularly fraught in situations where, as my own conversations with FARC and civilians evidenced, legacies of violence produce lasting suspicion. DDR programmes must therefore aim to redress these conflictive dynamics, for example by matching ex-combatants' economic assistance with additional support for communities (Felbab-Brown, 2020: 21).

Collective livelihoods

The emphasis on social dynamics in second-generation DDR coincided with research questioning the orthodoxy that armed groups must be disbanded altogether. Demobilisation was formerly considered 'the single most important factor determining the success of peace operations' (UN, 2004: 61). This typically occurred in cantonment sites, a 'halfway house between a mobilized state and the dissolution of forces', with ex-combatants then expected to go home (Knight and Özerdem, 2004: 507). From a policy perspective, cantonment provides efficiency benefits, streamlining needs and skills assessments and the distribution of material and other assistance (ibid: 508). Cantonment might also be preferable for ex-combatants without a home to return to, or they may simply prefer (or be accustomed to) 'communitarian living' (Subedi, 2014: 244). Crucially, this desire often transcends the cantonment period; in the Ukraine, ex-combatants established an urban café, offering a positive atmosphere, facilitating peaceful dialogue with civilians and a space for mutual support (Uehling, 2019). Thus, it is argued, there may be important benefits to allowing – even encouraging – ex-combatants to maintain a degree of cohesion (Wiegink, 2015).

However, as evidence from Colombia demonstrates, continued cohesion can expose or even exacerbate the fraught power relations between actors, while also making visible ex-combatants' culturally specific preferences. FARC leadership, having explicitly rejected the individualism of standard DDR, negotiated the establishment of Collective Spaces of Reincorporation and Training (ETCRs in Spanish), with food, public services and security guaranteed for a transitionary two-year period. There they established agricultural, tourism and other collective livelihood projects, and though many left the spaces,[3] others expressed a desire to stay permanently. In one ETCR in the Caquetá region, residents aggregated their individual financial support packages and purchased the land (Forero Rueda, 2019) – when I visited, I was told they had done so without obtaining consent from the government or international monitors.[4] This decision is revealing on a number of grounds. Firstly, their desire to stay – and the attachment to the space they expressed – coincides with research in El Salvador where land grants to ex-combatants were perceived as 'something won' and symbolised 'alternative futures' (de Bremond, 2013). Secondly, albeit suggestive of a degree of trust between FARC and the landowner, the decision to forego government consent evidences ongoing mistrust despite the supposed social contract established upon disarmament. Instead of positioning this as necessarily a security threat, however, researchers might instead usefully draw on livelihood research exploring how local people appropriate development interventions to their own advantage (Long, 1992, cited in de Haan, 2017: 27).

Sustaining advances

As the goals expanded, DDR has faced significant institutional and political impediments to sustaining advances in the long-term. National states frequently lack the capacity to deliver comprehensive livelihoods assistance and other services, leading to a reduction in assistance when humanitarian and development actors discontinue their support (Maxwell et al., 2017: 19). Myriad non-state actors often have a lasting interest in the war economy, shaping and constraining livelihood options during and after conflict (ibid: 21). Lastly, seismic political shifts can lead to wholescale re-orientation; in Colombia the 2018 presidential election and subsequent introduction of a *peace with legality* framing has threatened ex-combatant and community livelihood security and undermined peace's transformative potential (Diaz et al., 2021).

Another concern – particularly with regard to local community livelihoods – relates to the assumption that economic growth and job creation necessarily contribute to long-term political stability. Prioritisation of largescale investments in agricultural and natural resource industries can lead to policies that overlook social and environmental justice and exacerbate uneven development (Hennings, 2019). Furthermore, even when justice issues are incorporated into post-conflict interventions, for example regulating extractive industries so that rents are more equally distributed (Roy, 2017), priority is usually given to formal sector employment in spite of the significance of informal sector work for ex-combatants, civilians and indeed Global South economies more generally. These justice concerns – and the broader theoretical debates from which they stem – point to the unavoidable role of power and politics in DDR.

Third-generation DDR: from programme to process

DDR programmes are currently underway in contexts involving multiple, fragmented armed groups, as well as 'terrorist actors' engaging in violent extremism across entire regions; in all instances, reaching a comprehensive peace agreement is less likely (Muggah and O'Donnell, 2015: 7). DDR has thus adapted into a 'dynamic political process' employed before peace agreements are reached, with reintegration taking place prior to or alongside disarmament and demobilisation (ibid: 6). Recognising the complexity of these new, modern conflicts, an updated Integrated DDR Standards (IDDRS) re-positions DDR as a multi-dimensional process with possible applications along the 'entire peace continuum' (United Nations, 2019: Module 2.10).[5] The guidelines also introduce a new menu of options for when conventional pre-conditions are not met, including DDR-related tools and reintegration support to help prevent first-time recruitment. This final section considers some of the implications of this latest shift.

Building an evidence base: aspirations, agency and the politics of livelihoods

The new IDDRS guidelines emphasise context-specificity and a robust evidence base. This is welcome, as there are notable areas in which policy and practice has been slow to catch up with research (Hauge, 2020; Shariff, 2018). For example, senior members of armed groups tend to be more highly educated on average, with distinct expectations and aspirations to the rank-and-file (Alden, 2002; Humphreys and Weinstein, 2007), yet mid-level commanders remain an overlooked, influential and potentially disruptive group if not incorporated into plans (Bultmann, 2018). In addition, recent studies draw attention to the lasting psychological harm induced by exposure to conflict, and the importance of psychological stability for envisioning 'a life without weapons', yet only minimal psychosocial support has been provided in Colombia and elsewhere (Reardon, 2018). To factor in such concerns – and the wider range of assistance they necessitate – requires accurate, multi-disciplinary data on armed group internal structures as well as individual vulnerabilities; a task complicated by limited access to reliable information in modern conflicts.

Analyses of third-generation DDR will need to revisit many fundamental assumptions, foremost among them that engagement is voluntary. In the Democratic Republic of Congo, the UN employed a 'forceful DDR' model where offers of support coincided with military intervention (Muggah and O'Donnell, 2015: 4). Though punitive measures were mostly limited to 'hardcore' fighters – in that instance about 10% of the total – in Somalia, as many as 40% of DDR participants in one facility had been captured (Felbab-Brown, 2015: 120). This, along with the ever-greater blur between combatant and civilian, will require a rethink of the design of entry and eligibility criteria. As a solution, Shariff (2018) suggests that benefits should only be offered to already deradicalised combatants, though this begs the question of how to contribute to deradicalisation without a credible offer of support. Particularly given that ex-combatants are 'cause-oriented' (Ouaiss and Rowayheb, 2017), further research is required to explore how ex-combatants' agency is enacted under these conditions.

One of the primary disjunctures in the research-policy-practice nexus concerns the long-held assumption that livelihood security contributes to peace, with a multitude of studies now drawing attention to how labour market experiences influence political organisation (Mallett and Slater, 2016: 241). Of course, not all political expression is equally antagonistic: in the Democratic Republic of Congo, ex-combatants formed a bicycle taxi union, granting collective bargaining power to members while also providing a much-needed public good (Carayannis and Pangburn, 2020). In Namibia, however, ex-combatants drew on their role in the country's liberation to appeal to the government for improved work conditions, which re-awakened old societal cleavages and conflict narratives (Metsola, 2010). A 'politics of livelihoods' approach which explores how (contentious) political claims are 'earned through labour', and which accounts for ex-combatants' political aspirations, could enrich analysis of these dynamics, including their likelihood of producing positive social outcomes (Menon and Sundar, 2018).

Integration across scales

The new IDDRS guidelines prioritise flexibility, integration with local and national institutional frameworks and support from the private sector (United Nations, 2019, Module 2.40). Yet contemporary conflict environments lacking a 'clear beginning or end in time or space' are often not constrained to national borders (Koopman, 2019: 209). Even localised armed groups have connections and/or aspirations beyond their immediate surroundings, requiring analysis of multiple 'scales of violence' (World Bank Group et al., 2018: 2–3). Furthermore, legal distinctions stemming from international law can directly impede reintegration on the ground; the US State Department's continued designation of the FARC as a terrorist organisation has prevented their participation in local development planning (Colombia Peace, 2020). These multi-scalar entanglements require moving beyond the local in favour of a 'hybrid' or 'everyday peace' framework which interrogates how power relations intersect across scales (Mac Ginty, 2010; Williams, 2015). Analysts might also draw on livelihoods/development scholarship offering a relational understanding of poverty, inequality and power (Mosse, 2010). Integrating across scales in this way will also allow DDR to deliver on its promise of flexibility, including in response to shocks. Already, there are promising signs of progress in this regard; in the Central African Republic, DDR was re-purposed to engage ex-combatants and communities in combating the COVID-19 outbreak (United Nations, 2020).

New sensitivities, exclusionary effects

Despite claims to neutrality, the international community has been 'very much part of the conflict' in contemporary environments (Özerdem, 2013: 228). The new IDDRS module, The Political Dynamics of DDR (Module 2.20), acknowledges this, pointing to the increased risk of doing harm in ongoing conflicts. However, the very notion of harm must be interrogated, and include the potentially exclusionary effects of interventions themselves – as occurred in South Sudan, where minority ethnic participants had notably lower levels of satisfaction (Phayal et al., 2015). Again, inspiration can be taken from livelihood interventions, such as that employed by Ethiopia's Social Protection Programme, praised for 'its determination not to ignore signals of exclusion, including that caused by its own operations' (de Haan, 2017: 31).

Lastly, the latest IDDRS (United Nations, 2019 – Module 2.10: 16) warns practitioners to be aware of the sensitivities around language including the use of particular words and labels, aligning with research on the significance of how inequalities are perceived and experienced (Cederman et al., 2011). This awareness must extend beyond direct contact with beneficiaries, to how findings are presented and framed by researchers and policymakers; as Koopman (2019: 209) reminds us, the labels 'post-conflict' and 'post-agreement' connote wholly distinct realities. On the other hand, the commonalities between ex-combatants and other groups to which labels have been ascribed, and about which significant bodies of research exist, should not be unduly avoided. As this chapter has aimed to demonstrate, ex-combatants are – and aspire to be – many things, and drawing on these literatures (including the rich body of work on livelihoods) can support the design of effective interventions that meet individual needs and aspirations while contributing to broader societal goals.

Key Points

- The underlying rationale of DDR interventions has evolved from a preoccupation with immediate security goals to establishing the conditions for long-term recovery and development. This reflects changes in the international peace, security and development agendas and the evolving nature of conflict.
- In first-generation DDR, establishing livelihoods security was considered the best way to avoid ex-combatants committing crime or remobilising. Over time, logistical challenges, macro-economic conditions, labour market demand and ex-combatants' preferences were all identified as relevant determinants of success.

- The second-generation of DDR emphasised community relations and long-term recovery and development, introducing new stakeholders and potential beneficiaries of livelihoods assistance. In this context, ex-combatants were re-imagined as possessing the capacity to productively contribute to peace alongside communities. Yet new challenges emerged, including the need to balance the conflicting agendas of multiple actors, to consider ex-combatants' collective motivations, and to sustain advances in the long-term.
- DDR's third-generation sees interventions re-imagined as a dynamic process with potential applicability during conflict. This will require considerable dialogue between research, policy and practice, in order to better cater for individual needs and vulnerabilities; understand how agency is enacted and constrained; how labour market experiences interrelate with political behaviour, and how multi-scalar institutional frameworks impact DDR processes on the ground. At the same time, it requires greater awareness of the potentially exclusionary effects of interventions, and sensitive use of language, both in practice and in academic and policy discourse.

The paradigm shift from negative to positive peace was a welcome change in DDR's underlying rationale. But to deliver on its many promises, particularly in contemporary complex conflicts, a more refined understanding of ex-combatants' livelihood strategies – and paying close attention to processes of social inclusion – will be essential. Livelihood studies is well-placed to make a valuable contribution.

Further reading

1. United Nations (2019) *Integrated disarmament, demobilization and reintegration standards* (iDDRS) (Modules 1.10, 2.10, 2.20 and 2.40).
2. Maxwell, D., Mazurana, D., Wagner, M. and Slater, R. (2017) *Livelihoods, conflict and recovery: Findings from the Secure Livelihood studies Consortium*, London: Secure Livelihood Research Consortium.
3. Shariff, S. (2018) 'A critical review of evidence from ex-combatant reintegration programs', *Politics of Return Working Paper No.2*, London: LSE.

Notes

1. Data provided to author by the United Nations Department of Peace Operations (DPO)/Office of Rule of Law and Security Institutions (OROLSI)/DDR Section (July 2020).
2. In 2019, 252 ex-combatants in the Mambere Kadei region, Central African Republic, received training as 'Ambassadors for Peace' (UNITAR, 2019).
3. While this was entirely permissible under the agreement, it was broadly perceived as evidence of failure, and exposed the group's internal fragmentation and disillusionment with leadership.
4. When I asked why this decision was taken, the response was illuminating: 'When you buy something for yourself, a mobile phone or a house, you don't tell the government. I mean, why should we tell them?' (Conversation with FARC ex-combatant, August 2018).
5. At the time of writing, the final output is under revision, but several of the core modules have been released, including Module 2.10 'The UN Approach to DDR' which outlines core aims and principles.

References

Alden, C. (2002) 'Making old soldiers fade away: Lessons from the reintegration of demobilized soldiers in Mozambique', *Security Dialogue*, 33(3): 341–356. https://doi.org/10.1177/0967010602033003008

Bultmann, D. (2018) 'Insurgent groups during post-conflict transformation: The case of military strongmen in Cambodia', *Civil Wars*, 20(1): 24–44. https://doi.org/10.1080/13698249.2018.1446113

Cain, A., Clover, J. and Cornwell, R. (2004) 'Supporting sustainable livelihoods. A critical review of assistance in post-conflict situations', *Institute of Security Studies Monographs*, 2004(102). https://hdl.handle.net/10520/EJC48647

Carayannis, T. and Pangburn, A. (2020) 'Home is where the heart is: Identity, return, and the Toleka Bicycle Taxi Union in Congo's grand equateur', *Journal of Refugee Studies*, 33(4): 706–726. https://doi.org/10.1093/jrs/fez105

Cederman, L-E, Weidmann, N.B. and Gleditsch, K.S. (2011) 'Horizontal inequalities and ethnonationalist civil war: A global comparison', *The American Political Science Review*, 105(3): 478–495. https://doi.org/10.1017/S0003055411000207

Clubb, G. and Tapley, M. (2018) 'Conceptualising de-radicalisation and former combatant re-integration in Nigeria', *Third World Quarterly*, 39(11): 2053–2068. https://doi.org/10.1080/01436597.2018.1458303

Colletta, N., Kostner, M. and Wiederhofer, U. (1996) 'Case studies in war-to-peace transition: the demobilization and reintegration of ex-combatants in Ethiopia, Namibia and Uganda', *World Bank Discussion Papers*, Washington, DC: World Bank. https://doi.org/10.1596/0-8213-3674-6

Collier, P. (1994) 'Demobilisation and insecurity: A study in the economics of the transition from war to peace', *Journal of International Development*, 6(3): 343–351. https://doi.org/10.1002/jid.3380060308

Colombia Peace (2020) *An outdated interpretation of counter-terror law has painted U.S. Colombia programming into a corner. The way out is simple*. Available at: https://colombiapeace.org/an-outdated-interpretation-of-counter-terror-law-has-painted-u-s-colombia-programming-into-a-corner-the-way-out-is-simple/ (Accessed 7 January 2021).

de Bremond, A. (2013) 'Regenerating conflicted landscapes in post-war El Salvador: Livelihoods, land policy, and land use change in the Cinquera Forest', *Journal of Political Ecology*, 20(1): 116–136. https://doi.org/10.2458/v20i1.21761

de Haan, L. (2017) 'Livelihoods in development', *Canadian Journal of Development Studies/Revue Canadienne D'études Du Développement*, 38(1): 22–38. https://doi.org/10.1080/02255189.2016.1171748

de Haan, L. and Zoomers, A. (2005) 'Exploring the frontier of livelihood research', *Development and Change*, 36(1): 27–47. https://doi.org/10.1111/j.0012-155X.2005.00401.x

Diaz, J.M., Staples, H., Kanai, J.M. and Lombard, M. (2021) 'Between pacification and dialogue: Critical lessons from Colombia's territorial peace', *Geoforum*, 118: 106–116. https://doi.org/10.1016/j.geoforum.2020.12.005

Eastmond, M. (2010) 'Introduction: Reconciliation, reconstruction, and everyday life in war-torn societies', *Focaal*, (57): 3–16. https://doi.org/10.3167/fcl.2010.570101

Esland, C.C. (2017) 'Defending homeland and regaining freedoms interpreting livelihoods among conflict-affected communities in Southern Lebanon', in de Haan, L. (ed.), *Livelihoods and development*, Leiden, The Netherlands: Brill. pp. 124–147. https://doi.org/10.1163/9789004347182_007

Felbab-Brown, V. (2015) 'DDR – A bridge not too far: A field report from Somalia', in Cockayne, J. and O'Neil, S. (eds.), *UN DDR in an era of violent extremism: Is it fit for purpose?* United Nations University Centre for Policy Research, pp. 104–137.

Felbab-Brown, V. (2020) 'Detoxifying Colombia's drug policy: Colombia's counternarcotics options and their impact on peace and state building', *Foreign Policy at Brookings*, Washington, DC: The Brookings Institution.

Forero Rueda, S. (2019) 'Agua Bonita (Caquetá), el primer centro poblado de excombatientes de las Farc', *El Espectador*, 15 July 2019. Available at: www.elespectador.com/colombia2020/territorio/agua-bonita-caqueta-el-primer-centro-poblado-de-excombatientes-de-las-farc-articulo-870933/ (Accessed 7 January 2021).

Hauge, W.I. (2020) 'Gender dimensions of DDR – beyond victimization and dehumanization: Tracking the thematic', *International Feminist Journal of Politics*, 22(2): 206–226. https://doi.org/10.1080/14616742.2019.1673669

Hennings, A. (2019) 'From bullets to banners and back again? The ambivalent role of ex-combatants in contested land deals in Sierra Leone', *Africa Spectrum*, 54(1): 22–43. https://doi.org/10.1177/0002039719848511

Humphreys, M. and Weinstein, J.M. (2007) 'Demobilization and reintegration', *The Journal of Conflict Resolution*, 51(4): 531–567. https://doi.org/10.1177/0022002707302790

Kilroy, W. and Basini, S.A. (2018) 'Social capital made explicit: The role of norms, networks, and trust in reintegrating ex-combatants and peacebuilding in Liberia', *International Peacekeeping*, 25(3): 349–372. https://doi.org/10.1080/13533312.2018.1461564

Knight, M. and Özerdem, A. (2004) 'Guns, camps and cash: Disarmament, demobilization and reinsertion of former combatants in transitions from war to peace', *Journal of Peace Research*, 41(4): 499–516. https://doi.org/10.1177/0022343304044479

Koopman, S. (2019) 'Peace: Keywords in radical geography', in Jazeel, T., Kent, A., McKittrick, K., Theodore, N., Chari, S., Chatterton, P., Gidwani, V., Heynen, N., Larner, W., Peck, J., Pickerill, J., Werner, M. and Wright, M.W. (eds.), *Antipode at 50*, Oxford: Wiley Blackwell, pp. 207–211. https://doi.org/10.1002/9781119558071.ch38

Kostner, M. (2001) *A technical note on the design and provision of transitional safety nets for demobilization and reintegration programs*, Washington, DC: World Bank.

Long, N. (1992) 'From paradigm lost to paradigm regained?', in Long, A. and Long, N. (ed.), *Battlefields of knowledge. The interlocking theory and practice in social research and development*, London: Routledge, pp. 16–43.

Mac Ginty, R. (2010) 'Hybrid peace: The interaction between top-down and bottom-up peace', *Security Dialogue*, 41(4): 391–412. https://doi.org/10.1177/0967010610374312

Mac Ginty, R. and Richmond, O. (2013) 'The local turn in peace building: A critical agenda for peace', *Third World Quarterly*, 34(5): 763–783. https://doi.org/10.1080/01436597.2013.800750

Mallett, R. and Slater, R. (2016) 'Livelihoods, conflict and aid programming: Is the evidence base good enough?' *Disasters*, 40: 226–245. https://doi.org/10.1111/disa.12142

Maxwell, D., Mazurana, D., Wagner, M. and Slater, R. (2017) *Livelihoods, conflict and recovery: Findings from the secure livelihood research consortium*, London: Secure Livelihood Studies Consortium.

McEvoy, K. and Shirlow, P. (2009) 'Re-imagining DDR', *Theoretical Criminology*, 13(1): 31–59. https://doi.org/10.1177/1362480608100172

Menon, G.A. and Sundar, A. (2018) 'Uncovering a politics of livelihoods: Analysing displacement and contention in contemporary India', *Globalizations*, 16(2): 186–200. https://doi.org/10.1080/14747731.2018.1479017

Metsola, L. (2010) 'The struggle continues? The spectre of liberation, memory politics and 'war veterans' in Namibia', *Development and Change*, 41(4): 589–613. https://doi.org/10.1111/j.1467-7660.2010.01651.x

Mosse, D. (2010) 'A relational approach to durable poverty, inequality and power', *Journal of Development Studies*, 46(7): 1156–1178. https://doi.org/10.1080/00220388.2010.487095

Muggah, R. and O'Donnell, C. (2015) 'Next generation disarmament, demobilization and reintegration', *Stability: International Journal of Security and Development*, 4(1): Art.30. https://doi.org/10.5334/sta.fs

Nygård, H.M. (2018) 'Inequality and conflict – some good news', *World Bank Blogs*. Available at: https://blogs.worldbank.org/dev4peace/inequality-and-conflict-some-good-news (Accessed 7 January 2021).

Ouaiss, M. and Rowayheb, M. (2017) 'Ex-combatants working for peace and the Lebanese civil society: A case study in non-communal reintegration', *Civil Wars*, 19(4): 448–469. https://doi.org/10.1080/13698249.2017.1393282

Özerdem, A. (2013) 'Disarmament, demobilization and reintegration', in Mac Ginty, R. (ed.), *Routledge handbook of peacebuilding*, New York: Routledge, pp. 225–236.

Peters, K. (2007) 'Reintegration support for young ex-combatants: A right or a privilege?', *International Migration*, 45(5): 35–59. https://doi.org/10.1111/j.1468-2435.2007.00426.x

Phayal, A., Khadkha, P.B. and Thyne, C.L. (2015) 'What makes an ex-combatant happy? A micro-analysis of disarmament, demobilization, and reintegration in South Sudan', *International Studies Quarterly*, 59(4): 654–668. https://doi.org/10.1111/isqu.12186

Porto, J.G. and Parsons, I. (2003) 'Sustaining the peace in Angola: An overview of current demobilisation, disarmament and reintegration', *Paper 27*, Bonn: Bonn International Center for Conversion; Pretoria: Institute for Security Studies.

Reardon, S. (2018) 'Colombia: After the violence', *Nature*. Available at: www.nature.com/immersive/d41586-018-04976-7/index.html (Accessed 7 January 2021).

Richmond, O. (2006) 'The problem of peace: Understanding the "liberal peace"', *Conflict, Security and Development*, 6(3): 291–314. https://doi.org/10.1080/14678800600933480

Roy, V. (2017) 'Stabilize, rebuild, prevent? An overview of post-conflict resource management tools', *The Extractive Industries and Society*, 4: 227–234. https://doi.org/10.1016/j.exis.2017.01.003

Serrat, O. (2017) *Knowledge solutions: Tools, methods, and approaches to drive organizational performance*, Singapore: Springer Open.

Shariff, S. (2018) 'A critical review of evidence from ex-combatant reintegration programs', *Politics of Return Working Paper No.2*, London: LSE.

Sommers, M. (2011) 'Governance, security and culture: Assessing Africa's youth bulge', *International Journal of Conflict and Violence*, 5(2): 292–303. https://doi.org/10.4119/ijcv-2874

Subedi, D.B. (2014) 'Conflict, combatants, and cash: Economic reintegration and livelihoods of ex-combatants in Nepal', *World Development*, 59: 238–250. https://doi.org/10.1016/j.worlddev.2014.01.025

Tajima, Y. (2010) *Understanding the livelihoods of former insurgents: Aceh, Indonesia*, Jakarta: World Bank. https://doi.org/10.1596/27784

Torjesen, S. (2013) 'Towards a theory of ex-combatant reintegration', *Stability: International Journal of Security and Development*, 2(3): Art. 63. https://doi.org/10.5334/sta.cx

Uehling, G.L. (2019) 'Working through warfare in Ukraine: Rethinking militarization in a Ukrainian theme café', *International Feminist Journal of Politics*, 22(3): 335–358. https://doi.org/10.1080/14616742.2019.1678393

United Nations (2004) *A more secure world: Our shared responsibility. report of the high-level panel on threats, challenges and change*, New York: United Nations. Available at: www.un.org/en/events/pastevents/a_more_secure_world.shtml (Accessed 7 January 2021).

United Nations (2006) *Integrated disarmament, demobilization and reintegration standards (IDDRS)*. Available at: https://digitallibrary.un.org/record/609144?ln=en (Accessed 7 January 2021).

United Nations (2019) *Integrated disarmament, demobilization and reintegration standards (IDDRS)*, (Modules 1.10, 2.10, 2.20 and 2.40). Available at: www.unddr.org/the-iddrs/ (Accessed 7 January 2021).

United Nations (2020) *From fighting with guns to fighting the pandemic*. Available at: www.un.org/africarenewal/news/coronavirus/fighting-guns-fighting-pandemic (Accessed 7 January 2021).

United Nations Institute for Training and Research (UNITAR) (2019) *Journey for Peace and Development – Evolution and Impact of DDR*. Available at: https://www.youtube.com/watch?v=sFjRXV5xnTQ&feature=youtu.be (Accessed 30 March 2022).

Wiegink, N. (2015) 'Former military networks a threat to peace? The demobilisation and remobilization of Renamo in central Mozambique', *Stability*, 4(1): Art. 56. https://doi.org/10.5334/sta.gk

Williams, P. (2015) *Everyday peace: Politics, citizenship and Muslim lives in India*, Chichester: Wiley Blackwell. https://doi.org/10.1002/9781118837764

Willibald, S. (2006) 'Does money work? Cash transfers to ex-combatants in disarmament, demobilisation and reintegration processes', *Disasters*, 30(3): 316–339. doi:10.1111/j.0361-3666.2005.00323.x.

World Bank (2006) 'CDD in the context of conflict-affected countries: Challenges and opportunities', *Report No. 36425 – GLB*, Washington DC: World Bank.

World Bank Group, UN Peacekeeping and Social Science Research Council (2018) *The changing landscape of armed groups: Doing DDR in new contexts*, Washington, DC: World Bank Group, UN Peacekeeping and Social Science Research Council.

45
LIVELIHOODS IN CONFLICT-AFFECTED SETTINGS

Ibrahim Bangura

Introduction

Conflict-ridden contexts are characterised by a plethora of socio-economic and political challenges, that undermine development and have devastating effects on the livelihoods of people (Gates et al., 2010; Ammons, 1996). These challenges undermine resilience and increase vulnerability of at-risk populations, particularly children, the elderly and the disabled (Snoubar and Duman, 2016). In addition to this, conflicts lead to increased competition for limited resources and livelihood opportunities (Gates et al., 2010).

Conflict has a direct negative impact on livelihoods as it leads to loss of assets and limits economic opportunities, thereby shifting the focus onto survival and overcoming the everyday consequences of the conflict. Insecurity created by contextual changes leads to the need for people to adapt, by drawing on different assets in order for them to survive. As indicated by the United States Agency for International Development (USAID, 2005: 2–3):

> Conflict restricts or blocks access to one or more of these assets [physical, natural, human, financial, social, and political assets]. When this happens, people try to find other ways of obtaining those resources, or compensate for the loss of one resource by intensifying their efforts to secure another. Understanding this dynamic is essential for managing and minimising conflict's impact on civilians.

Understanding the survival dynamics and ability to secure livelihood assets is crucial, for individuals and communities to deal with the everyday challenges they contend with. Lautze et al. (2003: 18) describe survival dynamics as 'the sum of means by which people get by overtime'.

Despite the vast body of literature on both livelihoods and conflicts, there remains a gap in understanding and examining the symbiotic relationship between the two. Additionally, there are very few studies in the literature that focus on understanding livelihood practices in conflict settings and how they enhance household, individual and community adaptability, resilience and survival. In seeking to address this gap, this chapter draws on examples from a wide range of countries to critically investigate how effective livelihood responses and coordination by national and international actors have helped strengthen resilience, reduce vulnerability and fragility of local communities, and transformed approaches to managing conflicts. The chapter further explores the complex relationship between conflict and livelihoods, including examining the nature and typologies of conflict and its impact on livelihoods.

As many different and varied types of conflict exist, all of which cannot be addressed by this chapter, the focus is on civil wars and conflict related to organised criminal activities such as piracy. These types of conflict were chosen given the frequency of their occurrence especially in Asia and Africa (ACLED, 2015; Bateman, 2010; Bowd and Chikwanha, 2010; Cilliers, 2018).

This chapter is divided into seven sections. Section two presents the characteristics of conflicts that affect livelihoods. Section three discusses the challenges, complexities and opportunities related to livelihoods in fragile and conflict settings. Section four analyses livelihood responses and their coordination, detailing lessons learnt, best practice and recommendations that may meaningfully contribute to the literature and social policy, vis-à-vis approaches to livelihoods as not only a means for survival, but also a tool for conflict transformation and peacebuilding. Section five concludes the chapter, followed by key points in section six and recommendations for further reading in section seven.

Characteristics of conflicts affecting livelihoods

This section identifies key factors related to civil war and piracy that affect livelihoods. It is important to note that inasmuch as some conflicts are violent, the degree of violence and the dynamics of conflicts vary and depend largely on the context and actors involved. Violence could take many forms as it could be sporadic, long lived or interspersed depending on the actors involved, changes in the contexts and the kind of peacebuilding efforts that are initiated (Douma, 2003; Watts et al., 2017). The shift in conflict characteristics is observed by Richards (2003: 22) who stated that conflicts are 'never static; they evolve over time, taking on many forms and parameters'.

Conflicts usually lead to insecurity, with implications for livelihood activities. Factors such as the intensity and length of the conflict have direct impact on livelihoods. Activities such as trading, farming and even basic household chores such as collecting water from boreholes can become risky. As indicated by Mueller and Tobias (2016: 2), 'insecurity can disrupt economic activity through a number of channels, and the effects can be large and long lasting. Fear resulting from violence and destruction can hinder economic activity directly'.

The situation described above could be worse in contexts where armed groups also directly target livelihood assets by destroying them and looting goods and assets, thereby undermining livelihood strategies and food security. Such strategies are 'often intentionally aimed at undermining livelihoods' (Jaspars and Maxwell, 2009: 1), demoralising and impoverishing communities. Asset appropriation is often perpetrated in conjunction with sexual violence and the abduction of community members (Coulter, 2009; Specht, 2006).

Another form of conflict common in both Asia and Africa is piracy (Bateman, 2010; Whitman and Suarez, 2012), which is a transitional criminal activity with negative implications for livelihood activities in affected communities. Piracy is usually rooted in factors such as grievances against a state, corruption, limited state capacity, regional disputes and the relatively high chances of making substantial income by those involved in it (Whitman and Suarez, 2012). Box 45.1 presents a case study of Somalia, where decades of civil war have led to abuse and exploitation of marine resources by external actors. Consequently, this exploitation has led to the emergence of piracy, which has also negatively affected the livelihood activities of local communities in Somalia.

> **Box 45.1 External exploitation of livelihoods as a catalyst of piracy in Somalian waters**
>
> The exploitation of marine resources by foreign fishing ships and the dumping of nuclear and toxic waste by industrialised countries within the territorial waters of Somalia (Hussein, 2010) led to confrontation between Somali fishermen and these foreign actors. As such, piracy started off as a desire to punish outsiders carrying out illegal and unauthorised fishing but evolved into survivalism, opportunism and exploitation (Karawita, 2019).
>
> The Somali fishermen gradually realised that they could make large sums from ransoms paid for ships and crews captured. The evolution of piracy in Somalia and the region gradually undermined small scale fishing activities as a result of the waters becoming insecure, with communities depending on the exploits of piracy (Karawita, 2019). Young, deprived and marginalised youth entered into enterprises with transnational criminal networks that emerged (Gilpin, 2009). These confrontations culminated in the global piracy crisis in the region (Kantharia, 2019). In addressing the crisis, there appears to be an unwillingness on the part of the international community to identify the socio-genesis of the problem and a reluctance to fairly address the exploitation of a major source of livelihood of people in a conflict-ridden setting, an exploitation that continues to date.

The causes and effects of piracy in Somalia bear semblance to those of the Niger Delta crisis in Nigeria. For decades, communities in the oil rich Niger Delta region were marginalised, leading to the highest levels of illiteracy and youth unemployment in Nigeria (Bamgboye et al., 2011). Decades of frustration and grievance against the government and multinational oil corporations resulted in youth using disruptive means to access resources that were solely in the hands of elite, including oil bunkering, piracy and waging civil war (Emmanuel et al., 2009). This struggle for access to state resources resulted in insecurity, with immense implications for the environment, the local population and their livelihood assets (Ihayere et al., 2014).

Conflict and the associated insecurity described above may lead to mass displacement, disruption of livelihood opportunities and a volatile security situation. This may result in changes in livelihood patterns, as people move from rural to urban settings or into refugee and Internally Displaced Persons camps. As observed by Maxwell, Mazurana et al. (2017: vii) 'protracted conflict has facilitated a transition for some from rural agricultural-based livelihoods to informal urban livelihoods'.

Livelihood strategies and challenges in conflict-affected settings

The eruption of conflict in any society disrupts livelihood patterns, having immediate and longer-term implications for those affected. In some cases, conflict leads to disruptive shocks and uncertainty, leading to the need for those affected to adapt to the context. However, the ability to adapt is largely dependent on factors such as the opportunities available, the coping and resilience mechanisms that a community possesses, the intensity and longevity of the conflict, and the conduct of conflicting parties and criminal networks (Mueller and Tobias, 2016; Odusote, 2016).

Some of the immediate concerns with the eruption of conflict include the protection of livelihood assets and adaptation to the new context. Like civilians, armed groups also need food and resources to fuel their war efforts and as such embark on mobilising them. For instance, in the Democratic Republic of Congo, belligerent factions have over the years been accused of being involved in the mining of minerals to finance the conflict and enrich themselves (Butcher, 2010; Stearns, 2011). Similarly, during the conflict in Sierra Leone the mining of diamonds and gold was widespread among the armed groups (Gberie, 2005). These practices by armed groups had devastating consequences for the environment (Alao, 2007; Fisher et al., 2018; Maconachie and Hilson, 2018). The consequent destruction of the environment and deforestation has been detrimental, in some cases it has been irreversible. In countries such as Colombia and Afghanistan, armed groups have depended on the sale of narcotics for their survival and sustenance of their war efforts (Mancini and Sati, 2017).

In some conflict settings, the need for food and other essential items on the part of the armed groups could reduce the intensity of violence against local communities. For instance, Koren and Bagozzi (2017: 351) suggest that 'prospects of repeated interactions compel armed actors to pursue a strategy of co-optation – to avoid the use of violence – during their efforts to obtain food access from civilians'. This relationship could also be based on a mutual desire to protect livelihood assets and family members, which has in some contexts led to the establishment of militias and vigilante groups. For example, in Sierra Leone, local militias such as the Kamajors, Tamaboros, Donsos and the Kapras were established with the goal of protecting communities and their livelihood assets during the conflict (1991–2002); these groups were largely dependent on the communities for food and other material support (Gberie, 2005; Koroma, 2004). Similar patterns have been observed in Somalia, where groups such as Al Shabaab depend heavily on livelihood opportunities available in the communities they operate in (Rembold et al., 2013).

As indicated above, the disruption of livelihood patterns and the intensity and longevity of conflicts affect livelihood options, with those affected adapting to potential sources of livelihood available in their communities. In the case of Somalia, with the insecurity related to fishing as a result of piratic activities and the limited livelihood options in coastal villages and towns in Somalia, local community leaders and members adapted to livelihood options tied to piracy. Local leaders demand 10% of docking fees from pirates and local merchants sell items to pirates at very high prices (Beloff, 2013). The financial resources from the pirates is needed for the survival of local communities and the pirates depend on the communities for supplies that are essential for their survival and the survival of those they kidnap (Beloff, 2013).

Another common livelihood activity that people adapt to in conflict settings is transactional sex, which intensifies as the level of vulnerability increases and resilience declines, usually as a result of the longevity of the conflict. The lack of livelihood options, limited available resources, alongside the existing insecurity, render women and girls in conflict settings and refugee or displaced persons camps much more vulnerable to sexual exploitation and abuse. Writing on the vulnerability of displaced women, Sider and Sissons (2016: 436) state that some engage in 'harmful coping strategies in camps, including child marriage and sexual exploitation, as a result of their limited resources'. Even peacekeepers have been implicated in allegations of transactional sex in countries in which they are deployed (Mudgway, 2017).

In some conflict-ridden communities, remittance transfers have been a major source of livelihood (Lindley, 2007). Remittances not only support the daily survival of the individual, but stimulate local economies, strengthening longer-term resilience in households and communities, overcoming the challenges presented by the disruption in livelihood activities as a result of conflicts (Gioli et al., 2013; Ncube and Gómez, 2011). In 2015 alone, remittances to Somalia were estimated to reach a total of US $1.4 billion, making up 23% of the country's Gross Domestic Product (GDP) (World Bank, 2016). The transfer of remittances in Somalia has been facilitated through the use of the Hawala system; local money transfer couriers are used to channel money from one location to another with relative ease (Faith, 2011). This service has been significantly improved over time and not only serves as a source of employment but also as a means of supporting a functioning informal economy in conflict ridden Somalia.

The challenges presented by conflicts lead to the need for affected groups to identify and adapt to measures that foster their survival. As indicated above, this need is not limited to civilians but also includes armed groups, pirates and others who need to survive and maintain their war or criminal enterprises. The longevity and intensity of the conflict also undermines resilience and influences the kind of adaptive measures that are embarked on.

Livelihood support and coordination

This section presents the risks and challenges in responding to and coordinating humanitarian efforts geared towards providing livelihood assistance in conflict contexts. These obstacles which include insecurity, denial of access by armed groups, poor road networks and lack of infrastructure in local communities, undermine the ability of national and international actors to effectively deliver assistance required by people trapped in such settings. In several conflict zones across the world, there have been reports of aid workers being targeted, killed or having their operations disrupted by armed groups and transnational criminal networks (Carmichael and Karamouzian, 2014; Hoelscher et al., 2015). This disruption to aid activities may lead to death, starvation and health-related complications amongst affected populations, particularly in contexts with very limited livelihood options and those prone to natural disasters.

To overcome challenges related to humanitarian efforts, the respective actors involved have sought to plan and effectively coordinate their different interventions. Such a strategy has been found to be useful in reducing duplication of efforts and the wastage of limited resources and enhances efficiency and effectiveness in the delivery of required assistance (Saavedra and Knox-Clarke, 2015; Stephenson, 2005). However, effective coordination of humanitarian efforts is usually not easy to achieve as there are several international and national actors involved, with varied interests. Saavedra and Knox-Clarke (2015: 13) observe that:

> Many of the challenges in the international coordination system spring from the fact that, while organisations might wish to commit to coordination in principle, they have differing organisational mandates and priorities and are often in competition for funds, which makes coordination extremely difficult in practice.

In spite of the challenges involved, there are numerous examples of efforts that have yielded positive results in coordinating and consolidating food and medical assistance for people trapped in conflict settings. For instance, the establishment of the Food Security and Livelihoods Working Group (FSLWG) in southern Turkey (Food Security Cluster, 2013) and the Somalia Food Security Cluster in 2012 (Bonsignore, 2013). Both initiatives were introduced with the aim of promoting effective coordination and complementarity of efforts among development partners working in these two countries. The food security cluster in Somalia has a membership of over 400 organisations, co-led by the Food and Agriculture Organization (FAO) and World Food Programme (WFP) and is co-coordinated with Save the Children (Bonsignore, 2013). According to both clusters, there has been significant progress in the provision of food and other essential supplies to victims of the conflict, with better coordination of efforts and reduction in challenges related to delivering livelihood assistance (Bonsignore, 2013; Food Security Cluster, 2013). However, it is worth mentioning that these conclusions are self-reported assessments of impact by the Clusters. Nonetheless, observers such as Humphries (2013: 1) argue that the 'cluster approach has increased the effectiveness of humanitarian action, suggesting that it is a worthwhile mechanism to pursue'.

Central to the discussion on livelihoods programming is the promotion of gender responsive approaches, to address the challenges that men and women face and how they could be supported to better adapt to conflict settings. As such, livelihood programming should seek to address the factors that render the respective genders vulnerable. As found by Sider and Sissons (2016: 436), in Iraq, women's vulnerability to violence is 'exacerbated by the depletion of assets, lack of livelihood opportunities, lack of privacy, and general uncertainty'. To overcome this challenge, organisations such as Oxfam with funds from the United Nations Development Programme focused on 'asset building and asset replacement in conflict affected communities, with a focus on women's economic empowerment' (Sider and Sissons, 2016: 428). Oxfam argues that such support gradually empowers those targeted, increasing their resilience (Sider and Sissons, 2016).

Common methods being employed to provide socio-economic empowerment for women in fragile contexts in Africa, Asia and Latin America include the provision of microcredit and the establishment of savings associations such as the Village Savings and Loan Association (VSLA), a model developed by Care International in Niger in 1991. The VSLA provides training and support to small groups of 15–25 people, most often women, who save together and take small, low-interest loans from those savings (Care International, 2011). Ngegba et al. (2016) argue that VSLA can increase savings capacity and income levels as well as contribute not only to household food security, but also to poverty reduction and community development. As such, the increase in savings and income levels are essential in helping VSLA members better deal with the challenges in conflict settings and also improve communities' adaptive capacity to shocks and stresses (Chinyoka, 2016).

For livelihood programming and coordination to succeed in conflict settings, understanding and addressing the needs of vulnerable groups should remain an important objective (Jaspars and Maxwell, 2009). Additionally, the success of livelihood interventions depends largely on the ability and willingness of the development actors to promote local buy-in, ownership and leadership of the process (Gingerich and Cohen, 2015). However, it is important to note that while effective livelihood programming could help to reduce vulnerabilities and strengthen resilience, they cannot alone stabilise conflict settings (Maxwell, Stites et al., 2017). Nonetheless, some

> livelihood interventions may intentionally and directly aim to address the root cause of the conflict, such as conflicts over water, land, or other resources. Overall, however, there continues to be a gap in the evidence as to the ultimate effectiveness of such efforts.
>
> *(Maxwell, Stites et al., 2017: 30–31)*

Conclusion

Based on the literature, it is concluded that conflict settings are often plagued with challenges that affect livelihood activities. The types of conflict discussed in this chapter exacerbate tensions related to livelihoods due to the ensuing scarcity, insecurity and stiff competition for limited available resources. Furthermore, the destruction of livelihood assets has been used as a means of demoralising and punishing communities that may be perceived to be at odds with armed groups.

The chapter concludes that the desire to survive creates the need for conflict affected groups to adapt to available livelihood opportunities and coping mechanisms. The approaches adopted by the respective groups will depend largely on the context, the opportunities available, the nature, longevity and the intensity of the conflict. However, to better deliver on livelihood programming, they have to be designed to be context specific and better planned and coordinated to reduce the existing challenges in the field. Livelihood programming should be gender responsive to be effective and appropriate. Furthermore, the use of people-centred approaches, promoting local buy-in, ownership and leadership, are also crucial for the success of livelihood interventions in conflict settings.

Key Points

- Violent conflicts undermine resilience and increase vulnerability, particularly with the destruction of livelihood assets. The extent to which this occurs depends on the persistence and intensity of the conflict as well as the presence and actions of armed groups.
- The challenges presented by conflict lead to the adoption of survival and adaptation strategies by populations affected by conflict. These options/strategies are largely dependent on the context, the opportunities available, the nature of the conflict and the intentions of the respective actors.
- Humanitarian efforts are usually ad-hoc and ill-coordinated which leads to duplication of efforts.
- Central to the discussion on livelihoods programming is gender responsiveness and the need to design and implement livelihood interventions that specifically target and empower women.
- Livelihood interventions should be context specific and should be tailored to promote local buy-in, ownership and leadership.

Further reading

Gingerich, T.R. and Cohen, M.J. (2015) *Turning the humanitarian system on its head: Saving lives and livelihoods by strengthening local capacity and shifting leadership to local actors*, New York: Oxfam Research Reports.

Jaspars, S. and Maxwell, D. (2009) 'Food security and livelihoods programming in conflict: A review', *Network Paper* 65, London: Humanitarian Practice Network, Overseas Development Institute.

References

ACLED – Armed Conflict Location and Event Data Project (2015) *Conflict dynamics within and across Africa and Asia*. Available at: https://acleddata.com/2015/03/25/conflict-dynamics-within-and-across-africa-and-asia/ (Accessed 24 July 2020).

Alao, A. (2007) *Natural resources and conflict in Africa: The tragedy of endowment Rochester studies in African history and the diaspora volume: 29*, Rochester: Boydell & Brewer, University of Rochester Press.

Ammons, L. (1996) 'Consequences of war on African countries' social and economic development', *African Studies Review*, 39(1): 67–82. https://doi.org/10.2307/524669

Bamgboye, V., Shiras, P., Oliver, D. and Mendie, M. (2011) 'A report on Niger delta region youth assessment', *Abuja*, Nigeria: Foundation for Partnership Initiatives in the Niger Delta (PIND).

Bateman, S. (2010) 'Tackling piracy in Asia: The current situation and outlook', *Global Asia*, 5(4): 32–35.

Beloff, J.R. (2013) 'How piracy is affecting economic development in Puntland, how piracy is affecting economic development in Puntland, Somalia', *Journal of Strategic Security*, 6(1): 47–54. http://dx.doi.org/10.5038/1944-0472.6.1.4

Bonsignore, V. (2013) *Lessons learned & good practices in the Somalia food security cluster. WFP/FAO co-led global food security cluster*, Somalia, Rome: World Food Programme.

Bowd, R. and Chikwanha, A.B. (eds.) (2010) 'Understanding Africa's contemporary conflicts origins, challenges and peacebuilding', *ISS Monograph 173*, Pretoria: Africa Human Security Initiative.

Butcher, T. (2010) *Chasing the devil: On foot through Africa's killing fields*, London: Vintage Books.
CARE International (2011) 'Introducing the village savings and loan model', *Join My Village Blog*. Available at: www.care.org/our-impact/joinmyvillage/blog/2011/02/introducing-village-savings-and-loan-model (Accessed 7 July 2020).
Carmichael, J. and Karamouzian, M. (2014) 'Deadly professions: Violent Attacks against aid-workers and the health implications for local populations', *International Journal of Health Policy and Management*, 2: 65–67. https://doi.org/10.15171/ijhpm.2014.16.
Chinyoka, E.W. (2016) *An investigation of the contribution of village savings and loans associations to household resilience: The case of Twic county, South Sudan*, Published Thesis, Bulawayo: National University of Science and Technology.
Cilliers, J. (2018) 'Violence in Africa trends, drivers and prospects to 2023', *Africa Report 12*, Pretoria: Institute of Security Studies.
Coulter, C. (2009) *Bush wives and girl soldiers: Women's lives through the war and peace in Sierra Leone*, New York: Cornell University Press.
Douma, P.S. (2003) *The origin of contemporary conflict: A comparison of violence in the three regions*, The Hague: Netherlands Institute of International Relations, Clingendael.
Emmanuel, A.O., Olayiwola, J.J. and Babatunde, A.W. (2009) 'Poverty, oil exploration and Niger delta crisis: The response of the youth', *African Journal of Political Science and International Relations*, 3(5): 224–232.
Faith, D.C. (2011) 'The hawala system', *Global Security Studies*, 2(1): 23–33.
Fisher, F., Bavinck, M. and Amsalu, A. (2018) 'Transforming asymmetrical conflicts over natural resources in the Global South', *Ecology and Society*, 23(4): 28. https://doi.org/10.5751/ES-10386-230428
Food Security Cluster (2013) *Lessons learned: The food security and livelihoods working group, South Turkey*, Rome: World Food Programme.
Gates, S., Hegre, H., Nygård, H.M. and Strand, H. (2010) 'Consequences of civil conflict', *World Development Report 2011 Background Paper*, Washington, DC: World Bank. https://openknowledge.worldbank.org/handle/10986/27502
Gberie, L. (2005) *A dirty war in West Africa: The RUF and the destruction of Sierra Leone*, London: Hurst.
Gilpin, R. (2009) *Counting the cost of Somali piracy*, Washington, DC: United States Institute for Peace Center for Sustainable Economies.
Gingerich, T.R. and Cohen, M.J. (2015) *Turning the humanitarian system on its head: Saving lives and livelihoods by strengthening local capacity and shifting leadership to local actors*, New York: Oxfam Research Reports.
Gioli, G., Khan, T. and Scheffran, J. (2013) 'Remittances and community resilience to conflict and environmental hazards in northwestern Pakistan', in Rodima-Taylor, D. and Estey, N. (eds.), *Remittance flows to post-conflict states: Perspectives on human security and development*, Boston: The Frederick S. Pardee Center for the Study of the Longer-Range Future, Center for Finance, Law & Policy, Taskforce Report, University of Boston, USA.
Hoelscher, K., Miklian, J. and Nygård, H.M. (2015) 'Understanding violent attacks against humanitarian aid workers', *Conflict Trends 06: 2015*, Oslo: Peace Research Institute Oslo (PRIO).
Humphries, V. (2013) 'Improving humanitarian coordination: Common challenges and lessons learned from the cluster approach', *The Journal of Humanitarian Assistance*. Available at: http://sites.tufts.edu/jha/archives/1976 (Accessed 7 July 2020).
Hussein, B.M. (2010) *The evidence of toxic and radioactive wastes dumping in Somalia and its impact on the enjoyment of human rights: A case study*. Paper presented at the United Nations Human Rights Council's 14th Session Panel discussion on Toxic Wastes, Geneva, Switzerland.
Ihayere, C., Ogeleka, D.F. and Ataine, T.I. (2014) 'The effects of the Niger delta oil crisis on women folks', *Journal of African Studies and Development*, 6(1): 14–21. http://doi.org/10.5897/JASD11.078
Jaspars, S. and Maxwell, D. (2009) 'Food security and livelihoods programming in conflict: A review', *Network Paper 65*, London: Humanitarian Practice Network, Overseas Development Institute.
Kantharia, R. (2019) *Causes of maritime piracy in Somalia waters*, Marine Insights. Available at: www.marineinsight.com/marine-piracy-marine/causes-of-piracy-in-somalia-waters/ (Accessed 7 July 2020).
Karawita, A.K. (2019) 'Piracy in Somalia: An analysis of the challenges faced by the international community', *Journal Ilmu Sosial dan Ilmu Politik*, 23(2): 102–119. http://doi.org/10.22146/jsp.37855.
Koren, O. and Bagozzi, B.E. (2017) 'Living off the land: The Connection between cropland, food security, and violence against civilians', *Journal of Peace Research*, 54(3): 351–364. http://doi.org/10.1177/0022343316684543.
Koroma, A.K. (2004) *Crisis and intervention in Sierra Leone*, Freetown: Andromeda.
Lautze, S., Aklilu, Y., Raven Roberts, A., Young, H., Kebede, G. and Leaning, J. (2003) 'Risk and vulnerability in Ethiopia: Learning from the past, responding to the present, preparing for the Future', *A Report*

for the United States Agency for International Development (USAID), Boston, MA: Feinstein International Center, Tufts University.

Lindley, A. (2007) 'Remittances in fragile settings: A Somali case study', *HiCN Working Paper 27*, Brighton: Households in Conflict Network, Institute of Development Studies.

Maconachie, R. and Hilson, G. (2018) 'The war whose bullets you don't see: Diamond digging, resilience and Ebola in Sierra Leone', *Journal of Rural Studies*, 61: 110–122. https://doi.org/10.1016/j.jrurstud.2018.03.009

Mancini, A. and Sati, S. (2017) 'Afghanistan and Colombia: A common struggle against narcotics', *Working Paper No. 65*, Milan: Italian Institute for International Political Studies.

Maxwell, D., Mazurana, D., Wagner, M. and Slater, R. (2017) *'Livelihoods, conflict and recovery' findings from the secure livelihoods research consortium*, London: Secure Livelihoods Research Consortium.

Maxwell, D., Stites, E., Robillard, S.C. and Wagner, M. (2017) *Conflict and resilience: A synthesis of Feinstein International Center work on building resilience and protecting livelihoods in conflict-related crises*, Boston: Feinstein International Center, Tufts University.

Mudgway, C. (2017) 'Sexual exploitation by UN peacekeepers: The 'survival sex' gap in international human rights law', *The International Journal of Human Rights*, 21(9): 1453–1476. https://doi.org/10.1080/13642987.2017.1348720.

Mueller, H. and Tobias, J. (2016) 'The cost of violence: Estimating the economic impact of conflict', *IGC Growth Brief Series 007*, London: International Growth Centre.

Ncube, G. and Gómez, G.M. (2011) 'Local economic development and migrant remittances in rural Zimbabwe: Building on sand or solid ground?', *ISS Working Paper Series/General Series, 523*, The Hague: Institute of Social Studies.

Ngegba, M.P., Kassoh, T.L. and Sesay, M. (2016) 'Impact of village saving and loan association (VSLA) on farm productivity in lower Banta Chiefdom, Southern Sierra Leone', *International Research Journal of Social Sciences and Humanities*, 1: 29–32.

Odusote, A. (2016) 'Nigeria: The matrix between fragility of livelihoods and conflict', *Journal of Global Initiatives*, 10(2): 9–30. https://digitalcommons.kennesaw.edu/jgi/vol10/iss2/3

Rembold, F., Oduori, S.M., Gadain, H. and Toselli, P. (2013) 'Mapping charcoal driven forest degradation during the main period of Al Shabaab control in southern Somalia', *Energy for Sustainable Development*, 17(5): 510–514. http://dx.doi.org/10.1016/j.esd.2013.07.001

Richards, P. (2003) 'The political economy of internal conflict in Sierra Leone', *Working Paper 21*, The Hague: Netherlands Institute of International Relations Clingendael.

Saavedra, L. and Knox-Clarke P. (2015) 'Working together in the field for effective humanitarian response', *ALNAP Working Paper*, London: ALNAP/ODI.

Sider, R. and Sissons, C. (2016) 'Researching livelihoods recovery and support for vulnerable conflict-affected women in Iraq', *Gender and Development*, 24(3): 427–441. http://dx.doi.org/10.1080/13552074.2016.1233671.

Snoubar, Y. and Duman, N. (2016) 'Impact of wars and conflicts on women and children in Middle East: Health, psychological, educational and social crisis', *European Journal of Social Sciences Education and Research*, 6(2): 211–215. http://doi.org/10.26417/ejser.v6i2.p211-215.

Specht, I. (2006) *Red shoes: Experience of girl-combatants in Liberia*, Geneva: International Labour Organization.

Stearns, J. (2011) *Dancing in the glory of monsters: The collapse of the congo and the great war of Africa*, New York: Public Affairs.

Stephenson, M. (2005) 'Making humanitarian relief networks more effective: Operational coordination, trust and sense making', *Disasters*, 29(4): 337–350. https://doi.org/10.1111/j.0361-3666.2005.00296.x.

United States Agency for International Development (USAID) (2005) *Livelihoods and conflicts: A toolkit for intervention*, Washington, DC: Office of Conflict Management and Mitigation.

Watts, S., Kavanagh, J., Frederick, B., Norlen, T.C., O'Mahony, A., Voorhies, P. and Szayna, T.S. (2017) *Understanding conflict trends: A review of the social science literature on the causes of conflict*, Research Report for the United States Army, Santa Monica: RAND Corporation.

Whitman, S. and Saurez, C. (2012) 'The root causes and true costs of marine piracy. Dalhousie marine piracy project', *Marine Affairs Program Technical Report #1*, Halifax: Marine Affairs Program, Dalhousie University.

World Bank. (2016) *World bank makes progress to support remittance flows to Somalia*. Available at: www.worldbank.org/en/news/press-release/2016/06/10/world-bank-makes-progress-to-support-remittance-flows-to-somalia (Accessed 17 May 2020).

46
LAND TENURE TRANSFORMATIONS IN THE GLOBAL SOUTH

Privatisation, marketisation and dispossession in contemporary rural Asia

Lam Minh Chau

Introduction

In the last four decades, land tenure systems across the Global South have undergone unprecedented changes as the result of massive political and economic transformations characteristic of the post-Cold War era: the transition from socialist command economy to market-based economy in former Eastern European and Asian socialist states (Burawoy and Verdery, 1999; Hann, 2002), the rapid introduction of market liberalisation policies and structural adjustment programmes (SAPs) in Asia, Latin America and Africa under the advice and pressure from international financial institutions such as the World Bank and the International Monetary Fund (Deininger, 2003), and the 'land rush' led by finance-rich but resource-poor countries, transnational corporations and private investors, many of which have focused predominantly on the Global South as sources of food crops, mineral deposits, energy materials and massive windfall profit from transactions of land (Borras et al., 2012).

This chapter examines three of the most profound changes in land tenure systems in the Global South that those processes have engendered. The first is privatisation, which is the transfer of land management and possession from state collectives to individual households. The second is marketisation, which involves the transfer of land from small-holding 'peasant' households to large-scale farmers and from subsistence-oriented to commercial purposes. The third is dispossession, which is the process through which governments, private developers and transnational corporations use coercive measures to seize arable land from rural households against their will.

This chapter overviews the literature on the actual impact of those changes in land tenure systems on rural livelihoods in the Global South, with a particular focus on Asian contexts: China, India, Vietnam, and selected former Soviet socialist republics. It is in those countries that the impact of the three changes have played out in the most complex and unpredicted ways. This does not mean, however, that the impact I describe is limited to Asian settings. In fact what I am discussing has also been observed elsewhere across the Global South: in Latin American contexts that have experienced the tremendous destabilising impact of land marketisation (Borras et al., 2012; Zoomers and van der Haar, 2000), as well as in African countries, where thousands of small cultivators in Sudan, Tanzania and Madagascar have fallen victim to land dispossession (GRAIN, 2008; Zoomers, 2010).

Privatisation

If the transition from socialist command economy to market-based economy is one of the greatest transformations of the 20th century, then the privatisation of land is no doubt one of the most dramatic features of this transformation. Since the collapse of socialism in various Eastern Europe and Asian countries and the introduction of market reform policies in China and Vietnam, extensive programmes of land privatisation, also known as decollectivisation, have been launched with the aim of transferring the management and possession of land from former socialist collectives to individual households (Hann et al., 2002; Verdery, 2004).

Behind the privatisation policy is an idea that has been widely advocated by neoliberal commentators and adherents of structural adjustments in the Global South: that land would be used much more effectively if placed in the hands of private and individual users rather than states and collectives (Lipton and Sachs, 1990). The crisis in agricultural production across socialist command economies in the pre-1990s period was largely attributed to the lack of motivation on the part of farmers to invest in increasing productivity because they could not control how land was used and how the fruit from land was distributed (Kerkvliet, 2005). It is thus widely assumed that if households and individuals are given the authority of management and long-term possession over land, they will be much more motivated to invest in the land, purchase mechanised equipment and farm using modern methods. Thus by removing the state from land management, the potential of land will be unleashed with a positive impact on rural livelihoods, particularly on agricultural production.

In practice, the effects of privatisation have substantially varied. Amongst countries where privatisation has been launched, Vietnam has been widely praised as a distinctive example of success. In the late 1980s, arable land hitherto under centralised management of socialist cooperatives was distributed to rural households across the country. With the 1993 Land Law, rural households were granted title over new individual holdings, which they could transfer, give to others and cultivate at will. Privatisation effected immediate positive changes. Farmers across the country responded to the new opportunity to act as land holders with great enthusiasm, investing resources and labour on the land to increase productivity and generate surplus for selling at the market in ways rarely seen during the pre-1990s collectivised period (Lam, 2019b). From a country that experienced serious food shortage in the late-1980s, Vietnam has become a world-leading rice exporter and achieved a ten-fold increase in GDP per capita between 1990 and 2008 (Do and Iyer, 2008; Watts, 2002).

In China, land was also distributed to households on a long-term basis in the 1980s. While Chinese farmers originally benefited substantially from privatisation, the state retained the monopoly of the sale of the most important agricultural inputs, as well as a mandatory procurement system that required farmers to sell a production quota to the state at a low state-fixed price. As a consequence, Chinese households quickly lost the incentive to increase output in the 1990s (Luong and Unger, 1998). The China case shows that privatisation of land does not necessarily lead to an improvement in agricultural production if other factors of production are still centrally controlled by the state (also Kerkvliet and Selden, 1998).

While privatisation in China did not achieve a sustainable boost of agricultural production because the state still retained too much control over the supply of inputs and distribution of agricultural products, in other contexts the stagnation of agriculture has been the result of land privatisation itself. Post-socialist Mongolia is a case in point. Sneath (2004) shows that in the early 1990s, as Soviet-style socialism collapsed, the Mongolian state launched an ambitious privatisation scheme that dismantled pastoral collectives and state farms and distributed pastoral lands and livestock to individual households. Yet the results were disastrous. In the winter and spring of 1999–2000, Mongolia lost 3 million livestock, 10% of the national herd. Sneath (2004) points out that in Mongolia before privatisation, land had been owned and managed collectively, in ways that allowed herders much flexibility to move their livestock across vast open grazing spaces, to utilise different climatic conditions and to get the best from pastures in different seasons. This collective management system was maintained under the socialist centrally planned economy, where state collectives and farms both provided herders with mechanised transportation, such as motorised vehicles and fuel, and ensured access to pasture for everyone. Privatisation, however, disrupted what Sneath called the 'sociotechnical' systems (2004). Because land is now privately owned by those who exert exclusive rights and prevent others from accessing it, livestock can no longer be flexibly moved from localities seriously affected by harsh weather to better locations. And without the support from collectives, relocating livestock across long distance becomes much more difficult. The Mongolian case shows that state management of land is not necessarily ineffective and growth-inhibiting as neoliberal adherents presume, but in specific contexts can be a much more sustainable model of land management than systems based on private management.

Marketisation

If privatisation seeks to increase land use efficiency by transferring the management of land from state and collectives to private actors and individual households, marketisation seeks to maximise the efficiency of land use by fostering a 'natural transition' (Sneath, 2004), the result of which would be the concentration of land in the hands of those who can use it most effectively (Li, 2009; World Bank, 2007). Proponents of land marketisation such as Deininger (2003) and De Soto (2000) believe that the most effective way to foster such a transition is to turn land into a marketable asset, which can be easily marketised and transferred. Those unable to farm effectively can easily relinquish their land to more capable farmers, a move commonly described as 'exiting' agriculture (Li, 2009: 630). As long as the transfer is done by market mechanisms, land will be allocated to those who can and are willing to offer the best price (Akram-Lodhi, 2005; Deininger and Jin, 2003; also see Bernstein and Byres, 2001 and Borras, 2009).

Proponents of land marketisation hold that marketisation would rapidly lead to the replacement of small-scale, subsistence production with large-scale capital-intensive farming, as enterprise-minded farmers would quickly acquire arable holdings from less successful ones to create large, consolidated fields and practice commercial farming in pursuit of profit maximisation. Furthermore, this will be a harmonious, win-win process that benefits even the less successful farmers, because it provides them with an opportunity to 'exit' agriculture, i.e. to cash out their unproductive land so as to relocate to other commercial livelihood options, such as craft production or industrial waged employment (World Bank, 2007).

The results again have been substantially mixed and uneven across rural Asia. In various contexts where land marketisation has been fostered, there has been no 'natural' transition featuring the consolidation of land and the emergence of large-scale mechanised farming. For example, in Vietnam, China and various parts of India, small-scale subsistence farming remains the dominant way of using arable land (Lam, 2019b; Patnaik, 2007).

Amongst reasons that prevent a 'natural' transition, economists have focused largely on policy and legal constraints. In China, for example, land throughout the 1980s and 1990s was legally owned by rural collectives and hence difficult to transfer or mortgage (Guo, 2001). Compared to the Chinese counterpart, the Vietnamese government has implemented stronger marketisation policies. As Ravallion and Walle (2008) argue, in Vietnam, individual households have the rights to exchange, transfer, rent, mortgage and inherit land. Do and Iyer (2008) suggest that Vietnamese households have obtained the *de facto* private property right to land. However, for the economists, those policies were still not enough to create a sense of long-term security for farmers to accumulate land and invest in large-scale production. Under Vietnamese law, all land belongs to 'the entire people' and is managed by the state. The state can take back the land anytime under the pretext of national defence, public interest and development (Kerkvliet, 2014).

Cultural perceptions of land have been another factor that inhibits the transition. As Polanyi argues, land is a 'fictitious' commodity because it is valued in remarkably different ways and in many situations cannot be reduced to a marketable asset (Polanyi, 2001: 187). In Vietnam, China and India, land is not only a means of livelihood, but also a source of security with which farmers are very reluctant to part (for Vietnam see Kerkvliet, 2014; Labbé, 2016; for China see Guo, 2001; Yep and Fong, 2009; for India see Levien, 2012). The reluctance can also be attributed to a sentimental bond between people and the land they have long been attached to, which cannot be reduced to economic calculations, as observed amongst the *Adivasi* (indigenous people) in India (Ong, 2020; Walker, 2009).

Yet the biggest factor that prevents small farmers from relinquishing their land has been their concerns about the absence of promising non-farm economic opportunities. As Li (2009) argues, many rural communities in China and India reject an exit from agriculture not because of a 'sentimental' attachment to land and to locally-oriented production on small family farms. What they reject is actually the terms of their inclusion in off-farm economies. In India for example, the low level of education of rural people means that it would be impossible for them to find jobs in the booming service sector, or in high-tech Special Economic Zones (SEZs) that only offer white-collar jobs available exclusively to well-educated middle-class citizens (Levien, 2012).

This has been shown to be the case in situations where there has indeed been a trend of land consolidation. As many have shown, the transfer of land has been far from a harmonious, mutually beneficial process as proponents of land marketisation expect. The transition has actually benefited only a small group of rich, politically well-connected farmers (Akram-Lodhi, 2005). In contrast, for the majority, the transfer of land is a last resort. The literature on southern Vietnam has documented growing landlessness amongst the rural population, who have no choice but to sell their arable land because they are unable to settle the debt incurred due to output failures, climatic shocks, and ill health (86). Here Ellis's (2000) distinction between diversification out of necessity and out of choice becomes useful. Those small farmers do not sell land as an actively made decision so as to relocate to more productive livelihoods, but because they have been squeezed out by richer farmers and become landless proletariats with no reliable alternative livelihoods in store.

India is another example of rising landlessness as the result of land marketisation policies. The abrupt launch of neoliberal reforms in the 1990s has resulted in the withdrawal of state support for poor farmers, the removal of state subsidies for fertilisers and other agricultural inputs, and growing pressure on peasant production under the effect of WTO-mandated imports of cheap agricultural products from First World countries. These changes significantly reduce the viability and profitability of being small farmers. Poor peasants are driven into the arms of moneylenders who provide loans at usurious rates (Walker, 2009). Many are plunged into indebtedness, thus have to sell their land, a situation detected as early as the 1990s (Agarwal, 1990) and rapidly escalates in the last two decades (Walker, 2009). The selling of land, again, is far from an active choice, but a desperate decision. Between 1993 and 2003, roughly 100,000 Indian peasants took their own lives, a trend engendered mostly by growing indebtedness and landlessness amongst the country's poorest people (Mohanty, 2005).

Dispossession

While the focus of privatisation and marketisation has been on the better utilisation of land by means of reducing state intervention and fostering a 'natural' transition using voluntary market principles, in the last decades there has been an equally remarkable change in rural Asia that involves the transfer of land in the opposite direction: dispossession. This term refers to the acquisition of land by state authorities, who use administrative power to seize land at a low price and transfer to private investors, who put it up for sale or utilise the resources from it for a massive profit. In contrast to privatisation and marketisation, dispossession is a transformation in land tenure that involves strong state intervention and does not follow market logics.

This phenomenon has been far from confined to Asia. Dispossession has become widespread across the Global South, and has been referred to by various terms, such as land seizure, expropriation and land grabbing (Borras et al., 2012: 210; World Bank, 2010). But Asian settings, notably China, Vietnam and India, have become the most prominent sites of dispossession. Whereas land grabbing in Latin America, the Caribbean and Africa has been motivated largely by food security initiatives and fuel security ventures, in China, Vietnam and India, the seizure of land has been driven mostly by development purposes (Borras et al., 2012). In China, it is estimated that between 1987 and 2010, about 52 million farmers were displaced by land acquisition by the state (cf. Ong, 2020). In India, between 20 and 60 million people were affected by land seizure between 1947 and 2007 (Chakravorty, 2013). In Vietnam, official sources reported that between 2001 and 2005 alone, 366,000 hectares of agricultural land were converted into urban and industrial land by the government. This amount accounted for 4% of the total area of Vietnam's agricultural land. 2.5 million farmers were affected by the process (Suu, 2009).

Amongst the reasons that give rise to land seizure in Asia specifically, scholars have highlighted the adoption of neoliberal economic doctrines by governments in Asian marketised contexts as a principal means to achieve growth. Neoliberal doctrines not only promote prioritisation of GDP growth at all cost, but also entail economic liberalisation and fiscal decentralisation, which create both incentives for and pressure on lower-level governments to increase revenue (Ong, 2020; Walker, 2009; Yep and Fong, 2009).

In their quest for increasing revenue, Asian state authorities have found land expropriation a highly effective tool, if not the most lucrative. In China and Vietnam, because legally the state still holds the ultimate power over how land should be used, state authorities have the right to expropriate land from title holders under the excuse of public interests. While farmers are entitled to compensation, the rate of compensation is fixed by the state, and is normally much lower than the actual market value of the land, particularly if the land is later transferred to private developers, increasing its value hundred-fold in the process (Ho and Lin, 2004; Truong and Perera, 2010).

While the state in India does not hold such power as the ultimate manager of land, it has long been able to wield 'the power of the eminent domain' (Kim, 2009) to expropriate land under the guise of 'public purposes'. Because what counts as 'public purposes' has been vaguely defined, land taken for economic development, the construction of industrial zone, and urbanisation can all be justified by this excuse (Ong, 2020). Even when land is taken by private developers, such as for the establishment of Special Economic Zones, private developers still prefer to have the state requisition land on their behalf, because the developers would face high transaction costs to negotiate directly with thousands of farmers (Levien, 2012).

The most visible impact of dispossession has been deprivation of the rural poor. In India, the majority of those whose lands have been taken have not received sufficient compensation, and in many cases no compensation at all because they do not possess written titles to the land (Walker, 2009). Levien (2012) shows that farmers dispossessed by SEZs have found only low-paid jobs such as security guards, janitors and drivers because of their low level of education and English fluency. Yet even those jobs are too limited compared to the total number of the dispossessed. In Vietnam, it is estimated that merely 3–5% of people whose land has been requisitioned have found new jobs in the service and industrial sectors, mostly because their limited social and cultural capital makes them unqualified employees (Suu, 2009). And in China, for many dispossessed peasants, being forced to give up their lands means 'no land to farm, no job to take, no right to pension' (Yep, 2012).

Land dispossession has thus become the most contested form of market changes and neoliberal reforms in the Global South and in Asia particularly. In China, in the 2010s, land-related conflicts have been the single largest source of protest (Ong, 2020). In India, land seizure has been regarded as the country's 'single most contentious issue' (cf. Walker, 2009: 590). In Vietnam, in the last two decades, the country well-known as a case of successful and peaceful marketisation has become a site of dramatic protests against land appropriation projects initiated by state authorities to establish industrial parks in rural areas and residential complexes in urban contexts, with 70% of civil cases lodged with district and provincial courts involving land appropriation. In most cases, the disputes have engendered violent confrontations between the state and affected communities (Kerkvliet, 2014).

The dramatic rise in the number and magnitude of land-related conflicts in rural Asia can be attributed to several factors. The most common cause has been inadequate compensation. As mentioned above, many farmers in India receive minimal and even no compensation for the land taken (Walker, 2009). In Vietnam and China, the compensation rate is usually fixed by state authorities based on the agricultural value of the land, rather than its actual market value once the land is converted into commercial and private use. It is this sense of injustice that drives farmers to protest as soon as they realise the vast difference between the compensation they receive and the profit that is generated from land transactions (for Vietnam see Kerkvliet, 2014; for China see Cai, 2003; Guo, 2001). In Vietnam, China and India, resistance has also been driven by villagers' outrage over corruption and collusion between state officials and private developers, particularly in China where local governments hold monopoly power in land acquisition (Cai, 2003; Ong, 2020). Another cause of rural agitation against dispossession is the lack of sustainable alternative off-farm opportunities for those whose land has been taken. This situation is particularly severe in India, where land has been widely taken for the construction of SEZs that focus mostly on the high-tech service sector with little capacity to absorb the labour ejected from agriculture (Levien, 2012).

Yet dispossession should not always be seen as a threat for the farmers, and resistance should not always be seen as a desperate act of those seeking to retain land at all costs or cling onto a life of basic subsistence. There are situations in which land seizure can become an opportunity, and farmers who protest are actually seeking to maximise their gain from land transactions. Yep (2012) notes that many land-related protests break out in prosperous regions of China, where rural households have gained access to multiple off-farm activities and thus are no longer dependent on the land for subsistence. Farmers in those regions are enthusiastic about surrendering their land. For them, the resistance against land dispossession is not a desperate effort to retain their land, but an active, carefully calculated choice to negotiate a fairer share of the premiums from land transactions (for Vietnam see Lam, 2019a).

Key points

In the last four decades, land tenure systems in the Global South generally and in Asia particularly have experienced three profound transformations with tremendous impact on rural livelihoods: privatisation, marketisation, and dispossession. Key points about the impact of those processes on livelihoods in rural Asia include:

- Privatisation of land in some contexts can produce a take-off in agricultural production, but in others can lead to disastrous consequences.
- Marketisation does not necessarily foster a 'natural' transition from small-scale subsistence farming to large-scale commercial farming. When a transition does occur, it is rarely a harmonious, mutually benefiting process, but often results in rising inequality, landlessness, and proletarianisation.
- Dispossession, while generally a threat to rural households' livelihoods, can in certain circumstances be an opportunity for those in better-off regions to maximise the gain from land transactions.
- The impact of land tenure changes on rural livelihoods in Asia specifically and in the Global South more broadly has been shaped by various factors: state policies, legal frameworks governing land ownership and transaction, cultural perceptions of land, the availability of off-farm employment opportunities and alternative livelihoods, and the economic conditions and choices of those affected.

Further reading

Bernstein, H. and Byres, T. (2001) 'From peasant studies to agrarian change', *Journal of Agrarian Change*, 1(1): 1–56. https://doi.org/10.1111/1471-0366.00002

Borras, S.M. Jr., Franco, J., Gómez, S., Kay, C. and Spoor, M. (2012) 'Land grabbing in Latin America and the Caribbean', *Journal of Peasant Studies*, 39(3–4): 845–872. https://doi.org/10.1080/03066150.2012.679931

Humphrey, C. and Verdery, K. (eds.) (2004) *Property in question: Value transformation in the global economy*, Oxford: Berg.

Ong, L. (2020) '"Land grabbing" in an autocracy and a multi-party democracy: China and India compared', *Journal of Contemporary Asia*, 50(3): 361–379. https://doi.org/10.1080/00472336.2019.1569253

References

Agarwal, B. (1990) 'Social security and the family: Coping with seasonality and calamity in rural India', *Journal of Peasant Studies*, 17(3): 341–412. https://doi.org/10.1080/03066159008438426

Akram-Lodhi, A. (2005) 'Vietnam's agriculture: Processes of rich peasant accumulation and mechanisms of social differentiation', *Journal of Agrarian Change*, 5(1): 73–116. https://doi.org/10.1111/j.1471-0366.2004.00095.x

Bernstein, H. and Byres, T. (2001) 'From peasant studies to agrarian change', *Journal of Agrarian Change*, 1(1): 1–56. https://doi.org/10.1111/1471-0366.00002

Borras, S. (2009) 'Agrarian change and peasant studies: Changes, continuities and challenges – an introduction', *Journal of Peasant Studies*, 36(1): 5–31. https://doi.org/10.1080/03066150902820297

Borras, S.M. Jr., Franco, J., Gómez, S., Kay, C. and Spoor, M. (2012) 'Land grabbing in Latin America and the Caribbean', *Journal of Peasant Studies*, 39(3–4): 845–872. https://doi.org/10.1080/03066150.2012.679931

Burawoy, M. and Verdery, K. (eds.) (1999) *Uncertain transition: Ethnographies of change in the postsocialist world*, New York: Rowman and Littlefield.

Cai, Y. (2003) 'Collective ownership or cadres' ownership? The non-agricultural use of farmland in China', *The China Quarterly*, 175: 662–680. https://doi.org/10.1017/S0305741003000890

Chakravorty, S. (2013) *The price of land: Acquisition, conflict, consequence*, New Delhi: Oxford University Press.
De Soto, H. (2000) *The mystery of capital: Why capitalism triumphs in the West and fails everywhere else*, New York: Basic Books.
Deininger, K. (2003) *Land policies for growth and poverty reduction. A world bank policy research report*, Washington, DC and Oxford: World Bank, Oxford University Press.
Deininger, K. and Jin, S. (2003) 'Land sales and rental markets in transition: Evidence from rural Vietnam', *Policy Research Working Paper No. 3013*, Washington, DC: World Bank.
Do, T. and Iyer, L. (2008) 'Land titling and rural transition in Vietnam', *Economic Development and Cultural Change*, 56(3): 531–579. https://doi.org/10.1086/533549
Ellis, F. (2000) 'The determinants of rural livelihood diversification in developing countries', *Journal of Agricultural Economics*, 51(2): 289–302. https://doi.org/10.1111/j.1477-9552.2000.tb01229.x
GRAIN (2008) 'SEIZED! The 2008 land grab for food and financial security', *GRAIN Briefing*. Available at: http://farmlandgrab.org (Accessed October 2008).
Guo, X. (2001) 'Land expropriation and rural conflicts in China', *The China Quarterly*, 166: 422–439. https://doi.org/10.1017/S0009443901000201
Hann, C. (ed.) (2002) *The postsocialist agrarian question*, Münster: Lit Verlag.
Ho, S.P.S. and Lin, G.C.S. (2004) 'The state, land system, and land development processes in contemporary China', *Annals of the Association of American Geographers*, 95(2): 411–436. https://doi.org/10.1111/j.1467-8306.2005.00467.x
Kerkvliet, B. (2005) *The power of everyday politics: How Vietnamese peasants transformed national policy*, Ithaca: Cornell University Press.
Kerkvliet, B. (2014) 'Protests over land in Vietnam: Rightful resistance and more', *Journal of Vietnamese Studies*, 9(3): 19–54. https://doi.org/10.1525/vs.2014.9.3.19
Kerkvliet, B. and Selden, M. (1998) 'Agrarian transformation in China and Vietnam', *The China Journal*, 40: 37–58. https://doi.org/10.2307/2667453
Kim, A. (2009) 'Land taking in the private interest: Comparisons of urban land development controversies in the United States, China and Vietnam', *Cityscape: A Journal of Policy Development and Research*, 11(1): 19–32.
Labbé, D. (2016) 'Critical reflections on land appropriation and alternative urbanization trajectories in periurban Vietnam', *Cities*, 53: 150–155. https://doi.org/10.1016/j.cities.2015.11.003
Lam, M. (2019a) '"Extremely rightful" resistance: Land appropriation and rural agitation in contemporary Vietnam', *Journal of Contemporary Asia*, 49(3): 343–364. https://doi.org/10.1080/00472336.2018.1517896
Lam, M. (2019b) 'Negotiating uncertainty in late-socialist Vietnam: Households and livelihood options in the marketizing countryside', *Modern Asian Studies*, 53(6): 1701–1735. https://doi.org/10.1017/S0026749X17000993
Levien, M. (2012) 'The land question: Special economic zones and the political economy of dispossession in India', *Journal of Peasant Studies*, 39(3–4): 933–969. https://doi.org/10.1080/03066150.2012.656268
Li, T.M. (2009) 'Exit from agriculture: A step forward or a step backward for the rural poor?', *Journal of Peasant Studies*, 36(3): 629–636. https://doi.org/10.1080/03066150903142998
Lipton, D. and Sachs, J. (1990) 'Privatization in eastern Europe: The case of Poland', *Brookings Papers on Economic Activity*, 2(1990): 293–341.
Luong, H.V. and Unger, J. (1998) 'Wealth, power, and poverty in the transition to market economies: The process of socio-economic differentiation in rural China and northern Vietnam', *The China Journal*, 40: 61–93. https://doi.org/10.2307/2667454
Mohanty, B. (2005) 'We are like the living dead: Farmer suicides in Maharashtra, Western India', *Journal of Peasant Studies*, 32(2): 243–276. https://doi.org/10.1080/03066150500094485
Ong, L. (2020) '"Land grabbing" in an autocracy and a multi-party democracy: China and India compared', *Journal of Contemporary Asia*, 50(3): 361–379. https://doi.org/10.1080/00472336.2019.1569253
Patnaik, U. (2007) 'New data on the arrested development of capitalism in Indian agriculture', *Social Scientist*, 35(7–8): 4–23.
Polanyi, K. (2001) *The great transformation: The political-economic origins of our time*, Boston: Beacon Press.
Ravallion, M. and van de Walle, D. (2008) *Land in transition: Reform and poverty in rural Vietnam*, Washington, DC: Palgrave Macmillan and the World Bank.
Sneath, D. (2004) 'Proprietary regimes and sociotechnical systems: Rights over land in Mongolia's "age of the market"', in Humphrey, C. and Verdery, K. (eds.), *Property in question: Value transformation in the global economy*, Oxford: Berg, pp. 139–160. https://doi.org/10.4324/9781003086451

Suu, N. (2009) 'Agricultural land conversion and its effects on Vietnamese farmers', *Focaal – European Journal of Anthropology*, 54(2): 106–113. https://doi.org/10.3167/fcl.2009.540109

Truong, T.T. and Perera, R. (2010) 'Consequences of the two-price system for land in the land and housing market in Ho Chi Minh City, Vietnam', *Habitat International*, 35: 30–39. https://doi.org/10.1016/j.habitatint.2010.03.005

Verdery, K. (2004) 'The obligations of ownership: Restoring rights to land in postsocialist Transylvania', in Humphrey, C. and Verdery, K. (eds.), *Property in question: Value transformation in the global economy*, Oxford: Berg, pp. 139–160. https://doi.org/10.4324/9781003086451

Walker, K. (2009) 'Neoliberalism on the ground in rural India: Predatory growth, Agrarian crisis, internal colonization, and the intensification of class struggle', *Journal of Peasant Studies*, 35(4): 557–620. https://doi.org/10.1080/03066150802681963

Watts, M. (2002) 'Agrarian Thermidor: State, decollectivization, and the peasant question in Vietnam', in Szelényi, I. (ed.), *Privatizing the land: Rural political economy in post-communist societies*, London: Routledge, pp. 149–190.

World Bank (2007) *World development report 2008: Agriculture for development*, Washington, DC: World Bank.

World Bank (2010) *Rising global interest in farmland: Can it yield sustainable and equitable benefits?* Washington, DC: World Bank.

Yep, R. (2012) 'Containing land grabs: A misguided response to rural conflicts over land', *Journal of Contemporary China*, 22(80): 273–291. https://doi.org/10.1080/10670564.2012.734082

Yep, R. and Fong, C. (2009) 'Land conflicts, rural finance and capacity of the Chinese state', *Public Administration and Development*, 29: 69–78. https://doi.org/10.1002/pad.498

Zoomers, A. (2010) 'Globalisation and the foreignisation of space: Seven processes driving the current global land grab', *Journal of Peasant Studies*, 37(2): 429–447. https://doi.org/10.1080/03066151003595325

Zoomers, A. and Van de Haar, G. (eds.) (2000) *Current land policy in Latin America: Regulating land tenure under neoliberalism*, Amsterdam: KIT Press.

INDEX

Note: Page numbers in *italic* indicate a figure or table and page numbers in **bold** indicate a box on the corresponding page.

accountability *see* social accountability
activity choice approach 97–98
actor-oriented approach 471, 476–477
adaptation 481–487; hiding vulnerability behind 48–50; pastoralist 306–307; *see also* livelihood adaptation
agency 20–28, 31–34, 37–39, 45–52, 160–161, 361–363, 495; youth protagonism **232**
agrarian questions 284–285, 294–296; and agroecological alternatives 285–290; and rural development policy 290–294
agrarian reforms 207–208, 439
agricultural livelihoods *see* livelihoods, agricultural
agricultural policies 47, 184–185
agroecological practices 174, 285–290, 295
artisanal and small-scale mining (ASM) 345–353; employment estimates *348*; global trends *347*
aspirations **231–232**, 495–496
assets 14–15, 24, *94*; asset approach 96–97; and capabilities 24
asylum seeker 372; definition of **372**

Bangladesh 24–27, **26**, 363–365, 369–371, 463–465
behaviour *see* human behaviour
Bolivia 93–98, 209–212; neoliberalism in **211**
Bolsa Familia (BF) 445–446, **446**
bonding capital 59–60, 452
Botswana 184–188, 302–303; government response to drought in **188**
Brazil 134–136, 142–143, 327–329; PV with recycling cooperatives in **136**; social protection in 445–447
bridging capital 59–60

Burkina Faso 134–143; participatory diagramming with disabled people in *94*; training in PV with disabled people in **136, 138**, *140*
Bus Rapid Transit project 425–429

Cambodia 260–261; variations in absolute and relative environmental household-level income patterns *261*
capabilities **21**; and assets 24; collective 451; in existing livelihood approaches 23–24; *see also* Capability Approach (CA); collective capabilities framework; household capabilities
Capability Approach (CA) 20–28, *22*; key concepts **21**
capability theories 20–28
caste 326–328, **328**, 331
Chile 291–295; rural development projects *293*
choice experiment approach 98
choice of livelihood *see* livelihood, choice of
choice variables 95–98
citizenship 371–372
class 325–328, **327**, 330–331
climate change 44–45, 288–289, 295–296, 307–308, 462–463, 481–487; FPE in the context of 175–176
Cochabamba water wars **211**
collective action 240–241, 450–456, *451*, 486
collective capabilities framework 451, *451*
collective livelihoods *see* livelihoods, collective
collective organisations 450–456
colonial planning 336–337
colonisation 335–341
communities 250–253; heterogeneity of forest use within 274–275; lower-income 337–338

519

community-based development 246–254
community-based rehabilitation (CBR) 246, 250–254, *251*
complexity 36–40, 104–111, 324–331, 417–418
compulsory labour **376**
conflict-affected settings 501–507
conservation efforts 276; and pastoral livelihoods 305–306
consumption 62–64, 259–263, 398–399, 408–409, 440–442
coordination, livelihood 505–506
coping responses 317
co-production 94, 98, 242–243, 290
crisis, sustainability 285–290, 296
Critical Institutionalism (CI) 31, 34–40
cultural rights 69–70, 206, 208, 211–213
cultures 114–115, 212–213; interpretation and writing of 119–121

decent work *see* work, decent
Delhi **328**, 330, **331**
demobilisation *see* disarmament, demobilisation and reintegration (DDR)
democratic politics 181–190
Deonar dumping ground **163**, *164*
Department for International Development (DFID) 4; Sustainable Livelihoods Framework (SLF) 6, 10–17, *13*, 23–24, 36, 493
dependency 419–421
development 246–254; key features and implications for 69–70; rights-based approach to 68–69; rural projects *293*; social protection in 437–438, **439**
development, agricultural 285–287, 290–292, 295–296, 387–388
development, economic 210–211, 382–390
development, human capital 382, 387–390
development, inclusive 246–254, 389, 490–497
development, rural 284–285, *293*; and agroecological alternatives 285–290; and new agrarian questions 294–296; policy 290–294
disability 246–254; participatory diagramming *94*; training in PV **136, 138**, *140*
Disabled People's Organisations (DPOs) 247–248, 250–251, 253–254
disarmament, demobilisation and reintegration (DDR) 490–497; key terms *491*
disaster vulnerabilities 460–466
diversifying livelihoods 364–365
dominant income source approach 95, 99
drought 187–189, **188**, 262–263, 303–306, 442–445, 482–485
drylands 302–308

economic development *see* development, economic
economics, new 361–362

economy, informal 237–243; *see also* livelihoods, informal
eco-territorial turn 208–209
education 217–224; Bolsa Familia's impact on **446**; indicators *199*
education-livelihoods relationship 7, 217–218, 222–224
Egypt **25**, 292, 384–386, 404
elite-oriented political settlements 182–184
embodied engagements 82, 85–87, 90
emotional engagements 82, 85–87
empowering approach 247
empowerment 71, 159–161, 166–167, 246–247, 250–252, 407–409, 440–442; Mobile Money and women empowerment 398
entrepreneurship 384, 389–390, 407–409
environmental change 51; pastoralists adaptations to 306–307
environmental income 259–266, *261–262*
environmental rights 69, 206, 209–210, 213
epistemologies 82–84, **83**, 171–172
ethics 153–154
Ethiopia 292–295, 446–447, 452–453; rural development projects *293*; social protection in 443–445
ethnography 109–111, 114–122, **119**
exploitation 171–173, 371–372, 377–378, 419–421, 502–504, **503**
extractivism 209–210

families 142–143, 250–252, 369–370, 387–388, 398–399
famine relief 187–189
farmers *see* smallholder farmers
feminisation: of agricultural tasks 288; of migration 370–371
feminism **88–89**, 170–176, 369–370
Feminist Political Ecology (FPE) 170, 172; in the context of climate change 175–176; Latin American FPE (LAFPE) 170, 173–175; urban 174–175
fieldnotes 117–121
financial inclusion 393–400, 408–409
financial institutions (FIs) 394–396, 405–409
financial services 393–400, 403–409, 452–453
fisheries: and poverty 315–317; *see also* fisheries management; livelihoods, fisheries
fisheries management 38, 313, 316, 318, 454
fishing 38, **39**, **241–242**, 312–319, 465–466, **503**
food security 174–175, 398–399, 443–445, 461–462, 465–466, 505–506
forced labour 317, 376–377; definition of **376**
forests and groves 271–280, **276–277**; *see also* livelihoods, forest
formalising 324–331; safeguarding livelihoods through 350–352

Index

frameworks 10–18, 20–27, 35–36, 68–71, 159–167, 349–352, 450–452, 461–466
freedom 20–22, **21**, 375–376
functionings 20–25, **21**
future aspirations **231**

gender 15–18, 26, 38, 71, 73; and agricultural livelihoods 288–289, 482, 485–487; and artisanal mining 350; and collective organisations 454–455; complexity and heterogeneity in the informal economy of waste 326–331; and conflict-affected settings 506–507; feminist political ecology 170–176; fisheries livelihoods 312–319; and forests 275; and the informal economy of waste **327**; and migration 362, 370–371; and religion 475–477; and research 89, 107, 118; and social protection 442–443; and vulnerability **364**; and youth livelihoods 228–233
generations 33, 72, 110, 227–228
global markets (GMs) 211, 237, 415–421
Global South 3–8
gold miners *346*, 350–351
governance 25, 237–243; *see also* social governance
government response to drought in Botswana **188**
GPS technology: contemporary GPS tracking 148; contextualising data 149–150; to develop livelihood mobility research 148–149; in research practice 147–148; *see also* participatory GPS methods
Greater Serengeti-Mara Ecosystem *see* Serengeti-Mara Ecosystem
green revolution **49**, 288
Guyana 134, 139, 254; PV with community members on natural resource management in **139**

health 387–388; indicators *199*; and Mobile Money 399; *see also* primary health care (PHC)
higher-productivity livelihoods 185–187
'horizontal' processes 162–167
household: household approaches 361–362; household capabilities 62–64, *62*
human behaviour 31–37
human capital development 382, 387–390
human capital theory (HCT) 217–219, 222–224
human rights 68–74, 195–196, 369–378
Hyderabad **200**, 200–201, 425

immigration policy 371–372; *see also* migration
immobility **26**, 231; *see also* mobility
inclusion *see* development, inclusive; financial inclusion; social inclusion
inclusive community-based development 246–254; *see also* development, inclusive
inclusivity 339, 431, 455
income: dominant income source approach 99; income approach 95–97; variations in absolute and relative environmental household-level income patterns *261*; variations in income composition *262*; *see also* environmental income
India *326*, 368–369, 378–379; dynamics of gender, class and caste in informal waste work **327**; experiences of abuse and slavery like situations 376–378; human rights and international migration 372; implications of heterogeneity for worker organising in **331**; international migrant rights conventions 373–374; livelihood strategies **221**; migrant workers' rights 374; migration, rights and livelihoods 369–371; overview of rights 371–372; politics of representation in **328**; purpose, reform and analysis of the Kafala system 374–376; social protection in 441–443; structure of the informal economy of waste in **326**
informal economy *see* economy, informal; informal waste sector; livelihoods, informal
informality 162–163, 237–241, 337–341, 352–353
informal livelihoods *see* informal waste sector; livelihoods, informal
informal waste sector **163**, 324–331, *326*; dynamics of gender, class and caste in **327**; politics of representation **328**; structure of **326**
institutions and institutionalisation 31–40, 44–47, 57–61, 70–72; contextualising livelihoods 451–456, 485–486; enabling livelihoods 394–397, 403–409; generating livelihoods 272–279, 304–307; institutional arrangements 36, 455–456; negotiating livelihoods 159–167, 194–203, 208–212, 239–241
intergenerationality 227–233, 228–230
international migrant rights conventions 373–374
international migration 368–369, 378–379; experiences of abuse and slavery like situations 376–378; and human rights 372; international migrant rights conventions 373–374; migrant workers' rights 374; migration, rights and livelihoods 369–371; overview of rights 371–372; purpose, reform and analysis of the Kafala system 374–376; *see also* migration
interpretation of cultures 119–121
irregular migrants 360, 374; definition of **360**, **372**; *see also* migrants

Kafala 368–369, 378–379; experiences of abuse and slavery like situations 376–378; human rights and international migration 372; international migrant rights conventions 373–374; migrant workers' rights 374; migration, rights and livelihoods 369–371; origin and meaning of **374**; overview of rights 371–372; purpose, reform and analysis of 374–376

Index

labour: abuses 377–378; exploitation 377–378; forced or compulsory 376, **376**; and sustainability crisis 285–290
land 284–285; and agroecological alternatives 285–290; and new agrarian questions 294–296; and rural development policy 290–294
land management 307, 337, 445, 510–512
land tenure 304–305, 510–516
Latin American FPE (LAFPE) 170, 173–175; *see also* Feminist Political Ecology (FPE)
Latin American political ecology 171–172; *see also* political ecology
legal framework of human rights 372–376
linking capital 59–60
livelihood, choice of 473–474, 477
livelihood activities 93–100, 106–107, 110–111, 397–400, 502–506
livelihood adaptation 23, 26–28, 45, 49
livelihood definitions **11**
livelihood elements *94*
livelihood mobility 148–149, 155
livelihood outcomes 95–96, 105–111, 397–399
livelihood pathways **443–444**; contested 50–51
livelihoods 3–8; and artisanal mining 345–353; and bus rapid transit 425–429; and the Capability Approach 20–28; and collective organisations 450–456; concepts and frameworks 10–18; in conflict-affected settings 501–507; contested pathways 50–51; and critical understanding 81–90; and DDR 490–497; and democratic politics 181–190; and education 217–224; exploitation of **503**; and gender *364*; and governance 237–243; higher-productivity 185–187; and institutions 31–40; linked origins of vulnerability and 45–46; and longitudinal research 104–111; and migration 359–365, 369–371; and Mobile Money 393–400; networks for 61–62; in the occupied Palestinian territory **165**; and participatory GPS methods 147–155; and pastoralism 302–308; and power 159–167; and social accountability 194–203; and social movements 206–213; and social protection 437–447; and vulnerability **364**
livelihoods, agricultural 284–285, 481–487; and agroecological alternatives 285–290; and new agrarian questions 294–296; and rural development policy 290–294
livelihoods, collective 493–494
livelihoods, diversifying 16, 364–365
livelihoods, fisheries 312–319
livelihoods, forest 271–280, **273**, **277**, 289
livelihoods, informal **163**, 237–242; governance of 242–243; social governance of **241**; *see also* informal waste sector
livelihoods, mediating 36, 403–409; terminology 414
livelihoods, migrant 359, 362–363
livelihoods, pastoralist **25**, 302–308
livelihoods, politics of 182, 190, 495
livelihoods, rebuilding 460–466
livelihoods, rural *63*, 93–101, **221**, 259–266; India **221**; key PRA tools for researching **126**; rural livelihood strategies models *63*; South Africa **221**
livelihoods, spaces for 26, *164*
livelihoods, sustainable 68–74, 246–254, 335–341; ASM as a driver of 349–350; *see also* sustainable livelihoods approach (SLA)
livelihoods, urban 20, 26, 190, 228, 335–341, 503; and bus rapid transit 428–430
livelihoods, women's **26**, 149, 174, 476
livelihoods, youth 107, 227–233, **231**
livelihoods analysis 17–18, *17*, 44–46, 48, 51–52; unmasking vulnerability with 46–47
livelihoods approach 10–12, 16–17, 24, 47, 464–466, 471–473, 477; *see also* Sustainable Livelihoods Approach (SLA)
livelihoods policy 40; PV in 142–143
livelihoods practices **165**, **241**; PV in 142–143
livelihoods research *see* research
livelihoods studies 95, 110, 470–478
livelihood strategies 93–101; in conflict-affected settings 503–505; India **221**; rural *63*, **221**; South Africa **221**
livelihood support 316, 491, 505–506
localised approach to planning 340
longitudinal research 104–111; *see also* research

Mahatma Gandhi National Rural Employment Guarantee Act (MGNREGA) 441–443
Mainstream Institutionalism (MI) 34–37, 39–40
marketisation 176, 510–516
markets *see* global markets (GMs)
mediating livelihoods *see* livelihoods, mediating
Mexico 383–384, 405–406, 418, 425; network as household capability for survival in 62–64
micro-enterprises 247, 407–408
microfinance 403–409; terminology 414; types of institutions *406*
migrants 362–363, 368–369, 378–379; experiences of abuse and slavery like situations 376–378; irregular **372**; meaning of **371**; rights and livelihoods 369–376; in South-South migration 359–360, **360**
migration 359–365, 368–369, 378–379; experiences of abuse and slavery like situations 376–378; human rights and international migration 372; international migrant rights conventions 373–374; migrant workers' rights 374; overview of rights 371–372; purpose, reform and analysis of the Kafala system 374–376; and rights and livelihoods 369–371; *see also* migrants; South–South migration (SSM)

militant engagement **120**
mining *see* artisanal and small-scale mining (ASM); gold miners
Mobile Money (MM) 393–400
mobility **26**, 147–149, 153–155, 230–231, 303–307, 370–371, 425–427
money *see* Mobile Money (MM)
Morocco **25**, 362, 365
multiculturalism 208, 212–213
Mumbai: spaces for livelihoods in *164*; waste pickers in non-recognised slums **163–164**

natural resource management 124, 134, **139**, 142, 304–305; community-based (CBNRM) 306, 450, 453–455
neoliberalism 4–5, 208–213, **211**, 289–291, 295–296, 416–419, 511–514
Nepal **86–87**, 97–98, 263–265, 275–276, 370–371; Forest User Groups in 454; PV with marginalised women in **139**, 142
networks of social exchange (NSE) 56–57, 59–65; *see also* social networks
new economics *see* economics, new
New Rural Paradigm 289, 291, *291*

occupation-based approach 99
occupied Palestinian territory (oPt) **165**, *166*
Organisation for Economic Co-operation and Development (OECD) *293*; New Rural Paradigm 289, 291, *291*

Palestine *see* occupied Palestinian territory (oPt)
participant observation 115–121, **118**
participation: case study from Tanzania 151–152
participatory diagramming *94*, **136**, *137*
participatory GPS methods 147–155
participatory methods 110–111, 118–119, 121, 124, 135
Participatory Rural Appraisal (PRA) 124, 129–130; genealogy of 125–126; heterogeneity and local perspectives 127–128; key tools for researching rural livelihoods *126*; as a package of methods 126–127
participatory video (PV) 134–144; with community members on natural resource management in Guyana **139**; with marginalised women in Nepal **139**; with persons with disabilities in Burkina Faso **136**, **138**, *140*; with recycling cooperatives in Brazil **136**
pastoralism 302–308
pastoralist livelihoods *see* livelihoods, pastoralist
piracy 502–504, **503**
planning for sustainable urban livelihoods 335–341
policies 284–285; agricultural 184–185; and agroecological alternatives 285–290; forest 277–279; immigration 371–372; and new agrarian questions 294–296; pastoralists adaptations to policy change 306–307; rural development policy 290–294
political commitment 338, 340–341
political ecology 170–176, 284–285; and agroecological alternatives 285–290; Latin American 171–172; and new agrarian questions 294–296; and rural development policy 290–294
political economy of livelihoods 16–17
political settlements, elite-oriented 182–184
politics *see* democratic politics; livelihoods, politics of; political commitment; political ecology; political economy of livelihoods; political settlements, elite-oriented; politics of representation
politics of representation 85, **328–329**
positionalities 81, 84–90
post-disaster policies **464**
poverty 12–13, *264*, 384–387; and disability 249; and fisheries 315–317; and forests 272–274; graduation out of **441**, **443**; *see also* poverty reduction; Productive Safety Net Program (PNSP)
poverty reduction 263–265
power 159–167; in livelihoods research 82–88
primary health care (PHC) 201, **202**, 250
privatisation 210–211, 304–306, 324–325, 329–330, 510–516
producer organisations 450, 452–456
Productive Safety Net Program (PNSP) 437, 443–445, **443–445**
productivity 183–184, 305–308, 384–388; higher-productivity livelihoods 185–187
protagonism **232**, 233
Pune **328**, 330, **331**, 425

qualitative approaches 108–110
quantitative approaches 93–101, 106–108

race 335–341
rapid transit 425–431
rebuilding livelihoods *see* livelihoods, rebuilding
reciprocity 56–65, 115–116
recycling cooperatives 134, **136**, 142, 143
reflexivity 85–90, **88**, 119–120, 142–144
reforms, agrarian 207–208, 439
refugees 372, 503–504; definition of **360**, **372**
regulation 185–186, 237–243, 303–305, 405–406, 408–409
reintegration *see* disarmament, demobilisation and reintegration (DDR)
religion 232–233, 470–478
remittances 382–390, *383*; diversifying livelihoods through 364–365
representation: in livelihoods research 87–88, **88**; politics of 85, **328–329**

research 104–111, 114–122, 124–130, 134–144; critical understanding 81–82; epistemologies **83**; GPS in 147–149; key PRA tools for researching rural livelihoods *126*
researchers 82–88
research practices 81–88, 90, 121, 148
resilience 44–52, **49**; socio-economic 388–389
resource-user committees (RUCs) 453–454
rights 369–376; cultural 211–212; environmental 209–210; human rights 372–376; territorial 207; *see also* rights-based approach (RBA)
rights-based approach (RBA) 68–71; complementarities between a SLA and 71–72; nexus of SLA and 72–74
risk 44–49, 118–119, 219–220, 336–340, 417–421; disaster risk 460–466
Rufiji River floodplain 38, **39**
rural appraisal *see* participatory rural appraisal
rural development *see* development, rural
rural livelihoods *see* livelihoods, rural; livelihood strategies
rural livelihood strategies models *63*; *see also* livelihoods, rural; livelihood strategies
Rwanda 49, **49–50**, 73–74, 96, 107–108, 428, 453

safety nets 64–65, 186–187, 262–263, 364–365, 437–439
sanctions 32–33, 37–38, 60–61, 194–203
savings and credit groups 8, 450–456
scarcity 114–115, 338–339
schooling **446**; *see also* education; school systems
school systems 219–220
security 15–16, 490–497; food security 174–175, 398–399, 443–445, 461–462, 465–466, 505–506; land security 208; water security 209
sedentarisation 304–305
Sen, Amartya 20–27, 69, 187, 451, 473; *see also* Capability Approach (CA)
Serengeti-Mara Ecosystem: variations in absolute and relative environmental household-level income patterns *261*; variations in income composition *262*
Sierra Leone 253, 349–353, 504; gold miners in *346*, *351*; social governance of informal livelihoods in **241**
slavery 369, 376–377, 476; definition of **376**
slums **163–164**, *164*, 228–230
smallholder farmers 47, 481–487
small-scale fisheries 313; *see also* livelihoods, fisheries
social accountability 194–203
social capital 56–65, 451–452
social governance 240–241, **241**, 243
social groups 50–51, 61, 159–161, 217–219
social inclusion 399
social movements 206–213
social networks 56–65, 230–231, 360–362, 398–399, 452–453

social protection 187–189, **439–440**, 446–447; in Brazil 445–446; as a development strategy 437–439; in Ethiopia 443–445; in India 441–443; and livelihoods 439–441; rationales for **442**
social structures 22–26, 56–58, 61, 166–167
socio-economic resilience 388–389
Somalia 404, 462, 495, 502–505, **503**
sources of vulnerability 14, 313, 317, 319; *see also* vulnerability
South Africa 47, 99, 181–183, 187–189, 218–220, 336–340, 404–408; livelihood strategies **221**
South-South migration (SSM) 359, **360**, 363–365
spaces for livelihoods *see* livelihoods, spaces for
state regulation 237–243, 352; *see also* regulation
strategies 15–16, *94*; *see also* livelihood strategies
structural change 185–187
support, livelihood *see* livelihood support
sustainability crisis 285–290, 296
sustainable approach 248; *see also* Sustainable Livelihoods Approach (SLA)
sustainable livelihoods *see* livelihoods, sustainable; Sustainable Livelihoods Approach (SLA); Sustainable Livelihoods Framework (SLF)
Sustainable Livelihoods Approach (SLA) 247–248; complementarities between a RBA and 71–72; nexus of RBA and 72–74
Sustainable Livelihoods Framework (SLF) 6, 10–17, *13*, 23–24, 36, 493

Tanzania 98, 181–184, 188–190, 262–263; fishing 38, **39**; increasing participation 151–152; variations in absolute and relative environmental household-level income patterns *261*; variations in income composition *262*
technology *see* GPS technology
tenure *see* land tenure
territorial rights 206–209, 213
territorial turn 207–209
trafficking 376; debates 377; definition of **376**
transforming structures and processes 10, 13–16
transient poor *264*
transport *see* urban transport systems
trans-scalar connections 415–421
trust 57–61, 452–456

United Nations (UN) 68, 212, **228**, 353, 370, 373, 490–491, 506; Convention on the Rights of Persons with Disabilities (CRPD) 246–249; Food and Agriculture Organization (FAO) 260, 271, 312–313, 505; International Strategy for Disaster Reduction (UNISDR) 460–461; Sustainable Development Goals (SDGs) 48, 198, 219, 223, 345, 389–390, 428
urban FPE 174–175; *see also* Feminist Political Ecology (FPE)
urban livelihoods *see* livelihoods, urban
urban transport systems 425–429

variables, choice 95–100
video *see* participatory video
vulnerability 13–14, 44–52, **49**, **364**, 417–419; and migrant livelihoods 362–364; sources and coping responses 317; *see also* disaster vulnerabilities

WASH (water, sanitation, and hygiene) 194–195, 198, *199*, 202, 340
waste *see* informal waste sector; wastelands
wastelands 277, **277**
water **200**, 210–211, 284–285; and agroecological alternatives 285–290; Cochabamba water wars **211**; and new agrarian questions 294–296; and rural development policy 290–294
wellbeing 12–17, 20–27, 104–111, 239–243
women 172–173, 368–369, 378–379; experiences of abuse and slavery like situations 376–378; human rights and international migration 372; international migrant rights conventions 373–374; and microfinance 408–409; migrant workers' rights 374; migration, rights and livelihoods 369–371; Mobile Money and empowerment 398; overview of rights 371–372; purpose, reform and analysis of the Kafala system 374–376; PV with **139**; *see also* feminism; livelihoods, women's
women's livelihoods *see* livelihoods, women's
work: complexity and heterogeneity of 326–328; decent 237–243; and migrant livelihoods 362–363
workers: complexity and heterogeneity of 326–328; migrant workers' rights 374; *see also* organising

young people 227–228, *228*, 230–231; protagonism **232**, 233; *see also* livelihoods, young people

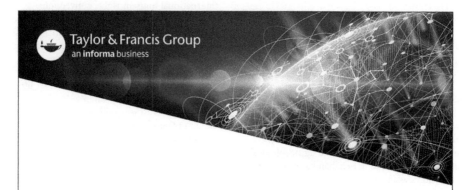

Milton Keynes UK
Ingram Content Group UK Ltd.
UKHW020616210624
444299UK00001B/1